RECEPTORS IN THE DEVELOPING NERVOUS SYSTEM
VOLUME 1

RECEPTORS IN THE DEVELOPING NERVOUS SYSTEM

VOLUME 1
Growth factors and hormones

Edited by

Ian S. Zagon and
Patricia J. McLaughlin

Department of Neuroscience and Anatomy
The Milton S. Hershey Medical Center
The Pennsylvania State University
Hershey, Pennsylvania, USA

CHAPMAN & HALL
London · Glasgow · New York · Tokyo · Melbourne · Madras

Published by Chapman & Hall, 2–6 Boundary Row, London SE1 8HN

Chapman & Hall, 2–6 Boundary Row, London SE1 8HN, UK

Blackie Academic & Professional, Wester Cleddens Road, Bishopbriggs, Glasgow G64 2NZ, UK

Chapman & Hall Inc. , 29 West 35th Street, New York NY10001, USA

Chapman & Hall Japan, Thomson Publishing Japan, Hirakawacho Nemoto Building, 6F, 1–7–11 Hirakawa-cho, Chiyoda-ku, Tokyo 102, Japan

Chapman & Hall Australia, Thomas Nelson Australia, 102 Dodds Street, South Melbourne, Victoria 3205, Australia

Chapman & Hall India, R. Seshadri, 32 Second Main Road, CIT East, Madras 600 035, India

First edition 1993

© 1993 Chapman & Hall

Typeset in 10/12 Palatino by EXPO Holdings, Malaysia
Printed in Great Britain at the University Press, Cambridge

ISBN 0 412 45240 5   Vols. 1 and 2 (Set) ISBN 0 412 54520 9

A catalogue record for this book is available from the British Library

Library of Congress Cataloging-in-Publication data

Receptors in the developing nervous system / edited by Ian S. Zagon and
    Patricia J. McLaughlin.—1st ed.
        p.   cm.
    Includes bibliographical references and index.
    Contents: v. 1. Receptors related to growth factors and hormones —
    v. 2. Neurotransmitters.
    ISBN 0-412-54520-9 (set: alk. paper). — ISBN 0-412-49400-0 (hb: v. 2:
alk. paper)
    1. Developmental neurology. 2. Neurotransmitter receptors. I. Zagon,
Ian S. II. McLaughlin, Patricia J.
    [DNLM: 1. Growth Substances. 2. Nervous System—embryology.
3. Neuroregulators. 4. Receptors, Endogenous Substances.
5. Receptors, Sensory. WL 101 R295]
QP363. 5.R43 1993
599' .0333—dc20
DNLM/DLC   92-49100
for Library of Congress   CIP

For Eileen
And my parents,
Beatrice and Benjamin
I. S. Z.

For my parents,
Katharine and Charles
P. J. M.

CONTENTS

CONTRIBUTORS

MARTIN ADAMO
Section on Molecular and Cellular
    Physiology, Diabetes Branch, National
    Institute of Diabetes and Digestive and
    Kidney Diseases, National Institutes of
    Health, Bethesda, MD 20892, USA

SEEMA BHATNAGAR
Developmental Neuroendocrinology
    Laboratory, Douglas Hospital Research
    Centre, Departments of Psychiatry and
    Neurology and Neurosurgery, McGill
    University, Montreal, Canada H4H 1R3

GERARD J. BOER
Netherlands Institute for Brain Research,
    Meibergdreef 33, 1105AZ Amsterdam Z0,
    The Netherlands

CAROLYN A. BONDY
Developmental Endocrinology Branch,
    National Institute of Child Health and
    Human Development, National Institutes
    of Health, Bethesda, MD 20892, USA

JEAN-GUY CHABOT
Douglas Hospital Research Centre and
    Department of Psychiatry, Faculty of
    Medicine, McGill University, Verdun,
    Québec, Canada H4H 1R3

KWEN-JEN CHANG
Department of Anesthesiology and
    Pharmacology, Duke University Medical
    Center, and Division of Cell Biology,
    Wellcome Research Laboratories,
    Burroughs Wellcome Co.
    Research Triangle Park, NC 27709, USA

MOSES V. CHAO
Department of Cell Biology and Anatomy,
    Hematology/Oncology Division, Cornell
    University Medical College, 1300 York
    Avenue, New York, NY 10021, USA

ERROL B. DE SOUZA
Central Nervous System Diseases Research,
    The DuPont Merck Pharmaceutical Co.,
    Wilmington, DE 19880, USA

JEAN DE VELLIS
Departments of Anatomy and Psychiatry,
    Mental Retardation Research Center, Brain
    Research Institute, Laboratory of
    Biomedical and Environmental Sciences,
    UCLA School of Medicine, University of
    California, Los Angeles, CA 90024, USA

ARACELI ESPINOSA DE LOS MONTEROS
Departments of Anatomy and Psychiatry,
    Mental Retardation Research Center, Brain
    Research Institute, Laboratory of
    Biomedical and Environmental Sciences,
    UCLA School of Medicine, University of
    California, Los Angeles, CA 90024, USA

DIMITRI E. GRIGORIADIS
Central Nervous System Diseases Research,
    The DuPont Merck Pharmaceutical Co.,
    Wilmington, DE 19880, USA

JEFFREY A. HEROUX
Central Nervous System Diseases Research,
    The DuPont Merck Pharmaceutical Co.,
    Wilmington, DE 19880, USA

THOMAS R. INSEL
Laboratory of Neurophysiology, NIMH,
    Poolesville, MD 20837, USA

SATYABRATA KAR
Douglas Hospital Research Centre and
    Department of Psychiatry, Faculty of Medi-
    cine, McGill University, Verdun, Québec,
    Canada H4H 1R3

STEPHEN L. KINSMAN
The Kennedy Krieger Institute, 707 North
    Broadway, Baltimore, MD 21205, USA

DEREK LEROITH
Section of Molecular and Cellular
    Physiology, Diabetes Branch, National
    Institute of Diabetes and Digestive and
    Kidney Diseases, National Institutes of
    Health, Bethesda, MD 20892, USA

PATRICIA J. MCLAUGHLIN
Department of Neuroscience and Anatomy,
    The Pennsylvania State University, The
    M.S. Hershey Medical Center, Hershey, PA
    17033, USA

MICHAEL J. MEANEY
Developmental Neuroendocrinology
    Laboratory, Douglas Hospital Research
    Centre, Departments of Psychiatry and
    Neurology and Neurosurgery, McGill
    University, Montreal, Canada H4H 1R3

DAJAN O'DONNELL
Developmental Neuroendocrinology
    Laboratory, Douglas Hospital Research
    Centre, Departments of Psychiatry and
    Neurology and Neurosurgery, McGill
    University, Montreal, Canada H4H 1R3

LUIS F. PARADA
Molecular Embryology Group, ABL-Basic
    Research Program, NCI-Frederick Cancer
    Research, P O Box B, Frederick, MD 21701,
    USA

DANIEL H. POLK
UCLA School of Medicine, Perinatal
    Laboratories, Harbor-UCLA Medical
    Center, Torrance, CA 90509, USA

RÉMI QUIRION
Douglas Hospital Research Centre and
    Department of Psychiatry, Faculty of
    Medicine, McGill University, Verdun,
    Québec, Canada H4H 1R3

MOHAN K. RAIZADA
Department of Physiology, University of
    Florida, Gainesville, FL 32610, USA

CHARLES T. ROBERTS
Section on Molecular and Cellular
    Physiology, Diabetes Branch, National
    Institute of Diabetes and Digestive and
    Kidney Diseases, National Institutes of
    Health, Bethesda, MD 20892, USA

ALAIN SARRIEAU
Developmental Neuroendocrinology
    Laboratory, Douglas Hospital Research
    Centre, Departments of Psychiatry and
    Neurology and Neurosurgery, McGill
    University, Montreal, Canada H4H 1R3

NOLA SHANKS
Developmental Neuroendocrinology
    Laboratory, Douglas Hospital Research
    Centre, Departments of Psychiatry and
    Neurology and Neurosurgery, McGill
    University, Montreal, Canada H4H 1R3

SAMUEL A. SHOLL
Wisconsin Regional Primate Research Center,
    University of Wisconsin, Madison,
    WI 53715–1299, USA

JAMES SMYTHE
Developmental Neuroendocrinology
    Laboratory, Douglas Hospital Research
    Centre, Departments of Psychiatry and
    Neurology and Neurosurgery, McGill
    University, Montreal, Canada H4H 1R3

YING-FU SU
Department of Anesthesiology and Pharmacology, Duke University Medical Center, Durham, NC 27710, USA

VICTOR VIAU
Developmental Neuroendocrinology Laboratory, Douglas Hospital Research Centre, Departments of Psychiatry and Neurology and Neurosurgery, McGill University, Montreal, Canada H4H 1R3

CLAIRE-DOMINIQUE WALKER
Department of Physiology, University of California at San Francisco, San Francisco, CA 94143, USA

HAIM WERNER
Section on Molecular and Cellular Physiology, Diabetes Branch, National Institute of Diabetes and Digestive and Kidney Diseases, National Institutes of Health, Bethesda, MD 20892, USA

IAN S. ZAGON
Department of Neuroscience and Anatomy, The Pennsylvania State University, The M.S. Hershey Medical Center, Hershey, PA 17033, USA

CONTENTS OF VOLUME TWO

CONTRIBUTORS TO VOLUME TWO

JOHN D. ALVARO
Laboratory of Molecular Psychiatry,
Department of Psychiatry and Program in
Neuroscience, Abraham Ribicoff Research
Facility, Yale University School of
Medicine, 34 Park St, New Haven, CT
06508, USA.

LUCIO G. COSTA
Department of Environmental Health, SC-34,
University of Washington, Seattle, WA
98195, USA.

THAN-VINH DAM
Douglas Hospital Research Centre and
Department of Psychiatry, Faculty of
Medicine, McGill University, Verdun,
Quebec, Canada H4H IR3.

ANGEL LUIS DE BLAS
Division of Molecular Biology and
Biochemistry, School of Biological Sciences,
University of Missouri-Kansas City,
Kansas City, MO 64110-2499, USA.

RONALD S. DUMAN
Laboratory of Molecular Psychiatry,
Department of Psychiatry and Program in
Neuroscience, Abraham Ribicoff Research
Facility, Yale University School of
Medicine, 34 Park St, New Haven, CT
06508, USA.

F. JAVIER GARCIA-LADONA
Institute of Pathology, Department of
Neuropathology, University of Basel,
Switzerland.

GAIL E. HANDELMANN
Department of Pharmacology and
Toxicology, University of Utah, Salt Lake
City, UT 84112, USA.

FRANCES M. LESLIE
Department of Pharmacology, University of
California at Irvine, Irvine, CA 92717, USA.

EDYTHE D. LONDON
Addiction Research Center, National
Institute on Drug Abuse, Baltimore, MD
21224, USA.

SANDRA E. LOUGHLIN
Department of Anatomy and Neurobiology,
University of California at Irvine, Irvine,
CA 92717, USA.

GUADALUPE MENGOD
Department of Neurochemistry, Centro
Investigacion y Desarrollo, Consejo
Superior de Investigaciones Cientificas
(CSIC), Jordi Girona, 18–26, Barcelona,
Spain.

TIMOTHY H. MORAN
Department of Psychiatry, Johns Hopkins
University School of Medicine, Baltimore,
MD 21205, USA.

JOSÉ M. PALACIOS
Department of Neurochemistry, Centro
Investigacion y Desarrollo, Consejo
Superior de Investigaciones Cientificas
(CSIC), Jordi Girona, 18–26, Barcelona,
Spain, and Research Institute, Laboratorios
Almirall, Barcelona, Spain.

REBECCA M. PRUSS
Marion Merrell Dow Research Institute, 2110 E.
Galbraith Road, Cincinnati, OH 45215, USA.

RÉMI QUIRION
Douglas Hospital Research Centre and
Department of Psychiatry, Faculty of
Medicine, McGill University, Verdun,
Quebec, Canada H4H 1R3.

PAUL H. ROBINSON
Westminster Hospital, London SW1, UK.

THOMAS ROTHE
University of Leipzig, Paul Flechsig Institute
for Brain Research, Department of
Neurochemistry, Leipzig, Germany.

REINHARD SCHLIEBS
University of Leipzig, Paul Flechsig Institute
for Brain Research, Department of
Neurochemistry, Leipzig, Germany.

JOHN D. STEPHENSON
Institute of Psychiatry, London SE5, UK.

ANN TEMPEL
Hillside Hospital, Division of Long Island
Jewish Medical Center, The Long Island
Campus for the Albert Einstein College of
Medicine, Glen Oaks, New York 11004,
USA.

PATRICIA M. WHITAKER-AZMITIA
Department of Psychiatry, State University
of New York, Stony Brook, NY 11794,
USA.

STEPHEN R. ZUKIN
Albert Einstein College of Medicine, Yeshiva
University, Bronx, New York, NY 10461,
USA.

PREFACE

Receptors for cell hormones, growth factors, and neurotransmitters are involved in the control and modulation of an enormous array of biological processes. The development of these receptors has distinct spatial and temporal arrangements, and alterations in this pattern during embryogenesis can have significant consequences for the well-being of the fetus, infant, child and adult. The developing nervous system is particularly dependent on receptors because its period of structural and functional organization extends through both prenatal and postnatal phases. Moreover, receptors are a key element in neural communication in both the developing and adult organism, so that the ontogeny of receptors is crucial in determining the myriad connections forming the circuitry of the nervous system.

The purpose of these two volumes is to provide a comprehensive review of receptors in the developing nervous system, placing basic and clinical information into perspective in order to formulate future scientific inquiry. A number of themes are maintained throughout the books. First, the receptors discussed in these books are part of a unit, and both the receptor and neurotransmitter, hormone, or growth factor must be considered. Therefore, the authors have spent some time in each chapter introducing the compound(s) that interacts with each receptor. Second, some receptors appear transiently and are responsible for participating in a biological process related to ontogeny, disappearing by maturation. Third, a receptor in the process of evolving into a structure important in the mature animal may also be vital in establishing the framework of particular pathways or the architecture of the brain.

Fourth, alterations in the development of neural receptors may have profound implications for the structure and function of the organism. As much as possible, the repercussions of disrupting the orchestration of receptor development in the nervous system are discussed. In many instances, however, we are just beginning to learn about some receptors and the authors may not be in a position to discuss the consequences of receptor dysfunction.

In designing these two volumes, we have asked major figures in each field to review the literature, to apprise the audience of their latest findings, and to provide a perspective on the role of receptors in the developing nervous system. These books are intended to summarize not only where we have been, but also to map directions for future research efforts, with the ultimate goals of understanding processes of normal development and the prevention or treatment of receptor dysfunction. The book is intended to be part of a continuing dialog about an important and emerging field. Given the broad scope of the subject matter, it is our expectation that the information provided by these experts will be of interest to basic and clinical researchers and graduate students in the field of developmental neurobiology, as well as to individuals in psychology, cellular and molecular biology, endocrinology, embryology, neuroscience, pharmacology, and in clinical professions such as pediatrics, neonatology, and neurology. The style adopted by the contributors should also ensure that the contents of the two volumes will be accessible to undergraduate students enrolled in advanced courses concerned with neuroscience and cell and molecular biology.

The subject matter with respect to receptors and the developing nervous system has been divided into two volumes. The editors have attempted to group central themes and topics within three major areas: receptors related to growth factors and hormones (Volume I) and to neurotransmitters (Volume II). The reader should be aware that these are rather arbitrary divisions, and that a particular compound may have multiple functions. For example, opioid receptors known to be involved in neurotransmission may also have a trophic influence on development, and could also be considered a growth factor. An introductory chapter on the biology of receptors has been included in Volume I. For convenience, an index of the subject matter in each volume is included. Finally, since Volumes I and II should be viewed as a single, integral entity, we have included the chapter titles and authors of both volumes in each book.

We thank all the authors for their thoughtful and stimulating chapters, and for the prompt attention shown to editorial requests. We have learned a great deal from reviewing these chapters and trust that the reader will have an equally enjoyable experience.

Ian S. Zagon
Patricia J. McLaughlin

INTRODUCTION: THE BIOLOGY OF RECEPTORS

1

Ying-Fu Su and Kwen-Jen Chang

1.1 INTRODUCTION

Receptors were originally defined as unknown entities of the cell that specifically recognize neurotransmitters, hormones, drugs, growth factors, or extracellular ligands and transmit signals from outside the cell to elicit physiological responses. The molecular structures of many receptors were unraveled in the last decade; enormous advances have been made toward understanding the structures and functions of those receptors. Many of their genes have been cloned and primary protein structures have been deduced from cDNA sequences. Investigations continue in an effort to comprehend the underlying biochemical and molecular mechanisms. Based on the structures and functions of the cloned receptors, receptors have been classified into six super families. These are: (a) G protein associated receptors, (b) tyrosine kinase receptors, (c) ionotropic receptors, (d) nuclear receptors – transcription regulating factors, (e) other enzyme receptors, and (f) cytokine receptors. The general features of each superfamily of receptors are discussed in this chapter.

Molecular biology technology has also revealed far more types or subtypes of receptors for a given ligand than were originally proposed from physiological and pharmacological studies. For instance, dopamine receptors have expanded from two subtypes to five subtypes. Similarly, muscarinic receptors were expanded from two to at least five subtypes of receptors based on their cloned genes. The numbers of α and β adrenergic receptors also continue to expand. The physiological significance of some of the newly proposed receptor subtypes is not clear at the present time, but the receptor diversity provides an opportunity for pharmacological intervention with more selective drugs. The possibility of developing receptor-specific drugs with fewer side effects is now more real than imaginary.

In the past, receptor structures were deduced from purified receptor proteins and subsequent molecular cloning of specific full-length cDNAs. Receptors studied this way include nicotinic acetylcholine (Noda *et al.*, 1983), muscarinic (Kubo *et al.*, 1986; Peralta *et al.*, 1987a), α and β adrenergic (Dixon *et al.*, 1986; Kobilka *et al.*, 1987a, b), and γ-aminobutyric acid (GABA) receptors (Schofield *et al.*, 1987). Now nearly all receptor genes are cloned using new molecular biology techniques. The most frequently used new techniques are: (a) low stringency hybridization screening, (b) polymerase chain reaction (PCR) technique, (c) COS cell transient expression system, (d) affinity panning technique, and (e) oocyte expression system (Table 1.1). How these techniques were used to clone different receptors will be discussed in detail in individual chapters. Some principles of these techniques are briefly discussed here.

Based on homology of nucleotide and

Receptors in the Developing Nervous System Vol. 1: *Growth factors and hormones* Edited by Ian S. Zagon and Patricia J. McLaughlin. Published in 1993 by Chapman & Hall. ISBN 0 412 45240 5. Vols. 1 and 2 (set) ISBN 0 412 54520 9.

Table 1.1.   Methods of cloning receptor genes

1. Purification of receptor proteins by affinity chromatography
2. Low stringency hybridization with homologous oligonucleotides
3. PCR amplification with homologous primers
4. Expression cloning – receptor autoradiography with ^{125}I-labeled ligands
5. Expression cloning – ligand affinity panning
6. Xenopus oocyte expression

amino acid sequences of receptor genes and proteins within a family of receptors, two molecular biology techniques were developed to clone new receptor genes (Fig. 1.1): (a) low stringency hybridization screening method to identify homologous new clones with a probe, and (b) PCR to amplify a cDNA fragment with degenerate oligonucleotide primers specifically designed for regions homologous to a family of receptors. The full-length clone can then be identified with the specific probes obtained from the hybridization screening or the PCR product under high stringency conditions. This full-length clone is inserted into a suitable mammalian expression vector and transfected into a cell line to establish a permanent cell line, or is transfected into a COS cell transient expression system. The identity of the cloned receptor can be verified with receptor binding or functional assays with specific agonists and antagonists. These two methods often result in unexpected clones, which then require thorough screening with a large number of receptor ligands to identify the receptor types. This may be assisted by knowing the tissue and brain

HYBRIDIZATION SCREENING   PCR AMPLIFICATION

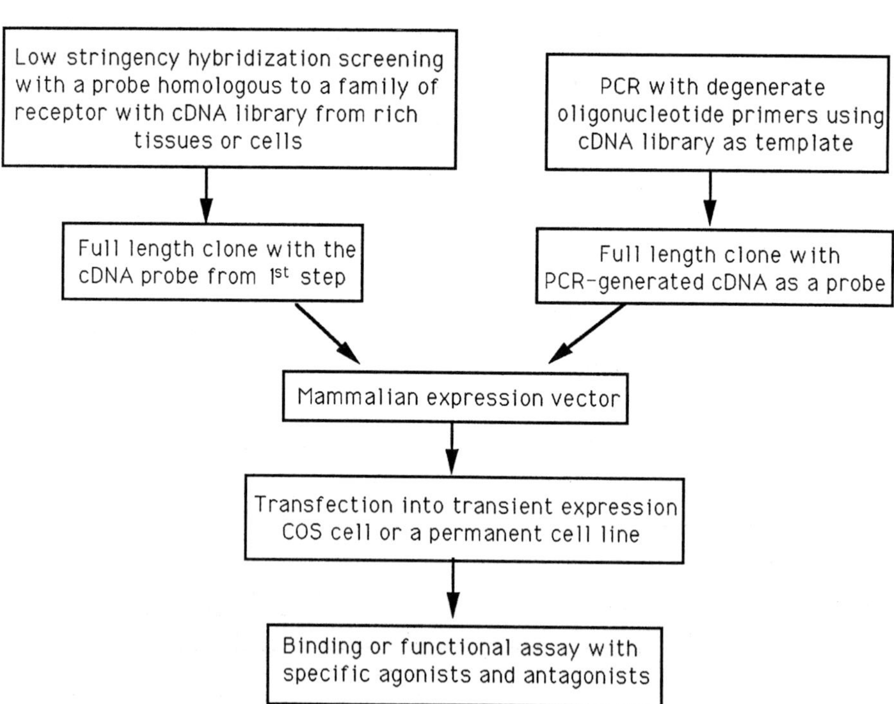

Fig. 1.1.   Summary of the procedures of molecular cloning of receptor cDNA by low stringency hybridization and PCR amplification methods.

regional distribution of the expressed mRNA of a known receptor. The low stringency hybridization method is an extremely powerful way to identify subtypes of α and β adrenergic receptors (Frielle *et al.*, 1987; Cotecchia *et al.*, 1988; Lefkowitz and Caron, 1988; Emorine *et al.*, 1989; Lorenz *et al.*, 1990; Schwinn *et al.*, 1990), muscarinic receptors (Peralta *et al.*, 1987a, b), and dopamine receptors (Bunzow *et al.*, 1988; Dearry *et al.*, 1990; Sokoloff *et al.*, 1990; Van Tol *et al.*, 1991; Sunahara *et al.*, 1991), and is used in cloning cannabinoid receptors (Matsuda *et al.*, 1990). Adenosine A_1 (Mahan *et al.*, 1991), dopamine D_1 (Zhou *et al.*, 1990), H_2 histamine (Gantz *et al.*, 1991) and odorant (Buck and Axel, 1991) receptors were cloned by PCR amplification techniques.

The third and fourth methods for receptor gene cloning use a COS cell transient expression system (Fig. 1.2). This technique involves the construction of a cDNA library into a mammalian expression vector from poly $(A)^+$ mRNA isolated from a receptor-rich cell line or tissue. Plasmid is transfected into a COS cell transient expression system. The transfected cells with the correct gene express receptors at a high level, and these can be either identified directly with a ^{125}I-labeled ligand and autoradiography or separated from other cells by affinity ligand panning. In the former, a repeated screening of a subpool of the positive cDNA pool results in a single clone. Examples of receptor genes cloned this way are endothelin B (Sakurai *et al.*, 1990), angiotensin II type-1 (Sasaki *et al.*, 1991; Murphy *et al.*, 1991), secretin (Ishihara *et al.*, 1991) and parathyroid hormone (Juppner *et al.*, 1991) receptors.

In the ligand panning method, cells expressing a specific receptor on the surface bind to specific ligand which is tagged with a

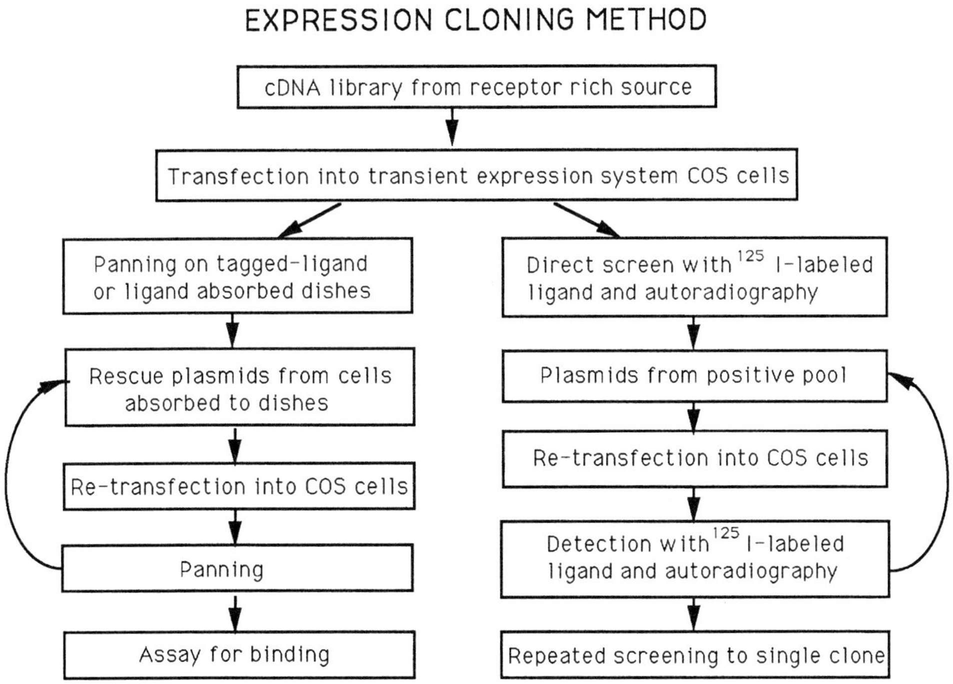

EXPRESSION CLONING METHOD

Fig. 1.2. Summary of the procedures of expression cloning method by either ^{125}I-labeling or ligand panning techniques.

small molecule. The tagged ligand is absorbed specifically onto the tag-specific antibody preabsorbed plates (Wysocki and Sato, 1978). The cells can also be directly absorbed onto a plate coated with a specific ligand. Plasmid vector is rescued from cells absorbed on to the plate and re-transfected into COS cells for repeated screening until a single clone is obtained. The typical example is the cloning of the receptor for ciliary neurotrophic factor (Davis *et al.*, 1991).

The fifth method, oocyte expression system (Fig. 1.3), has been used very successfully for receptors coupled to phospholipase C, which involves the Ca^{2+}/inositol polyphosphate second messenger system. In this system, capped poly(A)$^+$mRNA is first prepared by an *in vitro* transcription method with a cDNA library and then microinjected into stage VI Xenopus oocytes to express receptor molecules. The positive clones are identified by the electrophysiology techniques that measure Cl$^-$conductance. The Cl$^-$channels in oocytes

can be activated by ligands using the inositol phosphate/Ca^{2+} signaling pathway. The specific ligand-induced activation of Cl$^-$ conductance is used to obtain a desired clone by repeated cloning and screening. Examples of receptor genes cloned this way are receptors of substance K (Masu *et al.*, 1987), 5HT$_{1c}$ (Julius *et al.*, 1988), B$_2$ bradykinin (McEachern *et al.*, 1991), endothelin A (Arai *et al.*, 1990), and a metabotropic glutamate receptor (Masu *et al.*, 1991).

1.2   G PROTEIN ASSOCIATED RECEPTORS

For a large number of neurotransmitters and hormones the transduction of signals from receptors on the cell surface to intracellular effectors occurs via a family of coupling proteins called guanine nucleotide binding regulatory proteins, or G proteins. This signal transduction system regulated by G proteins was first discovered by Rodbell *et al.* (1971) who observed the effect of guanine nucleotides on cAMP production by hepatic adenylate cyclase. The subsequent identification and characterization of the system to include receptor, G protein, and effector protein was carried out in the early 1980s. Simultaneous with the studies of G protein-associated hormone receptor systems were studies on the visual transduction by rhodopsin in the retinal rod outer segment, which led to the discovery of transducin, a protein with biochemical properties virtually identical to the stimulatory G proteins (Stryer *et al.*, 1981). G proteins were found to couple receptors to a number of enzymes and effector proteins. Both positive and negative signals are transmitted by separate classes of G proteins. It is now recognized that the known intracellular second messengers – cAMP, phosphoinositides, diacylglycerol, Ca^{2+} – are all regulated by G proteins. In addition, there is accumulating evidence that ion channels (e.g., Ca^{2+} and K^+-channels) that respond to neurotransmitters and hormones are also directly regulated by G proteins (Brown, 1991). It is estimated that

XENOPUS OOCYTE EXPRESSION SYSTEM

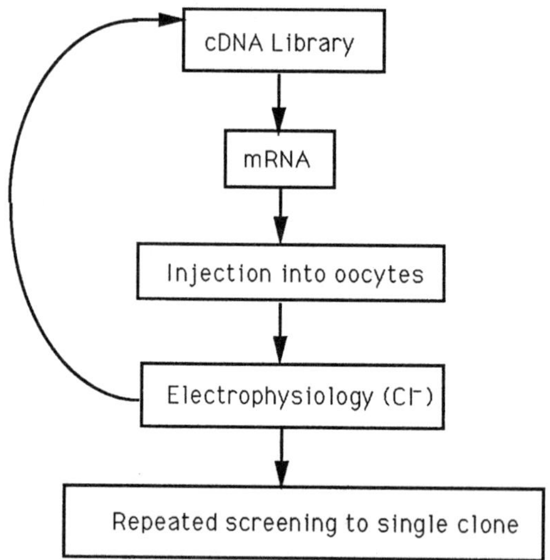

Fig. 1.3.   Summary of the procedures of molecular cloning of receptor cDNA by the oocyte expression system.

more than 40 different agonists, including all known hormones and neurotransmitters as well as some growth modulators and factors, transmit their messages through receptors coupled to G proteins. The list of receptors in this class (Table 1.2) underscores the importance of G proteins in the network of cell-to-cell communication in higher organisms. This subject has been the focus of a number of excellent review articles (Rodbell, 1980; Gilman, 1984; Lefkowitz *et al.*, 1983; Stryer and Bourne, 1986; Gilman, 1987; Birnbaumer *et al.*, 1990).

Receptors that interact with G proteins share common structural features despite the fact that there is a great diversity of receptor types, ranging from receptors for classical neurotransmitters such as norepinephrine, dopamine, histamine, and 5-hydroxytryptamine (5-HT) to those for hormones and neuromodulators such as somatostatin and neurokinin. Pharmacological and biochemical studies in the past have clearly established the existence of subtypes of receptor within a receptor type – the β_1 and β_2 adrenergic receptors, for example. Recent cloning studies with some of the known receptors confirmed the existence of multiple receptor subtypes. In many cases new subclasses of receptors with yet to be defined functions were uncovered by cloning technology. For example, after the β adrenergic receptor was cloned (Dixon *et al.*, 1986) a new β_3 receptor was discovered (Emorine *et al.*, 1989). Five species of muscarinic acetylcholine receptors, five dopamine receptors, and three serotonin receptors are now known (Dohlman *et al.*, 1991). The total number of receptors associated with G proteins has grown to more than a hundred (Birnbaumer *et al.*, 1990).

Despite the multiplicity of receptor types, all G protein coupled receptors seem to have basically the same transmembrane topological structure (Fig. 1.4). Analysis of the amino acid composition, deduced from cDNAs, of receptors cloned so far reveals that these receptors belong to a superfamily of proteins of approximately 40–55 kDa (350–500 amino acids) that span the plasma membrane seven times. The seven transmembrane α helices of hydrophobic sequences are connected by intracellular and extracellular loops. The extracellular *N*-terminal ends are glycosylated and the intracellular *C*-termini have several serine and threonine residues that may be the sites for phosphorylation. The ligand-binding sites have not been deciphered for all receptors, but evidence suggests that part of the transmembrane regions might form a ligand-binding pocket buried deep within the membrane. The loop that links the fifth and sixth transmembrane regions and part of the *C*-terminal tail may be involved in the interaction of receptors with G proteins. This information was mostly generated by site-directed mutagenesis studies of expressed receptor molecules, such as studies of the β adrenergic receptors in which deletion or substitution of amino acid residues altered the functional activities of the receptor to activate adenylate cyclase or to couple with G proteins (Dixon *et al.*, 1987; Lefkowitz *et al.*, 1988). Another powerful tool is the construction of chimeric receptors for the detailed analysis of the structure–function relationship of receptors. Since α and β adrenergic receptors have distinct and clearly defined receptor binding as well as functional specificity, hybrid receptors with peptide sequences from α and β receptors could yield information about domains critical for ligand binding and G protein interaction (Kobilka *et al.*, 1988; Frielle *et al.*, 1988). Similar genetic manipulation of other receptors and various subtypes will undoubtedly bring about understanding of the fundamental mechanisms of receptor activation and signal transduction in the near future.

Through the work of Gilman and others, a number of G proteins were identified and characterized. G proteins are a family of heterotrimeric proteins each composed of an α, β, and γ subunit; on activation the subunits dissociate into an α and a $\beta\gamma$ dimer. In contrast to the α subunit, the β and γ subunits are

Table 1.2.   G protein regulated receptors and effectors

Receptor	G Protein	Effector
Adrenergic		
$\alpha_{1\ (A,B,C)}$	Gp	PLC ↑
$\alpha_{2\ (A,B,C)}$	Gi, Go	AC ↓, K^+ channel ↑, Ca^{2+} channel ↓
$\beta_{\ (1,2,3)}$	Gs	AC ↑
Muscarinic		
$M_{\ (1,3,5)}$	Gp	PLC ↑
$M_{\ (2,4)}$	Gi	AC ↓, K^+ channel ↑
Dopamine		
$D_{\ (1,5)}$	Gs	AC ↑
$D_{\ (2,3,4)}$	Gi, Go	AC ↓, K^+ channel ↑, Ca^{2+} channel ↓
Serotonin		
$5\text{-HT}_{\ (1C,2)}$	Gp	PLC ↑
$5\text{-HT}_{\ (1A,1D)}$	Gi	AC ↓, K^+ channel ↑
Histamine		
H_1	Gp	PLC ↑
H_2	Gs	AC ↑
Adenosine		
A_1	Gi, Go	AC ↓, K^+ channel ↑, Ca^{2+} channel ↓
A_2	Gs	AC ↑
$GABA_B$	Gi, Go	AC ↓, K^+ channel ↑, Ca^{2+} channel ↓
Prostaglandin (D_1, E_2), Prostacyclin	Gs	AC ↑
Prostaglandin (E_1, F), Thromboxane	Gp	PLC ↑
LT (B_4, D_4)	Gs	PLC ↑
Purinoceptors (P_{2y})	Gp	PLC ↑
Vasopressin		
$V_{(1A,1B)}$	Gp	PLC ↑
V_2	Gs	AC ↑
Oxytocin	Gp	PLC ↑
Glucagon	Gs	AC ↑
ACTH	Gs	AC ↑
LH	Gs	AC ↑
FSH	Gs	AC ↑
MSH	Gs	AC ↑
TSH	Gs	AC ↑
VIP	Gs	AC ↑
Opioid	Gi, Go	AC ↓, K^+ channel ↑, Ca^{2+} channel ↓
Somatostatin	Gi, Go	AC ↓, K^+ channel ↑, Ca^{2+} channel ↓
$NK_{1,2,3}$ (Sub P, K and Neurokinin B)	Gp	PLC ↑
$CCK_{\ (A,B)}$	Gp	PLC ↑
Bradykinin $_{(B1,B2)}$	Gp	PLC ↑
PAF	Gp	PLC ↑
Et $_{(A,B)}$	Gp	PLC ↑
Neurotensin	Gp, Gi	PLC ↑, AC ↓
AgII	Gp, Gi	PLC ↑, AC ↓
NPY	Gi	AC ↓, Ca^{2+} channel ↓
Cannabinoid	Gi	AC ↓

In most cases, the type of G protein involved in the activation of PLC is unclear. Therefore, Gp is used to indicate putative G protein or PLC associated G protein. PLC is phospholipase C; AC is adenylyl cyclase.

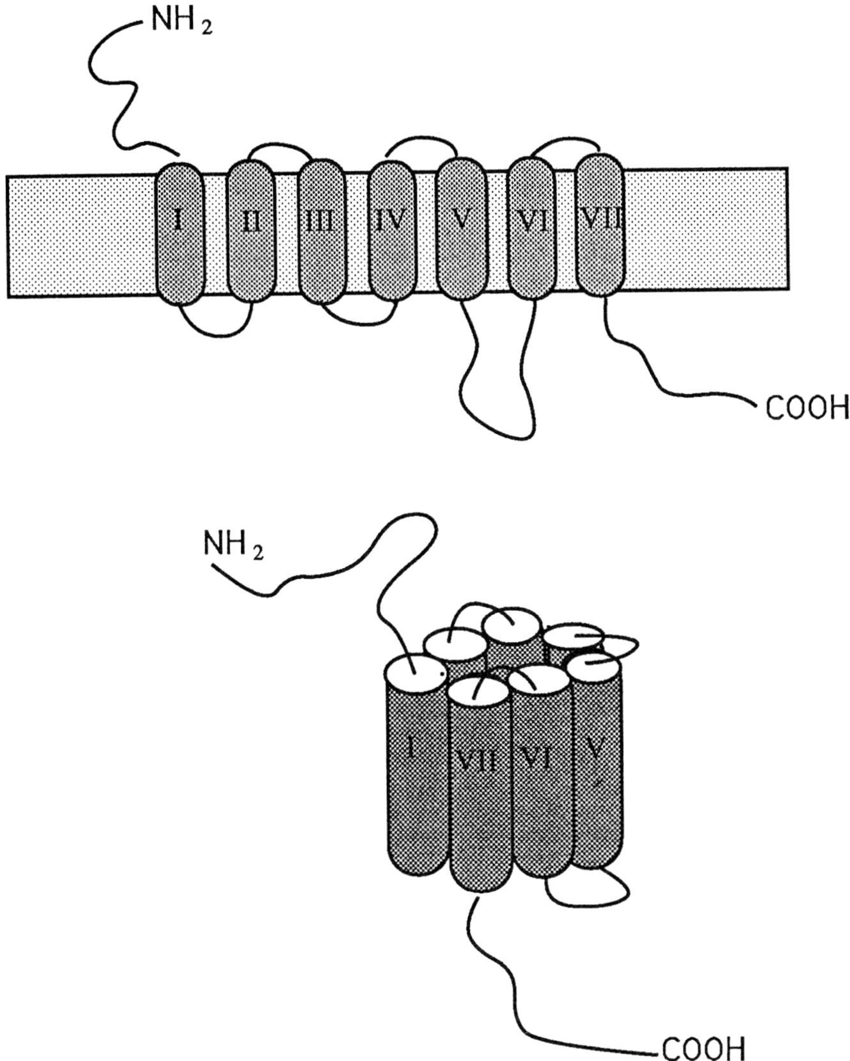

Fig. 1.4. Model of the transmembrane arrangement of G protein coupled receptors. Hydropathic analysis of amino acid composition from cDNA clones revealed that about half the residues form seven transmembrane helices connected by intra- and extracellular loops. The extracellular *N*-terminal end has several glycosylation sites, and the *C*-terminal end has several serine and threonine residues for phosphorylation.

relatively conserved and exhibit fewer isotypes. In the natural state, β and γ subunits are always associated as a dimer that is ready to interact with an α subunit to form a haloprotein. The α subunit (39–55 kDa) binds guanine nucleotides and activates effector proteins. It is also the unit that shows most structural diversity among the subunits. It is this subunit that defines the receptor and effector specificity of a given G protein. Purification and molecular cloning of α subunits from various tissue sources have resulted in identification of more than a dozen distinct types to date. Earlier, G proteins were grouped into four main categories – the stimulatory (Gs), the inhibitory (Gi), the transducin (T), and the G protein with unknown function (Go). The profusion of subtypes has resulted in new classification. Based on their functional activities, the Gα subunits have been classified into Gαs, Gαi, Gαo, GαT, Gαolf, Gαplc, Gαk, and others. Gαs activates adenylate cyclase and increases cellular cAMP levels. The Gαolf subunit is expressed in olfactory cilia, where it also stimulates adenylate cyclase and is closely related to Gαs. Both the Gαs and Gαolf subunits contain a site for ADP-ribosylation by cholera toxin. The GαT1 and GαT2 proteins, present in rods and cones of the retina, mediate the activation of cGMP-dependent phosphodiesterase. The proteins of GαT are susceptible to inhibitory actions by both cholera toxin and pertussis toxin. The Gαi subunit functions in inhibiting adenylate cyclase activity as well as regulating K$^+$ channels. The Gαo protein, the most abundant G protein subunit in brain tissue, is associated with K$^+$ and Ca^{2+} channel regulation. Recently, cloning and amino acid sequence analysis have revealed that G proteins belong to a gene superfamily consisting of at least four main groups: Gs, Gi, Gq, and G12 (Table 1.3) (Simon *et al.*, 1991). This new classification, based on amino acid sequence identity, grouped subunits Gαs and Gαolf into the Gs class. The α subunit of the Gi class consists of several species each of Gαi, Gαo, and GαT. Cloning experiments also uncovered two new classes of Gα subunit, G12 and Gq, each with several subspecies. Both of the subunits lack the cysteine residue at the C-terminus that is necessary for pertussis toxin-catalyzed ADP-ribosylation. One of the Gq subunits may be the subunit of Gplc, coupling phospholipase C to receptors. There are other Gα subunits already identified by cloning experiments but their functions remain to be identified.

The third component of the G protein regulated signal transduction system, the effectors, includes enzymes such as adenylate cyclase, cGMP activated phosphodiesterase, phosphoinositide specific phospholipase C and phospholipase A$_2$, and ion channels that gate K$^+$, Ca^{2+}, and Na$^+$. The molecular structures of these effectors are only beginning to be revealed. For example, adenylate cyclase was recently cloned (Krupinski, 1991); cyclic GMP dependent phosphodiesterase was also purified and cloned (Pittler and Baehr, 1991).

Table 1.3. Gα subunit superfamily classified according to amino acid sequence identity

Gα subtypes	Toxin sensitivity	Effector
Gs		
Gαs	Cholera toxin	AC,$\uparrow$ Ca^{2+} channel $\uparrow$
Gαolf	Cholera toxin	AC $\uparrow$
Gi		
Gαi 1,2,3	Pertussis toxin	AC, K$^+$ Channel $\uparrow$
Gαo A,B	Pertussis toxin	K$^+$, Ca^{2+} Channels $\downarrow$
GαT 1,2	Cholera and pertussin toxin	cGMP-Phosphodiesterase $\uparrow$
Gαz		
Gq		
Gαq	–	PLC $\uparrow$
Gα 11	–	PLC $\uparrow$
Gα 14,15,16		
G12		
Gα 12		
Gα 13		

Research on regulation of ion channels by G proteins is still at an early stage and little is known about the interaction between G proteins and these ion channels at the molecular level (Brown, 1991). However, the protein structure of ion channels is beginning to be revealed (Krueger, 1989; Montal, 1990).

Much is known about the activation of G protein-coupled transduction, and detailed molecular mechanisms of this transduction are the subject of numerous reviews (e.g., Stryer and Bourne, 1986; Gilman, 1987). To briefly summarize here (Fig. 1.5), signal transduction is initiated by the binding of ligand to receptor. This ligand–receptor interaction alters the conformational state of the receptor protein, which in turn leads to activation of the G protein. At resting state, the trimeric G protein has GDP bound to its Gα subunit with a slow rate of dissociation. The activated receptor catalyzes the dissociation of GDP and replaces GDP with GTP. The GTP-occupied G protein

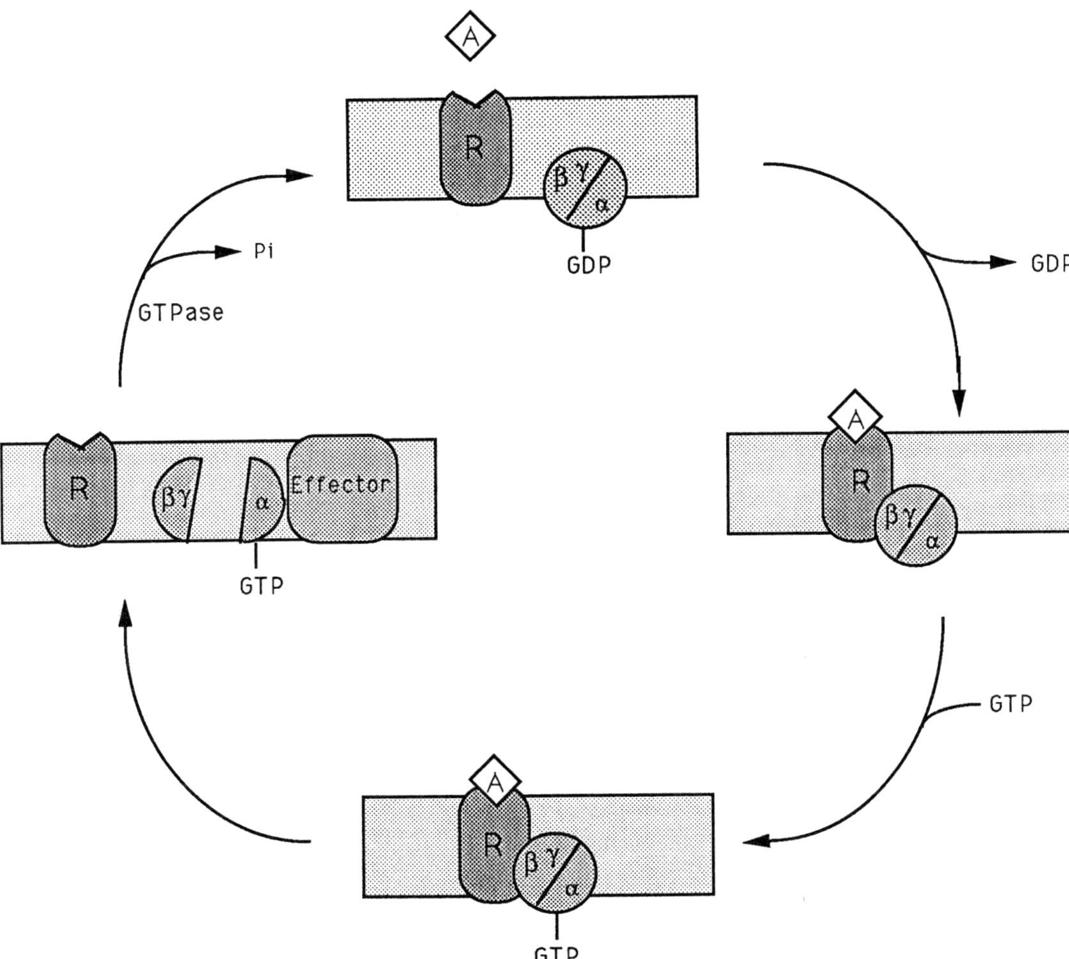

Fig. 1.5. G protein-mediated receptor signal transduction. Receptor (R) associates with a specific agonist (A) to form a receptor–G protein complex which triggers exchange of GTP for GDP at the α subunit. The α subunit is thus freed from the complex and activates effector protein. The intrinsic GTPase hydrolyzes GTP and completes the G protein cycle.

is transitory in nature and the receptor–agonist–G protein complex promptly dissociates. The G protein breaks down into an α monomer and βγ dimer. The dissociated Gα subunit is the active form that interacts with specific effector proteins to bring about cellular responses. The high intrinsic GTPase activity on the α subunit quickly hydrolyzes GTP and inactivates the Gα. The GDP–ligand Gα reassociates with the βγ dimer to complete the cycle. The G protein thus oscillates between active and inactive forms depending on the species of guanine nucleotide on the Gα. Binding of GTP also promotes dissociation of receptor from G protein, resulting in a lowering of the receptor's affinity for ligand. Agonists bind more tightly to the receptor when the receptor conformation is the one associated with Gαβγ in the absence of guanine nucleotides. High affinity receptors are thus limited by the availability of trimeric G protein and the rate of dissociation of GDP from the trimer. The high affinity of a receptor invites interaction with a ligand. The ligand–receptor complex, in turn, promotes GTP/GDP exchange with subsequent lowering of receptor afffinity. Thus, GTP serves not only to activate G proteins but also affects receptor affinity for ligands.

Continuous receptor stimulation by ligand results in diminishing cellular responses. This phenomenon of desensitization seems to be a general adaptive property exhibited by all receptor-related responses. Frequently observed receptor changes associated with desensitization in various cells and tissues includes receptor down-regulation and receptor uncoupling. Little is known about the mechanism involved in receptor down-regulation. Desensitization of receptor responses has been studied in many receptor systems and is especially well characterized in the G protein-associated β adrenergic receptor system (see review by Dohlman *et al.*, 1991). In the β receptor system, desensitization of adenylate cyclase is differentiated into homologous and heterologous. Homologous desensitization is

defined as attenuation of cellular responses to the desensitizing agent but not to other activators. Heterologous desensitization is the loss of responsiveness to a ligand due to prior exposure of the cell to other ligands. In β adrenergic receptors both homologous and heterologous desensitization are thought to be the result of receptor phosphorylation and receptor uncoupling from G proteins. Heterologous desensitization appears to be mediated by phosphorylation of the receptor by cAMP-dependent protein kinase PKA (Stadel *et al.*, 1983) as well as by protein kinase C through activation of receptor linked to the phosphatidylinositol system (Sibley *et al.*, 1984). Homologous desensitization of β adrenergic receptors is believed to be the result of receptor phosphorylation carried out by a non-cAMP dependent protein kinase, the β adrenergic receptors kinase (βARK) (Benovic *et al.*, 1986). The covalent modification of β receptors leads to uncoupling from G proteins and sequestration of β receptors (Benovic *et al.*, 1985), but the mechanism of receptor phosphorylation-induced uncoupling is still unknown.

Phosphorylation may also be the desensitization mechanism of other G protein-linked receptors. Muscarinic receptors and α-adrenergic receptors are phosphorylated by agonists, and the phosphorylation parallels desensitization (Kwatra and Hosey, 1986; Leeb-Lundberg *et al.*, 1987). The identity of the kinases mediating the phosphorylation reaction in these systems is still unknown. Judging from the structural and functional similarity, it is conceivable that receptor phosphorylation may be one of the molecular mechanisms of desensitization for all G protein-linked receptors. Whether a family of βARK-homologous kinases exists for each type of receptors remains to be seen.

1.3   TYROSINE KINASE RECEPTORS

In multicellular organisms, the regulation and coordination of cellular development is under the control of a variety of polypeptide factors such as epidermal growth factor (EGF), insulin-like growth factor (IGF), platelet-derived growth factor (PDGF), nerve growth factor (NGF), and many others. These polypeptide factors bind to cell surface receptors and trigger a series of complex and yet to be determined cellular reactions leading to cell division and differentiation. The receptors for these factors share a common feature: they possess intrinsic tyrosine kinase activity (Yarden and Ullrich, 1988; Ullrich and Schlessinger, 1990). Binding of growth factors to a receptor leads to activation of the receptor's tyrosine kinase and to phosphoryla-tion of tyrosine residues of various cellular protein substrates, including those on the receptor protein itself. Although the phosphorylation of cellular proteins by these receptor protein kinases is recognized as an important initial step, little is known about the details of the cascade of biochemical events that follow.

Structurally, all growth factor receptors have similar topological features including an extra-cellular ligand binding domain, a hydrophobic transmembrane domain, and an intracellular domain where kinase activity is localized (Fig. 1.6). On the basis of amino acid sequence

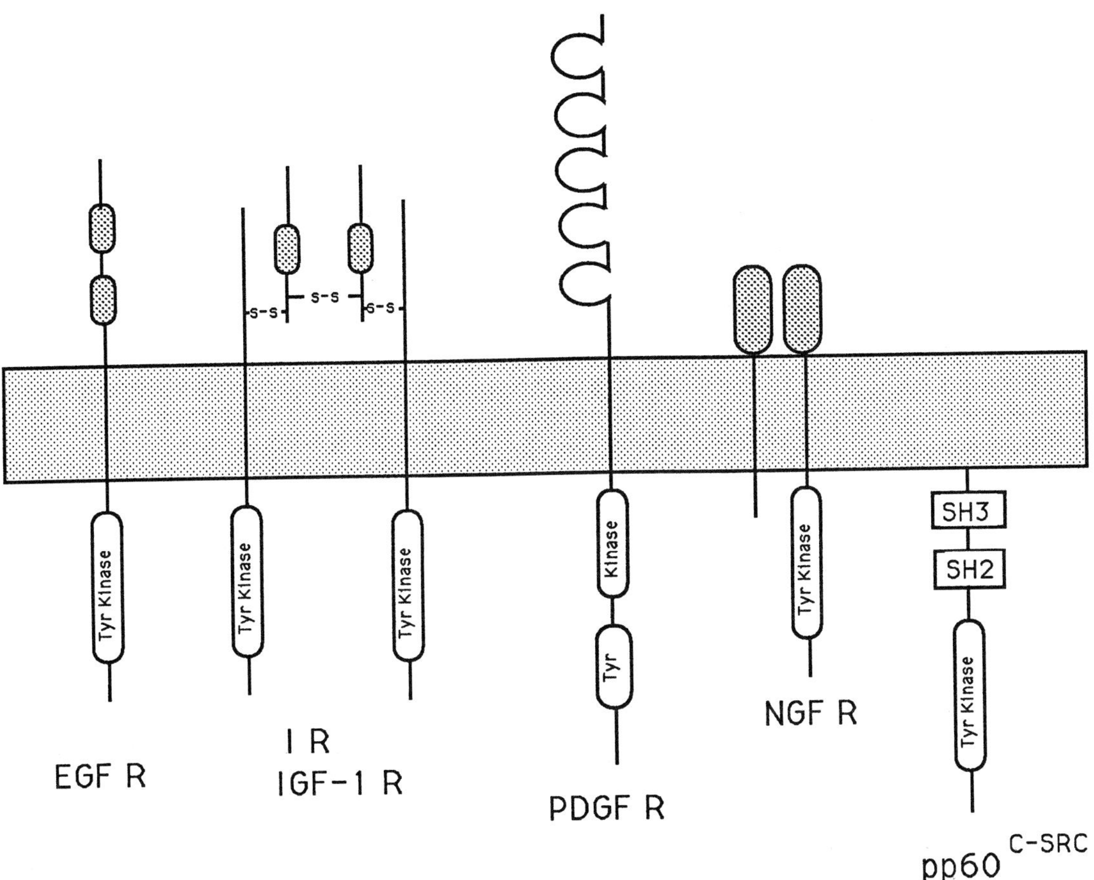

Fig. 1.6. Schematic representation of receptor tyrosine kinase subclasses. The extracellular domains contain common cysteine-rich repeat regions (shaded circles), except for the PDGF R class where repeated conserved regions are featured. The intracellular domains contain tyrosine kinase as well as tyrosine residues for phosphorylation.

homology and structural characteristics, these receptors can be classified into three subclasses – the EGF receptor, the insulin receptor, and the PDGF receptor. The receptor for EGF is monomeric and has two cysteine-rich repeat sequences in the extracellular domain. Insulin and insulin-like growth factor-1 (IGF-1) receptors consist of two α and two β subunits that are linked by disulfide bonds. The α subunits form the ligand binding domain and the β subunits form the kinase domain. The PDGF receptor has five immunoglobulin-like repeat sequences in its extracellular domain. For all tyrosine kinase receptors, the tyrosine kinase region resides in the cytoplasmic domains. This domain is the most conserved portion of all tyrosine kinase receptors and its activity is indispensable for cellular responses. The length of each peptide chain, its molecular weight, and the functional activities of the domains of each receptor have been well characterized with available cDNA clones and genetic studies (Yarden and Ullrich, 1988).

One of the early events in growth factor–receptor interaction appears to be activation of tyrosine kinase activity and autophosphorylation of tyrosine residues on the receptor protein. Evidence suggests that for all growth factor receptors, ligand binding triggers aggregation of receptors (Ullrich and Schlessinger, 1990). Dimerization and oligomerization of receptors seems to be a

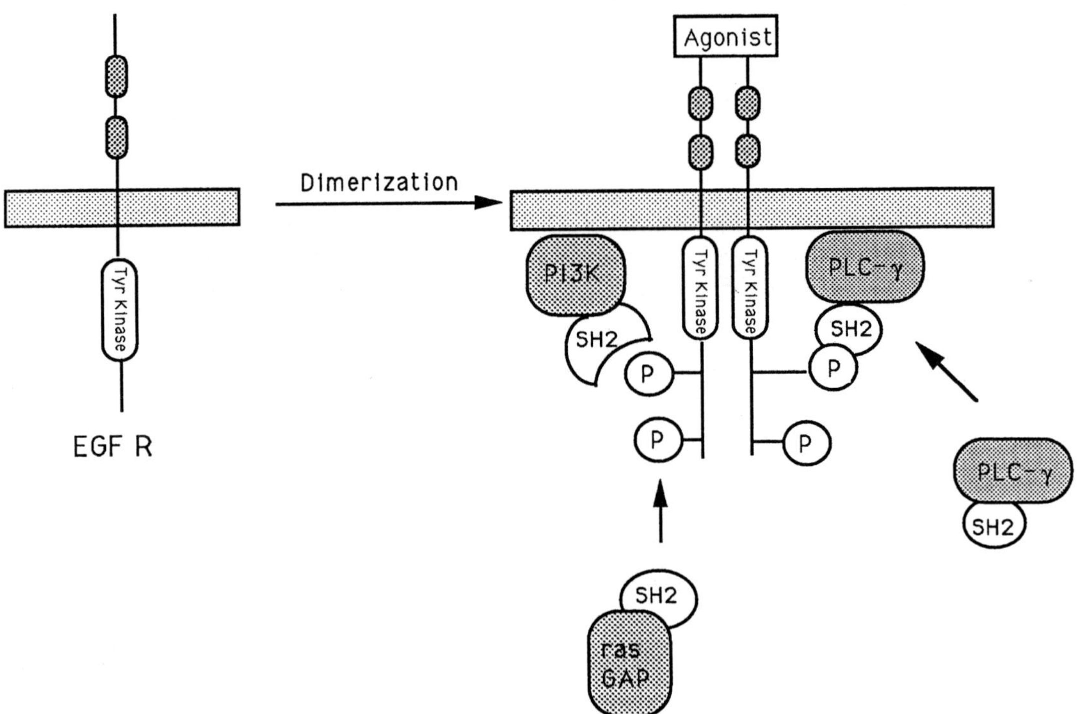

Fig. 1.7.   Models of protein tyrosine kinase activation. Ligand binding induces dimerization of receptor and cross-phosphorylation of tyrosine residues on the intracellular subunit. It is speculated that the SH2 domains of several cellular enzymes are the binding sites for phosphorylated tyrosine. Once these enzymes are attracted and bound, they become substrates for receptor protein kinase. Phosphorylation of these substrate enzymes results in modulation of their enzymatic activities. Recruitment of the enzymes also draws them close to membranes, where lipids and membrane proteins are likely targets for these enzymes.

prerequisite event that precedes kinase activation (Fig. 1.7). Apparently, the process allows adjacent receptors to cross-phosphorylate each other on tyrosine residues in the cytosolic domains. The autophosphorylation of receptors causes conformational changes that lead to increased receptor kinase activity toward other cellular substrates.

The search for receptor tyrosine kinase substrates has been the focal point of research in recent years and several candidates have been proposed. They include phospholipase C-γ (PLC-γ), phosphoinositol 3-kinase (PI3K), the proto-oncogene product c-*raf* (pp74 c-*raf*), and *ras* GTPase-activating protein (GAP). The stimulation of PLC-γ leads to the generation of diacyl glycerol and phosphoinositol metabolites such as inositol 1,4,5-triphosphate. Intracellular Ca^{2+} is elevated by inositol phosphate, and PKC is also activated by diacyl glycerol (Nishizuka, 1988). Several serine and threonine residues of cellular proteins were found to be phosphorylated by addition of growth factors. It is believed that PKC is involved in the secondary phosphorylation of these proteins, including threonine residues on the receptor itself (Boulton *et al.*, 1991). It is assumed that the phosphorylation of cellular substrates by PKC and possibly other kinases triggers cellular responses that lead to cell growth. Of the growth factors, only PDGF is capable of activating all the above-mentioned enzymes and stimulating cell division in the absence of other growth factors. Other receptors have more limited substrate targets. The pathways of signal transduction for other receptor kinase substrates – P13K, c-*raf* (pp74 c-*raf*) and GAP – are still unknown.

A number of oncogene products of the tyrosine kinase family are remarkably homogeneous in structure to the receptors for growth hormones (Table 1.4). For instance, the chicken v-*erb* B oncogene product is similar in structure to human EGF receptor but lacks the extracellular domain and part of the C-terminus. The neu/neu[*] products also closely resemble EGF receptors. The oncogene products of v-*fms* and c-*kit* are closely related to the growth factor receptors for colony stimulating factor (M-CSF) and PDGF, respectively. All of these oncogene products display tyrosine kinase activity and the study of these proteins will greatly expand our understanding of the structure and function of receptor protein kinases (Hanks *et al.*, 1988; Cantley *et al.*, 1991). The c-*kit* gene product, structurally related to the PDGF-receptor, was recently found to be a receptor for a new growth factor termed multipotential growth factor (MGF), which is thought to be important in the regulation of hemopoeisis (Huang *et al.*, 1990; Williams *et al.*, 1990; Zsebo *et al.*, 1990).

Yet another class of protein kinase that is associated with membranes but lacks the extracellular receptor domain is the *src*-type oncogene product. These non-receptor protein

Table 1.4.   Classification of growth factor receptors

Receptor	Protein tyrosine kinase	Substrates
EGF	HER/neu, *erb* B	PLCγ, GAP, PI3K, C-*src*
Insulin		PI3k, c-*raf*
IGF-1		PI3k
PDGF		PLCγ, PI3k, GAP, c-*raf*, C-*src*
M-CSF	c-*fms*	PI3K, GAP, c-*raf*
MGF	c-*kit*	
NGF	*trk*	PLCγ, PI3K
BDNF, NT-3	*trkB*	PLCγ, PI3K

GAP, GTPase activating protein; PI3K, phosphatidyl inositol-3-kinase; PLCγ, γ-type phospholipase C.

tyrosine kinases, have unique regions of amino acid sequence with $pp60^{c\text{-}src}$ as the prototype. These kinases are anchored to membranes via a short myristoylated peptide and two domains named src homology 2 and 3 (SH2 and SH3). The non-receptor protein kinases were found to have the same target substrates as receptor protein tyrosine kinases – phosphoinositol 3-kinase, phospholipase C-γ, *ras* GAP and c-*raf*. Significantly, all these tyrosine kinase substrates also have SH2 and SH3 domains within their peptide structures (Cantley *et al.*, 1991; Koch *et al.*, 1991). The functional role of SH2 was recently defined as a binding site for phosphotyrosine (Fig. 1.7). Studies on the activation of $pp60^{c\text{-}src}$ tyrosine kinase revealed that the kinase is inactive when a tyrosine residue near the C-terminal end (Tyr-527) is phosphorylated; this phosphotyrosine binds to the SH2 domain to form a self-associated conformation that apparently hinders the expression of kinase activity. Activation of $pp60^{c\text{-}src}$ tyrosine kinase might involve prevention of this self-association, perhaps through competitive binding to Tyr-527 by other proteins with SH2 domains. For receptor tyrosine kinase activation, it is suggested that ligand-induced autophosphorylation of tyrosine residues on receptors results in regions with specific binding sites for SH2-containing cytosolic substrates. Thus the SH2 domains serve an important function in bringing substrate proteins to receptor tyrosine kinases. Once bound to kinase, the enzymatic activity of substrate proteins is subsequently modified by phosphorylation of tyrosine residues. In addition, the recruitment of these proteins to the inner surface of the cell membranes as the result of phosphotyrosine-SH2 association might also be an important step in expressing the activity of these proteins, since substrates for phospholipase C-γ and phosphoinositol 3-kinase are membrane lipids and the target protein for GAP also resides at the membrane. The function of the SH3 domain is not known. It often accompanies SH2 and is

thought to participate, with SH2, in the interaction of substrate proteins with the cytoskeleton and membranes.

The biochemical events following ligand binding to growth receptors are beginning to come to light. The cloning and structural analysis of receptor protein kinase represent tremendous breakthroughs in our understanding of the receptor structure–function relationship. Much remains to be learned to link the initial events of autophosphorylation and recruitment of cellular substrates via phosphotyrosine–SH2 association to cellular responses such as gene activation, DNA synthesis and transformation. The convergence of our knowledge on signal transduction by oncogene products and growth receptors will allow the elucidation of the complex biochemical mechanisms involved in mammalian cell growth and transformation.

The development and maintenance of the mammalian nervous system depends on the presence of peptide growth factors. The vital role of these factors is perhaps best illustrated by the effect of nerve growth factor (NGF) on the proliferation and differentiation of neurons (Levi-Montalcini, 1987). This factor is required for the survival and function of neurons in sympathetic and dorsal root ganglia as well as in basal forebrain. Intensive efforts at isolating neurotrophic factors resulted in identification and cloning of NGF as well as homologous brain-derived neurotrophic factor (BDNF) and neurotrophin-3 (NT-3) (Leibrock *et al.*, 1989; Hohn *et al.*, 1990; Jones and Reichardt, 1990). These factors so far constitute the family of neurotrophic hormones with similar biochemical characteristics, as indicated by amino acid homology, cysteine residue conservation, and isoelectric points. The biological properties of these neurotrophic peptides are also similar in that they all support the survival of dorsal root ganglion neurons in culture. The precise physiological functions of BDNF and NT-3 in development of the embryonic nervous system and the survival of mature neurons are still largely unknown.

The receptors and signal transduction pathways utilized by neurotrophic factors are being explored. Recent studies showed that the receptors for nerve growth factor consist of two components: a glycoprotein of 75 kDa and a proto-oncogene product of 140–145 kDa (Hempstead *et al.*, 1991; Kaplan *et al.*, 1991a,b; Klein *et al.*, 1991). The 75-kDa protein has been cloned and characterized as a transmembrane protein with a glycosylated cysteine-rich extracellular domain; it shares homology with the tumor-necrosis factor receptor (Smith *et al.*, 1990). All three neurotrophins bind to this 75-kDa membrane protein albeit with varying affinities. The second receptor component for NGF was recently found to be the protein encoded by *trk* proto-oncogene. It is also a transmembrane glycoprotein of 140 kDa that is expressed preferentially within the nervous system. The protein has similar structural features to those of a receptor tyrosine kinase, and is phosphorylated on tyrosine residues in response to NGF. A closely related gene, the *trk*B, encodes a receptor protein for BDNF and NT-3 (Soppet *et al.*, 1991; Squinto *et al.*, 1991). The *trk*B product is also a receptor tyrosine kinase with a molecular weight of 145 kDa. Apparently, simultaneous binding to both the 75-kDa and the 140/145-kDa proteins results in high affinity binding, which is necessary for the neurotrophins to elicit biological responses. The initial event, as with many other growth factor receptors, seems to be autophosphorylation of tyrosines on the receptor proteins. Addition of NGF to responsive cells results in tyrosine kinase activation and phosphorylation of phospholipase C-γ and phosphoinositol 3-kinase (Vetter *et al.*, 1991; Kim *et al.*, 1991). A similar substrate profile is expected for the *trk*B tyrosine kinase. Again, the signal-tranducing pathways of these and other, yet to be discovered, neurotrophic factor receptors wait to be explored.

1.4 THE LIGAND-GATED ION CHANNEL RECEPTORS

There are two classes of ion channels on neurons that can be distinguished by their physiological functions – ligand-gated and voltage-gated ion channels. The voltage-gated ion channels include Na^+, K^+, and Ca^{2+} channels which account for the electric excitability of neurons and muscle cells. Signal transmission at neuronal synapses on the other hand, is accomplished by activation of specific neurotransmitter receptor–ion channel complexes that trigger alteration of membrane potentialities. Sodium ion influx into neurons results in excitation of neuronal firing, whereas chloride and potassium ion influxes inhibit firing. The neurotransmitters associated with regulation of ion channels are grouped into the so-called amino acid receptor family, which includes amino acids glutamate, glycine, acetylcholine, and others. The receptors for these amino acids are widely distributed in the brain and constitute the major transmitter system in the mammalian central nervous system. Glutamate is believed to be the major excitatory neurotransmitter, and glycine and γ-aminobutyric acid (GABA) are the inhibitory neurotransmitters (Table 1.5). Based on pharmacological and electrophysiological studies, the glutamate-gated excitatory ion channel is currently subdivided into four major subtypes: the *N*-methyl-D-aspartate (NMDA), kainate, quisqualate, and L-2-amino-4-phosphonobutyrate (APB) receptors (Foster and Fagg, 1984). The inhibitory ion channel is similarly subdivided into $GABA_A$, $GABA_B$, and glycine receptors (Dingledine *et al.*, 1990; Betz, 1990).

$GABA_A$ receptors are selectively activated by muscimol and isoquvacine and are selectively antagonized by bicuculline, whereas $GABA_B$ receptors are selectively activated by β-(*P*-chlorophenyl) GABA (baclofen) and antagonized by phaclofen (β-(*P*-chlorophenyl-3-amino propyl phosphonic

Table 1.5.   Classification of amino acid receptors

Receptor	Agonist	Modulator
Excitatory	Glutamate	
NMDA	NMDA, Aspartate	Glycine, PCP, Ketamine
Kainate	Kainate	
Quisqualate/AMPA	Quisqualate, AMPA	
APB		
ACPD	ACPD, Qisqualate	
Inhibitory	GABA	
GABA$_A$	Muscimol, Isoquvacine	Benzodiazepine, Barbiturates
GABA$_B$	Baclofen	
Glycine	Glycine	

acid). GABA$_A$ receptors are the classical receptors mediating rapid 'ionotropic' transmission through activation of chloride channels, whereas GABA$_B$ receptors do not contain an integral ion channel and are believed to affect Ca^{2+} and K^+ conductance through G protein-associated effector complexes. Conductance through the GABA$_A$-gated chloride channel can be blocked by picrotoxin, which binds directly to the chloride channel domain of the receptor complex, thus preventing ion movement.

GABA$_A$ receptors are also modulated by a number of drugs such as benzodiazepines and barbiturates. Investigation of the effects of benzodiazepines on neurotransmission led to the recognition that these drugs act as modulators of GABA$_A$-receptor gated chloride channel activity. Receptor binding studies suggest that benzodiazepine receptors are allosterically linked to GABA$_A$ receptors. GABA$_A$ agonists increase the affinity of benzodiazepine binding and vice versa. Barbiturates also enhance GABA binding and the barbiturate modulatory site at or near chloride channel is probably separate from the benzodiazepine site. The GABA$_A$ receptor is also the primary site of action for alcohol and certain convulsant toxins. The exact molecular mechanism of GABA$_A$ receptor activation and regulation by benzodiazepines, barbiturates, and other drugs is still not very clear.

Solubilization and purification of the GABA$_A$ receptor showed that it consists of several subunits. By cDNA cloning, four different classes of subunits (α, β, γ and δ) have been identified to date. Multiple variants of each class also exist. Each subunit has a molecular mass of approximately 50–60 kDa and the native molecular mass of the receptor is estimated to be 220–350 kDa. Cloning of the subunit (Schofield *et al.*, 1987; Grenningloh *et al.*, 1987; Barnard *et al.*, 1987) revealed a similar structural organization and significant amino acid homology between subunits of the GABA$_A$ receptor and the nicotinic acetycholine receptor (nAChR). Hydropathic analysis of primary protein sequences reveals several structural features common to all cloned ligand-gated ion channel proteins (Fig.1.8). Each subunit is composed of an *N*-terminal hydrophobic signal sequence and a long hydrophilic extracellular *N*-terminal domain, followed by four hydrophobic membrane-spanning domains (M_1–M_4) (Olsen and Tobin, 1990; Burt and Kamatchi, 1991). The transmembrane regions present the highest homology among all identified receptor clones. The M_2 region is especially conserved and is proposed to form part of the channel pore in ion-channel receptors. The well-characterized ion channel protein nACh receptor (Changeux *et al.*, 1987; McCarthy *et al.*, 1986; Steinbach, 1989) is a pentameric oligomer of approximately

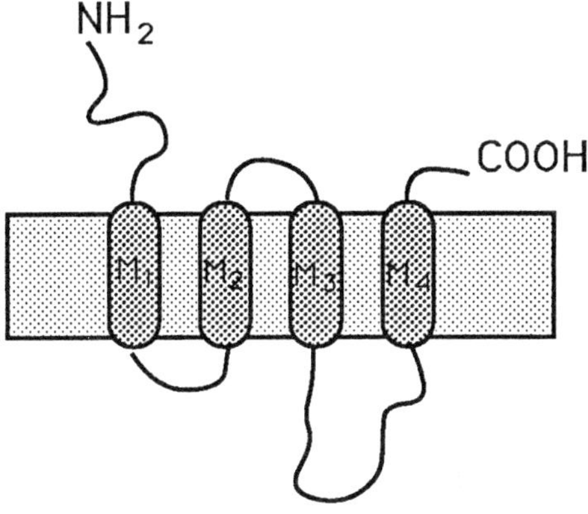

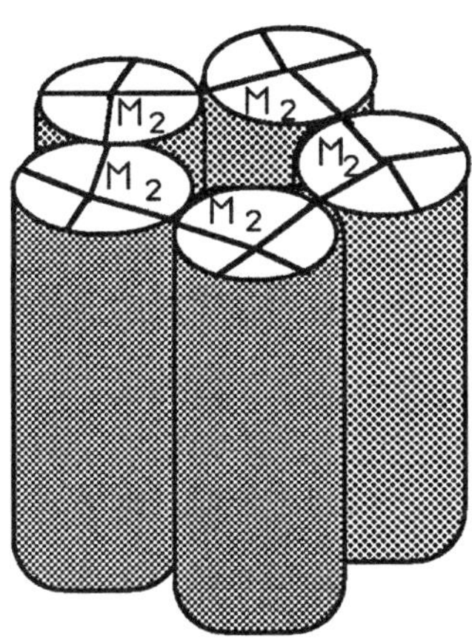

Fig. 1.8. Proposed subunit structure and the pentameric structure of a ligand-gated ion channel receptor complex. Four putative membrane-spanning α helices (M_1–M_4) are the main structural features of these ion channel subunit proteins. The ion channel is believed to consist of five such subunits in various combinations, with the M_2 lining the pore of the channel in analogy to the nicotinic receptor.

250 kDa, consisting of four types of subunits in the form of $\alpha_2\beta\gamma\delta$. The subunits are arranged pseudosymmetrically to form a roughly cylindrical structure around a central pore. The amino acid homology and common transmembrane topology among the subunit proteins of

ligand-gated ion channel receptors has promoted the notion that all ligand-gated ion channel receptors have the five-subunit stoichiometry of the nAChR. Whether this is true remains to be seen. The subunit composition and stoichiometries of these receptors are still unknown. Further complicating the issue is the fact that many variants of subunit proteins have been uncovered by molecular biology techniques. So far, six different α and three β subunits of $GABA_A$ receptors have been described, and more are anticipated. The combination of these variants could result potentially in a tremendously diverse set of distinct receptors. Functional expression in oocytes or mammalian cells with subunit variants has been studied (Schofield *et al.*, 1987), and the distribution of subunit transcripts in various brain regions has been examined (Wisden *et al.*, 1988). However, the composition of naturally occurring $GABA_A$R is still unknown.

Comparatively little is known about $GABA_B$ receptors. The pharmacologic and physiologic actions of $GABA_B$ receptor activation have just begun to be understood (Bowery, 1989; Ogata, 1990). It appears that the $GABA_B$ receptor is associated with a number of effects – neurotransmitter release and epileptiform activity, for example. The effects may be mediated by receptors linked to a G protein. Activation of $GABA_B$ results in increased K^+ conductance or decreased Ca^{2+} conductance. It is speculated that this receptor site also exists in multiple forms, but evidence for subclassification is still lacking.

The inhibitory glycine receptor has a similar molecular structure (Grenningloh *et al.*, 1987). Glycine-gated Cl^- channels contain two types of subunits (48 kDa and 58 kDa) that belong to the ligand-gated ion channel receptor superfamily. The glycine receptor is believed to have a pentameric arrangement similar to nAChR (Langosch *et al.*, 1988).

The amino acid glutamate mediates excitatory synaptic transmission in the brain. The diverse physiological responses elicited by glutamate have led to the postulation of multiple glutamate receptor classes. Pharmacological and electrophysiological studies have defined at least five classes: the *N*-methyl-D-aspartate (NMDA) receptor, the kainate receptor, the quisqualate receptor/ AMPA (α-amino-3-hydroxy-5-methyl-isoazole-4-proprionate) receptor, the APB (L-2-amino-4-phosphonobutyrate) receptor, and a receptor that is specifically stimulated by ACPD (1-amino-cyclopentyl-1,3-dicarboxylate) and is linked to a pertussis toxin-sensitive G protein that regulates formation of inositol phosphates. Among the receptor subtypes, the NMDA receptor has been the focus of studies for its role in memory, ischemia, epilepsy, and other physiological functions. The NMDA receptor is also a macromolecular complex containing the NMDA binding site, the strychnine-insensitive glycine binding site, the phencyclidine binding site, and a cation channel protein permeable to K^+ and Ca^{2+}. The receptor also possesses a binding site for Mg^{2+} and Zn^{2+}. Thus the receptor is under modulation by a multitude of factors and the exact molecular mechanism of ion channel regulation is still unknown. The biochemistry, pharmacology, and function of other glutamate receptors are even less understood. The NMDA receptor has an apparent molecular mass in the range of 196–209 kDa. However, the subunit structure of the receptor is still unsolved.

1.5   NUCLEAR HORMONE RECEPTORS

The ligands for the receptors mentioned above can not penetrate cell plasma membrane. They act indirectly by binding to cell surface receptors, which trigger a change in the membrane's physical properties, such as ion permeability or a change of concentration of intracellular second messengers, by modulating enzymatic activities as discussed above. Lipophilic hormones, such as steroids including vitamin D and thyroxine, and retinoic acids including vitamin A, interact with cytosolic or nuclear receptors to form complexes that accumulate finally in the nucleus. These lipophilic ligands activate nuclear receptors and then bind to their cognate hormone-responsive DNA elements (HRE) and modulate the activity of adjacent genes. Nuclear receptors are essentially transcription factors and lipophilic hormones are regulators of these transcription factors. Functionally, these receptors behave very similarly, although genes regulated by each hormone may be very different. This family of hormones coordinates complex events involving development, differentiation, growth, and physiological responses to diverse stimuli. Our understanding of the mode of action of lipophilic hormones at molecular level has been greatly advanced by successful isolation of genes of this superfamily of nuclear receptors.

The isolation of cDNAs for many nuclear hormone receptors has revealed the existence of a superfamily of related genes whose products are ligand-responsive transcription factors with a common structure (see Evans, 1988; Wahli and Martinez, 1991; Schwabe and Rhodes, 1991). Comparative analysis of these structures has revealed six homologous domains which are designated as A to F from *N*-terminus to *C*-terminus (Fig. 1.9). Domain C, composed of about 70 amino acids, is the most conserved among members of the superfamily. This domain also encodes DNA-binding activity through two zinc finger motifs. Domain E forms the hormone-binding domain, which is the second most-conserved region of receptors. This domain also participates in other functions, including receptor dimerization, nuclear translocation, and transcriptional activation, as well as binding of heat-shock protein (hsp90) that associates selectively with unliganded receptor molecules of glucocorticoid, progesterone, and estrogen receptors. Domain D may participate in the nuclear localization of glucocorticoid receptor. Domains A/B and F are highly variable, and the function of the latter is unknown. The length of the A/B domain

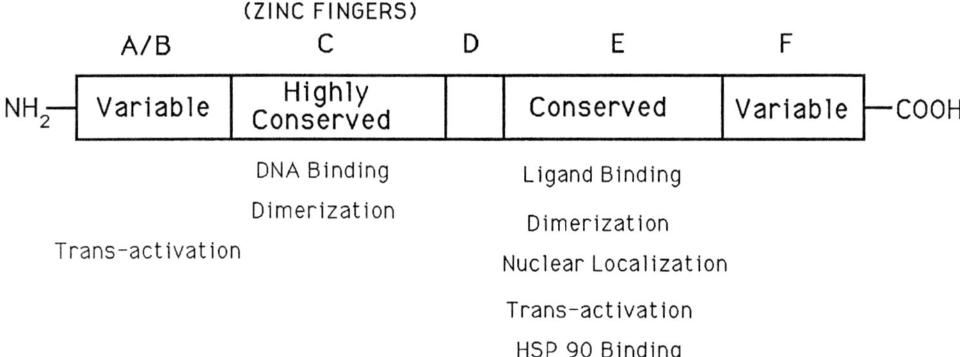

Fig. 1.9. The structural organization and functional assignment for the family of nuclear hormone receptors. Within a given type of nuclear receptor there are many subtypes which are generated by alternative promoters for the same gene by alternative splicing of RNA precursor, or by gene multiplicity. Many potential transcriptional factors were found to belong to this superfamily, but their putative ligands remain to be identified.

differs enormously among members of the nuclear receptor superfamily and the amino acid sequence of this domain is poorly conserved. This domain confers the function of transcriptional activation and the specificity of cell-type and gene under regulation.

Like membrane-localized receptors, there are many subtypes of receptor within a given type of nuclear receptor, and they are generated by alternative promoters for the same gene by alternative splicing of RNA precursor and by gene multiplicity. Many other potential transcriptional factors (orphan receptors) belong to this superfamily but their putative ligands have yet to be identified.

1.6   OTHER ENZYME RECEPTORS

Guanylate cyclase is now recognized as a receptor system with a diverse array of cellular functions. There are two classes of guanylate cyclase: the membrane-bound and the cytoplasmic. The receptors associated with each type of enzyme have been identified. Atrial natriuretic peptide was the first peptide found to stimulate cyclic GMP production through activation of membrane-bound guanylate cyclase. Other recently discovered peptide ligands for membrane guanylate cyclase are brain natriuretic peptide and C-type natriuretic peptide, both of which are found in brain tissue (Saper *et al.*, 1985; Minamino *et al.*, 1990). The ligand for cytoplasmic guanylate cyclase is proposed to be endothelium-derived relaxing factor, the factor has been tentatively identified as nitric oxide (Ignarro *et al.*, 1987; Bredt and Snyder, 1990). Although cyclic GMP is known to be the intracellular mediator in phototransduction and a regulator of vascular smooth muscle and renal functions, increasing evidence suggests that cyclic GMP is also an important second messenger system in regulating ion channels in nerve tissue. Nitric oxide synthesis has been demonstrated in brain slices (Bredt and Snyder, 1989; Bult *et al.*, 1990). The physiological function of brain natriuretic peptide is still unknown. The regulation by cyclic GMP of Na^+–Ca^{2+} channels in the phototransduction process, and of Ca^{2+} levels in smooth muscle cells, suggests that this receptor system is linked to these membrane ion channels in the neurons.

Several clones of guanylate cyclase genes were recently identified; they constitute a unique gene family commonly referred to as the guanylate cyclase gene family.

Cytoplasmic cyclase is apparently a heterodimer of 70 kDa and 73–82 kDa subunits with considerable homology at specific regions (Nakane *et al.*, 1988; Koesling *et al.*, 1990). The subunit structure of membrane-associated guanylate cyclase has not yet been determined, but the enzyme is currently believed to be a homomultimer of polypeptide (130–160 kDa) (Schulz *et al.*, 1989; Garbers, 1990). Membrane-associated guanylate cyclase consists of an extracellular peptide ligand binding domain, a transmembrane domain, and the intracellular domain. High sequence homology at the catalytic domain was demonstrated between the cytoplasmic and the membrane-associated guanylate cyclases. It is apparent that additional membrane cyclase species remain to be identified. Expression of mutant genes with altered sequences should help in defining functional activity of structural domains. Further physiological studies will establish the roles of this receptor system in the function of cells and neurons.

As discussed, the cytosolic region of many growth factor receptors has intrinsic tyrosine kinase activity, catalyzing the phosphorylation of tyrosine residues of many cellular substrates as well as the autophosphorylation of the receptor molecules. In contrast to the family of tyrosine kinase receptors, a new family of phosphotyrosine phosphatases (PTP) was uncovered and found to be localized in the plasma membrane (Trowbridge, 1991; Fischer *et al.*, 1991). These PTPs have long, extracellular *N*-terminal regions that contain ligand binding motifs such as cysteine-rich or fibronectin-like and/or immunoglobulin-like domains. However, despite the receptor-like structure of PTPs, the ligands for them remain to be identified. The fibronectin-like extracellular domain of PTPs suggests that they may participate in the process of cell adhesion, thereby controlling growth, development, and differentiation in response to cell–cell or cell–matrix interaction. They may also take part in processes such as contact inhibition by opposing the effects of the tyrosine protein kinases. It is interesting to note that a recently isolated PTP1c contains two SH2 domains, suggesting that a cytosolic PTP may be modulated through phosphotyrosine–SH2 type protein–protein interaction (Shen *et al.*, 1991). By opposing the tyrosine kinase activity, PTPs may function as anti-oncogenes.

The prototype of transmembrane PTPs is CD45, which is found only on hematopoietic cells. It consists of an external segment of a ligand binding motif, a single transmembrane stretch, and an intracellular segment that contains two tandem PTP-like domains. CD45 is essential for the antigen-induced proliferative response of T cells, and appears to be necessary for signaling of the B cell antigen receptors (Trowbridge, 1991). The mechanism by which CD45 regulates antigen response appears to be dephosphorylation of tyrosine residues on p56^{lck} or other *src*-related tyrosine kinases, although this has not been firmly established. It is exciting to look forward to the identification of the natural ligands for PTPs and an understanding of their underlying mechanism of growth regulation.

1.7   CYTOKINE RECEPTOR SUPERFAMILY

The final group of receptors belong to the cytokines, such as interleukins, erythropoietin, granulocyte-macrophage colony stimulating factor, granulocyte colony-stimulating factor, prolactin, and growth hormone. This superfamily of receptors has been referred to as cytokine receptor superfamily and is the subject of several reviews (Cosman *et al.*, 1990; Hall and Rao, 1992). This family of receptors is now believed to require two subunits (α and β subunit) to achieve high affinity binding for eliciting responses at physiological concentrations. The α subunit seems to confer the specificity and low affinity binding of cytokines, and the β subunit provides the high affinity-converting   and   signal-transducing component. The two subunits share several structural motifs at their extracellular region: four cysteine residue-domains located at the *N*-terminus, and a Trp-Ser-x-Trp-Ser motif located just outside the membrane-spanning domain (except the growth hormone receptor). There is not much sequence similarity between the cytoplasmic domains of these receptors. The α subunits have a short cytoplasmic domain. The β subunits have a long cytoplasmic domain that is rich in proline and serine residues. Both subunits have a single transmembrane domain. It appears that all cytokine receptors have both α and β subunits, although some receptors have not yet been isolated. So far, no enzymatic activity was observed on the cytoplasmic domain of these receptors. The underlying mechanism of signal transduction is unknown.

Two recently isolated neurokines, ciliary neurotrophic factor (CNTF) and cholinergic differentiation factor (CDF), which is identical to LIF (leukemia inhibitory factor), were shown to contain cytokine-like structure and influence neuronal survival and differentiation. Receptors for both neurokines have been cloned and found to belong to the cytokine receptor superfamily. The receptor of CNTF is similar to the α subunit of neurokine receptor (Davis *et al.*, 1991), whereas the receptor of CDF is similar to the β subunit of neurokine receptor (Gearing *et al.*, 1991). However, in each case, the affinity of the cloned receptor for the ligand is an order of magnitude below the K_d value of the native high affinity receptor. This suggests that, similar to other cytokine receptors, an additional subunit is required for full receptor function. Because of the similarity between neurokines and cytokines, and their receptors, plus the well-known function of cytokines in hemopoeisis, it is conceivable that cytokines or cytokine-like factors may play important roles in neuronal differentiation during development as well as neuronal survival upon aging.

1.8 CONCLUSION

Our understanding of receptor biology is rapidly advancing on two fronts. Superfamilies of receptor genes are being identified; each family encodes a large number of receptors and their variants. Transcriptional and translational regulation of many receptors can now be examined in detail during different stages of development. Parallel to this development is the rapid progress in the elucidation of biochemical mechanisms involving signal transduction for each receptor family. Various components participating in the signal-transduction pathways are being identified, studied, and their genes cloned, thus becoming amenable for developmental studies. We are anticipating an information explosion in the coming decade in the area of developmental studies of receptors and signaling.

REFERENCES

The citations listed here are not meant to be comprehensive. These references are intended only as starting points to various topics concerning receptor biology.

Arai, H., Hori, S., Aramori, I. *et al.* (1990) Cloning and expression of a cDNA encoding an endothelin receptor. *Nature*, **348**, 730–2.

Barnard, E. A., Darlison, M. G. and Seeburg, P. (1987) Molecular biology of the GABA$_A$ receptor: the receptor/channel superfamily. *Trends Neurosci.*, **10**, 502–9.

Benovic, J. L., Pike, L. J., Cerione, R. A. *et al.* (1985) Phosphorylation of the mammalian β-adrenergic receptor by cyclic AMP-dependent protein kinase. *J. Biol. Chem.*, **260**, 7094–101.

Benovic, J. L., Strasser, R. H., Caron, M. G. and Lefkowwitz, R. J. (1986) β-adrenergic receptor kinase: identification of a novel protein kinase that phosphorylates the agonist-occupied form of the receptor. *Proc. Natl. Acad. Sci. USA*, **83**, 2797–801.

Betz, H. (1990) Ligand-gated ion channels in the brain: the amino acid receptor superfamily. *Neuron*, **5**, 383–92.

Birnbaumer, L., Abramowitz, J. and Brown, A. M. (1990) Receptor–effector coupling by G proteins. *Biochim. Biophys. Acta*, **1031**, 163–224.

Boulton, T. G., Nye, S. H., Robbins, D. J. *et al.* (1991) ERKS: a family of protein serine/threonine kinases that are activated and tyrosine phosphorylated in response to insulin and NGF. *Cell*, **65**, 663–75.

Bowery, N. (1989) GABA$_B$ receptors and their significance in mammalian pharmacology. *Trends Pharmacol. Sci.*, **10**, 401–7.

Bredt, D. S. and Snyder, S. H. (1989) Nitric oxide mediates glutamate-linked enhancement of cGMP levels in the cerebellum. *Proc. Natl. Acad. Sci. USA*, **86**, 9030–3.

Bredt, D. S. and Snyder, S. H. (1990) Isolation of nitric oxide synthetase, a calmodulin-requiring enzyme. *Proc. Natl. Acad. Sci. USA*, **87**, 682–5.

Brown, A. M. (1991) A cellular logic for G protein-coupled ion channel pathways. *FASEB J.*, **5**, 2175–9.

Buck, L. and Axel, R. (1991) A novel multigene family may encode odorant receptors: a molecular basis for odor recognition. *Cell*, **65**, 175–87.

Bult, H., Boeckxstaens, G. E., Pelckmans, P. A. *et al.* (1990) Nitric oxide as an inhibitory non-adrenergic non-cholinergic neurotransmitter. *Nature*, **345**, 346–7.

Bunzow, J. R., Van Tol, H. H. M., Grandy, D. K. *et al.* (1988) Cloning and expression of a rat D2 dopamine receptor cDNA. *Nature*, **336**, 783–7.

Burt, D. R. and Kamatchi, G. L. (1991) GABA$_A$ receptor subtypes: from pharmacology to molecular biology. *FASEB J.*, **5**, 2916–23.

Cantley, L. C., Auger, K. R., Carpenter, C. *et al.* (1991) Oncogenes and signal transduction. *Cell*, **64**, 281–302.

Changeux, J. P., Giradat, J. and Dennis, M. (1987) The nicotinic acetylcholine receptor: molecular architecture of a ligand-regulated ion channel. *Trends Biochem. Sci.*, **8**, 459–65.

Cosman, D., Lyman, S. D., Idzerda, R. L. *et al.* (1990) A new cytokine receptor superfamily. *Trends Biochem. Sci.*, **15**, 265–70.

Cotecchia, S., Schwinn, D. A., Randall, R. R. *et al.* (1988) Molecular cloning and expression of the cDNA for the hamster α1-adrenergic receptor. *Proc. Natl. Acad. Sci. USA*, **85**, 7159–63.

Davis, S., Aldrich, T. H., Valenzuela, D. M. *et al.* (1991) The receptor for ciliary neurotrophic factor. *Science*, **253**, 59–63.

Dearry, A., Gingrich, J. A., Falardeau, P. *et al.* (1990) Molecular cloning and expression of the gene for a human D$_1$ dopamine receptor. *Nature*, **347**, 72–6.

Dingledine, R., Myers, S. J. and Nicholas, R. A. (1990) Molecular biology of mammalian amino acid receptors. *FASEB J.*, **4**, 2636–45.

Dixon, R. A. F., Kobilka, B. K., Strader, D. J. *et al.* (1986) Cloning of gene and cDNA for mammalian β-adrenergic receptor and homology with rhodopsin. *Nature*, **321**, 75–9.

Dixon, R. A. F., Sigal, I. S., Rands, E. *et al.* (1987) Ligand binding to the β-adrenergic receptor involves its rhodopsin-like core. *Nature*, **326**, 73–7.

Dohlman, H. G., Thorner, J., Caron, M. G. and Lefkowitz, R. J, (1991) Model systems for the study of seven-transmembrane-segment receptors. *Annu. Rev. Biochem.*, **60**, 653–88.

Emorine, L. J., Marullo, S., Briend-Sutren, M. -M. *et al.* (1989) Molecular characterization of the human β$_3$-adrenergic receptor. *Science*, **245**, 1118–21.

Evans, R. M. (1988) The steroid and thyroid hormone receptor superfamily. *Science*, **240**, 889–95.

Fischer, E. H., Charbonneau, H. and Tonks, N. K. (1991) Protein tyrosine phosphatases: a diverse family of intracellular and transmembrane enzymes. *Science*, **253**, 401–6.

Foster, A. and Fagg, G. E. (1984) Acidic amino binding sites in mammalian neuronal membranes: their characteristics and relationship to synaptic receptors. *Brain Res. Rev.*, **7**, 103–64.

Frielle, T., Collins, K. W., Daniel, M. G. *et al.* (1987) Cloning of the cDNA for the human β1-adrenergic receptor. *Proc. Natl. Acad. Sci. USA*, **84**, 7920–4.

Frielle, T., Daniel, S., Caron, K. W. and Lefkowitz, R. J. (1988) Structural basis of β-adrenergic receptor subtype specificity studied with chimeric β$_1$/β$_2$-adrenergic receptors. *Proc. Natl. Acad. Sci. USA*, **185**, 9494–8.

Gantz, I., Schaffer, M., DelValle, J. *et al.* (1991) Molecular cloning of a gene encoding the histamine H2 receptor. *Proc. Natl. Acad. Sci. USA*, **88**, 429–33.

Garbers, D. L. (1990) The guanylyl cyclase receptor family. *New Biol.*, **2**, 499–504.

Gearing, D. P., Thut, C. J., VandeenBos, T. *et al.* (1991) Leukemia inhibitory factor is structurally related to the IL-6 signal transducer, gp130. *EMBO J.*, **10**, 2839–48.

Gilman, A. G. (1984) G proteins and dual control of adenylate cyclase. *Cell*, **36**, 577–9.

Gilman, A. G. (1987) G proteins: transducers of receptor-generated signals. *Annu. Rev. Biochem.*, **56**, 615–49.

Grenningloh, G., Rienitz, A., Schmitt, B. *et al.*

(1987) The strychnine-binding subunit of the glycine receptor shows homology with nicotinic acetylcholine receptors. *Nature,* **328**, 215–20.

Hall, A. K. and Rao, M. (1992) Cytokines and neurokines: related ligand and related receptors. *Trends Neurosci.,* **15**, 35–7.

Hanks, S. K., Quinn, A. M. and Hunter, T. (1988) The protein kinase family: conserved features and deduced phylogeny of the catalytic domains. *Science,* **241**, 42–52.

Hempstead, B. L., Martin-Zanca, D., Kaplan, D. R. *et al.* (1991) High-affinity NGF binding requires coexpression of the *trk* proto-oncogene and the low-affinity NGF receptor. *Nature,* **350**, 678–83.

Hohn, A., Leibrock, J., Bailey, K. and Barde, Y. A. (1990) Identification and characterization of a novel member of the nerve growth factor/brain-derived neurotrophic factor family. *Nature,* **344**, 339–41.

Huang, E., Nocka, K., Beier, D. R. *et al.* (1990) The hematopoietic growth factor KL is encoded by the SI locus and is the ligand of the c-kit receptor, the gene product of the W locus. *Cell,* **63**, 225–33.

Ignarro, L. J., Buga, G. M., Wood, K. S. *et al.* (1987) Endothelin-derived relaxing factor produced and released from artery and vein is nitric oxide. *Proc. Natl. Acad. Sci. USA,* **84**, 9265–9.

Ishihara, T., Nakamura, S., Kaziro, Y. *et al.* (1991) Molecular cloning and expression of a cDNA encoding the secretin receptor. *EMBO J.,* **10**, 1635–41.

Jones, K. R. and Reichardt, L. F. (1990). Molecular cloning of a human gene that is a member of the nerve growth factor family. *Proc. Natl. Acad. Sci. USA,* **87**, 8060–4.

Julius, D., MacDermott, A. B., Axel, R. and Jessell, T. M. (1988) Molecular characterization of a functional cDNA encoding the serotonin 1c receptor. *Science,* **241**, 558–64.

Juppner, H., Abdul-Samra, A. -B., Freeman, M. *et al.* (1991) A G protein-linked receptor for parathyroid hormone and parathyroid hormone-related peptide. *Science,* **254**, 1024–6.

Kaplan, D. R., Martin-Zanca, D. and Parada, L. F. (1991a) Tyrosine phosphorylation and tyrosine kinase activity of the *trk*-proto-oncogene product induced by NGF. *Nature,* **350**, 158–60.

Kaplan, D. R., Hempstead, B. L., Martin-Zanca, D. *et al.* (1991b) The *trk* proto-oncogene product: a signal transducing receptor for nerve growth factor. *Science,* **252**, 554–8.

Kim, U. H., Fink, D., Kim, H. S. *et al.* (1991) Nerve growth factor stimulates phosphorylation of phospholipase C-γ in PC12 cells. *J. Biol. Chem.,* **266**, 1359–62.

Klein, R., Jing, S., Nanduri, V. *et al.* (1991) The *trk* proto-oncogene encodes a receptor for nerve growth factor. *Cell,* **65**, 189–97.

Kobilka, B. K., Matsui, H., Kobilka, T. S. *et al.* (1987a). Cloning, sequencing and expression of the gene coding for the human platelet α_2-adrenergic receptor. *Science,* **238**, 650–6.

Kobilka, B. K., Dixon, R. A. F., Frielle, T. *et al.* (1987b) cDNA for the human β_2-adrenergic receptor: a protein with multiple membrane spanning domains and encoded by a gene whose chromosomal location is shared with that of the receptor for platelet-derived growth factor. *Proc. Natl. Acad. Sci. USA,* **84**, 46–50.

Kobilka, B. K., Kobilka, T. S., Daniel, K. *et al.* (1988) Chimeric α-2, β-2-adrenergic receptors: delineation of domains involved in effector coupling and ligand binding specificity. *Science,* **240**, 1310–16.

Koch, C. A., Anderson, D., Moran, M. F. *et al.* (1991) SH2 and SH3 domains: elements that control interactions of cytoplasmic signaling proteins. *Science,* **252**, 668–74.

Koesling, D., Harteneck, C., Humbert, P. *et al.* (1990) The primary structure of the larger subunit of soluble guanylyl cyclase from bovine lung. Homology between the two subunits of the enzyme. *FEBS Lett.,* **266**, 128–32.

Krueger, B. K. (1989) Toward an understanding of structure and function of ion channels. *FASEB J.,* **3**, 1906–14.

Krupinski, J. (1991) The adenyl cyclase family. *Mol. Cell. Biochem.,* **104**, 73–9.

Kwatra, M. M. and Hosey, M. M. (1986) Phosphorylation of the cardiac muscarinic receptor in intact chick heart and its regulation by a muscarinic agonist. *J. Biol. Chem.,* **261**, 12429–32.

Kubo, T., Fukuda, K., Mikami, A. *et al.* (1986) Cloning, sequencing and expression of complementary DNA encoding the muscarinic acetylcholine receptor. *Nature,* **323**, 411–16.

Langosch, G., Thomas, L. and Betz, H. (1988) Conserved quaternary structure of ligand-gated ion channels: the post-synaptic glycine receptor is a pentamer. *Proc. Natl. Acad. Sci. USA,* **85**, 7394–8.

Leeb-Lundberg, L. M., Cotecchia, S., DeBlasi, A. *et al.* (1987) Regulation of adrenergic receptor function by phosphorylation: I. Agonist-promoted desensitization and phosphorylation of α_1-adrenergic receptors coupled to inositol phospholipid metabolism in DDT1, and MF-2 smooth muscle cells. *J. Biol. Chem.*, **262**, 3098–105.

Lefkowitz, R. J. and Caron, M. G. (1988) Adrenergic receptors: models for the study of receptor to guanine nucleotide regulatory proteins. *J. Biol. Chem.*, **263**, 4993–6.

Lefkowitz, R. J., Stadel, J. M. and Caron, M. G. (1983) Adenylate cyclase-coupled β-adrenergic receptors: structure and mechanisms of activation and desensitization. *Annu. Rev. Biochem.*, **52**, 159–86.

Lefkowitz, R. J., Kobilka, B. K., Benovic, J. L. *et al.* (1988) Molecular biology of adrenergic receptors. *Cold Spring Harbor Symp. Quant. Biol.*, **53**, 507–14.

Leibrock, J., Lottspeich, F., Hohn, A. *et al.* (1989) Molecular cloning and expression of brain-derived neurotrophic factor. *Nature*, **341**, 149–52.

Levi-Montalcini, R. (1987) The nerve growth factor 35 years later. *Science*, **237**, 1154–62.

Lorenz, W., Lomasney, J. W., Collins, S. *et al.* (1990) Expression of three α_2-adrenergic receptor subtypes in rat tissues: implications for α_2 receptor classification. *Mol. Pharmacol.*, **38**, 599–603.

Mahan, L. C., McVittie, L. D., Smyk-Randall, E. M. *et al.* (1991) Cloning and expression of an A_1 adenosine receptor from rat brain. *Mol. Pharmacol.*, **40**, 1–7.

Masu, M., Tanabe, Y., Tsuchida, K. *et al.* (1991) Sequence and expression of a metabotropic glutamate receptor. *Nature*, **349**, 760–5.

Masu, Y., Nakayama, K., Tamaki, H. *et al.* (1987) cDNA cloning of bovine substance-K receptor through oocyte expression system. *Nature*, **329**, 836–8.

Matsuda, L. A., Lolait, S. J., Brownstein, M. J. *et al.* (1990) Structure of a cannabinoid receptor and functional expression of the cloned cDNA. *Nature*, **346**, 561–4.

McCarthy, M. P., Earnest, J. P., Young, E. F. *et al.* (1986) The molecular neurobiology of the acetylcholine receptor. *Annu. Rev. Neurosci.*, **9**, 383–413.

McEachern, A. E., Shelton, E. R., Bhakta, S. *et al.* (1991) Expression cloning of a rat B_2 bradykinin receptor. *Proc. Natl. Acad. Sci. USA*, **88**, 7724–8.

Minamino, N., Kangawa, K. and Matsuo, H. (1990) N-terminally extended form of C-type natriuretic peptide (CNP-53) identified in porcine brain. *Biochem. Biophys. Res. Commun.*, **170**, 973–9.

Montal, M. (1990) Molecular anatomy and molecular design of channel proteins. *FASEB J.*, **4**, 2623–35.

Murphy, T. J., Alexander, R. W., Griendling, K. K.

et al. (1991) Isolation of a cDNA encoding the vascular type-1 angiotensin II receptor. *Nature*, **351**, 233–6.

Nakane, M., Saheki, S., Kuno, T. *et al.* (1988) Molecular cloning of a cDNA coding for 70 kilodalton subunit of soluble guanylate cyclase from rat lung. *Biochem. Biophys. Res. Commun.*, **157**, 1139–47.

Nishizuka, Y. (1988) The molecular heterogeneity of protein kinase C and its implications for cellular regulation. *Nature*, **334**, 661–5.

Noda, M., Takahashi, H., Tanabe, T. *et al.* (1983) Structural homology of *Torpedo californica* acetylcholine receptor subunits. *Nature*, **302**, 528–32.

Ogata, N. (1990) Pharmacology and physiology of $GABA_B$ receptors. *Gen. Pharmacol.*, **21**, 395–402.

Olsen, R. W. and Tobin, A. J. (1990) Molecular biology of $GABA_A$ receptors. *FASEB J.*, **4**, 1469–80.

Peralta, E. G., Winslow, J. W., Peterson, G. L. *et al.* (1987a) Primary structure and biochemical properties of an M2 muscarinic receptor. *Science*, **236**, 600–5.

Peralta, E. G., Ashkenazi, A., Winslow, J. W. *et al.* (1987b) Distinct primary structure, ligand-binding properties and tissue-specific expression of four human muscarinic acetylcholine receptors. *EMBO J.*, **6**, 3923–9.

Pittler, S. J. and Baehr, W. (1991) The molecular genetics of retinal photoreceptor proteins involved in cGMP metabolism. *Prog. Clin. Biol. Res.*, **362**, 33–66.

Rodbell, M. (1980) The role of hormone receptors and GTP-regulatory proteins in membrane transduction. *Nature*, **284**, 17–22.

Rodbell, M., Birnbaumer, L., Pohl, S. L. and Kraus, H. M. (1971) The glucagon-sensitive adenyl cyclase system in plasma membranes of rat liver. V. An obligatory role of guanyl nucleotides in glucagon action. *J. Biol. Chem.*, **246**, 1877–82.

Sakurai, T., Yanagisawa, M., Takuwa, Y. *et al.* (1990) Cloning of a cDNA encoding a non-isopeptide-selective subtype of the endothelin receptor. *Nature*, **348**, 732–5.

Saper, C. B., Standaent, D. G., Cunnic, M. G. *et al.* (1985) Atriopeptin-immunoreactive neurons in brain: presence in cardiovascular regulatory areas. *Science*, **227**, 1047–9.

Sasaki, K., Yamano, Y., Bardhan, S. *et al.* (1991) Cloning and expression of a complementary DNA encoding a bovine adrenal angiotensin II type-1 receptor. *Nature*, **351**, 230–3.

Schofield, P. R., Darlison, M. G., Fujita, N. *et al.* (1987) Sequence and function expression of the

GABA$_A$ receptor shows a ligand-gated receptor superfamily. *Nature*, **328**, 221–7.

Schulz, S., Chinkers, M. and Garbers, D. L. (1989) The guanylate cyclase/receptor family of proteins. *FASEB J.*, **3**, 2026–35.

Schwabe, J. W. R. and Rhodes, D. (1991) Beyond zinc fingers: steroid hormone receptors have a novel structural motif for DNA recognition. *Trends. Biochem. Sci.*, **16**, 291–6.

Schwinn, D. A., Lomasney, J. W., Lornez, W. *et al.* (1990) Molecular cloning and expression of the cDNA for a novel α_1-adrenergic receptor subtype. *J. Biol. Chem.*, **265**, 8183–9.

Shen, S. -H., Bastien, L., Posner, B. I. and Chretien, P. (1991) A protein-tyrosine phosphatase with sequence similarity to the SH2 domain of the protein-tyrosine kinase. *Nature*, **352**, 736–9.

Sibley, D. R., Nambi, P., Peters, J. R. and Lefkowitz, R. J, (1984) Phorbol diesters promote β-adrenergic receptor phosphorylation and adenylate cyclase desensitization in duck erythrocytes. *Biochem. Biophys. Res. Commun.*, **121**, 973–9.

Simon, M. I., Strathmann, M. P. and Gautam, N. (1991) Diversity of G protein in signal transduction. *Science*, **252**, 807–8.

Smith, C. A., Davis, T., Anderson, D. *et al.* (1990) A receptor for tumor necrosis factor defines an unusual family of cellular and viral proteins. *Science*, **248**, 1019–23.

Sokoloff, P., Giros, B., Martres, M. -P. *et al.* (1990) Molecular cloning and characterization of a novel dopamine receptor (D$_3$) as a target for neuroleptics. *Nature*, **347**, 146–51.

Soppet, D., Escandon, E., Maragos, J. *et al.* (1991) The neurotrophic factors brain-derived neurotrophic factor and neurotrophin-3 are ligands for the *trk* B tyrosine kinase receptor. *Cell*, **65**, 895–903.

Squinto, S. P., Stitt, T. N., Aldrich, T. H. *et al.* (1991) *trk* B encodes a functional receptor for brain-derived neurotrophic factor and neurotrophin-3 but not nerve growth factor. *Cell*, **65**, 885–93.

Stadel, J. M., Nambi, P., Shorr, R. G. L. *et al.* (1983) Catecholamine-induced desensitization of turkey erythrocyte adenylate cyclase is associated with phosphorylation of the β-adrenergic receptor. *Proc. Natl. Acad. Sci. USA*, **80**, 3173–7.

Steinbach, J. H. (1989) Structure and functional diversity in vertebrate skeletal muscle nicotinic acetylcholine receptors. *Annu. Rev. Physiol.*, **51**, 353–65.

Stryer, L., Hurley, J. B. and Fung, B. K. - K. (1981) Transducin: an amplifier protein in vision. *Trends Biochem. Sci.*, **6**, 245–7.

Stryer, L. and Bourne, H. R. (1986) G proteins: a family of signal transducers. *Annu. Rev. Cell Biol.*, **2**, 391–419.

Sunahara, R. K., Guan, H. -C., O'Dowd, B. F. *et al.* (1991) Cloning of the gene for a human dopamine D$_5$ receptor with higher affinity for dopamine than D$_1$. *Nature*, **350**, 614–19.

Trowbridge, I. S. (1991) CD45: a prototype for transmembrane protein tyrosine phosphatases. *J. Biol. Chem.*, **266**, 23517–20.

Ullrich, A. and Schlessinger, J. (1990) Signal transduction by receptors with tyrosine kinase activity. *Cell*, **61**, 203–12.

Van Tol, H. H. M., Bunzow, J. R., Guan, H. -C. *et al.* (1991) Cloning of the gene for a human dopamine D$_4$ receptor with high affinity for the antipsychotic clozapine. *Nature*, **350**, 610–14.

Vetter, M. L., Martin-Zanca, D., Parada, L. F. *et al.* (1991) Nerve growth factor rapidly stimulates tyrosine phosphorylation of phospholipase C-γ1 by a kinase activity associated with the product of the *trk* proto-oncogene. *Proc. Natl. Acad. Sci. USA*, **88**, 5650–4.

Wahli, W. and Martinez, E. (1991) Superfamily of steroid nuclear receptors: positive and negative regulators of gene expression. *FASEB J.*, **5**, 2243–9.

Williams, D. E., Eisenman, J., Baird, A. *et al.* (1990) Identification of a ligand for the c-kit proto-oncogene. *Cell*, **63**, 167–74.

Wisden, W., Morris, B. J., Darlison, M. G. *et al.* (1988) Localization of GABA$_A$ receptor α-subunit mRNAs in relation to receptor subtypes. *Mol. Brain Res.*, **5**, 305–10.

Wysocki, L. J. and Sato, V. L. (1978). 'Panning' for lymphocytes: a method for cell selection. *Proc. Natl. Acad. Sci. USA*, **75**, 2844–8.

Yarden, Y. and Ullrich, A. (1988) Growth factor receptor tyrosine kinases. *Annu. Rev. Biochem.*, **57**, 443–78.

Zhou, Q. -Y, Grandy, D. K., Thambi, L. *et al.* (1990) Cloning and expression of human and rat D$_1$ dopamine receptors. *Nature*, **347**, 76–80.

Zsebo, K. M., Wypych, J., McNiece, I. K. *et al.* (1990) Identification, purification and biological characterization of hematopoietic stem cell factor from Buffalo rat liver-conditioned medium. *Cell*, **63**, 195–201.

Moses V. Chao and Luis F. Parada

2.1 INTRODUCTION

Neurotrophic factors are necessary during development for the survival and proper maintenance of selective neuronal populations. The classic studies on nerve growth factor (NGF) have revealed its essential role as a survival and differentiation factor in both the peripheral and central nervous systems (Thoenen *et al.*, 1987; Levi-Montalcini, 1987). The restricted cellular targets for NGF have led to the search and identification of a family of related neurotrophic factors. Brain-derived neurotrophic factor (BDNF, Barde *et al.*, 1982; Leibrock *et al.*, 1989) and other closely related neurotrophin (NT-3, NT-4, NT-5) factors (Maisonpierre *et al.*, 1990; Hohn *et al.*, 1990; Hallbook *et al.*, 1991; Berkemeier *et al.*, 1991) display similar biological effects on distinct but overlapping populations of neuronal cells. The identification of these additional family members and their receptors will provide many insights into the mechanisms of cell survival and differentiation in the developing nervous system.

The biological actions of NGF include survival of embryonic and adult neurons and induction of neurite outgrowth and neuro-transmitter enzymes. The primary targets in the peripheral nervous system are sympathetic and populations of sensory neurons derived from the neural crest. These neurons require NGF for their survival and proper differentiation. A number of differentiative properties are induced by NGF, including increased tyrosine hydroxylase, acetyl-cholinesterase, and choline acetyltransferase activities, as well as many neuron specific proteins such as synapsin, tau, microtubule associated proteins, neurofilaments, and sodium channels (Chao, 1990).

In the central nervous system NGF is capable of influencing cholinergic neurons in the basal forebrain by the induction of choline acetyltransferase (ChAT) activity (Mobley *et al.*, 1985; Martinez *et al.*, 1985), and reducing, or even preventing, degeneration of basal forebrain neurons. Infusion of NGF into the ventricle adjacent to lesioned fimbria fornix of rodents prevented degeneration and cell death of cholinergic cell bodies (Williams *et al.*, 1986). Administration of NGF to aged rats increased the size of these neurons and also improved certain cognitive tasks (Fischer *et al.*, 1987).

An essential prerequisite for the actions of NGF is the presence of functional receptor sites on responsive cells. Two NGF receptors exist that are represented by transmembrane glycoproteins of 75 kDa ($p75^{NGFR}$) and 140 kDa ($p140^{trk}$), the product of the protooncogene *trk*, a receptor tyrosine kinase. Both receptor proteins bind NGF, and are expressed in the traditional target cells for NGF. Since all actions of NGF are initiated by binding to its cell surface receptors, expression of the two

Receptors in the Developing Nervous System Vol 1: *Growth factors and hormones* Edited by Ian S. Zagon and Patricia J. McLaughlin. Published in 1993 by Chapman & Hall. ISBN 0 412 45240 5. Vols. 1 and 2 (set) ISBN 0 412 54520 9.

NGF receptors would be expected to be tightly regulated during development and restricted to specific cell types in the nervous system. The discovery that the product of the proto-oncogene *trk* encodes a signalling receptor for NGF (Kaplan *et al.*, 1991b; Klein *et al.*, 1991a) has increased our understanding of the signal transduction mechanism for neurotrophic factors and provided an opportunity for understanding how specificity of action of neurotrophin factors are encoded in their receptors.

The interaction of NGF with its two receptors provides a unique growth factor-mediated mechanism for carrying out the crucial processes of cell maturation and differentiation in the nervous system. This chapter discusses the structural aspects and expression of the two NGF receptors with respect to the developing nervous system, and the functional roles that have been ascribed to each receptor. The neurotrophic field is a rapidly emerging area in which many new related factors and related receptors will eventually be discovered. The finding of multiple factors and two different types of neurotrophin receptors raises a challenging series of signal transduction questions and regulatory mechanisms, from which will emerge common unifying principles.

2.2   p75 NEUROTROPHIN RECEPTOR

The biologically active form of NGF is a dimer of two identical 118 amino acid polypeptides, that exists primarily as a β-stranded structure (McDonald *et al.*, 1991). The dimer is formed as a result of hydrophobic interactions. The 3-dimensional structure indicates that NGF has a long flat sheet-like structure, in which the three disulfide bonds are in close proximity. The structure resembles a 'hand', in which the fingers and the base represent hypervariable regions of amino acids that differ from other family members, including BDNF and the other neurotrophins. The structure presents many possibilities for specific interactions with NGF receptors.

In order to develop probes for the NGF receptor, monoclonal antibodies to the receptor were first generated against human melanoma cells (Ross *et al.*, 1984) and PC12 cells (Chandler *et al.*, 1984) that inhibited or altered NGF binding to these cells. These antibodies were specific either for the rat receptor (IgG-192 and RAN-1), or the human receptor (ME20.4), and recognized a glycoprotein of 75–80 kDa daltons (called p75^{NGFR}). The IgG-192 is capable of increasing binding of ^{125}I-labeled NGF to p75^{NGFR}, whereas the ME20.4 is effective at blocking binding. In addition, the IgG-192 antibody has been used to demonstrate that p75 is retrogradely transported in sciatic nerve and in neurons found in the basal forebrain (Johnson *et al.*, 1987). The IgG-192 has been shown to inhibit neurite regeneration and induction of the c-*fos* gene (Chandler *et al.*, 1984; Milbrandt, 1986), suggesting that p75^{NGFR} may be involved in multiple NGF processes.

The cloning of the p75 gene was accomplished by using total genomic DNA in gene transfer experiments, with the herpes virus thymidine kinase gene to generate transfected fibroblasts expressing the p75^{NGFR} (Chao *et al.*, 1986; Radeke *et al.*, 1987). ME20.4 and IgG-192 antibodies were used to identify positive cells by rosette analysis or the fluorescence-activated cell sorter, and the receptor genes were isolated by either subtractive hybridization (Radeke *et al.*, 1987) or by using Alu repetitive sequences as a tag for the receptor sequence (Chao *et al.*, 1986) (Table 2.1).

The p75 receptor protein, binds ^{125}I-NGF with low affinity ($K_d = 10^{-9}$ M) in transfected fibroblast cells, and is a glycosylated transmembrane protein, rich in cysteine residues and negatively charged. Of 28 cysteine residues in the predicted amino acid sequence, 24 cysteines are found in four canonical repeats in the *N*-terminal region (Fig. 2.1). Similar cysteine-rich repeats have been

Table 2.1.   Cloning of p75[NGFR] by gene transfer. Genomic DNA was transfected into mouse Ltk⁻ cells with the herpes virus thymidine kinase gene, and stable HAT-resistant clones were selected. Positive clones were identified with fluorescence-activated cell sorting (FACS, Radeke *et al.*, 1987), or red blood cell rosetting (Chao *et al.*, 1986) using monoclonal antibodies against the p75 NGF receptor

DNA	Selection	Screen	Monoclonal Antibody	Source
Rat	HAT	FACS	IgG-192	PC12
Human	HAT	Rosette	ME20.4	Melanoma

detected in the extracellular domains of a number of proteins including TNF receptors, the B cell activation molecule, CDw40, and the T cell molecule, OX40 (Smith *et al.*, 1990). A regular spacing of cysteine residues is observed in repeats found in this family of cell surface molecules. No similarities in sequence for the transmembrane and cytoplasmic domains have been detected to date. In the p75[NGFR] receptor, the cysteine repeats have been directly implicated in NGF binding (Welcher *et al.*, 1991; Yan and Chao, 1991a), and it is likely that similar domains in related cell surface molecules are used for ligand binding.

Equilibrium binding of ^{125}I-NGF to cells revealed two distinct affinity binding sites representing the NGF receptor (Sutter *et al.*, 1979; Landreth and Shooter, 1980; Schechter and Bothwell, 1981). In most responsive cells, such as sympathetic neurons or pheochromocytoma PC12 cells, approximately 10–15% of the receptors display high affinity binding with a K_d of 10^{-11} M, with the remainder of the receptors possessing a low affinity K_d of 10^{-9} M. Biological responsiveness to NGF depends on interactions with the high affinity form (Green *et al.*, 1986).

In transfected fibroblast cells, Schwann cells or melanoma cells, the p75 receptor binds ^{125}I-NGF with an equilibrium dissociation constant of 10^{-9} M (Radeke *et al.*, 1987; Hempstead *et al.*, 1989; DiStefano and Johnson, 1988). The p75 receptor does not appear to have any direct signal transduction capability in these cells. However, transfection of p75[NGFR] cDNA in cells of neural crest origin display both high and low affinity sites, and results in NGF responsiveness, such as increased tyrosine phosphorylation and induction of the c-*fos* gene (Hempstead *et al.*, 1989; Berg *et al.*, 1991). These studies indicated that the p75 receptor is a low affinity receptor that participates in high affinity sites, but only in a proper cellular context. These experiments also suggested that the p75 receptor is involved in NGF signal transduction and in the formation of high affinity NGF receptor sites, but that other signalling molecules were necessary to obtain biological responsiveness.

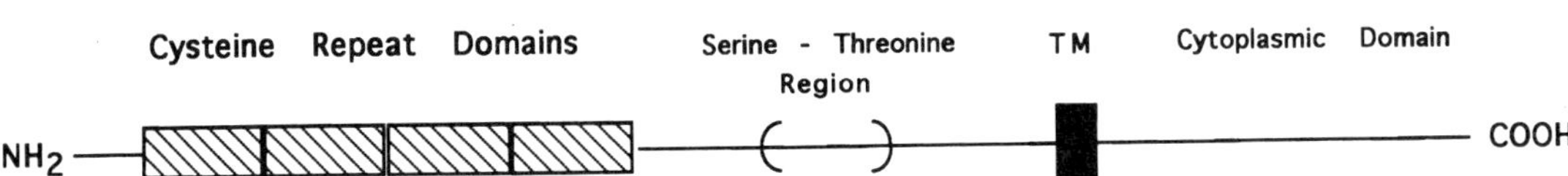

Fig. 2.1.   Structure of the p75[NGFR]. A signal peptide precedes the *N*-terminal sequence. Each cysteine repeat is negatively charged and contains 6 cysteines in invariant positions. Multiple serine and threonine residues are sites of *0*-linked glycosylation (Grob *et al.*, 1985).

The *trk* proto-oncogene, a tyrosine kinase receptor for NGF (see below), has fulfilled this role.

2.2.1   EXPRESSION OF p75[NGFR]

The differential *in vivo* pattern of p75[NGFR] gene expression has been well documented. Measurements of p75 NGF receptor levels in the peripheral and central nervous system by immunohistochemistry (Yan and Johnson, 1988; Schatteman *et al.*, 1988), hybridization to receptor mRNA (Buck *et al.*, 1987, 1988; Ernfors *et al.*, 1988; Large *et al.*, 1989; Escandon and Chao, 1989; Heuer *et al.*, 1990), and ^{125}I-NGF binding (Raivich *et al.*, 1987; Cohen-Cory *et al.*, 1989) indicate that the p75 receptor is expressed over a wide developmental time period.

An important consideration for potential functions of the p75 receptor is its distribution in responsive cells. The primary target cells for NGF in the vertebrate nervous system include neural crest derived sensory neurons, sympathetic neurons, and cholinergic neurons of the basal forebrain. The expression of the p75 receptor mRNA has been extensively studied in these regions (Buck *et al.*, 1987, 1988) and found to correlate with the period of responsiveness for each neuronal population. For example, sensory ganglia increase steady-state levels of p75 mRNA during embryonic periods, when NGF is crucially required for survival. Similarly, sympathetic ganglia depend on NGF after embryonic development, when p75 expression can be most readily detected (Buck *et al.*, 1987). Expression of receptors is highest in neuronal populations that are dependent on NGF for survival.

In the central nervous system, basal forebrain neurons express p75 receptor mRNA and protein (Hefti *et al.*, 1986; Gibbs *et al.*, 1989). Levels of receptor appear to increase during development, and can be correlated with the cholinergic phenotype by co-staining with antibodies against choline acetyltransferase. This population of neurons has been shown to respond to exogenous treatment with an increase in choline acetyltransferase activity. Similar results have been observed in the caudate putamen (Mobley *et al.*, 1985), where cholinergic neurons also undergo differentiation by neurotrophin factors.

In addition to appropriate expression of p75[NGFR] in all NGF-responsive cells, including PC12 pheochromocytoma cells, many other unexpected cell types express this cell surface receptor (Bothwell, 1990). p75[NGFR] is observed on glial cells such as Schwann cells of peripheral nerves, and in several prominent neuronal populations, such as cerebellar Purkinje cells (Yan and Johnson, 1988; Cohen-Cory *et al.*, 1989) and motor neurons (Ernfors *et al.*, 1989). A wide number of nerve sheath tumors such as neurofibromas, melanomas, and gliomas also express this receptor (Garin-Chesa *et al.*, 1988). This observation presents a paradox since NGF has no known biological effects in the majority of cases.

One explanation for the diverse pattern of expression for the p75 receptor may lie in the ability of multiple neurotrophin factors to interact with this receptor. Rodriguez-Tebar *et al.* (1990) first demonstrated that other neurotrophins such as BDNF could bind specifically to fibroblast cells expressing the p75 receptor. Other neurotrophin factors such as NT-3 can be directly crosslinked to the p75 receptor (Squinto *et al.*, 1991). Hence, p75[NGFR] may act as a receptor for all neurotrophin factors in the family. A direct prediction from neurotrophin binding to this receptor is the possibility that p75 receptor expression may reflect the site of action for other neurotrophin factors aside from NGF. This hypothesis has yet to be tested or verified.

Alternatively, the expression of p75 receptors may be indicative of another role, to act functionally as a presentation receptor (Johnson *et al.*, 1988). The expression of high levels of p75[NGFR] in Schwann cells after axotomy (Taniuchi *et al.*, 1986a) has suggested the possibility that Schwann cells used the receptor to present NGF to responsive neurons. It is unclear in this model which concentrations of receptors and NGF are necessary for this action to take place, and what role the p75 receptor would fulfill in the responsive neurons. A direct prediction would be that $p75^{NGFR}$ may be a component in a chemotatic response by NGF. This model for receptor function by $p75^{NGFR}$ has yet to be formally tested.

Other experiments have suggested that p75 may have additional roles in NGF signal transduction. Disruption of the $p75^{NGFR}$ gene by homologous recombination resulted in severe sensory deficits, including loss of CGRP, substance P fibers and heat sensitivity (Lee *et al.*, 1992). Also, expression of a chimeric receptor consisting of the extracellular domain of the EGF receptor and the transmembrane and intracellular domains of $p75^{NGFR}$ in PC12 cells gives a morphological response (Yan *et al.*, 1991). These results support transfection experiments in which introduction of p75 cDNAs in cells of neuronal origin gave rise to biological responses and high affinity binding (Hempstead *et al.*, 1989; Pleasure *et al.*, 1990; Matsushima and Bogenmann, 1990). Taken together, these results suggest that the $p75^{NGFR}$ molecule participates in signal transduction by NGF during cellular differentiation.

2.2.2   CROSSLINKING OF NGF RECEPTORS

Two crosslinking reagents have historically been used to characterize NGF receptors. The most reactive and widely used reagents have been carbodiimide compounds, such as ethyl-3-dimethylaminopropyl-carbodiimide (EDAC), which induces crosslinking of ^{125}I-NGF with the p75 receptor through amide formation to produce a species with a M_r of 100 000 (Grob *et al.*, 1983; Green and Greene, 1986; Taniuchi *et al.*, 1986b). Mouse NGF has been used for most studies of this kind since human, rat, bovine and chicken NGF receptors

bind mouse NGF specifically. The ability of mouse NGF to bind specifically to receptors of different species has been supported by the high degree of homology of the extracellular portions of the chicken (Escandon and Chao, 1989; Large *et al.*, 1989), rat (Radeke *et al.*, 1987), and human p75 receptors (Johnson *et al.*, 1986). Similarly, between chicken, mouse and human NGF genes, there is more than 80% amino acid homology with the mouse and human NGF genes.

In addition to EDAC, the use of another crosslinking agent, hydroxysuccinimidyl-4-azidobenzoate (HSAB), a lipid-soluble reagent, gave two crosslinked species; the 100 000 M_r species and an additional 160 000 M_r crosslinking species can be detected in PC12 cells after treatment (Massague *et al.*, 1981; Hosang and Shooter, 1985). This species was originally thought to represent the p75 receptor crosslinked to ^{125}I-NGF and an auxiliary

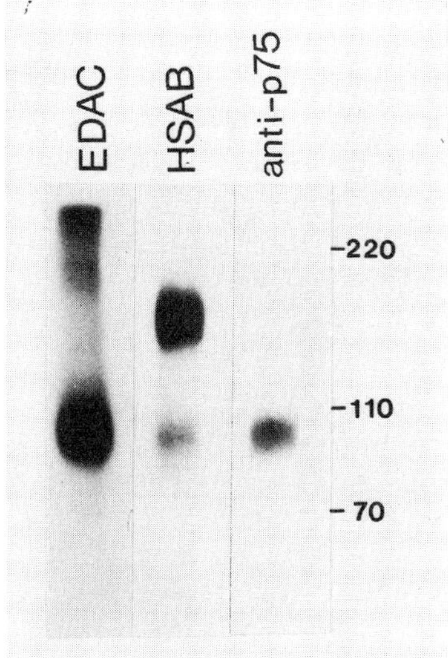

Fig. 2.2. Differential crosslinking of NGF receptors from PC12 cells. ^{125}I-NGF was incubated with PC12 cells and directly crosslinked with EDAC or HSAB (Hempstead *et al.*, 1989). Each reaction was immunoprecipitated with anti-NGF antibodies (Collaborative), except the third lane which was immunoprecipitated with anti-p75 antibodies. The 100 kDa species represents the $p75^{NGFR}$ crosslinked to NGF, and the 160 kDa species is p140^{prototrk} crosslinked to NGF.

protein of smaller size. However, antibodies made against the p75 receptor did not recognize the 160 kDa species (Fig. 2.2), indicating that the 160 kDa species represented an independent protein binding complex (Meakin and Shooter, 1991). The dramatic difference in the crosslinking pattern between EDAC and HSAB is probably due to the length of the spacer arm and the different mechanism of the crosslinking reaction. EDAC creates protein–protein crosslinks without a spacer arm, but HSAB can crosslink at a distance of 8 Å. These experiments indicate that NGF is capable of making different contacts with receptors on the cell surface.

2.3   THE NGF RECEPTOR *trk*

The higher-molecular-weight crosslinked species of 135–158 kDa identified by earlier HSAB affinity crosslinking studies is the product of the proto-oncogene *trk*. This crosslinked species had been observed only on cells which were responsive to NGF. The identity of this crosslinked species was obscure until the proto-oncogene *trk* was found to be specifically and rapidly phosphorylated on tyrosine residues by NGF treatment of PC12 cells (Kaplan *et al.*, 1991a). No other growth factors or treatment induced *trk* autophosphorylation in PC12 cells. Immunoprecipitation of the 160 kDa HSAB crosslinked species with anti-*trk* polyclonal antibodies firmly established the identity of the

crosslinked product (Kaplan *et al.*, 1991b). Experiments using anti-phosphotyrosine immunoblotting also detected *trk* in embryonic dorsal root ganglion cells. These observations led to the verification of the *trk* proto-oncogene as a signalling receptor for NGF that is activated after ligand binding.

The *trk* proto-oncogene was originally identified as a rearranged oncogene from a colon carcinoma (Martin-Zanca *et al.*, 1986), and additional *trk* oncogenes have been discovered in papillary thyroid carcinomas (Bongarzone *et al.*, 1989). *trk* encodes a receptor tyrosine kinase that exists as a transmembrane glycoprotein of mol.wt. 140 000 (p140$^{proto trk}$), capable of binding to NGF directly. The extracellular sequence of p140$^{proto trk}$ differs substantially from the p75 receptor ligand binding domain. In particular, multiple cysteine repeats are absent in p140$^{proto trk}$, and affinity crosslinking agents such as EDAC that crosslink ^{125}I-NGF efficiently to p75^{NGFR}, do not crosslink ^{125}I-labeled NGF to p140$^{proto trk}$ (Fig. 2.3). Therefore, different protein–protein interactions are made between NGF and each of the two receptors.

The product of the proto-oncogene product *trk* contains tyrosine kinase catalytic activity in its cytoplasmic domain. The *trk* gene was originally identified and cloned from NIH 3T3 cells in gene transfer experiments in which cells were transfected with high-molecular-weight DNA from human tumor cell DNA. The *trk* gene was discovered by its virtue of

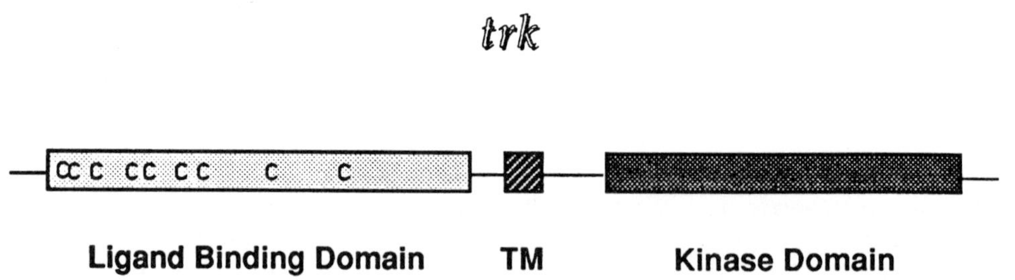

Fig. 2.3.   Structure of p140$^{proto trk}$. The *N*-terminal region consists of several leucine rich motifs and two C2 immunoglobulin-like repeats (Schneider and Schweiger, 1991), and the cytoplasmic domain contains a tyrosine kinase domain, a consensus ATP binding site and a short kinase insert (Martin-Zanca *et al.*, 1989).

inducing focus formation, and was found to be the result of a genomic rearrangement wherein tropomyosin sequences were aberrantly fused onto the transmembrane and cytoplasmic domains of *trk* (hence, the name 'tropomyosin receptor kinase'), thus producing an oncogene with malignant properties. The cellular homology of *trk* contains an extracellular domain similar in size to other growth factor receptors. The rearranged truncated *trk* oncogene product presumably unleashed its phosphorylation activities to create the transformed phenotype.

Because NGF triggers the tyrosine kinase activity of p140^{trk} in PC12 cells and in isolated embryonic dorsal root ganglia (Kaplan *et al.*, 1991b), a mechanism for initial signal transmission for NGF is provided (Maher, 1988). The discovery that a member of the tyrosine kinase family is the signal transducer for NGF simultaneously solved a number of questions that has previously stymied the neurotrophic field, including the signalling mechanism for NGF, and the way in which specificity of action is generated for the neurotrophic family of factors, including NGF and its relatives, BDNF (brain derived neurotrophic factor), and other neurotrophins (NT-3 and NT-4).

Molecules related to *trk*, such as *trkB* and *trkC*, which bind to other neurotrophic factors (BDNF and NT-3) but not NGF, have been identified (Soppet *et al.*, 1991; Squinto *et al.*, 1991; Klein *et al.*, 1991b; Cordon-Cardo *et al.*, 1991; Lamballe *et al.*, 1991). These *trk* family members bear considerable resemblance to the p140^{prototrk} molecule, not only in the consensus tyrosine kinase domain, but also in the extracellular domain. Cysteine residues are conserved among *trk*, *trkB*, and *trkC*, and there exists more than 50% conservation in the ligand binding domains. The catalytic tyrosine kinase domains contain a consensus ATP binding site and are characterized by dual tyrosine residues at the putative autophosphorylation site, and a short *C*-terminal tail. Among the family of tyrosine

kinases (Hanks *et al.*, 1988), the *trk* family shows the most similarity to the insulin receptor. Consistent with the identity of BDNF and NT-3 as ligands for these receptors is the highly specific expression of *trkB* (Klein *et al.*, 1989) and *trkC* (Lamballe *et al.*, 1991) in the central nervous system. The *trk* family of tyrosine kinase receptors are distinguished as a whole by their extremely specific expression in the developing nervous system, a trait not yet observed with any other receptor tyrosine kinases.

2.3.1 EXPRESSION OF THE *trk* PROTO-ONCOGENE

Though the *trk* gene was originally identified in a colon carcinoma tumor, the mRNA for the *trk* proto-oncogene is expressed primarily in the nervous system. RNA blot analysis of a wide variety of tissues indicated that *trk* mRNA was present at exceedingly low levels. Little expression has been noted in any other tissue, including the colon. *In situ* hybridization analysis of the developing mouse embryo showed that *trk* expression was exquisitely confined to dorsal root ganglia and also other sensory ganglia such as superior, jugular and trigeminal. Transcripts for *trk* were observed as early as embryonic day 9.5 and persisted throughout embryonic development. Ganglia not known to be responsive to NGF, such as nodose, petrosal, acoustic and facial did not express the *trk* gene during early development.

In addition to sensory ganglia, neonatal sympathetic ganglia (Schecterson and Bothwell, 1992), and neurons in the mature basal forebrain (Holtzmann *et al.*, 1992) also display *trk* mRNA expression. In the case of sympathetic neurons, very limited *trk* expression is observed during embryonic development. Increased expression has been noted in sympathetic ganglia RNA isolated at postnatal periods. This regulation in *trk* mRNA expression in sympathetic ganglia is consistent with the developmental history of the superior cervical ganglia. Sympathetic neuroblast mitosis ends approximately at E18 in the rodent, a period when there is little *trk* mRNA detected (Martin-Zanca *et al.*, 1990). Interestingly, the period of NGF dependence begins about this time, as well as higher expression of the *trk* tyrosine kinase.

Hence there is a correlation between *trk* expression and the primary neuronal targets that are known to be responsive to NGF. The expression of p140^{prototrk} in NGF responsive neurons verifies its essential part in NGF action. The requirement for *trk* in the signal transduction pathway of NGF in neuronal cells was confirmed by the introduction of the *trk* gene in mutant PC12 cells that are non-responsive to NGF. These PC12^{nnr} cells have a much lower p140^{prototrk} level, but the cells can differentiate with NGF when p140^{prototrk} is expressed after gene transfer. Therefore *trk* expression results in the recovery of morphological differentiation in these cells, showing that the expression of the *trk* molecule is essential for maintaining the neuronal phenotype (Loeb *et al.*, 1991).

2.4   NGF RECEPTOR BINDING

The finding that NGF binds directly to the *trk* tyrosine kinase to increase tyrosine phosphorylation raises an important question concerning the affinity of NGF binding to this receptor. Previous affinity crosslinking studies predicted that the higher molecular weight species constituting NGF bound to *trk* (M_r 160 kDa) represented the high affinity binding complex (Hosang and Shooter, 1985; Meakin and Shooter, 1991).

However, equilibrium binding studies with membranes isolated from fibroblast cells expressing *trk* alone demonstrate that p140^{prototrk} binds NGF with a K_d of 10^{-9} M, a value similar to the p75 NGF receptor (Kaplan *et al.*, 1991b; Hempstead *et al.*, 1991). Similar K_d values of 10^{-9} M to 10^{-10} M were found in whole cell binding studies of ^{125}I-NGF binding to *trk* at 4°C, but higher affinity binding (K_d =

10^{-11} M) was observed at 37°C (Klein *et al.*, 1991a). Since p140^{prototrk} is significantly internalized in fibroblasts at 37°C, the higher affinity binding constant measured may be due to internalized ^{125}I-NGF. Taken together, the binding studies indicate that the overwhelming majority of binding sites for NGF on p140^{prototrk} display a low affinity interaction. This observation has been extended to the *trkB* tyrosine kinase, which interacts with both BDNF and NT-3 with a K_d of 10^{-9}–10^{-10} M (Soppet *et al.*, 1991) by equilibrium binding analysis. Hence the *trk* family of tyrosine kinases appears to act as low affinity receptors for the neurotrophins.

Another method for analyzing the binding behavior is to measure binding by kinetic parameters. Analysis of time courses of ^{125}I-NGF binding to cells expressing either p75^{NGFR} or p140^{prototrk} at 4°C is an independent means of obtaining equilibrium dissociation constants. Interestingly, the on- and off-rates of ^{125}I-NGF binding differ dramatically for the two receptors, with p140^{prototrk} exhibiting a much slower off-rate than p75^{NGFR} (Meakin *et al.*, 1992). Likewise, the on-rate for NGF binding to *trk* is far slower than the binding to p75^{NGFR}.

Since the K_d is a ratio of the rate constants for dissociation and association, the measurement of these kinetic parameters verifies that p140^{prototrk} interacts with NGF with low affinity but with comparatively slow on- and off-rates (Hempstead *et al.*, unpublished results). Therefore, both p140^{prototrk} and p75^{NGFR} bind ^{125}I-NGF with a similar low affinity (10^{-9} M) K_d, implying that high affinity binding requires co-expression of both receptors.

Reconstitution studies by membrane fusion and transfection of cloned cDNAs in concert with equilibrium binding analysis indicated that expression of both receptors, p75^{NGFR} and p140^{prototrk}, is required for high affinity binding (Hempstead *et al.*, 1991) and biological function. If high affinity binding is taken as a criterion for biological action for NGF, then these findings suggest that both receptors will be co-expressed in the same cells for functional responses. Other studies, however, have indicated that NGF-induced *trk* or BDNF/NT-3-induced *trkB* tyrosine kinase activities were sufficient for cell proliferation in fibroblasts (Cordon-Cardo *et al.*, 1991; Glass *et al.*, 1991), suggesting that the biological actions mediated by neurotrophins are carried out solely by interactions with an appropriate *trk* tyrosine kinase family member.

The major biological functions of NGF in the developing nervous system, however, are to mediate cell survival and maintenance in postmitotic neuronal populations (Levi-Montalcini, 1987). That NGF exerts mitogenic properties *in vivo* is a widely held misconception; though its name implied that it acted to stimulate neuronal cell division, NGF does not actually do this (Black *et al.*, 1990).

Functional responses of *trk* family members in fibroblast cells have depended on the tyrosine kinase activities of the *trk* receptors to initiate increased cell division in heterologous cells, but have not addressed activities normally associated in cells that are postmitotic and undergoing terminal differentiation. Also, multiple neurotrophin factors are known to interact with each *trk* tyrosine kinase (Chao, 1992). Autophosphorylation of $p140^{prototrk}$ in fibroblast cells is induced by not only NGF, but also NT-3 and NT-5 (Berkemeier *et al.*, 1991). The promiscuity of neurotrophin action in fibroblasts expressing *trk* or *trkB* (Soppet *et al.*, 1991; Cordon-Cardo *et al.*, 1991; Squinto *et al.*, 1991) suggest that neurotrophin action must be examined more closely in neuronal cell environments. Hence the cellular context in which the *trk* tyrosine kinase is expressed is probably a major parameter in determining biological functions of the neurotrophin factor family.

2.5 CONCLUSIONS

The presence of two receptors in NGF responsive neurons poses new questions concerning the function of each receptor.

Why is there a need for expression of both receptors? The role of the *trk* tyrosine kinase is to act as the signalling transducing receptor for NGF. Initiation of NGF action depends on increased tyrosine phosphorylation by $p140^{prototrk}$. Lack of expression of this receptor leads to NGF unresponsiveness, indicating that $p140^{prototrk}$ function is essential for NGF signal transduction (Loeb *et al.*, 1991). The p75 receptor can bind to all neurotrophin molecules tested, and should be more appropriately considered as a neurotrophin receptor. Its functions are to establish the high affinity binding site for NGF, to participate in retrograde transport, and to modulate the tyrosine kinase activities of *trk*.

The description of *trk* expression in the developing nervous system (Martin-Zanca *et al.*, 1990) has allowed identification of the ligands for the *trk* subfamily of tyrosine kinases and defined the mechanism of NGF receptor-mediated signal transduction. The $p75^{NGFR}$ is widespread throughout development in a variety of diverse cell types (Bothwell, 1990), suggesting that this receptor must be acting in a wider manner as a receptor for the family of neurotrophins. A major unanswered question is how specificity of neurotrophin factors is determined by expression of these receptors. It is anticipated that this question will ultimately be answered by examining the pattern of differential expression of the p75 receptor and *trk* family members during development and maturity, and by determining how specificity is built into each neurotrophin–receptor interaction.

REFERENCES

Barde, Y. -A., Edgar, D. and Thoenen, H. (1982) Purification of a new neurotrophic factor from mammalian brain. *EMBO J.*, **1**, 549–53.

Berg, M. M., Sternberg, D., Hempstead, B. L. and Chao, M. V. (1991) The low affinity p75 nerve growth factor (NGF) receptor mediates NGF-induced tyrosine phosphorylation. *Proc. Natl. Acad. Sci. USA*, **88**, 7106–10.

Berkemeier, L. R., Winslow, J. W., Kaplan, D. R. *et al.*

(1991) Neurotrophin: a novel neurotrophic factor that activates *trk* and *trkB*. *Neuron*, **7**, 857–66.

Black, I. B., DiCicco-Bloom, E. and Dreyfus, C. (1990) Nerve growth factor and the issue of mitosis in the nervous system. *Curr. Topics Dev. Biol.*, **24**, 161–92.

Bongarzone, I., Pierotti, M. A., Monzini, N. *et al.* (1989) High frequency of activation of tyrosine kinase oncogenes in human papillary thyroid carcinoma. *Oncogene*, **4**, 1457–62.

Bothwell, M. A. (1990) Tissue localization of nerve growth factor and nerve growth factor receptors. *Curr. Topics Microbiol. Immunol.*, **165**, 55–70.

Bothwell, M. (1991) Keep track of neurotrophin receptors. *Cell*, **65**, 915–18.

Buck, C. R., Martinez, H., Black, I. B. and Chao, M. V. (1987) Developmentally regulated expression of the nerve growth factor receptor gene in the periphery and brain. *Proc. Natl. Acad. Sci. USA*, **84**, 3060–3.

Buck, C. R., Martinez, H., Chao, M. V. and Black, I. B. (1988) Differential expression of the nerve growth factor receptor gene in multiple brain areas. *Dev. Brain Res.*, **44**, 259–68.

Chandler, C. E., Parsons, L. M., Hosang, M. and Shooter, E. M. (1984) A monoclonal antibody modulates the interaction of nerve growth factor with PC12 cells. *J. Biol. Chem.*, **259**, 6882–9.

Chao, M. V. (1990) Nerve growth factor, in *Peptide Growth Factors and Their Receptors* (eds M. B. Sporn and A. B. Roberts), Springer-Verlag, Berlin, pp. 135–65.

Chao, M. V. (1992) Neurotrophin receptors: a window into neuronal differentiation. *Neuron*, **9**, 583–93.

Chao, M. V., Bothwell, M. A., Ross, A. H. *et al.* (1986) Gene transfer and molecular cloning of the human NGF receptor. *Science*, **232**, 418–21.

Cohen-Cory, S., Dreyfus, C. F. and Black, I. B. (1989) Expression of high- and low-affinity nerve growth factor receptors by Purkinje cells in the developing rat cerebellum. *Exp. Neurol.*, **105**, 104–9.

Cordon-Cardo, C., Tapley, P., Jing, S. *et al.* (1991) The *trk* tyrosine protein kinase mediates the mitogenic properties of nerve growth factor and neurotrophin-3. *Cell*, **66**, 173–83.

DiStefano, P. S. and Johnson, E. M. (1988) Nerve growth factor receptors on cultured rat Schwann cells. *J. Neurosci.*, **8**, 231–41.

Ernfors, P., Hallbook, F., Ebendal, T. *et al.* (1988) Developmental and regional expression of β-nerve growth factor receptor mRNA in the chick and rat. *Neuron*, **1**, 983–96.

Ernfors, P., Henschen, A., Olson, L. and Persson, H. (1989) Expression of nerve growth factor receptor mRNA is developmentally regulated and increased after axotomy in rat spinal cord motorneurons. *Neuron*, **2**, 1605–13.

Escandon, E. and Chao, M. V. (1989) Developmental expression of the chicken nerve growth factor receptor gene during brain morphogenesis. *Dev. Brain Res.*, **47**, 187–96.

Fischer, W., Wictorin, K., Bjorklund, A. *et al.* (1987) Amelioration of cholinergic neuron atrophy and spatial memory impairment in aged rats by nerve growth factor. *Nature*, **329**, 65–8.

Garin Chesa, P., Rettig, W. J., Thomson, T. M. *et al.* (1988) Immunohistochemical analysis of nerve growth factor expression in normal and malignant human tissues. *J. Histochem. Cytochem.*, **36**, 383–9.

Gibbs, R. B., McCabe, J. T., Buck, C. R. *et al.* (1989) Expression of the NGF receptor in the rat forebrain detected with *in situ* hybridization and immunohistochemistry. *Mol. Brain Res.*, **6**, 275–87.

Glass, D. J., Nye, S. H., Hantzopoulos, P. *et al.* (1991) *trkB* mediates BDNF/NT-3-dependent survival and proliferation in fibroblasts lacking the low affinity NGF receptor. *Cell*, **66**, 405–13.

Green, S. H. and Greene, L. A. (1986) A single M_r=103,000 ^{125}I-β-nerve growth-factor-affinity-labeled species represents both the low and high affinity forms of the nerve growth factor receptor. *J. Biol. Chem.*, **261**, 15316–26.

Green, S. H., Rydel, R. E., Connolly, J. L. and Greene, L. A. (1986) PC12 cell mutants that possess low- but not high-affinity nerve growth factor receptors neither respond to nor internalize nerve growth factor. *J. Cell Biol.*, **2**, 830–43.

Grob, P. M., Berlot, C. H. and Bothwell, M. A. (1983) Affinity labeling and partial purification of nerve growth factor receptors from rat pheochromocytoma and human melanoma cells. *Proc. Natl. Acad. Sci. USA*, **80**, 6819–23.

Grob, P. M., Ross, A. H., Koprowski, H. and Bothwell, M. A. (1985) Characterization of the human melanoma nerve growth factor receptor. *J. Biol. Chem.*, **260**, 8044–9.

Hallbook, F., Ibanez, C. F. and Persson, H. (1991) Evolutionary studies of the nerve growth factor family reveal a novel member abundantly expressed in *Xenopus* ovary. *Neuron*, **6**, 845–58.

Hanks, S. K., Quinn, A. M. and Hunter, T. (1988) The protein kinase family: conserved features and deduced phylogeny of the catalytic domains. *Science*, **241**, 42–52.

Hefti, F., Hartikka, J., Salvatierra, A. *et al.* (1986) Localization of nerve growth factor receptors on cholinergic neurons of the human basal forebrain. *Neurosci. Lett.*, **69**, 37–41.

Hempstead, B. L., Schleifer, L. S. and Chao, M. V. (1989) Expression of functional nerve growth factor receptors after gene transfer. *Science*, **243**, 373–5.

Hempstead, B. L., Martin-Zanca, D., Kaplan, D. *et al.* (1991) High affinity NGF binding requires co-expression of the *trk* proto-oncogene and the low affinity NGF receptor. *Nature*, **350**, 678–83.

Heuer, J. G., Fatemie-Nainie, S., Wheeler, E. F. and Bothwell, M. (1990) Structure and developmental expression of the chicken NGF receptor. *Dev. Biol.*, **137**, 287–304.

Hohn, A., Leibrock, J., Bailey, K. and Barde, Y. -A. (1990) Identification and characterization of a novel member of the nerve growth factor/brain-derived neurotrophic factor family. *Nature*, **334**, 339–41.

Holtzman, D. M., Li,Y., Parada, L. F. *et al.* (1992) p140^{trk} mRNA marks NGF-responsive forebrain neurons: evidence that *trk* gene expression is induced by NGF. *Neuron*, **9**, 465–78.

Hosang, M. and Shooter, E. M. (1985) Molecular characteristics of nerve growth factor receptors on PC12 cells. *J. Biol. Chem.*, **260**, 655–62.

Johnson, D., Lanahan, A., Buck, C. R. *et al.* (1986) Expression and structure of the human NGF receptor. *Cell*, **47**, 545–54.

Johnson, E. M., Taniuchi, M., Clark, H. B. *et al.* (1987) Demonstration of the retrograde transport of nerve growth factor receptor in the peripheral and central nervous system. *J. Neurosci.*, **7**, 920–8.

Johnson, E. M., Taniuchi, M. and DiStefano, P. S. (1988) Expression and possible function of nerve growth factor receptors on Schwann cells. *Trends Neurosci.*, **11**, 299–304.

Kaplan, D. R., Martin-Zanca, D. and Parada, L. F. (1991a) Tyrosine phosphorylation and tyrosine kinase activity of the *trk* proto-oncogene product induced by NGF. *Nature*, **350**, 158–60.

Kaplan, D. R., Hempstead, B. L., Martin-Zanca, D. *et al.* (1991b) The *trk* proto-oncogene product: a signal transducing receptor for nerve growth factor. *Science*, **252**, 554–558.

Klein, R., Parada, L. F., Coulier, F. and Barbacid, M. (1989) *trkB*, a novel tyrosine protein kinase receptor expressed during mouse neural development. *EMBO J.*, **8**, 3701–9.

Klein, R., Jing, S., Nanduri, V. *et al.* (1991a) The *trk* proto-oncogene encodes a receptor for nerve growth factor. *Cell*, **65**, 189–97.

Klein, R., Nanduri, V., Jing, S. *et al.* (1991b) The *trkB* tyrosine protein kinase is a receptor for brain-derived neurotrophic factor and neurotrophin-3. *Cell*, **66**, 395–403.

Lamballe, F., Klein, R. and Barbacid, M. (1991) *trkC*, a new member of the *trk* family of tyrosine protein kinases, is a receptor for neurotrophin-3. *Cell*, **66**, 967–79.

Landreth, G. E. and Shooter, E. M. (1980) Nerve growth factor receptors on PC12 cells: ligand-induced conversion from low- to high-affinity states. *Proc. Natl. Acad. Sci. USA*, **77**, 4751–5.

Large, T. H., Weskamp, G., Helder, J. C. *et al.* (1989) Structure and developmental expression of the nerve growth factor in the chicken nervous system. *Neuron*, **2**, 1123–34.

Lee, K. F., Li, A., Huber, L. *et al.* (1992) Targeted mutation of the gene encoding the low affinity NGF receptor p75 leads to deficits in the peripheral nervous system. *Cell*, **69**, 737–49.

Leibrock, J., Lottspeich, F., Hohn, A. *et al.* (1989) Molecular cloning and expression of brain-derived neurotrophic factor. *Nature*, **341**, 149–52.

Levi-Montalcini, R. (1987) The nerve growth factor thirty-five years later. *Science*, **237**, 1154–62.

Loeb, D., Martin-Zanca, D., Chao, M. V. *et al.* (1991) The *trk* proto-oncogene rescues NGF responsiveness in mutant NGF-nonresponsive PC12 cell lines. *Cell*, **66**, 961–6.

Maher, P. A. (1988) Nerve growth factor induces protein-tyrosine phosphorylation. *Proc. Natl. Acad. Sci. USA*, **85**, 6788–91.

Maisonpierre, P. C., Belluscio, L., Squinto, S. *et al.* (1990) Neurotrophin-3: a neurotrophic factor related to NGF and BDNF. *Science*, **247**, 1446–51.

Martinez, H. J., Dreyfus, C. F., Jonakait, G. M. and Black, I. B. (1985) Nerve growth factor selectively increases cholinergic markers but not neuropeptides in rat basal forebrain in culture. *Brain Res.*, **412**, 295–301.

Martin-Zanca, D., Hughes, S. H. and Barbacid, M. (1986) A human oncogene formed by the fusion of truncated tropomyosin and protein tyrosine kinase sequences. *Nature*, **319**, 743–77.

Martin-Zanca, D., Oskam, R., Mitra, G. *et al.* (1989) Molecular and biochemical characterization of the human *trk* proto-oncogene. *Mol. Cell. Biol.*, **9**, 24–33.

Martin-Zanca, D., Barbacid, M. and Parada, L. F. (1990) Expression of the *trk* proto-oncogene is restricted to the sensory cranial and spinal ganglia of neural crest origin in mouse development. *Genes Dev.*, **4**, 683–8.

Massague, J., Guillette, B. J., Czech, M. P. *et al.* (1981) Identification of a nerve growth factor receptor protein in sympathetic ganglia membranes by affinity labeling. *J. Biol. Chem.*, **256**, 9419–24.

Matsushima, H. and Bogenmann, E. (1990) NGF induces neuronal differentiation in neuroblastoma cells transfected with the NGF receptor cDNA. *Mol. Cell. Biol.* **10**, 5015–20.

McDonald, N. Q., Lapatto, R., Murray-Rust, J. *et al.* (1991) A new protein fold revealed by a 2.3 Å resolution crystal structure of nerve growth factor. *Nature*, **354**, 411–14.

Meakin, S. O. and Shooter, E. M. (1991) Molecular investigation on the high affinity nerve growth factor receptor. *Neuron*, **6**, 153–63.

Meakin, S. O., Suter, U., Drinkwater, C. C. *et al* (1992) The rat *trk* proto-oncogene product exhibits properties characteristic of the slow NGF receptor. *Proc. Natl. Acad. Sci. USA*, **89**, 2374–8.

Milbrandt, J. (1986) Nerve growth factor rapidly induces c-fos mRNA in PC12 rate pheochromocytoma cells. *Proc. Natl. Acad. Sci. USA*, **83**, 4789–93.

Mobley, W. C., Rutkowski, J. L., Tennekoon, G.I. *et al.* (1985) Choline acetyltransferase activity in striatum of neonatal rats increased by nerve growth factor. *Science*, **229**, 284–7.

Pleasure, S. J., Reddy, U. R., Venkatakrishnan, G. *et al.* (1990) Introduction of NGF receptors in a medulloblastoma cell line results in expression of high- and low-affinity NGF receptors but not NGF-mediated differentiation. *Proc. Natl. Acad. Sci. USA*, **87**, 8496–500.

Radeke, M. J., Misko, T. P., Hsu, C. *et al.* (1987) Gene transfer and molecular cloning of the rat nerve growth factor receptor. *Nature*, **325**, 593–7.

Raivich, G., Zimmermann, A. and Sutter, A. (1987) Nerve growth factor (NGF) receptor expression in chicken cranial development. *J. Comp. Neurol.*, **256**, 229–45.

Rodriguez-Tebar, A., Dechant, G. and Barde, Y. -A. (1990) Binding of brain-derived neurotrophic factor to the nerve growth factor receptor. *Neuron*, **4**, 187–92.

Ross, A. H., Grob, P., Bothwell, M. A. *et al.* (1984) Characterization of nerve growth factor receptor in neural crest tumors using monoclonal antibodies. *Proc. Natl. Acad. Sci. USA*, **81**, 6681–5.

Schatteman, G. C., Gibbs, L., Lanahan, A. A. *et al.* (1988) Expression of NGF receptor in the developing and adult primate central nervous system. *J. Neurosci.*, **8**, 860–73.

Schechter, A. L. and Bothwell, M. A. (1981) Nerve growth factor receptors on PC12 cells: evidence for two receptor classes with differing cytoskeletal association. *Cell*, **24**, 867–74.

Schecterson, L. C. and Bothwell, M. A. (1992) Novel roles for neurotrophins are suggested by BDNF and NT-3 mRNA expression in developing neurons. *Neuron*, **9**, 867–74.

Schneider, R. and Schweiger, M. (1991) A novel modular mosaic of cell adhesion motifs in the extracellular domains of the neurogenic *trk* and *trkB* tyrosine kinase receptors. *Oncogene*, **6**, 1807–11.

Smith, C. A., Davis, T., Anderson, D. *et al.* (1990) A receptor for tumor necrosis factor defines an unusual family of cellular and viral proteins. *Science*, **248**, 1019–23.

Soppet, D., Escandon, E., Maragos, J. *et al.* (1991) The neurotrophic factors, brain-derived neurotrophin factor and neurotrophin-3 are ligands for the *trkB* tyrosine kinase receptor. *Cell*, **65**, 895–903.

Squinto, S. P., Stitt, T. N., Aldrich, T.H. *et al.* (1991) *trkB* encodes functional receptor for brain-derived neurotrophic factor and neurotrophin-3 but not nerve growth factor. *Cell*, **65**, 885–93.

Sutter, A., Riopelle, R. J., Harris-Warrick, R. M. and Shooter, E. M. (1979) NGF receptors: characterization of two distinct classes of binding sites on chick embryo sensory ganglia cells. *J. Biol. Chem.*, **254**, 5972–82.

Taniuchi, M., Clark, H. B. and Johnson, E. M. (1986a) Induction of nerve growth factor in Schwann cells after axotomy. *Proc. Natl. Acad. Sci. USA*, **83**, 4094–8.

Taniuchi, M., Schweitzer, J. B. and Johnson, E. M. (1986b) Nerve growth factor receptor molecules in rat brain. *Proc. Natl. Acad. Sci. USA*, **83**, 1950–4.

Thoenen, H., Bandtlow, C. and Huemann, R. (1987) The physiological function of nerve growth factor in the central nervous system: comparison with the periphery. *Rev. Physiol. Biochem. Pharmacol.*, **109**, 145–78.

Welcher, A. A., Bitler, C. M., Radeke, M. J. and Shooter, E. M. (1991) Nerve growth factor binding domain of the nerve growth factor receptor. *Proc. Natl. Acad. Sci. USA*, **88**, 159–63.

Williams, L. R., Varon, S., Peterson, G. M. *et al.* (1986) Continuous infusion of nerve growth factor prevents basal forebrain neuronal death after fimbria fornix transection. *Proc. Natl. Acad. Sci. USA*, **83**, 9231–5.

Yan, H. and Chao, M. V. (1991) Disruption of cysteine-rich repeats of the p75 nerve growth factor receptor leads to loss of ligand binding. *J. Biol. Chem.*, **266**, 12099–104.

Yan, H. and Johnson, E. M. (1988) An immunohistochemical study of the nerve growth factor receptor in developing rats. *J. Neurosci.*, **8**, 3481–98.

Yan, H., Schlessinger, J. and Chao, M. V. (1991) Chimeric NGF and EGF receptors define domains responsible for neuronal differentiation. *Science*, **252**, 561–3.

OPIOID GROWTH FACTOR RECEPTOR IN THE DEVELOPING NERVOUS SYSTEM

Ian S. Zagon and Patricia J. McLaughlin

3.1 INTRODUCTION

An endogenous opioid system involved in the growth of normal (Zagon and McLaughlin, 1983a, 1983b, 1984a) and abnormal (Zagon and McLaughlin, 1983c, 1984b) cells and tissues was first discerned in the early 1980s. In recent years, identification of the opioid growth factor involved with development of the nervous system (Zagon and McLaughlin, 1991), as well as with the biology of neural cancers (Zagon and McLaughlin, 1989a), has been established. The receptor associated with the opioid growth factor has been identified in both the developing nervous system (Zagon *et al.*, 1991a) and in neural tumor cells (Zagon *et al.*, 1989a, 1990a). This receptor has been characterized by function, biochemistry, and subcellular location as a unique opioid receptor termed zeta (ζ). The major focus of this chapter is to review information about the zeta receptor in the developing nervous system. The history and function of the opioid growth factor has been included in order to provide a more complete picture of the zeta receptor.

3.2 OPIOID GROWTH FACTOR

3.2.1 ESTABLISHMENT OF THE PRINCIPLES OF OPIOID PEPTIDE ACTION ON NERVOUS SYSTEM DEVELOPMENT: OPIOID ANTAGONISM PARADIGMS

Opioids have been important in medicine, as well as in the social setting, for thousands of years (Blum, 1969; Terry and Pellens, 1970; Musto, 1973). The discovery of opioid receptors in the nervous system by three independent laboratories in 1973 (Pert and Snyder, 1973; Simon *et al.*, 1973; Terenius, 1973), and the subsequent finding in 1975 of endogenous opioids (Hughes, 1975; Hughes *et al.*, 1975), have stimulated considerable activity in understanding the role of opioids and opioid receptors in neurobiology. In the early 1980s, our research group posed the question of whether endogenous opioids and opioid receptors are related to the growth of the nervous system (Zagon and McLaughlin, 1983a,b, 1984a). Two approaches were contemplated to explore the potential role of endogenous opioid systems in ontogeny. The first was to subject developing animals to opioid peptides and examine the repercussions using defined developmental parameters. Since opioid peptides have a short half-life in plasma, and we also had no information as to which opioid peptide to utilize, this approach was not pursued initially. An alternative paradigm was designed in our laboratory which involved the interruption of endogenous opioid–opioid receptor interaction with opioid antagonists. If opioids are related to development, act tonically, and interface with a receptor, a biological effect on growth should be revealed when the receptor is occupied by an opioid antagonist. Both naloxone and naltrexone are opioid antagonists (Blumberg and Dayton, 1974) which have considerable

Receptors in the Developing Nervous System Vol. 1: *Growth factors and hormones* Edited by Ian S. Zagon and Patricia J. McLaughlin. Published in 1993 by Chapman & Hall. ISBN 0 412 45240 5. Vols. 1 and 2 (set) ISBN 0 412 54520 9.

affinity for opioid receptors and induce what is referred to as an 'opioid receptor blockade'. We chose to utilize the latter antagonist because naltrexone, in particular, is a long-acting and extremely potent opioid antagonist which has been regarded as a 'pure' antagonist to the many biological actions of opioid substances and is devoid of significant intrinsic activity (Blumberg and Dayton 1974; Sawynok *et al.*, 1979). These compounds have been very useful in investigating the biology of opioid peptides and receptors. In postnatal rat pups which grow from 6 g to 45 g in the first three postnatal weeks, a daily injection of naltrexone capable of blocking opioid receptors for a 24 h period (yet only 2% of the LD_{50}) produced a dramatic acceleration in somatic and neurobiological development (e.g., Zagon and McLaughlin, 1983a, b, 1984a). For example, at weaning (day 21), animals given a total opioid receptor blockade weighed 15–20% more than control rats (Zagon and McLaughlin, 1984a). The interpretation of these results along with other studies on brain and organ development was that endogenous opioids must be involved with growth, and that one or more opioid peptides serve as negative regulators. Moreover, these investigations inferred that opioid peptides related to growth must operate in a tonic fashion, since interruption of the interaction of opioids and receptors stimulated development. To examine whether drug dosage rather than the blockade of receptors was associated with opioid action on growth, an experiment was designed that utilized low dosages of naltrexone given multiple times a day in order to achieve a continuous occupation of opioid receptors (Zagon and McLaughlin, 1984a). The results of these experiments showed that the duration of the opioid receptor blockade, rather than the dosage of the opioid antagonist, determined the course of developmental events. In order to fully secure a foundation of opioid receptor participation in growth, the stereospecific nature of opioid action required documentation. A classic pharmacological test of opioid receptor involvement has been to demonstrate stereospecificity of opioid activity (e.g., Pert and Snyder, 1973; Sawynok *et al.*, 1979). Although isomeric forms of naltrexone were not available, stereoisomers of another opioid antagonist, naloxone, were obtained. Using the active (–) and inactive (+) isomers of naloxone, the stereospecific mechanism of opioid-related influences on growth was revealed (Zagon and McLaughlin, 1989b). The (–) isomer of naloxone, but not the (+) isomer, was found to alter the course of development. In addition to confirming the pharmacological property of stereospecificity, experiments with naloxone also showed that the effects of naltrexone seen in the initial reports were not peculiar to naltrexone itself, but were related to opioid antagonism.

With the hypothesis articulated that endogenous opioid systems participate in growth, and evidence in hand that firmly supported this postulate, studies could be designed that employed the opioid antagonist paradigm to explore the role of opioid–opioid receptor interaction in development, with special attention directed towards the nervous system. Investigations concerning the ramifications of complete opioid receptor blockade showed that not only was body weight increased by this treatment, but so were the weights of most organs in these rats (Zagon and McLaughlin, 1983b, 1985a, 1989b). Calculations of relative organ weight (i.e., ratio of organ to body weight) indicated changes in organ wet weights of naltrexone-treated animals that were generally proportional to body weight changes (Zagon and McLaughlin, 1985a). Opioid antagonist administration was found to influence the ontogeny of physical characteristics and behavioral development (Zagon and McLaughlin, 1983b, 1985b). For example, on the day that all the animals given a complete opioid receptor blockade opened their eyes (i.e., day 15), only one-half of the control animals had their eyes open. Yet another example concerned the development of walking, considered to be a behavioral milestone. Walking occurred 2 days earlier in the group subjected to a total opioid receptor blockade than in controls.

A series of histological, morphometric, and ultrastructural studies were conducted to establish more fully the influence of endogenous opioid systems in brain development (Hess and Zagon, 1988; Zagon and McLaughlin, 1986a, b). By using neural tissues from 21-day-old rats that received an opioid receptor blockade (i.e., naltrexone) for the first 3 weeks of postnatal life, detailed analyses of the cerebellum, cerebral cortex, and hippocampus were undertaken. Highlights of the findings included: (a) opioid modulation is not restricted to a particular brain region; (b) cellular differentiation of both prenatally and postnatally derived neural elements are influenced by opioids during the preweaning period; (c) opioid modulation during the preweaning period may affect the total population of cells derived postnatally, but not prenatally; (d) the morphogenesis of both neurons and glia is governed by opioid peptides; and (e) endogenous opioids serve as an inhibitory factor in developmental processes. For example, in the cerebellum, a brain region that undergoes dramatic postnatal development, morphometric, histological, histochemical, and quantitative analysis of sections revealed that a complete receptor blockade during the preweaning period significantly increased macroscopic and areal dimensions, the number and size of some neural cells, and histotypic organization. Companion studies (Hauser *et al.*, 1987, 1989) in which neural tissues were impregnated with silver and dendrite elaboration and spine formation examined, showed a number of remarkable morphogenetic events influenced by opioid peptides. Observation of pyramidal cells from frontoparietal cortex (layer III) and hippocampal field CA1, and cerebellar Purkinje cells, indicated region-dependent alterations in dendrite complexity and/or spine concentrations in all regions, with complete opioid receptor blockade stimulating

ontogeny. In the hippocampus total basilar dendritic length increased 52% in animals given an appropriate naltrexone regimen with respect to control levels, and increases of 33% and 31% in mean segment length and number of branches, respectively, were observed in 10-day-old animals treated with opioid antagonists. Moreover, the concentration of basilar and oblique spines in 10-day-old animals subjected to a total opioid receptor blockade from birth were 69% and 83% greater than control levels, respectively. These data indicate that endogenous opioid peptides are critical regulators of neuronal differentiation, and control growth through inhibitory channels. Whether the influence on differentiation was direct, or secondary due to repercussions because of changes in the population of germinative brain cells, is a question that remains unanswered. The profound developmental effects in rats chronically subjected to naltrexone were not the result of changes in the ultrastructure of the cells (Hess and Zagon, 1988). Thus, the regulatory abilities of endogenous opioids utilize mechanisms that do not induce overt anatomical changes but rather employ basic developmental processes (e.g., cell proliferation).

The greater number of cells in developing animals receiving complete opioid receptor blockade during the preweaning period suggested that opioids were involved in controlling the proliferation of neural cells. Closer examination of this mechanism using thymidine incorporation to monitor DNA synthesis was in order. Utilizing the external germinal layer of the cerebellum in 6-day-old rats, opioid receptor blockade produced a significant elevation in the number of radio-labeled cells (labeling index) in this region of highly proliferative neural cells (Zagon and McLaughlin, 1987a). These data indicated that opioids must actively inhibit cell proliferation in the developing brain since interference of opioid–receptor interaction results in an increase in the labeling index. An interesting observation in these experiments was that

DNA synthesis was significantly altered within an hour of opioid antagonist administration, suggesting the intimate and vital control of cell proliferation exerted by opioid peptides. Up to this time, our experiments relied on opioid antagonist paradigms to understand the influence of opioids on developmental processes. In a first attempt to inspect the effects of endogenous opioid peptides on growth, 6-day-old rats were subjected to 80 μg/kg of [Met^5]-enkephalin, one of the first endogenous opioid peptides discovered (Hughes, 1975; Hughes *et al.*, 1975). This peptide significantly depressed cell proliferation of external germinal cells in the cerebellum. To investigate whether [Met^5]-enkephalin operated at the level of the opioid receptor with respect to cell replication, some animals received injections of both the opioid peptide and an opioid antagonist, naloxone. No changes in the labeling index were recorded between control animals and those given both the peptide and antagonist or the antagonist alone, indicating that opioid receptors mediated opioid action.

A number of conclusions could be drawn from these experiments with the opioid antagonist paradigm and the developing nervous system. Firstly, blockade of endogenous opioids from opioid receptors showed that one or more of these peptides participated in the regulation of developmental processes in the brain. Secondly, opioid peptide action obeyed pharmacological principles of stereospecificity and involvement of opioid receptors. Thirdly, opioid peptides appeared to be associated with cell proliferation as well as cell differentiation.

3.2.2   IDENTIFICATION OF THE OPIOID GROWTH FACTOR

An important question that needed to be addressed was which opioid peptide(s) is(are) related to neural ontogeny. To examine this matter directly, an assay system was devised which took advantage of the fact that opioid peptides have a profound effect on replicating cells. Using the external germinal layer of 6-day-old rats and thymidine uptake combined with autoradiography, the effects of a wide variety of natural and synthetic opioids were investigated (Zagon and McLaughlin, 1991). A key element in these experiments was to administer the peptide acutely and examine the repercussions on DNA synthesis within a few hours so that degradation products and/or untoward reactions of the compounds were not confounding influences. The results (Fig. 3.1) showed that concentrations as low as 0.1 mg/kg of the naturally occurring pentapeptide [Met^5]-enkephalin, as well as the longer molecule and related opioid, peptide F, were extremely potent in depressing cell replication. Experiments with a synthetic peptide in which methionine was omitted, [Des-Met^5]-enkephalin, also showed a reduction in DNA synthesis, suggesting that even a breakdown product of [Met^5]-enkephalin had an effect on thymidine incorporation. Interestingly, a wide variety of other opioids, including many that were selective for the μ, δ, κ, σ and ε receptors, did not influence DNA synthesis. The lack of effect of μ-related compounds on DNA synthesis was not surprising in view of earlier findings using β-funaltrexamine (β-FNA), a highly selective and irreversible μ opioid receptor antagonist (Zagon and McLaughlin, 1986c). Animals given β-FNA from birth to day 21 did not show any alterations in body weight, brain and cerebellar weights, macroscopic dimensions of the brain, the area of the cerebellum, or in organ weight.

Further dose–response experiments showed that dosages of [Met^5]-enkephalin as low as 80 μg/kg altered DNA synthesis. In order to determine whether the opioid peptide(s) involved with proliferation of neuronal cells also governed the generation of glial cells, a population of glial cells (astrocytes and oligodendrocytes) in the medullary layer of the 6-day-old rat cerebellum was assessed for labeled thymidine incorporation (Zagon and McLaughlin, 1991) (Fig. 3.1). The profile of

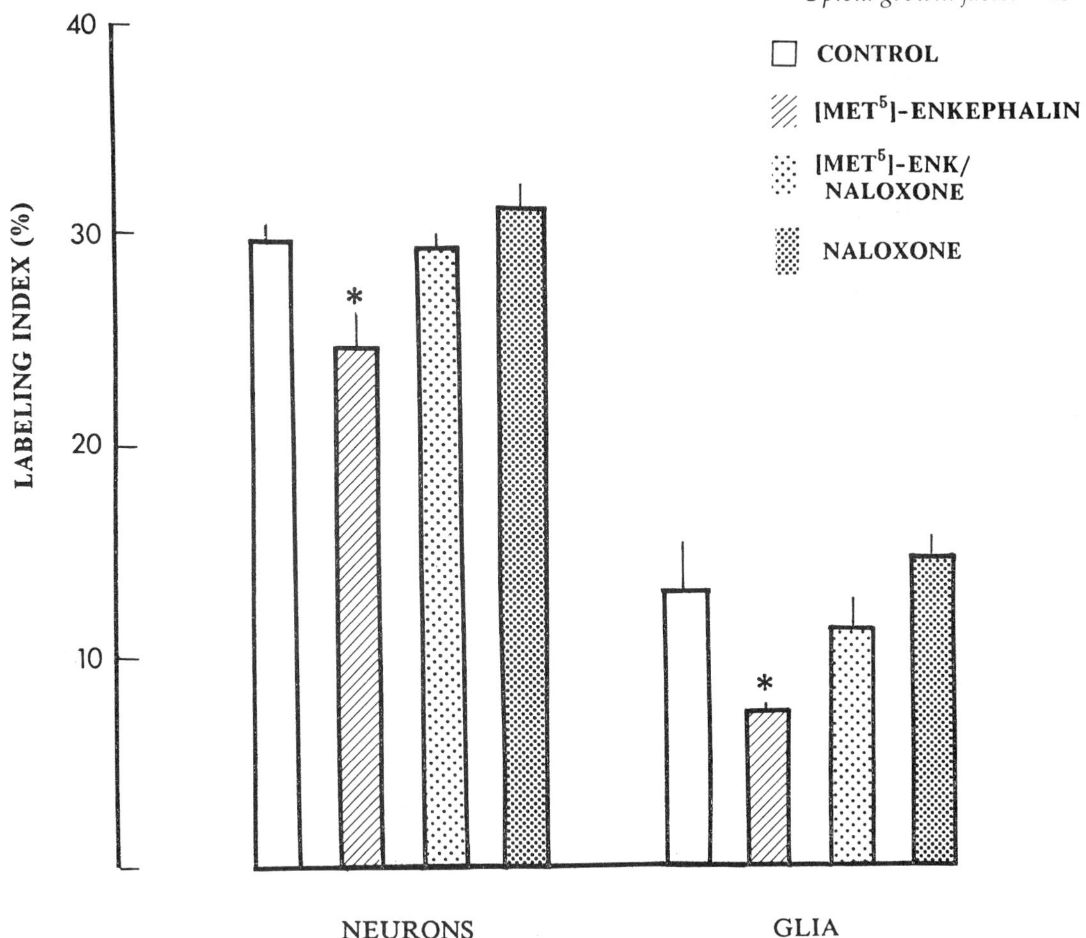

Fig. 3.1. The effects of [Met⁵]-enkephalin (100 µg/kg), [Met⁵]-enkephalin (100 µg/kg) and naloxone (1 mg/kg), or naloxone (1 mg/kg) on the labeling index of neural cells in the external germinal (granule) layer and medullary layer of the 6-day-old rat cerebellum. Rats were injected with drugs and given an injection of [³H]thymidine 30 min before being killed. Tissues were harvested at 4 h after injection of the opioid. Controls received sterile water. Bars = SEM. Significantly different from controls at *$P < 0.05$. (Reproduced from Zagon and McLaughlin, 1991, with permission of the publisher.)

compounds that reduced cell proliferation in glial precursors was similar to that observed earlier in replicating cells destined to be neurons. However, the magnitude of depression in labeling index for glia (43% of control values) was greater than that recorded for neurons (24% of control levels). In experiments exploring DNA synthesis in both neurons and glia, concomitant administration with naloxone eliminated the inhibitory effects of [Met⁵]-enkephalin (Fig. 3.1). Consistent with the results obtained using opioid antagonists, DNA synthesis was extremely sensitive to opioids, with [Met⁵]-enkephalin significantly depressing thymidine incorporation within 2 h of drug administration. Finally, the action of [Met⁵]-enkephalin was not restricted to germinative cells of the cerebellum. Similar results showing that this opioid peptide could depress DNA synthesis in the neonatal rat retina, and in a naloxone-reversible manner, were reported by Isayama *et al.* (1991).

3.2.3 LOCALIZATION OF THE OPIOID GROWTH FACTOR

Interpretation of these data suggested that the opioid growth factor targeted proliferative cells, since DNA synthesis was dramatically altered by this compound. The distribution of the opioid growth factor in the developing brain required elucidation, and studies localizing the opioid growth factor in germinal cells of a variety of tissues have been reported (Zagon *et al.*, 1985, 1986; Zagon and McLaughlin, 1990; Isayama *et al.*, 1991). Using a polyclonal antibody to [Met^5]-enkephalin, light and electron microscopy were employed to examine immunoreactivity. Peptide was associated with proliferating and differentiating cells of the cerebellum (Zagon *et al.*, 1985; Zagon and McLaughlin, 1990) (Fig. 3.2), as

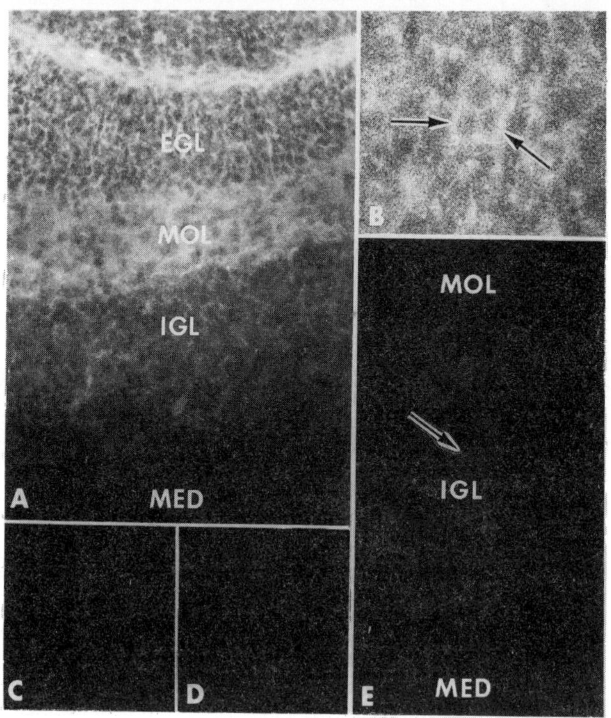

Fig. 3.2.   Photomicrographs showing the location of [Met^5]-enkephalin immunoreactivity, using fluorescent secondary antibodies, in the cerebellum of 10- and 35-day-old rats. (A) The cerebellum of a 10-day-old rat. Note the bright staining of the external germinal (granule) layer (EGL), and the low to moderate staining of the molecular layer (MOL), internal granule layer (IGL), and medullary layer (MED). Blood vessels in the pia are often extremely fluorescent. (B) A higher magnification photomicrograph of external germinal cells in the EGL of 10-day-old rat cerebellum showing an intensely stained cortical cytoplasm (arrows) encircling a non-fluorescent nucleus. (C) and (D) Control sections of the 10-day-old rat cerebellum stained with [Met^5]-enkephalin antiserum absorbed with [Met^5]-enkephalin (C) or processed with serum from unimmunized animals (D). Both (C) and (D) are of the EGL (identified by phase optics) and times of exposure and printing correspond to those in (A). (E) Cerebellum of 35-day-old rat. Except for lightly fluorescent endothelial cells of blood vessels (arrow), little staining was observed. The times of exposure and printing are similar to those in (A). (Reproduced from Zagon *et al.*, 1985, with permission of the publisher.)

well as the retina (Isayama *et al.*, 1991). [Met⁵]-enkephalin immunoreactivity, however, was not found in the cerebellum or retina of adult animals. In the developing cerebellar cortex the peptide was associated with external germinal (granule) cells, being located in the cortical cytoplasm as a mesh-work throughout the cell, but not in the cell nucleus. With electron microscopy (Zagon and McLaughlin, 1990), immunoreactivity was found to be concentrated in the cytoplasm of external germinal cells of the developing rat cerebellum. Within the cytoplasm, immunostaining aggregates were seen subjacent to the plasma membrane, and associated with the surface of the nuclear envelope. Staining was prominent in Purkinje neurons. In addition to cytoplasmic staining, immunoreactivity was also found within vesicles and associated with 20–26 nm microtubule-like elements. Reaction product was prominent in the synaptic spines of the dendrites, where it was often located near postsynaptic densities. No immunostaining product was associated with cell nuclei. The cerebellar cortex of 45-day-old rats exhibited reduced immunoreactivity, with only the soma of basket/stellate cells and Purkinje neurons retaining reaction product. In adulthood no immunoreactivity was detected in the cerebellum. These results are consistent with the suggestion that an opioid growth factor regulates cellular replication. The finding that opioid growth factor was also located in a macroneuron, the Purkinje cell, which had its birthday prenatally in the rat, suggested that this growth factor might function earlier than the postnatal period.

3.2.4 GENE EXPRESSION OF OPIOID GROWTH FACTOR PROHORMONE

Given the significance that an opioid growth factor controlled some features of neural development, the source of this opioid peptide remained an important question. [Met⁵]-enkephalin is derived from proenkephalin A (Comb *et al.*, 1982; Gubler *et al.*, 1982; Noda *et al.*, 1982). An experiment was designed using the neonatal rat retina and an oligonucleotide probe for preproenkephalin A (Isayama and Zagon, 1991). Examination of *in situ* hybridization preparations showed that pre-proenkephalin mRNA was found in neuroblast cells, precursors to the cells in the retina, as well as ganglion cells which were derived prenatally (Fig. 3.3). These results indicated that both proliferating cells and large macroneurons (i.e., ganglion cells) produced the mRNA for the prohormone responsible for the opioid growth factor. Studies described earlier using immunocytochemistry demonstrated that [Met⁵]-enkephalin was associated with both cell types, indicating that the mRNA was, indeed, translated. Interpretation of these data suggested that the source of the opioid growth factor in developing tissues such as the retina appears to be autocrine (e.g., retinal neuroblasts) and paracrine (e.g., ganglion cells) in nature.

3.2.5 OTHER STUDIES CORROBORATING AND EXTENDING THE CONCEPT OF AN OPIOID GROWTH FACTOR

The observations about the influence of opioid antagonists on neural growth, as well as the action of [Met⁵]-enkephalin on neural developmental events, have been corroborated and extended by a number of scientists. In some cases the investigators were in pursuit of compounds related to growth; in other instances their observations were secondary to different intentions (e.g., studying the ontogeny of a neurotransmission function). Using tissue culture paradigms, [Met⁵]-enkephalin has been found to be inhibitory to the proliferation of glia (Stiene-Martin and Hauser, 1990, 1991; Hauser and Stiene-Martin, 1991), whereas [Leu⁵]-enkephalin has been reported to be a negative neuronal growth regulatory factor in cultured serotonergic neurons (Davila-Garcia and Azmitia, 1989). These studies, along with an extensive literature about the influence of

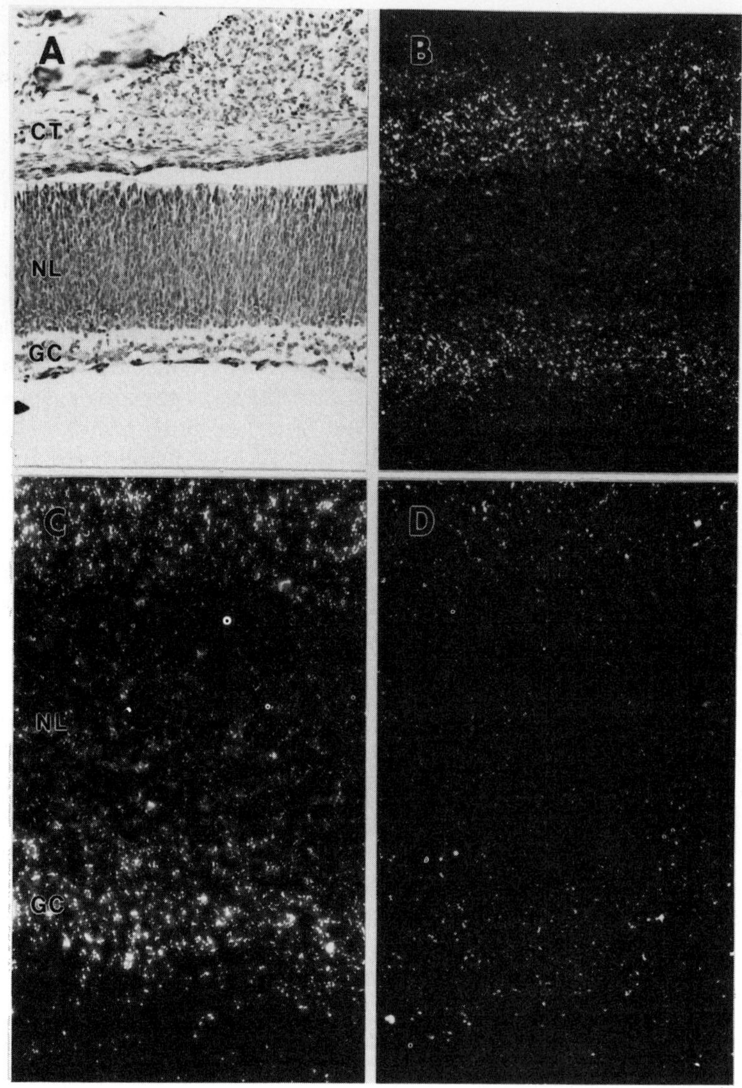

Fig. 3.3. An *in situ* hybridization study of the 1-day-old rat retina using an oligonucleotide probe to preproenkephalin A (PPE). (A) Bright-field photomicrograph indicating the different layers of the neonatal retina. GC = ganglion cell layer; NL = neuroblast layer; CT = connective tissue. (B) Dark-field micrograph of emulsion-coated autoradiogram showing the labeling in the GC, NL, and CT. Note that the concentration of PPE message found associated with the neuroblasts of the NL appears to be less abundant than that related to the ganglion cells. (C) Higher magnification micrograph of the retina showing the abundance of PPE mRNA in the GC and the CT. (D) Dark-field micrograph of an RNase A-pretreated control eye displaying a lack of labeling. (Reproduced from Isayama and Zagon, 1991, with permission of the publisher.)

opioid antagonists and agonists on growth of neural and non-neural tumor cells in tissue culture (see Zagon *et al.*, 1982, 1984, 1989a, b) support the concept from observations *in vivo* about the direct action of these compounds on developmental processes. It should be mentioned that in contradistinction to the majority of other findings, Ilyinsky *et al.* (1987) have reported a growth-stimulating effect in the organotypic cultures of neural tissue using a tissue culture system; however, these effects occurred in a non-opioid receptor manner (i.e., opioid antagonists did not block the growth-related influence of opioid peptides). Investigating the possibility that neurotransmitter compounds may have a regulatory role in the neurogenesis of the brain, Vertes *et al.* (1982) discovered that naloxone produced an increased rate of [^{3}H]thymidine incorporation in the forebrain and hypothalamus at 1 and 3 h after drug administration in rats, with a significant reduction at 9 and 12 h postdrug injection; no effect was observed in the cerebellum. In a second part of this study, [D-Met^2,Pro^5]-enkephalamide, a long-acting synthetic enkephalin analog, reduced DNA synthesis in the cerebral cortex and hypothalamus during a 12 h postinjection period, but only influenced the cerebellum for the first hour after drug administration. Schmahl *et al.* (1989) examined the effects of the opioid antagonist, naltrexone, on cell proliferation in the developing rat brain and found a long-lasting increase in the mitotic rate of the forebrain ependymal layer. These workers have also discovered that naltrexone given during early brain development has reparative growth effects on the injured nervous system (Schmahl *et al.*, 1987). Finally, the effects of opioid antagonists on neurobehavioral development in rats have been examined using naloxone (Vorhees, 1981; Najam and Panskepp, 1989) or naltrexone (Paul *et al.*, 1978; Harry and Rosecrans, 1979).

The presence and distribution of the opioid growth factor in proliferative neural cells has been widely documented. Enkephalin immunoreactivity has been reported in a number of tissue culture studies, including those on mouse spinal cord (Neale *et al.*, 1978), rat brain (Knodel and Richelson, 1980; Weyhenmeyer *et al.*, 1980), and astrocytes in organotypic explants of rat cerebellum

(Hauser *et al.*, 1990). Tsang *et al.* (1982b) have carefully studied the ontogenesis of enkephalin in brain by radioimmunoassay. Moreover, gene expression of proenkephalin A, the prohormone that the opioid growth factor is derived from, has been examined in the developing nervous system by a number of investigators using Northern analysis or *in situ* hybridization; these include Spruce *et al.* (1990), Vilijn *et al.* (1988), Schwartz and Simantov (1988), Melner *et al.* (1990), Hauser *et al.* (1990), Di Scala-Guenot *et al.* (1990), Laurent-Huck *et al.* (1991), Rosen and Polakiewicz (1989), Osborne *et al.* (1991), Tecott *et al.* (1989), Morita *et al.* (1990), and Shinoda *et al.* (1989).

3.2.6 OTHER ISSUES CONCERNING THE OPIOID GROWTH FACTOR

A number of issues and concerns related to the opioid growth factor should be mentioned. The identification of [Met^5]-enkephalin as the opioid growth factor does not mean that this peptide cannot serve in a different capacity (e.g., neurotransmitter) nor does it preclude that other opioid peptides may have an action on developmental events. Hypothetically, the ontogeny of synapses related to an opioid receptor could be radically altered if an abundance or lack of a related opioid peptide important to its function is present during synaptogenesis. Thus, disturbances in the ontogeny of other opioid receptors, and associated peptides, may have ramifications that signify a modulatory role in developmental neurobiology. However, it should be kept in mind that the opioid growth factor is an intrinsic force that is present only during growth, and exerts a profound influence on ontogeny.

Another important question to be addressed is whether there is more than one opioid growth factor. Some reports have mentioned β-endorphin (Bartolome *et al.*, 1986, 1987, 1990; Lorber *et al.*, 1990) and other opioid peptides (Stiene-Martin and Hauser,

1991) as potential factors in determining growth-related events. The relationship of these compounds to endogenous opioid–opioid receptor mechanisms, and to criteria for inclusion as an opioid growth factor must be approached cautiously and necessitates documentation of a variety of criteria. Some of these criteria include: (a) stereospecificity, (b) sensitivity to opioid antagonists, (c) activity at physiologically relevant concentrations, (d) effects demonstrated within a relatively short time if cell replication is implicated, (e) cross-reactivity to other receptor-selective compounds, (f) relationship to a naturally occurring opioid peptide, (g) *in vivo* activity, (h) spatial and temporal distribution consistent with specific growth-related effects, (i) lack of cross-reactivity to known opioid growth factor(s), (j) effectiveness by acute, as well as chronic, application, (k) elimination of confounding variables such as physical dependence, tolerance, and/or withdrawal, and (l) presence in the tissue of a related opioid receptor. To date, reports of candidates other than [Met^5]-enkephalin have fallen short of qualifying as an opioid growth factor that is mediated by an opioid receptor. Rigorous, systematic, and extensive experiments will be needed to support such claims.

3.2.7   OPIOID GROWTH FACTOR: SUMMARY

In summary, studies from a number of laboratories provide compelling evidence that an opioid growth factor is present in, and important to, the development of the nervous system. This opioid growth factor exerts an inhibitory influence, operates at physiologically relevant concentrations, and is especially targeted to cell proliferative events. The generation of both neurons and glia are governed by this peptide. Opioid action insofar as growth is concerned operates through classical opioid receptor mechanisms, being stereospecific and blocked by opioid antagonists (e.g., naloxone). Moreover, the growth-related activity of opioids is under

tonic control since receptor blockade by potent opioid antagonists such as naltrexone results in an elevation of the number of cells undergoing DNA synthesis. The opioid peptide involved with growth has been identified as [Met^5]-enkephalin; opioids selective for μ, δ, κ, σ, and ε receptors do not influence cell proliferation after acute exposure. This opioid growth factor is a naturally occurring pentapeptide, derived from proenkephalin A, and produced by replicating cells as well as neighboring macroneurons. The spatial and temporal distribution of opioid growth factor is transient in appearance, with a distinct relationship to proliferating and, to a lesser degree, differentiating cells. The opioid growth factor governs developmental processes under *in vivo* and *in vitro* conditions

3.3   THE OPIOID GROWTH FACTOR RECEPTOR: ZETA

Endogenous opioid systems have two major components: endogenous opioid peptides and opioid receptors. With the identification of the opioid growth factor, [Met^5]-enkephalin, and evidence that this peptide interfaced with an opioid receptor, determination of which opioid receptor was involved with proliferation of neural cells was pursued. The ontogeny of opioid receptors (e.g., μ, δ, κ) has been examined in the developing nervous system by a number of authors (e.g., Clendeninn *et al.*, 1976; Coyle and Pert, 1976; Tsang and Ng, 1980; Leslie *et al.*, 1982; Tsang *et al.*, 1982a, b; Spain *et al.*, 1985; Tavani *et al.*, 1985; Kornblum *et al.*, 1987; Petrillo *et al.*, 1987; Barg *et al.*, 1989; Barg and Simantov, 1989, 1991; Kinney *et al.*, 1990; Zagon *et al.*, 1990b; Kinney and White, 1991; Rius *et al.*, 1991; see also the review by McDowell and Kitchen, 1987, and bibliographies by Zagon *et al.*, 1982, 1984, 1989b). These earlier studies were focused on understanding how opioid receptors, related to neurotransmission, develop in the brain. Few studies mentioned the role of opioid receptors in the regulation of growth. In one of the earliest investigations about opioid receptors and the developing nervous system, Tsang and Ng (1980) used radiolabeled [Met[5]]-enkephalin to probe developing rat brain. This study was performed at a time when ligands selective for different opioid receptors were not available and these workers chose [Met[5]]-enkephalin because it was a naturally occurring opioid peptide (Hughes *et al.*, 1975). Moreover, since the identification of [Met[5]]-enkephalin as an opioid growth factor was not reported until recently, presumably Tsang and Ng had no indication that it regulated growth and governed cell replication. Using preparations of the rat cerebellum, these investigators found that binding to [Met[5]]-enkephalin was transiently expressed, with the greatest binding occurring within the first week of life and declining to low levels thereafter. At that time the reason for transient expression of binding in the rat cerebellum was unclear, but the information was provocative.

With this background in mind, and with the knowledge that ligands selective for known opioid receptors such as μ, δ, κ, ε and σ do not alter growth of cells either *in vivo* (Zagon and McLaughlin, 1991) or *in vitro* (Zagon and McLaughlin, 1989a), our strategy was to use the opioid growth factor, [Met[5]]-enkephalin, as a probe to study the binding site associated with brain growth (Zagon *et al.*, 1991a). Initially, we chose to study the developing rat cerebellum, a region of the brain that enjoyed attention earlier with respect to defining how opioid peptides regulate growth. Not only is the developmental pattern of the cerebellum well known, but over 90% of the cerebellar cells originate in the preweaning period (Zagon and McLaughlin, 1979), thereby providing a rich source of proliferative cells that contain opioid growth factor receptors. The approach to conducting binding assays was systematic, since assay conditions for the radiolabeled opioid growth factor needed to be established. Utilizing the cerebellum of 6-day-

old rats, studies on tissue concentration and [3H][Met[5]]-enkephalin indicated that binding to cerebellar homogenates was linear for protein concentrations from 0.2 to 1 mg/ml (Fig. 3.4). Binding was time dependent, with equilibrium occurring at 60 min (Fig. 3.5). The interactions of radiolabeled [Met[5]]-enkephalin were found to be a function of temperature, with 22°C being optimal (Fig. 3.5). Boiling of cerebellar homogenates for 20 min reduced binding to negligible levels. Binding of [3H][Met[5]]-enkephalin was dependent on pH, with a pH of approximately 7.4 being optimal (Fig. 3.6). To achieve optimal binding of [3H][Met[5]]-enkephalin, a cocktail of selective agents was required (Table 3.1). Omission of these agents (EGTA, thiorphan, leupeptin, phenylmethylsulfonyl fluoride (PMSF)) resulted in approximately 78% reduction in binding compared to optimal conditions. No differences in binding activity between fresh tissue and specimens frozen up to 30 days were detected.

Under optimal conditions in terms of time, protein concentration, temperature, and pH, [3H][Met[5]]-enkephalin binding to cerebellar homogenates was found to be saturable and

Table 3.1. The effects of selective agents on the binding of [3H][Met[5]]-enkephalin to cerebellar homogenates of 6-day-old rats. The binding of [3H][Met[5]] - enkephalin to (1.0 nM) was assayed with EGTA (1 mM), thiorphan (6 nM), leupeptin (1 μg/ml), and/or PMSF (3.5 mM). Values are means ± SEM for two experiments. (Reproduced from Zagon *et al.*, 1991a, with permission of the publisher.)

Inhibitor(s)	Binding (%)
EGTA + PMSF + thiorphan + leupeptin	100
EGTA + PMSF + thiorphan	89 ± 3
EGTA + PMSF	75 ± 6
Thiorphan + leupeptin	59 ± 8
Thiorphan + PMSF	54 ± 4
PMSF + leupeptin	54 ± 7
EGTA + thiorphan	31 ± 5
None of the above	22 ± 4

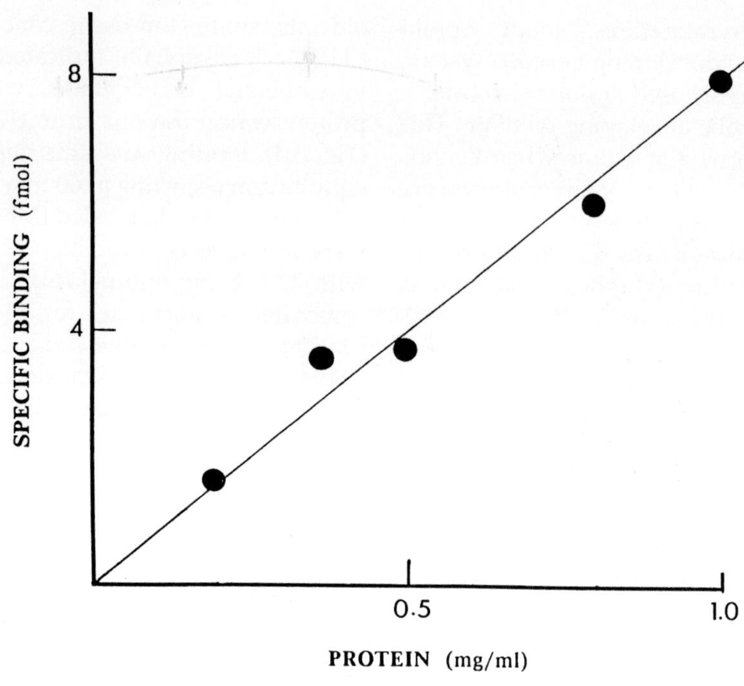

Fig. 3.4.   The dependence of specific [³H][Met⁵]-enkephalin binding to cerebellar homogenates of 6-day-old rats on the concentration of protein. The reaction was initiated by incubating various concentrations of cerebellar homogenate protein with 0.4 nM-[³H][Met⁵]-enkephalin for 60 min at 22°C (pH 7.4) followed by filtration and determination of radioactivity. Points (means ± SEM) represent duplicate determinations from each of three assays. Non-specific binding has been subtracted. (Reproduced from Zagon *et al.*, 1991a, with permission of the publisher.)

to have a high affinity (Fig. 3.7). Computer analysis of binding showed that the data best fit a one-site model. The equilibrium dissociation constant (K_d) for [³H][Met⁵]-enkephalin was 2.2 ± 0.3 nM and the binding capacity (B_{max}) was 22.3 ± 1.4 fmol/mg protein. Analysis of cerebella from male and female rats revealed no differences in binding affinity or capacity for [³H][Met⁵]-enkephalin.

A number of studies were performed that provided further information about the nature of the opioid growth factor receptor. Incubation of homogenates with 1M KCl indicated that all the binding was associated with the insoluble fraction, suggesting that the binding site to [Met⁵]-enkephalin was an integral membrane protein rather than a peripheral membrane protein. Studies on the effects of proteolysis of cerebellar homo-genates were conducted by incubating prep-arations with trypsin, terminating the reaction with trypsin inhibitor, and perform-ing binding assays. The results showed that trypsin decreased specific binding of [³H][Met⁵]-enkephalin by 67% of control levels; trypsin inhibitor alone had no influence on binding. These studies suggested that the opioid growth factor receptor was proteinaceous in character.

To examine the specificity and relationship of radiolabeled [Met⁵]-enkephalin to its bind-ing site, competition experiments utilizing a range of opioid agonists and antagonists were performed (Table 3.2). [Met⁵]-enkephalin exhibited the greatest potency of any of the compounds tested. The binding site was recog-nized by opioid antagonists such as naltrexone and (–)-naloxone. Competition studies also

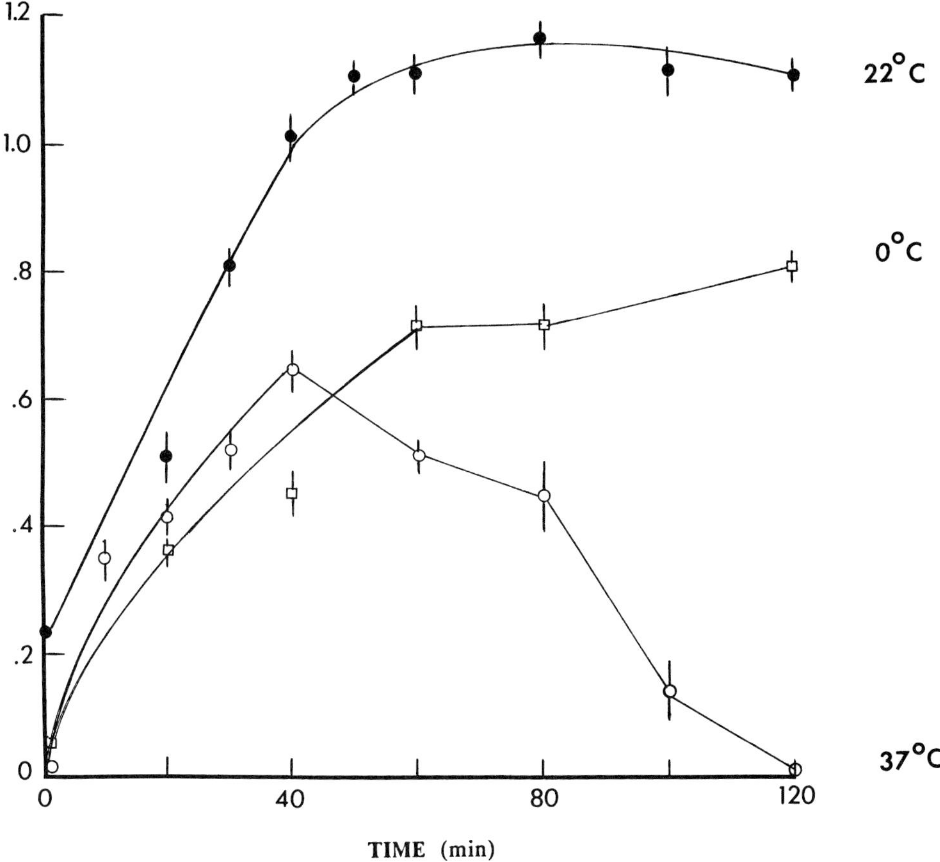

Fig. 3.5.   The dependence of [^{3}H][Met^5]-enkephalin binding to cerebellar homogenates of 6-day-old rats on the time and temperature of incubation. The reaction was initiated by adding cerebellar homogenate protein to 0.4 nM-[^{3}H][Met^5]-enkephalin at either 0, 22 or 37°C for varying periods of time, followed by filtration and determination of radioactivity. Points (means ± SEM) represent duplicate determinations from each of two assays. Non-specific binding has been subtracted. (Reproduced from Zagon *et al.*, 1991a, with permission of the publisher.)

showed that this binding site distinguished stereoisomers, with (−)-naloxone being about 1000-fold more potent than (+)-naloxone in displacing [^{3}H][Met^5]-enkephalin. Ligands selective for μ ([D-Ala^2, MePhe^4, Gly-ol^5]-enkephalin (DAGO), β-FNA, morphine), δ ([D-Pen^{2,5}]-enkephalin (DPDPE), ICI 174,864), κ (U50,488, U69,593, ethylketocyclazocine (EKC), dynorphin A$_{1-13}$), ∈ (β-endorphin), and σ and related (SKF-10,047, haloperidol) binding sites exhibited no distinct pattern of potency. Individually, however, some of these compounds (e.g., DAGO, morphine) were quite potent in displacing [^{3}H][Met^5]-enkephalin. Interestingly, some proenkephalin A products (e.g., [Met^5]-enkephalin, [Met^5, Arg^6, Phe^7]-enkephalin, [Met^5, Arg^6, Gly^7, Leu^8]-enkephalin) had considerable ability to displace [^{3}H][Met^5]-enkephalin, whereas others (e.g., BAM12P, peptide F) did not. These studies revealed that a variety of compounds had some affinity to the [Met^5]-enkephalin binding site. Placed together with earlier data about the effects of opioid

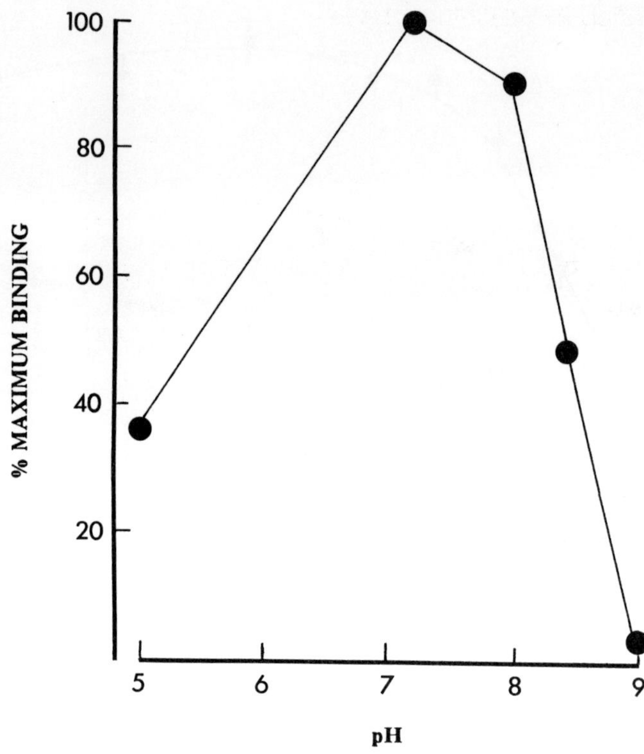

Fig. 3.6.   The dependence of $[^3H][Met^5]$-enkephalin binding on pH of cerebellar homogenates of 6-day-old rats. The reaction was initiated by adding cerebellar homogenate protein to 1 nM-$[^3H][Met^5]$-enkephalin at 22°C for 60 min. Points (means ± SEM) represent duplicate determinations from each of three assays. Non-specific binding has been subtracted.

peptides on cell proliferation, it is interesting to note that there is no special correlation between peptides that actively competed with $[Met^5]$-enkephalin and those that influenced growth. For example, peptide F significantly depressed DNA synthesis at concentrations as low as 100 μg/kg, but its K_i was 36 nM compared to dynorphin A (1–13), which had an estimated K_i of 4.7 nM yet did not change DNA synthesis at concentrations as great as 10 mg/kg. Of course, these *in vivo* studies which identified the opioid growth factor were conducted 4 h after drug administration, with the animals studied shortly (30 min) after $[^3H]$thymidine administration. This paradigm was chosen to minimize toxicity and confounding interpretations by lengthy or multiple exposure to thymidine, and to

avoid chronic exposure which introduces the effects of physical dependence and withdrawal and may confuse interpretation of the results. The effects of chronic opioid use would be interesting to examine from the standpoint of cell replication and interaction with the $[Met^5]$-enkephalin binding site as well as other opioid receptors. However, one must be extremely cautious in interpreting data that suggest an opioid peptide influences growth only when long-term paradigms of drug exposure are employed.

In order to elucidate further the relationship of $[Met^5]$-enkephalin binding to other opioid receptor types in the developing rat cerebellum, binding assays using radiolabeled ligands selective for the μ (DAGO), δ (DPDPE), and κ (U69,593) receptors were conducted

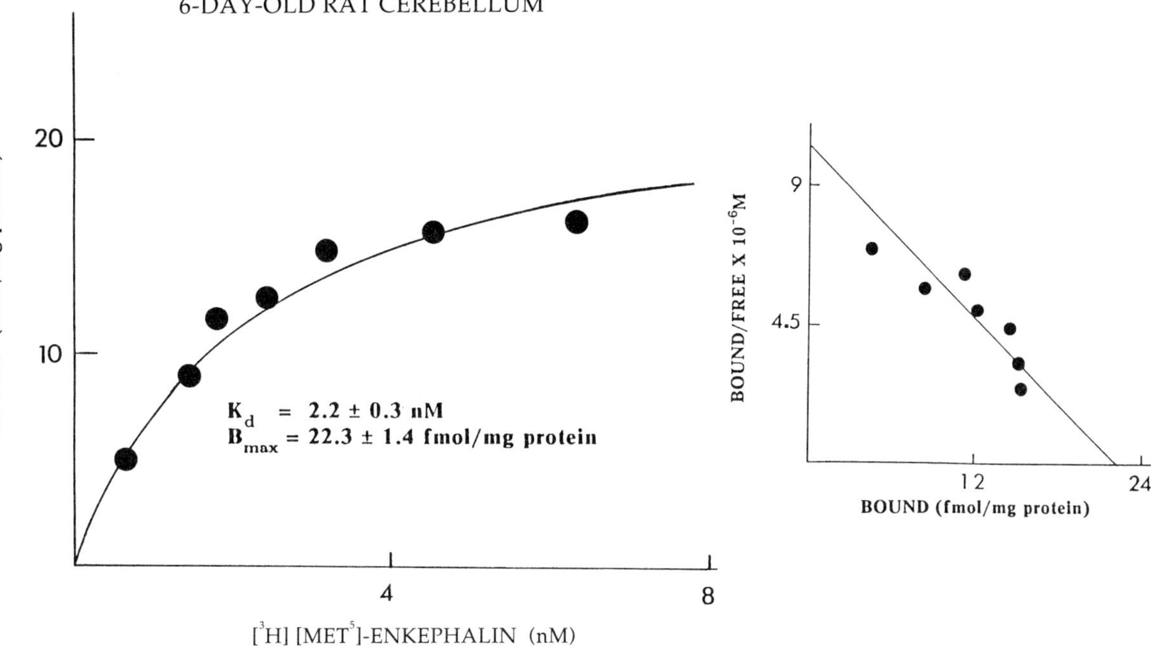

Fig. 3.7.   Representative saturation binding isotherm and Scatchard plot (inset) of the specific binding of [^{3}H][Met^5]-enkephalin to cerebellar homogenates of 6-day-old rats. (Reproduced from Zagon *et al.*, 1991a, with permission of the publisher.)

with homogenates of 6-day-old rat cerebellum. The results of these binding assays are presented in Table 3.3 and indicate specific and saturable binding to [^{3}H]DAGO and [^{3}H]-labeled U69,593, but not to [^{3}H]DPDPE. A one-site model of binding was delineated by computer analysis for both [^{3}H]DAGO and [^{3}H]U69,593. Competition studies using unlabeled [Met^5]-enkephalin indicated a K_i of 7.3 ± 3.6 nM for [^{3}H]DAGO and >10^{-5}M for [^{3}H]U69,593. These data would suggest that if binding assays for the opioid growth receptor are performed with a ligand concentration below 5 nM, there would be little cross-reactivity with κ receptors and minimal interaction with μ receptors. It is interesting to note that in earlier studies by Barg and Simantov (1989) using [^{3}H]diprenorphine and unlabeled DAGO, DPDPE, or U50,488 to examine μ, δ, and κ receptors, respectively, the greatest binding in the cerebellum was

recorded during the first week of postnatal life, although an increase in binding per mg of protein was noted throughout ontogeny. Therefore, both our study and that of Barg and Simantov concur that μ and κ binding exists in the cerebellum of animals approximately 1-week old. We did not find δ receptors in the developing cerebellum, which can probably be explained by differences in binding conditions and the radiolabeled ligands utilized.

Evidence from functional identification studies suggested that [Met^5]-enkephalin was the opioid growth factor because other peptides selective for opioid receptors such as μ, δ, κ, ε, and σ did not influence growth. This would imply that μ, δ, κ, ε, and σ receptors may not play a role in governing growth. Binding assays by our group (Zagon *et al.*, 1991a), as well as Tsang and Ng (1980) and Tsang *et al.* (1982a, b), indicated that [Met^5]-enkephalin binding sites could be

Table 3.2. Potency of opioid and non-opioid compounds to compete for the binding of [³H][Met⁵]-enkephalin (2 nM). Values represent the means ± SEM from at least two experiments. (Reproduced from Zagon *et al.*, 1991a, with permission of the publisher.)

Ligand	K_i (nM)
[Met⁵]-enkephalin	3.2 ± 0.7
Morphine	4.2 ± 0.7
DAGO	4.6 ± 0.5
[Met⁵ (O)]-enkephalin	4.7 ± 0.5
Dynorphin A_{1-13}	4.7 ± 0.5
[Met⁵, Arg⁶, Phe⁷]-enkephalin	4.9 ± 2.0
[Met⁵, Arg⁶, Gly⁷, Leu⁸]-enkephalin	8.7 ± 3.1
β-endorphin	12 ± 6
(–)-naloxone	21 ± 0
Naltrexone	21 ± 0
β-funaltrexamine	24 ± 12
Methadone	27 ± 2
Ethylketocyclazocine	31 ± 4
[Des-Tyr¹, Met⁵]-enkephalin	35 ± 9
Peptide F	36 ± 15
[Leu⁵]-enkephalin	42 ± 5
ICI 174, 864	42 ± 4
BAM12P	47 ± 5
[D-Ala², D-Leu⁵]-enkephalin	49 ± 3
[Des-Met⁵]-enkephalin	52 ± 12
U69, 593	169 ± 35
Heroin	195 ± 117
[D-Pen²,⁵]-enkephalin	331 ± 86
SKF-10, 047	$>10^{-6}$ M
U50, 488	$>10^{-6}$ M
(+)-naloxone	$>10^{-5}$ M
Haloperidol	$>10^{-5}$ M

detected in developing tissues. However, our studies and those of Barg and Simantov (1989) did demonstrate that μ and κ, and perhaps δ, receptors (according to Barg and Simantov), are present in cerebellum of 6-day-old animals. Our competition studies also revealed that ligands selective for other opioid receptors could interact with the [Met⁵]-enkephalin binding site. Conceivably, it could be argued that the binding of [Met⁵]-enkephalin was to one of these known receptors, especially the μ receptor, since [Met⁵]-enkephalin had an affinity of 7.3 nM.

Thus, it would be difficult to design experiments to quench binding to some of these receptors and perform a [Met⁵]-enkephalin binding assay to examine for sites related to [Met⁵]-enkephalin as the opioid growth factor, since quenching compounds also cross-react to the [Met⁵]-enkephalin binding site. This would have the effect of occupying most, if not all, of the binding sites for [Met⁵]-enkephalin, even when concentrations as low as 10 nM were used (e.g., μ ligands).

To resolve this problem, we decided to examine the subcellular location of the [Met⁵]-enkephalin binding site. Previous subcellular fractionation studies (Pert and Snyder, 1973; Pert *et al.*, 1974) have demonstrated that other opioid receptors were associated predominantly with the membrane fraction. In work to this point, the cerebellar homogenate examined was prepared by centrifugation at moderate speeds (e.g., 39 000 *g*), a procedure commonly employed in opioid receptor analysis. Our laboratory had been previously engaged in studies examining the nuclei of brain cells (Zagon and McLaughlin, 1979). At that time, we had utilized a scheme of subcellular fractionation that capitalized on differentiation of cellular elements by sucrose gradients. Discontinuous sucrose gradients were also used in earlier studies on determining the subcellular location of the opioid

Table 3.3. The binding affinity (K_d) and binding capacity (B_{max}) of [³H]DAGO, [³H]DPDPE, [³H] U69, 593, and [³H][Met⁵]-enkephalin ([³H]MET) to cerebellar homogenates of 6-day-old rats. Values represent means ± SEM from at least two experiments. (Reproduced from Zagon *et al.*, 1991a, with permission of the publisher.)

Receptor	Ligand	K_d (nM)	B_{max} (fmol/mg protein)
μ	[³H]DAGO	0.6 ± 0.1	15.8 ± 1.8
δ	[³H]DPDPE	***	***
κ	[³H]U69, 593	4.3 ± 0.7	24.3 ± 2.9
ζ	[³H]MET	2.2 ± 0.3	22.3 ± 1.4

*** No specific and saturable binding detected.

receptor (Pert and Snyder, 1973; Pert *et al.*, 1974). Applying this methodology to our binding assays for opioid receptors, homogenates (including protease inhibitors) of 6-day-old and adult rat cerebellum were layered over a 0.32M/1.4 M sucrose gradient and centrifuged at 2200 *g* (the pellet of this spin was termed P1); the supernatant was centrifuged at 39 000 *g* (pellet = P2). Phase microscopy was used to examine the purity of the preparations. Binding assays with [^{3}H][Met^5]-enkephalin and [^{3}H]DAGO were conducted on each fraction (Table 3.4). [^{3}H][Met^5]-enkephalin binding was located in the P1 fraction, whereas [^{3}H]DAGO binding was detected in the P2 fraction. Competition assays showed that DAGO had a K_i of 33.0 ± 7.6 nM for [^{3}H][Met^5]-enkephalin in the P1, in contrast to the 4.6 nM K_i calculated in the 39 000 *g* homogenate used earlier.

Another characteristic that should distinguish the opioid growth factor receptor from other opioid receptors would be the demonstration of temporal expression. Thus, experiments were designed to investigate [Met^5]-enkephalin binding in developing cerebellum, as well as in adult cerebellum. We also examined the binding of ligands selective for μ, δ, and κ receptors for comparison to our work in the developing cerebellum. Our assays of the binding of [^{3}H]DAGO, [^{3}H]DPDPE, and [^{3}H] U69, 593 revealed specific and saturable binding for each ligand in the adult cerebellum (Table 3.5); computer evaluation of the data showed a one-site model of binding for each ligand. No differences in binding affinity or capacity between males and females could be detected for any ligand.

Binding assays of adult cerebellum with [^{3}H][Met^5]-enkephalin showed a K_d of 2.4 ± 0.4 nM and a B_{max} of 5.5 ± 0.4 fmol/mg protein. Subcellular fractionation studies of adult rat cerebellum revealed that [^{3}H][Met^5]-enkephalin, as well as [^{3}H]DAGO binding was located in the P2 fraction (Table 3.4). Further evaluation showed that [^{3}H][Met^5]-enkephalin binding was not stereospecific, with both (+)- and (−)-naloxone having a K_i of >10^{-6}M and not differing from one another. The binding of [^{3}H][Met^5]-enkephalin could not be altered by the addition of a cocktail of 10 nM DAGO, DPDPE, and U69, 593. However, [^{3}H][Met^5]-enkephalin binding was notably displaced by β-endorphin (K_i = 3.0 ± 0.1 nM).

The detection of μ, δ, and κ receptors in the

Table 3.4. The subcellular fractionation of [^{3}H][Met^5]-enkephalin ([^{3}H]MET) and [^{3}H]DAGO binding in the 6-day-old and adult rat cerebellum. Cerebellar homogenates from 6-day-old and adult rats were fractionated through sucrose, with a first pellet obtained at 2200 *g* (P1) and a second pellet obtained at 39 000 *g* (P2). Values represent the means ± SEM from at least two experiments. (Reproduced from Zagon *et al.*, 1991a, with permission of the publisher.)

	[^{3}H]MET	[^{3}H]DAGO
6-days		
P1	23.5 ± 2.6	***
P2	***	23.7 ± 0.7
Adult		
P1	***	***
P2	5.2 ± 0.2[†]	3.7 ± 0.2

*** No specific and saturable binding detected.
[†] Non-opioid receptor binding.

Table 3.5. The binding affinity (K_d) and binding capacity (B_{max}) of [^{3}H]DAGO, [^{3}H]DPDPE, [^{3}H]U69, 593, and [^{3}H][Met^5]-enkephalin ([^{3}H]MET) to cerebellar homogenates of adult rats. Values represent means ± SEM from at least two experiments. (Reproduced from Zagon *et al.*, 1991a, with permission of the publisher.)

Receptor	Ligand	K_d (nM)	B_{max} (fmol/mg protein)
μ	[^{3}H]DAGO	4.5 ± 0.9	13.2 ± 3.1
δ	[^{3}H]DPDPE	2.8 ± 0.4	11.1 ± 4.4
κ	[^{3}H]-U69, 593	3.7 ± 1.1	11.3 ± 2.4
ζ	[^{3}H]MET	***	***

*** Binding was unrelated to opioid receptors.

adult rat cerebellum support Barg and Simantov's (1989) findings utilizing different binding conditions and radiolabeled ligands, and reveals that opioid receptors are in the adult cerebellum. With respect to the binding of [³H][Met⁵]-enkephalin in the adult cerebellum, it is interesting to note that this binding was located in the P2 fraction whereas binding of this ligand in developing cerebellum was in the P1 fraction. Since it is known that [Met⁵]-enkephalin in adults recognizes the μ and δ receptors, the binding in P2 is probably related to these opioid receptors. [³H][Met⁵]-enkephalin was found to bind in a non-specific and non-saturable manner in P2 fractions, and this binding was neither stereospecific nor quenched by the addition of a cocktail of ligands specific for other opioid receptors. It was interesting that β-endorphin was an avid displacer of the [³H][Met⁵]-enkephalin binding. Non-opioid receptors related to β-endorphin have been reported (Hazum *et al.*, 1979; Schweigerer *et al.*, 1982), and the relationship between non-opioid receptors associated with β-endorphin and [Met⁵]-enkephalin requires clarification.

The expression of [Met⁵]-enkephalin binding during brain and cerebellar development has been explored (Zagon *et al.*, 1991b). In recent studies the ontogeny of [³H][Met⁵]-enkephalin binding was examined in homogenates of the whole rat brain and the cerebellum (Fig. 3.8). These studies show that binding occurs in the embryonic period and during the first 3 weeks after birth – a time of

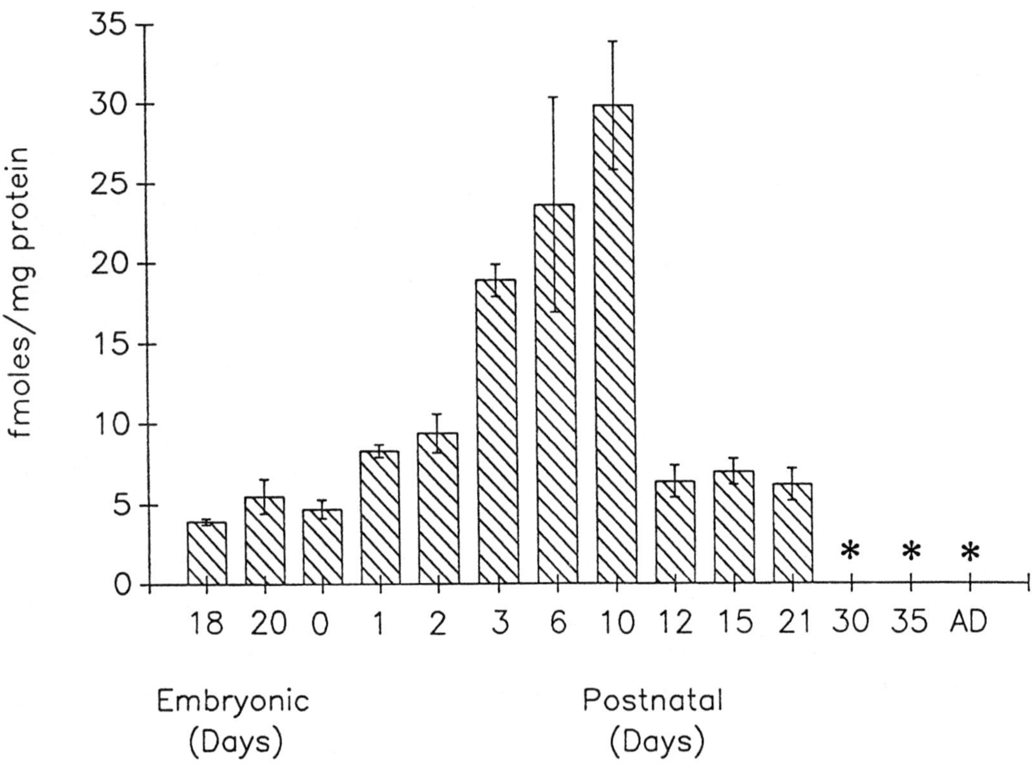

Fig. 3.8.   Specific and saturable binding of [³H][Met⁵]-enkephalin in homogenates of rat cerebellum from animals of various embryonic and postnatal ages. Binding affinities (K_d) at each age were approximately 2.0 ± 0.1 nM; no significant differences in mean K_d existed between ages. * = No specific and saturable binding was detected.

major cell proliferation in both the brain and cerebellum. A particularly strong correlation between [^{3}H][Met^5]-enkephalin binding and cell proliferation can be recorded in cerebellar preparations, with the number of germinative cerebellar cells rising during early postnatal life and diminishing by weaning (Altman, 1972; Zagon, 1972; Zagon and McLaughlin, 1987b); no binding of [^{3}H][Met^5]-enkephalin could be determined after the third postnatal week. The [^{3}H][Met^5]-enkephalin binding sites could also be influenced by opioid receptor blockade, since animals receiving naltrexone for the first 6 days of life had an elevation in binding capacity. Whether this increase was due to more cells (preliminary analysis showed an increase in the number of layers of germinative cells), or to an increase in the number of receptors per cell needs to be explored. However, the increase in receptors is consistent with earlier observations of more cerebellar cells seen at weaning.

To begin to address the question of where the binding sites were located in the developing brain, radiolabeled [Met^5]-enkephalin can be employed in *in vitro* autoradiography experiments. When this technique is applied to the neonatal retina (Fig. 3.9) one can detect radiolabeled [Met^5]-enkephalin associated with the retina, particularly the neuroblasts, suggesting that the binding sites for this peptide are related to proliferating and differentiating cell populations. Further work in which emulsion-coated slides are used, along with studies using an antibody to the ζ receptor, will be helpful in determining the

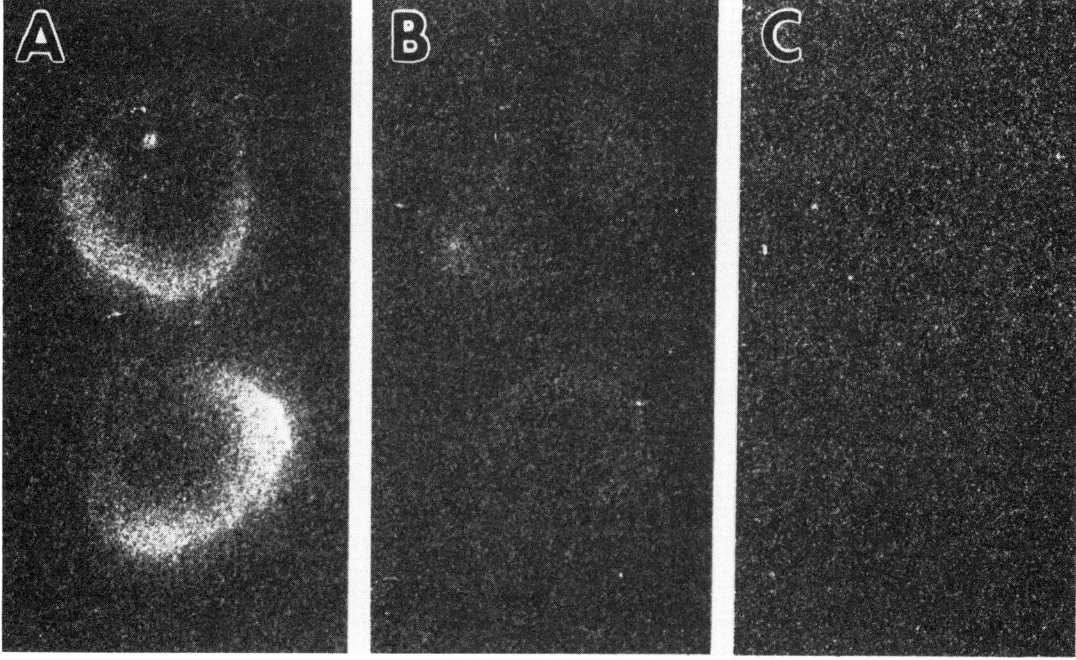

Fig. 3.9.   Localization of radiolabeled [Met^5]-enkephalin in the rat retina using *in vitro* autoradiography. One-day-old (A,B) or adult (C) rat retina incubated with either 1.5 nM [^{125}I][Met^5]-enkephalin (A,C) or [^{125}I][Met^5]-enkephalin and 1 µM naloxone (B). (A) Note that the radiolabel is associated with the retina. (B) Labeling with [^{125}I][Met^5]-enkephalin is blocked by concomitant exposure to naloxone. (C) No labeling of the adult rat retina with [^{125}I][Met^5]-enkephalin can be discerned. (Reproduced from Isayama *et al.*, 1991, with permission of the publisher.)

location of the opioid growth factor receptor in the developing nervous system.

An important extension of this work would be if the [Met⁵]-enkephalin binding occurred in the developing human cerebellum. Indeed, in such studies (Zagon *et al.*, 1990b) specific and saturable binding of [³H][Met⁵]-enkephalin was detected in infant cerebellum, at a level 20-fold greater in subjects 2–19 days of age than in newborns (Fig. 3.10). The adult human cerebellum did display minor [Met⁵]-enkephalin binding, but this binding was quenched by the addition of small concentrations of DAGO, suggesting that in the adult [Met⁵]-enkephalin recognized a μ receptor. Another interesting feature of [Met⁵]-enkephalin in the human brain was also recently reported. In the cerebellum from an individual with metastatic adenocarcinoma of the cerebellum, a high level of [³H][Met⁵]-enkephalin binding was observed, indicating that when cell replication occurs this binding site can be found (Zagon *et al.*, 1990c).

3.4 CONCLUSIONS

Evidence from a number of morphological, pharmacological, biochemical, and behavioral studies points to the role of an opioid growth factor in the developing nervous system. The finding that [Met⁵]-enkephalin is an opioid growth factor involved in the ontogeny of the nervous system, particularly with regard to cell replication, and that other opioids selective for known opioid receptors (μ, δ, κ, ∈, σ) do not influence growth, indicates a distinct place for [Met⁵]-enkephalin in neural development. Yet, the actions of [Met⁵]-

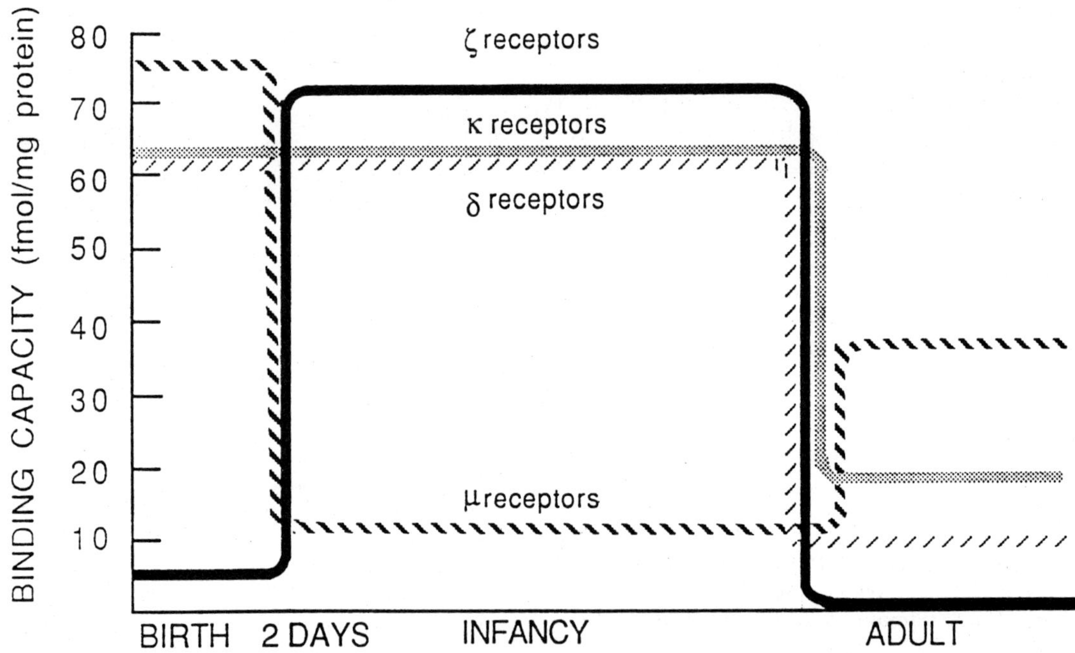

Fig. 3.10. A schematic depicting the ontogeny of opioid receptors in the developing and adult human cerebellum. At birth, μ, δ, and κ receptors are relatively abundant, but only a limited number of ζ receptors are recorded. Within 2 days after birth, assays revealed that both the binding affinity and capacity of μ receptors had decreased substantially. However, the concentration of ζ receptors increased markedly shortly after birth. In adults, all opioid receptors were decreased in number and the ζ receptor was not detected. (Reproduced from Zagon *et al.*, 1990b, with permission of the publisher.)

enkephalin on developmental processes obey the pharmacological rules of an opioid receptor-mediated event: displacement by an opioid antagonist and stereospecificity. The temporal and spatial expression of [Met⁵]-enkephalin also supports a unique role for this opioid peptide during brain development. The identification and characterization of the [Met⁵]-enkephalin binding site in cerebellum and brain provides further support of a novel opioid growth factor receptor. This binding site was sensitive to opioid antagonists, displayed stereospecificity, and exhibited a transient appearance consistent with brain development. In the retina, studies with radiolabeled [Met⁵]-enkephalin and autoradiography also demonstrated a spatial and temporal expression. The distinct location (i.e., nuclear rather than cytoplasmic) of the [Met⁵]-enkephalin binding site corroborates its unique nature. Taken together, these data suggest that [Met⁵]-enkephalin is an opioid growth factor, and that the receptor involved is different from those previously recognized. This receptor has been termed zeta (ζ), the sixth letter of the Greek alphabet, and named for the Greek word 'zoe' (life).

ACKNOWLEDGEMENTS

This work was supported by NIH grant NS-20500.

REFERENCES

Altman, J. (1972) Postnatal development of the cerebellar cortex in the rat. I. The external germinal layer and the transitional molecular layer. *J. Comp. Neurol.*, **136**, 353–98.

Barg, J. and Simantov, R. (1989) Developmental profile of kappa, mu and delta opioid receptors in the rat and guinea pig cerebellum. *Dev. Neurosci.*, **11**, 428–34.

Barg, J. and Simantov, R. (1991) Transient expression of opioid receptors in defined regions of developing brain: are embryonic receptors selective? *J. Neurochem.*, **57**, 1978–84.

Barg, J., Levy, R. and Simantov, R. (1989) Expression of the three opioid receptor subtypes μ, δ, and κ in guinea pig and rat brain cell cultures and *in vivo*. *Int. J. Dev. Neurosci.*, **7**, 173–9.

Bartolome, J. V., Bartolome, M. B., Daltner, L. A. *et al.* (1986) Effects of beta-endorphin on ornithine decarboxylase in tissues of developing rats: a potential role for this endogenous neuropeptide in the modulation of tissue growth. *Life Sci.*, **38**, 2355–62.

Bartolome, J. V., Bartolome, M. B., Harris, E. B. *et al.* (1987) N-alpha-acetyl-beta-endorphin stimulates ornithine decarboxylase activity in preweanling rat pups: opioid- and non-opioid-mediated mechanisms. *J. Pharmacol. Exp. Ther.*, **240**, 895–9.

Bartolome, J. V., Bartolome, M. B., Lorber, B. A. *et al.* (1990) Effects of central administration of beta endorphin on brain and liver DNA synthesis in preweanling rats. *Neuroscience*, **40**, 289–94.

Blum, R. H. (1969) A history of opium, in *Society and Drugs*. I. *Social and Cultural Observations* (eds R. H. Blum and Associates), Jossey-Bass, San Francisco, CA, pp. 45–58.

Blumberg, H. and Dayton, H. B. (1974) Naloxone, naltrexone, and related noroxymorphones, in *Narcotic Antagonists* (eds M. C. Braude, L. C. Harris, E. L. May *et al.*), Raven Press, New York, pp. 33–43.

Clendeninn, N. J., Petraitis, M. and Simon, E. J. (1976) Ontological development of opiate receptors in rodent brain. *Brain Res.*, **118**, 157–60.

Comb, M., Seeburg, P., Adelman, J. *et al.* (1982) Primary structure of the human met- and leu-enkephalin precursor and its mRNA. *Nature*, **295**, 663–6.

Coyle, J. T. and Pert, C. B. (1976) Ontogenetic development of [³H]naloxone binding in rat brain. *Neuropharmacology*, **15**, 555–60.

Davila-Garcia, M. I. and Azmitia, E. C. (1989) Effects of acute and chronic administration of leu-enkephalin on cultured serotonergic neurons: evidence for opioids as inhibitory neuronal growth factors. *Dev. Brain Res.*, **49**, 97–103.

Di Scala-Guenot, D., Strosser, M. T., Felix, J. M. *et al.* (1990) Expression of vasopressin and opiates but not of oxytocin genes studied by *in situ* hybridization in embryonic rat brain primary cultures. *Dev. Brain Res.*, **56**, 35–9.

Gubler, U., Seeburg, P., Hoffman, B. J. *et al.* (1982) Molecular cloning establishes proenkephalin as precursor of enkephalin-containing peptides. *Nature*, **295**, 206–9.

Harry, G. J. and Rosecrans, J. A. (1979) Behavioral effects of perinatal naltrexone exposure: a preliminary investigation. *Pharmacol. Biochem. Behav.*, **11**, 19–22.

Hauser, K. F. and Stiene-Martin, A. (1991) Characterization of opioid-dependent glial development in dissociated and organotypic cultures of mouse central nervous system: critical periods and target specificity. *Dev. Brain Res.*, **62**, 245–55.

Hauser, K. F., McLaughlin, P. J. and Zagon, I. S. (1987) Endogenous opioids regulate dendritic growth and spine formation in developing rat brain. *Brain Res.*, **416**, 157–61.

Hauser, K. F., McLaughlin, P. J. and Zagon, I. S. (1989) Endogenous opioid systems and the regulation of dendritic growth and spine formation. *J. Comp. Neurol.*, **281**, 13–22.

Hauser, K. F., Osborne, J. G., Stiene-Martin, A. *et al.* (1990) Cellular localization of proenkephalin mRNA and enkephalin peptide products in cultured astrocytes. *Brain Res.*, **522**, 347–53.

Hazum, E., Chang, K. J. and Cuatrecasas, P. (1979) Specific non-opiate receptors for β-endorphin. *Science*, **205**, 1033–5.

Hess, G. D. and Zagon, I. S. (1988) Endogenous opioid systems and neural development: ultrastructural studies in the cerebellar cortex of infant and weanling rats. *Brain Res. Bull.*, **20**, 473–8.

Hughes, J. (1975) Isolation of an endogenous compound from the brain with pharmacological properties similar to morphine. *Brain Res.*, **88**, 195–208.

Hughes, J., Smith, T. W., Kosterlitz, H. W. *et al.* (1975) Identification of two pentapeptides from the brain with potent opiate agonist activity. *Nature*, **258**, 577–9.

Ilyinsky, O. B., Kozlova, M. V., Kondrikova, E. S. *et al.* (1987) Effects of opioid peptides and naloxone on nervous tissue in culture. *Neuroscience*, **22**, 719–35.

Isayama, T. and Zagon, I. S. (1991) Localization of preproenkephalin A mRNA in the neonatal rat retina. *Brain Res. Bull.*, **27**, 805–8.

Isayama, T., McLaughlin, P. J. and Zagon, I. S. (1991) Endogenous opioids regulate cell proliferation in the retina of developing rat. *Brain Res.*, **544**, 79–85.

Kinney, H. C. and White, W. F. (1991) Opioid receptors localize to the external granular cell layer of the developing human cerebellum. *Neuroscience*, **45**, 13–25.

Kinney, H. C., Ottoson, C. K. and White, W. F. (1990) Three-dimensional distribution of ^{3}H-naloxone binding to opiate receptors in the human fetal and infant brainstem. *J. Comp. Neurol.*, **291**, 55–78.

Knodel, E. L. and Richelson, E. (1980) Methionine-

enkephalin immunoreactivity in fetal rat brain cells in aggregating culture and in mouse neuroblastoma cells. *Brain Res.*, **197**, 565–70.

Kornblum, H. I., Hurlbut, D. E. and Leslie, F. M. (1987) Postnatal development of multiple opioid receptors in rat brain. *Dev. Brain Res.*, **37**, 21–41.

Laurent-Huck, F. M., Stoeckel, M. E. and Felix, J. M. (1991) Ontogeny of proenkephalin gene expression in the rat hypothalamus. *Dev. Brain Res.*, **62**, 33–43.

Leslie, F. M., Tso, S. and Hurlbut, D. E. (1982) Differential appearance of opiate receptor subtypes in neonatal rat brain. *Life Sci.*, **31**, 1393–6.

Lorber, B. A., Freitag, S. K. and Bartolome, J. V. (1990) Effects of beta-endorphin on DNA synthesis in brain regions of preweanling rats. *Brain Res.*, **531**, 329–32.

McDowell, J. and Kitchen, I. (1987) Development of opioid systems: peptides, receptors and pharmacology. *Brain Res. Rev.*, **12**, 397–421.

Melner, M. H., Low, K. G., Allen, R. G. *et al.* (1990) The regulation of proenkephalin expression in a distinct population of glial cells. *EMBO J.*, **9**, 791–6.

Morita, Y., Zhang, J. H., Hironaka, T. *et al.* (1990) Postnatal development of preproenkephalin mRNA containing neurons in the rat lower brainstem. *J. Comp. Neurol.*, **292**, 193–213.

Musto, D. F. (1973) *The American Disease*, Yale University Press, New Haven, CT.

Najam, N. and Panksepp, J. (1989) Effect of chronic neonatal morphine and naloxone on sensorimotor and social development of young rats. *Pharmacol. Biochem. and Behav.*, **33**, 539–44.

Neale, J. H., Barker, J. L., Uhl, G. R. *et al.* (1978) Enkephalin-containing neurons visualized in spinal cord cell cultures. *Science*, **201**, 467–9.

Noda, M., Furutani, Y., Takahashi, H. *et al.* (1982) Cloning and sequence analysis of cDNA for bovine adrenal preproenkephalin. *Nature*, **295**, 202–6.

Osborne, J. G., Kindy, M. S. and Hauser, K. F. (1991) Expression of proenkephalin mRNA in developing cerebellar cortex of the rat: expression levels coincide with maturational gradients in Purkinje cells. *Dev. Brain Res.*, **63**, 63–9.

Paul, L., Diaz, J. and Bailey, B. (1978) Behavioral effects of chronic narcotic antagonist administration to infant rats. *Neuropharmacology*, **17**, 655–7.

Pert, C. B. and Snyder, S. H. (1973) Opiate receptor: demonstration in nervous tissue. *Science*, **179**, 1011–14.

Pert, C. B., Snowman, A. M. and Snyder, S. H.

(1974) Localization of opiate receptor binding in synaptic membranes of rat brain. *Brain Res.*, **70**, 184–8.

Petrillo, P., Tavani, A., Verotta, D. *et al.* (1987) Differential postnatal development of μ-, δ- and κ-opioid binding sites in rat brain. *Dev. Brain Res.*, **31**, 53–8.

Rius, R. A., Barg, J., Bem, W. T. *et al.* (1991) The prenatal developmental profile of expression of opioid peptides and receptors in the mouse brain. *Dev. Brain Res.*, **58**, 237–41.

Rosen, H. and Polakiewicz, R. (1989) Postnatal expression of opioid genes in rat brain. *Dev. Brain Res.*, **46**, 123–9.

Sawynok, J., Pinsky, C. and LaBella, F. S. (1979) Minireview on the specificity of naloxone as an opiate antagonist. *Life Sci.*, **25**, 1621–2.

Schmahl, W., Plendl, J., Reinohl-Kompa, S. (1987) Effects of naltrexone in postnatal rats on the recovery of disturbed brain and lymphatic tissues after X-irradiation or ethylnitrosourea treatment *in utero*. *Res. Commun. Chem. Pathol. Pharmacol.*, **55**, 89–99.

Schmahl, W., Funk, R., Miaskowski, U. *et al.* (1989) Long-lasting effects of naltrexone, an opioid antagonist, on cell proliferation in developing rat forebrain. *Brain Res.*, **486**, 297–300.

Schwartz, J. P. and Simantov, R. (1988) Developmental expression of proenkephalin mRNA in rat striatum and in striatal cultures. *Dev. Brain Res.*, **40**, 311–14.

Schweigerer, L., Bhakdi, S. and Teschemacher, H. (1982) Specific nonopiate binding sites for human β-endorphin on the terminal complex of human complement. *Nature*, **296**, 572–4.

Shinoda, H., Marini, A. M., Cosi, C. *et al.* (1989) Brain region and gene specificity of neuropeptide gene expression in cultured astrocytes. *Science*, **245**, 415–17.

Simon, E. J., Miller, J. M. and Edelman, I. (1973) Stereospecific binding of the potent narcotic analgesic [^{3}H] etorphine to rat brain homogenate. *Proc. Natl. Acad. Sci. USA*, **70**, 1947–9.

Spain, J. W., Roth, B. L. and Coscia, C. J. (1985) Differential ontogeny of multiple opioid receptors (μ, δ, and κ). *J. Neurosci.*, **5**, 584–8.

Spruce, B. A., Curtis, R., Wilkin, G. P. *et al.* (1990) A neuropeptide precursor in cerebellum: proenkephalin exists in subpopulations of both neurons and astrocytes. *EMBO J.*, **9**, 1787–95.

Stiene-Martin, A. and Hauser, K. F. (1990) Opioid-dependent growth of glial cultures: suppression of astrocyte DNA synthesis by met-enkephalin. *Life Sci.*, **46**, 91–8.

Stiene-Martin, A. and Hauser, K. F. (1991) Glial growth is regulated by agonists selective for multiple opioid receptor types *in vitro*. *J. Neurosci. Res.*, **29**, 538–48.

Tavani, A., Robson, L. E. and Kosterlitz, H. W. (1985) Differential postnatal development of μ-, δ- and κ-opioid binding sites in mouse brain. *Dev. Brain Res.*, **23**, 306–9.

Tecott, L. H., Rubenstein, J. L. R., Paxinos, G. *et al.* (1989) Developmental expression of pro-enkephalin mRNA and peptides in rat striatum. *Dev. Brain Res.*, **49**, 75–86.

Terenius, L. (1973) Stereospecific interaction between narcotic analgesics and a synaptic plasma membrane fraction of rat cerebral cortex. *Acta Pharmacol. Toxicol.*, **32**, 307–14.

Terry, C. E. and Pellens, M. (1970) *The Opium Problem*, Patterson Smith, Montclair, New Jersey (originally published in 1928 by the Bureau of Social Hygiene, Inc.).

Tsang, D. and Ng, S. C. (1980) Effect of antenatal exposure to opiates on the development of opiate receptors in rat brain. *Brain Res.*, **188**, 199–206.

Tsang, D., Ng, S. C. and Ho, K. P. (1982a) Development of methionine-enkephalin and naloxone binding sites in regions of rat brain. *Dev. Brain Res.*, **3**, 637–44.

Tsang, D., Ng, S. C., Ho, K. P. *et al.* (1982b) Ontogenesis of opiate binding sites and radio-immunoassayable β-endorphin and enkephalin in regions of rat brain. *Dev. Brain Res.*, **5**, 257–61.

Vertes, Z., Melegh, G., Vertes, M. *et al.* (1982) Effect of naloxone and D-met^2-pro^5-enkephalinamide treatment on the DNA synthesis in the developing rat brain. *Life Sci.*, **31**, 119–26.

Vilijn, M.-H., Vaysse, P. J., Zukin, R. S. *et al.* (1988) Expression of preproenkephalin mRNA by cultured astrocytes and neurons. *Proc. Natl. Acad. Sci. USA*, **85**, 6551–5.

Vorhees, C. V. (1981) Effects of prenatal naloxone exposure on postnatal behavioral development of rats. *Neurobehav. Toxicol. Teratol.*, **3**, 295–301.

Weyhenmeyer, J. A., Raizada, M. K., Miller, R. J. *et al.* (1980) Leucine-enkephalin immunoreactivity: localization in rat brain cells in culture. *Brain Res.*, **181**, 465–9.

Zagon, I. S. (1972) Light and electron microscopic observations on the postnatal development of the rat cerebellum. Doctoral Thesis, University of Colorado Medical Center.

Zagon, I. S. and McLaughlin, P. J. (1979) Morphological identification and biochemical characterization of isolated brain cell nuclei from the developing rat cerebellum. *Brain Res.*, **170**, 443–57.

Zagon, I. S. and McLaughlin, P. J. (1983a) Increased brain size and cellular content in infant rats treated with an opiate antagonist. *Science*, **221**, 1179–80.

Zagon, I. S. and McLaughlin, P. J. (1983b) Naltrexone modulates growth in infant rats. *Life Sci.*, **33**, 2449–54.

Zagon, I. S. and McLaughlin, P. J. (1983c) Naltrexone modulates tumor response in mice with neuroblastoma. *Science*, **221**, 671–3.

Zagon, I. S. and McLaughlin, P. J. (1984a) Naltrexone modulates body and brain development in rats: a role for endogenous opioid systems in growth. *Life Sci.*, **35**, 2057–64.

Zagon, I. S. and McLaughlin, P. J. (1984b) Duration of opiate receptor blockade determines tumorigenic response in mice with neuroblastoma: a role for endogenous opioid systems in cancer. *Life Sci.*, **35**, 409–16.

Zagon, I. S. and McLaughlin, P. J. (1985a) Opioid antagonist-induced regulation of organ development. *Physiol. Behav.*, **34**, 507–11.

Zagon, I. S. and McLaughlin, P. J. (1985b) Naltrexone's influence on neurobehavioral development. *Pharmacol. Biochem. Behav.*, **22**, 441–8.

Zagon, I. S. and McLaughlin, P. J. (1986a) Opioid antagonist (naltrexone) modulation of cerebellar development: histological and morphometric studies. *J. Neurosci.*, **6**, 1424–32.

Zagon, I. S. and McLaughlin, P. J. (1986b) Opioid antagonist-induced modulation of cerebral and hippocampal development: histological and morphometric studies. *Dev. Brain Res.*, **28**, 233–46.

Zagon, I. S. and McLaughlin, P. J. (1986c) β-funaltrexamine (β-FNA) and the regulation of body and brain development in rats. *Brain Res. Bull.*, **17**, 5–9.

Zagon, I. S. and McLaughlin, P. J. (1987a) Endogenous opioid systems regulate cell proliferation in the developing rat brain. *Brain Res.*, **412**, 68–72.

Zagon, I. S. and McLaughlin, P. J. (1987b) The location and orientation of mitotic figures during histogenesis of the rat cerebellar cortex. *Brain Res. Bull.*, **18**, 325–36.

Zagon, I. S. and McLaughlin, P. J. (1989a) Endogenous opioid systems regulate growth of neural tumor cells in culture. *Brain Res.*, **490**, 14–25.

Zagon, I. S. and McLaughlin, P. J. (1989b) Naloxone modulates body and organ growth of rats: dependency on the duration of opioid receptor blockade and stereospecificity. *Pharmacol. Biochem. Behav.*, **33**, 325–8.

Zagon, I. S. and McLaughlin, P. J. (1990) Ultrastructural localization of enkephalin-like immunoreactivity in developing rat cerebellum. *Neuroscience*, **34**, 479–89.

Zagon, I. S. and McLaughlin, P. J. (1991) Identification of opioid peptides regulating proliferation of neurons and glia in the developing nervous system. *Brain Res.*, **542**, 318–23.

Zagon, I. S., McLaughlin, P. J., Weaver, D. J. and Zagon, E. (1982) Opiates, endorphins, and the developing organism: a comprehensive bibliography. *Neurosci. Biobehav. Rev.*, **6**, 439–79.

Zagon, I. S., McLaughlin, P. J. and Zagon, E. (1984) Opiates, endorphins, and the developing organism: a comprehensive bibliography, 1982–1983. *Neurosci. Biobehav. Rev.*, **8**, 387–403.

Zagon, I. S., Rhodes, R. E. and McLaughlin, P. J. (1985) Distribution of enkephalin immunoreactivity in germinative cells of developing rat cerebellum. *Science*, **227**, 1049–51.

Zagon, I. S., Rhodes, R. E. and McLaughlin, P. J. (1986) Localization of enkephalin immunoreactivity in diverse tissues and cells of the developing and adult rat. *Cell Tissue Res.*, **246**, 561–5.

Zagon, I. S., Goodman, S. R. and McLaughlin, P. J. (1989a) Characterization of zeta (ζ): a new opioid receptor involved in growth. *Brain Res.*, **482**, 297–305.

Zagon, I. S., Zagon, E. and McLaughlin, P. J. (1989b) Opioids and the developing organism: a comprehensive bibliography, 1984–1988. *Neurosci. Biobehav. Rev.*, **13**, 207–35.

Zagon, I. S., Goodman, S. R. and McLaughlin, P. J. (1990a) Demonstration and characterization of zeta (ζ), a growth-related opioid receptor, in a neuroblastoma cell line. *Brain Res.*, **511**, 181–6.

Zagon, I. S., Gibo, D. M. and McLaughlin, P. J. (1990b) Adult and developing human cerebella exhibit different profiles of opioid binding sites. *Brain Res.*, **523**, 62–8.

Zagon, I. S., Gibo, D. and McLaughlin, P. J. (1990c) Expression of zeta (ζ), a growth-related opioid receptor, in metastatic adenocarcinoma of the human cerebellum. *J. Natl. Cancer Inst.*, **82**, 325–7.

Zagon, I. S., Gibo, D. M. and McLaughlin, P. J. (1991a) Zeta (ζ), a growth-related opioid receptor in developing rat cerebellum: identification and characterization. *Brain Res.*, **551**, 28–35.

Zagon, I. S., Gibo, D. M. and McLaughlin, P. J. (1991b) Ontogeny of the zeta (ζ) receptor in developing rat cerebellum. *Soc. Neurosci. Abstr.*, **17**, 596.

TRANSFERRIN AND THE DEVELOPING NERVOUS SYSTEM 4

Araceli Espinosa de los Monteros and Jean de Vellis

4.1 INTRODUCTION

Iron is essential for life and plays a key role in oxidative metabolism. Iron is necessary for the activity of a variety of enzymes that interact with oxygen directly. These include dioxygenases, cytochrome monoxidases, mitochrondrial non-heme iron monooxygenases, fatty acid desaturase, peroxidases and catalase (for a review see Cammack *et al.*, 1990). Because of the toxicity of free iron radicals, special proteins for iron transport, uptake and storage are required for iron utilization in the mammalian system.

The brain functions by communication among neurons and between neurons and glial cells. The classic method of neuronal communication is chemical synaptic transmission. However, between neurons and glial cells and between different types of neuroglia, signalling consists of the release of a substance (e.g. growth factors, neurotransmitters, peptides) and its diffusion and interaction with specific receptors present on neighboring cell membranes, as well as receptors expressed on the membrane of the same cell. Many drugs used for the treatment of neurological and psychiatric disorders exert their effects by acting on specific receptors. Exogenous growth factors, hormones and other signalling agents are also known to regulate intracellular processes via interactions with receptors. The iron carrier glycoprotein transferrin (Tf) as well as the Tf-like proteins that belong to the Tf family act via receptors in all organs that have been studied. (Crichton and Charloteaux-Wauters, 1987). Briefly, Tf secreted into the extracellular environment binds two iron molecules at the N- and C- terminals (Chasteen, 1983). The iron-loaded transferrin interacts with receptors (R) expressed on the cell surface, forming a Tf–Tf-R complex. This complex is internalized into the cytoplasm and iron is delivered into acidic endosomes. The apotransferrin coupled to the Tf receptor is recycled to the cell surface. This mechanism allows Tf and Tf-R to be recycled several times.

At the beginning of each section a brief description of the general characteristics of mammalian proteins involved in iron transport and storage will be given. Transferrin, transferrin receptor and ferritin act in concert to maintain iron homeostasis necessary for cell function. Most of our understanding of Tf and its receptor derives from studies performed on liver and blood cells. However, all mammalian non-transformed cells of the nervous system also use a 'Tf–Tf-R' system to transfer iron from the extracellular environment into the cell (Espinosa *et al.*, 1989).

4.2 IRON METABOLISM WITHIN THE CNS: WHAT DO WE KNOW?

The brain, like the rest of the body, needs iron for its normal development and function.

Receptors in the Developing Nervous System Vol. 1: *Growth factors and hormones* Edited by Ian S. Zagon and Patricia J. McLaughlin. Published in 1993 by Chapman & Hall. ISBN 0 412 45240 5. Vols. 1 and 2 (set) ISBN 0 412 54520 9.

During the last two decades, various investigators have contributed to the characterization of the distribution of iron, transferrin, Tf-R receptor and ferritin within the CNS. This information has recently been reviewed (Espinosa *et al.*, 1989). A second review focusing mainly on the regional and subcellular localization of iron, Tf-R, and ferritin in the brain has also been published (Hill, 1990). Here we discuss recent findings on Tf and Tf-R in the CNS from our research group and other laboratories. We also speculate about future directions based on research on Tf and Tf-R gene regulation in various other organs and cells.

The presence of iron in the human brain was detected almost a century earlier (Zaleski, 1887) than Tf and its receptor. However, the biochemical processes in which iron takes part during normal brain development and in the adult brain have just begun to be elucidated. Iron distribution in the adult brain of several species has been well characterized. In the adult rat brain iron is visualized at the level of the neuropil, oligodendrocytes and myelin in the globus pallidus, while in the lateral septum the cells that are iron-positive are neurons (Hill, 1990). Many other areas of the brain display iron staining at different intensities. The comparison of gamma aminobutyric acid (GABA) distribution with iron distribution shows an overlapping pattern, suggesting that iron is involved in GABA metabolism (Hill *et al.*, 1985). Preventing GABA degradation by injecting the irreversible inhibitor of GABA-transaminase, gamma vinyl GABA (GVG), which has no direct effects on GAD activity, results in the elevation of GABA levels (Schechter *et al.*, 1977; Loscher, 1980; Gale and Casu, 1981). GVG injected into the striatum and globus pallidus significantly reduced iron accumulation in the ventral pallidus, globus pallidus and substantia nigra, all of which are target sites for GABAergic efferents. The exact mechanism is still unknown, but it has been proposed that iron reduction may involve changes at the level of other neurotransmitter(s) besides GABA itself (Hill, 1990). LeVine and Macklin (1990) have reanalyzed iron distribution within the adult rat brain. They found that most oligodendrocytes are enriched in iron without any regional restriction, as was previously described (François *et al.*, 1981; Hill and Switzer, 1984). On the basis of these results LeVine and Macklin questioned the previously proposed relationship between iron and GABA distribution. They found that brain tissue needs to undergo extensive permeabilization with sodium dodecyl sulfate or Triton X-100 in order to visualize iron-enriched oligodendroglia. We have used the same permeabilization technique in an attempt to compare iron distribution to Tf-immunoreactivity in paraffin and vibratome sections from 5–25-day-old rat brains. Iron was not detectable in either paraffin or vibratome sections (unpublished observations). It is possible that iron in young tissues is somewhat loosely bound and that with the LeVine and Macklin (1990) technique iron gets washed out.

4.3 IRON TRANSPORT PROTEIN: MOLECULAR AND CELLULAR PROPERTIES OF TRANSFERRIN

The transferrin gene family has at least six members. There are three forms of Tf that are structurally and functionally similar; ovotransferrin, lactoferrin and serotransferrin (transferrin contained in the serum component of the blood). A fourth molecule, P97 or melanotransferrin, is a cell surface molecule found in human melanomas, which binds iron and shares 39% homology with human Tf (Brown *et al.*, 1982; Rose *et al.*, 1986). A fifth member of the Tf gene family is a factor responsible for transformation in chicken β-cell lymphomas *ChBlym*-1 (or Hu Blym-1, Goubin *et al.*, 1983), or in human Burkitt's lymphoma (Diamond *et al.*, 1983). The sixth Tf-like molecule, TBG41, a repeated

DNA sequence in the human KprI gene family, is associated with pseudogenes near the β globin gene (Hattori *et al.*, 1985).

Transferrin and the 'transferrin-like' molecules function as extracellular iron-carrier molecules; their molecular weight is around 80 kDa. Human transferrin is a glycoprotein encoded by 2.3 kb cDNA (Uzan *et al.*, 1984; Yang *et al.*, 1984). The genomic sequence is contained within multiple exons: 17 distinct exons for chick ovotransferrin (Jeltsch *et al.*, 1987) and a similar exon organization for human Tf (Park *et al.*, 1985). These iron carrier proteins have two conserved domains at the *N*- and *C*- terminus, each site containing a binding site for ferric iron atoms and a single glycosylation site (Metz-Boutigue *et al.*, 1984). Both sites present 1–6×10^{-22}M affinity for iron at physiological conditions.

4.3.1   TRANSFERRIN WITHIN THE CNS

In contrast to the poor characterization of iron content within the CNS during development an extensive characterization of Tf accumulation within the brain of various species has been performed (for details see Espinosa *et al.*, 1989). Transferrin can be considered as a 'transient' marker of cell development, since it is expressed by different cell types at different ages. A similar pattern has been found in every species investigated to date. In the chick, transferrin is first detectable during embryonic life; neurons are Tf+ as early as embryonic day 7 (ED) (Oh *et al.*, 1986). The number of Tf+ cells increases significantly between ED10 and ED16, when all neurons are strongly positive, then the Tf signal starts to fade in neurons while it begins to show up in capillary endothelium. Later, at ED19, some glial cells start showing Tf immunoreactivity. In the mouse brain, neurons are Tf+ between ED18 to postnatal day (PD) 8. However, some Tf+ neurons have also been detected in the adult brain (Toran-Allerand, 1980). In the rat brain Tf appears in neurons at ED18, peaking at PD1 in the spinal cord

and PD6 in the brain (Dion *et al.*, 1988). During postnatal life the cells that are Tf+ are oligodendrocytes appearing at PD5 in white matter and in gray matter around PD5 (Connor and Fine, 1986, 1987). In the sheep brain a brief Tf+ period is detected in diencephalic nuclei at ED34, but it disappears permanently by ED60 (Dziegielewska *et al.*, 1981; Reynolds and Mollgard, 1985). In the pig, Tf is detectable in neurons, meninges and choroid plexus by ED31 (Cavanagh and Mollgard, 1985). In the human neocortex, Tf is localized in the ventricular zone around ED43 but it disappears at later stages (Mollgard and Jacobsen, 1984). In general, in the brains of adult animals Tf-immunoreactivity in neurons is either absent or quite weak, whereas Tf is present in oligodendrocytes, choroid plexus, epithelial cells and endothelial cells of the chick brain (Oh and Markelonis, 1982). In the rat, Tf is found in oligodendrocytes and choroid plexus (Bloch *et al.*, 1985, 1987; Connor and Fine, 1986, 1987). There is no restriction for Tf expression with respect to the topographical localization of oligodendrocytes: all three types, interfascicular, perivascular and perineuronal or satellite oligodendrocytes have been found to be Tf+ (Espinosa *et al.*, 1988). Neurons are Tf negative in the normal adult brain (Oh *et al.*, 1986; Espinosa *et al.*, unpublished observation). We have explored postnatal rat brain from newborn, PD1 to 1-year-old animals searching for Tf+/GFAP+ astrocytes; we have not found such a cell phenotype within the intact normal brain. The embryonic rat brain has not yet been investigated.

The sequential but transient accumulation of Tf-immunoreactivity in the developing CNS appears to be related to a given 'cell stage' more than to a given 'cell type'. Early in embryogenesis neuroblast proliferation and cell migration correlate with the early Tf expression by neuronal cells. Subsequently neuroblasts stop dividing and at birth precursor cells for oligodendroglia start to develop, this seems to correlate with the appearance of Tf in oligodendroblasts and choroid plexus epithelial cells (Scully *et al.*, unpublished results). Subsequently when the blood–brain barrier (BBB) begins to form, Tf-immunoreactivity is detectable in the capillary endothelium and the percentage of Tf+ oligo-dendrocytes decreases but remains positive throughout the lifespan of the animal.

4.3.2   DEVELOPMENTAL EXPRESSION OF TRANSFERRIN *IN VITRO*

The developmental expression of Tf has been investigated in our laboratory in primary glial cultures containing oligodendrocytes and their progenitors as well as astrocytes, or in pure populations of astrocytes, oligoden-drocytes and cerebral neurons. We have shown that transferrin is detectable in pri-mary glial cultures by immunocytochemistry in oligodendrocytes and their progenitors but not in astrocytes (Espinosa *et al.*, 1988). The number of Tf+ oligodendrocytes in culture increases with time, peaks by 15 days *in vitro* (DIV) and decreases later on. Double immunocytochemical studies demonstrate that all oligodendrocytes express Tf during development, but some lose Tf immunore-activity concomitantly with the appearance of myelin basic protein (MBP). Three pheno-types of oligodendrocytes were observed: Tf^+/MBP^-, Tf^+/MBP^+ and Tf^-/MBP^+. These subpopulations exist throughout oligoden-droglial maturation, and persist in older mature cultures (Espinosa and de Vellis, 1988). These findings have also been con-firmed in studies of normal rat brains from PD 0 to 1-year-old animals (Espinosa and de Vellis, 1990). This finding of segregation of Tf and MBP expression among some oligoden-drocytes implies that within the mature brain these cells may perform a key role in iron transport and storage (Espinosa *et al.*, 1989). We also tested the hypothesis that oligoden-drocytes synthesize and release Tf in the CNS (aside from choroid plexus). We analyzed pure cultures of oligodendrocytes, astrocytes

and neurons; all three neural cell types express Tf mRNA, however, oligodendro-cytes and astrocytes synthesize and secrete a Tf molecule immunoprecipable by an anti-rat Tf antibody (Espinosa *et al.*, 1990) but neurons do not. Other cell types such as fibroblasts and C6 glioma cells do not express Tf mRNA. *In situ* hybridization for Tf mRNA in brain tissue sections shows a distribution pattern similar to that for MBP mRNA, although less intense, and coincides with the immunocytochemical analysis. The differ-ences between *in vivo* and *in vitro* Tf gene expression suggest the existence of epigenetic factors which regulate this gene. Among possible factors we have found that hydrocortisone decreases Tf mRNA levels in cultured astrocytes (Espinosa et al., 1990), a result contrary to that found in liver cells where Tf levels increased in the presence of hydrocortisone (Morgan, 1981). Transferrin itself appears to have a regulatory effect on the Tf gene. When astrocytes are cultured in the presence of a Tf-containing medium, the levels of newly synthesized protein are increased (Fig. 4.1). Besides the effect of hormones and exogenous transferrin on the Tf gene in cultured astrocytes, we have found that the initial plating cell density has a long-term effect on the expression of Tf mRNA (Condorelli *et al.*, unpublished results). When astrocytes are plated at low density, the Tf

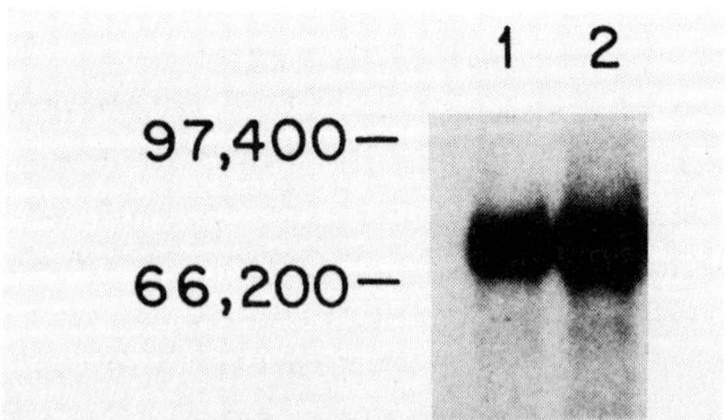

Fig. 4.1.   Fluorograph of radiolabeled transferrin (Tf) synthesized and secreted by 14-day-old primary astroglial cultures. Line 1 incubated in Tf-depleted medium; line 2 incubated in medium supplemented with transferrin 100µg/ml.

mRNA levels are lower than those found in high density plated astrocytes, even if both cultures are confluent. These findings indicate that environmental conditions can have a long-term influence on the regulation of the Tf gene. Recently, DNA regulatory elements controlling tissue specific expression of Tf have been identified in the 5' upstream region of the Tf gene (Schaeffer *et al.*, 1987; Brunel *et al.*, 1988). The plasticity of Tf expression in neurons and astrocytes suggests that the Tf gene may be expressed by these cells during development or under pathological conditions, due to injury and other degenerative conditions.

4.4 CELLULAR IRON UPTAKE: MOLECULAR AND CELLULAR FEATURES OF THE Tf RECEPTOR

The transferrin receptor (Tf-R) was first purified and characterized from human placenta (Seligman *et al.*, 1979; Wada *et al.*, 1979). The isolation and analysis of this receptor has been performed by the use of monoclonal antibodies (Trowbridge and Omary, 1981; for review see Crichton and Charloteaux-Wauters, 1987). Cells obtain iron-loaded Tf through Tf-specific surface receptors; this is the unique general mechanism, and there is no other iron transport system utilized by mammalian cells under normal conditions (Hamilton *et al.*, 1979; Omary *et al.*, 1980; Morgan, 1981; Newman *et al.*, 1982). In other words Tf receptors (Tf-Rs) are expressed by all normal mammalian cell types. However, the number of Tf receptors expressed by a given cell type may vary depending on the physiological and developmental stage of the cell. For instance proliferating cells express a larger number of Tf receptors than resting cells (Hamilton *et al.*, 1979; Larrick and Cresswell, 1979; Galbraith *et al.*, 1980).

The Tf-R is a 180 kDa membrane glycoprotein. It consists of two 90 kDa identical subunits that form a disulfide-linked dimer with two Tf binding sites. The bulk of the molecule is located in the extracellular domain. The iron-loaded Tf–Tf-R complex is internalized in endosomes where iron is liberated and then bound by ferritin. (For review see Crichton and Charloteaux-Wauters, 1987).

4.4.1   Tf-R WITHIN THE CNS

The presence of iron and Tf in the brain implicates the existence of Tf receptors within the CNS. In fact, Tf receptors are present in the brain at the luminal surface of capillary endothelial cells, further suggesting that circulating serotransferrin participates in iron-transport into the CNS via receptor-mediated transport across the blood–brain barrier (Jefferies *et al.*, 1984; Pardridge *et al.*, 1986).

Experiments to investigate the transcytosis of circulating transferrin into the brain have shown that after perfusing rat brain with ^{125}I-Tf there is a rapid receptor-mediated uptake of the radiolabeled protein into capillary endothelium of the brain (Fishman *et al.*, 1987). Approximately 50% of the internalized ^{125}I-Tf was found in the brain. Similar experiments were performed to show permeability of the BBB to radiolabeled proteins including Tf in developing sheep. These labeled proteins penetrated into the brain and the brain/plasma ratios increased with time. *In vitro* studies performed with isolated human brain capillaries confirmed these *in vivo* findings. The kinetics of endocytosis of ^{125}I -labeled human holotransferrin was temperature dependent. The affinity and capacity of the BBB Tf-R are within the same order of magnitude as those for insulin receptors and insulin growth factor I and II receptors (Pardridge *et al.*, 1987).

The use of immunocytochemistry and autoradiography have allowed investigators to characterize Tf-R distribution within the CNS. Systemic injections of monoclonal antibodies against rat and human Tf-Rs exclusively labeled brain capillaries but not capillaries in other tissues. In these sections of brain tissue, scattered cells having the morphology of neurons were also stained (Jefferies *et al.*, 1984). Monoclonal antibodies against Tf-R, B 3/25 (Barclay, 1981) and OKT₉ (Trowbridge and Omary, 1981) stained capillaries in the human visual cerebral cortex, basal ganglia, cerebellum, mid-brain and spinal cord.

Immunocytochemistry performed on various species has shown that early in development Tf-R+ cells appear at ED5 on the surface of germinal cells in the chick brain. At ED6, Tf-R+ cells are present throughout the walls of developing cerebral hemispheres (Markelonis *et al.*, 1985b). The inner layer of the ventricular zone strongly expresses the Tf-R (Oh *et al.*, 1986). The Tf-R can also be visualized on the surface of membranes of dorsal root ganglion neurons in culture (Markelonis *et al.*, 1985a).

In the rat brain the patterns of Tf-R distribution, visualized both by using ^{125}I-Tf and by immunocytochemistry with an anti Tf-R monoclonal antibody, are almost identical (Hill *et al.*, 1985). A report (Hill, 1990) shows that only at the level of the interpeduncular nucleus is there a small overlap between iron distribution and Tf-R. The areas rich in Tf-R show little or no iron staining. However, interpretation of these findings has been discussed in a hypothetical form; it has been proposed that since most areas with a high concentration of iron have input from sites with high levels of Tf-R, and that brain areas accumulating iron are efferent to areas of high Tf-R density, these sites receive iron through axonal transport. Furthermore, Hill (1990) suggested that 'transferrin is involved in myelinogenesis in addition to its role in iron uptake'. The same suggestion has been made by Connor and Fine (1986, 1987) who found that Tf immunostaining is absent in the brain of myelin-deficient (md) rat mutants. Although this hypothesis remains to be demonstrated, the specific mechanisms by which oligodendrocyte Tf may be involved in the process of myelination could be through Tf's main function in iron transport or its action as a trophic/differentiation factor, as

was previously proposed (Espinosa *et al.*, 1989). Although the myelin-deficient rat is highly deficient in Tf mRNA (Scully *et al.*, unpublished results) as well as Tf itself, it seems that oligodendrocytes are affected at the maturation level as a consequence of the point PLP mutation (Boison and Stoffel, 1989). This impairment affects not only myelin-associated genes but also oligodendroglial genes that do not directly relate to myelinogenesis (Kumar *et al.*, 1988, 1990; Espinosa and de Vellis, 1990). At present, it seems premature to link the myelination process directly to Tf gene expression. Iron is required for respiratory enzymes, protein synthetic mechanisms and lipid desaturation. It is clear that iron metabolism is linked to normal cell development and function for all cell types, not solely oligodendroglia. Extensive work needs to be done to elucidate whether the mechanisms of myelinogenesis are triggered or regulated by Tf through its iron transport function or by Tf itself independently of iron.

The transferrin receptor has been characterized in pure cultures of oligodendrocytes (Espinosa and Foucaud, 1987). The availability of pure cultures of oligodendrocytes opened several lines of investigation for myelinogenesis and remyelination. Also of note was the fact that oligodendrocytes, but not astrocytes, survive in cell suspension and mature further when replated and maintained under adequate conditions (Espinosa *et al.*, 1986).

Pure oligodendrocytes in cell culture can survive at relatively low iron concentrations (0.1–0.3 μM) in the absence of Tf. When both exogenous Tf and iron at micromolar concentrations are present in the culture medium a toxic effect is seen. The time frame to observe the toxic effect depends on the concentration of iron (Espinosa and Foucaud, 1987). Oligodendroglial progenitors are more sensitive to iron toxicity than mature myelinating cells. Both iron and Tf have a marked effect on cultured oligodendroglia maturation and myelin-like membrane formation (Espinosa, unpublished observations). On the basis of these findings the question of the occurrence of iron uptake by these cells was investigated. Since the expression of the Tf-R is regulated in response to extracellular stimulation (Neckers *et al.*, 1984; Hamilton, 1985), maturation (Harding *et al.*, 1984) or Tf itself (Rao *et al.*, 1986), we determined the binding constants at different ages in culture, corresponding to the stages of proliferation, maturation, and myelin-like membrane formation by oligodendrocytes. The results indicated that pure cultures of oligodendrocytes have Tf-Rs with high affinity binding sites (K_D=0.2–0.6 nM). Most of the published K_D values for the Tf-R in other cell types are in the nanomolar range or higher; for example, human fibroblast K_D values are approximately 1.5 nM and liver cells 89–100 nM. In fact, the values found for cultured oligodendrocytes are the highest values ever reported for the Tf-R. During the time course of the culture, several events occur: proliferation at 5 to 6 days post-plating; expression of myelin markers together with morphological changes; maturation to premyelinating and myelinating stages where a high synthesis of myelin components takes place; and finally cells reach the state where myelin components are organized as *in vivo* into a myelin-like membrane (Espinosa, 1987). From 3 to 20 days *in vitro* there was no significant changes in K_D values. Furthermore, in contrast to other cell types, the number of receptors per cell was at least two orders of magnitude lower than in most other cell types. When oligodendrocytes proliferate $1.1–3.6 \times 10^3$ receptors per cell are expressed. During maturation and myelin-like membrane synthesis, the number of binding sites remain unchanged despite the large increase in the membrane surface area. These results may suggest that Tf-R remains restricted to the oligodendrocyte cell body. The small and steady number of Tf-Rs throughout oligodendrocyte development suggest that oligodendrocyte iron requirements do not vary in a significant manner over time and that the iron Tf-R is involved in regulating iron utilization by other cells and/or iron storage within the oligodendrocyte. Oligodendrocytes also produce and secrete Tf (see preceding section). Therefore, the high affinity surface Tf-Rs of oligodendrocytes, which are expressed in small numbers, may very well function as an autoreceptor mediated regulation for Tf synthesis within the cell.

There have been two other reports concerning high affinity Tf-R expression and distribution (Mash *et al.*, 1990; Roskams and Connor, 1990), in which saturation binding assays revealed an apparent single class of sites with a K_D in the range 2–4 nM. These findings confirm the report by Espinosa and Foucaud (1987) of a single high affinity receptor expressed on cultured oligodendrocytes. The variation in terms of the K_D could be due to the fact that either isolated brain cell membranes (Mash *et al.*, 1990) or total homogenized brain tissue (Roskams and Connor, 1990) were used for Tf-R binding assays, whereas Tf-R binding on cultured oligodendrocytes was measured on intact live cells (Espinosa and Foucaud, 1987). The procedures for membrane preparation may have slightly altered or denatured the Tf-R, whereas its structure remains intact in cultured cells.

A complete developmental study of changes in ^{125}I- and ^{59}Fe-Tf uptake in the rat brain (Taylor and Morgan, 1990) showed that iron transport and the number of Tf-Rs increased rapidly during the first 15 days of life. The number peaked at 15 days and thereafter declined. A similar pattern occurred in the uptake of ^{125}I-Tf. The turnover of radiolabelled iron and Tf in the brain was determined at 15 days of age. A time course was done by measuring radioactivity in the brain at various times postinjection from 15 min to 13 days. The amount of ^{59}Fe-Tf in the brain increased over the first 4h and remained constant thereafter. In contrast, ^{125}I-Tf values increased rapidly during the first 15 min, peaked and remained constant for 6h, and then declined. This uptake pattern was compared to those found in liver and femur. Tf and iron uptake by the brain through capillaries is maximal at the time of rapid brain growth. The pattern of iron uptake was found to be a unidirectional process, whereas the uptake of Tf through the blood–brain barrier involves two components, one associated with a large pool of Tf recycling to the circulation and another retained by the endothelium or transported into the brain parenchyma. The Tf kept in the brain parenchyma seems to be a smaller pool compared to that recycled to the systemic circulation. It also seems to be much less in developing (15-day-old) animals compared to the previously reported values for adult rat brain (Fishman *et al.*, 1987).

4.5   IRON STORAGE: FERRITIN

Intracellular iron metabolism involves iron uptake and storage, as well as iron mobilization and distribution within the cell, either for synthesis of iron-containing proteins or secretion from the cell.

Two major compounds have been identified for iron storage: hemosiderin and ferritin. Hemosiderin was isolated from horse spleen by Cook in 1929 and ferritin was extracted from horse spleen by Laufberger in 1935 (for detailed review see Laufberger, 1937; Crichton and Charloteaux-Wauters, 1987; Harrison *et al.*, 1990). This section summarizes the general features of ferritin. Ferritin is present as insoluble granules within secondary lysosomes and consists of a chemically inert 'iron core' within a large protein structure. The iron contained in the core varies from 0 to 4500 atoms (Fishbach and Anderbergg, 1965). The 476 kDa protein structure is composed of 24 subunits, organized in a 24 symmetrical form.

The apoferritin shell can give dimers, trimers, etc. by aggregation. In SDS-polyacrylamide gel electrophoresis subunits of 19 kDa and 21 kDa are frequently observed; these two subunits have been designated light (L) and heavy (H) subunits respectively. The ratio of H/L subunits has been found to be both species and tissue specific. The molecular biology of the ferritin gene and the biochemistry of iron storage and mobilization has been reviewed by Harrison *et al.* (1990) and Munro *et al.* (1990). It is well known that iron excess can cause extensive damage. The damage occurs mainly by the formation of free radicals promoted by iron (O'Connell *et al.*, 1985, 1986).

In vitro studies have shown that ferritin incubated with ascorbate in the presence of oxygen (Bienfait and Van den Briel, 1981) at pH 4.5 can give rise to peptides of 19 and 20 kDa (Boyd *et al.*, 1985). However, the mechanism of iron mobilization still remains an enigma. Various compounds have been postulated for mobilization of iron from ferritin, for example, xanthine oxidase (Mazur *et al.*, 1958) and mononucleotides (Jones *et al.*, 1978), which may act by penetrating into the ferritin shell; other more controversial reports have also been published (Funk *et al.*, 1985). Molecules such as dithionate, which can completely mobilize iron from ferritin *in vitro*, are not present in the *in vivo* environment. In summary, only ascorbic acid can be considered a strong candidate for iron mobilization *in vivo*. (For review of clinical studies see Bridges, 1990.)

4.5.1   FERRITIN WITHIN THE CNS

Ferritin distribution in the rat brain has been determined by radioimmunocytochemistry and immunocytochemistry (for review see Hill, 1990). The patterns of distribution found for ferritin are similar to those described for iron, although density is lower for ferritin when compared to iron. Both ferritin and iron are most abundant in substantia nigra, globus pallidus, ventral pallidus, mossy fibers and granule cell layers of the cerebellar cortex. Differences in distribution and abundance for ferritin and iron have been detected in hippocampus and lateral septum. Ferritin is abundant in mossy fibers and throughout the septal ganglia, whereas iron is abundant in the pyramidal cell layer of the hippocampus, in the dorsolateral septal nucleus and granule cell layer of the dentate gyrus.

Martin *et al.* (1985) have found ferritin localized by immunocytochemistry exclusively in neurons in various regions of the brain such as the cerebral cortex and globus pallidus. Ferritin immunoreactivity has been found in oligodendroglia of adult human brain (Gerber and Connor, 1989). We have also performed an immunocytochemical analysis of the expression of MBP, Tf and ferritin in cultured oligodendrocytes, as well as in primary glial cultures containing both astrocytes and oligodendrocytes. In both types of cultures ferritin was found in oligodendrocytes. The immunocytochemical pattern is similar to the distribution of MBP+/Tf+ subpopulations (Espinosa and de Vellis, 1988). Ferritin was clearly visualized inside oligodendroglia as bright granular structures. Double immunocytochemistry was also performed for both iron-related proteins, Tf and ferritin. These two proteins were found to co-localize within oligodendrocytes. Astrocytes identified as GFAP+ cells were not ferritin positive. Similar phenotypes were found to be ferritin positive in 17-day-old rat brain sections (Espinosa *et al.*, unpublished observations). Quantitative studies of ferritin and Tf expression in primary glial cell cultures, described above, were performed in relation to MBP developmental profile (Fig. 4.2). The results show that ferritin appears first before BBB develops, followed sequentially by Tf and MBP.

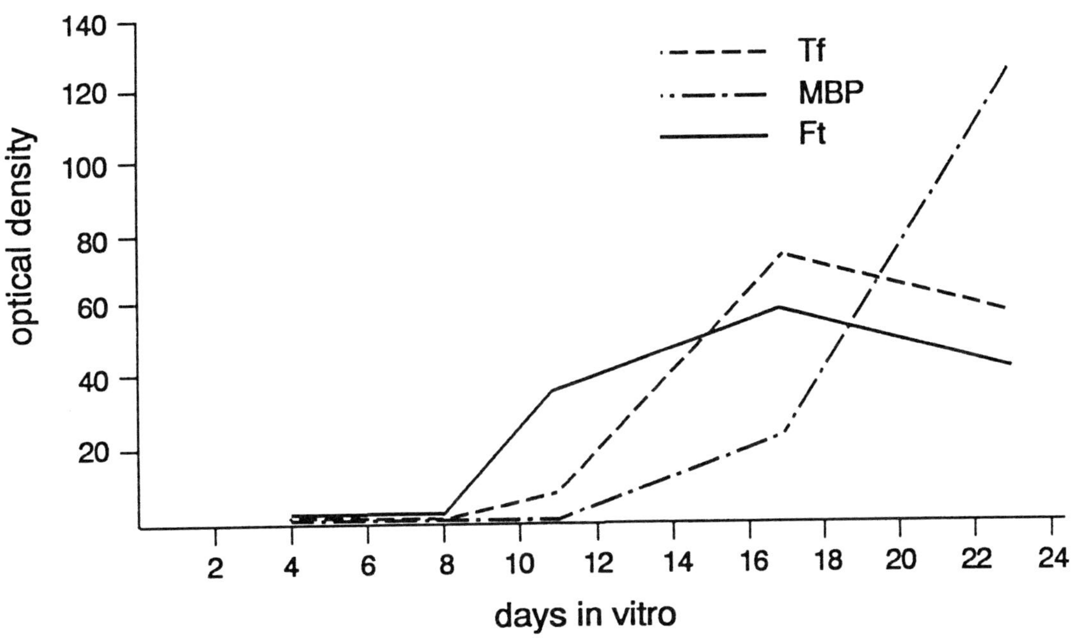

Fig. 4.2.   Time course determination (Elisa) of Tf and ferritin (Ft) in primary glial cultures (containing astrocytes and oligodendrocytes). The developmental pattern of iron-related proteins (Tf and Ft) was compared to the oligodendrocyte and myelin marker, myelin basic protein (MBP). Tf and ferritin display a similar developmental pattern, appearing earlier than MBP. Values represent the mean of absorbance units at 405nm from three experiments composed of triplicates. SD fluctuated between ± 0.007 and ± 0.1.

4.6   THE Tf-R IN ABNORMAL AND PATHOLOGICAL CONDITIONS

4.6.1   THE Tf-R AS A MARKER FOR ALTERED IRON METABOLISM

A number of reports indicate that in addition to the membrane-associated Tf-R, a soluble form of the Tf-R (Sol Tf-R) can be detected in human serum or plasma (Kohgo, 1986). The levels of Sol Tf-R are elevated in patients with iron-deficiency, anemia, autoimmune hemolytic anemia and polycythemia. Sol Tf-R values are lower in patients with anaplastic anemia. The exact biochemical nature for this Sol Tf-R is not fully defined, but it is recognized by the monoclonal antibodies OKT9 and B3/25. The antibodies were developed against intact cells (Reinherz *et al.*, 1980; Schneider *et al.*, 1982) and may react mainly with the extracellular portion of the Tf-R molecule (Flowers *et al.*, 1989). It is not known if Sol Tf-R is present in the cerebrospinal fluid under some abnormal conditions.

4.6.2 Tf-R IN MYELIN DEFICIENCY

Lin and Connor (1989) have published a descriptive immunocytochemical study of the developmental pattern for Tf and Tf-R which they compared to the oligodendroglial/myelin markers, myelin basic protein (MBP) and GC, and to the astrocyte marker GFAP, both in the normal and myelin-deficient (md) rat. In the normal rat optic nerve, Tf-R appears earlier than Tf itself. They also confirmed our data (Espinosa *et al.*, 1987) concerning the early appearance of Tf preceding the synthesis of myelin components. GC and MBP appeared simultaneously at PD10, whereas in corpus callosum and cerebellum they appeared at PD6. The authors did not find any correlation between Tf-Tf-R expression and GFAP.

In md rat mutants Tf and Tf-R immunoreactivity was found only in endothelial cells by Lin and Connor (1989), who also reported an absence of Tf-R, Tf, MBP and GC expression. These findings confirm our previous reports

(Kumar *et al.*, 1988, 1990; Gordon *et al.*, 1989; Espinosa and de Vellis, 1990) that have shown a marked reduction of all oligodendrocytes/myelin markers. Astrogliosis has also been described in md rats by Dentinger *et al.* (1985) and confirmed by Lin and Connor (1989). The myelin-deficient rat mutants as well as all other proteolipid protein mutants, led to an inhibition of differentiation and maturation of the oligodendroglial cell, resulting in the almost total absence of myelin-related and oligodendroglial (myelin independent) markers, including Tf. However, a low number of receptors should exist in order to keep CNS cells alive in the md rat. It is possible that the immunocytochemical technique used by Lin and Connor (1989) was not sensitive enough to detect low levels of the Tf receptor. Despite the temporal association between myelinogenesis and Tf accumulation in oligodendrocytes in the normal rat brain, and the negative results obtained for oligodendroglial markers as well as for Tf accumulation in the oligodendroglial cells of the md rat, it is not possible to affirm that both phenomena, myelinogenesis and Tf synthesis are directly linked; more work needs to be done to ascertain whether Tf is a causal factor in myelinogenesis, in addition to its well-known function as an iron transport protein.

4.6.3 THE Tf-R IN REGENERATING CELLS

After injury, cells show a number of metabolic and protein changes that occur in response to the altered microenvironment, besides the response to epigenetic factors transiently present in the post-injured system. Neuronal cells seem to have an inherent neuronal regeneration program which eventually leads to restitution of function. The regeneration mechanisms have not yet been characterized. However, Graeber *et al.* (1989) have shown that axotomy of motor neurons results in a strong but transient increase in Tf-Rs. This large amount of Tf-R expression is associated with an elevated uptake of exogenous iron.

This finding supports the idea of neuron activation for repair. For instance, iron is an essential co-factor in mitochondrial oxidative enzyme metabolism. During nervous system development the mitochondrial form of malate dehydrogenase (mMDH) occurs coincidentally with the onset of oxidative metabolism, and the rise of mMDH follows a peak of intra-neuronal Tf immunoreactivity (Dion *et al.*, 1988). Neurons may recapitulate to some extent developmental stages where serum is present. Iron is also necessary for the enzyme ribonucleotide reductase, which exhibits strong activity in correlation to DNA synthesis (Laskey *et al.*, 1988).

Remyelination is poor within the CNS, in contrast to the PNS where total remyelination occurs. Remyelination can be taken as an example of glial regeneration. Preliminary data from our laboratory indicate that cultured myelinating oligodendrocytes that are $MBP^+/Tf^-/MBP^+/Tf^+$ or MBP^-/Tf^+ can be injured by pH changes, exposure to high oxygen and freeze–thaw techniques. After injury all three populations of oligodendroglial cells lose cell processes and a concomitant transient expression of Tf within the cell body is observed. After 6 to 8 days, some cells start losing Tf immunoreactivity and express MBP at the same time as new cell processes appear and eventually myelin-like membrane formation occurs. These regenerating phenomena are dependent on the severity of injury and age in culture. This partial recovery mimics the poor remyelination displayed *in vivo* within the adult and aged brain. The Tf accumulation observed in our injury models suggests that oligodendroglia, as well as neurons, exhibit a high requirement of iron for new protein synthesis.

Recent studies (Graeber *et al.*, 1989) of the increase in Tf-R and iron uptake by regenerating motorneurons, as well as the recapitulation of transient Tf accumulation by oligodendrocytes after injury reported above, suggest that Tf-R may play a key role in neural cell regeneration.

4.7. REGULATION OF IRON METABOLISM AT THE MOLECULAR LEVEL

The uptake and storage of iron is regulated by the expression of Tf-R and iron content. The mediation of iron transport and storage involves the iron carrier glycoprotein Tf.

4.7.1   REGULATION OF FERRITIN SYNTHESIS

The correlation of ferritin levels with iron levels was reported by Granick (1946). The regulation of ferritin biosynthesis by iron was proposed by Hentze *et al.* (1986). Deletion experiments in the 5' upstream region of the ferritin gene led to the identification of sequences involved in iron regulation. The full mRNA contains a 5' untranslated region (5' UTR) that can form 'stem loop' structures and is responsible for the iron-dependent translational control of the ferritin gene (for review see Klausner, 1988). This 5' UTR is present in both heavy and light ferritin chains in human, rat, rabbit, chicken and bullfrog. An iron-responsive element (IRE) within the 5' UTR (Rouault *et al.*, 1988; Leibold and Munro, 1988) provides the recognition site for binding of protein(s) which may act as a repressor of translation. Leibold and Munro (1988) and Rouault *et al.* (1988) have identified such cytoplasmic proteins responsible for the regulation of Tf synthesis. There is also a 3' untranslated region of bullfrog mRNA that is essential for translation regulation (Dickey *et al.*, 1988). The existence of sequences similar to the metal regulating elements in the 5' flanking sequences of metallothionein have been reported (Drysdale, 1988) (for review see Worwood, 1989).

4.7.2   REGULATION OF Tf-R SYNTHESIS

Iron also regulates Tf-R synthesis and Klausner and colleagues (1988) have thoroughly elucidated the regions of the Tf-R gene which are implicated in the iron-dependent regulation. In contrast to the situation with ferritin, the levels of Tf-R synthesis in response to iron correlate with the levels of Tf-R mRNA.

A 5' flanking region and part of a 3' untranslated region are in charge of the regulation for the Tf-R synthesis. However, most of the regulation resides in a number of IRE stem loop structures similar to those found in the 5' UTR of ferritin mRNA (Casey *et al.*, 1988).

It seems that an IRE at the 5' terminal region of the ferritin protein sequence inhibits translation if iron is absent, whereas several IREs located at the 3' end confer mRNA protection by binding a protein (in the absence of iron) or by avoiding protein binding if iron is present, leading to mRNA degradation (Klausner, 1988).

4.7.3 REGULATION OF Tf SYNTHESIS

Tf synthesis is tissue-specific and varies during development. There is a complex system of regulation. Iron also regulates Tf synthesis through metal regulatory elements (MREs) located at the 5' end of the human Tf gene (Adrian *et al.*, 1986). These MREs are similar to enhancer sequences found in the promoter region of the metallothionein gene which respond to copper and zinc. Iron inhibits Tf synthesis when bound to MREs (for review see Funk and MacGillivray, 1990).

4.8 CONCLUDING REMARKS AND FUTURE DIRECTIONS

The differences in iron transport and metabolism between the developing and adult brain are illustrated diagramatically in Fig.4.3. It seems that because of the blood–brain barrier and despite some transcytosis that may allow some peripheral iron-loaded Tf into the brain, the CNS is an independent system that exhibits all the machinery for iron homeostasis. Under normal conditions the oligodendrocytes synthesize and secrete Tf, and this gene seems to have a particular regulatory mechanism that differs to some extent from liver Tf gene regulation. The synthesis of ferritin is also performed mainly by the oligodendroglial cell and some neurons. This step ensures the safety of the environment by storage of non-utilized iron. The coordinated interaction of iron-related proteins necessary for iron homeostasis and normal CNS function is controlled by cell–cell interactions, trophic factors, and hormones.

During development iron requirements are larger than in adulthood, since developmental activities include cell proliferation, growth, migration and differentiation. The CNS also displays some capability to protect itself from an adverse environment. During abnormal conditions, such as pathological alterations and injury, the main cellular components of the CNS, oligodendrocytes, astrocytes and neurons, seem to be the machinery for the synthesis of iron-related protein. This response appears to recapitulate in part some developmental stages that are required to recover their normal functions. However, regeneration within the CNS generally remains incomplete and poor when compared to the peripheral nervous system.

Many factors affect the regeneration process, including blood–brain barrier disruption leading to entry of circulating substances into the brain, tissue degeneration, myelin disruption and macrophage invasion. These pathological reactions lead to iron liberation from dying cells and myelin breakdown. Free iron leads to free radical formation and lipid peroxidation.

Although it seems that CNS cells have the ability to regenerate, the disruptive events in the initial insult occur faster than the regenerating process can begin. The elucidation (or identification) of time course reactions immediately post-trauma or injury can lead us to identify the factors that favor regeneration and those factors that may (a) prevent further damage and (b) retard or prevent cell recovery. Various approaches, including the characterization of *in vivo* and *in vitro* models that mimic the response of nervous tissue to injury, are being studied in our laboratory. Initially,

EARLY POST NATAL STAGES

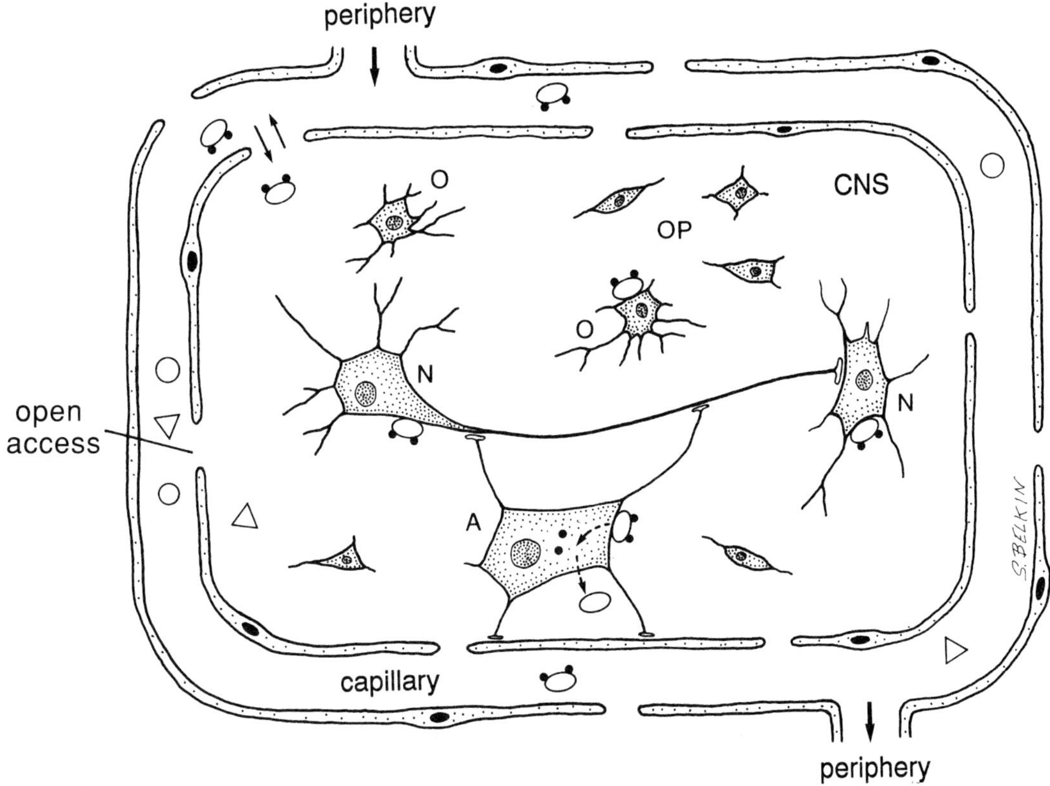

Fig. 4.3 (a). Diagrammatic representation of CNS parenchyma in relation to its blood supply during early postnatal (P0 to approx P10) stages. CNS parenchyma during embryonic life is readily accessible to influences from the periphery since it lacks a blood–brain barrier. Molecules and proteins contained within plasma have free access to the CNS and vice versa. Among such molecules is the iron-carrier glycoprotein, Tf. Tf is synthesized in the liver and circulates in the plasma providing iron to all tissues including CNS cells. This free access also exists throughout the first week of postnatal life. As the blood–brain barrier develops many other phenomena occur in concert within the CNS. Synaptogenesis takes place among non-myelinated neurons (N). Oligodendroglial progenitors (OP) proliferate, migrate, and start to differentiate, undergoing typical morphological changes from bipolar or tripolar flat cells with short cell processes into round multipolar oligodendrocytes (O), with thin long cell processes. Together with these morphological changes the expression of oligodendroglial and myelin markers occurs. At the same time astrocytes (A) send out cell processes inducing endothelial cells to form the blood–brain barrier as well as interacting with neurons.

we are focusing on the early events that occur immediately after injury. Time courses of receptor expression, including Tf-R, and cell markers as well as morphological changes are being followed in various model systems.

Because of the newly discovered function of oligodendrocytes in iron homeostasis discussed in this chapter, it will be important to assess the impact of this function on myelination, response of nervous tissue to injury,

NORMAL YOUNG AND ADULT STAGES

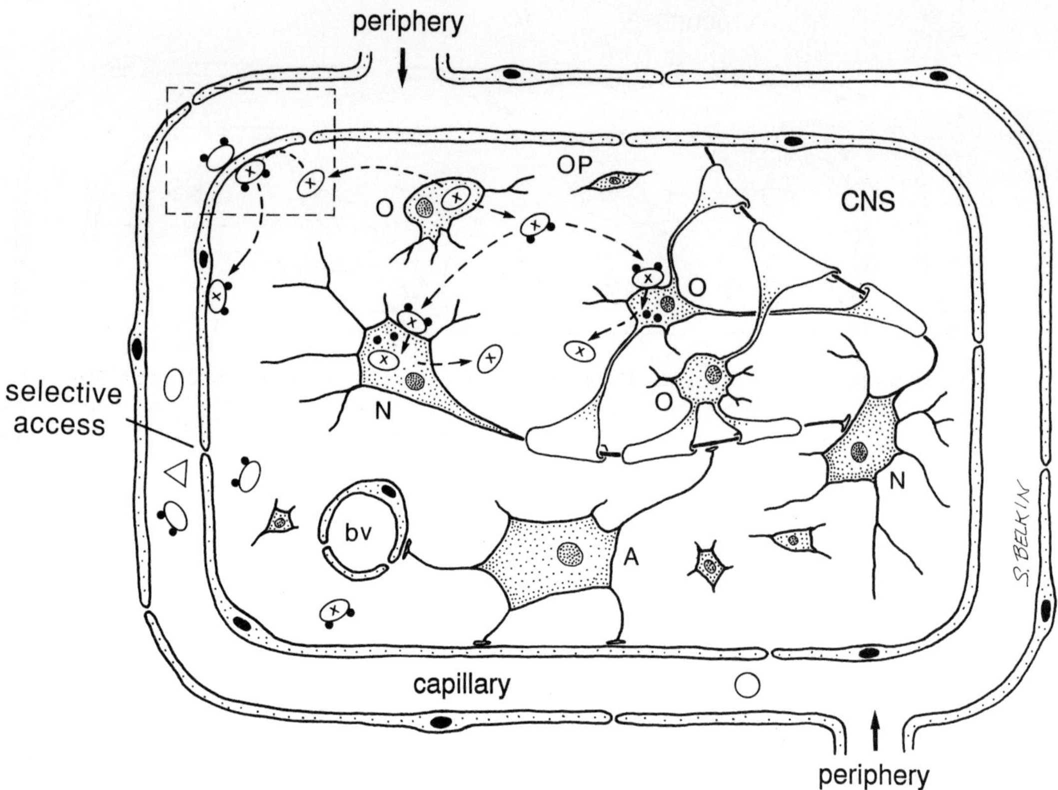

Fig. 4.3. (b).   Diagrammatic representation of normal CNS parenchyma once the blood–brain barrier has been established in young and adult stages. At the time the blood–brain barrier develops oligodendrocytes (O) are actively myelinating. Once myelination is completed the role of the oligodendrocytes is to maintain the myelin sheath. Here we propose a second role for oligodendrocytes; the BBB decreases the amount of available sero-Tf within the CNS. This could be the signal for oligodendrocytes to synthesize and secrete CNS-Tf. This newly synthesized CNS-Tf can reach either other myelinating oligodendrocytes, astrocytes and neurons acting as a paracrine trophic factor, or it could also be used as an autocrine factor providing iron to the same cell (not shown on the diagram).

remyelination and other aspects of regeneration in the CNS. Obviously, the ability of oligodendrocytes to maintain iron homeostasis in a dramatically changing environment involves mechanisms to regulate iron-related protein expression. The molecular biology pertinent to this regulation has just begun to be studied by a few researchers. They will be guided by the rapid advances that have been made in other organs, resulting in the identification of several transcriptional regulatory mechanisms. Some similarities in regulation between the CNS and other tissues will certainly exist. However, one can expect to find tissue-specific mechanisms.

ACKNOWLEDGEMENTS

We thank Nancy Wainwright for assisting with the preparation of the manuscript. We

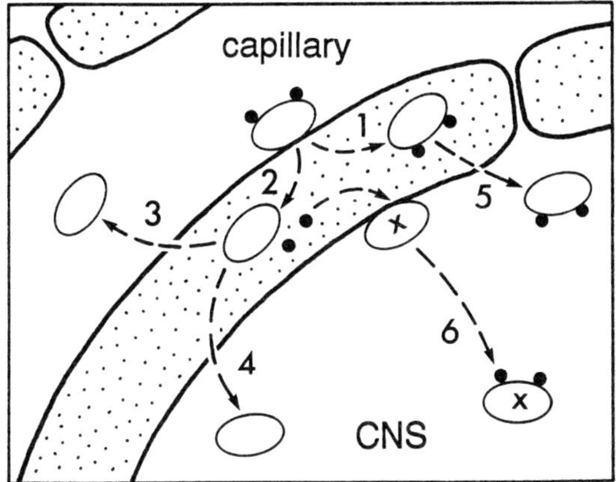
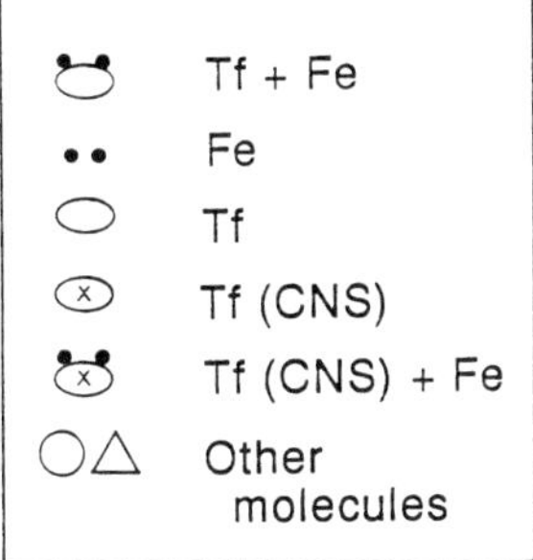

Fig. 4.3. (c). Diagrammatic representation of alternative possible pathways for sero-Tf and CNS-Tf. It has been shown that sero-Tf can go into the CNS by transcytosis through the endothelium. (1) In this case sero-diferric Tf would bind to the Tf-R on the surface of the endothelial cell. Once inside the cell, the same molecule would either deliver its iron (2) and recycle to the bloodstream (3) as apo-Tf; or deliver its iron and enter the CNS parenchyma. This apo-Tf may bind two more iron molecules and deliver them to a CNS cell (4). Another likely possibility is that iron delivered by sero-Tf would be mobilized out of the capillary cell and captured by CNS-Tf, this diferric CNS-Tf could deliver its iron to a CNS cell (6) and continue the Tf-cycle within the CNS. A fourth, more remote possibility, would be that sero diferric-Tf could go through the capillary wall and itself deliver iron into the CNS environment or a CNS cell (5).

also thank Sharon Belkin and Carol Gray of the MRRC Media Unit for assistance in preparing the illustrations. This work was supported by NIH Grant HD-06576 and DOE Contract DE-FC03-87-ER60615.

REFERENCES

Adrian, G. S., Korinek, B. M. and Yang, F. (1986) The human transferrin gene: 5' region contains conserved sequences which match the control elements regulated by heavy metals glucocorticoids and acute phase reaction. *Gene*, **49**, 167–75.

Barclay, A. N. (1981) The localization of populations of lymphocytes defined by monoclonal antibodies in rat lymphoid tissues. *Immunology*, **42**, 915–27.

Bienfait, H. F. and Van den Briel, M. L. (1981) Rapid mobilization of ferritin iron by ascorbate in the presence of oxygen. *Biochim. Biophys. Acta*, **631**, 507–10.

Bloch, B., Popovici, T., Levin, M. J. *et al.* (1985) Transferrin gene expression visualized in oligodendrocytes of the rat brain by using *in situ* hybridization and immunohistochemistry. *Proc. Natl. Acad. Sci. USA*, **82**, 6706–10.

Bloch, B., Popovichi, T., Chahman, S. *et al.* (1987) Transferrin gene expression in choroid plexus of the adult rat brain. *Brain Res. Bull.*, **18**, 573–6.

Boison, D. and Stoffel, W. (1989) Myelin-deficient rat: a point mutation in exon III (A–C, Thr 75–Pro) of the myelin proteolipid protein causes dysmyelination and oligodendrocyte death. *EMBO J.*, **8**, 3295–302.

Boyd, D., Vecoli, C., Belcher, D. M. *et al.* (1985) Structural and functional relationships of human ferritin H- and L-chains deduced from cDNA clones. *J. Biol. Chem.*, **26**, 11755–61.

Bridges, K. R. (1990) Modulation of the availability of intracellular iron, in *Iron Transport and Storage* (eds P. Ponka, H. M. Schulman, and R. C. Woodworth), CRC Press, Boston, pp. 297–314.

Brown, P. J., Hewichk, R. M., Hellström, I. *et al.* (1982) Human melanoma-associated antigen p97 is structurally and functionally related to transferrin. *Nature*, **296**, 171–3.

Brunel, F., Ochoa, A., Schaeffer, E. *et al.* (1988) Interactions of DNA-binding proteins with the

5′ region of the human transferrin gene. *J. Biol. Chem.*, **263**, 10180–5.

Cammack, R., Wrigglesworth, J. M. and Baum, H. (1990) Iron-dependent enzymes in mammalian systems, in *Iron Transport and Storage* (eds P. Ponka, H. M. Schulman and R. C. Woodworth), CRC Press, Boston, pp. 17–39.

Casey, J. L., Hentze, M. W., Koeller, D. M., *et al.* (1988) Iron-responsive elements: regulatory RNA sequences that control mRNA levels and translation. *Science*, **240**, 924–8.

Cavanagh, M. E. and Mollgard, K. (1985) An immunological study of the distribution of some plasma proteins within the developing forebrain of the pig with special reference to the neocortex. *Brain Res.*, **349**, 182–94.

Chasteen, N. D. (1983) Identification of the probable locus of iron and anion binding in the transferrins. *Trends Biochem. Sci.*, **8**, 272–5.

Connor, J. R. and Fine, R. (1986) The distribution of transferrin immunoreactivity in the rat nervous system. *Brain Res.*, **368**, 319–28.

Connor, J. R. and Fine, R. (1987) Development of transferrin-positive oligodendrocytes in the rat CNS. *J. Neurosci. Res.*, **17**, 51–9.

Cook, S. F. (1929) Structure and composition of hemosiderin. *J. Biol. Chem.*, **82**, 595–609.

Crichton, R. R. and Charloteaux-Wauters, M. (1987) Iron transport and storage. *Eur. J. Biochem.*, **164**, 485–506.

Dentinger, M. P., Barron, K. D. and Csiza, C. K. (1985) Glial and axonal development in optic nerve of myelin deficient rat. *Brain Res.*, **344**, 255–66.

Diamond, A., Cooper, G. M., Ritz, J. and Lane, M. A. (1983) Identification and molecular cloning of the human Blym gene activated in Burkitts lymphomas. *Nature*, **305**, 112–16.

Dickey, L. F., Wang, Y. H, Shull, G. E. *et al.* (1988) The importance of the 3′-untranslated region in the translational control of ferritin mRNA. *J. Biol. Chem.*, **263**, 3071–4.

Dion, T. L., Markelonis, G. J., Oh, T. H. *et al.* (1988) Immunocytochemical localization of transferrin and mitochondrial malate dehydrogenase in the developing nervous system of the rat. *Dev. Neurosci.*, **10**, 152–64.

Drysdale, J. W. (1988) Human ferritin gene expression. *Prog. Nucleic Acid Res. Mol. Biol.*, **35**, 127–55.

Dziegielewska, K. M., Evans, C. A., Lorscheider, F. L. *et al.* (1981) Plasma proteins in fetal sheep brain: blood brain barrier and intracerebral distribution. *J. Physiol.*, **318**, 239–50.

Espinosa de los Monteros, A. (1987) Contribution a l'etude des oligodendrocytes *in vitro*. University Louis Pasteur, Strasbourg. Thesis for PhD.

Espinosa de los Monteros, A. and de Vellis, J. (1988) Myelin basic protein and transferrin characterize different sub-populations of oligodendrocytes in rat primary glial cultures. *J. Neurosci. Res.*, **21**, 181–7.

Espinosa de los Monteros, A. and de Vellis, J. (1990) Oligodendrocyte differentiation: developmental and functional subpopulations, in *Cellular and Molecular Biology of Myelination* (eds G. Jeserich, H. H. Althaus, T. V. Waehneldt), Springer-Verlag, Berlin, pp. 33–45.

Espinosa de los Monteros, A. and Foucaud, B. (1987) Effect of iron and transferrin on pure oligodendrocytes in culture: characterization of a high-affinity transferrin receptor at different ages. *Dev. Brain Res.*, **35**, 123–30.

Espinosa de los Monteros, A., Roussel, G., Neskovic, N. and Mussbaum, J. -L (1986) A study *in vitro* of premyelination oligodendrocytes. *Biochem. Soc. Trans.*, **14**, 648–50.

Espinosa de los Monteros, A., Chiappeli, F., Fisher, R. S. and de Vellis, J. (1988) Transferrin: an early marker of oligodendrocytes in culture. *Int. J. Dev. Neurosci.*, **6**, 167–75.

Espinosa de los Monteros, A., Peña, L. A. and de Vellis, J. (1989) Does transferrin have a special role in the nervous system? *J. Neurosci. Res.*, **24**, 125–36.

Espinosa de los Monteros, A., Kumar, S., Scully, S. *et al.* (1990) Transferrin gene expression and secretion by rat brain cells *in vitro*. *J. Neurosci. Res.*, **25**, 576–80.

Fischbach, F. A. and Anderbergg, J. W. (1965) An X-ray scattering study of ferritin and apoferritin. *J. Mol. Biol.*, **14**, 458–73.

Fishman, J. B., Rubin, J. B., Handrahan, J. V. *et al.* (1987) Receptor-mediated transcytosis of transferrin across the blood–brain barrier. *J. Neurosci. Res.*, **18**, 299–304.

Flowers, C. H., Skikne, B. S., Covell, A. M. and Cook, J. D. (1989) The clinical measurement of serum transferrin receptor. *J. Lab. Clin. Med.*, **114**, 368–77.

François, C., Nguyen-Legros, J. and Percheron, G. (1981) Topographical and cytological localization of iron in rat and monkey brain. *Brain Res.*, **215**, 317–22.

Funk, F., Lenders, J. P., Crichton, R. R. and Schneider, W. (1985) Reductive mobilisation of ferritin iron. *Eur. J. Biochem.*, **152**, 167–72.

Funk, W. D. and MacGillivray, T. A. (1990) Molecular biology of the transferrins, in *Iron*

Transport and Storage (eds P. Ponka, H. M. Schulman and R. C. Woodworth), CRC Press, Boston, pp. 103–18.

Galbraith, R. M., Werner, P., Arnaud, P. and Gillian M. P. (1980) Transferrin binding to peripheral blood lymphocytes activated by phytohaemoglutinin involves a specific receptor: ligand interaction. *J. Clin. Invest.*, **66**, 1135–43.

Gale, K. and Casu, M. (1981) Dynamic utilization of GABA in substantia nigra: regulation by dopamine and GABA in the striatum, and its clinical and behavioral implications. *Mol. Cell Biochem.*, **39**, 369–405.

Gerber, M. R. and Connor, J. R. (1989) Do oligodendrocytes mediate iron regulation in the human brain? *Neurology*, **26**, 95–8.

Gordon, M. N., Kumar, S., Espinosa de los Monteros, A. *et al.* (1989) Developmental regulation of myelin-associated genes in the normal and the myelin-deficient mutant rat, in *Molecular Aspects of Development and Aging of the Nervous System* (eds J. Lauder, A. Privat, A. E. Giacobini *et al.*) Plenum, New York, pp. 11–22.

Goubin, G., Goldman, D. S., Luce, J. *et al.* (1983) Molecular cloning and nucleotide sequence of a transforming gene detected by transfection of chicken B-cell lymphoma DNA. *Nature*, **302**, 114–19.

Graeber, M. B., Raivich, G. and Kreutzberg, G. W. (1989) Increase of transferrin receptors and iron uptake in regenerating motor neurons. *J. Neurosci. Res.*, **23**, 342–5.

Granick S. (1946) Ferritin increase of the protein apoferritin in the gastrointestinal mucosa as a direct response to iron feeding. The function of ferritin in the regulation of iron absorption. *J. Biol. Chem.*, **164**, 737–46.

Hamilton, T. A. (1985) Regulation of transferrin receptor expression in concanavalin-A stimulated and Gross virus transformed rat lymphoblasts. *J. Cell Physiol.*, **113**, 40–6.

Hamilton, T. A., Wada, H. G. and Susman, H. H. (1979) Identification of transferrin receptors on the surface of human cultured cells. *Proc. Natl. Acad. Sci. USA*, **76**, 6406–10.

Harding, C., Heuser, J. and Stahl, P. (1984) Endocytosis and intracellular processing of transferrin and colloidal gold-transferrin in rat reticulocytes: demonstration of a pathway for receptor shedding. *Eur. J. Cell Biol.*, **35**, 256–63.

Harrison, P. M., Andrews, S. C., Artymiuk, P. J. *et al.* (1990) Ferritin, in *Iron Transport and Storage* (eds P. Ponka, H. M. Schulman and R. C. Woodworth), CRC Press, Boston, pp. 81–101.

Hattori, M., Hidaka, S. and Sakaki, Y. (1985) Sequence analysis of a KpnI family member near the 3' end of human β-globin gene. *Nucleic Acids Res.*, **13**, 7813–27.

Hentze, M., Keim, S., Hartford, J. B. *et al.* (1986) Characterization, expression, and chromosomal localization of a human ferritin heavy chain gene. *Proc. Natl. Acad. Sci. USA*, **83**, 7226–30.

Hill, J. M. (1985) Iron concentration reduced in ventral pallidum, globus pallidus and substantia nigra by GABA-transaminase inhibitor, *gamma*-vinyl GABA. *Brain Res.*, **342**, 18–25.

Hill, J. M. (1990) Iron and proteins of iron metabolism in the central nervous system, in *Iron Transport and Storage* (eds P. Ponka, H. M. Schulman and R. C. Woodworth), CRC Press, Boston, pp. 315–30.

Hill, J. M. and Switzer, R. D. (1984) The regional distribution and cellular localization of iron in the rat brain. *Neuroscience*, **11**, 595–603.

Hill, J. M., Ruff, M. R., Weber, R. J. and Pert, C. B. (1985) Transferrin receptors in rat brain: neuropeptide-like pattern and relationship to iron distribution. *Proc. Natl. Acad. Sci. USA*, **82**, 4553–7.

Jefferies, W. A., Brandon, M. R., Hunt, S. U. *et al.* (1984) Transferrin receptor on endothelium of brain capillaries. *Nature*, **312**, 162–3.

Jeltsch, J. -M., Hen, R., Maroteaux, L. *et al.* (1987) Sequence of the chicken ovotransferrin gene. *Nucleic Acids Res.*, **15**, 7643–5.

Jones, T., Spencer, R. and Walsh, C. (1978) Mechanism and kinetics of iron release from ferritin by dihydroflavins and dihydroflavin analogues. *Biochemistry*, **17**, 4011–17.

Klausner, R. D. (1988) From receptors to genes – insights from molecular iron metabolism. *Clin. Res.*, **36**, 494–500.

Kohgo, Y. (1986) Structure of transferrin and transferrin receptor. *Acta Haematol. Jpn*, **49**, 1627–34.

Kumar, S., Gordon, M. N., Espinosa de los Monteros, A. and de Vellis, J. (1988) Developmental expression of neural cell type-specific mRNA markers in the myelin-deficient mutant rat brain: inhibition of oligodendrocyte differentiation. *J. Neurosci. Res.*, **21**, 268–74.

Kumar, S., Macklin, W. B., Gordon, M. N. *et al.* (1990) Transcriptional regulation studies of myelin-associated genes in myelin-deficient mutant rats. *Dev. Neurosci.*, **12**, 316–25.

Larrick, J. W. and Cresswell, P. (1979) Modulation

of cell surface iron transferrin receptors by cellular density and state of activation. *J. Supramol. Struct.*, **11**, 579–86.

Laskey, J., Webb, I., Schulman, H. M. and Ponka, P. (1988) Evidence that transferrin supports cell proliferation by supplying iron for DNA synthesis. *Exp. Cell Res.*, **176**, 87–95.

Laufberger, V. (1937) Sur la cristallisation de la ferritine. *Bull. Soc. Chim. Biol.*, **19**, 1575–82.

Leibold, E. A. and Munro, H. N. (1988) A cytoplasmic protein binds *in vitro* to a highly conserved sequence in the 5'-untranslated region of ferritin H and L subunit mRNAs. *Proc. Natl. Acad. Sci. USA*, **88**, 2171–5.

LeVine, S. M. and Macklin, W. B. (1990) Iron-enriched oligodendrocytes: a reexamination of their spatial distribution. *J. Neurosci. Res.*, **26**, 508–12.

Lin, H. H. and Connor, J. R. (1989) The development of the transferrin–transferrin receptor system in relation to astrocytes, MBP and galactocerebroside in normal and myelin-deficient rat optic nerves. *Dev. Brain Res.*, **49**, 281–93.

Loscher, W. (1980) A comparative study of the pharmacology of inhibitors of GABA metabolism. *Naunyn Schmeidebergs Arch. Pharmacol.*, **315**, 119–28.

Markelonis, G. J., Oh, T. H., Park, L. P. *et al.* (1985a) Receptor-mediated uptake of labeled transferrin by embryonic chicken dorsal root ganglion neurons in culture. *Int. J. Dev. Neurosci.*, **3**, 257–66.

Markelonis, G. J., Oh, T. H., Park, L. P. *et al.* (1985b) Synthesis of the transferrin receptor by cultures of embryonic chicken spinal neurons. *J. Cell Biol.*, **100**, 8–17.

Martin, T. L., Switzer, P. C., Joshi, J. and Zimmerman, A. (1985) Ferritin localization in rat brain by immunocytochemistry. *Soc. Neurosci. Abstr.*, **11**, 502.

Mash, D. C., Pablo, J., Flynn, D. D. *et al.* (1990) Characterization and distribution of transferrin receptors in the rat brain. *J. Neurochem.*, **55**, 1972–9.

Mazur, A., Green, S., Salia, A. and Charleton, A. (1958) Mechanism of release of ferritin iron *in vitro* by xanthine oxidase. *J. Clin. Invest.*, **37**, 1809–17.

Metz-Boutigue, M. -H., Jolles, J., Mazurier, J. *et al.* (1984) Human lactotransferrin: amino acid sequence and structural comparisons with other transferrins. *Eur. J. Biochem.*, **145**, 659–76.

Mollgard, K. and Jacobsen, M. (1984) Immunohistochemical identification of some plasma proteins in human embryonic and fetal forebrain with particular reference to the development of the neocortex. *Dev. Brain Res.*, **13**, 49–63.

Morgan, E. H. (1981) Transferrin: biochemistry, physiology and clinical significance. *Mol. Aspects Med.*, **4**, 1–123.

Munro, H. N., Leibold, E. A., Aziz, N. *et al.* (1990) Ferritin gene structure and expression, in *Iron Transport and Storage*. CRC Press, Boca Raton, FL, pp. 133–48.

Neckers, L. M., Yenokida, G. and James, S. P. (1984) The role of the transferrin receptor in human B-lymphocyte activation. *J. Immunol.*, **133**, 2437–41.

Newman, R., Schneider, C., Sukerland, R. *et al.* (1982) The transferrin receptor. *Trends Biochem. Sci.*, **7**, 397–400.

O'Connell, M. J., Ward, R. J., Baum, H. and Peters, T. J. (1985) The role of iron in ferritin- and haemosiderin-mediated lipid peroxidation in liposomes. *Biochem. J.*, **229**, 135–9.

O'Connell, M., Halliwell, B., Moorhouse, C. P. *et al.* (1986) Formation of hydroxyl radicals in the presence of ferritin and haemosiderin. *Biochem. J.*, **234**, 727–31.

Oh, T. H. and Markelonis, G. J. (1982) Chicken serum transferrin duplicates the myotrophic effects of sciatin on cultured muscle cells. *J. Neurosci. Res.*, **8**, 535–45.

Oh, T. H., Markelonis, G. J., Royal, G. M. and Bregman, B. S. (1986) Immunocytochemical distribution of transferrin and its receptor in the developing chicken nervous systems. *Dev. Brain Res.*, **30**, 207–20.

Omary, M. B., Trowbridge, I. S. and Minowada, J. (1980) Human cell surface glycoprotein with unusual properties. *Nature*, **286**, 888–91.

Pardridge, W. M., Eisenberg, J. and Yang, J. (1987) Human blood–brain barrier transferrin receptor metabolism. *Metabolism*, **36**, 892–5.

Park, I., Schaeffer, I., Sidoli, A. *et al.* (1985) Organization of the human transferrin gene: direct evidence that it originated by gene duplication. *Proc. Natl. Acad. Sci. USA*, **82**, 3149–53.

Rao, K. K., Hartford, J., Rouault, T. *et al.* (1986) Transcriptional regulation by iron of the gene for the transferrin receptor. *Cell Mol. Biol.*, **6**, 236–40.

Reinherz, E., Kung, P., Goldstein, G. *et al.* (1980) Discrete stages of human intrathymic differentiation: analysis of normal thymocytes and leukemic lymphoblasts of T-cell lineage. *Proc. Natl. Acad. Sci. USA*, **77**, 1588–92.

Reynolds, M. L. and Mollgard, K. (1985) The distribution of plasma proteins in the neocortex and early allocortex of the developing sheep brain. *Anat. Embryol.*, **171**, 41–60.

Rose, T. M., Plowman, G., Teplow, D. *et al.* (1986) Primary structure of the human melanoma-associated antigen p97 (melanotransferrin) deduced from mRNA sequence. *Proc. Natl. Acad. Sci. USA*, **83**, 1261–5.

Roskams, A. J. and Connor, J. R. (1989) Alterations in transferrin and its receptor in the brain of myelin-deficient rats. *Ann. NY Acad. Sci.*, **19**, 391–3.

Roskams, A. J. and Connor, J. R. (1990) Aluminum access to the brain: a role for transferrin and its receptor. *Proc. Natl. Acad. Sci. USA*, **87**, 9024–7.

Rouault, T. A., Hentze, M. W., Caughman, S. W. *et al.* (1988) Binding of a cytosolic protein to the iron-responsive element of human ferritin messenger RNA. *Science*, **241**, 1207–10.

Schaeffer, E., Lucero, M. A., Py, M. -C. *et al.* (1987) Complete structure of the human transferrin gene. Comparision with analogous chicken gene and human pseudogene. *Gene*, **56**, 109–16.

Schechter, P. J., Tranier, Y., Jung, M. J. and Bohlen, P. (1977) Audiogenic seizure protection by elevated brain GABA concentration in mice: affects of γ–acetylenic GABA and γ-vinyl GABA, two irreversible GABA-T inhibitors. *Eur. J. Pharmacol.*, **45**, 319–28.

Schneider, C., Sutherland, R., Newman, R. and Greaves, M. (1982) Structural features of the cell surface receptor for transferrin that is recognized by the monoclonal antibody OKT9. *J. Biol. Chem.*, **257**, 8516–22.

Seligman, P. A., Schleicher, R. B. and Allen, R. H. (1979) Isolation and characterization of the transferrin receptor from human placenta. *J. Biol. Chem.*, **254**, 9943–6.

Taylor, E. M. and Morgan, E. H. (1990) Developmental changes in transferrin and iron uptake by the brain in the rat. *Dev. Brain Res.*, **55**, 35–40.

Toran-Allerand, C. D. (1980) Coexistence of α-feto protein, albumin and transferrin immunoreactivity in neurones of the developing mouse brain. *Nature*, **286**, 733–5.

Trowbridge, I. S. and Omary, M. B. (1981) Human cell surface glycoprotein related to cell proliferation is the receptor for transferrin. *Proc. Natl. Acad. Sci. USA*, **78**, 3039–43.

Uzan, G., Frain, M., Park, I. *et al.* (1984) Molecular cloning and sequence analysis of cDNA for human transferrin. *Biochem. Biophys. Res. Commun.*, **119**, 273–81.

Wada, H. G., Hass, P. E. and Sussman, H. H. (1979) Transferrin receptor in human placental brush border membranes. *J. Biol. Chem.*, **254**, 12629–35.

Worwood, M. (1989) An overview of iron metabolism at a molecular level. *J. Intern. Med.*, **226**, 381–91.

Yang, F., Lum, J. B., McGill, S. J. R. *et al.* (1984) Human transferrin: cDNA characterization and chromosomal localization. *Proc. Natl. Acad. Sci. USA*, **81**, 2752–6.

Zaleski, S. (1887) Das Eisen de Organe bei morhus maculosus werlhofii. *Arch. Exp. Pathol. Pharmakol.*, **23**, 77.

ONTOGENIC PROFILE OF EPIDERMAL GROWTH FACTOR RECEPTORS IN RAT BRAIN

5

Jean-Guy Chabot, Satyabrata Kar and Rémi Quirion

5.1 INTRODUCTION

Recently, there has been an increase in research activities on growth factors (GF) and their receptors in the brain (for reviews see Thoenen and Edgar, 1985; Crutcher, 1986; Hefti *et al.*, 1989; Araujo *et al.*, 1990; Perry, 1990; Plata-Salamán, 1991). Emphasis has focused on the role of neurotrophin-related GF, such as nerve growth factor (NGF) and brain-derived nerve growth factor (BDNF), in brain functions (Thoenen and Edgar, 1985; Hefti and Weiner, 1986; Levi-Montalcini and Calissano, 1986; Hefti *et al.*, 1989; Perry, 1990). However, it is clear that many other classes of GF are expressed, and are biologically active, in the central nervous system (CNS) (LeRoith *et al.*, 1987; Baskin *et al.*, 1988; Araujo *et al.*, 1990; Plata-Salamán, 1991). Over the past few years, our attention has focused on the characterization and biological functions of epidermal (EGF) and insulin (IGF)-like GFs in the developing and mature brain (Chabot *et al.*, 1988; Quirion *et al.*, 1988; Araujo *et al.*, 1989a, b, 1990; Kar *et al.*, 1991, 1992). This chapter briefly reviews recent findings concerning the ontogenic profile of EGF receptor binding sites in the rat brain using receptor autoradiography. The possible functional significance of these findings is discussed.

5.2 METHODOLOGICAL CONSIDERATIONS

5.2.1 *IN VITRO* RECEPTOR AUTORADIOGRAPHY

As it has been rather difficult to visualize brain EGF receptor sites using *in vitro* receptor autoradiographic methods alternative approaches including immunohistochemistry have been used (Loy *et al.*, 1987; Gomez-Pinilla *et al.*, 1988; Nieto-Sampedro *et al.*, 1988; Werner *et al.*, 1988b; Styren *et al.*, 1990; Birecree *et al.*, 1991).

We succeeded in developing a protocol (Table 5.1) which allowed the use of *in vitro* receptor autoradiography; relatively good signal-to-noise ratio was achieved especially in late prenatal and early postnatal brains (Chabot *et al.*, 1988; Quirion *et al.*, 1988). Fetal, neonatal and adult Sprague–Dawley rats, obtained from Charles River, Canada were used for all experiments. Brain sections were prepared according to the method of Herkenham and Pert (1982) in order to ensure adherence of sections to slides. Briefly, animals were decapitated and their brains removed and frozen in 2-methylbutane at $-40°C$. The tissues were then serially cut (20 μm) using a cryostat at $-17°C$ and thaw-mounted on precleaned chrome–alum gelatin-

Receptors in the Developing Nervous System Vol.1: *Growth factors and hormones* Edited by Ian S. Zagon and Patricia J. McLaughlin. Published in 1993 by Chapman & Hall. ISBN 0 412 45240 5. Vols. 1 and 2 (set) ISBN 0 412 54520 9.

Table 5.1   Effect of different incubation conditions on the binding of [^{125}I]EGF in neonatal and adult rat brain sections using *in vitro* receptor autoradiography

Buffer composition	Specific binding	
	Neonate	Adult
Tris–HCl (25mM, pH 7.4) + MgCl$_2$(10mM) + 0.1% BSA	+++	–
Tris–HCl (170mM, pH 7.4) + 0.2% BSA + protease inhibitors	n.d.	±
Tris–HCl (170mM, pH 7.4) + 0.2% BSA + EDTA (1mM) + protease inhibitors	n.d.	+++
HEPES (10mM, pH 7.6) + 0.5% BSA + 0.0125% NEM + protease inhibitors	n.d.	–
Krebs–phosphate buffer (pH 7.4) + 0.1% BSA + protease inhibitors	n.d.	–

Incubation for 2 h at room temperature in the presence of 200 pM-[^{125}I]EGF. Non-specific binding was defined in the presence of an excess of unlabeled EGF (100–200 nM). The minus sign (–) indicates that signal-to-noise ratio was very poor. The (+++) icon indicates a good signal-to-noise ratio. n.d., not determined.

coated slides, dehydrated overnight at 4°C under vacuum and kept at –80°C until use.

For the labelling of [^{125}I]EGF binding sites, frozen sections were allowed to acclimate for one hour on ice. In our earlier studies, fetal and postnatal rat brain sections were preincubated in Tris–HCl buffer (25 mM, pH 7.4) containing 0.1% bovine serum albumin (BSA) for 30 min at room temperature (22°C). Incubation was then carried out for 2 h at room temperature in the same buffer containing MgCl$_2$ (10 mM) and 200 pM[^{125}I]EGF. Non-specific binding was evaluated on adjacent sections incubated under the same conditions and in the presence of 100 nM unlabelled mouse EGF (ICN Biomedicals Canada Ltd). At the end of the incubation time, sections were washed four times (2 min each) in ice-cold incubation buffer, dipped in cold deionized water to remove salts before being rapidly dried and exposed to films (Quirion *et al.*, 1988). Under these conditions, a good signal was obtained in developing but not adult rat brain tissues (Table 5.1). It is especially critical to use the highest quality of sections and a fresh batch of radioligand. Thus, we routinely obtained only minimal quantities of freshly prepared [^{125}I]EGF (1000 Ci/mmol; ICN Biomedicals Canada Ltd) in order to maximize signal-to-noise ratio. The use of relatively old batches of [^{125}I]EGF is

certainly not advisable, the ligand needing repurification before its use.

In order to optimize conditions for adult rat brain tissues, incubation with 200 pM-[^{125}I]EGF (ICN Biomedicals Canada Ltd) was carried out using four different buffers for 2 h at room temperature (Table 5.1). Non-specific binding was defined by incubating adjacent sections in the presence of 200 nM unlabelled mouse EGF. The composition of each buffer was as follows: (a) Tris–HCl (170 mM, pH 7.4) containing 0.2% BSA and protease inhibitors (0.025% bacitracin and 100 K.I.U./ml of aprotinin); (b) same as (a) plus 1 mM-tetrasodium ethylenediamine tetra-acetate (EDTA); (c) HEPES buffer (*N*-2-hydroxylethylpiperazone-*N*'-2-ethanesulfonic acid) (10 mM, pH 7.6) containing 0.5% BSA, 0.0125% NEM (*N*-ethylmaleimide) and protease inhibitors; and (d) Krebs–phosphate buffer (pH 7.4) containing 0.1% BSA and protease inhibitors. After incubation, slides were rinsed three times (1 min each) in ice-cold Tris–HCl buffer (50 mM, pH 7.4), briefly dipped in cold distilled water, rapidly dried under a stream of cold air and juxtaposed tightly against highly sensitive film (Hyperfilm, Amersham, Canada) for 2–3 weeks along with standard strips (Amersham, Canada) (Chabot *et al.*, 1988; Quirion *et al.*, 1981, 1988). Table 5.1 shows that only buffer

(b) gave positive results in the adult rat brain.

5.2.2   IMMUNOHISTOCHEMICAL APPROACHES

For the immunohistochemical detection of EGF receptors, various antibodies have been used (Loy *et al.*, 1987; Gomez-Pinilla *et al.*, 1988; Nieto-Sampedro *et al.*, 1988; Shiurba *et al.*, 1988; Werner *et al.*, 1988b; Styren *et al.*, 1990; Birecree *et al.*, 1991). We have recently used a mouse monoclonal antibody to human EGF receptor (EGFR: E-2760) obtained from Sigma Chemical Co., St Louis, MO, USA. In our laboratories the protocol used for the immunohistochemical detection of EGFR was as follows. Adult Sprague–Dawley rats were anesthetized with pentobarbital (Nembutal, 50 mg/kg, i.p.) and perfused through a cardiac cannula with 100 ml of cold saline, followed by 300 ml of a cold solution of 4% paraformaldehyde in 0.1 M phosphate-buffered saline (PBS, pH 7.4). The brains were removed and immersed in the 4% paraformaldehyde for 1 h at 4°C. Following fixation, the brains were cryoprotected in 20% PBS–sucrose solution (overnight) at 4°C before sectioning, and free-floating sections (30 μm) cut using a cryostat at –20°C. Sections were first incubated in 50mM Tris–HCl saline buffer (TBS, pH 7.6) containing 0.3% v/v Triton X-100 for 30 min. To saturate non-specific tissue binding sites, all sections were then incubated in 1% normal horse serum in TBS for 30 min. This was followed by incubation for 48 h at 4°C with the mouse monoclonal antibody to human EGFR diluted 1:500 in TBS containing 1% normal horse serum. After three washes in TBS, staining was performed according to the Vectastain ABC protocol using a biotinylated horse antimouse IgG as secondary antibody (Vector Labs., Burlingame, CA). After a final wash, sections were incubated with 0.025% w/v 3,3'-diaminobenzidine tetrahydrochloride, 0.01% v/v hydrogen peroxide in TBS for 15 min. The sections were rinsed thoroughly with TBS, mounted on precleaned slides, rinsed in distilled water, dehydrated in ethanol and mounted with glass coverslips before examination. Furthermore, for each immunostaining, 'negative' control sections were included in which the primary antibody was omitted or substituted by normal mouse serum (see below).

5.3   ONTOGENIC PROFILE OF BRAIN EGF BINDING SITES

EGF receptors, detected using quantitative autoradiography, undergo major redistribution during the development of the rat CNS. Figure 5.1A shows that specific [^{125}I]EGF binding sites can be detected in 20-day-old rat embryos with a high density of sites located in various areas including the cortex and the olfactory system. Earlier embryonic stages were not investigated here for technical reasons but it has already been reported that EGF receptor sites are detectable as early as embryonic day 15 in membrane preparations of mouse brain using binding assays (Adamson and Meek, 1984). Thus, it appears that EGF receptor sites are present in the rat brain at very early embryonic stages.

Soon after birth it appears that the expression of [^{125}I]EGF binding sites is markedly increased. As shown in Fig. 5.1, the distribution of [^{125}I]EGF sites is more extensive on day 3 (PD3) postnatally, as compared to the situation reported in prenatal tissues. At P3, [^{125}I]EGF binding sites are most abundant in various superficial cortical laminae (Fig. 5.1B, C) and in the cingulate cortex (Fig. 5.1C). High densities of sites are also present in various subcortical areas including the nucleus accumbens (Fig. 5.1B), the olfactory tubercule (Fig. 5.1B), the septal area, especially in its lateral portion (Fig. 5.1C) and in the striatum (Fig. 5.1C). Much lower levels of binding are detected at the level of the anterior hypothalamus (Fig. 5.1C). White matter areas were generally devoid of specific labelling (Fig. 5.1).

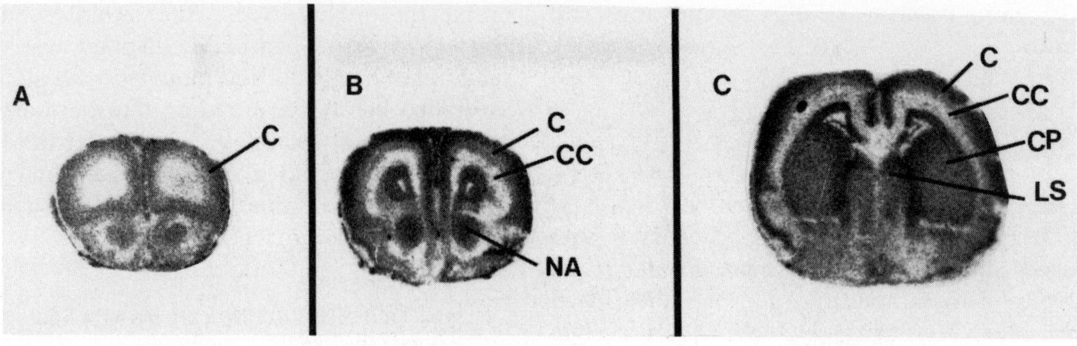

Fig. 5.1.   Autoradiograms representing the distribution of [^{125}I]EGF receptor binding sites at the level of the striatum in rats at 1 day before birth (A) and 3 days after birth (B,C). Coronal brain sections were incubated in Tris–HCI buffer (25 mM, pH 7.4) containing 10 mM MgCI$_2$, 0.1% bovine serum albumin (BSA) and 200 pM [^{125}I]EGF. C, cerebral cortex; CC, corpus callosum; CP, caudate-putamen; LS, lateral septum; NA, nucleus accumbens.

An earlier study (Quirion *et al.*, 1988) failed to detect the presence of specific [^{125}I]EGF binding in adult brain using *in vitro* receptor autoradiography. However, the development of optimized incubation conditions (Table 5.1) allowed for the detection of [^{125}I]EGF sites in various regions of the adult rat brain. Although only low densities of specific [^{125}I]EGF labeling can be detected in most rostral areas of the adult rat brain (in contrast to pre- and early postnatal tissues), moderate levels of [^{125}I]EGF binding are seen in more caudal regions (Figs 5.2–5.5). For example, a discrete pattern of labeling is observed at the level of the posterior hypothalamus with rather high densities of [^{125}I]EGF sites surrounding the medial portions of the dorso- and ventromedial nuclei, and to a certain extent the dorsal hypothalamic area. Discrete labeling is also detected in the area of the arcuate nucleus (Fig. 5.2). At the level of the dorsal hippocampal formation, limited quantities of specific [^{125}I]EGF binding are seen in most areas with some enrichment present in the posterior cingulate cortex, the superficial laminae of the parietal cortex and certain thalamic nuclei (Fig. 5.3). More caudally, it is of interest to note that moderate levels of specific [^{125}I]EGF binding are detected in the substantia nigra pars compacta and the ventral tegmental area (Fig. 5.4), two brain regions enriched in dopaminergic cell perikarya. The interpeduncular nucleus is not enriched with specific [^{125}I]EGF binding sites (Fig. 5.4). Relatively high amounts of specific [^{125}I]EGF label are also detected in the granular layer of the adult cerebellum (Fig. 5.5) whereas only very limited quantities of sites are present in this structure at early postnatal stages (not shown). Wiedermann *et al.* (1988) reported a similar distribution of [^{125}I]EGF binding sites in the adult rodent brain. Finally, specific [^{125}I]EGF sites are also present in the pituitary gland with higher levels located in the pars intermedia and, to a lesser extent, the pars distalis (Fig. 5.6). The pars nervosa is mostly devoid of specific labeling (Fig. 5.6).

We have also used a monoclonal EGFR antiserum to show that immunoreactive EGFR are detectable in the adult rat central nervous system. Figure 5.7A shows prominent EGFR-like immunostaining in cerebellar Purkinje cells. Moreover, neurons of laminae IV and V of the cerebral cortex are also EGFR-immunostained (Fig. 5.7C). A similar pattern of immunostaining has been reported (Gomez-Pinilla *et al.*, 1988).

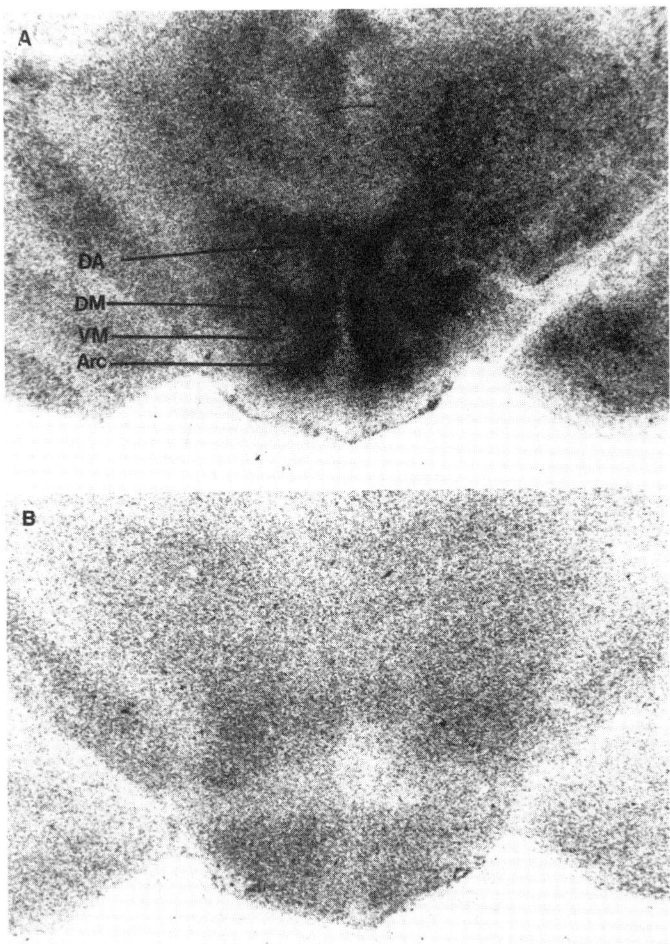

Fig. 5.2.   Autoradiograms representing the distribution of [^{125}I]EGF binding sites in the adult (3 months) rat brain at the level of the posterior hypothalamus. Coronal brain sections were incubated in Tris–HCI buffer (170 mM, pH 7.4) containing 0.1% BSA, 1 mMEDTA and either (A) 200 pM[^{125}I]EGF or (B) 200 pM[^{125}I]EGF plus 200 nM unlabelled EGF to evaluate the level of specific binding, which in this case is above 60–70%. Arc, arcuate nucleus; DA, dorsal hypothalamic area; DM, dorsomedial hypothalamic nucleus; VM, ventromedial hypothalamic nucleus.

5.4   POSSIBLE FUNCTIONAL SIGNIFICANCE

It is clear that the distribution and expression of EGF receptors undergo major modifications during postnatal development and brain maturation. However, the functional significance of these observations remains to be fully established as limited information is currently available on the role of EGF during brain maturation and in adulthood. We briefly review here the known characteristics of EGF and EGF receptors and their reported biological effects, in an attempt to assess the possible relevance of the differential expression of EGF receptors occurring during the development of the CNS.

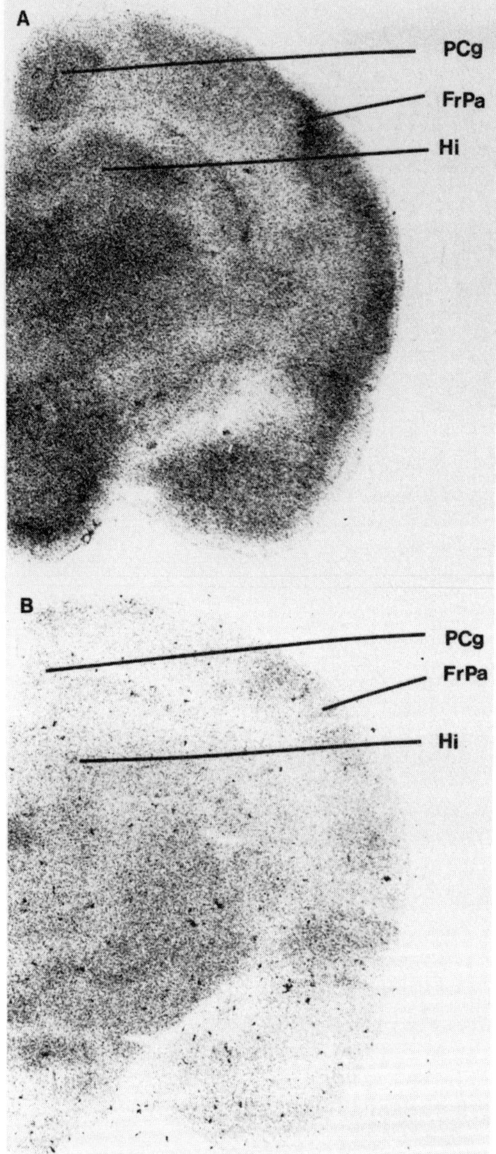

Fig. 5.3. Autoradiograms representing the distribution of [^{125}I]EGF binding sites in the adult (3 months) rat brain at the level of the dorsal hippocampus. Coronal brain sections were incubated in Tris–HCI buffer (170 mM, pH 7.4) containing 0.1% BSA, 1 mM EDTA and either (A) 200 pM [^{125}I]EGF or (B) 200 pM[^{125}I]EGF plus 200 nM unlabelled EGF to evaluate the level of specific binding, which in this case represents approximately 85–90% of the binding. FrPa, frontoparietal cortex; Hi, hippocampus; PCg, posterior cingulate cortex.

5.4.1   EGF-LIKE PEPTIDES IN THE CNS

EGF has not been widely studied in mammalian brain despite the evidence for its presence in this tissue (Fallon *et al.*, 1984; Araujo *et al.*, 1990). EGF was first isolated from the mouse submaxillary gland (Cohen, 1962). In view of its similarity to the homeotic loci *lin-12* in *Caenorhabditis elegans* and *Notch* in *Drosophila* (Wharton *et al.*, 1985; Knust *et al.*, 1987), two molecules which play crucial roles as morphogenetic and differentiation signals in non-mammals, it was proposed that EGF can have similar functions in higher species. Moreover, the EGF receptor, a proto-oncogene product, could play a critical role in the regulation of growth and differentiation (Adamson, 1987). In addition, some evidence suggests that EGF can also act as a direct neuromodulator on certain neuronal populations (Section 5.4.3).

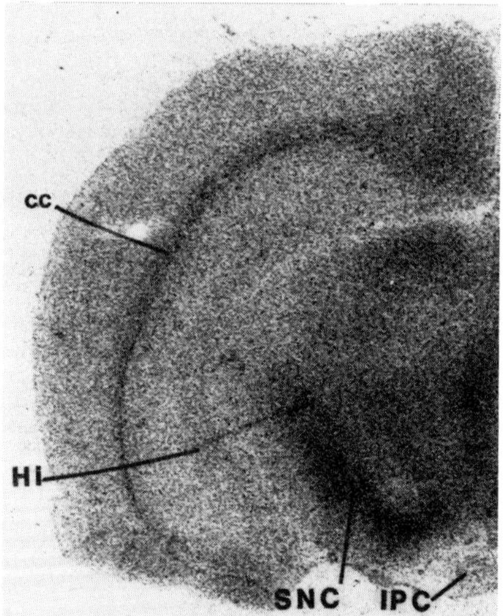

Fig. 5.4. Autoradiogram representing the distribution of [^{125}I]EGF binding sites in the adult (3 months) rat brain at the level of the substantia nigra. Coronal brain sections were incubated in Tris–HCI buffer (170 mM, pH 7.4) containing 0.1% BSA, 1 mMEDTA and 200 pM[^{125}I]EGF. Note that EGF binding is concentrated in the substantia nigra and the ventral tegmental area. cc, corpus callosum; HI, hippocampus; IPC, interpeduncular nucleus, caudal subnucleus; SNC, substantia nigra, pars compacta.

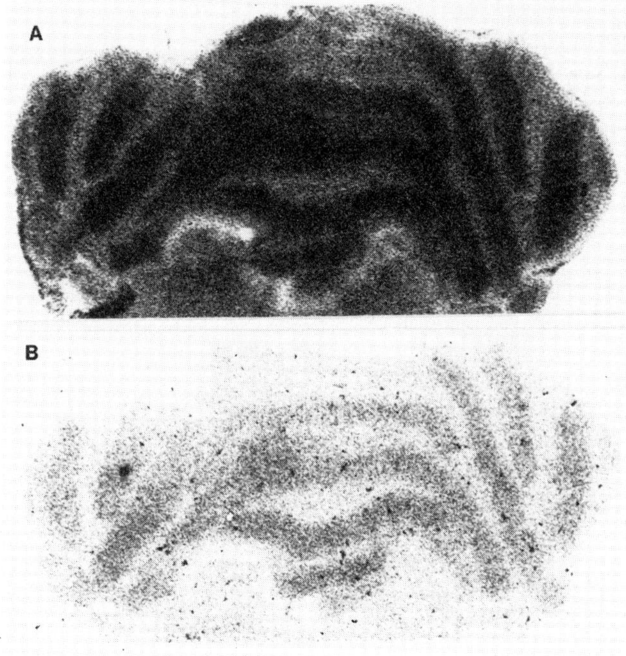

Fig. 5.5. Autoradiograms representing the distribution of [^{125}I]EGF binding sites in the adult (3 months) rat brain at the level of the cerebellum. Coronal brain sections were incubated in Tris–HCl buffer (170 mM, pH 7.4) containing 0.1% BSA, 1 mMEDTA and either (A) 200 pM[^{125}I]EGF or (B) 200 pM[^{125}I]EGF plus 200 nM unlabelled EGF to evaluate the level of specific binding, which in this case represents approximately 60% of the binding.

The presence of EGF-like peptides has been detailed in embryonic, postnatal and adult brain (Fallon *et al.*, 1984; Lakshmanan *et al.*, 1986; Lazar *et al.*, 1988; Schaudies *et al.*, 1989; Birecree *et al.*, 1991; Kajikawa *et al.*, 1991) and human cerebrospinal fluid (Hirata *et al.*, 1982). However, it appears that EGF itself is not synthesized in the rodent brain before the second postnatal week (Rall *et al.*, 1985; Popliker *et al.*, 1987; Lazar *et al.*, 1988). Since we have shown the presence of specific EGF receptor at much earlier stages (ED20 and PD3; Fig. 5.1), EGF-like material required to act on these sites could be of maternal origin. Alternatively, a structurally related factor, transforming growth factor α (TGFα), could activate EGF receptors in embryonic and neonatal rat brain.

TGFα is a member of a family of peptides implicated in the control of cell growth and differentiation. TGFα, and a homologue, TGFβ (Massague, 1987; Morstyn and Burgess, 1988) are present with their respective mRNAs at very early stages in mammalian CNS development (Roberts *et al.*, 1981; Marquardt *et al.*, 1984; Lee *et al.*, 1985; Clark and Bressler, 1988; Kakucska *et al.*, 1988; Kudlow *et al.*, 1989). In neonatal, as well as adult brain, TGFα is predominantly of neuronal origin (Wilcox and Derynck, 1988) with high concentrations of TGFα mRNA present in areas such as the caudate putamen and the hippocampal formation. TGFβ is apparently detected only at embryonic states and during brain development, but not in adulthood (Wilcox and Derynck, 1988).

It is well known that TGFα binds to EGF receptor sites (Massague, 1987; Morstyn and Burgess, 1988) and thus induces biological effects which are similar to those reported for EGF. It is thus possible that, at embryonic stages and during early postnatal develop-

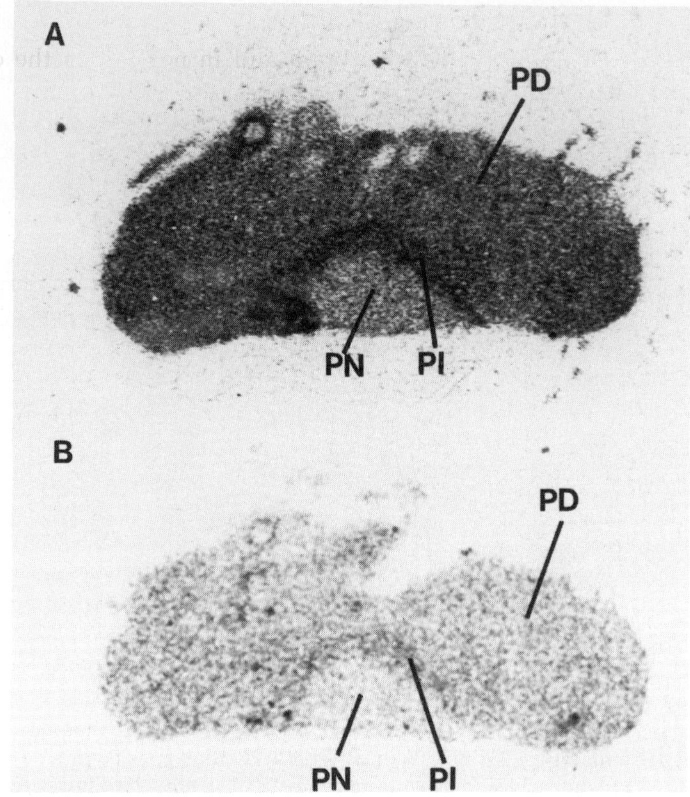

Fig. 5.6.   Autoradiograms representing the distribution of [^{125}I]EGF binding sites in the adult (3 months) rat pituitary gland. Coronal sections were incubated in Tris–HCI buffer (170 mM, pH 7.4) containing 0.1% BSA, 1 mMEDTA and either (A) 200 pM-[^{125}I]EGF or (B) 200 pM-[^{125}I]EGF plus 200 nM unlabelled EGF to determine the level of specific binding, which represents approximately 80% in the intermediate lobe. PD, pars distalis; PI, pars intermedia; PN, pars nervosa.

ment, TGFα acts as the endogenous ligand of the EGF receptor. During adulthood both EGF and TGFα could act as ligands for this receptor class, which is broadly distributed in multiple regions of the CNS.

5.4.2   EGF RECEPTORS IN THE CNS

As shown in section 5.3, EGF receptors are widely distributed in the neonatal and mature rodent brain. The presence of EGF binding sites have been reported in homogenates of rat (Hiramatsu *et al.*, 1988), rabbit (Sadiq *et al.*, 1985) and mouse (Nexo *et al.*, 1980; Adamson *et al.*, 1981; Adamson and Warshaw, 1982; Adamson and Meek, 1984) brains. These sites

are present in a variety of cell types including neurons, astrocytes, oligodendrocytes and several glial cell lines (Leutz and Schachner, 1981, 1982; Simpson *et al.*, 1982; Sagen and Pappas, 1987; Wang *et al.*, 1989) as well as in human brain tumors (Libermann *et al.*, 1984; Meyers *et al.*, 1988; Shiurba *et al.*, 1988; Werner *et al.*, 1988a; Arita *et al.*, 1989; Reubi *et al.*, 1989; Maruno *et al.*, 1991; Torp *et al.*, 1991). Immunohistochemically, the existence of EGF receptors has been demonstrated in the brain of a variety of species (Loy *et al.*, 1987; Birecree *et al.*, 1988, 1991; Gomez-Pinilla *et al.*, 1988; Shiurba *et al.*, 1988; Werner *et al.*, 1988b; Arita *et al.*, 1989; Samuels *et al.*, 1989; Torp *et al.*, 1991). EGF-receptor   immunoreactivity   is especially enriched in astroglia in neonatal rat brain and in neurons of the cerebral cortex, cerebellar astrocytes and Purkinje cells in adult CNS (Gomez-Pinilla *et al.*, 1988). In aged rats, EGF-receptor staining is most prominent in glial fibrillary acid protein (GFAP)-reactive astrocytes (Nieto-Sampedro *et al.*, 1988). EGF-receptor immunoreactivity also increases in reactive astrocytes in response to brain injury (Nieto-Sampedro *et al.*, 1988). Moreover, EGF-receptor staining is increased in Alzheimer's disease (AD) (Styren *et al.*, 1990) especially in association to neuritic plaques (Nanney *et al.*, 1986; Birecree *et al.*, 1988). This raises the possibility that EGF may be involved in the formation of these elements in the aged, pathological human brain.

In relation to the present study, Adamson and Meek (1984) have reported on the ontogenic profile of EGF receptors in the mouse brain. These authors observed the appearance of EGF sites as early as day 15 of gestation. Their densities increase during late gestation and decrease in the last phase of postnatal development. Although we did not investigate the presence of [^{125}I]EGF receptors at very early embryonic stages, the high levels of [^{125}I]EGF sites detected one day before birth also support their early appearance in the rat brain.

5.4.3 BIOLOGICAL EFFECTS OF EGF IN THE CNS

(a) Trophic and mitotic activity

EGF and related peptides induce multiple biological effects in the CNS. Those have been extensively detailed in an excellent recent review by Plata-Salamán (1991). Thus, we shall concentrate here on a few aspects which are particularly relevant to the possible role of EGF and its receptor during brain maturation.

The presence of high densities of EGF receptor sites at embryonic and early post-natal stages, as well as their proliferation following mechanical or excitotoxic lesions,

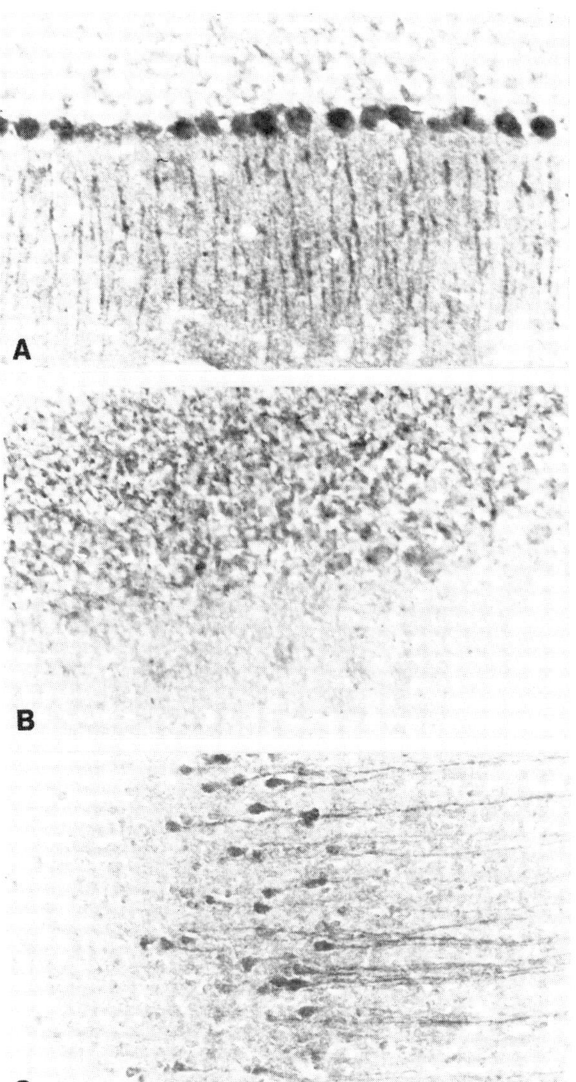

Fig. 5.7. Cellular distribution of EGF receptor-like immunoreactive material in the neocortex and cerebellum of the adult rat brain. Free-floating coronal sections (30 µm) of 3-months-old rat brain were stained for EGF receptors using a monoclonal antibody and the avidin–biotin method. (A) In the cerebellum, note the immunostaining in the perikarya and dendrites of Purkinje cells. (B) Control staining of the cerebellum with normal mouse serum. This reveals the specificity of the immunostaining observed in (A). (C) In the frontoparietal cortex, note the immunostaining localization in cells having the morphology of basket neurons in layers IV and V.

suggest that this GF probably has important roles in CNS growth and differentiation. EGF has been shown to possess mitogenic effects in glial cells (Leutz and Schachner, 1981, 1982; Simpson *et al.*, 1982; Raff *et al.*, 1983; Fischer, 1984; Patel *et al.*, 1988; Westermark, 1988; Loret *et al.*, 1989; Kenigsberg *et al.*, 1989; Kenigsberg and Mazzoni, 1990) in addition to inducing their differentiation (Guentert-Lauber and Honegger, 1983; Honegger and Guentert-Lauber, 1983; Almazan *et al.*, 1985; Monnet-Tschudi and Honegger, 1989; Loret *et al.*, 1989). In embryonic rat astroglial cell cultures, EGF, in a dose-dependent manner, increases mRNA and protein syntheses, as well as the labelling of cytoskeletal proteins (Avola *et al.*, 1988a, b; Oleszak *et al.*, 1988). Similarly, EGF also stimulates DNA synthesis of astrocytes in cerebellar cultures (Leutz and Schachner, 1981), and the activity of ornithine decarboxylase, a known marker of cellular differentiation in brain cells (Almazan *et al.*, 1985). It has also been shown that low concentrations of EGF stimulate the differentiation of astrocytes, whereas higher amounts induce mitogenesis *in vitro* (Guentert-Lauber and Honegger, 1983). Additionally, EGF activates the transcription of a number of immediate–early genes including *c-fos* and *c-myc*, which are involved in the cell growth cycle and differentiation (Kudlow *et al.*, 1989). These various biological effects induced by EGF suggest that a good proportion of [^{125}I]EGF binding sites detected during embryogenesis and early postnatal life are located on various types of glial cells. This could also explain the apparent decrease in the density of [^{125}I]EGF binding observed at later stages of brain maturation, as the ratio between glia and neuron decreases during these periods.

On the other hand, it is also clear that EGF can act as a neurotrophic, survival and maintenance factor on various neuronal populations. For example, EGF, in a concentration-dependent manner, enhances neurite outgrowth and survival of neonatal neocortical (Kornblum *et al.*, 1990), subneocortical (Morrison *et al.*, 1987; Abe *et al.*, 1990) and cerebellar (Morrison *et al.*, 1988; Abe *et al.*, 1990, 1991a) neurons. Interestingly, EGF has also been shown to stimulate the growth of septal pontine cholinergic and mesencephalic dopaminergic neurons in primary cultures (Casper *et al.*, 1989; Knusel *et al.*, 1990). This is of special interest in view of the recent demonstration that EGF decreases choline acetyltransferase activity (ChAT) and intracellular acetylcholine stores in primary medial septum cultures (Kenigsberg *et al.*, 1989; Kenigsberg and Mazzoni, 1990). This is in contrast to the apparent lack of effect of EGF on acetylcholine release in neonatal hippocampal slices (Araujo *et al.*, 1989a). However, in mature hippocampus, EGF induced a dose-dependent inhibition of acetylcholine release (Araujo *et al.*, 1989a). This apparent discrepancy could be explained by the possible mediation by glial cells of the effects of EGF on cholinergic markers in septal cultures (Kenigsberg and Mazzoni, 1990). In any case, it appears that under certain conditions, EGF can act directly on neurons to modulate their growth and differentiation (Casper *et al.*, 1989), this hypothesis being supported by the existence of EGF receptors on these various neuronal populations during brain development as well as in adulthood (Figs 5.1–5.5).

(b)   EGF as a neuromodulator/ neurotransmitter

In addition to their better trophic and mitotic actions, EGF-like peptides also act as modulators on various neuronal populations. These effects are probably related to the presence of relatively high densities of [^{125}I]EGF receptor sites in various areas of the normal adult brain (section 5.3). For example, EGF increases the dopamine uptake site on mesencephalic neurons in a glia-dependent manner (Knusel *et al.*, 1990; Hadjiconstantinou *et al.*, 1991) and stimulates the magnitude of long-term potentiation in the hippocampal formation (Terlau and Seifert, 1989; Abe *et al.*, 1991b). In addition, we have shown that EGF can inhibit the release of acetylcholine in adult hippocampal slices (Araujo *et al.*, 1989a).

Moreover, EGF modulates various hypothalamic and neuroendocrine functions. This is in good correlation with the presence of [^{125}I]EGF binding sites in certain hypothalamic nuclei and in the pituitary gland (Figs 5.2 and 5.6). In that regard, it is known that EGF can stimulate the release of corticotropin-releasing factor (CRF) (Luger *et al.*, 1988) and thus indirectly increase plasma ACTH levels. EGF can also stimulate the secretion of luteinizing hormone-releasing hormone (LHRH) (Ojeda *et al.*, 1989) as well as the release of growth hormone (Chabot *et al.*, 1986), ACTH and prolactin (Polk *et al.*, 1987) from the pituitary gland. Finally, EGF inhibits food intake, probably by a direct action at the level of the hypothalamus (Panaretto *et al.*, 1982; Plata-Salamán, 1989, 1991) in accordance with the presence of EGF receptors in this region of the adult rat brain. It thus appears that brain EGF is a neuromodulator on a variety of neurotransmitter and hormonal systems.

5.5 CONCLUSION

In summary, EGF receptors are clearly expressed in various brain regions during embryogenesis, early postnatal life and in adulthood. Their differential expression and redistribution during brain maturation suggest a possible modification of their role at various stages of development. Although probably acting as trophic, differentiation and mitogenic factors during pre- and postnatal ontogeny, EGF and related peptides probably also possess various neuromodulatory roles in the mature brain. Hopefully, the development of potent EGF receptor antagonists in the near future will facilitate a better understanding of the direct physiological relevance of this GF during brain development and maturation.

ACKNOWLEDGEMENTS

Support from the Medical Research Council of Canada and the Canadian Parkinson Foundation is acknowledged. R. Quirion is a 'Chercheur-Boursier' Senior of the 'Fonds de la Recherche en Santé du Quebec'. The expert secretarial assistance of Mrs J. Currie is acknowledged.

REFERENCES

Abe, K., Takayanagi, M. and Saito, H. (1990) A comparison of neurotrophic effects of epidermal growth factor and basic fibroblast growth factor in primary cultured neurons from various regions of fetal rat brain. *Jpn. J. Pharmacol.*, **54**, 45–51.

Abe, K., Takayanagi, M. and Saito, H. (1991a) Basic fibroblast growth factor and epidermal growth factor promote survival of primary cultured cerebellar neurons from neonatal rats. *Jpn. J. Pharmacol.*, **56**, 113–16.

Abe, K., Xie, F. and Saito, H. (1991b) Epidermal growth factor enhances short-term potentiation and facilitates induction of long-term potentiation in rat hippocampal slices. *Brain Res.*, **547**, 171–4.

Adamson, E.D. (1987) Oncogenes in development. *Development*, **99**, 449–71.

Adamson, E.D., Deller, M.J. and Warshaw, J.B. (1981) Functional EGF receptors are present on mouse embryo tissues. *Nature*, **291**, 656–9.

Adamson, E.D. and Meek, J. (1984) The ontogeny of epidermal growth factor receptors during mouse development. *Dev. Biol.*, **103**, 62–70.

Adamson, E.D. and Warshaw, J.B. (1982) Downregulation of epidermal growth factor receptors in mouse embryos. *Dev. Biol.*, **90**, 430–4.

Almazan, G., Honegger, P., Matthieu, J.-M. and Guentert-Lauber, B. (1985) Epidermal growth factor and bovine growth factor stimulate differentiation and myelination of brain cell aggregates in culture. *Dev. Brain Res.*, **21**, 257–64.

Araujo, D.M., Lapchak, P.A., Chabot, J.-G. *et al.* (1989a) Characterization and possible role of growth factor and lymphokine receptors in the regulation of cholinergic function in the mammalian brain, in *Alzheimer's Disease and Related Disorders* (eds K. Iqbal, H.M. Wisniewski and B. Winblad), Alan R. Liss, New York, pp. 423–36.

Araujo, D.M., Lapchak, P.A., Collier, B. *et al.* (1989b) Insulin-like growth factor-1 (somatomedin-C) receptors in the rat brain: distribution and interaction with the hippocampal cholinergic system. *Brain Res.*, **484**, 130–8.

Araujo, D.M., Chabot, J.-G. and Quirion, R. (1990) Potential neurotrophic factors in the mammalian central nervous system: functional significance in the developing and aging brain, in *International Review of Neurobiology*, Vol. 32 (eds J.R. Symthies and R.J. Bradley), Academic Press, New York, pp. 141–74.

Arita, N., Hayakawa, T., Izumoto, S. *et al.* (1989) Epidermal growth factor receptor in human glioma. *J. Neurosurg.*, **70**, 916–19.

Avola, R., Condorelli, D.F., Surrentino, S. *et al.* (1988a) Effect of epidermal growth factor and insulin on DNA, RNA and cytoskeletal protein labeling in primary rat astroglial cell cultures. *J. Neurosci. Res.*, **19**, 230–8.

Avola, R., Condorelli, D.F., Turpeenoja, L. *et al.* (1988b) Effect of epidermal growth factor on the labeling of the various RNA species and of nuclear proteins in primary rat astroglial cell cultures. *J. Neurosci. Res.*, **20**, 54–63.

Baskin, D.G., Wilcox, B.J., Figlewicz, D.P. and Dorsa, D.M. (1988) Insulin and insulin-like growth factors in the CNS. *Trends Neurosci.*, **11**, 107–11.

Birecree, E., Whetsell, W.O., Stoscheck, C.M. *et al.* (1988) Immunoreactive epidermal growth factor receptor in neuritic plaques from patients with Alzheimer's disease. *Exp. Neurol.*, **47**, 549–60.

Birecree, E., King, L.E., and Nanney, L.B. (1991) Epidermal growth factor and its receptor in the developing human nervous system. *Dev. Brain Res.*, **60**, 145–54.

Casper, D., Blum, M. and Mytilineou, C. (1989) Epidermal growth factor is neurotrophic in rat embryo mesencephalic primary culture. *Soc. Neurosci. Abstr.*, **15**, 708.

Chabot, J.-G., Walker, P. and Pelletier, G. (1986) Distribution of epidermal growth factor binding sites in the adult rat anterior pituitary gland. *Peptides*, **7**, 45–50.

Chabot, J.-G., Araujo, D.M. and Quirion, R. (1988) Epidermal growth factor (EGF) binding sites in adult rat brain and pituitary gland. An *in vitro* autoradiographic study. *Soc. Neurosci. Abstr.*, **14**, 1073.

Clark, W.C. and Bressler, J. (1988) Transforming growth factor-β-like activity in tumours of the central nervous system. *J. Neurosurg.*, **68**, 920–4.

Cohen, S. (1962) Isolation of a submaxillary gland protein accelerating incisor eruption and eyelid opening in the newborn animal. *J. Biol. Chem.*, **137**, 1555–62.

Crutcher, K.A. (1986) The role of growth factors in neuronal development and plasticity. *CRC Crit. Rev. Clin. Neurobiol.*, **2**, 297–333.

Fallon, J.H., Seroogy, K.B., Loughlin, S.E. *et al.* (1984) Epidermal growth factor immunoreactive material in the central nervous system: location and development. *Science*, **224**, 1107–9.

Fischer, G. (1984) Growth requirements of immature astrocytes in serum-free hormonally defined media. *J. Neurosci. Res.*, **12**, 543–52.

Gomez-Pinilla, F., Knauer, D.J., and Nieto-Sampedro, M. (1988) Epidermal growth factor receptor immunoreactivity in rat brain. Development and cellular localization. *Brain Res.*, **438**, 385–90.

Guentert-Lauber, B. and Honegger, P. (1983) Epidermal growth factor (EGF) stimulation of cultured brain cells. II. Increased production of extracellular soluble proteins. *Dev. Brain Res.*, **11**, 253–60.

Hadjiconstantinou, M., Fitkin, J.G., Dalia, A. and Neff, N.H. (1991) Epidermal growth factor enhances striatal dopaminergic parameters in the 1-methyl-4-phenyl-1,2,3,6-tetrahydropyridine-treated mouse. *J. Neurochem.*, **57**, 479–82.

Hefti, F., Hartikka, J. and Knusel, B. (1989) Function of neurotrophic factors in the adult and aging brain and their possible use in the treatment of neurodegenerative diseases. *Neurobiol. Aging*, **10**, 515–33.

Hefti, F. and Weiner, W.J. (1986) Nerve growth factor and Alzheimer's disease. *Ann. Neurol.*, **20**, 275–81.

Herkenham, M. and Pert, C.B. (1982) Light microscopic localization of brain opiate receptors: a general autoradiographic method which preserves tissue quality. *J. Neurosci.*, **2**, 1129–49.

Hiramatsu, M., Kashimata, M., Sato, A. *et al.* (1988) Influence of age on epidermal growth factor receptor level in the rat brain. *Experimentia*, **44**, 23–5.

Hirata, Y., Uchihashi, M., Nakajima, H. *et al.* (1982) Presence of epidermal growth factor in human cerebrospinal fluid. *J. Clin. Endocrinol. Metab.*, **55**, 1174–7.

Honegger, P. and Guentert-Lauber, B. (1983) Epidermal growth factor (EGF) stimulation of cultured brain cells. I. Enhancement of the developmental increase in glial enzymatic activity. *Dev. Brain Res.*, **11**, 245–51.

Jorgensen, P.E., Poulsen, S.S. and Nexo, E. (1988) Distribution of i.v. administered epidermal growth factor in the rat. *Regul. Pept.*, **23**, 161–9.

Kajikawa, K., Yasui, W., Sumiyoshi, H. *et al.* (1991) Expression of epidermal growth factor in human tissues. Immunohistochemical and biochemical analysis. *Virchows Arch. [A]*, **418**, 27–32.

Kakucska, I., Tappaz, M.L., Gaal, G. *et al.* (1988) GABAergic innervation of somatostatin-containing neurosecretory cells of the anterior periventricular hypothalamic area: a light and electron microscopy double immunolabelling study. *Neuroscience*, **25**, 585–93.

Kar, S., Chabot, J.-G. and Quirion, R. (1991) Ontogenic profiles of insulin, insulin-like growth factors I and II binding sites in rat brain. *Third IBRO World Congress of Neuroscience*, Montreal, Canada, p. 327.

Kar, S., Quirion, R. and Chabot, J.-G. (1992) Distribution of receptor sites for insulin, insulin-like growth factor-I (IGF-I) and insulin-like growth factor-II during development and adulthood in rat brain. (Submitted)

Kenigsberg, R.L. and Mazzoni, I.E. (1990) Astroglial mitogens affect septal cholinergic cell expression differentially. *Soc. Neurosci. abst.*, **16**, 998.

Kenigsberg, R.L., Lapchak, P., Mazzoni, I.E. *et al.* (1989) Differential effects of nerve growth factor and epidermal growth factor on dissociated septal cells in culture: interactions with mono-sialoganglioside GM1. *J. Neurochem. Suppl.*, **52**, Abstr. no. S192D.

Knusel, B., Michel, P.P., Schwaber, J.S. and Hefti, F. (1990) Selective and nonselective stimulation of central cholinergic and dopaminergic development *in vitro* by nerve growth factor, basic fibroblast growth factor, epidermal growth factor, insulin and the insulin-like growth factors I and II. *J. Neurosci.*, **10**, 558–70.

Knust, E., Dietrich, U., Tepass, U. *et al.* (1987) EGF homologous sequences encoded in the genome of *Drosophila melanogaster* and their relation to neurogenic genes. *EMBO J.*, **6**, 761–6.

Kornblum, H.I., Raymon, H.K., Morrison, R.S. *et al.* (1990) Epidermal growth factor and basic fibroblast growth factor: effects on an overlapping population of neocortical neurons *in vitro*. *Brain Res.*, **535**, 255–63.

Kudlow, J.E., Leung, A.W.C., Kobrin, M.S., *et al.* (1989) Transforming growth factor-α in the mammalian brain. Immunohistochemical detection in neurons and characterization of its mRNA *J. Biol. Chem.*, **264**, 3880–3.

Lakshmanan, J., Weischel, M.E. and Fisher, D.A. (1986) Epidermal growth factor in synaptosomal fractions of mouse cerebral cortex. *J. Neurochem.*, **46**, 1081–5.

Lazar, L.M., Roberts, J.L. and Blum, M. (1988) Regional distribution of epidermal growth factor mRNA in the mammalian central nervous system. *Soc. Neurosci. abstr.*, **15**, 1162.

Lee, D.C., Rochford, R., Todaro, G.J. and Villareal, L.P. (1985) Developmental expression of rat transforming growth factor-α mRNA. *Mol. Cell. Biol.*, **5**, 3644–6.

LeRoith, D., Lowe, W.L. and Roberts, C.T., (1987) Evolution of insulin and insulin receptors, in *Insulin, Insulin-like Growth Factors, and Their Receptors in the Central Nervous System* (eds M.K. Raizada, M.I. Phillips and D. LeRoith), Plenum Press, New York and London, pp. 107–20.

Leutz, A. and Schachner, M. (1981) Epidermal growth factor stimulates DNA synthesis of astrocytes in primary cerebellar cultures. *Cell Tissue Res.*, **220**, 393–400.

Leutz, A. and Schachner, M. (1982) Cell type-specificity of epidermal growth factor (EGF) binding in primary cultures of early postnatal mouse cerebellum. *Neurosci. Lett.*, **30**, 179–82.

Levi-Montalcini, R. and Calissano, P. (1986) Nerve growth factor as a paradigm for other polypeptide growth factor. *Trends Neurosci.*, **9**, 473–7.

Libermann, T.A., Razon, N., Bartal, A.D. *et al.* (1984) Expression of epidermal growth factor receptors in human brain tumors. *Cancer Res.*, **44**, 753–60.

Loret, C., Sensenbrenner, M. and Labourdette, G. (1989) Differential phenotypic expression induced in cultured rat astroblasts by acidic fibroblast growth factor, epidermal growth factor, and thrombin. *J. Biol. Chem.*, **264**, 8319–27.

Loy, R., Springer, J.F. and Koh, S. (1987) Localization of cells immunoreactive for epidermal growth factor (EGF) receptor in the adult and developing rat forebrain. *Soc. Neurosci. abstr.*, **13**, 575.

Luger, A., Calogera, A.E., Kalogeras, K. *et al.* (1988) Interaction of epidermal growth factor with hypothalamic–pituitary–adrenal axis: potential physiologic relevance. *J. Clin. Endocrinol. Metab.*, **66**, 334–7.

Marquardt, H., Hunkapiller, M.W., Hood, L.E. and Todaro, G.J. (1984) Rat transforming growth type 1: structure and relation to epidermal growth factor. *Science*, **223**, 1079–82.

Maruno, M., Kovach, J.S., Kelly, P.J. and Yanagihara, T. (1991) Transforming growth factor-α epidermal growth factor receptor, and proliferating potential in benign and malignant gliomas. *J. Neurosurg.*, **75**, 97–102.

Massague, J. (1987) The transforming growth factors, in *Oncogenes and Growth Factors* (eds R.A. Bradshaw and S. Prentis), Elsevier Science Publishers, New York, pp. 157–63.

Meyers, M.B., Shen, W.P.V., Spengler, B.A. *et al.* (1988) Increased epidermal growth factor receptor in multidrug-resistant human neuroblastoma cells. *J. Cell. Biochem.*, **38**, 87–97.

Monnet-Tschudi, F. and Honegger, P. (1989) Influence of epidermal growth factor on the maturation of fetal rat brain cells in aggregate culture. An immunocytochemical study. *Dev. Neurosci.*, **11**, 30–40.

Morrison, R.S., Kornblum, H.I., Leslie, F.M. and Bradshaw, R.A. (1987) Trophic stimulation of cultured neurons from neonatal rat brain by epidermal growth factor. *Science*, **238**, 72–5.

Morrison, R.S., Keating, R.F. and Moskal, J.R. (1988) Basic fibroblast growth factor and epidermal growth factor exert differential trophic effects on CNS neurons. *J. Neurosci. Res.*, **21**, 71–9.

Morstyn, G. and Burgess, A.W. (1988) Hemopoietic growth factors: a review. *Cancer Res.*, **48**, 5624–37.

Nanney, L.B., Werner, M., Stoscheck, C. *et al.* (1986) Epidermal growth factor-receptor (EGF-R) detected in normal brain and Alzheimer's disease. *J. Neuropathol. Exp. Neurol.*, **45**, 339 (Abstract).

Nexo, E., Hollenberg, M.D., Figueroa, A. and Pratt, R.M. (1980) Detection of epidermal growth factor-urogastrone and its receptor during fetal mouse development. *Proc. Natl. Acad. Sci. USA*, **77**, 2782–5.

Nieto-Sampedro, M., Gómez-Pinilla, F., Knauer, D.J. and Broderick, J.T. (1988) Epidermal growth factor receptor immunoreactivity in rat brain astrocytes. Response to injury. *Neurosci. Lett.*, **91**, 276–82.

Ojeda, S.R., Urbanski, H.F., Costa, M.E. *et al.* (1989) Transforming growth factor-α (TGF-α) mRNA is expressed in the developing hypothalamus and TGF-α stimulates luteinizing hormone-releasing hormone (LHRH) release. *Soc. Neurosci. abstr.*, **15**, 1086.

Oleszak, E.L., Murdoch, G., Manuelidis, L. and Manuelidis, E.E. (1988) Growth factor production by Creutzfeldt– Jakob disease cell lines. *J. Virol.*, **62**, 3103–8.

Panaretto, B.A., Moore, G.P.M. and Robertson, D.M. (1982) Plasma concentrations and urinary excretion of mouse epidermal growth factor associated with the inhibition of food consumption and of wool growth in Merino wethers. *J. Endocrinol.*, **94**, 191–202.

Patel, A.J., Seaton, P. and Hunt, A. (1988) A novel way of removing quiescent astrocytes in a culture of subcortical neurons grown in a chemically defined medium. *Dev. Brain Res.*, **42**, 283–8.

Perry, E. (1990) Nerve growth factors and the basal forebrain cholinergic systems: a link in the etiopathology of neurodegenerative dementias. *Alzheimer Dis. Assoc. Disord.*, **4**, 1–13.

Plata-Salamán, C.R. (1989) Growth factors, feeding regulation and the nervous system. *Life Sci.*, **45**, 1207–17.

Plata-Salamán, C.R. (1991) Epidermal growth factor and the nervous system. *Peptides*, **12**, 653–63.

Polk, D.H., Ervin, M.G., Padbury, J.F. *et al.* (1987) Epidermal growth factor acts as a corticotropin-releasing factor in chronically catheterized fetal lambs. *J. Clin. Invest.*, **79**, 984–8.

Popliker, M., Shatz, A., Avivi, A. *et al.* (1987) Onset of endogenous synthesis of epidermal growth factor in neonatal mice. *Dev. Biol.*, **119**, 38–42.

Quirion, R., Hammer, R.P., Herkenham, M. and Pert, C.B. (1981) Phenylcyclidine (angel dust) σ 'opiate' receptor: visualization by tritium sensitive film. *Proc. Natl. Acad. Sci. USA*, **78**, 5881–5.

Quirion, R., Araujo, D., Nair, N.P.V. and Chabot, J.C. (1988) Visualization of growth factor receptor sites in rat forebrain. *Synapse*, **2**, 212–18.

Raff, M., Abney, E., Cohen, J. *et al.* (1983) Two types of astrocytes in cultures of developing rat white matter. Differences in morphology, surface gangliosides, and growth characteristics. *J. Neurosci.*, **3**, 1289–300.

Rall, L.B., Scott, J., Bell, G.I. *et al.* (1985) Mouse prepro-epidermal growth factor synthesis by the kidney and other tissues. *Nature*, **313**, 228–31.

Reubi, J.C., Horisberger, U., Lang, W. *et al.* (1989) Coincidence of EGF receptors and somatostatin receptors in meningiomas but inverse, differentiation-dependent relationship in glial tumors. *Am. J. Pathol.*, **134**, 337–44.

Roberts, A.B., Anzano, M.A., Lamb, L.C. *et al.* (1981) New class of transforming growth factor: a bifunctional regulator of cellular growth. *Proc. Natl. Acad. Sci. USA*, **78**, 5339–43.

Sadiq, H.F., Chechani, V., Devaskar, S.U. and Devaskar, U.P. (1985) Thyroid-dependent maturation of neonatal brain but not lung epidermal growth factor receptors. *Dev. Pharmacol. Ther.*, **8**, 292–301.

Sagen, J. and Pappas, G.D. (1987) Morphological and functional correlates of chromaffin cell transplants in CNS pain modulatory regions. *Ann. NY Acad. Sci.*, **495**, 306–33.

Samuels, V., Barrett, J.M., Bockman, S. *et al.* (1989) Immunocytochemical study of transforming growth factor expression in benign and malignant gliomas. *Am. J. Pathol.*, **134**, 895–902.

Schaudies, R.P., Christian, E.L. and Savage, C.R., (1989) Epidermal growth factor immunoreactive material in the rat brain. Localization and identification of multiple species. *J. Biol. Chem.*, **264**, 10447–50.

Shiurba, R.A., Eng, L.F., Vogel, H. *et al.* (1988) Epidermal growth factor receptor in meningiomas is expressed predominantly on endothelial cells. *Cancer*, **62**, 2139–44.

Simpson, D.L., Morrison, R., de Vellis, J. and Herschman, H.R. (1982) Epidermal growth factor binding and mitogenic activity on purified populations of cells from the central nervous system. *J. Neurosci.*, **8**, 453–62.

Styren, S.D., Mufson, E.J., Styren, G.C. *et al.* (1990) Epidermal growth factor receptor expression in demented and aged human brain. *Brain Res.*, **512**, 347–52.

Terlau, H. and Seifert, W. (1989) Influence of epidermal growth factor on long-term potentiation in the hippocampal slice. *Brain Res.*, **484**, 352–6.

Thoenen, H. and Edgar, D. (1985) Neurotrophic factors. *Science*, **229**, 238–42.

Torp, S.H., Helseth, E., Dalen, A. and Unsgaard, C. (1991) Epidermal growth factor receptor expression in human gliomas. *Cancer Immunol. Immunother.*, **33**, 61–4.

Wang, S.-L., Shiverick, K.T., Ogilvie, S. *et al.* (1989) Characterization of epidermal growth factor receptors in astrocytic glial and neuronal cells in primary culture. *Endocrinology*, **124**, 240–7.

Werner, M.H., Humphrey, P.A., Bigner, D.D. and Bigner, S.H. (1988a) Growth effects of epidermal growth factor (EGF) and a monoclonal antibody against the EGF receptor on four glioma cell lines. *Acta Neuropathol. (Berl.)*, **77**, 196–201.

Werner, M.H., Nanney, L.B., Stoscheck, C.M. and King, L.E. (1988b) Localization of immunoreactive epidermal growth factor receptors in human nervous system. *J. Histochem. Cytochem.*, **36**, 81–6.

Westermark, B. (1988) Density dependent proliferation of human glia cells stimulated by epidermal growth factor. *Biochem. Biophys. Res. Commun.*, **76**, 304–10.

Wharton, K.A., Johansen, K.M., Xu, T. and Artavonis-Tsakonis, S. (1985) Nucleotide sequence from the neurogenic locus notch implies a gene product that shares homology with proteins containing EGF-like repeats. *Cell*, **43**, 567–81.

Wiedermann, C.J., Jelesof, N.J., Pert, C.B. *et al.* (1988) Neuromodulation by polypeptide growth factors: preliminary results on the distribution of epidermal growth factor receptors in adult brain. *Wien. Klin Wochenshcr.*, **100**, 760–3.

Wilcox, J.N. and Derynck, R. (1988) Localization of cells synthesizing transforming growth factor-α mRNA in the mouse brain. *J. Neurosci.*, **8**, 1901–4.

SOMATOSTATIN RECEPTORS IN THE DEVELOPING NERVOUS SYSTEM

6

Stephen L. Kinsman

6.1 INTRODUCTION

Somatostatin, a 14 amino acid peptide, was identified in hypothalamic extracts for its ability to inhibit growth hormone secretion almost twenty years ago. It is now clear that somatostatin is a family of peptides derived from a precursor propeptide of 92 amino acids. Several lines of investigation reveal the importance of this somatostatin peptide system not only to the regulation of pituitary function, but also to the nervous system, pancreas and the gut. Pharmacologic and biochemical studies of somatostatin receptors reveal multiple subtypes, as well as several different roles for this receptor in the regulation of signal transduction. Studies of somatostatin and its receptors during ontogeny make it clear that the somato-statinergic system exhibits dynamic patterns of expression during the development of the nervous system. These patterns allow us to hypothesize about the potential biologic roles somatostatin is playing during development. Somatostatin is now implicated in the regulation of such developmental processes as cell proliferation and/or differentiation, neurite extension and trophic interactions. The future understanding of the role of somatostatinergic systems during develop-ment will require more direct testing of these hypotheses.

6.2 SOMATOSTATIN AND THE DEVELOPING NERVOUS SYSTEM

Somatostatin was originally isolated and characterized as a somatotropin-release-inhibiting factor (SRIF) in 1973 (Brazeau *et al.*, 1973). The purification of this tetradecapeptide allowed the subsequent production of specific antisera, which were then used to study the distribution and content of SRIF in various tissues of the nervous system and other organs (Brownstein *et al.*, 1975; Patel and Reichlin, 1978; Finley *et al.*, 1981; Johansson *et al.*, 1984; Vincent *et al.*, 1985). These studies made it clear that somatostatin had a widespread but specific distribution throughout the nervous system, showing the peptide to be present in significant amounts not only in the hypothalamus, but also in the neocortex, hippocampus, striatum and retina of adult animals. Several other organ systems, most prominently the pancreas and gastrointestinal tract, were also found to contain somatostatin-ergic cell systems. Cloning of a cDNA for the preprosomatostatin message, extracted from rat medullary carcinoma of the thyroid cells in 1982 (Goodman *et al.*, 1982), indicated that somatostatin-14 occupies the C-terminus of a prepropeptide precursor molecule. Immuno-logical and chromatographic methods were also used to investigate the processing of the preprosomatostatin peptide (Rorstad *et al.*,

Receptors in the Developing Nervous System Vol. 1: *Growth factors and hormones* Edited by Ian S. Zagon and Patricia J. McLaughlin. Published in 1993 by Chapman & Hall. ISBN 0 412 45240 5. Vols. 1 and 2 (set) ISBN 0 412 54520 9.

SS 28

H-Ser-Ala-Asn-Ser-Asn-Pro-Ala-Met-Ala-Pro-Arg-Glu-Arg-Lys-
Gly-Cys-Lys-Asn-Phe-Phe-Trp-Lys-Thr-Phe-Thr-Ser-Cys-OH

SS 14

H-Ala-Gly-Cys-Lys-Asn-Phe-Phe-Trp-Lys-Thr-Phe-Thr-Ser-Cys-OH

SS 28 (1–12)

H-Ser-Ala-Asn-Ser-Asn-Pro-Ala-Met-Ala-Pro-Arg-Glu-OH

SMS 201–995

H-(D)Phe-Cys-Phe-(D)Trp-Lys-Thr-Cys-Thr(ol)

N-Ahep(7–10) S14 Bzl

cyclo 7-aminoheptanoyl-Phe-D-Trp-Lys-Thr(Bzl)

CGP 23996

des-Ala-Gly-[desamino-Cys,Tyr]-3, 14 dicarbosomatostatin

MK 678

cyclo[*N*-Me-Ala-Tyr-D-Trp-Lys-Val-Phe]

Fig. 6.1.   Somatostatin peptides and analogs.

1979; Patel *et al.*, 1981; Benoit *et al.*, 1982). From this work it became clear that rat brain contained not only somatostatin-14 (SS-14) but SS-28, SS-28(1–12), and at least four other peptides with larger molecular weights (Fig. 6.1). This work suggested that somatostatin is actually a family of prosomatostatin-derived peptides with the potential for multiple biologic functions. As noted, the first discovered function of somatostatin was as a somatotropin-release-inhibiting factor (SRIF). Studies of the distribution of somatostatin outside the hypothalamus suggested that the peptide probably also functions in the nervous system as either a neurotransmitter or neuro-modulator (Epelbaum, 1986). Studies of the potential functions in other organ systems suggested that it is also important in inhibiting the release of hormones other than somato-tropin, such as glucagon in the pancreas and gastrin in the stomach (Reichlin, 1987).

The notion that a family of peptides was being used by several unrelated physiologic systems within the body led researchers to investigate whether somatostatin was present during the development of an organism. Particular attention was focused on whether somatostatin was present in the developing central nervous system. The earliest studies to investigate the presence of somatostatin in the developing nervous system resulted in three major findings. First, the somatostatin expression in several brain regions began prenatally in rat development, at least as early as gestational day 14. Second, over the next several days of gestation, the number of somatostatin-positive cells increased drama-tically. Many of the regions studied reached a plateau in cell number during the perinatal stage and then more or less maintained their immunoreactivity into adulthood. Third, sev-eral populations of somatostatin-positive cells showed a decrease in number after the peri-natal stage and actually reached a point where none or only a few somatostatin-positive cells were still identifiable in the corresponding adult regions (Shiosaka *et al.*, 1981, 1982; Inagaki *et al.*, 1982). McGregor *et al.* (1982) confirmed the transient elevations of somato-statin immunoreactivity in some of these regions of developing nervous system using a radioimmunoassay for somatostatin. They also noted that the cerebellum was particularly noteworthy in that it showed rather high contents of somatostatin during perinatal development and had little or no detectable somatostatin in the adult state. These findings led both sets of investigators to hypothesize that somatostatin must be playing a developmental role during these stages.

Recent studies have focused on a detailed analysis of the distribution of somatostatin expression during development of the nervous system to better understand its potential functions (Laemle *et al.*, 1982; Eadie *et al.*, 1987; Lowe *et al.*, 1987; Cavanagh and Parnavelas, 1988; Naus *et al.*, 1988 a,b; Villar *et al.*, 1989; Naus, 1990; Naus and Durand, 1990; McElligott *et al.*, 1991). These studies have used molecular biology techniques to look directly at cells expressing the preprosomatostatin gene and have also used antisera, which is more specific for the various peptides derived from preprosomatostatin. Most of this work has been done in the rat, although there has been some work done in the mouse and in humans (Chigr *et al.*, 1989; Mitrofanis *et al.*, 1989; Najimi *et al.*, 1989; Bendotti *et al.*, 1990; Forloni *et al.*, 1990). Most brain regions have been investigated, but for present purposes only selected areas will be discussed. These regions include neocortex, the developing visual system and cerebellum.

The work of Naus *et al.* (1988b) is particularly noteworthy in terms of recent studies of the development of rat somatostatin containing neurons in the nervous system. Northern (RNA) blotting showed increases in neocortex message levels for preprosomatostatin postnatally, with a peak between postnatal day (PD)9 and PD15, with decreasing levels in the adults, suggesting a period of increased expression of preprosomatostatin either in the number of somatostatinergic cells or in the level of expression per cell. Hybridization studies *in situ*, using a probe specific for somatostatin mRNA, were done to define the distribution of somatostatin-expressing neurons during the developmental period (Naus *et al.*, 1988b). These studies of neocortex show the presence of two populations of somatostatin cells. The first wave of cells is present from PD1 to PD11 below the cortical plate in the developing infragranular layers. It has been hypothesized that these are so-called subplate cells. These peptidergic cells have been implicated by some authors as being important for cortical morphogenesis (Chun *et al.*, 1987). A shift occurs by PD12, when large supragranular populations of somatostatin-positive neurons appear. Using antibodies to SS-28 and SS-28(1–12) Naus *et al.* were able to show that these two populations of somatostatinergic cells have some immunochemical differences. The early appearing somatostatinergic cells in the infragranular layers preferentially reacted with the antibody to SS-28(1–12). The later-appearing cells of the supragranular layers were more reactive with the SS-28 antibodies. The authors give two potential explanations for this predominance of SS-28(1–12) reactive cells during early brain development. The first is that these cells preferentially contain the processed form of the preprosomatostatin molecule. The other possibility is that this SS-28(1–12) antibody recognized larger molecular weight forms of the preprosomatostatin molecule present in the infragranular cells. The authors favor the latter explanation. To date there has been no evidence for biologic function of these larger preprosomatostatin-derived molecules. The final point these authors make regarding neocortical expression of somatostatin is that there is a rather dramatic decrease in the number of neurons containing either somatostatin mRNA or somatostatin-28 immunoreactivity between PD12 and adulthood. First, as noted above, the infragranular somatostatinergic cells are no longer present. Also, whereas at around PD12 the somatostatinergic neurons appear to be present throughout all layers of the developing cortex, in adulthood there is a bilaminar distribution of somatostatinergic neurons with a predominance in layers 2/3 and 5/6.

Studies on the developmental appearance of somatostatin in the rat cerebellum show a pattern of somatostatinergic cell appearance quite different from that in neocortical development (Naus, 1990). *In situ* hybridization showed that many cells were positive for somatostatin mRNA at early postnatal stages and that progressively fewer cells were labeled after PD 20. In adulthood, very few cells could be identified as positive for somatostatin mRNA. Immunohistochemical analysis indicated that in the first postnatal week cells in the granule cell layer were stained, as were fibers in this area. During the second postnatal week there was a decrease in this granule cell layer staining with some Purkinje cells displaying immunoreactivity. Some recent work suggests that some of these cells could be derived from the glia. There was then a decrease in the immunoreactive staining of these cell populations over the next two weeks. In the adult cerebellum there was very little immunoreactivity.

These studies point out the prominent expression of somatostatin during both pre-natal and early postnatal brain development. However, they do not provide enough infor-mation to understand the biologic functions of these peptides during development of the mammalian nervous system. In order to better understand the role of somatostatin and its receptors during development, some investi-gators have focused their attention on studying the distribution and time course of somatostatin receptor expression during brain development. In order to understand these studies, it is first necessary to review work looking at the function of somatostatin recep-tors and their distribution in the adult brain.

6.3 SOMATOSTATIN RECEPTORS

The characterization of a functional receptor for somatostatin was first accomplished using a clonal pituitary cell line, GH4C1 cells, by Schonbrunn and Tashjian (1978). Several studies then followed which characterized the regional distribution of somatostatin receptors in adult rat brain and pituitary (Leroux *et al.*, 1985; Reubi and Maurer, 1985; Tran *et al.*, 1985; Patel *et al.*, 1986; Whitford *et al.*, 1986; Heiman *et al.*, 1987; Whitford *et al.*, 1987; Leroux *et al.*, 1988; Barone *et al.*, 1989; Epelbaum *et al.*, 1989; Krantic *et al.*, 1989; Heidet *et al.*, 1990; Krantic *et al.*, 1990). Iodinated-tyrosine-substituted somatostatin analogs were used as the ligand in these studies. Saturation studies with these radiolabeled analogs show a single population of binding sites with a dissociation constant in the nanomolar range. Displacement studies done in pituitary preparations show a mono-phasic curve with several analogs such as SMS 201–995. In rat brain, however, displacement studies using the same somatostatin analogs show a biphasic curve in such areas as the cortex, hippocampus and striatum (Tran *et al.*, 1985). Thus, analogs such as SMS 201–995 are able to differentiate between two classes of somatostatin receptors in the brain. These receptors have been called SSA, with nano-molar binding concentration, and SSB, with micromolar affinity for the agonist.

Once appropriate radiolabeled ligands were defined, studies to investigate the developmental distribution of somatostatin receptors were started. To date, most of this work has focused on development of the central nervous system. A group of investi-gators using the iodinated somatostatin analog[^{125}I]Tyr^{0}DTrp^{8}SS14 has published several papers analyzing the distribution and pharmacology of somatostatin receptors dur-ing ontogeny in several brain regions. The most important of these studies is the consistent finding that somatostatin receptor distribution is quite dynamic during develop-ment and that the observed changes in its distribution and time course of expression are consistent with potential developmental functions of somatostatin peptides.

First, these studies support the hypothesis

that somatostatin acts as a neurotransmitter/ neuromodulator in the adult nervous system by demonstrating the emergence of somatostatin receptors in their adult distribution during synaptogenesis (Gonzalez *et al.*, 1991). What is more interesting for this review is the earlier and mostly transient expression of somatostatin receptors in several parts of the developing nervous system. The presence of these transient distributions again suggests a potential role for the somatostatin neuropeptide family in such developmental processes as cell differentiation, control of neuroblast proliferation, and trophic interactions.

The work by Gonzalez *et al.* (1991) on somatostatin receptor expression patterns during the development of the somatosensory cortex in rat examines the potential role that somatostatin might play in a differentiating neuroblast population. They show the presence of somatostatin receptors as early as embryonic day 16 in the developing rat neocortex using receptor autoradiography. During this time, the signal from bound iodinated somatostatin ligand is intense in the intermediate zone of the developing cortex, a region rich in migrating neuroblasts. As development progresses prenatally, there is an increase in this signal and the beginning of the appearance of receptors in the developing cortical plate. As the newborn period is approached, there are marked decreases, approximately 2–4-fold, in somatostatin receptor signal from the cortical plate and intermediate zone. These results suggest that somatostatin receptors are associated with migrating neuroblasts. This prenatal increase in somatostatin receptor density in these areas of neuroblast migration also corresponds, temporally and spatially, to increased somatostatin gene expression and peptide immunoreactivity in these areas.

Other work on somatostatin receptor distribution during the ontogeny of the cerebellum supports the hypothesis that somatostatinergic systems may regulate cell proliferation in this brain region during development (Gonzalez *et al.*, 1988, 1990 a,b). By using membrane binding preparations, it was shown that developing cerebellar tissue possesses somatostatin receptors that exhibit functions similar to receptors in adult nervous tissue. Using receptor autoradiography, it was found that somatostatin receptors are transiently expressed in the external granular cell layer of the developing rat cerebellum, a region that is a germinative lamina. During this transient expression of receptors, there is also evidence for expression of somatostatin immunoreactivity in cerebellum (Inagaki *et al.*, 1982; Villar *et al.*, 1989; Naus, 1990). Unlike cortex, however, the cellular location of somatostatin receptors does not overlap with the distribution of somatostatin-like immunoreactivity. The immunoreactivity is seen in ascending fibers from the brain stem, developing granule cells and some Purkinje cells. It should be emphasized that these investigators found an extremely high density of somatostatin receptors in this external granular cell layer, some five times the density of somatostatin receptors in adult neocortex. The authors hypothesize that somatostatin may be functioning either to regulate the proliferation of these germinative cells and/or to regulate their early differentiation. More direct studies to test this hypothesis remain to be carried out.

Finally, studies by Bodenant *et al.* (1991) in the developing visual system of the rat, illustrate another potential role of somatostatinergic systems during development, the role of a trophic factor. Somatostatin receptors were found in relatively high concentrations in optic nerve, optic chiasm and optic tract during prenatal and early postnatal visual system development. Around PD21 this receptor expression was completely undetectable. The authors identified a situation where there is a high concentration of somatostatin peptide and receptors in the lateral geniculate nucleus, an area which contains target cells of these tracts. The authors suggest the possibility that somatostatinergic peptides may be acting as a retrogradely transported growth factor, much as nerve growth factor and brain-derived neurotrophic factor act. They could not, however, eliminate the possibility that myelination at later stages of development may be masking the presence of somatostatin receptors in these systems in adulthood. If somatostatin is present in these fiber tract systems in the adult, then it is not clear what functional role it is playing. Somatostatin receptors have also been identified in the developing white matter tracts of the human pyramidal tracts (Charnay *et al.*, 1988).

In summary, these studies emphasize that somatostatinergic systems probably play an important role during early brain development. Studies of somatostatin receptor distribution during development implicate this system in the regulation of various developmental processes in selective brain regions. The presence of somatostatinergic systems serving potentially different developmental functions in several different areas suggests that this peptide family plays a multifaceted role in ontogeny. This would not be particularly surprising in view of the many roles of somatostatinergic systems in the adult organism.

6.4   FUTURE DIRECTIONS

A better understanding of the roles of somatostatin in the developing brain will improve our understanding of both normal and abnormal brain development. To understand further the potential role of somatostatinergic systems in brain development, several future directions will need to be pursued. Three areas will be discussed here: (a) the further biochemical characterization and purification of the somatostatin receptor(s) leading to cloning of a somatostatin receptor gene or genes; (b) pharmacologic investigations to identify and characterize somatostatin receptor subtypes and to develop potential antagonists for these

receptors; and (c) experiments designed to manipulate endogenous levels of somatostatin peptides during brain development.

Biochemical purification of somatostatin receptors is being actively pursued by several groups (Knuhtsen *et al.*, 1988; He *et al.*, 1989; Murphy *et al.*, 1989; Reyl *et al.*, 1989; Brown *et al.*, 1990; Knuhtsen *et al.*, 1990; Lewin and Reyl, 1990). To date, these groups have isolated relatively pure preparations of somatostatin receptors and are in the process of trying to sequence parts of this protein. Using sequences of peptide fragments of these isolated receptors, cloning of the gene or genes for this receptor will be possible. Other investigators are trying to express receptors for this neuropeptide in frog oocyte expression systems (White and Reisine, 1990). The cloning of somatostatin receptor genes will allow further study of how the receptor is genetically regulated during development. It will also permit investigators to study the environmental factors that affect somatostatin receptor expression in adulthood and development.

A large selection of potent somatostatin analogs is being utilized to identify and study subtypes of the somatostatin receptor. As noted earlier, one such analog is SMS 201–995, which was used to define the somatostatin receptor subtypes SSA and SSB. More recent work using newer analogs CGP 23996 and MK 678 (Figure 6.1) has revealed even more heterogeneity in SS receptor subclasses at both the pharmacologic and anatomic levels (Martin *et al.*, 1991). The significance that these receptor subtypes hold for the function of somatostatin in development is not yet clear. These analogs have also been shown to exhibit different signal transduction characteristics, some linked to adenylate cyclase inhibition and some not, again providing evidence for the diversity of somatostatin's biologic actions (Raynor and Reisine, 1989).

The search for potential antagonists of somatostatin has not been particularly successful to date. In 1982, Fries *et al.* reported the identification of a somatostatin antagonist analog, cyclo[7-aminoheptanoyl-Phe-D-Trp-Lys-Thr-(Bzl)], that increased growth hormone, insulin and glucagon release in rat. Subsequent studies, however, have not shown this putative antagonist to be very effective in pharmacologic inhibitory binding experiments (Gonzalez *et al.*, 1990a, b). There is still a great need for reliable, pharmacologically well-defined antagonists of somatostatin binding.

Finally, manipulation of endogenous levels of somatostatin peptides during ontogeny is a direct way of testing hypotheses about the role of somatostatin during development, but this approach has been hampered by technical difficulties. Studies of the function of somatostatin in the adult nervous system have been aided by the use of cysteamine to deplete neurons of their somatostatin content (Palkovits *et al.*, 1982). The use of this substance has not been particularly fruitful in either early postnatal development or prenatal development because of problems with administration and toxicity. The use of antibodies to decrease circulating somatostatin levels during development has not been reported. Finally, work is ongoing to use the technology of transgenesis to produce mice with increased levels of endogenous somatostatin. Low *et al.* (1985) reported the production of somatostatin transgenic mice that carried copies of a metallothionein promotor–somatostatin construct. These transgenic mice had high circulating levels of immunoreactive somatostatin but did not have overexpression of somatostatin in the nervous system. By producing mice that carry extra copies of the functioning mouse preprosomatostatin gene, using constructs that include the endogenous promoter elements, it has been possible to produce mice that overexpress somatostatin in a tissue-appropriate distribution during the development of the nervous system (Kinsman *et al.*, 1990). These mice are being used to study whether an increased endogenous level of somatostatin has any effect on the development of the cerebellum, hippocampus, neocortex or visual system by disturbing one or more of its roles in the developmental organization of these regions. The regulation of somatostatin receptors in the developing brain in response to increased levels of the peptide can also be studied.

Recently, Yamada *et al.* (1992) reported the cloning and functional characterization of a family of human and mouse somatostatin receptors expressed in brain, gastrointestinal tract and kidney. Southern blot analysis of genomic DNA probed with these two very different receptor clones indicated that a larger family of somatostatin receptors exists.

REFERENCES

Barone, D., Ferrara, B., Fracchia, S. and Eshkol, A. (1989) Binding affinity profile of somatostatin (SS-14) to rat CNS receptors. *Pharmacol. Res.*, **1**, 77–8.

Bendotti, C., Hohmann, C., Forioni, G. *et al.* (1990) Developmental expression of somatostatin in mouse brain. II. *In situ* hybridization. *Dev. Brain Res.*, **53**, 26–39.

Benoit, R., Ling, N., Alford, B. and Guillemm, R. (1982). Seven peptides derived from pro-somatostatin in rat brain. *Biochem. Biophys. Res. Commun.*, **107**, 944–50.

Bodenant, C., Leroux, P., Gonzalez, B. J. and Vandry, H. (1991) Transient expression of somatostatin receptors in the rat visual system during development. *Neuroscience*, **41**, 595–606.

Brazeau, P., Vale, W., Burgus, R. *et al.* (1973) Hypothalamic polypeptide that inhibits the secretion of immunoreactive pituitary growth hormone. *Science*, **129**, 77–9.

Brown, P. J., Lee, A. B., Norman, M. G. *et al.* (1990) Identification of somatostatin receptors by covalent labeling with a novel photoreactive somatostatin analog. *J. Biol. Chem.*, **265**, 17995–8004.

Brownstein, M. A., Amimura, M., Sato, A. *et al.* (1975). The regional distribution of somatostatin in the rat brain. *Endocrinology*, **96**, 1456–61.

Cavanagh, M. E. and Parnavelas, J. G. (1988) Development of somatostatin immunoreactive neurons in the rat occipital cortex: a combined immunocytochemical-autoradiographic study. *J. Comp. Neurol.*, **268**, 1–12.

Charnay, Y., Leroux, P., Epelbaum, J. *et al.* (1988)

Displaceable somatostatin binding sites in the gray matter and pyramidal paths of the human developing spinal cord. *Neurosci. Lett.*, **84**, 245–50.

Chigr, F., Najimi, M., Leduque, P. *et al.* (1989) Anatomical distribution of somatostatin immunoreactivity in the infant brainstem. *Neuroscience*, **29**, 615–28.

Chun, J. J. M., Nakamura, M. J. and Shatz, C. J. (1987) Transient cells of the developing mammalian telencephalon are peptide-immunoreactive neurons. *Nature*, **325**, 617–20.

Eadie, L. A., Parnavelas, J. G. and Franke, E. (1987) Development of the ultrastructural features of somatostatin-immunoreactive neurons in the rat visual cortex. *J. Neurocytol.*, **16**, 445–59.

Epelbaum, J. (1986) Somatostatin in the central nervous system: physiology and pathological modification. *Progr. Neurobiol.*, **22**, 63–100.

Epelbaum, J., Agid, F., Agid, Y. *et al.* (1989) Somatostatin receptors in brain and pituitary. *Horm. Res.*, **31**, (1–2), 45–50.

Finley, J. C. W., Maderdrut, J. L., Roger, L. J. and Petrusz, P. (1981) The immunocytochemical localization of somatostatin-containing neurons in the rat central nervous system. *Neuroscience, 6*, 2173–92.

Forloni, G., Hohmann, G. and Coyle, J. T. (1990) Developmental expression of somatostatin in mouse brain. I. Immunocytochemical studies. *Dev. Brain Res.*, **53**, 6–25.

Fries, J. L., Murphy, W. A., Sueiras, D. J. and Coy, D. H. (1982) Somatostatin antagonist analog increases GH, insulin and glucagon release in the rat. *Peptides*, **3**, 811–14.

Gonzalez, B. J., Leroux, P., Laquerriere, H. *et al.* (1988) Transient expression of somatostatin receptors in the rat cerebellum during development. *Brain Res.*, **468**, 154–7.

Gonzalez, B. J., Leroux, P., Bodenant, C. *et al.* (1990a) Pharmacological characterization of somatostatin receptors in the rat cerebellum during development. *J. Neurochem.*, **55**, 729–37.

Gonzalez, B. J., Leroux, P., Boer, G. J. and Vandry, H. (1990b) Expression of somatostatin receptors is impaired in the cerebellum of developing Brattleboro rats. *Brain Res.*, **532**, 115–9.

Gonzalez, B. J., Leroux, P., Bodenant, C. and Vandry, H. (1991) Ontogeny of somatostatin receptors in the rat somatosensory cortex. *J. Comp. Neurol.*, **305**, 177–88.

Goodman, R. H., Jacobs, J. W., Dee, P. C. and Habener, J. F. (1982) Somatostatin-28 encoded in a cloned cDNA obtained from a rat medullary thyroid carcinoma. *J. Biol. Chem.*, **257**, 1156–9.

He, H. T., Johnson, K., Thermos, K. and Reisine, T. (1989) Purification of a putative brain somatostatin receptor. *Proc. Natl. Acad. Sci. USA*, **86**, 1480–4.

Heidet, V., Faivre-Bauman, A., Kordon, C. *et al.* (1990) Functional maturation of somatostatin neurons and somatostatin receptors during development of mouse hypothalamus *in vivo* and *in vitro*. *Dev. Brain Res.*, **57**, 85–92.

Heiman, M. L., Murphy, W. A. and Coy, D. H. (1987). Differential binding of somatostatin agonists to somatostatin receptors in brain and adenohypophysis. *Neuroendocrinology*, **45**, 429–36.

Inagaki, S., Shiosaka, S., Takatsuki, K. *et al.* (1982) Ontogeny of somatostatin-containing neuron system of the rat cerebellum including its fiber connections: an experimental and immunohistochemical analysis. *Dev. Brain Res.*, **3**, 509–27.

Johansson, O., Hökfelt, T. and Elde, R. P. (1984) Immunohistochemical distribution of somatostatin-like immunoreactivity in the central nervous system of the adult rat. *Neuroscience*, **13**, 265–339.

Kinsman, S., Moran, T., Reeves, R. *et al.* (1990) Somatostatin transgenic mice: neurobiologic consequence of neuropeptide gene expression. *Ann. Neurol.*, **28**, 442.

Knuhtsen, S., Esteve, J. P., Bernadet, B. *et al.* (1988) Molecular characterization of the solubilized receptor of somatostatin from rat pancreatic acinar membranes. *Biochem. J.*, **254**, 641–7.

Knuhtsen, S., Esteve, J. P., Cambillan, C. *et al.* (1990) Solubilization and characterization of active somatostatin receptors from rat pancreas. *J. Biol. Chem.*, **265**, 1129–33.

Krantic, S., Martel, J.-C., Weissman, D. and Quirion, R. (1989). Radioautographic analysis of somatostatin receptor sub-type in rat hypothalamus. *Brain Res.*, **498**, 267–78.

Krantic, S., Martel, J. C., Weissman, D. *et al.* (1990) Quantitative radioautographic study of somatostatin receptors heterogeneity in the rat extra-hypothalamic brain. *Neuroscience*, **39**, 127–37.

Laemle, L. K., Feldman, S. C. and Lichtenstein, E. (1982) Somatostatin-like immunoreactivity in the central visual pathway of the prenatal rat. *Brain Res.*, **251**, 365–70.

Leroux, P., Quirion, R. and Pelletier, G. (1985) Localization and characterization of brain somatostatin receptors as studied with somatostatin-14 and somatostatin-28 receptor radioautography. *Brain Res.*, **347**, 74–84.

Leroux, P., Gonzalez, B. J., Laquerriere, A. *et al.* (1988) Autoradiographic study of somatostatin receptors in the rat hypothalamus: validation of a GTP-induced desaturation procedure. *Neuroendocrinology*, **47**, 533–44.

Lewin, M. J. and Reyl, D. F. (1990) Molecular characterization of a purified human gastric somatostatin receptor. *Metabolism*, **39**, 74–7.

Low, M. J., Hammer, R. E., Goodman, R. H. *et al.* (1985) Tissue-specific posttranslational processing of pre-prosomatostatin encoded by a metallothionein-somatostatin fusion gene in transgenic mice. *Cell*, **41**, 211–19.

Lowe, W. L., Schaffner, A. E., Roberts, C. T. and LeRoith, D. (1987) Developmental regulation of somatostatin gene expression in the brain is region specific. *Mol. Endocrinol.*, **1**, 181–7.

Martin, J. -L., Chesselet, M. -F., Raynor, K. *et al.* (1991) Differential distribution of somatostatin receptor subtypes in rat brain revealed by newly developed somatostatin analogs. *Neuroscience*, **41**, 581–93.

McElligott, S. F., Card, J. P., O'Kane, T. M. and Baldino, F. (1991) Ontogeny of somatostatin mRNA-containing perikarya in the rat central nervous system. *Synapse*, **7**, 123–34.

McGregor, G. P., Woodhams, P. L., O'Shaughnessy, D. J. *et al.* (1982) Developmental changes in bombesin, substance P, somatostatin and vasoactive intestinal polypeptide in the rat brain. *Neurosci. Lett.*, **28**, 21–7.

Mitrofanis, J., Robinson, S. R. and Provis, J. M. (1989) Somatostatinergic neurones of the developing human and cat retinae. *Neurosci. Lett.*, **104**, 209–16.

Murphy, K. K., Srikant, C. B. and Patel, Y. C. (1989) Evidence for multiple protein constituents of the somatostatin receptor in pituitary tumor cells: affinity cross-linking and molecular characterization. *Endocrinology*, **125**, 948–56.

Najimi, M., Chigr, F., Luduque, P. *et al.* (1989) Immunohistochemical distribution of somatostatin in the infant hypothalamus. *Brain Res.*, **483**, 205–20.

Naus, C. C. G. (1990) Developmental appearance of somatostatin in the rat cerebellum: *in situ* hybridization and immunohistochemistry. *Brain Res. Bull.*, **24**, 583–92.

Naus, C. C. G. and Durand, M. M. (1990) Ontogeny of somatostatin expression in primary cultures of cortical neurons parallels that seen *in vivo*. *Brain Res. Bull.*, **25**, 749–54.

Naus, C. C. G., Miller, F. D., Morrison, J. H. and Bloom, F. E. (1988a) Immunohistochemical and *in situ* hybridization analysis of the development of the rat somatostatin-containing neocortical neuronal system. *J. Comp. Neurol.*, **269**, 448–63.

Naus, C. C. G., Morrison, J. H. and Bloom, F. E. (1988b) Development of somatostatin-containing neurons and fibers in the rat hippocampus. *Dev. Brain Res.*, **40**, 113–21.

Palkovits, M., Brownstein, M. A., Eiden, L. E. *et al.* (1982) Selective depletion of somatostatin in rat brain by cysteamine. *Brain Res.*, **240**, 178–80.

Patel, Y. C. and Reichlin, S. (1978) Somatostatin in hypothalamus, extrahypothalamic brain, and peripheral tissues of the rat. *Endocrinology*, **102**, 523–30.

Patel, Y. C., Wheatley, T. and Ning, C. (1981) Multiple forms of immunoreactive somatostatin: comparison of distribution in neural and nonneural tissues and portal plasma of the rat. *Endocrinology*, **109**, 1943–9.

Patel, Y. C., Baquiran, G., Srikant, C. B. and Posner, B. I. (1986) Quantitative *in vivo* autoradiographic localization of [^{125}I-Tyr11]somatostatin-14 and [Leu8,D-Trp22–^{125}I-Tyr25]somatostatin-28-binding sites in rat brain. *Endocrinology*, **119**, 2262–9.

Raynor, K. and Reisine, T. (1989) Analogs of somatostatin selectively label distinct subtypes of somatostatin receptors in rat brain. *J. Pharmacol. Exp . Ther.*, **251**, 510–7.

Reichlin, S. (1987) Secretion of somatostatin and its physiologic function. *J. Lab. Clin. Med.*, **109**, 320–6.

Reubi, J. C. and Maurer, R. (1985) Autoradiographic mapping of somatostatin receptors in the rat central nervous system and pituitary. *Neuroscience*, **15**, 1183–93.

Reyl, D. F., Le, R. S., Linard, C. *et al.* (1989) Solubilization and immunopurification of a somatostatin receptor from the human gastric tumoral cell line HGT-1. *J. Biol. Chem.*, **264**, 18789–95.

Rorstad, O. P., Epelbaum, J., Brazeau, P. and Martin, J. B. (1979) Chromatographic and biological properties of immunoreactive somatostatin in hypothalamic and extrahypothalamic brain regions of the rat. *Endocrinology*, **105**, 1083–92.

Schonbrunn, A. and Tashjian, A. (1978) Characterization of functional receptors for somatostatin in rat pituitary cells in culture. *J. Biol. Chem.*, **253**, 6473–83.

Shiosaka, S., Takatsuki, K., Sakanaka, M. *et al.* (1981) Ontogeny of somatostatin-containing neuron system of the rat: immunohistochemical observations. I. Lower brainstem. *J. Comp. Neurol.*, **203**, 173–88.

Shiosaka, S., Takatsuki, K., Sakanaka, M. *et al.* (1982) Ontogeny of somatostatin-containing neuron system of the rat: immunohistochemical analysis. II. Forebrain and diencephalon. *J. Comp. Neurol.*, **204**, 211–24.

Tran, V. T., Beal, M. F. and Martin, J. B. (1985) Two types of somatostatin receptors differentiated by cyclic somatostatin analogs. *Science*, **228**, 492–5.

Villar, M. J., Hökfelt, T. and Brown, J. C. (1989) Somatostatin expression in the cerebellar cortex during postnatal development. *Anat. Embryol.*, **179**, 257–67.

Vincent, S. R., McIntosh, C. H., Buchan, A. M. J. and Brown, J. C. (1985) Central somatostatin systems revealed with monoclonal antibodies. *J. Comp. Neurol.*, **238**, 169–86.

White, M. M. and Reisine, T. (1990) Expression of functional pituitary somatostatin receptors in Xenopus oocytes. *Proc. Natl. Acad. Sci. USA*, **87**, 133–6.

Whitford, C. A., Bloxham, C. A., Snell, C. R. *et al.* (1986) Regional distribution of high-affinity [³H]somatostatin binding sites in the human brain. *Brain Res.*, **398**, 141–7.

Whitford, C. A., Candy, J. M., Snell, C. R. *et al.* (1987) Autoradiographic visualization of binding sites for [³H]somatostatin in the rat brain. *Eur. J. Pharmacol.*, **138**, 327–33.

Yamada, Y., Post, S. R., Wang, K. *et al.* (1992) Cloning and functional characterization of a family of human and mouse somatostatin receptors expressed in brain, gastrointestinal tract and kidney. *Proc. Natl. Acad. Sci. USA*, **89**, 251–5.

DEVELOPMENTAL REGULATION OF THE INSULIN AND INSULIN-LIKE GROWTH FACTOR RECEPTORS IN THE CENTRAL NERVOUS SYSTEM

Haim Werner, Charles T. Roberts, Jr, Mohan K. Raizada, Carolyn A. Bondy, Martin Adamo and Derek LeRoith

7.1 INTRODUCTION

The ontogeny of the mammalian nervous system is probably one of the most complicated and intriguing biological processes. Full development of a mature brain, starting from a microscopic embryo composed of relatively undifferentiated cells, requires the orderly completion of a myriad of interrelated biochemical and cellular events. This ontogenic process is controlled by the concerted action of many different neurohormones and growth factors.

One family of regulatory substances whose importance in the differentiation and development of the nervous system has been well established are the insulin-like growth factors (IGFs) (Hepler and Lund, 1990; LeRoith *et al.*, 1992) (Fig. 7.1). IGF-I and IGF-II are multifunctional polypeptides structurally related to insulin. According to the classical view, which is currently under scrutiny, insulin is mainly responsible for the control of metabolic processes such as the regulation of carbohydrate metabolism, whereas the IGFs are preferentially devoted to more long-term growth processes (Rechler and Nissley, 1985; Daughaday and Rotwein, 1989). Although most of the circulating IGFs are synthesized in the liver, many tissues express these genes and have measurable levels of mRNA and peptide (D'Ercole *et al.*, 1984; Adamo *et al.*, 1989a). At the local level, the IGFs are believed to act in a paracrine or autocrine fashion by interacting with receptors located, respectively, in a neighboring cell or in the same IGF-producing cell.

The insulin and the IGF receptors are widely distributed membrane proteins, and this ubiquity reflects their physiological importance (Werner *et al.*, 1991; Taylor *et al.*, 1991; Nissley *et al.*, 1991). The insulin and IGF-I receptors are structurally very similar (Fig. 7.2). In humans, they are each the product of a single gene. The gene encoding the insulin receptor is located on chromosome 19 and the gene encoding the IGF-I receptor on chromosome 15 (Ebina *et al.*, 1985; Ullrich *et al.*, 1985, 1986). These genes each extend over more than 150 kb, contain 22 exons, and have similar exon/intron organization (Seino *et al.*, 1989; Abbott *et al.*, 1991).

Each receptor is synthesized as a single-chain polypeptide precursor, which then becomes glycosylated and cleaved into α and β subunits. The mature receptor is a heterotetramer, composed of two α and two β subunits. The α subunits are entirely extra-

Receptors in the Developing Nervous System Vol. 1: *Growth factors and hormones* Edited by Ian S. Zagon and Patricia J. McLaughlin. Published in 1993 by Chapman & Hall. ISBN 0 412 45240 5. Vols. 1 and 2 (set) ISBN 0 412 54520 9.

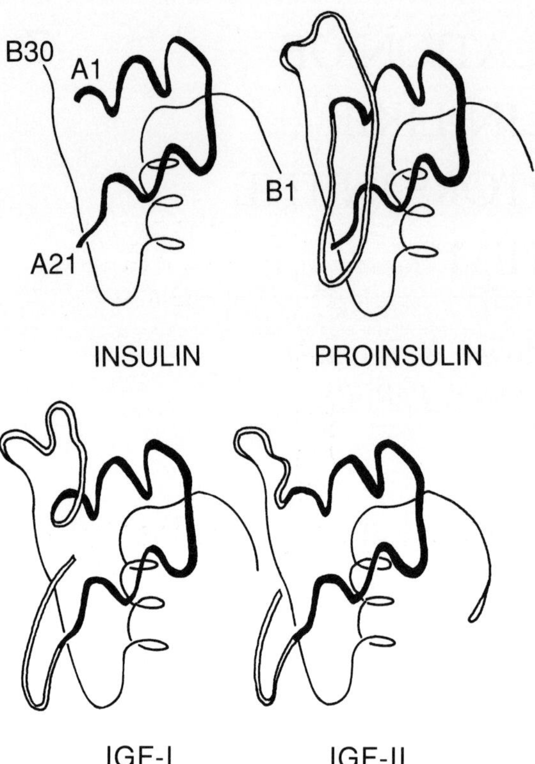

Fig. 7.1. The insulin-like growth factors are a family of multifunctional peptides structurally and phylogenetically related to insulin. According to the classical concept, insulin is mainly involved in the regulation of metabolic processes whereas the IGFs are mitogenic factors in a number of growth processes.

cellular and are mainly responsible for ligand binding, whereas the β subunits are transmembrane polypeptides which contain a tyrosine kinase domain in their intracellular portion. As with other growth factor receptors of the tyrosine kinase family, mediation of a biological effect by these receptors is believed to depend primarily on activation of this enzymatic domain (Yarden and Ullrich, 1988).

It has been proposed that insulin and IGF-I receptors can form hybrid tetramers on the cell surface, composed of an insulin α/β hemireceptor and an IGF-I α/β hemireceptor (Moxham *et al.*, 1989; Soos and Siddle, 1989). The physiological significance of these hybrids, both in general and in the nervous system in particular, is currently under investigation.

The IGF-II receptor is structurally and functionally dissimilar to the insulin and IGF-I receptors (Morgan *et al.*, 1987). It consists of a long extracellular chain, composed of 15 repeat sequences, and a short *C*-terminal cytoplasmic domain (Fig. 7.2). Cloning and sequencing of cDNAs encoding this receptor have revealed that it is identical to the cation-independent mannose 6-phosphate (M-6-P) receptor. The receptor has been shown to be involved in the intracellular targeting of lysosomal proteins which contain M-6-P residues. As the IGF-II receptor lacks tyrosine kinase activity, its role in signal transduction has been questioned. However, recent studies suggest that the IGF-II/M-6-P receptor is coupled to a calcium channel, through a G protein which is sensitive to pertussis toxin (Okamoto *et al.*, 1990).

Although each ligand of the insulin-like family binds preferentially to its specific receptor, there is a certain degree of cross-recognition (Rechler *et al.*, 1980). Thus, the K_d of the IGF-I receptor for IGF-I is ~ 1 nM, 2–10-fold lower for IGF-II and 100–500-fold lower for insulin. In contrast, the K_d of the insulin receptor for insulin is ~ 1 nM, 10–50-fold lower for IGF-II and 100–500-fold lower for IGF-I. The IGF-II/M-6-P receptor binds IGF-I with medium to high affinity but does not bind insulin.

The third component of the IGF system comprises the IGF binding proteins (BPs) (Table 7.1) (Clemmons, 1991). These proteins were initially found in the circulation, where they deliver IGFs to their target tissues and protect them from degradation. In addition to being an integral component of the IGF endocrine system, BPs have also been found to be produced locally by many tissues, including neural tissues. At this level, BPs were shown to be important in the modulation of the paracrine and autocrine actions of IGFs. Six BPs have been identified and characterized to date. In general, they inhibit IGF effects

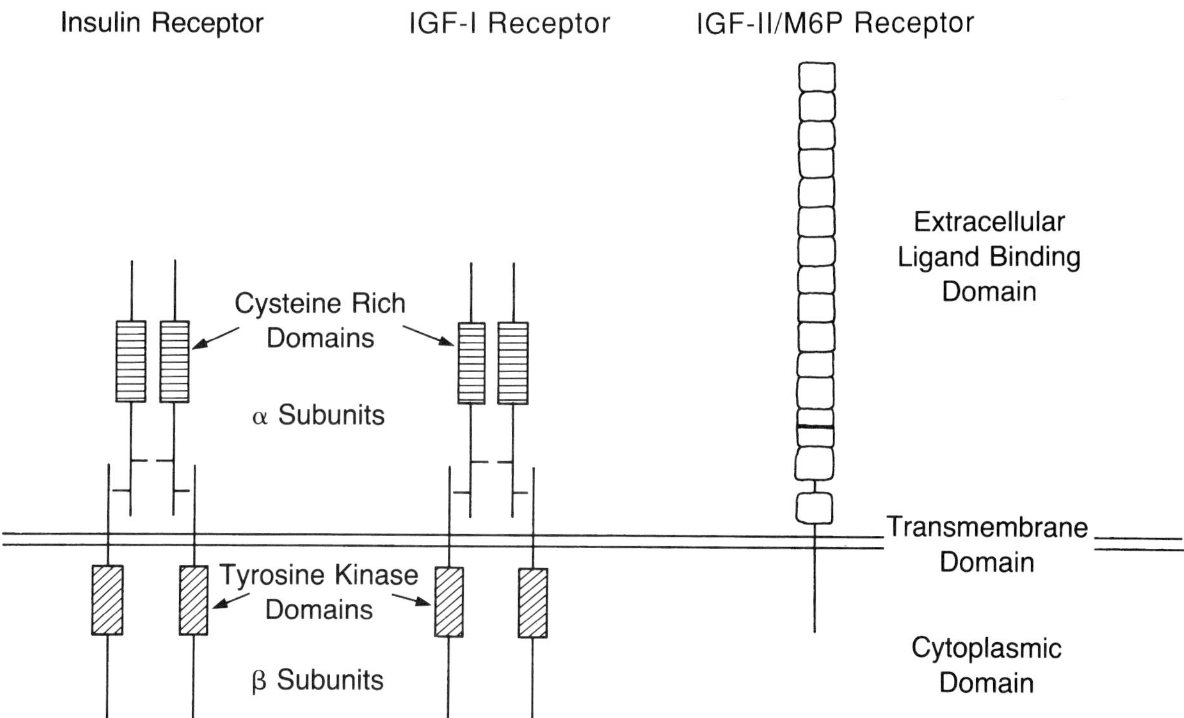

Fig. 7.2. The biological actions of insulin and the IGFs are intitiated by their interaction with specific cell-surface receptors. Insulin and IGF-I receptors are tetrameric proteins composed of two extracellular α subunits involved in ligand binding, and two transmembrane β subunits which contain a tyrosine kinase domain in their intracellular portion. The IGF-II/M-6-P receptor is a single-chain polypeptide composed of 15 repeat sequences and a short cytoplasmic domain.

although in certain cases, they may enhance the biological effects of these growth factors.

This review will focus on the regulation of insulin and IGF receptors in the developing nervous system. It is important, however, to keep in mind that the regulation of these receptors is intimately connected to that of the ligands and, in the case of IGF receptors,

Table 7.1.   IGF binding proteins (BP)

	BP1	BP2	BP3	BP4	BP5	BP6
Mass (Da)	25 000	33 000	53 000	24 000	32 000	31 000
Source for purification	Amniotic fluid	BRL-3A cell conditioned media	Blood	Many types of cultured cells	C2 myoblast conditioned media	Cerebrospinal fluid
Relative binding affinities for IGFs	IGF-II= IGF-I	IGF-II> IGF-I	IGF-II> IGF-I	IGF-I= IGF-II	IGF-I= IGF-II	IGF-II>> IGF-I

to that of the binding proteins. Therefore, all the biological activities of these receptors at any developmental stage must be analyzed in the context of their interactions with ligands and binding proteins.

7.2 DEVELOPMENTAL REGULATION OF INSULIN AND IGF RECEPTORS IN THE CNS

7.2.1 REGULATION OF INSULIN RECEPTORS IN THE DEVELOPING NERVOUS SYSTEM

Although the source of insulin in the nervous system is still open to speculation, the wide distribution of specific insulin receptors in brain suggests a significant role for this hormone/growth factor in nervous tissue differentiation and development (Havrankova and Roth, 1978; Adamo *et al.*, 1989b; Unger *et al.*, 1991).

Phylogenetically, the lowest organism in which the insulin receptor mRNA was found to be expressed in nervous tissue is *Drosophila*. Using *in situ* hybridization, insulin receptor mRNA was found in eggs and early embryos. However, after midembryogenesis, the highest levels of expression were localized in the developing nervous system. Consistent with a role for the insulin receptor in neural growth, expression of the insulin receptor gene in nervous tissue was also very high at the larval stage and persisted in the adult, at which stage the cortex and ganglion cells were heavily labeled (Petruzzelli *et al.*, 1986; Garofalo and Rosen, 1988).

Characterization of brain insulin receptors in vertebrates by affinity cross-linking followed by polyacrylamide gel electrophoresis (PAGE) showed that the size of the α subunit is smaller than in liver and other peripheral non-neural tissues (Fig. 7.3). This structural difference is evolutionarily conserved. Thus, the size of the frog, alligator, and lizard brain insulin receptor α subunit is ~ 120 kDa, whereas the size of the α subunit in the liver of each species is

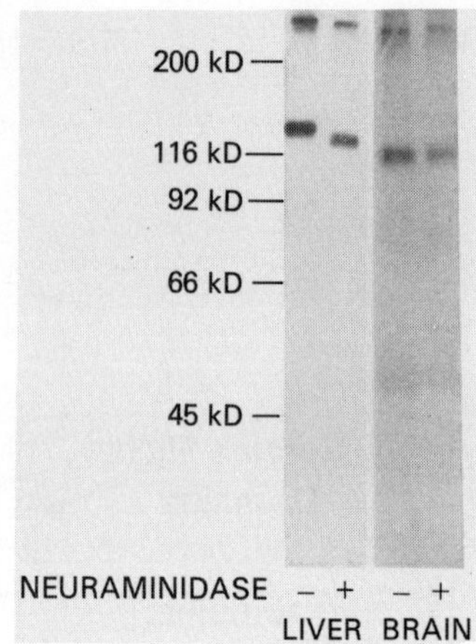

Fig. 7.3. Brain insulin receptors are smaller than their counterparts in liver and other peripheral tissues. Treatment of guinea pig insulin receptors with neuraminidase increased the electrophoretic mobility of the liver receptor but had no effect on the brain receptor, suggesting that the difference between brain and peripheral receptors was due primarily to differences in glycosylation. (Adapted from LeRoith *et al.*, 1988.)

~ 135 kDa (Hart *et al.*, 1987; Shemer *et al.*, 1986, 1987a, 1987b). Similarly, the α subunit of the human brain insulin receptor is significantly smaller than that in placenta and other tissues (Roth *et al.*, 1986). This difference between brain and peripheral non-neural insulin receptors was found to be due primarily to differences in glycosylation. Thus, exposure to neuraminidase increased the electrophoretic mobility of the liver receptor, but had no effect on the brain insulin receptor (Hendricks *et al.*, 1984; Lowe *et al.*, 1986; LeRoith *et al.*, 1988; McElduff *et al.*, 1988). The physiological significance of these differences in carbohydrate moieties is unknown. Gammeltoft *et al.* (1984) found that brain insulin receptors are functionally different from those of other tissues with respect to insulin specificity and the absence of negative cooperativity, and it is possible that differences in glycosylation of the receptor may, at least partially, explain the observed functional differences. The size of the α subunit was also found to be conserved during ontogeny, in addition to phylogeny. Thus, Lowe *et al.* (1986) found that cross-linked α subunits from 19-day-old fetuses, 1-day-old neonates and adult rats migrated identically on SDS-PAGE. The ~ 10 kDa difference between brain and peripheral non-neural insulin receptor α subunits has been observed not only at the adult stage but also during embryonic development (Devaskar *et al.*, 1988). When different brain cells were isolated, the neuronal (125 kDa) and microvascular (125 kDa) insulin receptors were not sensitive to neuraminidase action, thus conferring neuraminidase-insensitivity to the whole brain, whereas the glial cell insulin receptor (128 kDa) was sensitive to neuraminidase. One single study, however, showed somewhat contradictory results by reporting that development of the rat brain was associated with a progressive decrease in the apparent molecular weight of the insulin receptor α subunit: 130 kDa on day 16 of gestation, 126 kDa at birth and 120 kDa in the adult (Brennan, 1988). This developmental decrease in brain α subunit size was proposed to be due to a reduction in sialic acid content. However, the reason for these discrepant results remains unclear.

Differences in sialic acid content are, apparently, not the only differences between brain and peripheral non-neural insulin receptors. Thus, whereas the purified human brain insulin receptor was found to be identical to the placental insulin receptor in the amount of neuraminidase-sensitive sialic acid and in its reaction with three monoclonal antibodies to the β subunit, a monoclonal antibody directed against the insulin binding site recognized the placental receptor with an affinity ~ 300 times higher than that for the brain receptor (Roth *et al.*, 1986).

One model in which the ontogenic regulation of insulin binding in the nervous system was carefully analyzed is the chicken embryo (Bassas *et al.*, 1985, 1987, 1988, 1989; DePablo *et al.*, 1990). Insulin receptors are expressed as early as gastrulation (day 1 post-laying), and during neurulation (day 2) they are widely distributed throughout the nervous system. Using autoradiographic techniques, ^{125}I-labeled insulin was localized to the neural folds, neural tube and optic vesicle at this early stage of development (Girbau *et al.*, 1989). In young chicken embryos (day 6) the highest binding of ^{125}I-labeled insulin was to the retina and lateral encephalic vesicles, with a similar pattern described for [^{125}I]IGF-I. At late ontogenic stages (day 18, i.e., 3 days before hatching), insulin binding was mainly found in the paleostriatum augmentatum and in the molecular layers of the cerebellum (Bassas *et al.*, 1989).

In the rat brain, highest levels of binding were found around day 20 of gestation (Brennan, 1988) (364 ± 42 fmol insulin/mg protein). Two days after birth, insulin binding to brain membranes decreased by 60% and subsequently decreased to the adult level of 63 ± 15 fmol/mg. These fluctuations in binding resulted in similar changes in insulin-stimulated β subunit autophosphorylation. Peaks in insulin binding in the perinatal period were also reported by other groups (Kappy *et al.*, 1984; Lowe *et al.*, 1986).

To study the expression of the insulin receptor gene in rat brain at various fetal and postnatal stages an S1 nuclease protection assay was employed with a 48-base synthetic oligonucleotide probe complementary to the human insulin receptor mRNA sequence. Insulin receptor mRNA was expressed at very low levels at embryonic days 13 and 15, and was most abundant at embryonic day 20 and the day of birth. The levels of receptor transcript were then drastically reduced to adult values. Thus, the ontogenic changes observed in insulin binding and receptor number are associated with parallel changes in gene expression (Baron-Van Evercooren *et al.*, 1991).

A trend similar to the one described in rat was observed in the rabbit brain (Devaskar *et al.*, 1986). Specific insulin binding and receptor number increased from day 20 of gestation to postnatal day 1 , and attained adult values by postnatal day 6. Both binding and receptor number in the adult rabbit brain were less than in fetal and neonatal brain. In addition, the affinity of the receptors increased during early fetal development and remained constant in the postnatal period.

In human cerebral cortex, insulin binding was measured in synaptosomal membrane fractions obtained at different stages of development. Receptor concentrations were 3.0 ± 0.8 pmol/mg protein in pre-term up to 30 weeks of gestation, 0.6 ± 0.14 pmol/mg in full-term and newborns and 0.2 ± 0.024 pmol/mg from one year to adulthood (Potau *et al.*, 1991). Insulin receptors were also measured in plasma membranes from whole human fetal brain (Sara *et al.*, 1983). However, in comparison with the liver, specific insulin binding to brain membrane was 5 times lower.

In summary, the concentrations of insulin receptors in the brains of several different vertebrate species have been shown to fluctuate during ontogeny, with peaks at the critical periods of cell migration and differentiation which, in rodents, occurs around birth. Levels of insulin binding and insulin receptor mRNA subsequently decrease to the low levels characteristic of the adult brain.

7.2.2 REGULATION OF IGF RECEPTORS IN THE DEVELOPING NERVOUS SYSTEM

A role for the IGF-I receptor at early stages of development is suggested by the fact that both IGF-I binding and IGF-I receptor mRNA can be detected in oocytes and unfertilized eggs from *Xenopus laevis* (Scavo *et al.*, 1991). IGF-I receptors were much more abundant than insulin receptors in full-grown oocytes, and unfertilized eggs expressed almost only the IGF-I receptor. No information is, however, available on the specific expression of these receptors during the development of *Xenopus* nervous system.

As with the insulin receptor, the electrophoretic mobility of the α subunit of the brain IGF-I receptor in vertebrates was ~ 10 kDa smaller than in liver and other peripheral

organs (Gammeltoft *et al.*, 1985; Heidenreich *et al.*, 1986; McElduff *et al.*, 1988) and, again, this difference is evolutionarily conserved. Interestingly, the IGF-II/M-6-P receptor from rat brain is also of lower apparent molecular weight than that from rat liver (McElduff *et al.*, 1987).

IGF-I receptors are very abundant in the developing chicken nervous system. In fact, they dominate over insulin receptors at early embryonic stages (Bassas *et al.*, 1985, 1988, 1989), although the localization of binding was very similar for both receptors. At late embryonic stages, IGF-I binding was prominent in the hippocampus and neostriatum. This pattern of development suggests a possible overlap in the function of insulin and IGF-I in early brain development, whereas later in development it predicts a more specialized role for each peptide. Since the chicken M-6-P receptor lacks an IGF-II binding site, both ^{125}I-IGF-I and ^{125}I-IGF-II are presumably binding to IGF-I receptors (Bassas *et al.*, 1988).

In the rat brain, postnatal development was characterized by a decrease in the binding capacity for IGF-I (and insulin) from birth to 15 days of age, remaining stable thereafter. This decrease in binding resulted in a parallel decrease in tyrosine kinase activity of the receptor over the same period of time. Since the K_D value (~ 2 nM) did not change significantly over the period of time studied (from day 2 to day 37), it was suggested that the decrease in binding was due to a reduction in receptor number (Pomerance *et al.*, 1988). Consistent with these results, the expression of the IGF-I receptor gene in homogenized rat brain and in other tissues was especially high at perinatal stages, decreasing subsequently to low levels at adulthood. Thus, the steady-state levels of brain IGF-I receptor mRNA at postnatal day (PD) 50 were 5–6-fold lower than at embryonic day (ED) 20 (Werner *et al.*, 1989b) (Fig. 7.4). Similar results were recently reported by Baron-Van Evercooren *et al.* (1991). By using a synthetic oligonucleotide probe derived from the human IGF-I receptor

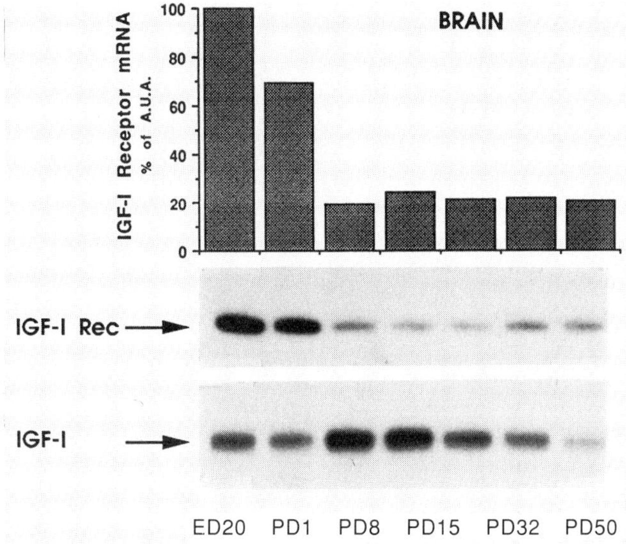

Fig. 7.4. Developmental regulation of IGF-I and IGF-I receptor mRNA in rat brain. Steady-state mRNA levels were measured by solution hybridization/RNase protection assay using specific antisense RNA probes and expressed as a percentage of the level at stage E20. (Werner *et al.*, 1989b). Highest levels of receptor mRNA were found at embryonic day 20 (ED20), whereas the highest levels of IGF-I mRNA were seen at postnatal day 8 (PD8). A.U.A.: arbitrary units of absorbance.

sequence, IGF-I receptor transcripts in rat brain were detected as early as ED13; however, maximal levels were seen at ED15–20.

To measure IGF-I and insulin receptor protein levels in membranes from fetal (14–21 days of gestation), neonatal and adult rat brain, three antipeptide antibodies against epitopes in the β subunit were employed. These antibodies (AbP2, AbP4 and AbP5) recognize both insulin and IGF-I receptors. Scanning densitometry of the immunoblots showed that receptor proteins were 4–10-fold more abundant in fetal than in adult brain. Maximal levels of protein expression were seen at 16–18 days of gestation, consistent with mRNA results (Garofalo and Rosen, 1989). Interestingly, whereas immunoblot analyses with AbP4 and AbP5 revealed a 92 kDa protein corresponding to the β subunit of both receptors, AbP2 revealed an additional protein of 97 kDa. This protein was phosphorylated in response to IGF-I and was not directly recognized by AbP4 or AbP5. These results are

interpreted by the authors to suggest that two distinct β subunits may be present in a single IGF-I receptor in brain. Alternatively, these results may be analyzed in the light of the insulin/IGF-I hybrid receptor theory. According to this concept, many cells may express on their surface tetrameric receptors composed of an insulin α/β hemireceptor and an IGF-I α/β hemireceptor (Fig. 7.5). The abundance and distribution of insulin/ IGF-I hybrid receptors would presumably be solely a result of the developmental regulation of insulin and IGF-I receptors separately.

Results similar to those obtained for the developmental regulation of the IGF-I receptor have been reported by Sklar *et al.* (1989) for the IGF-II/M-6-P receptor. Using quantitative immunoblotting, they found that receptor levels were high in brain and in other fetal tissues and declined dramatically in late gestation or in the early postnatal period.

During human ontogenesis, a dramatic change in IGF receptors occurs in the brain.

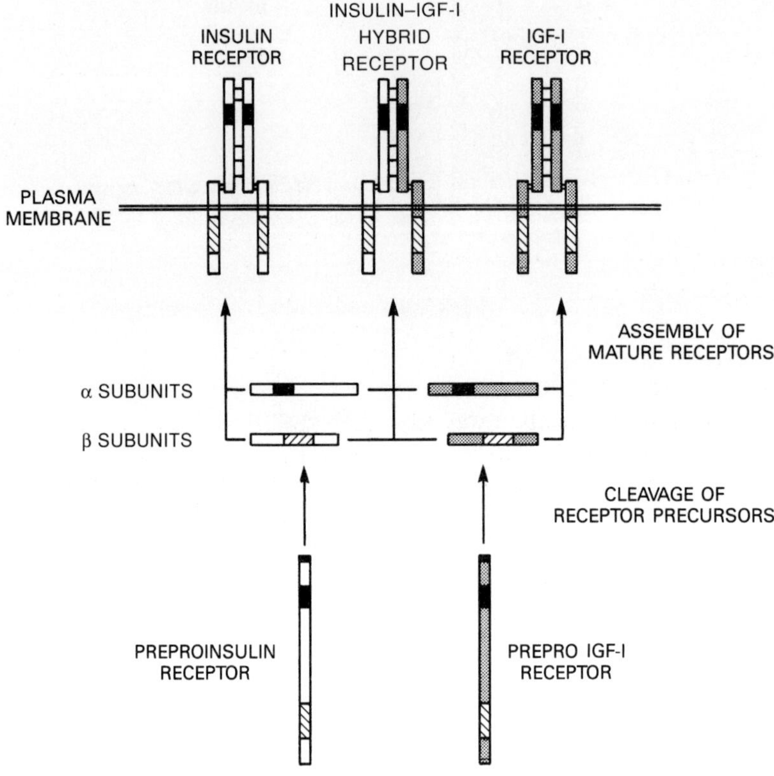

Fig. 7.5. Insulin–IGF-I hybrid receptors. It has been suggested that many cells, in addition to homodimer insulin and IGF-I receptors, may express heterodimer hybrids composed of an insulin receptor α/β subunit joined to an IGF-I receptor α/β subunit by disulfide bridges.

Thus, before the end of the first trimester of gestation, a low-affinity, high-capacity IGF-I receptor is present. Around week 17 of gestation this low-affinity receptor disappears, and a high-affinity IGF-I receptor takes its place. Interestingly, this increase in the affinity of the IGF-I receptor was accompanied by a concomitant decrease in the affinity of the insulin receptor. At week 25 of gestation, the affinity constant and pattern of cross-reaction of the IGF-I receptor are already identical to the adult receptor (Sara *et al.*, 1983). According to Sara *et al.*, the gestational period characterized by the low-affinity IGF-I receptor coincides with a rapid proliferative phase in fetal forebrain during which neuroblasts are formed. This suggests that this low-affinity receptor may be present on stem cells or immature neurons. After week 18 of gestation, the increase in cell number becomes more gradual. This late proliferative phase is dominated by glial cell formation and differentiation of immature neurons into mature ones. Accordingly, the low-affinity IGF-I receptor may recognize a specific fetal form of IGF originally termed human embryonic somatomedin. Structural analysis revealed that this variant IGF-I (des 1,3-IGF-I) lacks the first three *N*-terminal amino acids, and most probably arises from post-translational modification of the prohormone (Sara *et al.*, 1986).

Finally, the importance of IGF-I receptors in brain development is further strengthened by the finding that deletion of the distal long arm of chromosome 15 (q 26.1→ q ter) and loss of the IGF-I receptor gene in an infant was shown to be associated with microcephaly, neurodevelopmental delay and mental retardation (Roback *et al.*, 1991).

7.3 DISTRIBUTION OF INSULIN AND IGF RECEPTORS DURING DEVELOPMENT OF THE CNS

The original approach employed to determine the distribution of insulin and IGF receptors in the CNS consisted of measuring binding to membranes prepared from discrete brain regions. However, most of the recent information on the comparative localization of insulin and IGF receptors in the CNS was obtained by means of autoradiographic studies using ^{125}I-labeled insulin, IGF-I and IGF-II on brain slices. More recently, these ligand-binding studies were further extended with the introduction of *in situ* hybridization histochemistry, a powerful technique which allows the detection of as little as a few mRNA molecules in a single cell.

Autoradiographic localization of insulin receptors in adult rat brain showed a fairly even distribution with an especially high concentration in the external plexiform layer of the olfactory bulb (Table 7.2). Large numbers of binding sites were also found in the arcuate, dorsomedial and paraventricular nuclei of the hypothalamus. At the fetal stage, the distribution of insulin receptors in brain is more heterogeneous and is characterized by a waxing and waning, with maximal densities at 15 days of gestation (Young *et al.*, 1980; Baskin *et al.*, 1986, 1988; Hill *et al.*, 1986; Lesniak *et al.*, 1988).

The steady-state levels of insulin receptor mRNA were measured in discrete brain regions at ED20 (Baron-Van Evercooren *et al.*, 1991). Interestingly, at this stage, extremely low levels of receptor message were detected in the olfactory bulb, whereas highest levels were found in the diencephalon and spinal cortex, and medium levels in the cortex.

In a recent study, the cellular pattern of IGF-I receptor gene expression during matu-

Table 7.2. Regional localization of brain insulin and IGF-I and -II receptors as determined by autoradiography

	Insulin	IGF-I	IGF-II
Choroid plexus	++++[a]	++++	++++
Olfactory bulb	++++	++++	++[c]
Median eminence	+[d]	++++	+++
Hypothalamus	++	+++	++
Cerebral cortex	++	++	++
Cerebellum	+++[b]	+++	++
Amygdala	++	++	++
Hippocampus	++	+++	+++
Thalamus	++	+++	++
Anterior pituitary	−[e]	−	++++

Data compiled from Hill *et al.* (1986), Baskin *et al.* (1986), Lesniak *et al.* (1988), Adamo *et al.* (1989b).
[a] ++++ = Highest concentration.
[b] +++ = High.
[c] ++ = Moderate.
[d] + = Lower concentration.
[e] − = Extremely low or not present.

ration of rat brain was analyzed by *in situ* hybridization using antisense RNA probes (Bondy *et al.*, 1992). Two basic patterns of gene expression emerged (Figs 7.6 and 7.7). From early in ontogenesis, a fairly homogeneous pattern of expression of this receptor gene is observed throughout the nervous system, local variations being mainly a reflection of different cell densities. Thus, the apparent high level of expression in the adult cerebellar inner granule cell layer is most probably due to a constitutive expression by a population of densely packed cells rather than an especially high level of expression in each cell. This basal level of expression is present at fetal, perinatal and postnatal stages, and it was suggested that the function of this constitutively expressed receptor is mainly related to cell survival. Probably, both circulating and locally produced IGF-I and IGF-II (such as IGF-II secreted from the choroid plexus or circumventricular organs) can act as ligands for this receptor.

Superimposed on this basal pattern of expression, high levels of IGF-I receptor mRNA appear at localized areas during the

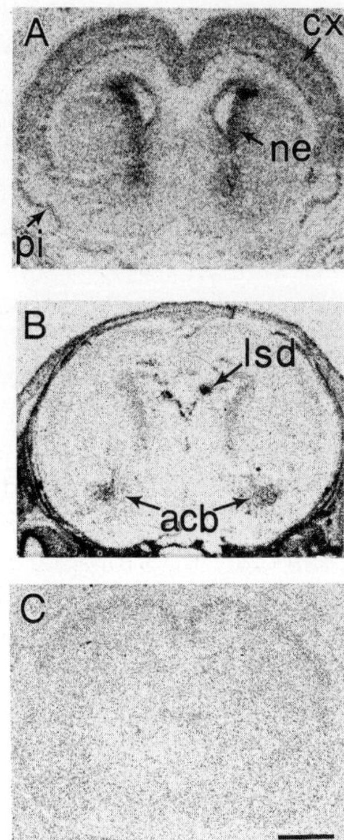

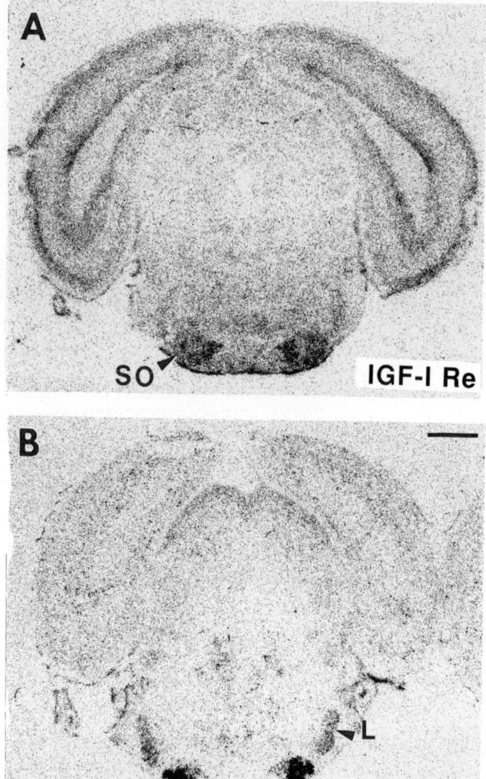

Fig. 7.6. Distribution of IGF-I receptor (A) and IGF-I (B) mRNAs in the rat brain on the day of birth, shown in film autoradiographs of *in situ* hybridization. (C) a section adjacent to (A) and (B) which was hybridized to an IGF-I receptor 'sense' probe. These are coronal sections through the forebrain at the level of the acumbens n. (acb). IGF-I receptor mRNA is abundant, particularly in the neocortex (cx), piriform cortex (pi) and neuroepithelium (ne). IGF-I mRNA is concentrated in the acumbens and lateral septal dorsal n. (lsd). Bar = 1mm.

Fig. 7.7. Distribution of IGF-I receptor (A) and IGF-I (B) mRNAs in the rat mid-brain on postnatal day 14. Both the receptor and IGF-I mRNAs are abundant in the superior olivary complex (SO) and, to a lesser extent, in the nuclei of the lateral lemniscus (L), both of which are auditory relay stations. Bar = 1mm.

course of development. Thus, during mid-embryogenesis (ED12–ED16), especially high concentrations of IGF-I receptor mRNA are found in the ventral floorplate of the hindbrain (Bondy *et al.*, 1990). It was postulated that high receptor levels in these specialized neuroepithelial cells may be involved in the process of axonal targeting. Receptor mRNA also accumulates in sensory and cerebellar relay systems, concomitant with a high local expression of IGF-I mRNA.

Coexistence of IGF-I and the IGF-I receptor in these relay systems supports the notion of local paracrine/autocrine interactions in the maintenance of these synaptic circuitries. Consistent with previous studies showing IGF-I binding sites in the glomerular and external plexiform layers of the olfactory bulb (Lesniak *et al.*, 1988; Bohannon *et al.*, 1988; Werther *et al.*, 1990), this study describes the presence of IGF-I receptor mRNA in mitral and tufted cell layers.

Autoradiographic analysis of the IGF-II binding sites shows co-localization with the IGF-I binding sites in the olfactory bulbs, granule cell layers of the cerebellum, hippocampus, choroid plexus and median eminence (Smith *et al.*, 1988; Matsuo *et al.*, 1989). However, the concentration of IGF-II binding sites in the anterior pituitary is several-fold higher than that of IGF-I (Goodyer *et al.*, 1984; Rosenfeld *et al.*, 1984; Ocrant *et al.*, 1988; Lesniak *et al.*, 1988).

In human brain, the highest densities of IGF-I receptors were found in the hippocampus, amygdala and parahippocampal gyrus. Intermediate densities were seen in the cerebellum, cerebral cortex and caudate nucleus, and low densities were found in the substantia nigra, red nucleus, white matter and cerebral pedunculus (Adem *et al.*, 1989).

7.4   CHARACTERIZATION OF INSULIN AND IGF RECEPTORS IN NEURAL-DERIVED TISSUES

Most studies on the binding and action of insulin and the IGFs in the nervous system have been performed on neural-derived cells. These cells can be divided into two major groups: primary cultures (including neurons, astroglia and oligodendrocytes) and established cell lines.

High-affinity insulin, IGF-I and IGF-II receptors were found in both neuronal and glial cells derived from rat brain (Lowe *et al.*, 1986; Masters *et al.*, 1987; Shemer *et al.*, 1987c; Ballotti *et al.*, 1987). Primary neuronal cultures from developing rat brain (ED15) showed cell-surface binding sites for IGF-I and IGF-II (Nielsen *et al.*, 1991). Neuronal synthesis of RNA and DNA is increased two-fold by IGF-I, with 10 times higher potency than IGF-II; however, an antibody specific against the IGF-II receptor has no effect on the DNA response to both IGF-I or IGF-II. Therefore, proliferation of neuronal precursor cells *in vitro* is presumably stimulated by IGF-I and IGF-II via activation of the IGF-I receptor.

In primary cultures of neuronal and glial cells from 1-day-old rat brains, the apparent molecular mass of the α subunit of the IGF-I receptor was 125 kDa in neuronal and 135 kDa in glial cells (Shemer *et al.*, 1987c). Similar differences between neuronal and glial cells were reported for the insulin receptor, suggesting that glial receptors are similar to the peripheral non-neural type, whereas the neuronal receptors are more representative of the brain type (Masters *et al.*, 1987). Differences in apparent molecular mass were also noted in the β subunit of the IGF-I receptor. Thus, the size of this subunit was 91 kDa in neurons and 95 kDa in glial cells, possibly reflecting minor glycosylation or protein differences (Shemer *et al.*, 1987c).

Extensive characterization of IGF receptors on neonate brain astrocytes in primary culture was reported by two groups (Ballotti *et al.*, 1987; Han *et al.*, 1987). IGF-I, which is synthesized by these cells, acts as a growth promoter by activation of the IGF-I receptor tyrosine kinase, although, according to one group (Han *et al.*,1987), the IGF-II receptor may also be responsible for the mitogenic effect of IGFs in these cells.

Interestingly, steady-state levels of IGF-I receptor mRNA were higher in astrocytes prepared from 21-day-old rat brains than in astrocytes prepared from neonates (Masters *et al.*, 1991a). Since the levels of receptor message in total brain were higher at the neonatal stage and decreased to much lower levels at 3 weeks of age, and because the contribution of astroglia to total brain mass is several orders of magnitude larger in adult than in neonate brain, it is possible that the trend observed *in vitro* represents either a culture artifact, or alternatively the total brain levels of mRNA more directly reflect expression by other cell types.

Primary cultures of oligodendrocytes also show developmental regulation of the IGF-I receptor message (Masters *et al.*, 1991b). Thus, IGF-I receptor mRNA levels were 180% higher after 8 days in culture than after 3 days. This regulation of IGF-I receptors is consistent with a role for this receptor in mediating IGF-I effects on the proliferation as well as the differentiation of oligodendrocytes.

Receptors for IGF-I and IGF-II were characterized in cultured brain microvessel endothelial cells, suggesting a role for these receptors as a transport system for the circulating peptides, particularly for IGF-II (Rosenfeld *et al.*, 1987; Duffy *et al.*, 1988). Insulin, IGF-I and IGF-II receptors were also demonstrated in many neuroblastoma cell lines, such as N18, NG108 and B104 and in the Y79 retinoblastoma and glioblastoma cell lines, including C6 cells and the cell line 1690 (Ota *et al.*, 1988a; Kiess *et al.*, 1989; Sturm *et al.*, 1989; Yorek *et al.*, 1989; McCusker *et al.*, 1990; Merrill and Edwards, 1990).

7.5 FUNCTION OF INSULIN AND IGF RECEPTORS IN THE DEVELOPING NERVOUS SYSTEM

Although early studies suggested that insulin receptors are mainly responsible for mediating short-term metabolic actions of insulin (e. g., glucose uptake), whereas IGF receptors mediate long-term effects of the IGFs (e.g., proliferation, differentiation), more recent studies, particularly on the artificial overexpression of insulin and IGF-I receptor cDNAs, suggested that both receptors have the inherent capability to mediate short-, medium-, and long-term effects (Steele-Perkins *et al.*, 1988). Thus, the involvement of one specific receptor in mediating a certain biological action will be dictated by the spatial and temporal expression of that receptor, i.e., its distribution and developmental regulation.

Most of the available information on the biological actions of insulin and IGFs in neural tissues was generated from cell culture studies, although valuable data were also obtained from intact *in vivo* models. For didactical purposes, these effects can be classified into metabolic types of action, neuromodulatory–neurotransmitter effects and long-term, growth/differentiation processes (Table 7.3).

Table 7.3. Biological actions of insulin and IGFs in neural tissues

---

A. *Metabolic actions*

Increase glucose uptake in glial cells
Increase Glut-1 mRNA in neuronal and glial cells
Activate pyruvate dehydrogenase
Stimulate protein, fatty acid, sulfolipid, and
  cholesterol synthesis
Central action to lower plasma glucose and free
fatty acid levels

B. *Neuromodulatory actions*

Alter neuronal firing rates
Promote electrical coupling
Inhibit norepinephrine reuptake
Stimulate serotonin biosynthesis and uptake
Maintain Na^+/K^+ pump in synaptosomes

C. *Growth/differentiation actions*

Stimulate synaptogenesis
Promote neurite outgrowth and survival
Mitogenesis in oligodendrocytes, astrocytes,
  neuronal cells and sympathetic neuroblasts
Promote glial cell development

D. *Neuroendocrine actions*

Inhibition of pituitary GH secretion

Among the metabolic actions mediated by insulin/IGF receptors, insulin was shown to stimulate 2-deoxy-D-glucose uptake in glial but not neuronal cells in culture (Clarke *et al.*, 1984), although the basal level of glucose uptake was much higher in neurons than in glia. Insulin (and IGF-I) induced an increase in the steady-state levels of glucose transporter 1 (Glut-1) mRNA and protein, suggesting that these hormones may be involved in the control of brain energy metabolism not only by translocating glucose transporters from an intracellular pool to the cell surface, but also by acting at the level of gene expression (Karnieli and Garvey, 1987; Werner *et al.*, 1989a). Interestingly, a recent study has shown that the stimulation of Glut-1 by IGF-I in brain astrocytes is developmentally regulated. Thus, in spite of a significantly higher level of IGF-I receptor mRNA and IGF-I binding in astrocytes from 21-day-old as compared to 1-day-old brains, IGF-I caused a dose-dependent stimulation of 2-deoxy-D-glucose uptake in the neonate-derived cultures but had no effect in adult-derived cells (Masters *et al.*, 1991a).

In vivo, brain insulin receptors mediate the hypoglycemic action of centrally administered insulin (Amir and Schechter, 1987). In addition, chronic i.c.v. injection of insulin in baboons resulted in a decrease in food intake and body weight (Woods *et al.*, 1979). IGF-I receptors in the hypothalamus have also been shown to be sensitive to the nutritional status of the animal. Thus, decreased caloric intake and low levels of circulating IGF-I significantly increased the number of receptors in the median eminence (Bohannon *et al.*, 1988).

Regulation of feeding behaviour by insulin and IGFs includes receptor-mediated transfer of information along pathways that connect the olfactory bulb with hypothalamic structures and the limbic system (Hill *et al.*, 1986; Unger *et al.*, 1991). Since these regions contain neurotransmitters and neuropeptides associated with feeding behavior, it is possible that this behavior is achieved by the interaction of the insulin/IGF-I system of peptides and receptors (and probably binding proteins) with classical neurotransmitters (Araujo *et al.*, 1989).

The distribution of insulin and IGF-I receptors in brain, with special enrichment in olfactory and limbic systems, is very similar to that of other neuropeptides. Thus, several investigators have suggested that insulin and IGFs may themselves act as neuromodulatory agents. However, this contention is supported only by studies *in vitro*. For example, insulin alters neuronal firing rates (Palovcik *et al.*, 1984; Sakaguchi and Bray, 1987), promotes electrical coupling between cultured sympathetic neurons (Wolinsky *et al.*, 1985), inhibits norepinephrine uptake in neuronal cells (Boyd *et al.*, 1985) and increases tryptophan

and serotonin biosynthesis and transport (Raizada *et al.*, 1987, 1988). In chromaffin cells, IGF-I was shown to enhance catecholamine secretion (Dahmer and Perlman, 1988). A role for the insulin receptor in signal transduction in neurons was suggested by the observation that insulin activates a cytosolic protein kinase that phosphorylates ribosomal SP6 and which is distinct from protein kinase C and cAMP-dependent protein kinase (Heidenreich and Toledo, 1989). Furthermore, by using immunocytochemical techniques, Moss *et al.* (1990) showed that the distribution of material containing phosphotyrosine (presumably reflecting endogenous substrates) in specific areas of the rat forebrain agreed well with the distribution of the insulin receptor.

An important neuroendocrine role of IGF-I is its inhibition of growth hormone secretion by the pituitary and hypothalamus (Berelowitz *et al.*, 1981; Rosenfeld *et al.*, 1984). In a recent report, cells overexpressing the transfected human IGF-I receptor were shown to be hyperresponsive to exogenous IGF-I, i.e., IGF-I treatment attenuated growth hormone secretion by 80% as compared to 40% attenuation in control cells (Yamasaki *et al.*, 1991). This indicates that IGF-I receptor number is important in mediating the negative feedback effect of IGF-I on growth hormone gene transcription.

Finally, IGFs fulfill important roles in cell growth and development. Thus, IGF-I stimulates DNA and mRNA synthesis in cultured brain cells (Lenoir and Honegger, 1983; Burgess *et al.*, 1987), supports survival of neuronal cells and induces neurite outgrowth (Recio-Pinto and Ishii, 1988; Svrzic and Schubert, 1990). Many of these growth-promoting actions are also seen with insulin, but at much higher concentrations, suggesting that the insulin effects are mediated by the IGF-I receptor (McMorris *et al.*, 1986). IGF-II may also have mitogenic actions in brain cells, but it is believed that these effects are mediated by interaction with IGF-I receptors (Ballotti *et al.*, 1987).

7.6   CONCLUDING REMARKS

The complexity of the nervous system precludes the postulation of oversimplified schemes to explain the role of insulin and IGF receptors in the developing brain. However, it is clear that this family of ligands, receptors and binding proteins has an important role in nervous system development and differentiation.

In this review we have presented evidence that insulin, IGF-I and IGF-II receptors are expressed in the developing central nervous system. IGF-I receptor mRNA is widely distributed from the very earliest developmental stages and throughout maturity and this constitutive level of expression may be necessary for neuronal survival. Superimposed on this basal level of IGF-I receptor gene expression are very high levels of receptor mRNA in certain regions at defined developmental stages.

Neural receptors are biochemically and immunologically distinct from peripheral receptors, and it is believed that these structural differences constitute the basis for the specialized functions mediated by this family of receptors in brain. Future work will define the specific functions of insulin and IGF receptors in the developing nervous system.

REFERENCES

Abbott, A.M., Bueno, R., Murray, J.G. and Smith, R. J. (1991) Structure of the human IGF-I receptor gene: comparison with insulin receptor gene structure. Abstract presented at the 73rd Annual Meeting of the Endocrine Society, Washington, DC, p. 31

Adamo, M., Lowe, W.L., LeRoith, D. and Roberts, C. T. (1989a) Insulin-like growth factor I messenger ribonucleic acids with alternative 5'-untranslated regions are differentially expressed during development of the rat. *Endocrinology*, **124**, 2737–42.

Adamo, M., Raizada, M.K. and LeRoith, D. (1989b) Insulin and insulin-like growth factor receptors in the nervous system. *Mol. Neurobiol.*, **3**, 71–100.

Adem, A., Jossan, S.S., d'Argy, R. *et al.* (1989) Insulin-like growth factor 1 (IGF-1) receptors in the human brain: quantitative autoradiographic localization. *Brain Res.*, **503**, 299–303.

Amir, S. and Schechter, Y. (1987) Centrally mediated hypoglycemic effects of insulin: apparent involvement of specific insulin receptors. *Brain Res.*, **418**, 152–6.

Araujo, D.M., Lapchak, P.A., Collier, B. *et al.* (1989) Insulin-like growth factor-1 (somatomedin-C) receptors in the rat brain: distribution and interaction with the hippocampal cholinergic system. *Brain Res.*, **484**, 130–8.

Ballotti, R., Nielsen, F.C., Pringle, N. *et al.* (1987) Insulin-like growth factor I in cultured rat astrocytes: expression of the gene and receptor tyrosine kinase. *EMBO J.*, **6**, 3633–9.

Baron-Van Evercooren, A., Olichon-Berthe, C., Kowalski, A. *et al.* (1991) Expression of IGF-I and insulin receptor genes in the rat central nervous system: a developmental, regional and cellular analysis. *J. Neurosci. Res.*, **28**, 244–53.

Baskin, D.G., Brewitt, B., Davidson, D.A. *et al.* (1986) Quantitative autoradiographic evidence for insulin receptors in the choroid plexus of the rat brain. *Diabetes*, **35**, 246–9.

Baskin, D.G., Wilcox, B.J., Figlewicz, D.P. and Dorsa, D. M. (1988) Insulin and insulin-like growth factors in the CNS. *Trends Neurosci.*, **11**, 107–11.

Bassas, L., DePablo, F., Lesniak, M.A. and Roth, J. (1985) Ontogeny of receptors for insulin-like peptides in chick embryo tissues: early dominance of insulin-like growth factor over insulin receptors in brain. *Endocrinology*, **117**, 2321–9.

Bassas, L., DePablo, F., Lesniak, M.A. and Roth, J. (1987) The insulin receptors of chick embryo show tissue-specific structural differences which parallel those of the insulin-like growth factor-I receptors. *Endocrinology*, **121**, 1468–76.

Bassas, L., Lesniak, M.A., Serrano, J. *et al.* (1988) Developmental regulation of insulin and type I insulin-like growth factor receptors and absence of type II receptors in chicken embryo tissues. *Diabetes*, **37**, 637–44.

Bassas, L., Girbau, M., Lesniak, M.A. *et al.* (1989) Development of receptors for insulin and insulin-like growth factor-I in head and brain of chick embryos: autoradiographic localization. *Endocrinology*, **125**, 2320–7.

Berelowitz, M., Szabo, M., Frohman, L.A. *et al.* (1981) Somatomedin-C mediates growth hormone negative feedback by effects on both the hypothalamus and the pituitary. *Science*, **212**, 1279–81.

Bohannon, N.J., Corp, E.S., Wilcox, B.J. *et al.* (1988) Characterization of insulin-like growth factor I receptors in the median eminence of the brain and their modulation by food restriction. *Endocrinology*, **122**, 1940–7.

Bondy, C.A., Werner, H., Roberts, C.T. and LeRoith, D. (1990) Cellular pattern of insulin-like growth factor-I (IGF-I) and Type I IGF receptor gene expression in early organogenesis: comparison with IGF-II gene expression. *Mol. Endocrinol.*, **4**, 1386–98.

Bondy, C.A., Werner, H., Roberts, C.T. and LeRoith, D. (1992) Cellular pattern of type I insulin-like growth factor receptor gene expression during maturation of the rat brain: comparison with insulin-like growth factors I and II. *Neuroscience*, **46**, 909–23.

Boyd, F.T., Clarke, D.W., Muther, T.F. and Raizada, M.K. (1985) Insulin receptors and insulin modulation of norepinephrine uptake in neuronal cultures from rat brain. *J. Biol. Chem.*, **260**, 15880–4.

Brennan, W.A. (1988) Developmental aspects of the rat brain insulin receptor: loss of sialic acid and fluctuation in number characterize fetal development. *Endocrinology*, **122**, 2364–70.

Burgess, S.K., Jacobs, S., Cuatrecasas, P. and Sahyoun, N. (1987) Characterization of a neuronal subtype of insulin-like growth factor I receptor. *J. Biol. Chem.*, **262**, 1618–22.

Clarke, D.W., Boyd, F.T., Kappy, M.S. and Raizada, M. K. (1984) Insulin binds to specific receptors and stimulates 2-deoxy-glucose uptake in cultured glial cells from rat brain. *J. Biol. Chem.*, **259**, 11672–5.

Clemmons, D.R. (1991) Insulin-like growth factor binding proteins, in *Insulin-like Growth Factors: Molecular and Cellular Aspects* (ed D. LeRoith), CRC Press, Boca Raton, FL, pp. 151–79.

Dahmer, M.K. and Perlman, R.L. (1988) Bovine chromaffin cells have insulin-like growth factor-I (IGF-I) receptors: IGF-I enhances catecholamine secretion. *J. Neurochem.*, **51**, 321–3.

Daughaday, W.H. and Rotwein, P. (1989) Insulin-like growth factors I and II. Peptide, messenger ribonucleic acid and gene structures, serum and tissue concentrations. *Endocrine Rev.*, **10**, 68–91.

DePablo, F., Scott, L.A. and Roth, J. (1990) Insulin and insulin-like growth factor I in early development: peptides, receptors and biological events. *Endocrine Rev.*, **11**, 558–77.

D'Ercole, A., Stiles, A. and Underwood, L. (1984) Tissue concentrations of somatomedin C: further evidence for multiple sites of synthesis, and paracrine or autocrine mechanisms of action. *Proc. Natl. Acad. Sci. USA*, **81**, 935–9.

Devaskar, S.U., Holekamp, N., Karycki, L. and Devaskar, U.P. (1986) Ontogenesis of the insulin receptor in the rabbit brain. *Hormone Res.*, **24**, 319–27.

Devaskar, S., Holtzclaw, L. and Sadiq, F. (1988) The heterogeneity of the developing brain insulin receptor. *Ped. Res.*, **24**, 683–92.

Duffy, K.R., Pardridge, W.M. and Rosenfeld, R.G. (1988) Human blood–brain barrier insulin-like growth factor receptor. *Metabolism*, **37**, 136–40.

Ebina, Y., Ellis, L., Jarnagin, K. *et al.* (1985) The human insulin receptor cDNA: the structural basis for hormone-activated transmembrane signalling. *Cell*, **40**, 747–58.

Gammeltoft, S., Staun-Olsen, P., Ottesen, B. and Fahrenkrug, J. (1984) Insulin receptors in rat brain cortex. Kinetic evidence for a receptor subtype in the central nervous system. *Peptides*, **5**, 937–44.

Gammeltoft, S., Haselbacher, G.K., Humbel, R.E. *et al.* (1985) Two types of receptor for insulin-like growth factors in mammalian brain. *EMBO J.*, **4**, 3407–12.

Garofalo, R.S. and Rosen, O.M. (1988) Tissue localization of *Drosophila melanogaster* insulin receptor transcripts during development. *Mol. Cell Biol.*, **8**, 1638–47.

Garofalo, R.S. and Rosen, O.M. (1989) Insulin and insulin-like growth factor 1 (IGF-1) receptors during central nervous system development: expression of two immunologically distinct IGF-1 receptor β subunits. *Mol. Cell. Biol.*, **9**, 2806–17.

Girbau, M., Bassas, L., Alemany, J. and dePablo, F. (1989) *In situ* autoradiography and ligand-dependent tyrosine kinase activity reveal insulin receptors and insulin-like growth factor I receptors in prepancreatic chicken embryos. *Proc. Natl. Acad. Sci. USA*, **86**, 5868–72

Goodyer, C.G., De Stephano, L., Lai, W.H. *et al.* (1984) Characterization of insulin-like growth factor receptors in rat anterior pituitary, hypothalamus, and brain. *Endocrinology*, **114**, 1187–94.

Han, V.K.M., Lauder, J.M. and D'Ercole, A.J. (1987) Characterization of somatomedin/insulin-like growth factor receptors and correlation with biologic action in cultured neonatal rat astroglial cells *J. Neurosci.*, **7**, 501–11.

Hart, C., Shemer, J., Penhos, J.C. *et al.* (1987) Frog brain and liver show evolutionary conservation of tissue-specific differences among insulin receptors. *Gen. Comp. Endocrinol.*, **68**, 170–8.

Havrankova, J. and Roth, J. (1978) Insulin receptors are widely distributed in the central nervous system of the rat. *Nature*, **272**, 827–9.

Heidenreich, K.A., Freidenberg, G.R., Figlewicz, D.P. and Climore, P. R. (1986) Evidence for a subtype of insulin-like growth factor I receptor in brain. *Regul. Pept.*, **15**, 301–10.

Heidenreich, K.A. and Toledo, S.P. (1989) Insulin receptors mediate growth effects in cultured fetal neurons. II. Activation of a protein kinase that phosphorylates ribosomal protein SP6. *Endocrinology*, **125**, 1458–63.

Hendricks, S.A., Agardh, C.D., Taylor, S.I. *et al.* (1984) Unique features of the insulin receptor in rat brain. *J. Neurochem.*, **43**, 1302–9.

Hepler, J.E. and Lund, P.K. (1990) Molecular biology of the insulin-like growth factors. Relevance to nervous system function, in *Molecular Neurobiology* (ed N. Bazan), Humana Press, Clifton, NJ, pp. 93–127.

Hill, J.M., Lesniak, M.A., Pert, C.B. and Roth, J. (1986) Autoradiographic localization of insulin receptors in rat brain: prominence in olfactory and limbic areas. *Neuroscience*, **17**, 1127–38.

Kappy, M.S., Sellinger, S. and Raizada, M.K. (1984) Insulin binding in four regions of the developing rat brain. *J. Neurochem.*, **42**, 198–203

Karnieli, E. and Garvey, W.T. (1987) Glucose transporters: overview and implications for the brain, in *Insulin, Insulin-like Growth Factors and their Receptors in the Central Nervous System* (eds M.K. Raizada, M.I. Phillips and D. LeRoith), Plenum Press, New York, pp. 71–92.

Kiess, W., Lee, L., Graham, D.E. *et al.* (1989) Rat C6 glial cells synthesize insulin-like growth factor I (IGF-I) and express IGF-I receptors and IGF-II/mannose 6-phosphate receptors. *Endocrinology*, **124**, 1727–36.

Lenoir, D. and Honegger, P. (1983) Insulin-like growth factor I (IGF-I) stimulates DNA synthesis in fetal rat brain cell cultures. *Dev. Brain Res.*, **7**, 205–13.

LeRoith, D., Lowe, W.L., Shemer, J. *et al.* (1988) Development of brain insulin receptors. *Int. J. Biochem.*, **20**, 225–30.

LeRoith, D., Roberts, C.T., Werner, H. *et al.* (1992) Insulin-like growth factors in the brain, in *Neurotrophic Factors* (eds J.H. Fallon and S.E. Loughlin), Academic Press, Orlando, FL, (in press).

Lesniak, M.A., Hill, J.M., Kiess, W. *et al.* (1988) Receptors for insulin-like growth factors I and II: autoradiographic localization in rat brain and comparison to receptors for insulin. *Endocrinology*, **123**, 2089–99.

Lowe, W.L., Boyd, F.T., Clarke, D.W. *et al.* (1986) Development of brain insulin receptors: structural and functional studies of insulin receptors from whole brain and primary cell cultures. *Endocrinology*, **119**, 25–35.

Masters, B.A., Shemer, J., Judkins, J.H. *et al.* (1987) Insulin receptors and insulin action in dissociated brain cells. *Brain Res.*, **417**, 247–56.

Masters, B.A., Werner, H., Roberts, C.T. *et al.* (1991a) Developmental regulation of insulin-like growth factor-I-stimulated glucose transporter in rat brain astrocytes. *Endocrinology*, **128**, 2548–57.

Masters, B.A., Werner, H., Roberts, C.T., *et al.* (1991b) Insulin-like growth factor I (IGF-I) receptors and IGF-I action in oligodendrocytes from rat brains. *Regul. Pept.*, **33**, 117–31.

Matsuo, K., Niwa, M., Kurihara, M. *et al.* (1989) Receptor autoradiographic analysis of insulin-like growth factor-I (IGF-I) binding sites in rat forebrain and pituitary gland. *Cell. Mol. Neurobiol.*, **9**, 357–67.

McCusker, R.H., Camacho-Hubner, C., Bayne, M.L. *et al.* (1990) Insulin-like growth factor (IGF) binding to human fibroblast and glioblastoma cells: the modulating effect of cell released IGF binding proteins (IGFBPs). *J. Cell. Physiol.*, **144**, 233–53.

McElduff, A., Poronnik, P. and Baxter, R.C. (1987) The insulin-like growth factor II (IGF-II) receptor from rat brain is of lower apparent molecular weight than the IGF -II receptor from rat liver. *Endocrinology*, **121**, 1306–11.

McElduff, A., Poronnik, P., Baxter, R.C. *et al.* (1988) A comparison of the insulin and insulin-like growth factor I receptors from rat brain and liver. *Endocrinology*, **122**, 1933–9.

McMorris, F.A., Smith, T.M., DeSalvo, S. and Furlanetto, R. W. (1986) Insulin-like growth factor I/somatomedin C: a potent inducer of oligodendrocyte development. *Proc. Natl. Acad. Sci. USA*, **83**, 822–6.

Merrill, M.J. and Edwards, N.A. (1990) Insulin-like growth factor-I receptors in human glial tumors. *J. Clin. Endocrinol. Metab.*, **71**, 199–209.

Morgan, D.O., Edman, J.C., Standring, D.N. *et al.* (1987) Insulin–like growth factor II receptor as a multifunctional binding protein. *Nature*, **329**, 301–7.

Moss, A.M., Unger, J.W., Moxley, R.T. and Livingston, J.N. (1990) Location of phosphotyrosine-containing proteins by immunocytochemistry in the rat forebrain corresponds to the distribution of the insulin receptor. *Proc. Natl. Acad. Sci. USA*, **87**, 4453–7.

Moxham, C.P., Duronio, V. and Jacobs, S. (1989) Insulin-like growth factor I receptor β subunit heterogeneity. *J. Biol. Chem.*, **264**, 13238–44.

Nielsen, F.C., Wang, E. and Gammeltoft, S. (1991) Receptor binding, endocytosis, and mitogenesis of insulin-like growth factors I and II in fetal rat brain neurons. *J. Neurochem.*, **56**, 12–21.

Nissley, S.P., Kiess, W. and Sklar, M.M. (1991) The insulin-like growth factor II/Mannose 6-phosphate receptor, in *Insulin-like Growth Factors: Molecular and Cellular Aspects* (ed D. LeRoith), CRC Press, Boca Raton, FL, pp. 111–50.

Ocrant, I., Valentino, K.L., Eng, L.F. *et al.* (1988) Structural and immunohistochemical characterization of insulin-like growth factor-I and II receptors in the murine central nervous system. *Endocrinology*, **123**, 1023–34.

Okamoto, T., Katada, T., Murayama, Y. *et al.* (1990) A simple structure encodes G protein-activating function of the IGF-II/mannose 6-phosphate receptor. *Cell*, **62**, 709–17.

Ota, A., Wilson, G.L. and LeRoith, D. (1988a) Insulin-like growth factor I receptors on mouse neuroblastoma cells. Two β subunits are derived from differences in glycosylation. *Eur. J. Biochem.*, **174**, 521–30.

Ota, A., Wilson, G.L., Spilberg, O. *et al.* (1988b) Functional insulin-like growth factor I receptors are expressed by neural-derived continuous cell lines. *Endocrinology*, **122**, 145–52.

Palovcik, R.A., Phillips, M.I., Kappy, M.S. and Raizada, M.K. (1984) Insulin inhibits pyramidal neurons in hippocampal slices. *Brain Res.*, **309**, 187–91.

Petruzzelli, L., Herrera, R., Arenas-Garcia, R. *et al.* (1986) Isolation of a *Drosophila* genomic sequence homologous to the kinase domain of the human insulin receptor and detection of the phosphorylated *Drosophila* receptor with an antipeptide antibody. *Proc. Natl. Acad. Sci. USA*, **83**, 4710–14.

Pomerance, M., Gavaret, J-M., Jacquemin, J.C. *et al.* (1988) Insulin and insulin-like growth factor I receptors during postnatal development of rat brain. *Dev. Brain Res.*, **42**, 77–83.

Potau, N., Escofet, M.A. and Martinez, M.D. (1991) Ontogenesis of insulin receptors in human cerebral cortex. *J. Endocrinol. Invest.*, **14**, 53–8.

Raizada, M.K., Boyd, F.T., Clarke, D.W. and LeRoith, D. (1987) Physiologically unique insulin receptors on neuronal cells, in *Insulin, Insulin-like Growth Factors and Their Receptors in the Central Nervous System* (ed M.K. Raizada, M.I. Phillips, and D. LeRoith), Plenum Press, New York, pp. 163–75.

Raizada, M.K., Shemer, J., Judkins, J.H. *et al.* (1988) Insulin receptors in the brain: structural and physiological characterization. *Neurochem. Res.*, **13**, 297–303.

Rechler, M.M. and Nissley, S.P. (1985) The nature and regulation of the receptors for insulin-like growth factors. *Annu. Rev. Physiol.*, **47**, 425–42.

Rechler, M.M., Zapf, J., Nissley, S.P. *et al.* (1980) Interactions of insulin-like growth factors I and II and multiplication-stimulating activity with receptors and serum carrier proteins. *Endocrinology*, **107**, 1451–9.

Recio-Pinto, E. and Ishii, D.N. (1988) Insulin and insulin-like growth factor receptors regulating neurite formation in cultured human neuroblastoma cells. *J. Neurosci. Res.*, **19**, 312–20.

Roback, E.W., Barakat, A.J., Dev, V.G. *et al.* (1991) An infant with deletion of the distal long arm of chromosone 15 (q26.1 → qter) and loss of insulin-like growth factor I receptor gene. *Am. J. Med. Genet.*, **38**, 74–9.

Rosenfeld, R.G., Ceda, G., Wilson, D.M. *et al.* (1984) Characterization of high affinity receptors for insulin-like growth factors I and II on rat anterior pituitary cells. *Endocrinology*, **114**, 1571–5.

Rosenfeld, R.G., Pham, H., Keller, B.T. *et al.* (1987) Demonstration and structural comparison of receptors for insulin-like growth factor-I and -II (IGF-I and -II) in brain and blood–brain barrier. *Biochem. Biophys. Res. Commun.*, **149**, 159–66.

Roth, R.A., Morgan, D.O., Beaudoin, J. and Sara, V. (1986) Purification and characterization of the human brain insulin receptor. *J. Biol. Chem.*, **261**, 3753–7.

Sakaguchi, T. and Bray, G.A. (1987) Intrahypothalamic injection of insulin decreases firing rate of sympathetic nerves. *Proc. Natl. Acad. Sci. USA*, **84**, 2012–14.

Sara, V.R., Hall, K., Misaki, M. *et al.* (1983) Ontogenesis of somatomedin and insulin receptors in the human fetus. *J. Clin. Invest.*, **71**, 1084–94.

Sara, V.R., Carlsson-Skwirut, C., Andersson, C. *et al.* (1986) Characterization of somatomedins from human fetal brain: identification of a variant form of insulin-like growth factor I. *Proc. Natl. Acad. Sci. USA*, **83**, 4904–7.

Scavo, L., Shuldiner, A.R., Serrano, J. *et al.* (1991) Genes encoding receptors for insulin and insulin-like growth factor I are expressed in *Xenopus* oocytes and embryos. *Proc. Natl. Acad. Sci. USA*, **88**, 6214–18.

Seino, S., Seino, M., Nishi, S. and Bell, G.I. (1989) Structure of the human insulin receptor gene and characterization of its promoter. *Proc. Natl. Acad. Sci. USA*, **86**, 114–18.

Shemer, J., Penhos, J.C. and LeRoith, D. (1986) Insulin receptors in lizard brain and liver: structural and functional studies of α and β subunits demonstrate evolutionary conservation. *Diabetologia*, **29**, 321–9.

Shemer, J., Perrotti, N., Roth, J. and LeRoith, D. (1987a) Characterization of an endogenous substrate related to insulin and insulin-like growth factor-I receptors in lizard brain. *J. Biol. Chem.*, **262**, 3436–9.

Shemer, J., Raizada, M. and LeRoith, D. (1987b) Structural and functional studies on insulin receptors from alligator brain and liver. *Comp. Biochem. Physiol.*, **86B**, 55–61.

Shemer, J., Raizada, M.K., Masters, B.A. *et al.* (1987c) Insulin-like growth factor I receptors in neuronal and glial cells. *J. Biol. Chem.*, **262**, 7693–9.

Sklar, M.M., Kiess, W., Thomas, C.L. and Nissley, S.P. (1989) Developmental expression of the tissue-insulin-like growth factor II/mannose 6-phosphate receptor in the rat. Measurement by quantitative immunoblotting. *J. Biol. Chem.*, **264**, 16733–8.

Smith, M., Clemens, J., Kerchner, G.A. and Mendelsohn, L.G. (1988) The insulin-like growth factor-II (IGF-II) receptor of rat brain: regional distribution visualized by autoradiography. *Brain Res.*, **445**, 241–6.

Soos, M.A. and Siddle, K. (1989) Immunological relationships between receptors for insulin and insulin-like growth factor I. *Biochem. J.*, **263**, 553–63.

Steele–Perkins, G., Turner, J., Edman, J.C. *et al.* (1988) Expression and characterization of a functional human insulin-like growth factor I receptor. *J. Biol. Chem.*, **263**, 11486–92.

Sturm, M.A., Conover, C.A., Pham, H. and Rosenfeld, R.G. (1989) Insulin-like growth factor receptors and binding protein in rat neuroblastoma cells. *Endocrinology*, **124**, 388–96.

Svrzic, D. and Schubert, D. (1990) Insulin-like growth factor 1 supports embryonic nerve cell survival. *Biochem. Biophys. Res. Commun.*, **172**, 54–60.

Taylor, S.I., Najjar, S., Cama, A. and Accili, D. (1991) Structure and function of the insulin receptors, in *Insulin-like Growth Factors: Molecular and Cellular Aspects* (ed D. LeRoith), CRC Press, Boca Raton, FL, pp. 221–44.

Ullrich, A., Bell, J.R., Chen, E.Y. *et al.* (1985) Human insulin receptor and its relationship to the tyrosine kinase family of oncogenes. *Nature*, **313**, 756–61.

Ullrich, A., Gray, A., Tam, A.W. *et al.* (1986) Insulin-like growth factor I receptor primary structure: comparison with insulin receptor suggests structural determinants that define functional specificity. *EMBO J.*, **5**, 2503–12.

Unger, J.W., Livingston, J.N. and Moss, A.M. (1991) Insulin receptors in the central nervous system: localization, signalling mechanisms and functional aspects. *Progr. Neurobiol.*, **36**, 343–62.

Werner, H., Raizada, M.K., Mudd, L. *et al.* (1989a) Regulation of rat brain/Hep G2 glucose transporter gene expression by insulin and insulin-like growth factor-I in primary cultures of neuronal and glial cells. *Endocrinology*, **125**, 314–20.

Werner, H., Woloschak, M., Adamo, M. *et al.* (1989b) Developmental regulation of the rat insulin-like growth factor I receptor gene. *Proc. Natl. Acad. Sci. USA*, **86**, 7451–5.

Werner, H., Woloschak, M., Stannard, B. *et al.* (1991) The insulin-like growth factor I receptor: molecular biology, heterogeneity and regulation, in *Insulin-like Growth Factors: Molecular and Cellular Aspects* (ed D. LeRoith), CRC Press, Boca Raton, FL, pp. 17–47.

Werther, G.A., Abate, M., Hogg, A. *et al.* (1990) Localization of insulin-like growth factor-I mRNA in rat brain by *in situ* hybridization – relationship to IGF-I receptors. *Mol. Endocrinol.*, **4**, 773–8.

Wolinsky, E.J., Patterson, D.H. and Willard, A.L. (1985) Insulin promotes electrical coupling between cultured sympathetic neurons. *J. Neurosci.*, **5**, 1675–9.

Woods, S.C., Lotter, E.C., McKay, L.D. and Porte, D. (1979) Chronic intracerebroventricular infusion of insulin reduces food intake and body weight of baboons. *Nature*, **282**, 503–5.

Yamasaki, H., Prager, D., Gebremedhin, S. and Melmed, S. (1991) Insulin-like growth factor I (IGF-I) attenuation of growth hormone is enhanced by overexpression of pituitary IGF-I receptors. *Mol. Endocrinol.*, **5,** 890–6.

Yarden, Y. and Ullrich, A. (1988) Growth factor receptor tyrosine kinases. *Annu. Rev. Biochem.*, **57**, 443–78.

Yorek, M., Leeney, E., Dunlap, J. and Ginsberg, B. (1989) Effect of fatty acid composition on insulin and IGF-I binding in retinoblastoma cells. *Invest. Ophthalmol. Vis. Sci.*, **30**, 2087–92.

Young, W.S., Kuhar, M.J., Roth, J. *et al.* (1980) Radiohistochemical localization of insulin receptors in the adult and developing rat brain. *Neuropeptides*, **1**, 15–22.

Daniel H. Polk

8.1 INTRODUCTION

The dependence of brain maturation on thyroid hormone has been well documented. However, the mechanism(s) of the effects of thyroid hormone on developing neural structures remain unclear. The recent cloning of thyroid hormone receptors provides an important new tool for such investigations. The focus of this review is to outline the biochemistry of nuclear thyroid hormone receptors, classification of receptor subtypes, the ontogeny and physiology of the relevant ligand triiodothyronine (T$_3$) and the impact of receptor–ligand coupling on post receptor phenomena. The unique relationship between occupancy of nuclear thyroid receptors and subsequent effects on brain maturation is responsible for the developmental focus of this review.

8.2 THYROID HORMONE RECEPTORS

The first experiments characterizing the subcellular localization of high-affinity, low-capacity binding sites for T$_3$ in cell nuclei were published by Oppenheimer *et al.* (1972) who used *in vivo* tracer quantities of radiolabeled T$_3$ and competing doses of unlabeled hormone which were injected into rats. These studies clearly demonstrated that liver and kidney expressed specific high-affinity, low-capacity T$_3$ binding sites, and suggested a similar phenomenon for the pituitary gland (Schadlow *et al.*, 1974). Subsequently, methods for the quantification of T$_3$ receptors both

in intact nuclei and extracted forms were developed (Samuels *et al.*, 1974; Surks *et al.*, 1975; Lindenburg *et al.*, 1978). Application of these techniques led to studies characterizing the binding affinity and maximal binding capacity of nuclear thyroid hormone receptors in a wide variety of tissues obtained from many species (for a review, see Oppenheimer and Schwartz, 1986). Taking the differences in tissue source (*in vitro* vs. *in vivo*) and technique (in whole nuclei vs. extract) into account, the consistency of relative binding affinities of various thyroid hormone analogs and of binding characteristics across many studies suggested a single class of thyroid hormone nuclear receptors.

Characterization of the thyroid nuclear hormone receptors has been hampered by the difficulties in obtaining pure receptor protein. Initial studies demonstrated that the nuclear thyroid hormone receptor protein had a molecular weight of approximately 50 000 and a sedimentation coefficient of 3.5 s (Apriletti *et al.*, 1981; Latham *et al.*, 1981); progressive enzymatic degradation suggested a non-histone class of chromatin-binding proteins (Surks *et al.*, 1973). Subtle differences in receptor characteristics were defined. By using the technique of photoaffinity labeling, the presence of two different-sized thyroid hormone receptor proteins was shown by Pascual *et al.* (1982), a finding confirmed by Oppenheimer *et al.* (1987). Work in our laboratory demonstrated a five-fold greater affinity for T$_3$ by receptors isolated from brain

Receptors in the Developing Nervous System Vol. 1: *Growth factors and hormones* Edited by Ian S. Zagon and Patricia J. McLaughlin. Published in 1993 by Chapman & Hall. ISBN 0 412 45240 5. Vols. 1 and 2 (set) ISBN 0 412 54520 9.

than for hepatic receptors in the developing ovine fetus, a pattern consistent with different receptor subtypes (Polk *et al.*, 1989). However, efforts to isolate and sequence the nuclear thyroid hormone receptor protein provided only partial purification and little sequence information.

8.3   THYROID HORMONE RECEPTORS AND THE *erbA* ONCOGENE

A major advance in our understanding of the structure of the nuclear thyroid hormone receptor came with the observation that the receptor was related to the erbA family of oncogenes. A malignant form of avian erythroleukemia is induced by an oncogenic retrovirus termed avian erythroblastosis virus (AEV). AEV infection of bone marrow in chickens results in the proliferation of immature erythroblasts which do not differentiate and do not respond to the growth factor erythropoietin (Graf and Beug, 1983). The oncogenic properties of the AEV arise from two loci within the viral genome, termed v-erbA and v-erbB. Studies involving the regulation of cellular differentiation have demonstrated that the protein product of the *v-erbB* gene is a 74 kDa transmembrane glycoprotein which resembles a truncated epidermal growth factor receptor. In contrast to its cellular homolog, the v-erbB protein constitutively expresses its intracellular tyrosine kinase activity, leading to unregulated phosphorylation of (unknown) intracellular substrates, presumably facilitating continued replication of the transformed cell line (Downward *et al.*, 1984). Expression of v-erbA results in the *C*-terminal portion of a 75 kDa fusion protein. Comparison of this protein sequence with other known regulatory protein structures led to an unanticipated finding: the v-erbA protein shared striking amino acid sequence homology with the steroid hormone nuclear receptors (Weinberger *et al.*, 1985). This family of nuclear proteins includes receptors for a

variety of ligands including the glucocorticoids, mineralocorticoids, estrogens, progesterone, vitamin D and retinoic acid (Evans, 1988). The receptors of this family share an overall basic structure of five domains, to termed A to E. In all the receptors and related receptor-like proteins described to date, the C region is the putative site of receptor-DNA binding. This is accomplished by means of the so-called 'zinc fingers'. These cystine-rich sequences appear to complex with zinc ions, forming loops of intervening amino acid residues which resemble dual finger-like projections. These fingers are thought to interdigitate within the major groove of the DNA helix and bind to specific nucleotide sequences (Evans *et al.*, 1988). This C region is highly conserved among members of the steroid–thyroid receptor family and represents the region of highest sequence homology with the v-erbA-related protein.

Two groups of workers (Sap *et al.*, 1986; Weinberger *et al.*, 1986) reported that c-erbA, the cellular homologue of v-erbA, encoded not one but a variety of different proteins which bound T₃ with high affinity. Sap *et al.* (1986) screened a chicken embryonic cDNA library with probes derived from the v-erbA sequence and detected 10 positive clones, two of which were subcloned and sequenced. Sequence analysis revealed an open reading frame of 1224 nucleotides, which would predict a peptide of 46 kDa, a size similar to that reported for the thyroid hormone receptor. Subsequent transcription and translation of this sequence yielded two protein products of 46 and 40 kDa. The 46 kDa protein was found to bind T₃ in a saturable and specific manner with a dissociation constant (K_d) between 0.2 and 0.3 nM, a value in agreement with that of the native T₃ nuclear receptor. Weinberger *et al.* (1986) used a similar approach but screened a human placental cDNA library instead of an avian-derived library. Again, two positive clones were identified, subcloned and sequenced. An open reading frame of 1368 nucleotides predicting a 52 kDa protein was identified. Expression of this sequence resulted in the isolation of three proteins with an M_r of 55, 52 and 35 kDa. The 55 and 52 dalton proteins were shown to bind T_3 with K_d of 5×10^{-11} M, a value somewhat higher than that observed by Sap, and suggesting that other nuclear factors might influence the binding of T_3 to its nuclear receptor. Subsequent to these reports, careful sequence analysis and restrictive enzyme mapping suggested the probability that multiple copies of the c-erbA related gene were encoded within the human genome.

8.4   MULTIPLE FORMS OF THE THYROID HORMONE RECEPTOR

After these initial observations, there were numerous reports concerning the protein products of the cellular homologs of the v-erbA sequence, and it is now clear that within a given species multiple forms of both thyroid receptor mRNAs and proteins are expressed. These multiple forms can be divided into two subtypes, α and β, based on site of expression, sequence homology and chromosomal location.

8.4.1 α THYROID HORMONE RECEPTORS

Thompson *et al.* (1987) first suggested the alpha nomenclature for the thyroid hormone receptor, which they localized to human chromosome 17. The alpha gene has been cloned in a variety of mammalian species including human (Nakai *et al.*, 1988; Miyajima *et al.*, 1989), rat (Thompson *et al.*, 1987; Murray *et al.*, 1988; Lazar *et al.*, 1988; Mitsuhashi *et al.*, 1988), chicken (Weinberger *et al.*, 1986), and mouse (Prost *et al.*, 1988). Given the importance of thyroid hormones during amphibian morphogenesis, it is not surprising that this gene also is expressed in lower vertebrates (Banker *et al.*, 1991). The

α thyroid receptor gene is complex, and transcription leads to a variety of mRNAs with subsequent translation to several different protein species. The genomic organization of the human thyroid hormone α receptor has been reported by Laudet *et al.* (1991). The gene consists of 10 exons encompassing 27 kb of DNA on chromosome 17. Alternate mRNA splicing produces two α receptor protein species. One receptor, consisting of 410 amino acids, is translated from a 5 kb mRNA which initiates at exon 1 and terminates at the 3' end of exon 9. This receptor protein, termed hTHR α(A)-1, binds T_3 and activates or inhibits transcription of thyroid hormone-responsive target genes. A second related receptor, 490 amino acids long, is derived from a 2.6 kb mRNA produced via alternate splicing between exons 9 and 10, replacing the C-terminal 40 amino acids in hTHR α-1 with 120 amino acids specific for hTHR α-2. This receptor, also known as hTHR variant I, binds DNA but does not bind T_3. It may thus negatively regulate the action of other thyroid receptors at the level of transcription (Koenig *et al.*, 1988). A third related thyroid hormone receptor variant termed hRev-ErbA α is translated from a 3 kb mRNA transcribed from the opposite (reverse) strand of DNA in the hTHR-α gene. Analysis of hRev-ErbA α cDNA predicts a 614 amino acid protein with sequence complimention to that of the other hTHR-α gene products (Miyajima *et al.*, 1989; Lazar *et al.*, 1990). This protein, like hTHRα-variant I, binds to DNA but does not bind thyroid hormone, suggesting that it, too, might participate in thyroid hormone nuclear receptor regulation. The amino acid homologies for these related thyroid hormone receptors in the rat and human species are shown in Fig. 8.1. In the rat, in addition to THR-α1 and the variants rTHR α–2 and rRev-ErbA α, a third variant has been described (Mitsuhashi *et al.*, 1988). This alpha variant, termed rTHR-α vII, is identical to rTR-α vI, lacking only the initial 40 amino acids of the ligand binding domain.

As expected, this variant also does not bind thyroid hormone.

8.4.2 β THYROID HORMONE RECEPTORS

The human placental thyroid hormone receptor originally characterized by Weinberger *et al.* (1986) and localized to human chromosome 3 is the prototypic β thyroid hormone receptor. This protein is translated from a 6.0 kb mRNA, resulting in a 456 amino acid peptide. The original reports of Weinberger *et al.* attributing translation of the β gene product to a 2.6 kb mRNA were probably the result of cross hybridization to the 2.6 hTHR α (variant I) mRNA due to the highly homologous sequence of the probe used (Santos *et al.*, 1988; Lazar and Chin, 1990). Another variant of the β receptor, termed THR β-2, has been isolated in the rat (Hodin *et al.*, 1989) and mouse (Lazar and Chin, 1990). This protein of 514 amino acids differs from rat THR β-1 in the initial N-terminal 47 amino acids but is identical to the structure of rTHR β-1 thereafter, thus

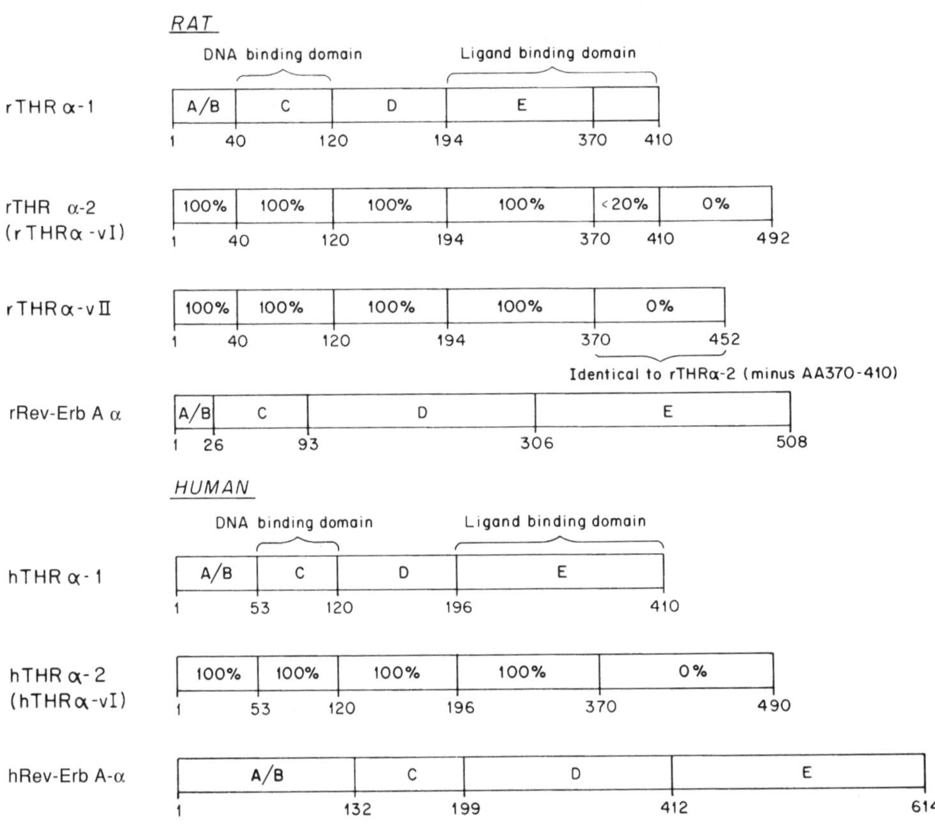

Fig. 8.1. Schematic representation of the amino acid sequences for the alpha subtype of nuclear thyroid receptors in the rat (upper panel) and human (lower panel). Numbers below boxes represent amino acid number deduced from cDNA sequence data. Letters within boxes refer to regions analogous to those described for other nuclear receptors (Evans, 1988). Numbers within boxes refer to sequence homology with the parent thyroid nuclear receptor protein. See text for details.

conserving both the DNA and ligand binding sequences. The structures of the β gene products for the rat and human species are schematically represented in Fig. 8.2. In general, the DNA binding domains are closely conserved among members of the α and β families, whereas there is less homology in this domain between the two subgroups. This suggests that these two classes of nuclear thyroid receptors might interact in different ways with thyroid responsive genes.

8.5   ONTOGENY OF THYROID HORMONES

The active ligand for the nuclear thyroid hormone receptor is triiodothyronine (T_3). Tetraiodothyronine or thyroxine (T_4) binds to the receptor 100–1000 times less avidly than T_3 (Oppenheimer, 1983). The biological activity of the receptor is dependent on its occupancy by ligand. The ontogeny of thyroid hormones during development has been well characterized. Circulating thyroid hormones are derived from thyroid secretion of T_4 and T_3 as well as peripheral tissue conversion of T_4 into T_3. Although thyroidal secretion is the only source of T_4, both thyroidal and peripheral sources contribute to circulating T_3 levels. In general, tissue levels of T_3 represent uptake of serum T_3 as well as enzymatically catalyzed, *in situ* monodeiodination of T_4 to T_3. Thus, both the availability of circulating thyroid hormones and the capacity of a particular tissue to deiodinate T_4 to T_3 modulate the levels of intracellular T_3 available for receptor occupancy.

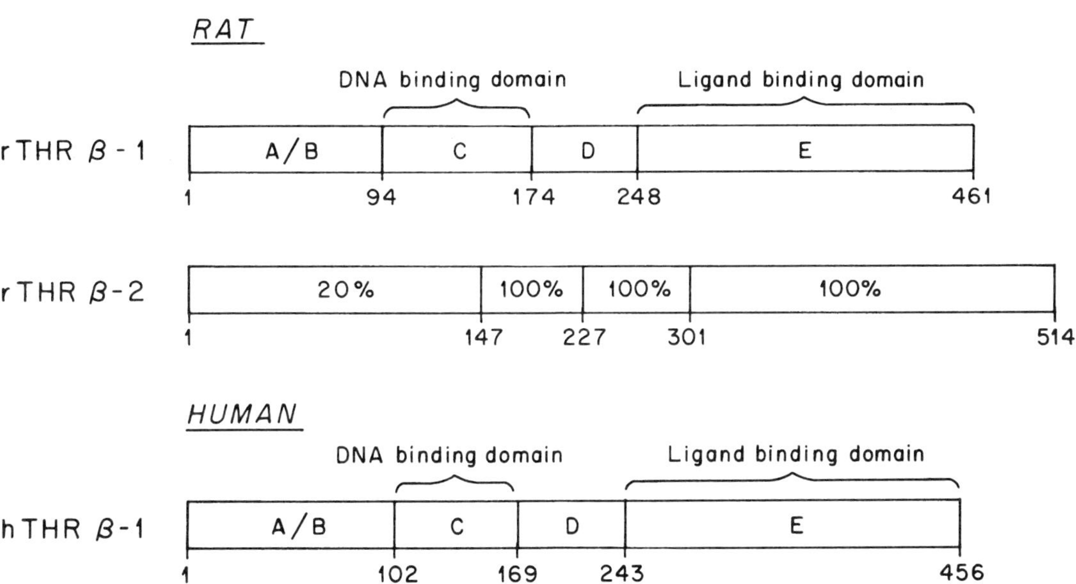

Fig. 8.2.   Schematic representation of the amino acid sequences for the beta subtype of nuclear thyroid receptors in the rat (upper panel) and human (lower panel). Numbers below boxes represent amino acid numbers deduced from cDNA sequence data. Letters within boxes refer to regions analogous to those described for other nuclear receptors (Evans, 1988). Numbers within boxes refer to sequence homology with the parent nuclear receptor protein. See text for details.

Monodeiodination of T_4 may occur at either the outer (β) or inner (α) ring of the molecule. This process is outlined in Fig. 8.3. Outer ring deiodination of T_4 results in T_3, the active form of thyroid hormone. Inner ring monodeiodination of T_4 produces 3,3'5'-triiodothyronine (rT_3) an inactive metabolite. Two types of outer-ring iodothyronine monodeiodinase (5'-MDI) have been described. Type I 5'-MDI, predominantly expressed in liver and kidney, is a high K_m enzyme inhibited by propylthiouracil. Expression of enzyme activity is augmented by thyroid hormone. A type II 5'-MDI, located principally in brain, pituitary and brown adipose tissue, is a low K_m enzyme, insensitive to propylthiouracil and its activity inhibited by thyroid hormone (Refetoff and Larsen, 1989). In adults, type I 5'-MDI activity in liver (and, to some degree, kidney and muscle) accounts for most of the peripheral deiodination of T_4 and maintains, to a large extent, the levels of T_3 in the circulation. The type II 5'-MDI acts primarily to increase local intracellular levels of T_3 in brain and pituitary tissues.

Although little is known about the ontogeny of these enzyme systems in the developing human, they have been extensively studied in fetal sheep. Type II 5'-MDI in brain is present by midgestation whereas hepatic type I activity is low at midgestation, increasing only near term (Wu *et al.*, 1990a). Both types of enzyme

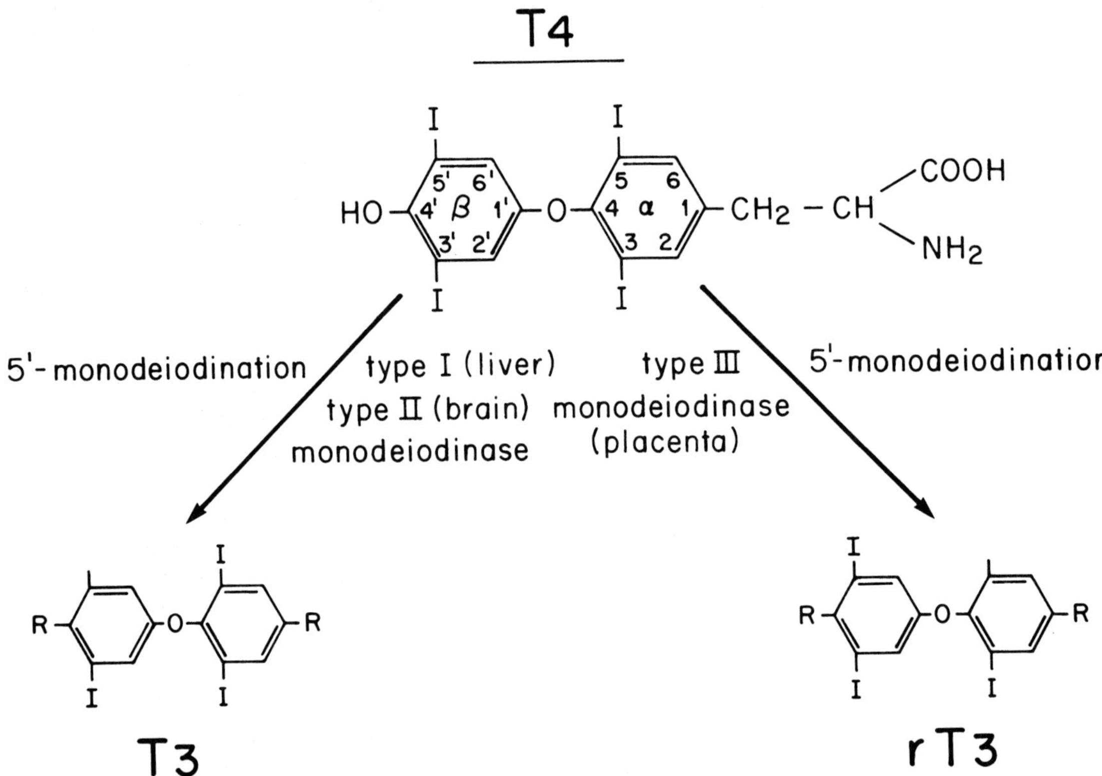

Fig. 8.3.   Pathways for T_4 monodeiodination. Abbreviations: T_4, thyroxine; T_3, triiodothyronine; rT_3, reverse triiodothyronine. Type I 5' monodeiodinase is expressed principally in hepatic tissues whereas type II 5' monodeiodinase occurs almost exclusively in brain and pituitary tissues. Type III 5'-monodeiodinase occurs in placental and other fetal tissues and is the principal pathway for T_4 deiodination in the developing fetus.

activity are responsive to fetal thyroid status. Brain type II 5'-MDI activity increases in response to hypothyroidism throughout the final third of gestation (Polk *et al.*, 1988). In contrast, hepatic type I 5'-MDI activity becomes thyroid hormone responsive (i.e., activity decreases with hypothyroidism) only during the final weeks of gestation. It is likely that the type II enzyme, particularly in the hypothyroid fetus, plays an important role in providing a source of intracellular T_3 to those tissues (brown fat, brain) dependent on T_3 during fetal life (Calvo *et al.*, 1990).

A third type of iodothyronine deiiodinase is particularly active and important during fetal development. Type III 5'-MDI has been characterized in several fetal tissues, including the placenta (Wu *et al.*, 1990a). This enzyme system catalyzes the conversion of T_4 to the inactive metabolite rT_3 (Fig. 8.3). Fetal thyroid hormone metabolism is characterized by a predominance of type III 5'-MDI activity, particularly in liver, kidney and placenta. This activity probably accounts for the increased levels of circulating rT_3 in the fetus. Placental type III 5'-MDI also contributes to amniotic fluid rT_3 levels and is an important pathway for inactivation of maternally derived thyroid hormones.

The patterns of tissue expression of the type I, II and III deiodinases and the maturation of thyroidal secretion of T_4 and T_3 are responsible for the ontogeny of circulating thyroid hormones in the fetus. The patterns of perinatal thyroid hormone secretion in the human and the rat are depicted in Fig. 8.4. Although the patterns are comparable there are important differences which must be taken into account when comparing the effects of thyroid hormone on brain maturation in the two species. Humans (as well as sheep, another important experimental model of fetal thyroid hormone effects) are a precocial species, born relatively mature, whereas the rodent is an altricial species, born quite immature and characterized by a great deal of postnatal organ maturation. Thus, the major-ity of thyroid system maturation in precocial species occurs prenatally, with birth as a late event, whereas in altricial species thyroid hormone system maturation occurs largely after birth.

Although the human fetal thyroid gland is able to concentrate iodide and synthesize thyroid hormones as early as 10 weeks gestation, circulating T_4 levels remain low until 18–20 weeks gestation. After this time, both total and free T_4 concentrations increase steadily until the final weeks of gestation (Fisher, 1985). This pattern of maturation of serum T_4 levels differs from that of circulating T_3 concentrations in the human fetus. Fetal T_3 serum concentrations remain low until 30 weeks gestation, increasing only modestly to levels of 50 µg/dl (0.77 nM) at term. This prenatal increase in T_3 is largely due to progressive maturation of hepatic type I 5'-MDI activity. Birth at term is associated with marked further changes in thyroid hormone production and metabolism. There is an acute increase in pituitary TSH secretion which peaks at 30 min and stimulates thyroidal iodine uptake and release of T_4 and T_3. Serum T_4 and T_3 levels peak at 36–48 h, decreasing thereafter to adult values by 4–5 weeks of life (Fisher *et al.*, 1977). Important changes in thyroid hormone metabolism occur simultaneously in the early postnatal period. These changes have been best characterized in the sheep, in which there is a progressive increase in type I 5'-MDI activity in hepatic, and to a lesser extent, renal and brown adipose tissues during the prenatal period (Polk *et al.*, 1986; Wu *et al.*, 1990b). In the early postnatal period, there is a further increase in hepatic type I 5'-MDI activity and a rapid increase in type II 5'-MDI activity in brown adipose tissue. Thus, whereas the neonatal TSH surge is brief, the further maturation of the monodeiodinase systems for T_3 production in several tissues accounts for the maintenance of high serum T_3 levels (in comparison to fetal levels) after birth.

In the rat, similar events occur except that parturition is a relatively early event during

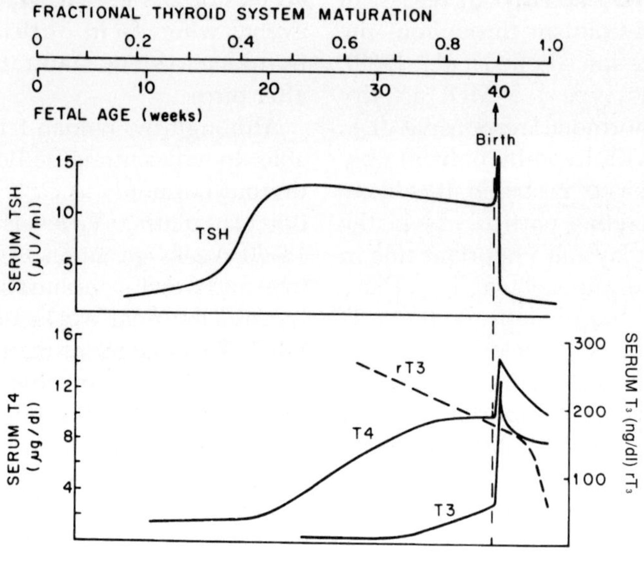

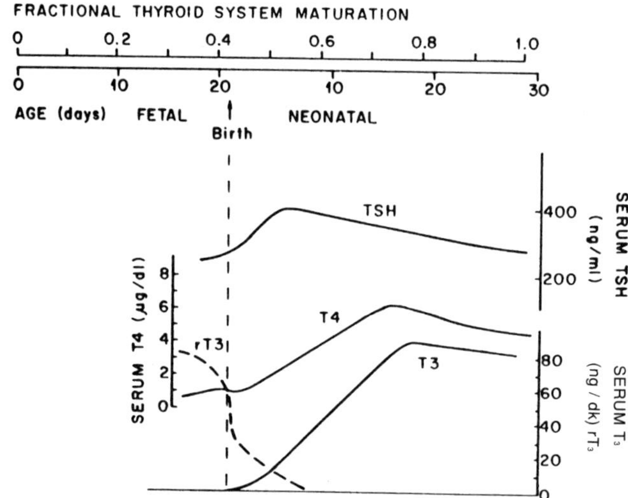

Fig. 8.4. Patterns of serum TSH, T₄, T₃ and rT₃ during development in the human (upper panel) and rat (lower panel). The fraction of total thyroid system maturation is indicated by the upper time line in both panels. Fetal and neonatal age is indicated by the lower time line in each panel. The time of birth is indicated by the dashed vertical line.

thyroid system maturation and there is no parturitionally associated surge in circulating thyroid hormones. Instead, T₄ rises slowly during the first postnatal week, followed by increases in circulating T₃ during the period of time that hepatic (as well as kidney and muscle) 5'-MDI activities increase (Fisher *et al.*, 1977).

8.6 NUCLEAR THYROID HORMONES: BRAIN DISTRIBUTION AND ONTOGENY

The demonstration of nuclear thyroid hormone receptors in brain has relied on three different techniques: immunocytochemistry, binding studies, and the most recent studies using techniques of molecular biology. In most instances the studies are complementary; however, the contribution of receptor subtypes to overall nuclear binding characteristics remains uncertain.

8.6.1 IMMUNOCYTOCHEMISTRY

Ruel *et al.* (1985), using monoclonal antibodies raised against hepatic T_3 nuclear receptors, described the distribution of thyroid nuclear receptors in adult rat brain. The distribution of receptors varied with cell type and brain region, with highest activities expressed in cortical, hippocampal and amygdaloid tissues and lowest activities in brainstem and cerebellar tissues. The principal expression of nuclear T_3 receptors occurred in neuronal cells; the maximal binding capacity in neuronal tissues *in vitro* was about three-fold that exhibited by astrocytes (Dussault and Ruel, 1987; Luo *et al.*, 1989). The ontogeny of nuclear T_3 receptors has been studied by immunohistochemical staining in the developing rodent. Weakly staining neurons confined to noncortical nuclei are demonstrable in the 16 day rat fetus. Staining intensity increases by 18 days gestation with cortical localization of receptor after this time. Immunostaining in rat brain peaks by the end of the first postnatal week, and is particularly prominent in the cingulate and superior frontal gyri of the cerebral cortex as well as in hypothalamic nuclei. Activity in these areas decreases by the end of the second postnatal week, with increases in staining in hippocampus and cerebellar cortex at that time (Dussault, 1989). Recall that the major increase in thyroid hormone secretion occurs during the second postnatal week in the developing rodent. No immunohistochemical studies of T_3 nuclear receptors are available from precocial species.

8.6.2 NUCLEAR BINDING STUDIES

The majority of studies characterizing the nuclear T_3 receptor in developing brain have relied on receptor binding characteristics. A variety of techniques have been described which use either nuclear extracts or intact nuclei as the source of nuclear T_3 receptors. Thus, binding affinities and maximal binding capacities often differ among reports. Nuclear T_3 receptor binding characteristics have been characterized in both altricial (developing rodent) and precocial species (the ovine fetus). Limited human data are also available. Bernal and Pekonen (1984) demonstrated saturable, specific, high-affinity binding sites for L-T_3 in human fetal cerebral cortex with binding affinities comparable to those in rodent cortex (K_a ~1–2 $\times$ 10^{10} M^{-1}). Maximal binding capacities were low at 10 weeks gestation and increased 10-fold by week 16 of gestation (Bernal and Pekonen, 1984). These data have been expanded by Su *et al.* (1989) who described a six-fold increase in brain cortical T_3 nuclear receptor binding capacity from 104 fmol T_3/mg DNA at 18 weeks gestation to 648 fmol T_3/mg DNA at 36 weeks gestation. These data suggest that during the critical periods of neuroblast proliferation and synaptogenesis there are progressive increases in the expression of nuclear T_3 receptors in the developing human brain. Estimations of T_3 receptor occupancy during this period average 25% (Ferreiro *et al.*, 1988).

In contrast to that reported for the developing human fetus, receptor binding in the rat occurs in two distinct phases. Receptor binding is demonstrable after day 15 in developing fetal rodent brain cortex; maximal binding capacities between fetal day 15 and the second day of life are comparable, averaging about 250 fmol T_3/mg DNA. After this period there is a rapid increase in brain T_3 receptor binding capacity to nearly 780 fmol T_3/mg DNA by day 5 of life (Perez-Castillo *et al.*, 1985). Maximal levels of receptor occupancy by T_3 occurred at day 15 of postnatal life, coincident with the elevated levels of plasma T_3 at that time (Ferreiro *et al.*, 1990). The increase in brain T_3 receptor binding during the first two weeks of postnatal life in the developing rat has been described by other investigators (Coulombe *et al.*, 1981; Schwartz and Oppenheimer, 1978). The significance of this observation is uncertain. Interestingly, the restoration of euthyroid status in neonatal rats rendered hypothyroid by a variety of methods is only minimally effective in reversing the resultant defects in brain development if the hypothyroidism is prolonged beyond this two-week period (Schwartz, 1983).

The ontogeny of nuclear thyroid hormone receptor binding characteristics has also been described in the developing sheep, a precocial species with thyroid system maturation analogous to that occurring in the human species (Fisher *et al.*, 1977). Ferreiro *et al.* (1987) demonstrated saturable high-affinity binding of L-T_3 in salt-extracted nuclei isolated from cerebral cortex at 50 days gestation (term = 145 days gestation). Receptor binding capacity increased from 190 ± 48 fmol/mg DNA to approximately 500 fmol T_3/mg DNA by 80 and 100 days gestation (Ferreiro *et al.*, 1987). Despite the low levels of circulating thyroid hormones present at this stage of fetal development (serum T_3 <20 ng/dl, T_4 2–3 μg/dl) receptor occupancy in cerebral cortex was estimated to be nearly 75%. This probably reflects *in situ* conversion of T_4 into T_3 by brain type II 5'-MDI as described earlier.

Work in our laboratory has extended these observations (Polk *et al.*, 1989). We characterized nuclear T_3 receptor binding characteristics in brain cortex of developing sheep from 80 days gestation through the first weeks of life. In agreement with previous reports, fetal brain T_3 nuclear receptor bind-

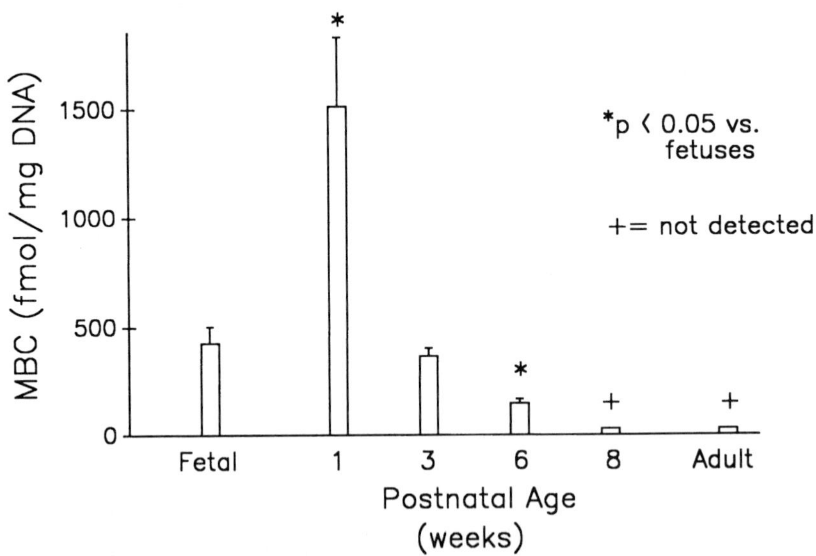

Fig. 8.5.   Brain nuclear thyroid receptor binding capacities (MBC) in developing sheep. Cortical binding capacities are comparable during the first third of gestation in the fetus, then increase nearly three-fold during the first postnatal week. Cortical binding capacity decreases to undetectable levels by the second month of life and this persists into adulthood. *$P<0.05$ vs. fetuses; + = not detected.

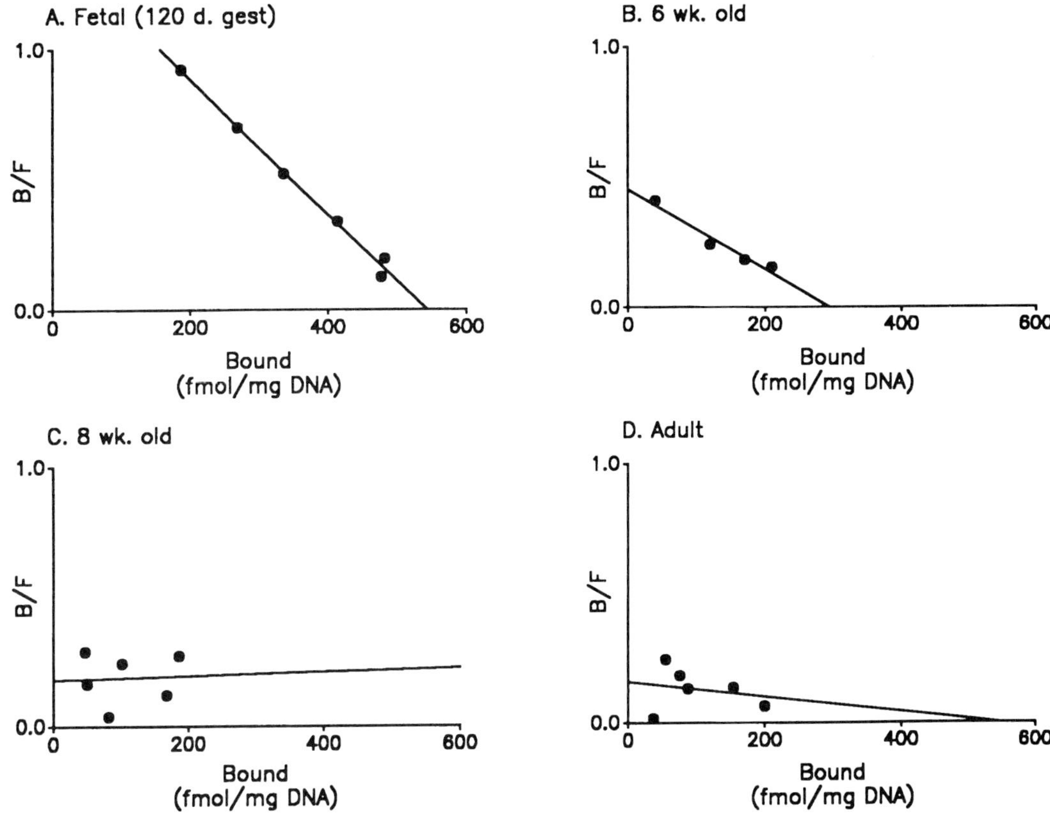

Fig. 8.6. Scatchard analysis of brain nuclear thyroid receptor binding activity in the developing sheep. (A) Data from the 120-day-old fetus (term = 145 days); (B) data from six-week-old animals; (C) data from 8-week-old animals; and (D) adult data.

ing capacity was comparable from 100 days gestation to term, averaging 410 fmol T_3/mg DNA. Interestingly, there was an almost four-fold increase in binding capacity of receptor isolated from cortex obtained from lambs 5–7 days after birth (1500 fmol T_3/mg DNA). Association constants were comparable during this period of time, indicating no change in receptor affinity. Thus, the new-born sheep also demonstrates a marked increase in brain cortical T_3 nuclear receptor binding capacity during the early neonatal period. This is associated with the thyroid hormone surge which occurs during the same period of time, a response analogous to that

demonstrated in the neonatal rodent during the first weeks of life. Continuing work in our laboratory has documented a further decline in T_3 nuclear receptor maximal binding capacities in cortical brain tissue over the first two months of life, leading to non-detectable binding which persists in the adult. These data are summarized in Figs 8.5 and 8.6. These observations contrast with data in hepatic tissues during this time. Nuclear T_3 receptor binding capacity in hepatic tissues increases progressively during fetal development, reaching adult levels near term. Thus, the paucity of thyroid hormone effects in adult brain in precocial species may be due in part to altered nuclear receptor binding characteristics in this tissue. These data are also consistent with the observation that the effects of congenital hypothyroidism on central nervous system function can only be reversed if replacement therapy is begun soon after birth; delayed therapy is associated with permanent intellectual sequelae despite restoration of normal function in other systems.

8.6.3   mRNA STUDIES OF BRAIN T_3 RECEPTORS

Given the previous data, it is not surprising that the distribution of the different species of mRNA encoding the α and β forms of the various thyroid hormone receptors differs in various sites within the developing brain and their expression is developmentally regulated. Data regarding these mRNAs are obtained from studies in the developing rodent; to date, no data from developing precocial species have been reported. Mitsuhashi and Nikodem (1989) reported low levels of mRNA encoding the rTHR α-1 receptor in total rat brain obtained from the 19-day-old fetus. These levels increase by 10 days after birth to values found in adult brain. However, the predominant mRNA species expressed during development is the mRNA encoding the rTHR α–2 variant (which does not bind thyroid hormone). We have observed a similar phenomenon in the developing rodent (North and Fisher, 1990) and have compared the relative abundance of the $\alpha 1$ and $\alpha 2$ mRNAs with previous data regarding T_3 receptor binding. The message corresponding to rTHR α-2 was the most abundant in the developing rat brain, and was evident from fetal day 19 through to postnatal day 50, with increased expression during the second week of life. Recall that this is the period of increased T_3 binding to isolated receptors from rat brain discussed earlier. Also noted was a small but significant increase in rTHR β–1 message which occurs in the two weeks after

birth in the rodent. Although this message is not abundant, it may also contribute to the overall T_3 binding capacity in developing brain during this period.

These results have been confirmed by Mellstrom *et al.* (1991), who, using probes specific to rTHR α and rTHR β subtypes and techniques of *in situ* hybridization, showed the extensive distribution of thyroid hormone receptor mRNAs in developing rodent brain. They confirmed the presence of α-1 mRNA as early as fetal day 14 and observed levels peaking during the first weeks after birth in cortex, amygdala, hippocampus, and cerebellum. The level of rTHR β-1 was low during fetal life and increased during the neonatal period to a level which persisted into adulthood. The first areas to express THR β–1 mRNA were the mitral cell layer in the olfactory bulb, followed by expression in the caudate nucleus, hippocampus (CA1 field) and nucleus accumbens. Cortical expression of rTHR β–1 mRNA was observed at birth in the intermediate layer followed by hybridization in the superficial and inner layers during the first weeks after birth. Again, expression of α gene transcripts including the non-thyroid hormone binding variants predominate early in development, followed by expression of β transcripts which persist into adulthood. This generalization is in agreement with other reports (Bradley *et al.*, 1989; Forrest *et al.*, 1990; Strait *et al.*, 1990).

In summary, these studies suggest that the overall effect of T_3 on developing brain is mediated through a balance of expression of the various alpha messages, which encode proteins with various capacities to bind thyroid hormone, as well as expression of THR beta mRNA, which in the rat persists into adulthood. These *in situ* and immunohistochemical techniques suggest that the effects of altered thyroid status on developing brain are not global, but rather site (and perhaps cell type) specific. The developmental expression of nuclear thyroid hormone receptors in various specific sites within the central nervous system is, at least in part, responsible for the sensitivity of specific brain structures to altered thyroid hormone status. Moreover, the observation that restoration of normal central nervous system function in response to abnormal thyroid status is achievable only during certain critical periods of development is probably due to developmentally regulated changes in abundance of specific thyroid hormone receptors during these periods.

8.7 EFFECTS OF ALTERED THYROID STATUS ON BRAIN MATURATION

The most striking effects of thyroid hormone deficiency during development involve the central nervous system. Although the dependency of developing brain structures on a euthyroid state has been known for over a century, the exact biochemical processes responsible for this effect remain largely obscure. The demonstration of nuclear thyroid hormone receptors as proteins capable of modifying gene transcription has prompted the reinvestigation of the effects of altered thyroid status on specific central nervous system gene expression.

8.7.1 HISTOLOGIC ALTERATIONS

Since the pioneering studies of Eayrs (1960) and Legrand (1967), it has been recognized that one of the most striking consequences of hypothyroidism is the reduction of the complexity of axonal and dendritic branching, particularly involving the pyramidal cells in the cortex and Purkinje cells in the cerebellum. These observations have been most extensively studied in the developing rodent. Ruiz-Marcos *et al.* (1983) described the ontogeny of these effects on acoustic cortical pyramidal cells. Neonatal rat pups underwent thyroidectomy at 10 days of life, and were studied after various periods of thyroid hormone replacement. Cortical pyramidal cells obtained from rats receiving thyroid replacement from day 12

of life were normal when studied at 60 days of age. However, delay of therapy until day 20 of life resulted in a severe reduction in neuropil. One hypothesis for these findings is that neuronal structural organelles may be dependent on thyroid hormone during development. In support of this view, Nunez *et al.* (1989) reported altered cortical tubulin mRNA in rats rendered hypothyroid during the first weeks after birth. A similar mechanism is proposed for the effect of hypothyroidism on cerebellar Purkinje cells (Legrand, 1979). Proof of this as a primary effect will ultimately require demonstration of thyroid receptor regulatory elements within the gene sequences of interest.

8.7.2 GROWTH FACTOR-MEDIATED BRAIN DEVELOPMENT AND THYROID HORMONE EFFECTS

It was noted earlier that some of the developmental abnormalities associated with hypothyroidism are indirectly mediated via thyroid hormone effects on other growth factors (for review see Fisher and Lakshmanan, 1990). Of particular interest is the relationship between thyroid status and expression of epidermal growth factor (EGF). EGF is a potent mitogen for a variety of ectodermal and mesodermal cell and tissue types. In the rat, both EGF and the EGF receptor have been isolated from brain tissues (Adamson and Meek, 1984; Lakshmanan *et al.*, 1986). Relatively high (in comparison to adult) levels of EGF are present in developing forebrain and midbrain neurons during the second week of life (Fallon *et al.*, 1984). EGF receptors, predominantly localized to glial cells, follow a similar pattern of maturation (Hiramatsu *et al.*, 1988; Gomez-Pinilla *et al.*, 1988). EGF is mitogenic for astrocytes, causes glial differentiation and promotes neurite outgrowth of telencephalic neurons in the developing rat (Monnet-Tschudi and Honegger, 1989; Morrison *et al.*, 1989). Sequence analysis of the mouse EGF gene suggests the presence of several consensus thyroid receptor binding elements (Pascall and Brown, 1988). Stein *et al.* (1989) have demonstrated in the mouse that thyroid hormone modulates expression of EGF mRNA during the same period of time (the first weeks of life) that normal central nervous system development is dependent on thyroid hormone. This observation is compatible with the hypothesis that thyroid hormone effects on developing brain are secondary to influences on specific growth factors.

In summary, the effect of thyroid hormone on brain development is complex and probably mediated by thyroid hormone modulation of specific gene products via the action of nuclear thyroid hormone receptors. These transcriptional events represent a balance among the interaction of various receptor subtypes with specific regulatory gene sequences. The extent and degree of these various interactions remain largely uncharacterized.

REFERENCES

Adamson, E. D. and Meek, J. (1984) Epidermal growth factor receptors during mouse development. *Dev. Biol.*, **193**, 62–70.

Apriletti, J. W., Eberhardt, N. L., Latham, K. R. and Baxter, J. D. (1981) Affinity chromatography of thyroid hormone receptors. *J. Biol. Chem.*, **256**, 12094–101.

Banker, D. E., Bigler, J. and Eisenman, R. N. (1991) The thyroid hormone receptor gene (c-erbA α) is expressed in advance of thyroid gland maturation during the early embryonic development of *Xenopus laevis*. *Mol. Cell Biol.*, **11**, 5079–89.

Bernal, J. and Pekonen, F. (1984) Ontogenesis of the nuclear T₃ receptor in the human fetal brain. *Endocrinology*, **114**, 677–9.

Bradley, D. J., Young, W. S. and Weinberger, C. (1989) Differential expression of α and β thyroid receptors in rat brain and pituitary. *Proc. Natl. Acad. Sci. USA*, **86**, 7250–4.

Calvo, R., Obregon, M. J., Ruiz de Ona, C. *et al.* (1990) Congenital hypothyroidism as studied in rats. *J. Clin. Invest.*, **86**, 889–99.

Coulombe, P., Ruel, R. and Dussault, J. H. (1981) Receptors nucleares de la T₃ dans le cerveau et le cervelet du rat au cours du developpement. *Union Med. Can.*, **110**, 658–61.

Downward, J., Yarden, Y., Mayes, E. *et al.* (1984) Close similarity of epidermal growth factor receptor and v-erb-β oncogene protein sequences. *Nature*, **307**, 521–7.

Dussault, J. H. (1989) Action of thyroid hormones on brain development, in *Research in Congenital Hypothyroidism* (eds F. Delange, D. A. Fisher and D. Glinoer) Plenum Press, New York, pp. 95–103.

Dussault, J. H. and Ruel, J. (1987) Thyroid hormones and brain development. *Annu. Rev. Physiol.*, **49**, 321–34.

Eayrs, J. T. (1960) Influence of the thyroid on the central nervous system. *Br. Med. Bull.*, **16**, 122–7.

Evans, R. M. (1988) The steroid and thyroid hormone receptor superfamily. *Science*, **240**, 889–95.

Evans, R. M. and Hollenberg, S. (1988) Zinc fingers: guilt by association. *Cell*, **52**, 1–3.

Fallon, J. H., Seroogy, K. B., Loughlin, S. E. *et al.* (1984) Epidermal growth factor immunoreactive material in the central nervous system: location and development. *Science*, **224**, 1107–9.

Ferreiro, B., Bernal, J. and Potter, B. J. (1987) Ontogenesis of thyroid hormone receptor in fetal lambs. *Acta Endocrinol.*, **116**, 205–10.

Ferreiro, B., Bernal, J., Goodyer, C. G. and Branchard, C. L. (1988) Estimation of nuclear thyroid hormone receptor saturation in human fetal brain and lung during early gestation. *J. Clin. Endocrinol. Metab.*, **67**, 853–6.

Ferreiro, B., Pastor, R. and Bernal, J. (1990) T₃ receptor occupancy and T₃ levels in plasma and cytosol during rat brain development. *Acta Endocrinol. (Copenh.)*, **123**, 95–9.

Fisher, D. A. (1985) Thyroid hormone and thyroglobulin synthesis and secretion, in *Pediatric Thyroidology* (eds F. Delange, D. A. Fisher and P. Malvaux), Karger, Basel, pp. 44–56.

Fisher, D. A. and Lakshmanan, J. (1990) Metabolism and effects of epidermal growth factor and related growth factors in mammals. *Endocr. Rev.*, **11**, 418–42.

Fisher, D. A., Dussault, J. H., Sack, J. and Chopra, I. J. (1977) Ontogenesis of hypothalamic–pituitary–thyroid function and metabolism in man, sheep and rat. *Recent Prog. Horm. Res.* **33**, 59–116.

Forrest, D., Sjoberg, M. and Vennstrom, B. (1990) Contrasting developmental and tissue-specific expression of α and β thyroid hormone receptor genes. *EMBO J.*, **9**, 1519–28.

Gomez-Pinilla, F., Knauer, D. J. and Nieto-Sampedro, M. (1988) Epidermal growth factor receptor immunoreactivity in rat brain. Development and cellular localization. *Brain Res.*, **438**, 385–90.

Graf, T. and Beug, H. (1983) Role of the v-erbA and v-erbB oncogenes of avian erythroblastosis virus in erythroid cell transformation. *Cell*, **34**, 7–9.

Hiramatsu, M., Kashimata, M., Sata, M. *et al.* (1988) Influences of age on epidermal growth factor receptor level in the rat brain. *Experientia*, **44**, 23–5.

Hodin, R., Lazar, M., Wintman, B. *et al.* (1989) Identification of a thyroid hormone receptor that is pituitary specific. *Science,* **244**, 76–9.

Koenig, R., Warne, R. L., Brent, G. A. *et al.* (1988) Isolation of a cDNA clone encoding a biologically active thyroid hormone receptor. *Proc. Natl. Acad. Sci. USA*, **85**, 5031–5.

Lakshmanan, J., Weichesel, M. E. and Fisher, D. A. (1986) Epidermal growth factor in synaptosomal fractions of mouse cerebral cortex. *J. Neurochem.*, **46**, 1081–5.

Latham, K. R., Apriletti, J. W., Eberhardt, N. L. and Baxter, J. D. (1981) Development of support matrices for affinity chromatography of thyroid hormone receptors. *J. Biol. Chem.*, **256**, 12088–94.

Laudet, V., Begue, A., Henry-Duthoit, C. *et al.* (1991) Genomic organisation of the human thyroid hormone receptor α (c-erbA-1) gene. *Nucleic Acid. Res.*, **19**, 1105–12.

Lazar, M. and Chin, W. W. (1990) Nuclear thyroid hormone receptors. *J. Clin. Invest.*, **86**, 1777–82.

Lazar, M. A., Hodin, R. A., Darling, D. S. and Chin, W. W. (1988) Identification of a rat c-erbA α-related protein which binds deoxyribonucleic acid but does not bind thyroid hormone. *Mol. Endocrinol.*, **2**, 893–901.

Lazar, M. A., Jones, K. and Chin, W. W. (1990) Isolation of a cDNA encoding human Rev-ErbAα: transcription from the noncoding DNA strand of a thyroid hormone receptor gene results in a related protein that does not bind thyroid hormone. *DNA Cell Biol.*, **9**, 77–83.

Legrand, J. (1967) Analyse de l'effect morphogenetique des hormones thyroidiennes sur le cervulet du jeune rat. *Arch. Anat. Microsc. Morphol. Exp.*, **56**, 205–9.

Legrand, J. (1979) Morphogenetic action of thyroid hormones. *Trends Neurosci.*, **2**, 234–41.

Lindenburg, J. A., Brehier, A. and Ballard, P. H. (1978) Triiodothyronine nuclear binding in fetal and adult rabbit lung and cultured lung cells. *Endocrinology*, **103**, 1725–31.

Luo, M., Puymirat, J. and Dussault, J. (1989) Immunocytochemical localization of nuclear 3, 5, 3'-triiodothyronine (L-T$_3$) receptors in astrocyte cultures. *Dev. Brain Res.*, **46**, 131–6.

Mellstrom, B., Naranjo, J. R., Santos, A. *et al.* (1991) Independent expression of the α and β c-erbA genes in developing rat brain. *Mol. Endocrinol.*, **5**, 1339–50.

Mitsuhashi, T. and Nikodem, V. M. (1989) Regulation of expression of the alternative mRNAs of the rat alpha thyroid hormone receptor gene. *J. Biol. Chem.*, **264**, 8900–4.

Mitsuhashi, T. G., Tennyson, G. E. and Nikodem, V. M. (1988) Alternative splicing generates messages encoding rat c-erbA proteins that do not bind thyroid hormone. *Proc. Natl. Acad. Sci. USA*, **65**, 5804–8.

Miyajima, N., Horiuchi, R., Shibuya, Y. *et al.* (1989) Two erb-A homologs encoding proteins with different T$_3$ binding capacities are transcribed from opposite DNA strands of the same genetic locus. *Cell*, **57**, 31–9.

Monnet-Tschudi, F. and Honegger, P. (1989) Influence of epidermal growth factor on the maturation of fetal rat brain cells in aggregate culture. An immunocytochemical study. *Dev. Neurosci.*, **11**, 30–40.

Morrison, R. S., Kornblum, H. I., Leslie, F. M. and Bradshaw, R. A. (1989) Trophic stimulation of cultured neurons from neonatal rat brain by epidermal growth factor. *Science*, **238**, 72–5.

Murray, M. B., Zilz, N. D., McCreary, N. L. *et al.* (1988) Isolation and characterization of rat cDNA clones from two distinct thyroid hormone receptors. *J. Biol. Chem.*, **263**, 12270–7.

Nakai, A., Sakarai, A., Bell, G. I. and deGroot, L. I. (1988) Characterization of a third human thyroid hormone receptor co-expressed with other thyroid hormone receptors in several tissues. *Mol. Endocrinol.*, **2**, 1087–92.

North, D. and Fisher, D. A. (1990) Thyroid hormone receptor and receptor-related RNA levels in developing rat brain. *Pediatr. Res.*, **28**, 622–5.

Nunez, J., Couchie, D. and Brion, J. P. (1989) Microtubule assembly: regulation by thyroid hormones, in *Iodine and the Brain* (eds G. R. Delong, J. Robbins and P. G. Condliffe), Plenum Press, New York, pp. 103–12.

Oppenheimer, J. H. (1983) The nuclear receptor–triiodothyronine complex: relationship to thyroid hormone distribution, metabolism and biological action, in *Molecular Basis of Thyroid Hormone Action* (eds J. H. Oppenheimer and H. H. Samuels), Academic Press, New York, pp. 1–35.

Oppenheimer, J. H. and Schwartz, H. L. (1986) Thyroid hormone action at the nuclear level, in *Thyroid Hormone Metabolism* (ed G. Hennemann), Marcel Dekker, New York, pp. 383–415.

Oppenheimer, J. H., Koemer, D., Schwartz, H. L. and Surks, M. I. (1972) Specific nuclear triiodothyronine binding sites in rat liver and kidney. *J. Clin. Endocrinol. Metab.*, **35**, 330–3.

Oppenheimer, J. H., Schwartz, H. L., Mariasu, C. N. *et al.* (1987) Advances in our understanding of thyroid hormone action at the cellular level. *Endoc. Rev.*, **8**, 288–308.

Pascall, J. C. and Brown, K. D. (1988) Structural analysis of the 5' flanking sequence of the mouse epidermal growth factor gene. *J. Mol. Endocrinol.*, **1**, 5–11.

Pascual, A., Casanova, J. and Samuels, H. H. (1982) Photoaffinity labeling of thyroid hormone receptors in intact cells. *J. Biol. Chem.*, **257**, 9640–7.

Perez-Castillo, A., Bernal, J., Ferreiro, B. and Pans, T. (1985) The early ontogenesis of thyroid hormone receptor in the rat fetus. *Endocrinology*, **117**, 2457–61.

Polk, D., Chermocha, D., Reviczky, A. and Fisher, D. A. (1989) Nuclear thyroid hormone receptors: ontogeny and thyroid hormone effects in sheep. *Am. J. Physiol.*, **256** (*Endocrinol. Metab.*, **19**), E543–9.

Polk, D. H., Wu, S. Y. and Fisher, D. A. (1986) Serum thyroid hormones and tissue 5' mono-deiodinase activity in acutely thyroidectomized newborn lambs. *Am. J. Physiol.* **251** (*Endocrinol. Metab.* **14**), 151–5.

Polk, D. H., Wu, S. Y., Wright, C. *et al.* (1988) Ontogeny of thyroid hormone effects on 5' monodeiodinase activity in fetal sheep. *Am. J. Physiol.* **254** (*Endocrinol. Metab.* **17**), 337–41.

Prost, E., Koenig, R. J, Moore, D. D. *et al.* (1988) Multiple sequences encoding potential thyroid hormone receptors isolated from mouse skeletal muscle cDNA libraries. *Nucleic Acids Res.*, **16**, 6248.

Refetoff, S. and Larsen, P. R. (1989) Transport, cellular uptake and metabolism of thyroid hormone, in *Endocrinology* (eds L.J. de Groot *et al.*) W. B. Saunders, Philadelphia, pp. 541–61.

Ruel, J., Faure, R. and Dussault, J. H. (1985) Regional distribution of nuclear T₃ receptors in rat brain and evidence for preferential local-ization in neurons. *J. Endocrinol. Invest.*, **8**, 343–8.

Ruiz-Marcos, A., Salas, J., Sanchez-Toscano, F. *et al.* (1983) Effects of neonatal and adult onset hypothyroidism on pyramidal cells of the rat auditory cortex. *Dev. Brain. Res.*, **9**, 205–13.

Samuels, H. H., Tsai, J. J. and Casanova, J. (1974) Thyroid hormone action: *in vitro* demonstration of putative receptors in isolated nuclei and soluble nuclear extracts. *Science*, **184**, 1188–91.

Santos, A., Freake, H. C., Rosenberg, M. *et al.* (1988) Triiodothyronine nuclear binding capacity in rat tissues correlates with a 6.0 kilobase (kb) and not a 2.6 kb messenger ribonucleic acid hybridization signal generated by a human c-erbA probe. *Mol. Endocrinol.*, **2**, 992–8.

Sapp, J., Munoz, A., Damm, K. *et al.* (1986) The c-erbA protein is a high affinity receptor for thyroid hormone. *Nature*, **324**, 635–40.

Schadlow, A. R., Surks, M. I., Schwartz, H. L. and Oppenheimer, J. H. (1974) Specific triio-dothyronine binding sites in the anterior pitui-tary of the rat. *Science*, **176**, 1252–4.

Schwartz, H. L. (1983) Effect of thyroid hor-mone on growth and development, in *Molecular Basis of Thyroid Hormone Action* (eds J. H. Oppenheimer and H. H. Samuels), Academic Press, New York, pp. 413–44.

Schwartz, H. L. and Oppenheimer, J. H. (1978) Nuclear triiodothyronine receptor sites in the brain: probable identity with hepatic receptors and regional distribution. *Endocrinology*, **103**, 267–73.

Stein, S. A., Shanklin, D. R., Adams, P. M. *et al.* (1989) Thyroid hormone regulation of specific mRNAs in the developing brain, in *Iodine and the Brain* (eds G. R. De Long, J. Robbins and P. G. Condliffe), Plenum Press, New York, pp. 59–78.

Strait, K. A., Schwartz, H. L., Perez-Castillo, A. and Oppenheimer, J. H. (1990) Relationship of c-erbA mRNA content to tissue triiodothyronine nuclear binding capacity and function in developing and adult rats. *J. Biol. Chem.*, **265**, 10514–21.

Su, H. L., Ling, P., Yang, R. K. and Chao, H. C. (1989) Ontogenesis of nuclear T₃ receptor in human fetal brain, in *Iodine and the Brain* (eds G. R. Delong, I. Robbins and P. G. Condliffe), Plenum Press, New York, p. 358.

Surks, M. I., Koemer, D., Dillmann, W. and Oppenheimer, J. H. (1973) Limited capacity binding sites for L-triiodothyronine in rat liver nuclei: localization to the chromatin and partial characterization of the L-triiodothyronine–chromatin complex. *J. Biol. Chem.*, **248**, 7066–72.

Surks, M. I., Koerner, D. H. and Oppenheimer, J. H. (1975) *In vitro* binding of L-triiodothyronine to receptors in rat liver nuclei: kinetics of binding, extraction properties and lack of requirement for cytosol proteins. *J. Clin. Invest.*, **55**, 50–60.

Thompson, C. C., Weinberger, C., Lebo, R. and Evans, R. M. (1987) Identification of a novel thyroid hormone receptor expressed in the mammalian central nervous system. *Science*, **237**, 1610–14.

Weinberger, C., Hollenberg, S. M., Rosenfeld, M. G. and Evans, R. M. (1985) Domain structure of human glucocorticoid receptor and its relationship to the v-erbA oncogene product. *Nature*, **318**, 670–2.

Weinberger, C., Thompson, C. C., Ong, E. S. *et al.* (1986) The c-erbA gene encodes a thyroid hormone receptor. *Nature*, **324**, 641–8.

Wu, S. Y., Fisher, D. A., Polk, D. H. and Chopra, I. (1990a) Maturation of thyroid hormone metabolism, in *Thyroid Hormone Metabolism* (ed S. Y. Wu), Blackwell Scientific Publications, Boston, pp. 293–320

Wu, S. Y., Merryfield, M. L., Polk, D. H. and Fisher, D. A. (1990b) Two pathways for thyroxine 5' monodeiodination in brown adipose tissue in fetal sheep: ontogenesis and divergent responses to hypothyroidism and 3, 5, 3'-triiodothyronine replacement. *Endocrinology*, **126**, 1950–8.

CORTICOTROPIN-RELEASING HORMONE RECEPTORS AND THE DEVELOPING NERVOUS SYSTEM

Dimitri E. Grigoriadis, Thomas R. Insel, Jeffrey A. Heroux and Errol B. De Souza

9.1 INTRODUCTION

Investigations into the ontogeny of receptor proteins for a variety of neurohormones have clearly demonstrated that the characteristics of these proteins, both in terms of appearance and function, can change over the course of development. Neuropeptides in the developing rat brain have an impact on everything from cell maturation (Bardo *et al.*, 1985) and protein synthesis (Bartolome *et al.*, 1986) to neuronal differentiation (Shyr *et al.*, 1986). Receptors for some of these neuropeptides demonstrate very distinct patterns of labeling during development and it has been suggested that certain peptides may play an organizational role in development.

With the advent of exquisitely sensitive techniques, such as immunocytochemistry, which enables the measurement of minute quantities of the hormone itself, or autoradiography, enabling the discrete localization of receptors in tissue sections, it has become increasingly possible to address the question of the development of these proteins and the consequences of their maturation (for reviews see Bugnon *et al.*, 1987; Ko *et al.*, 1989) For example, immunocytochemical techniques have been used to elucidate the development of the corticoliberin hypothalamic neuroglandular system in rats (Bugnon *et al.*, 1982) and humans (Bresson *et al.*, 1987), as well as to examine the levels of corticotropin-releasing hormone (CRH) in plasma during pregnancy in both maternal and fetal plasma (Goland *et al.*, 1986; Sasaki *et al.*, 1987). *In vitro* receptor autoradiographic techniques have been used extensively to study the appearance and distribution within the developing central nervous system of a variety of neurotransmitter or neurohormone receptor systems. Receptors for opiates (Kent *et al.*, 1982; Hammer, 1985), dopamine D-2 (Sales *et al.*, 1989), oxytocin (Shapiro and Insel, 1989), somatostatin (Gonzales *et al.*, 1991), neurohypophyseal peptides (Petracca *et al.*, 1986) neurotensin (Kiyama *et al.*, 1987) and substance P (Quirion and Dam, 1986) have all been shown by receptor autoradiography to appear transiently in discrete brain regions during development. The study of neuropeptides or hormones and the receptors with which they interact during ontogeny may provide some insight into the normal development of the organism as well as some possible clues into abnormalities that may occur later in life.

Receptors in the Developing Nervous System Vol. 1: *Growth factors and hormones* Edited by Ian S. Zagon and Patricia J. McLaughlin. Published in 1993 by Chapman & Hall. ISBN 0 412 45240 5. Vols. 1 and 2 (set) ISBN 0 412 54520 9.

9.1.1 ROLE OF CRH IN THE CENTRAL NERVOUS SYSTEM

Stress, or disruptions of homeostasis, can result from a variety of challenges: physical, psychological and immunological. Stress in the adult elicits a wide spectrum of changes within the nervous, endocrine, and immune systems. The activation of the pituitary adrenocortical axis resulting in the secretion of pituitary adrenocorticotropin (ACTH) and adrenal cortisol or corticosterone is the primary stress-related response in the endocrine system. CRH, a 41-residue peptide, has been shown to be the primary physiological regulator of synthesis and secretion of ACTH from the pituitary gland (Vale *et al.*, 1981; Rivier *et al.*, 1983). In the adult pituitary, CRH binding sites have been localized in the anterior and intermediate lobes of rat (De Souza *et al.*, 1984; Grigoriadis and De Souza, 1989), mouse (Webster *et al.*, 1991) and human (De Souza *et al.*, 1985b) anterior pituitary. CRH binding sites in a variety of species have demonstrated a wide distribution within the adult CNS as well as in peripheral tissues such as spleen (for review see De Souza *et al.*, 1990). The distribution pattern of CRH receptors within the pituitary supports the functional role of CRH as the primary physiological regulator of secretion of pro-opiomelanocortin (POMC)-derived peptide from the anterior and intermediate lobes of the pituitary. In addition to its established actions at the pituitary, immunohistochemical localization of CRH has demonstrated that the hormone has a broad extrahypothalamic distribution in the central nervous system and there is increasing evidence suggesting that it may act as a neurotransmitter or neuromodulator in the CNS, where it appears to have a major role in coordinating the organism's autonomic, behavioral and endocrine response to stress (Vale *et. al.*, 1983; De Souza *et al*, 1985a; Koob, 1985).

It is now well known, however, that during development there is a decrease in the ability of neonatal rats to secrete ACTH and glucocorticoids in response to stress or exogenously administered CRH from day 2 to day 12. This period has been designated the 'stress non-responsive period' (SNRP) (Sapolsky and Meaney, 1986). However, throughout this non-responsive period, stressors such as urethane or prolonged social isolation can elicit a robust hypothalamic–pituitary–adrenocortical (HPA) response (Walker *et al.*, 1986b; Stanton *et al.*, 1987). In addition, during this period there is a transient 'overshoot' of CRH receptors in the anterior pituitary (Walker *et al.*, 1986a).

The differences observed in the pituitary adrenocortical response to various stressors during development could be due to proximal influences on the pituitary in central CRH pathways. CRH has been detected in the median eminence as early as day 18 of fetal life (Bugnon *et al.*, 1982). During the first postnatal week, basal levels of CRH in the hypothalamus appear to be only about 20% of adult levels (Walker *et al.*, 1986a). Studies utilizing a bioassay for CRH demonstrated a decreased CRH response to stress during the first postnatal week. Thus during this stress non-responsive period, in addition to pituitary hyporesponsiveness, there appears to be decreases in the CRH content of the hypothalamus and a selective reduced release with respect to certain stressors. Extrahypothalamic CRH pathways, which are believed to coordinate the neural response to stress, have not yet been adequately examined in development. Therefore, in contrast to the abundant information on CRH and its effects on the pituitary during development, relatively little is known about the extrahypothalamic development of CRH and its receptor, although these pathways are clearly important in the regulation of behavior, autonomic function and the stress response. This chapter will very briefly describe some developmental features of the hormone CRH itself and then focus primarily on the development of the CRH receptor in the rat brain.

9.2   CRH IN DEVELOPMENT

During gestation, many studies utilizing several different species have demonstrated that the fetus is exposed to very high levels of CRH produced mainly by the placenta (Goland *et al.*, 1986; Sasaki *et al.*, 1987). The amount of these high levels of circulating CRH that actually get across the fetal blood–brain barrier is at present unknown but may involve some active transport mechanism. Such a mechanism has been described in the selective active transport of insulin into fetal brain. A similar mechanism may therefore exist for the selective active transport of CRH into the fetal brain. In addition to the maternal exposure of CRH during gestation, the fetal brain is capable of producing CRH on its own. Using immuno-histochemical techniques, CRH has been identified in rat median eminence as early as embryonic day 18 (ED18), with hypothalamic levels decreasing after birth (Bugnon *et al.*, 1982). In the human hypothalamus, immuno-reactive CRH has been detected as early as 16 weeks of gestation. Using *in situ* hybridization techniques, messenger RNA for CRH has been detected in extrahypothalamic sites by postnatal day 2 (PD2). Message is particularly dense in cingulate cortex and the bed nucleus of the stria terminalis early in postnatal life. Although it appears that the number of cerebral cortical cells expressing CRH mRNA decreases by PD28, this difference may in part reflect the concurrent increase in the neuropil so that the same number of cells are present in the total structure whereas fewer cells are present in any given unit of area. It is not yet clear whether the average number of grains per cell changes throughout development. These data suggest that CRH is present early on in development in both hypothalamic and extrahypothalamic sites. In addition it does not appear that CRH is selectively present in cells that are eliminated during postnatal growth (Insel, 1990).

9.2.1   RESPONSE TO EXOGENOUSLY APPLIED CRH

In adult animals, CRH elicits a variety of gross behavioral effects such as increases in rearing, grooming and locomotor activity in a familiar environment. These measures clearly cannot be assessed in neonatal rat pups, however, when rat pups are stressed, they emit ultrasonic vocalizations (USV). These vocalizations or distress calls are usually in the range of 35–45 kHz and are extremely potent stimuli for maternal retrieval. Anxio-lytic compounds, such as diazepam, decrease USV, whereas anxiogenic compounds, such as pentylenetetrazol, increase USV (Insel *et al.*, 1986). Furthermore, strains of rat bred for 'nervousness' emit more USV than normal rat pups when isolated for short periods of time (Insel and Hill, 1987). Since CRH has been described as 'anxiogenic', it would seem reasonable to expect that CRH administration to neonatal rat pups would increase the frequency of USV. Figure 9.1 illustrates the dose-related effect of centrally administered CRH to 5–6-day-old rat pups. Intracere-broventricular (i.c.v.) administration of CRH to rat pups clearly decreased the rate of isolation calls compared to animals injected with saline. This effect was not due to decreases in general arousal, thermoreg-ulatory capacity or peripheral actions of the peptide (Insel and Harbaugh, 1989).

9.3   CRH RECEPTORS IN DEVELOPMENT

As described, CRH receptors have been well characterized in the brain and pituitaries of adults in many different species. These receptors generally, though not exclusively, use the adenylate cyclase system as the second messenger (for a review see Webster *et al.*, 1991). It has also been observed that developing CRH receptors do not strictly coincide with CRH terminals (described immunocytochemically) suggesting that CRH may have its effect on extrahypothalamic

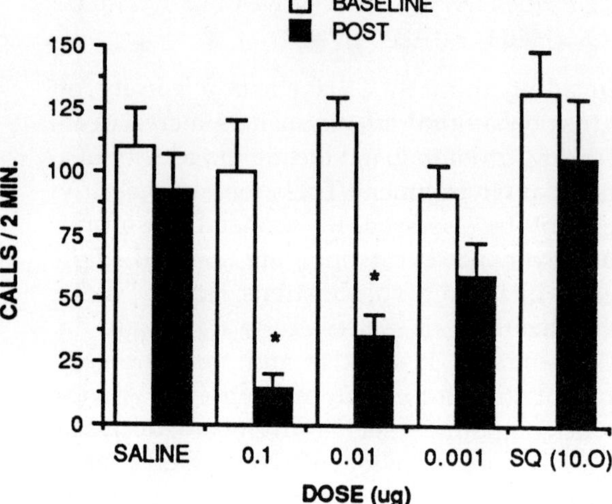

Fig. 9.1. Ultrasonic vocalization in 5–6-day-old rat pups isolated at room temperature for 2 min pre- and 30 min post- i.c.v. injection of oCRH at the doses indicated. Data represent means from between 6 and 12 pups at each dose and are analyzed by repeated measure ANOVA. In the presence of significant main effect for drug, * indicates change from pre- to post-significantly different ($P < 0.05$) from corresponding change with i.c.v. saline administration by *post hoc* one way ANOVA.

receptors via some non-synaptic pathway (Herkenham, 1987). Based on these observations, autoradiographic localization of CRH receptors in the developing brain would provide a better understanding of the functionality of this system as opposed to the immunocytochemical localization of the hormone itself. The following section will detail the various pharmacological and biochemical techniques used to study CRH receptors and their association with adenylate cyclase in the developing rat brain.

9.3.1 HOMOGENATE BINDING STUDIES

Radioligand binding is a very powerful tool in the determination of affinity, density and pharmacological profile of any receptor system. The use of iodinated radioligands allows for the determination of small quantities of receptors (in the femtomole range) in various tissues. The binding of ^{125}I-Tyr[0]-ovine CRH has been described in detail by De Souza (1987). Briefly, binding of ^{125}I-oCRH to membrane receptors was performed in 1.5 ml polypropylene microfuge tubes (Beckman Instruments) in 50 mM Tris, 10 mM $MgCl_2$, 2mM EGTA, 1.5% BSA, 0.15 mM bacitracin and 1.5% aprotinin (incubation buffer) in a final volume of 300 µl. Membrane suspension (100 µl), prepared as described above, was added to tubes containing 100 µl of an ^{125}I-oCRH solution in incubation buffer (final concentration approximately 100–200 pM) and 100 µl of the incubation buffer in the absence or presence of any competing peptides. Incubations were carried out at room temperature (22°C) for 2 h. The incubation reaction was terminated by centrifugation in a Beckman microfuge for 5 min. at 12 000 *g*. The resulting pellet was washed gently with 1.0 ml of ice-cold phosphate-buffered saline, pH 7.2, containing 0.01% Triton X-100 and re-centrifuged for 5 min at 12 000 *g*. The microfuge tubes were cut just above the pellet, placed into 12 × 75 mm polystyrene tubes and monitored for radioactivity in an LKB gamma counter at approximately 80% efficiency. The non-specific binding of ^{125}I-oCRH to membrane homogenates was defined in the presence of 1 µM unlabeled CRH.

By using this method, specific CRH binding was detected in whole rat brain homogenates as early as day 17 (ED17) of gestation in fetal rats, (i.e. 5 days before birth) at a level of approximately 40% of that seen at PD28 (adult). Receptor number was shown to increase steadily with the peak receptor density occurring at PD8. At this time CRH receptors measured 312% of normal adult values. To determine whether these changes reflected alterations in receptor affinity or receptor density, saturation studies were performed on whole brain homogenates at various times of gestation and postnatal life

(Table 9.1). These studies indicated that the affinity of CRH receptors remained constant at approximately 200 pM, whereas the changes observed reflected changes in the B_{max} (i.e. receptor density). Thus, following the peak increase of CRH receptors at PD8, the density of CRH receptors then decreases steadily, reaching normal adult levels by PD28. This observed overshoot of receptors was not due to a transient decrease in the total protein content in brain homogenates. In fact, the measured protein content increased steadily until PD21 (Fig. 9.2). It is worth mentioning that the increases observed represent an underestimate of the actual number of CRH receptors present since binding was performed in whole brain rather than in the discrete regions where these receptors are being expressed. However, these data clearly indicate that CRH receptors are present and active during the first few days of postnatal life. Discrete localization of brain receptors in fetal or early postnatal life cannot easily be performed using standard dissection techniques. The use of receptor autoradiography is a sensitive method for determining the precise localization of receptor proteins in discrete regions of the fetal brain on slide-mounted sections.

9.3.2 AUTORADIOGRAPHIC LOCALIZATION

The fundamental advantages of *in vitro* microscopic autoradiographic receptor mapping are increased anatomical resolution and greatly increased sensitivity of measurement when compared to the 'biochemical' approaches such as radioligand homogenate binding described above. The reason for the increased sensitivity of this technique is based on the fact that one can detect radiolabel binding to microscopically discrete areas of tissue sections. The increase in the sensitivity of this technique can thus be orders of magnitude greater than that found in biochemical studies. Using this method, receptor densities can be localized to areas of the brain which cannot be obtained even by microdissection techniques. Thus, receptor mapping will be useful in any situation which requires receptor measurement in small regions or where the overall quantities of receptors (but not necessarily the densities) are low.

Table 9.1. Ontogeny of CRH receptors and CRH-stimulated adenylate cyclase activity in homogenates of whole rat brain[a]

Age	CRH receptors (fmol/mg protein)	CRH-stimulated adenylate cyclase (pmol cAMP/ min/mg protein)	CRH activity (fmol cAMP/min/ fmol CRH receptor)
ED17	20.6±1.3	1.3±0.05	60.7
ED19	63.1±1.7	28.1±5.7	352
ED21	79.8±2.3	26.6±3.8	333
PD2	94.6±2.0	40.3±0.7	426
PD8	150.1±7.5	36.4±1.7	242
PD14	82.6±6.5	64.8±3.4	784
PD21	55.7±3.4	36.6±3.4	657
PD28	48.1±3.5	42.3±6.1	879

[a] CRH receptors were measured in homogenates of whole rat brain using 0.1 nM^{125}I-Tyr^0-oCRH and CRH-stimulated adenylate cyclase activity was determined using 1 μM rat/human CRH. Data represent the mean ± SEM from 2–4 animals at each time point. Specific CRH receptor binding was defined in the presence of 1 μM unlabeled rat/human CRH and represented approximately 80% of total binding whereas CRH-stimulated adenylate cyclase activity ranged from 10 to 35% above basal (+100 μM-GTP) activity. (Reproduced with permission from Insel *et al.*, 1988.)

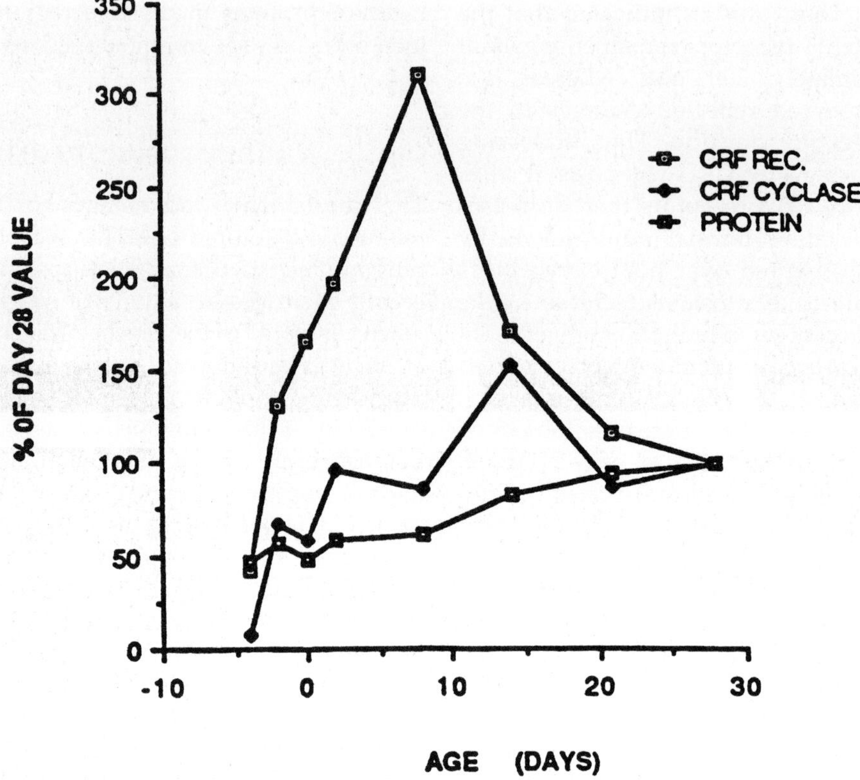

Fig. 9.2.   Development of ^{125}I-Tyr^0-oCRH specific binding, CRH stimulation of adenylate cyclase activity and protein concentration from ED17 (4 days before birth) to PD28 (28 days after birth) in whole rat brain homogenates. Data are expressed as percentages of the adult (PD28) values. Absolute values are given in Table 9.1. (Reproduced with permission from Insel *et al.*, 1988.)

Basically, *in vitro* labeling autoradiography requires that slide-mounted unfixed tissue sections, which have been cut from fresh frozen tissues, be incubated with a radio-labeled compound to equilibrium under very controlled conditions. Following receptor labeling, the slide-mounted tissue sections are rinsed, dried and apposed directly to tritium-sensitive Ultrofilm (LKB) or X-ray film. In this way, localization of receptors can be made to even microscopically discrete areas of tissues. Details on the methods used in receptor autoradiography are given by Kuhar (1981, 1985) and Kuhar *et al.* (1986).

To localize CRH receptors autoradio-graphically through ontogeny, 16 μm sections were cut from unfixed rat brain at various times throughout development. Binding was carried out at room temperature using 0.1 nM ^{125}I-Tyr^0-oCRH and using 1 μM unlabeled r/hCRH to define the non-specific binding in adjacent sections. CRH receptors were found to develop along the same time course as described using receptor binding in whole brain homogenates (Fig. 9.3). CRH receptors appear initially in the striatum, where they show the highest density from ED17 to PD8 (Fig. 9.3B–9.3F), after which they decrease markedly. The overshoot therefore observed using homogenate binding is largely due to the receptors concentrated in the striatum as well as across several layers of the neocortex (Fig. 9.3F). CRH receptors in the amyg-dala and claustrum, and the adult laminar

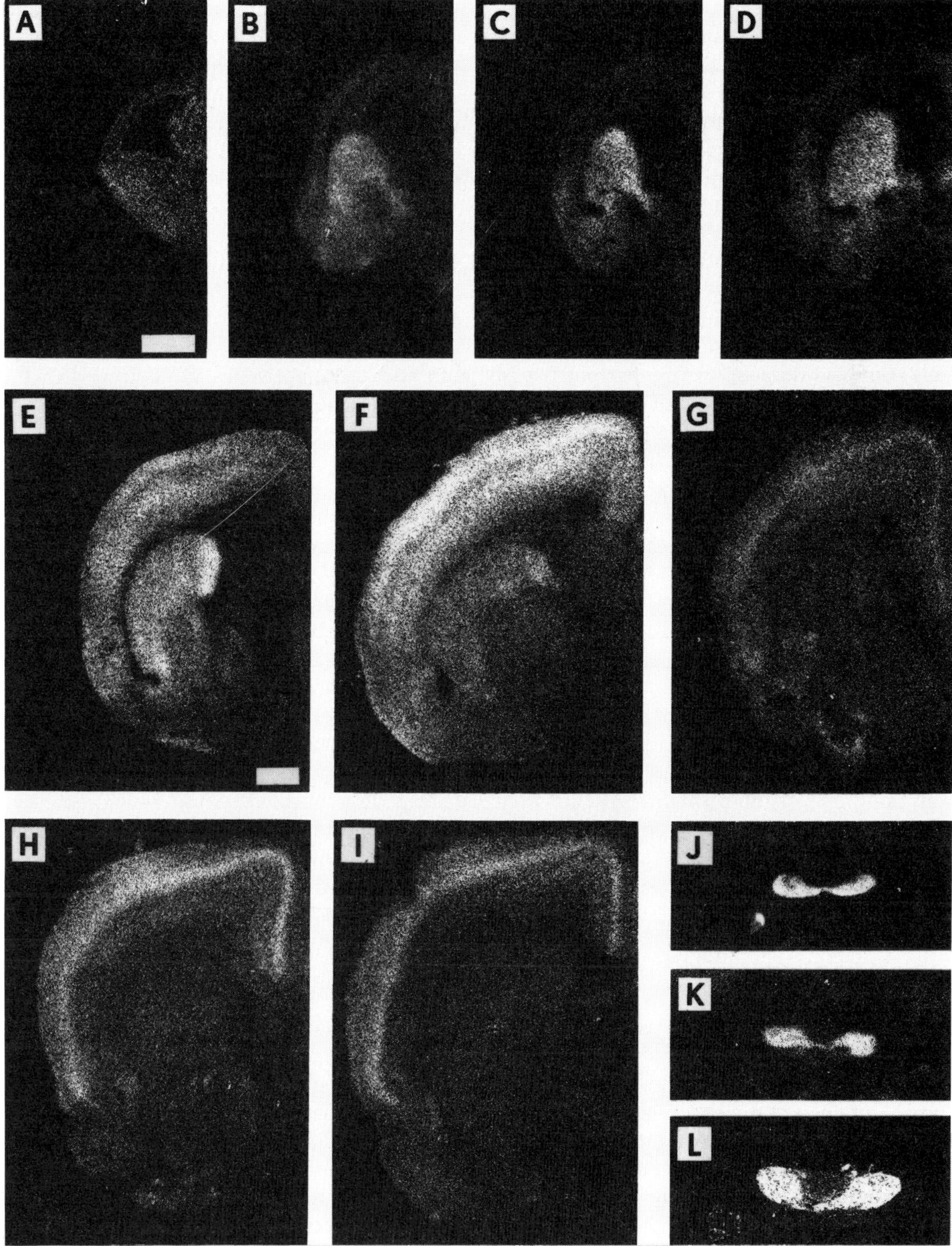

Fig. 9.3. Dark-field autoradiograms of ^{125}I-Tyr^0-oCRH binding to 16μm sections of rat brain and pituitary from different ages. A–D, ED15, ED17, ED19 and ED21 all photographed at the same magnification; E–I, PD2, PD8, PD14, PD21 and PD28 at the same magnification; J–L, pituitary sections of ED19, ED21 and PD14. The light areas indicate regions with the highest binding of ^{125}I-Tyr^0-oCRH. (Reproduced with permission from Insel *et al.*, 1988.)

distribution in cortex can be seen by PD14 (Fig. 9.3G). In comparison with brain receptors, the CRH receptors in the pituitary appear very dense by ED19 (Fig. 9.3J–9.3L) and experience a similar 'overshoot' during the first week of postnatal life (Walker *et al.*, 1986a).

The transient proliferation of receptors in the striatum is probably not a reflection of subsequent decreased density as previously reported with opiate receptors, where these receptors appear dense and homogeneous and then become patchy when nigrostriatal afferents arrive postnatally (Kent *et al.*, 1982). CRH receptors do not form similar patches in the striatum; in fact, these striatal CRH receptors are probably not on migrating neural elements since cortical neurons are formed earlier and migrate from a more medial, periventricular zone. It is conceivable that the transient CRH receptors in the striatum are labeling a neural or synaptic population that is eliminated between PD8 and PD21. It is also possible, however, that these receptors are expressed only transiently in cells that survive. In the light of these data, it becomes imperative that this pattern of CRH receptors be assessed for functionality during this phase of development. This can easily be achieved by examining one of the mechanisms by which the hormone CRH signals in cells, namely the adenylate cyclase system.

9.3.3 FUNCTIONAL LINKAGE WITH ADENYLATE CYCLASE

As described in the previous sections, the primary event in the action of CRH which initiates its effects on target cells is the interaction with cell-surface receptor proteins. These proteins are directly responsible for at least two important functions. First, the receptor must exhibit selectivity for the neurotransmitter (CRH) and second it must mediate a physiologic effect in response to a challenge by CRH. CRH stimulation of POMC-derived peptide secretion is mediated by the activation of adenylate cyclase in both anterior

(Giguere *et al.*, 1982) and neurointermediate (Aguilera *et al.*, 1983) lobes of the pituitary, is dose-related, of high affinity (nanomolar) and exhibits the appropriate pharmacology for CRH-related analogs (Labrie *et al.*, 1982; Aguilera *et al.*, 1983). In a way similar to the pituitary, CRH also stimulates adenylate cyclase activity in the adult rat brain (Chen *et al.*, 1986; Battaglia *et al.*, 1987) and mouse spleen (Webster *et al.*, 1989). The pharmacological potency of various CRH analogs in stimulating adenylate cyclase activity is also consistent with their affinities for CRH receptors and their potencies in eliciting POMC-derived peptide secretion from the pituitary. Based on these observations, there is increasing evidence that CRH initiates a cascade of enzymatic reactions in anterior pituitary cells, beginning with the receptor-mediated stimulation of adenylate cyclase which ultimately regulates POMC-peptide secretion and possibly synthesis. Therefore, during ontogeny, the functionality of CRH receptors can be assessed directly by measuring the ability of CRH receptors to stimulate the adenylate cyclase system.

The actual time at which receptors become linked to their second messenger system during development would be expected to vary depending on the type of receptor, the brain region and the functional significance of the system in development. As shown in Table 9.1, CRH receptors are capable, although at very low levels, of stimulating the production of cAMP from as early as ED17 with the adult levels of stimulation generated by PD2 (Fig. 9.2) suggesting that the system is functional very early in development. However, if the activity is expressed as a function of the large numbers of CRH receptors present during this period of time, the effect seems less dramatic. By expressing the results as a unit of cyclase generated per receptor, adult values of CRH-stimulated adenylate cyclase activity are not reached until PD14 (Fig. 9.2). Furthermore, studies have demonstrated that in the striatum, the region with the highest density of CRH receptors during this period, relatively little cyclase production is apparent until PD8 (Insel *et al.*, 1988). Thus, although the CRH receptors appear capable of stimulating cyclase activity as early as PD2, relatively few receptors actually demonstrate this ability. It is noteworthy that although the catalytic subunit of the adenylate cyclase system appears to be functional by ED17, the levels of basal cAMP production do not reach adult levels until PD21 (Insel *et al.*, 1988). In addition, at this early time, the enzyme can be directly stimulated by forskolin (a direct action at the catalytic unit) as well as by GTP and NaF, presumably through functional G proteins. During embryonic and postnatal development, both the basal enzyme activities and the stimulated enzyme activities increase in parallel, probably due to increases in the number of catalytic subunits. Thus it is possible that the discrepancy observed between the peak of CRH receptor density at PD8 and the peak of CRH-stimulated cAMP production at PD14 (Fig. 9.2) is due to the differential development of all the components required for a fully functional adenylate cyclase system. Further work is necessary in examining the ontogeny of the G proteins involved, since the studies described above can only assess the stimulation of adenylate cyclase which presumably requires a functioning G protein. These data therefore suggest that at least in striatum, the transiently appearing CRH receptors do not signal through the adenylate cyclase system. This is not to say, however, that they are not functional; CRH receptors have been hypothesized to signal through many other pathways in adult brain.

CRH has been shown to stimulate adenylate cyclase in the absence of Ca^{2+} and in the presence of EGTA in brain (Chen *et al.*, 1986; Battaglia *et al.*, 1987), pituitary (Labrie *et al.*, 1982; Holmes *et al.*, 1984) and spleen (Webster *et al.*, 1989) homogenates. This would suggest that CRH-receptor mediated stimulation of cAMP production does not involve a $Ca^{2+}/$ calmodulin mechanism. However, as has been demonstrated for a number of other stimulus–secretion hormone and transmitter systems, the stimulation of ACTH release by CRH and other agents requires Ca^{2+} (reviewed in Abou-Samra *et al.*, 1987). CRH appeared to elevate cytosolic Ca^{2+} in AtT-20 cells through a receptor-mediated mechanism linked to cAMP-dependent protein kinase (Luini *et al.*, 1985; Guild and Reisine, 1987). Other signal transduction mechanisms involving methyltransferases also appear to be linked to CRH receptors as CRH has been shown to increase ACTH secretion, protein carboxylmethylation, and phospholipid methylation in AtT-20 cells (Hook *et al.*, 1982; Heisler *et al.*, 1983). The stimulation of both protein carboxylmethylation and phospholipid methylation by CRH was dose-dependent and exhibited an appropriate pharmacology for the CRH receptor (Hook *et al.*, 1982; Heisler *et al.*, 1983). Preliminary evidence suggests that CRH may also regulate cellular responses through products of arachidonic acid metabolism (Abou-Samra *et al.*, 1986; 1987) where CRH increased the release of [^{3}H]arachidonic acid from prelabeled pituitary cells. Further evidence in anterior pituitary cells suggests that CRH does not directly regulate phosphatidylinositol turnover or protein kinase C activity (Abou-Samra *et al.*, 1987; Guillon *et al.*, 1987). However, stimulation of protein kinase C either directly or by specific ligands (vasopressin or angiotensin II) enhanced CRH-stimulated adenylate cyclase activity, ACTH release, and inhibited phosphodiesterase activity (Abou-Samra *et al.*, 1987; Bilezikjian and Vale, 1987; Labrie *et al.*, 1987). In turn, CRH suppressed arginine vasopressin-stimulated inositol triphosphate levels in rat anterior pituitary cells (Bilezikjian and Vale, 1987). Thus, the effects of CRH on anterior pituitary cells and possibly in neurons and other cell types expressing CRH receptors are likely to involve complex interactions among intracellular second messenger systems. In the light of these reports, further studies are necessary during ontogeny to determine which second messenger system (if any) is activated and at which specific times during prenatal and early postnatal life.

9.4   CRH RECEPTORS IN AGING

The effect of aging has been shown to have a dramatic effect on the normal functioning of the HPA axis in the rat (Sapolsky *et al.*, 1983). In the aged rat, there is a general increase in the activity of the HPA axis manifested by increased basal levels of corticosterone as well as a decrease in the sensitivity to hormonal negative feedback mechanisms. In addition, aged rats demonstrate a distinctive inability to recover from stress (Sapolsky *et al.*, 1983). For example, after exposure to a stressor, aged rats demonstrate an increase in the blood levels of gluococorticoids, which is not due to a decrease in the clearance of glucocorticoid, thus implying a loss in negative feedback control (McEwen *et al.*, 1987). Furthermore, the aged pituitary and adrenal glands exhibit a dampened response to CRH (Hylka *et al.*, 1984) and ACTH (Tang and Phillips, 1978) respectively. CRH has also been implicated in the manifestation of certain age-related dementias or neurodegenerative disorders. Decreases in CRH-like immunoreactivity accompanied by reciprocal changes in CRH receptor binding have been described in post-mortem cerebral cortex samples of Alzheimer's disease patients (Bissette *et al.*, 1985; De Souza *et al.*, 1986; Whitehouse *et al.*, 1987; De Souza, 1988). Additionally, alterations in the levels of CRH in other brain regions have been reported in Parkinson's disease, Huntington's chorea and progressive supranuclear palsy (Whitehouse *et al.*, 1987; De Souza, 1988). In view of the alterations in the HPA axis as a consequence of aging and the changes in CRH observed associated with several neurodegenerative disorders, CRH receptors are likewise affected by the aging process.

In the rat CNS, there is a selective decrease (approximately 30%) of CRH receptor binding activity in the hypothalamus of aged versus young animals (Fig. 9.4). The mechanism for this specific regional decrease is as yet unknown, however, studies have indicated the aged hypothalamus is in a state of hypersecretion of ACTH secretagogues (Scaccianoce *et al.*, 1990) and if the hormone itself is hypersecreted, this could account for the receptor down-regulation observed. On the other hand, a hypersecretion of hypothalamic factors and/or glucocorticoids could cause the death of certain hypothalamic cells containing CRH receptors. CRH receptors are not altered significantly in the cerebral cortex of aged rats confirming the data from post-mortem human tissues demonstrating that no significant relationship exists between age and CRH receptors in a variety of human cerebrocortical brain areas (De Souza *et al.*, 1986).

In the anterior pituitary, there is a very profound age-related change in CRH receptor binding activity with approximately a 60% decrease in receptors in aged rats. The rate of decline of CRH receptors is also age dependent (Heroux *et al.*, 1991). Saturation analyses clearly demonstrated that this change was not due to a alteration in the affinity of the receptor for the hormone but rather from a loss in the receptor density (Fig. 9.5). These data strongly support the observations that the pituitary gland of aged rats demonstrates a reduction in its ability to release ACTH in response to exogenously applied CRH (Hylka *et al.*, 1984). Several mechanisms may be involved in the age-related loss of CRH receptors. One explanation is that the decrease of CRH receptors observed may merely reflect a loss of corticotropes in the anterior pituitary of aged rats. Studies utilizing blot hybridization analysis have demonstrated that there is no difference in the levels of POMC mRNA expression between young and old rats (Heroux *et al.*, 1991). This observation is supported by studies demonstrating no

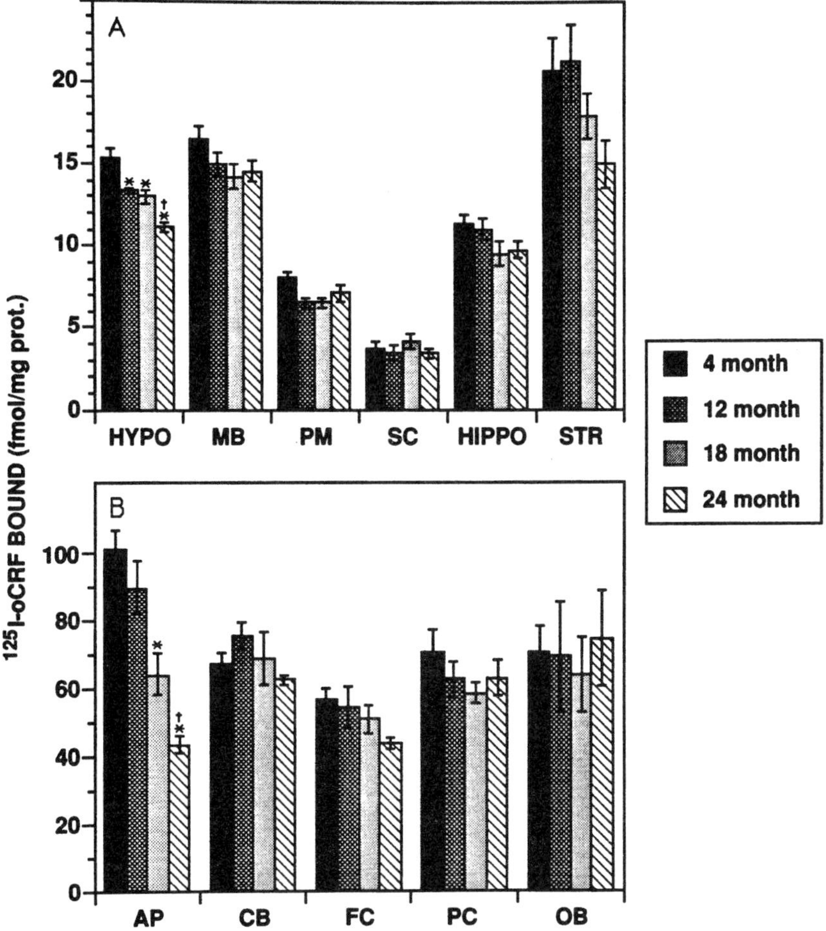

Fig. 9.4.   Effect of age on ^{125}I-Tyr^0-oCRH binding sites in discrete regions of rat brain and in anterior pituitary. Values represent the mean ± SEM; n = 6 rats/group. HYPO, hypothalamus; MB, midbrain; PM, pons medulla; SC, spinal cord; HIPPO, hippocampus; STR, striatum; AP, anterior pituitary; CB, cerebellum; FC, frontal cerebral cortex; PC, parietal cerebral cortex; OB, olfactory bulb. [*] indicates significant difference ($P < 0.01$) from corresponding 4-month-old rats; [†] significantly different at $P < 0.05$ and $P < 0.01$ from corresponding values in 12- and 18-month-old rats respectively. Data were analyzed for differences using one-way ANOVA and Duncan's multiple range test. (Reproduced with permission from Heroux *et al.*, 1991.)

differences in basal or stimulated ACTH release from pituitaries of young or aged rats (Scaccianoce *et al.*, 1990). If the assumption that an equal amount of POMC mRNA is transcribed per corticotrope is true, the data would suggest that there are no changes in the population of corticotropes in young or aged rats.

It is tempting to postulate that since the HPA axis is hyperactivated, hypothalamic hypersecretion of CRH in aged rats (Sapolsky *et al.*, 1983), as well as the elevated levels of circulating glucocorticoids (Scaccianoce *et al.*, 1990), will both contribute to the down-regulation of CRH receptors in the pituitary. Further studies are needed to determine the

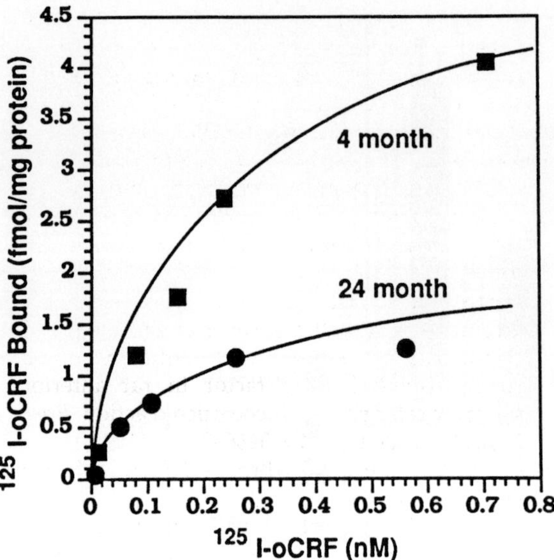

Fig. 9.5.   Saturation profile of ^{125}I-Tyr^0-oCRH binding sites in rat anterior pituitary membranes. The receptor densities (B_{max}) were 3.29 and 1.44 fmol/mg protein and the affinity (K_d) values were 155 and 126 pM for 4- and 24-month-old rats respectively. (Reproduced with permission from Heroux *et al.*, 1991.)

exact mechanism of the observed down-regulation of CRH receptors in both the hypothalamus and pituitary during aging and the consequences of this down-regulation to the aged individual.

9.5   SUMMARY AND CONCLUSIONS

This chapter has examined some major aspects of the ontogeny of brain CRH pathways. CRH receptors and CRH mRNA were found to develop by ED17 and the receptors appeared to show developmentally restricted expression early in postnatal life. CRH receptors appear to peak in density by PD8, where they exhibit a 300% increase in density compared to adult animals primarily due to an overexpression of CRH receptors in the developing striatum. These receptors, however, do not all appear to be linked to the second messenger system adenylate cyclase, as the peak activity for

CRH-stimulated adenylate cyclase does not occur until PD14. Exogenously applied CRH during early postnatal life was associated with a significant and dose-dependent decrease in ultrasonic vocalizations which was only elicited by central and not peripheral administration of the hormone. Although this observation is seen only in early postnatal life and is contrary to the established hypothesis that CRH is anxiogenic, it may reflect the manifestation of what has been termed the stress-non-responsive period. Finally, it is interesting to note that there is a selective decrease in CRH receptors in the hypothalamus and anterior pituitary of aged animals which supports the observation of a hyperactivation of the HPA axis during aging. Clearly there are many unanswered questions regarding the role of CRH during development and the changes that occur with increasing age. Further work should be focused on the relationship between CRH receptors and their association with various second messenger systems in order to gain some insight into the specific role that CRH plays in the developing organism.

REFERENCES

Abou-Samra, A.-B., Catt, K.J. and Aguilera, G. (1986) Role of arachidonic acid in the regulation of adrenocorticotropin release from rat anterior pituitary cells in culture. *Endocrinology*, **119**, 1427–31.

Abou-Samra, A.-B., Harwood, J.P., Catt, K.J. *et al.* (1987) Mechanisms of action of CRF and other regulators of ACTH release in pituitary corticotrophs. *Ann. NY Acad. Sci.*, **512**, 67–84.

Aguilera, G., Harwood, J.P., Wilson, J.X. *et al.* (1983) Mechanisms of action of corticotropin-releasing factor and other regulators of corticotropin release in rat pituitary cells. *J. Biol. Chem.* **258**, 8039–44.

Bardo, M.T., Schmidt, R.H. and Bhatnagar, R.K. (1985) Effects of morphine on sprouting of locus coeruleus fibers in the neonatal rat. *Dev. Brain Res.*, **22**, 161–8.

Bartolome, J.V., Bartolome, M.B., Daltner, L.A. *et al.* (1986) Effects of β-endorphin on ornithine decarboxylase in tissues in developing rats: a potential role for this endogenous neuropeptide in the modulation of tissue growth. *Life Sci.*, **38**, 2355–62.

Battaglia, G., Webster, E.L. and De Souza, E.B. (1987) Characterization of corticotropin-releasing factor receptor-mediated adenylate cyclase activity in the rat central nervous system. *Synapse*, **1**, 572–81.

Bilezikjian, L.M. and Vale, W.W. (1987) Regulation of ACTH secretion from corticotrophs: the interaction of vasopressin and CRF. *Ann. NY Acad. Sci.*, **512**, 85–96.

Bissette, G., Reynolds, G.P., Kilts, C.D. *et al.* (1985) Corticotropin-releasing factor-like immunoreactivity in senile dementia of the Alzheimer type. *J. Am. Med. Ass.*, **254**, 3067–9.

Bresson, J.L., Clavequin, M.C., Fellmann, D. *et al.* (1987) Human corticoliberin hypothalamic neuroglandular system: comparative immunocytochemical study with anti-rat and anti-ovine corticotropin-releasing factor sera in the early stages of development. *Dev. Brain Res.*, **32**, 241.

Bugnon, C., Fellmann, D., Gouget, A. *et al.* (1982) Ontogeny of the corticoliberin neuroglandular system in rat brain. *Nature*, **298**, 159–61.

Bugnon, C., Fellmann, D., Bloch, B. *et al.* (1987) Contribution of immonocytochemistry to the study of the development of neuroglandular peptidergic systems in the human fetal hypothalamus *Ann. Endocrinol. (Paris)*, **48**, 343–51.

Chen, F.M., Bilezikjian, L.M., Perrin, M.H. *et al.* (1986) CRF receptor mediated stimulation of adenylate cyclase activity in rat brain. *Brain Res.* **381**, 49–57.

De Souza, E.B. (1987) Corticotropin-releasing factor receptors in the rat central nervous system: characterization and regional distribution. *J. Neurosci.*, **7**, 88–100.

De Souza, E.B. (1988) CRH defects in Alzheimer's and other neurologic diseases. *Hosp. Pract.*, **23**, 59–71.

De Souza, E.B., Perrin, M.H., Rivier, J.E. *et al.* (1984) Corticotropin-releasing factor receptors in rat pituitary gland: autoradiographic localization. *Brain Res.*, **296**, 202–7.

De Souza, E.B., Insel, T.R., Perrin, M.H. *et al.* (1985a) Differential regulation of corticotropin-releasing factor receptors in anterior and intermediate lobes of pituitary and in brain following adrenalectomy in rats. *Neurosci. Lett.*, **56**, 121–8.

De Souza, E.B., Perrin, M.H., Whitehouse, P.J. *et al.* (1985b) Corticotropin-releasing factor receptors in human pituitary gland: autoradiographic localization. *Neuroendocrinology* **40**, 419–22.

De Souza, E.B., Whitehouse, P.J., Kuhar, M.J. *et al.* (1986) Reciprocal changes in corticotropin-releasing factor (CRF)-like immunoreactivity and CRF receptors in cerebral cortex of Alzheimer's disease. *Nature*, **319**, 593–5.

De Souza, E.B., Grigoriadis, D.E. and Webster, E.L. (1991) Role of brain, pituitary and spleen corticotropin-releasing factor receptors in the stress response. *Methods Achiev. Exp. Pathol.*, **14**, 23–44.

Giguere, V., Labrie, F., Cote, J. *et al.* (1982) Stimulation of cyclic AMP accumulation and corticotropin release by synthetic ovine corticotropin-releasing factor in rat anterior pituitary cells: site of glucocorticoid action. *Proc. Natl. Acad. Sci. USA*, **79**, 3466–9.

Goland, R.S., Starle, S.L., Brown, L.S. *et al.* (1986) High levels of CRF immunoreactivity in maternal and fetal plasma during pregnancy. *J. Clin. Endocrinol. Metab.*, **63**, 199–203.

Gonzales, B.J., Leroux, P., Bodenant, C. *et al.* (1991) Ontogeny of somatostatin receptors in the rat somatosensory cortex. *J. Comp. Neurol.* **305**, 177–88.

Grigoriadis, D.E. and De Souza, E.B. (1989) Corticotropin-releasing factor receptors in intermediate lobe of the pituitary: biochemical characterization and autoradiographic localization. *Peptides*, **10**, 179–88.

Guild, S. and Reisine, T. (1987) Molecular mechanisms of corticotropin-releasing factor stimulation of calcium mobilization and adrenocorticotropin release from anterior pituitary tumor cells. *J. Pharmacol. Exp. Ther.*, **241**, 125–30.

Guillon, G., Gaillard, R.L., Kehrer, P. *et al.* (1987) Vasopressin and angiotensin induce inositol lipid breakdown in rat adrenohypophysial cells in primary culture. *Regul. Pept.*, **18**, 119–29.

Hammer R.P. (1985) Ontogeny of opiate receptors in the rat medial preoptic area: critical periods in regional development. *Int. J. Dev. Neurosci.*, **3**, 541–8.

Heisler, S., Hook, V.Y.H. and Axelrod, J. (1983) Corticotropin-releasing factor stimulation of protein carboxylmethylation in mouse pituitary tumor cells. *Biochem. Pharmacol.*, **32**, 1295–9.

Herkenham, M. (1987) Mismatches between neurotransmitter and receptor localizations in brain: observations and implications. *Neuroscience*, **23**, 1–38.

Heroux, J.A., Grigoriadis, D.E. and De Souza, E.B. (1991) Age-related decreases in corticotropin-releasing factor (CRF) receptors in rat brain and anterior pituitary gland. *Brain Res.*, **542**, 155–8.

Holmes, M.C., Antoni, F.A. and Szentendrei, T. (1984) Pituitary receptors for corticotropin-releasing factor: no effect of vasopressin on binding or activation of adenylate cyclase. *Neuroendocrinology*, **39**, 162–9.

Hook, V.Y.H., Heisler, S. and Axelrod, J. (1982) Corticotropin-releasing factor stimulates phospholipid methylation and corticotropin secretion in mouse pituitary tumor cells. *Proc. Natl. Acad. Sci. USA*, **79**, 6220–4.

Hylka, V.W., Sonntag, W.E. and Meites, J. (1984) Reduced ability of old male rats to release ACTH and corticosterone in response to CRF administration. *Proc. Soc. Exp. Biol. Med.*, **175**, 1–4.

Insel, T.R. (1990) Brain corticotropin-releasing factor and development, in *Corticotropin-Releasing Factor: Basic and Clinical Studies of a Neuropeptide* (eds E.B. De Souza and C.B. Nemeroff), CRC Press, Boca Raton, FL, pp. 91–106.

Insel, T.R. and Harbaugh, C.R. (1989) Central administration of corticotropin-releasing factor alters rat pup isolation calls. *Pharmacol. Biochem. Behav.*, **32**, 197–201.

Insel, T.R. and Hill, J.L. (1987) Infant separation distress in genetically fearful rats. *Biol. Psychiatr.*, **22**, 783–9.

Insel, T.R., Hill, J.L. and Mayor, R.B. (1986) Rat pup isolation calls: possible mediation by the benzodiazepine receptor complex. *Pharmacol. Biochem. Behav.*, **24**, 1263–7.

Insel, T.R., Battaglia, G., Fairbanks, D.W. *et al.* (1988) The ontogeny of brain receptors for corticotropin-releasing factor and the development of their functional association with adenylate cyclase. *J. Neurosci.*, **8**, 4151–8.

Kent, J.L., Pert, C.B. and Herkenham, M. (1982) Ontogeny of opiate receptors in rat forebrain: visualization by *in vitro* autoradiography. *Dev. Brain Res.*, **2**, 487–504.

Kiyama, H., Inagaki, S., Kito, S. *et al.* (1987) Ontogeny of [³H]neurotensin binding sites in the rat cerebral cortex: autoradiographic study. *Dev. Brain Res.*, **31**, 303–6.

Ko, G.N., Wilkox, B.I., Petracca, F.M. *et al.* (1989) Localization and measurement of neurotransmitter receptors in rat and human brain by quantitative autoradiography. *Comput. Med. Imaging Graph*, **13**, 37–45.

Koob, G.F. (1985) Stress, corticotropin-releasing factor and behavior. *Perspect. Behav. Med.*, **2**, 39–52.

Kuhar, M.J. (1981) Autoradiographic localization of drug and neurotransmitter receptors in the brain. *Trends Neurosci.*, **4**, 60–4.

Kuhar, M.J. (1985) Receptor localization with the microscope, in *Neurotransmitter Receptor Binding* (eds H.I. Yamamura, S.J. Enna and M.J. Kuhar), Raven Press, New York, pp. 153–76.

Kuhar, M.J., De Souza, E.B. and Unnerstall, J.R. (1986) Neurotransmitter receptor mapping by autoradiography and other methods. *Annu. Rev. Neurosci.*, **9**, 27–59.

Labrie, F., Gagne, B. and Lefevre, G. (1982) Corticotropin-releasing factor stimulates adenylate cyclase activity in anterior pituitary. *Life Sci.*, **31**, 1117–21.

Labrie, F., Giguere, V., Meunier, H. *et al.* (1987) Multiple factors controlling ACTH secretion at the anterior pituitary level. *Ann. NY Acad. Sci.*, **512**, 97–114.

Luini, A., Lewis, D., Guild, S. *et al.* (1985) Hormone secretagogues increase cytosolic calcium by increasing cAMP in corticotropin-secreting cells. *Proc. Natl. Acad. Sci.*, **82**, 8034–8.

McEwen, B., Chao, H., Spencer, R. *et al.* (1987) Corticosteroid receptors in brain: relationship of receptors to effects in stress and aging. *Ann. NY Acad. Sci.*, **512**, 394–401.

Petracca, F.M., Baskin, D.G., Diaz, J. and Dorsa, D.M. (1968) Ontogenetic changes in vasopressin binding site distribution in rat brain: an autoradiographic study. *Dev. Brain Res.*, **28**, 63–8.

Quirion R. and Dam, T.V. (1986) Ontogeny of substance P receptor binding sites in rat brain. *J. Neurosci.*, **6**, 2187–99.

Rivier, J., Spiess, J. and Vale, W. (1983) Characterization of rat hypothalamic corticotropin-releasing factor. *Proc. Natl. Acad. Sci. USA*, **80**, 4851–5.

Sales, N., Martres, M.P., Bouthenet, M.L. *et al.* (1989) Ontogeny of dopaminergic D-2 receptor in the rat nervous system: characterization and detailed autoradiographic mapping with [¹²⁵I] iodosulpiride. *Neuroscience*, **28**, 673–700.

Sapolsky, R.M. and Meaney, K.J. (1986) Maturation of the adrenocortical response: neuroendocrine control mechanisms and the stress hyperresponsive period. *Brain Res. Rev.* **11**, 65–76.

Sapolsky, R.M., Krey, L.C. and McEwen, B.S. (1983) The adrenocortical stress response in the aged male rat: impairment of recovery from stress. *Exp. Gerontol.*, **18**, 55–64.

Sasaki, A., Shinkawa, O., Margioris, A.N. *et al.* (1987) Immunoreactive CRF in human plasma during pregnancy, labor and delivery. *J. Clin. Endocrinol. Metab.*, **64**, 224–9.

Scaccianoce, S., Di Sciullo, A. and Angelucci, L. (1990) Age-related changes in hypothalamo–

pituitary–adrenocortical axis activity in the rat. *Neuroendocrinology*, **52**, 150–5.

Shapiro, L.E. and Insel, T.R. (1989) Ontogeny of oxytocin receptors in rat forebrain: a quantitative study. *Synapse*, **4**, 259–66.

Shyr, S.W., Crowley, W.R. and Grosvenor, C.E. (1986) Effect of neonatal prolactin deficiency on prepubertal tuberoinfundibular and tuberohypophyseal dopaminergic neuronal activity. *Endocrinology*, **119**, 1217–21.

Stanton, M.E., Wallstrom, E.J. and Levine, S. (1987) Maternal contact inhibits pituitary adrenal stress responses in preweanling rats. *Dev. Psychobiol.*, **20**, 131–5.

Tang, G. and Phillips, R. (1978) Some age-related changes in pituitary adrenal function in the male laboratory rat. *J. Gerontol.*, **33**, 377–82.

Vale, W., Rivier, C., Brown, M.R. *et al.* (1983) Chemical and biological characterization of corticotropin-releasing factor. *Recent Prog. Horm. Res.*, **39**, 245–70.

Vale, W., Spiess, J., Rivier, C. *et al.* (1981) Characterization of a 41-residue ovine hypothalamic peptide that stimulates secretion of corticotropin and β-endorphin. *Science*, **213**, 1394–7.

Walker, C.D., Perrin, M., Vale, W. *et al.* (1986a) Ontogeny of the stress response in the rat: role of the pituitary and the hypothalamus. *Endocrinology*, **118**, 1445–51.

Walker, C.D., Sapolsky, R.M., Meaney, M.J. *et al.* (1986b) Increased pituitary sensitivity to glucocorticoid feedback during the stress non responsive period in the neonatal rat. *Endocrinology*, **119**, 1816–21.

Webster, E. L., Battaglia, G. and De Souza, E.B. (1989) Functional corticotropin-releasing factor (CRF) receptors in mouse spleen: evidence from adenylate cyclase studies. *Peptides*, **10**, 395–401.

Webster, E.L., Grigoriadis, D.E. and De Souza, E.B. (1991) Corticotropin-releasing factor receptors in the brain–pituitary–immune axis, in *Stress, Neuropeptides, and Systemic Disease* (eds J.A. McCubbin, P.G. Kaufmann and C.B. Nemeroff), Academic Press, San Diego, pp. 233–60.

Whitehouse, P.J., Vale, W.W., Zweig, R.M. *et al.* (1987) Reductions in corticotropin releasing factor-like immunoreactivity in cerebral cortex in Alzheimer's disease, Parkinsons's disease, and progressive supranuclear palsy. *Neurology*, **37**, 905–9.

CORTICOSTEROID RECEPTORS IN THE RAT BRAIN AND PITUITARY DURING DEVELOPMENT AND HYPOTHALAMIC–PITUITARY–ADRENAL FUNCTION

Michael J. Meaney, Dajan O'Donnell, Victor Viau, Seema Bhatnagar, Alain Sarrieau, James Smythe, Nola Shanks and Claire-Dominique Walker

With the integration of receptor theory into the neurosciences, it became clear that the regulation of neurotransmission can occur at the level of the receptor and receptor-coupled events, as well as at the level of transmitter synthesis and release. Likewise, signal transduction in endocrine systems is subject to regulation by changes in receptor function. This chapter summarizes the variations in corticosteroid receptor systems over development in relation to hypothalamic–pituitary–adrenal (HPA) function. Variations in corticosteroid receptor systems appear to underlie the complex pattern of HPA activity that is observed in the neonatal rat under both basal and stressful conditions (Fig. 10.1). Indeed, the complexity of HPA function in the neonate and the differences between the neonate and the adult actually deceived researchers into thinking that the HPA axis of the neonatal rat was unresponsive to stress during the first two weeks of life (e.g. Sapolsky and Meaney, 1986).

In addition, there is considerable plasticity in the corticosteroid receptor systems during the first weeks of life. This plasticity is reflected in changes in the development of corticosteroid receptors in specific brain regions that occur in response to environmental events. Thus, environmental stimuli can permanently alter corticosteroid receptor density in brain regions which regulate HPA responses to stress. These changes persist throughout the life of the animal, and are accompanied by altered endocrine responses to conditions which threaten homeostasis. Such developmental effects probably represent one way in which early life events can predispose an individual to pathology in later life (Meaney *et al.*, 1988a).

10.1 CORTICOSTEROID RECEPTOR ACTION

Corticosteroids exert their intracellular effects via soluble, high affinity corticosteroid receptors. As discussed later there are at least two corticosteroid receptor subtypes, the mineralocorticoid and glucocorticoid receptors, also

Receptors in the Developing Nervous System Vol. 1: *Growth factors and hormones* Edited by Ian S. Zagon and Patricia J. McLaughlin. Published in 1993 by Chapman & Hall. ISBN 0 412 45240 5. Vols. 1 and 2 (set) ISBN 0 412 54520 9.

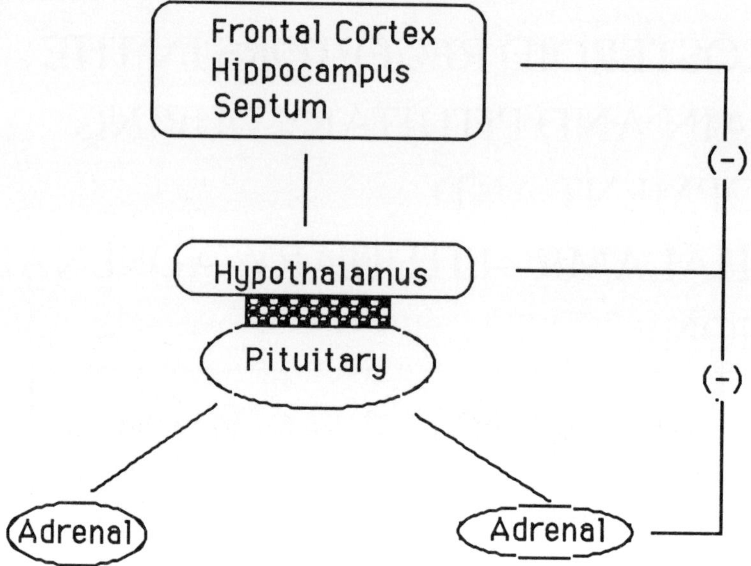

Fig. 10.1   A schema of the HPA axis depicting activational influences emerging from the hypothalamus in the form of ACTH secretagogues, such as CRH, AVP, OT etc., resulting in the release of ACTH from the pituitary, the adrenal glucocorticoids, and corticolimbic regions which are thought to mediate inhibitory glucocorticoid negative-feedback inhibition of HPA activity.

referred to as type I and type II corticosteroid receptors, respectively. The importance of corticosteroid action is indicated by the presence of corticosteroid receptors in virtually every cell type in the body (Burnstein and Cidlowski, 1989). The cloning of the human (Hollenberg *et al.*, 1985), rat (Miesfield *et al.*, 1984) and mouse (Danielson *et al.*, 1986) glucocorticoid receptors and the human mineralocorticoid receptor (Arriza *et al.*, 1988) marked a period of tremendous progress in our knowledge of steroid receptor action. Figure 10.2 provides a summary of the basic features of the prevailing understanding of steroid receptor action (see Carson-Jurica *et al.*, 1990). These receptors belong to a family of ligand-dependent transcription factors. Briefly, the steroid diffuses across the cell membrane where it binds to soluble, unoccupied receptor sites. The receptor is then 'activated', a process which corresponds to the dissociation of a 90 kDa heat-shock protein, into a form which is then capable of binding to specific DNA

sequences referred to as hormone regulatory (or responsive) elements (Pratt *et al.*, 1989). Receptors complexed with a heat-shock protein are unable to bind DNA. Corticoid effects on the transcription of target genes is then mediated by an interaction between the hormone–receptor complex and specific DNA sequences. The effect is then reflected in terms of changes in mRNA transcription and protein synthesis. It should be noted that this summary does not include any reference to the effects of various intracellular factors such as protein kinases and the early gene products, *cfos/jun*, which can regulate glucocorticoid receptor function. Indeed, there are a number of potential sites for the regulation of corticosteroid receptor action (Burnstein and Cidlowski, 1989).

Steroid receptors are composed of discrete, functional domains (see Evans and Arriza, 1989 for a review). For the glucocorticoid receptor, ligand binding is localized at the C-terminal portion of the receptor. The role for

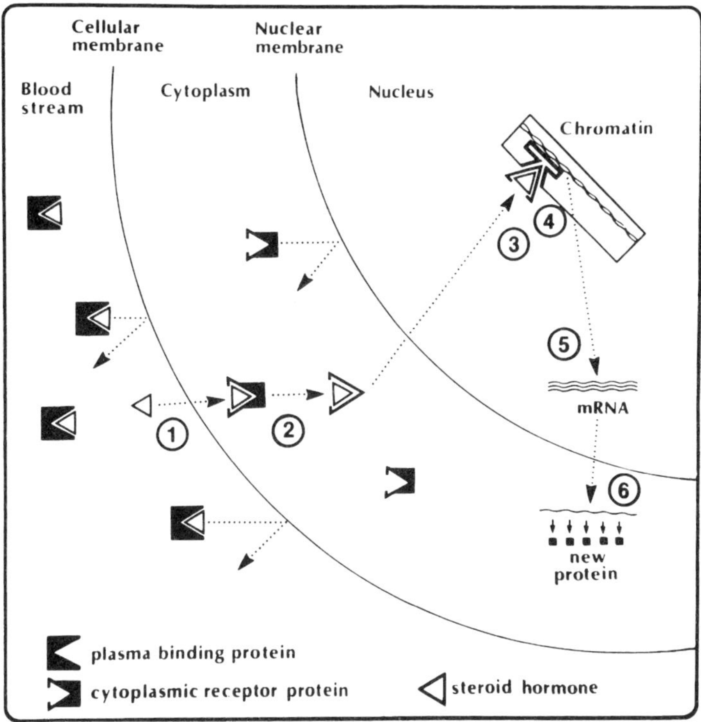

Fig. 10.2.   A simplified model of steroid receptor action (see text). (1) The free steroid (i.e. steroid not bound to plasma binding proteins) diffuses into the cell, (2) where it binds to an unoccupied, intracellular receptor (located either in the cytoplasmic or nuclear compartment) resulting in an 'activated' hormone–receptor complex (3–4). This complex then binds to specific DNA target sequences resulting in (5) the regulation of mRNA production and (6) protein synthesis.

the amino portion of the receptor is less clear. This region may serve a modulatory role in gene transcription. The DNA binding domain lies in the mid-portion of the receptor and is an especially well-conserved region. Indeed, the homology among steroid receptors has led to the speculation that they are derived from a common ancestral gene.

An essential feature of this model is that it predicts that receptor density (or binding capacity) is a major determinant of target cell sensitivity to glucocorticoids. In the adult animal this appears to be true in many, but not all cases (Burnstein and Cidlowski, 1989; Ringold, 1985). The presence of corticosteroid receptors in a cell is a prerequisite for a glucocorticoid response. Moreover, variations in corticosteroid receptor levels have often been found to predict the magnitude of cellular responses to glucocorticoids. These conclusions are also supported by studies identifying glucocorticoid receptor anomalies as a major source of naturally occurring syndromes of glucocorticoid resistance. However, it is not known at present whether the efficacy of the processes underlying receptor activation and DNA binding are the same in the neonate as in the adult. In the absence of relevant studies on this topic, it is assumed that variations in receptor density in neonates do indeed result in changes in tissue sensitivity to glucocorticoids.

10.2 THE HPA AXIS

The adrenal glucocorticoids, together with the medullary catecholamines, comprise a front-line of defense for mammalian species under conditions which threaten homeostasis (conditions commonly referred to as stress, Selye, 1950). These hormones serve as major endocrine regulators of carbohydrate and lipid metabolism, cardiovascular tone, muscle function, immunocompetence, and behavior. The adrenal glucocorticoids represent the endproduct of HPA activation. Under most conditions, this axis lies under the dominion of specific peptides secreted by neurons located in the paraventricular nuclei (PVN) of the hypothalamus. Most notable among these releasing factors is corticotropin-releasing hormone (CRH). CRH neurons from the parvocellular cells of the PVN project almost exclusively to the portal capillary zone of the median eminence, furnishing a potent excitatory signal for the synthesis and release of adrenocorticotropin (ACTH) from the anterior pituitary. These inputs, along with co-secretagogues such as arginine vasopression (AVP) and oxytocin (OT) provide the chemical signals by which neural inputs are coded into endocrine signals through their actions on pituitary corticotrophs (Plotsky and Vale, 1984; Antoni, 1986; Gibbs, 1986; Rivier and Plotsky, 1986; Plotsky, 1987). Dynamic changes in ACTH levels occurring under resting state conditions or in response to stress are associated with changes in the secretion of one or more of these secretagogues from hypothalamic neurons.

Circulating glucocorticoid levels rise and fall in concert with a circadian rhythm. Generally, glucocorticoid levels increase in anticipation of the locomotor activity and feeding associated with the foraging that follows awakening. In diurnal species, such as humans, this refers to the early AM period (or very, very late PM period for many graduate students). In nocturnal species, such as the rat, the peak in plasma glucocorticoid levels occurs in the early PM. This rhythm is pronounced; in most species, glucocorticoid levels during the active period of the day are about 2–3 times higher than during the inactive period, whereas peak to trough ratios are around 10 to 1. Hence, even under basal conditions, there is a very significant change in the glucocorticoid signal over the circadian cycle. In addition to the influence of light/dark cycles, the rhythm in HPA activity is also linked to food and water intake (e.g., Johnson and Levine, 1973; Kreiger, 1974; Kreiger and Hauser, 1978; Wilkinson *et al.*, 1979). This finding underscores the important relationship between the adrenal corticosteroids and metabolism.

The HPA is exquisitely sensitive to stress (Selye, 1950). The increased secretion of CRH and co-secretagogues such as AVP and OT during stress enhances the synthesis and release of pro-opiomelanocortin (POMC)-derived peptides, including ACTH and β-endorphin. Elevated plasma ACTH levels increase the synthesis and release of adrenal glucocorticoids. The magnitude of the glucocorticoid response varies greatly depending on the species and the intensity of the stressor. In humans, major stress responses generally involve about a three-fold increase and in the rat, as much as a 10-fold increase in circulating glucocorticoids (note, the principal glucocorticoid in humans is cortisol, whereas in the rat it is corticosterone). The highly catabolic glucocorticoids induce lipolysis, increasing free fatty acid and triglyceride levels, glycogenolysis, increasing blood glucose levels, and protein catabolism, which increases amino acid availability as substrates for gluconeogenesis, further increasing blood glucose (Munck *et al.*, 1984; Brindley and Rolland, 1989). Together, these actions assist the organism under stressful conditions by increasing the availability of energy substrates. The glucocorticoids also suppress immunological responses (Munck *et al.*, 1984), protecting against the occurrence of inflammation at a time when mobility may be important to the animal, such as situations requiring flight or fight responses.

However, prolonged exposure to elevated glucocorticoid levels can lead to a suppression of anabolic processes, muscle atrophy, decreased sensitivity to insulin and steroid-induced diabetes, hypertension, hyperlipidemia, hypercholesterolemia, arterial disease, amenorrhea, impotency, and the impairment of growth and tissue repair, as well as immunosuppression (Munck *et al.*, 1984, Baxter and Tyrrell, 1987; Brindley and Rolland, 1989). Thus, once the stressor is terminated, it is in the animal's interest to 'turn off' the HPA stress response and the efficacy of this process is determined by the ability of glucocorticoids to inhibit ACTH release (i.e. glucocorticoid negative-feedback).

Circulating glucocorticoids feedback into the pituitary and specific brain regions to inhibit secretory activity and thus synthesis and release of ACTH from the anterior pituitary (Keller-Wood and Dallman, 1984; Plotsky and Vale, 1984; Plotsky *et al.*, 1986; Dallman *et al.*, 1987). In addition to the more obvious target sites, such as the pituitary and the medial-basal hypothalamus, there is now considerable evidence for the importance of extrahypothalamic regions in the inhibition of HPA activity (Dallman *et al.*, 1987). Most notable among these regions is the hippocampus (Sapolsky *et al.*, 1986; Jacobson and Sapolsky, 1991 for reviews). Hippocampal lesions or ablations are associated with elevated corticosterone levels under both basal, stress, and post-stress conditions (e.g., Feldman and Conforti, 1976; Fischette *et al.*, 1980; Wilson *et al.*, 1980; Sapolsky *et al.*, 1984a). Moreover, hippocampectomized animals show reduced suppression of ACTH following exogenous glucocorticoid administration (Feldman and Conforti, 1980) and increased CRH and AVP mRNA levels in the PVN of the hypothalamus (Herman *et al.*, 1989). In adrenalectomized rats, the accompanying increase in plasma ACTH is attenuated by corticosterone implants into the hippocampus (Kovacs and Makara, 1988). These findings, together with the fact that the hippocampus is rich in corticosteroid receptors (McEwen *et al.*, 1986b), suggest that this structure is involved in the inhibitory influence of glucocorticoids over adrenocortical activity.

Evidence from a number of models indicates that a decrease in hippocampal corticosteroid receptor density is associated with a hypersecretion of corticosterone, both under basal conditions and following the termination of stress (i.e. less effective negative feedback). There are decreased levels of hippocampal corticosteroid receptors binding in the aged rat (Ritger *et al.*, 1984; Meaney *et al.*, 1988a, 1991, 1992; Reul *et al.*, 1988; Peiffer *et al.*, 1991) and lactating rats (Meaney *et al.*, 1989a), and these animals also hypersecrete corticosterone (Goldman *et al.*, 1978; Sapolsky *et al.* 1983a; Stern and Levine, 1974). Perhaps the most impressive evidence comes from studies with the vasopressin-deficient, Brattleboro rat (Sapolsky *et al.*, 1984b). These animals show a deficit in corticosteroid receptors in the hippocampus and pituitary and hypersecrete corticosterone following stress. The hippocampal receptor deficit is reversed with vasopressin treatment and, so long as the treatment is continued, receptor levels remain elevated and the animals exhibit normal corticosterone secretion following stress (Sapolsky *et al.*, 1984b). Thus, a decrease in corticosteroid receptor density results in a proportional loss in sensitivity to glucocorticoids. When this occurs in structures that mediate glucocorticoid negative-feedback regulation over HPA activity, there ensues an increase in the release of pituitary ACTH and adrenal glucocorticoids, which is particularly evident under stressful conditions.

10.3 CORTICOSTEROID RECEPTORS

The uptake of corticosterone in rat brain is associated with at least two distinct corticosteroid receptor subtypes (e.g., Veldhuis *et al.*, 1982; Beaumont and Fanestil, 1983; Reul and De Kloet, 1985, 1986; Emadian *et al.*, 1986; Funder and Sheppard, 1987; Reul *et al.*, 1987a, b; Sheppard and Funder, 1987; Sarrieau *et al.*, 1988a). The mineralocorticoid (or type I) receptor binds *in vitro* to both corticosterone and the mineralocorticoids, aldosterone, spironolactone, and RU 26752, with high affinity, and binds the synthetic glucocorticoid, RU 28362, with very low affinity. In the brain, mineralocorticoid receptor density is highest in the septal-hippocampal system. The glucocorticoid (or type II) receptor, which is far more diffusely distributed throughout the brain, binds corticosterone, dexamethasone, and RU 28362 with high affinity, and RU 26752 and aldosterone with lower affinity. Although both receptors bind corticosterone with high affinity, the K_d of the mineralocorticoid receptor for corticosterone (~0.5 nM) is lower than that of the glucocorticoid receptor (~2.0–5.0 nM) (Reul and De Kloet, 1985).

A physiological consequence of this difference in affinity for corticosterone is the fact that these receptors then show different rates of occupancy under basal corticosterone levels. About 80–90% of the mineralocorticoid sites are occupied under moderate basal corticosterone levels (~5 µg/dl; Reul and De Kloet, 1985; Reul *et al.*, 1987a, b). This renders the hippocampal mineralocorticoid receptor relatively insensitive to dynamic variations in corticosterone levels above those seen in the resting state. In contrast, the glucocorticoid receptor is highly responsive to dynamic changes in corticosterone titers, such as those occurring during the circadian peak or stress (Reul and De Kloet, 1985; Reul *et al.*, 1987a, b; Meaney *et al.*, 1988b). Under conditions of basal circulating (2–5 µg/dl) corticosterone only about 10–15% of the glucocorticoid receptors are occupied. Stress results in a dramatic increase in the hormone-receptor signal, such that immediately after a 20 min period of restraint about 75% of glucocorticoid receptors are occupied. Cortico-

sterone injections (15 mg/kg) that mimic the steroid levels seen during stress (~30 µg/dl) also result in about 75% occupancy of glucocorticoid receptors. These findings, together with the known negative-feedback efficacy of the synthetic corticoids such as dexamethasone, once thought to bind selectively to the glucocorticoid receptor, suggested that it was this site, and not the mineralocorticoid-like receptor, that underlies the negative-feedback actions of glucocorticoids.

However, evidence has been provided for the involvement of both hippocampal mineralocorticoid and glucocorticoid receptors in the regulation of ACTH levels in rats. In these studies hippocampal implants of both the glucocorticoid receptor antagonist, RU 38486, and the mineralocorticoid receptor antagonist, RU 26752, resulted in raised levels of plasma ACTH (Bradbury and Dallman, 1989). Moreover, Ratka *et al.* (1989) have found that systemic injections of either antagonist resulted in elevated post-stress corticosterone levels in intact rats. These findings suggest that glucocorticoid negative feedback may involve both mineralocorticoid and glucocorticoid receptor sites in the hippocampus. In addition, it should also be noted that although there is at present considerable emphasis on the role of the hippocampus and hippocampal corticosteroid receptors in glucocorticoid negative-feedback regulation of HPA function, several other cortico-limbic sites are likely to be relevant. Kovacs and Makara (1988) reported that glucocorticoid implants into the septum attenuated adrenalectomy-induced increases in plasma ACTH. Moreover, Diorio *et al.* (unpublished) have found that corticoid implants into the frontal cortex-cingulate gyrus dampen plasma ACTH responses to restraint stress. Interestingly, these regions contain both mineralocorticoid and glucocorticoid receptor sites (Diorio *et al.*, unpublished; Meaney and Aitken, 1985a).

Likewise, there is evidence for the importance of both mineralocorticoid and glucocorticoid receptor mediation of negative-feedback over basal HPA activity. Studies of the glucocorticoid regulation of ACTH release have revealed a shift in negative-feedback sensitivity to corticosterone between the AM (light) and PM (dark) phases of the cycle (Akana *et al.*, 1986; Dallman *et al.*, 1987, 1989). Thus, the IC_{50} of free corticosterone for the normalization of ACTH levels in adrenalectomized animals during the AM phase of the day is about 0.6 nM, a good approximation of the K_d of the mineralocorticoid receptor for corticosterone. Moreover, aldosterone, an endogenous, high affinity ligand for mineralocorticoid receptors, also normalizes ACTH in adrenalectomized animals (Dallman *et al.*, 1989). During the PM period there is a functional shift in negative-feedback sensitivity, such that higher concentrations of free corticosterone are required to normalize ACTH in adrenalectomized rats (Akana *et al.*, 1986; Dallman *et al.*, 1989). These concentrations (~4.0 nM) approximate the K_d of the glucocorticoid receptor for corticosterone. It is not known whether this shift is receptor mediated, or is simply a function of greater CNS drive on pituitary ACTH secretion in the PM phase. However, these data are also consistent with the findings of human studies showing that RU 38486, the glucocorticoid receptor antagonist, increases plasma cortisol levels during the AM phase of the cycle (the period of peak, basal HPA activity in humans), but not in the PM phase (Gaillard *et al.*, 1984). These data suggest that negative-feedback regulation of ACTH release during (and following) the peak in basal activity involves the glucocorticoid receptor (as suggested by Dallman *et al.*, 1987). This would not be surprising, since corticosterone levels during the PM peak are comparable to those occurring in response to a mild stressor (Dallman *et al.*, 1987).

A final consideration in this matter is the recent finding that glucocorticoid receptor density in the hippocampus is regulated by mineralocorticoid receptor activation (Luttge *et al.*, 1989). Thus, in the mouse brain aldosterone decreases glucocorticoid receptor density and the effect is blocked by RU 26752. In primary rat hippocampal cell cultures (O'Donnell and Meaney, 1991) aldosterone decreases glucocorticoid receptor density with an EC_{50} in the low nanomolar range and the effect is attenuated by spironolactone, an antagonist which is highly specific for the mineralocorticoid receptor site. Moreover, O'Donnell and Meaney (1991) found that the effect of aldosterone on glucocorticoid receptor density occurs in a neuronal-enriched cell culture preparation. Since the majority of hippocampal neurons express the genes for both the mineralocorticoid and glucocorticoid receptors (Arriza *et al.*, 1988; Evans and Arriza, 1989), it is possible that this regulation occurs within neurons. This interaction underscores the likelihood that glucocorticoid negative feedback involves both corticosteroid receptor subtypes.

In addition to these intracellular receptors, proteins also exist in plasma which bind to glucocorticoids. These plasma binding proteins bind glucocorticoids with varying efficacy, but one protein, corticosteroid-binding globulin (CBG), binds glucocorticoids with high affinity and very high capacity (Siiteri *et al.*, 1982; Hammond, 1990; Rosner, 1990). CBG serves a major regulatory role in HPA function. To the best of our knowledge, the entry of steroids into cells occurs via passive diffusion. This seems an appropriate route of entry considering the small size and lipophilic nature of these molecules. However, corticosteroids bound to CBG do not enter the cell, and are therefore biologically inactivated. Indeed, the biological effects of steroids are associated with the concentration of free steroid. In the rat under most physiological conditions about 90–95% of glucocorticoids are bound to CBG or to other, weaker binders, such as albumin; leaving approximately 5–10% of the hormone in the free state. Thus, changes in CBG alter the percentage of the steroid that is in the free, biologically active form. This is an important consideration for the study of corticosteroid effects on neural function.

Pardridge *et al.* (1983) have shown that the uptake of corticosterone in brain reflects the free portion of the steroid. In addition to these considerations, CBG also regulates the steroid metabolic clearance rate for corticosterone, as CBG-bound corticosteroid is protected from degradation.

The liver appears to be the major site of CBG synthesis (Wolf *et al.*, 1981) and hepatocytes from fetal animals synthesize and secrete CBG (Smith and Hammond, 1991). Moreover, CBG levels are down-regulated by corticosterone in both the adult (Levin *et al.*, 1987) and the neonate (Walker *et al.*, 1991). Thus, under conditions of elevated corticosterone the percentage of free corticosterone can rise as a function of corticoid inhibition of CBG production. This finding underscores the important interplay between CBG and corticosterone in determining the magnitude of the plasma corticoid signal. Indeed, there are very important changes in plasma CBG levels over development and, as will be described later, these changes are central to HPA function in the neonate.

CBG is found not only in plasma, but also in tissue (e.g., DeKloet and McEwen, 1976; Koch *et al.*, 1981; Kuhn *et al.*, 1986). The pituitary of the adult animal contains high levels of intracellular CBG, which appears to be loosely bound to membranes following internalization via membrane CBG receptors (Perrot-Applanat *et al.*, 1984; Hryb *et al.*, 1986; Singer *et al.*, 1988). Like plasma CBG, the intracellular CBG binds with high affinity to corticosterone and appears to act as a buffer, reducing the binding of corticosterone to classical corticosteroid receptors, and thus the hormone-receptor signal.

10.4 CORTICOSTEROID RECEPTORS DURING EARLY DEVELOPMENT

10.4.1 GLUCOCORTICOID RECEPTORS

The results of several receptor binding studies have shown that glucocorticoid recep-

tor density in rat brain is low, about 30% of adult values, during the first week of life. These data apply to virtually all regions within the cortico-limbic system, including the hypothalamus, amygdala, septum, hippocampus, olfactory bulbs, and cortex (Clayton *et al.*, 1977; Olpe and McEwen, 1976; Meaney *et al.*, 1985a; Pavlik and Buresova, 1984; Rosenfeld *et al.*, 1988a; Sarrieau *et al.*, 1988). Thereafter, receptor density in these regions increases, such that receptor levels approximate to those observed in the adult by about the third week of life. On the basis of these findings, and assuming a comparable level of receptor activation and DNA binding it is reasonable to believe that the forebrain of the rat is less sensitive to circulating glucocorticoids during the first two weeks of postnatal life. In contrast, glucocorticoid receptor density in the fetal (i.e. E16–E22) rat forebrain is comparable to that of the adult (Meaney *et al.*, 1985c). Receptor levels therefore decrease at birth and then increase again at about 12–14 days of age. Considering the data from the fetal rat, this two-week period stands out conspicuously. It is clearly a unique ontogenic pattern, and it begs the question of why glucocorticoid sensitivity in the brain should decrease after birth.

It is also interesting to note that the developmental pattern for the glucocorticoid receptor in the brain is different from that in several other tissues. For instance, glucocorticoid receptor density in regions such as the lung, intestine, and pituitary is actually higher during the first week of life than in the adult animal (e.g., Kalimi, 1984; Henning *et al.*, 1975; Olpe and McEwen, 1976; Sakly and Koch, 1981). These findings suggest that (a) there are tissue-specific differences in glucocorticoid receptor gene expression during development, and (b) these differences in glucocorticoid receptor number are probably associated with the different function subserved by glucocorticoids in these tissues. For instance, it is not hard to imagine the adaptive value of enhanced glucocorticoid sensitivity in the lung of the neonate, considering the role of glucocorticoids in the induction of lung surfactant and the critical importance of surfactant for pulmonary function.

The findings from receptor binding studies have generally been corroborated by recent work on glucocorticoid receptor gene expression. Glucocorticoid receptor mRNA is readily detectable in the neonate using *in situ* hybridization (van Eekelen *et al.*, 1991). A semi-quantitative analysis applied in the van Eekelen study suggested that mRNA levels are generally consistent with receptor binding studies. Thus, there was a general increase in mRNA levels over the first three weeks of life. To our knowledge, there are no studies of glucocorticoid receptor mRNA levels in the fetal rat brain.

The results of immunocytochemical analyses using various monoclonal antibodies to the glucocorticoid receptor are also of considerable interest. It should be noted, however, that the measures used in immunocytochemical studies often reflect receptor function in ways that differ from receptor binding assays. The strength of the binding assays is that they provide quantitative information on the status of functional receptors, albeit with little information on the distribution of receptors beyond that afforded by dissection techniques (although see Reul and De Kloet (1985) for a successful adaptation of microdissection techniques to corticosteroid receptor binding studies). In contrast, the immunocytochemical studies provide a wealth of information on developmental changes in the distribution of corticosteroid receptors. What is less obvious in these studies is the source of the immunoreactivity (-ir) and there is likely to be considerable variability across various antibodies in the nature of the labeled receptor. In many cases, because of the procedures used in receptor purification, the antibody was generated against the 'activated', DNA-binding form of the receptor, and

therefore the staining should, in part at least, reflect circulating levels of the endogenous ligand, corticosterone. Labeling with these antibodies is often decreased or lost entirely in adrenalectomized animals, and increases with steroid exposure. In other cases (e.g. Lawson *et al.*, 1991) the antibody seems to recognize both the activated and unactivated forms of the receptor.

Rosenfeld *et al.* (1988b), working with an antibody generated against the activated form of the glucocorticoid receptor, found that staining in the hippocampus was moderately high shortly after birth and then declined during the second week of life. Glucocorticoid receptor-ir then increased during the third week of life. This pattern corresponds well to the changes in circulating corticosterone. Thus, several authors have reported that plasma total corticosterone (total = the free component + steroid bound to plasma binders such as CBG) is high at birth and until about 2–3 days of age, and thereafter decreases and remains low until about the beginning of the third week of life. Similarly Rosenfeld *et al.* (1988b) reported intense glucocorticoid receptor-ir staining in a wide range of areas in the period shortly after birth, when plasma corticosterone levels are still high. Staining then declined until about day 16, when corticosterone levels begin to rise toward adult values. At this age glucocorticoid receptor-ir increases, reflecting the greater corticoid signal. Lawson *et al.* (1991) found a similar pattern of changes in glucocorticoid receptor-ir. Although these studies may not directly reflect changes in glucocorticoid receptor density, they provide a very valuable indication that, throughout development, the glucocorticoid receptor is activated in relation to the intensity of the hormonal signal.

Finally, there are reports of very different patterns of glucocorticoid receptor across development in specific structures. Van Eekelen *et al.* (1987) reported on the transient presence of glucocorticoid receptor-ir in the superchiasmatic nucleus of the postnatal rat. Glucocorticoid receptor immunostaining was detectable only during the first two weeks of life. The functional significance of this finding is not clear, but it is possibly related to the developmental changes in the diurnal rhythm of corticosterone that is seen over the first three weeks of life (see below). Likewise, Bohn *et al.* (1984) have reported on the presence of glucocorticoid receptors in the superior cervical ganglion (SCG) in the neonate. These receptors appear either in glia cells or in the small intensely fluorescent (SIF) cells. Interestingly, glucocorticoids have been found to increase phenylethanolamine N-methyltransferase (PNMT) activity in SIF cells, an effect that is restricted to fetal and early postnatal life. Bohn *et al.* (1984) reported that glucocorticoid receptor density decreases over the first two weeks of life, and that in the adult receptor density is < 30% of that seen in the neonate. It is possible that the decrease in receptor concentrations could underlie the loss of the PNMT response to glucocorticoids. Thus far, these examples are somewhat unique, but they serve to caution that changes in corticosteroid receptor density across development are likely to be varied, depending upon the CNS region and the function subserved by these receptors. The tissue-specificity of these events is a further alert to the importance of local regulators of steroid receptor gene expression. We will undoubtably hear much more about such events in the near future.

The pharmacological profile of the glucocorticoid receptor in the brain during early development appears to be similar to that in the adult. Competition studies have shown that the affinity of the rat forebrain receptor for a wide range of steroids is of the same order as in the adult. Moreover, the results of pulse–chase studies (Meaney, unpublished) have suggested that the K_d values for the hippocampal glucocorticoid receptor during the first 3 weeks of life are identical to those obtained with adults over a range of temperatures (i.e. 0, 25, or 37°C). Nevertheless, Rosenfeld *et al.* (1988a) have reported that the K_d of the glucocorticoid receptor for [^{3}H]RU28362 is slightly lower in rat hippocampus during the first week of life. This is a potentially important finding and it should be noted that a number of intracellular factors can regulate glucocorticoid receptor function, such as protein kinases (e.g., Kido *et al.*, 1987) and early gene products (e.g., Diamond *et al.*, 1990; Lian *et al.*, 1991). Changes over development in the biochemical milieu of the glucocorticoid receptor are likely to emerge as a major source of variation in receptor function with age. There are also reports of changes in the elution profile of the rat liver glucocorticoid receptor using various ion exchange resins (reviewed in Kalimi, 1984) and of differences in the heat-activation of glucocorticoid receptors (Sharma and Timiras, 1987). It is not clear at this time whether these differences exist in the brain glucocorticoid receptor, nor whether these differences correspond to changes in receptor function. This is an important topic that merits considerable attention.

10.4.2 MINERALOCORTICOID RECEPTORS

The results of both receptor binding and *in situ* hybridization studies suggest that neither mineralocorticoid receptor number nor mRNA levels vary greatly over early development. Thus, mineralocorticoid receptor density in hippocampus during the first week of life is not significantly different from that in the adult rat. To our knowledge, there is little information on developmental changes in mineralocorticoid receptor density in other brain regions. Likewise, van Eekelen *et al.* (1991) found that mineralocorticoid receptor mRNA levels in the hippocampus showed no apparent change between days 2–4 and the adult period. There is a paucity of evidence on the pharmacokinetics of the mineralocorticoid receptor during early life.

Some studies (Rosenfeld *et al.*, 1988a; Sarrieau *et al.*, 1988a) suggest that the affinity of the receptor for either aldosterone or corticosterone is stable during development.

In contrast, Lawson *et al.* (1991) found that mineralocorticoid receptor-ir in the hippocampus showed a pattern similar to that for the glucocorticoid receptor. Thus, staining was moderately high on postnatal day 5 and then declined during the second week of life. Mineralocorticoid receptor-ir then increased by day 15. Again, these data are roughly comparable to the changes seen in plasma corticosterone levels. However, in each hippocampal region studied the percentage of cells showing mineralocorticoid receptor-ir on days 5 and 10 was higher than that for the glucocorticoid receptor. This finding is consistent with the results of receptor bindings assays which suggest that the density of mineralocorticoid receptors during this period is greater than the density of glucocorticoid receptors. Note, that in the hippocampus of the adult rat, mineralocorticoid and glucocorticoid receptor density are roughly equivalent.

In the early 1980s a number of papers appeared derived from *in vivo* autoradiographic studies using tracer levels of [^{3}H]corticosterone as radioligand. These studies predated the clear enunciation of corticosteroid receptor subtypes, and were often performed under the assumption that these conditions allowed for the visualization of glucocorticoid receptor distribution. Considering the low doses of radioligand used in these studies, and the higher affinity of the mineralocorticoid receptor for corticosterone, it seems likely that these studies mostly reflect mineralocorticoid receptor binding (Reul and De Kloet, 1985). This, in part at least, explains why these studies were really only successful in visualizing corticosteroid receptors in the hippocampus. The results of one such study (Meaney *et al.*, 1985b) provide useful information on changes in mineralocorticoid receptor density within various regions of the rat hippocampus as a function of age. However, the pattern of neural development in the hippocampus also complicates estimates of silver grain density used in these studies. Nissl staining of the hippocampus during the first week of life reveals a diffuse band of cells that do not become compacted into the characteristically thin laminae of the cell bodies of Ammon's Horn in adults until the second week of life. There does not seem to be any major change in pyramidal neuron number over this period, rather, as neuritic outgrowth procedes, there occurs a 'packing' of cell bodies. This means that a normal density analysis reflects both the increased neuron density and changes in receptor density. Thus, the results of earlier studies using this analysis should be interpreted with caution. Nevertheless, with this caveat in mind, the results of these studies are consistent with the more recent data in suggesting that the majority of pyramidal neurons in the hippocampus are corticosterone-sensitive during early development, and there are interesting changes within the hippocampus itself. Meaney *et al.* (1985b) found that the uptake and retention of [^{3}H]corticosterone within the dentate gyrus peaked at day 15, and was substantially greater than in the adult. Other regions, such as the subiculum and the CA1–CA3 regions of Ammon's Horn, showed a very moderate but consistent increase that might well have been associated with the changes in neuron packing alluded to earlier. But the peak in [^{3}H]corticosterone uptake in the dentate gyrus at day 15 was striking.

One final problem with the study of mineralocorticoid receptors in the hippocampus concerns the differential labeling of these receptors with either aldosterone or corticosterone. The problem emerged in consideration of the question of how aldosterone could bind *in vivo* to the renal mineralocorticoid receptor if (a) corticosterone levels in plasma are about 100-fold higher than those of aldosterone, and (b) the affinity of the mineralocorticoid receptor for corticosterone is comparable to that for corticosterone aldosterone (e.g., Krozowski and Funder, 1983). Moreover, McEwen *et al.* (1986b) found that about 30% of the [^{3}H]aldosterone binding in the hippocampus was unaffected by prior injection of unlabeled corticosterone. This finding indicates that a subpopulation of hippocampal mineralocorticoid receptor sites are selective for aldosterone over corticosterone. Aldosterone binding in the brain regulates salt appetite and blood pressure, and these effects are not mimicked by glucocorticoids (Funder and Sheppard, 1987; McEwen *et al.*, 1986b for reviews). So how does aldosterone bind *in vivo* to mineralocorticoid receptor sites in the presence of excessive corticosterone levels? The problem has two resolutions. First, aldosterone, unlike corticosterone, does not bind with high affinity to CBG. Therefore, the competition at the level of the mineralocorticoid receptor is between aldosterone and free corticosterone. Recall that free corticosterone concentrations are only about 5–10% of those for total corticosterone. The second solution involves the intracellular metabolism of corticosterone by 11β-hydroxysteroid dehydrogenase (11β-OHSD), an enzyme which metabolizes corticosterone to the inactive 11-dehydrocorticosterone, a steroid which does not bind mineralocorticoid receptors (Edwards *et al.*, 1988; Funder *et al.*, 1988). High levels of 11β-OHSD activity and mRNA expression have been found not only in the kidney, but also in the hippocampus (Funder *et al.*, 1988; Moisen *et al.*, 1990; Lakshmi *et al.*, 1991) and in the corticolimbic system there is considerable colocalization between 11β-OHSD and mineralocorticoid receptor-immunoreactivity. Seckl *et al.* (1991) found that inhibition of 11β-OHSD activity with glycyrrhetinic acid results in increased corticosterone uptake in the hippocampus. This finding provides direct *in vivo* evidence that 11β-OHSD regulates the access of corticosterone to mineralocorticoid receptors.

Moisan *et al.* (1992) have recently shown that in the hippocampus and the cortex 11β-OHSD activity (measured by the conversion of corticosterone into 11β-dehydrocorticosterone) is high at the time of birth and declines substantially over the next few days. Enzyme activity decreases over the first two weeks of life, and then increases to reach adult (and day 1) values during the third week of life. The reduced level of enzyme activity in the neonate could lead to increased binding of corticosterone to mineralocorticoid receptors. Interestingly, 11β-OHSD levels are regulated by corticosterone. Adrenalectomy or treatment with the glucocorticoid receptor antagonist, RU 38486, decreases 11β-OHSD activity, whereas dexamethasone increases 11β-OHSD activity (Seckl *et al.*, 1991). This suggests another level of interaction between the two corticosteroid receptor subtypes. Moreover, the developmental changes in 11β-OHSD activity are also strikingly similar to those in plasma corticosterone.

10.4.3 CBG

Plasma levels of CBG show considerable variation during the period between the third week of gestation and puberty. In the fetal animal plasma CBG levels are in the adult range during the early portion of the third (and last) week of gestation; about 50–60% of those observed in maternal circulation (Smith and Hammond, 1991). However, at about the time of birth CBG levels fall markedly and are less than 1% of the levels observed in adults (D'Agostino and Henning, 1981; Henning, 1978; Koch, 1969; Meaney *et al.*, unpublished). Levels of the plasma binder increase during the second and third weeks of life (in concert with plasma corticosterone levels, see below) and are once again within the adult range by about the time of puberty. At this time the well-documented sex difference in plasma CBG and CBG mRNA levels appears in response to increased plasma levels of estrogen, which stimulates CBG production. After this time CBG levels are significantly higher in females than in males (Gala and Westphal, 1965).

In a series of papers Smith and Hammond (1991) have examined the ontogenetic changes in plasma CBG levels. Interestingly, these authors report that although CBG levels do not approximate to those seen in the adult until about 6 weeks of life, CBG mRNA levels are comparable to those observed in the adult as early as the third week of life. During the first week post-partum, both CBG binding capacity and CBG mRNA levels are extremely low. However, in the following weeks the increase in the levels of steady-state CBG mRNA is greater than for the protein itself. Smith and Hammond (1991) also found that the clearance rate for radiolabeled CBG was significantly faster in the neonate (~6.9 h) compared to the adult (14.5 h). The difference was due to the increased degradation rate in the neonate and not to differences in the composition of the protein. CBG isolated from adult serum was degraded at the same rate as that of CBG isolated from neonates when administered to infant rats.

Taken together, the results of these studies show that the decreased plasma levels of CBG in the neonate are due to (a) decreased CBG biosynthesis during the first few weeks of life, and (b) an increase in the clearance rate for CBG. The latter effect probably accounts for the discrepancy between CBG mRNA and CBG levels between the third and sixth week of life.

10.5 THE HPA AXIS DURING EARLY DEVELOPMENT

The early studies on the development of the HPA axis have been reviewed in detail by Sapolsky and Meaney (1986). In general there is intense HPA activity in the fetal rat, with high levels of corticosterone synthesis and secretion (Martin *et al.*, 1977) as well as dense CRH-ir staining in the mediobasal hypothalamus (Bugnon *et al.*, 1984). After parturition there is a profound alteration in HPA function due to changes at each level of the axis. This alteration corresponds to a period of apparent quiesence in HPA function, with steroidogenic activity increasing towards adult levels by week 3–4 of life. However, because of the important tissue-specific changes in the density of various corticosteroid receptors, as well as changes in both plasma and pituitary CBG, the corticoid signal varies considerably during early development. In certain regions, such as the pituitary, the pattern of changes in receptor density may actually result in an increased corticoid signal. This is also probably true in the lung and pancreas.

What we hope to show here is that whereas HPA function in the neonate differs from that of the adult, this is not a question of immaturity. Indeed, the concept of immaturity is one that we would shun. It is a notion that obstructs any clear understanding of neuroendocrine function during development, and underestimates the ability of the neonate to adapt to a profoundly different external environment. Rather, we believe that the nature of HPA function at this stage of life reflects the biological needs of the neonate, and as these needs come to approximate better those of the adult animal, so too does HPA function. The HPA axis of the neonate subserves the needs of the growing, maternally protected neonate, just as the HPA axis of the adult subserves the needs of the independent, self-regulating adult. Our understanding of neuroendocrine development is unlikely to be furthered by tacitly assuming that differences in function between the neonate and the adult are due to a deficit in the neurobiology of the neonate. As the animal grows physically larger and becomes independent of its caretakers, its CNS adapts to meet the changing demands placed on the system (Purves, 1988). It seems to us that this is what is meant by the term 'development'.

10.5.1 ADRENAL

Considerable morphological changes occur at the level of the adrenal in the neonate (Van Dorp and Deane, 1956; reviewed in Sapolsky and Meaney, 1986). There are also reports of decreased adrenal sensitivity to ACTH during the first weeks of life (e.g., Levine *et al.*, 1972; Guillet *et al.*, 1980; Witek-Janusek, 1988). There is also a decreased cAMP response to ACTH during the neonatal period (Coté and Yasumura, 1975). Chatelin *et al.* (1989) have reported a decreased density of ACTH receptor sites in the adrenal of the neonate, with no change in the affinity of the receptor for ACTH. Such a decrease in ACTH receptor sites could certainly constitute one of the mechanisms for dampened steroidogenic and second messenger responses to ACTH in the rat pup. At the same time, other studies have found little or no change in adrenal sensitivity to ACTH (Schroeder and Henning, 1989). These conflicting data are difficult to resolve, but it may be, at least for the *in vivo* experiments, that we have the wrong measures. As we shall see later the volume of distribution (V_d) for corticosterone in the neonate differs considerably from the adult, and therefore the measures of corticosterone levels in the plasma compartment may be a less reliable index of adrenal secretion.

10.5.2 PITUITARY

Pituitary corticotrophs show a very interesting pattern of change in the neonatal period. Childs *et al.* (1982; also see Childs, 1987) showed that the population of corticotrophs which stain only for ACTH decline dramatically (by 95%) after birth and remain at low levels until the third week of life. However, a second population of corticotrophs, which stain for ACTH as well as gonadotropins (LH and/or FSH) remain stable during this period and decrease in frequency in adulthood. The functional significance of these changes is not clear. It is

also unclear whether there are major changes in the number of corticotrophs, although there is considerable mitosis in the anterior pituitary through to the peripubertal period (Siperstein *et al.*, 1954).

POMC mRNA levels in the anterior pituitary increase during the late fetal period, peak at about E20, and then decline after birth (Grino *et al.*, 1989). During weeks 2–3 of life POMC mRNA levels increase dramatically (Grino *et al.*, 1989; Wang *et al.*, 1989). In general, this pattern is similar to results from studies examining the pituitary content of ACTH (Walker *et al.*, 1986a). Corticotroph POMC is processing in the neonate differs considerably from that in the adult (Hotta *et al.*, 1988; Mains and Eipper, 1990; Sato and Mains, 1985, 1988). First, only about 50% of POMC is processed in the neonate compared with the almost complete levels of processing seen in the adult. Second, there is considerable formation of α-melanocyte-stimulating hormone (MSH)-like material (determined as (ACTH$_{1-13}$)NH2), a condition that is unique to the neonate and underscores the considerable plasticity in POMC processing over development (Sato and Mains, 1985, 1988). This finding suggests that under basal conditions there is reduced formation of ACTH and this has been confirmed in a number of studies on pituitary ACTH content (Bartova, 1968; Walker *et al.*, 1986a). The altered pattern of POMC processing into ACTH contrasts with the high levels of β-endorphin in the anterior pituitary (Seizinger *et al.*, 1984) and plasma (Iny *et al.*, 1987) of the neonate. The β-endorphin found in the anterior pituitary is largely non-acetylated, indeed as in the adult little α-N-acetylation of POMC products occurs (Seizinger *et al.*, 1984; Sato and Mains, 1985), and the ratio of β-lipotrophin/β-endorphin resembled that found in the adult (Seizinger *et al.*, 1984). The changes with age in pituitary ACTH content and the absence of such changes in β-endorphin suggest that the increased formation of α-MSH-like material, in favor of ACTH, is of considerable importance.

Finally, Walker *et al.* (1986a) found that pituitary CRH receptor sites were actually higher on day 5 of life than in adults. Receptor levels decreased by day 10 to levels that were indistinguishable from adults. Both *in vivo* and *in vitro*, the pituitary corticotroph of the neonate shows a potent pituitary–adrenal response to CRH (Guillet and Michaelson, 1978; Hary *et al.*, 1984; Walker *et al.*, 1986a; Sato and Mains, 1986, 1988), and this response is sensitive to dexamethasone suppression (Sato and Mains, 1986; Walker *et al.*, 1986a, b).

At the same time, pituitary responsiveness to AVP appears to be decreased in the neonate (Koch and Lutz-Bucher, 1990). There is a decrease in AVP binding sites in the anterior pituitary of the neonate. Moreover, there appears to be less-efficient coupling between the AVP receptor and the inositol phosphate (IP) second messenger system. AVP exerts its effects on ACTH release through an activation of IP metabolism. This finding is of considerable importance, since the increase in ACTH release in response to stress is often associated with changes in hypophysial-portal AVP concentrations (see opening section).

In summary, there is increased formation of α-MSH-like material and decreased sensitivity to AVP. Together, these findings would suggest low levels of ACTH synthesis, and this is consistent with studies of pituitary ACTH content (Walker *et al.*, 1986a). Nevertheless, there is surprisingly little difference in basal ACTH levels between neonates and adults (Meaney *et al.*, unpublished), and adrenalectomized day 5 pups respond to exogenous CRH with a massive increase in plasma ACTH (Walker *et al.*, 1986a, 1990). Thus, despite the status of the neonatal pituitary, it is capable of maintaining a basal secretion rate that is in the same range as the adult. Moreover, the pituitary of the neonate shows larger increases in ACTH release when provided with (a) the appropriate stimulation in the form of hypophysiotrophic secretagogues, and (b) the absence of the exaggerated glucocorticoid negative-feedback that exists at the level of the pituitary in the neonate (Section 10.8). Thus, the status of the corticotroph and POMC processing do not preclude a very robust response to CRH. This is instructive in many ways. In one context, it underscores the pitfall of assuming that differences between the neonate and the adult in either anatomical or biochemical measures necessarily implies reduced HPA function.

10.5.3 HYPOTHALAMUS

CRH-like immunoreactivity (CRH-ir) in the fetal animal is readily apparent and staining is generally comparable with that in the adult. However, following birth there is a rapid decline in staining, and CRH-ir remains low during the next two weeks of life (Bugnon *et al.*, 1984). Likewise, hypothalamic CRH content is low (< 10% adult values) during the first week of life, and then increases steadily thereafter, with about 50% adult values during the second and third weeks of life. This pattern generally parallels changes in CRH gene expression. Steady-state levels of CRH mRNA in the hypothalamus decline following birth and remain low during the first week of life (Grino *et al.*, 1989b). Likewise, vasopressin mRNA levels are also low during the first weeks of the postnatal period (Rundle and Funder, 1988; Almazan *et al.*, 1989).

Thus, hypothalamic CRH content and mRNA levels are both comparatively low in the neonate. In a series of studies on the *in vitro* release of ACTH secretagogues in hypothalami from neonates, *in vitro* basal release of ACTH was found to be about 20–30% of the levels seen in adult hypothalami (Widmaier 1989; Widmaier *et al.*, 1990). Moreover, adult-like levels of CRH release were not apparent until about 35 days of age. Thus, the decreased CRH biosynthesis is accompanied by a comparable decrease in release in the neonate.

We do not know a great deal about the hypophysiotropic regulation of ACTH release in the neonate. However, in one study, Walker *et al.* (1986a) found that urethane stimulated ACTH release in pups on days 3, 5, and 10 of life, and that the response was attenuated by the administration of a CRH antiserum. These findings suggest that (a) CRH can mediate a pituitary stress response in the neonate, and (b) communication between the hypothalamus and the pituitary is sufficiently developed to support such a response. What is truely missing at this point is information on the status of the neural pathways which are involved in the transduction of the stress signal to the PVN of the hypothalamus (Adamson *et al.*, 1991).

10.6   BASAL HPA FUNCTION DURING EARLY DEVELOPMENT

Resting state levels of plasma corticosterone are generally elevated in the fetal rat pup, with a marked peak at about E19 (Dupouy *et al.*, 1975; Martin *et al.*, 1977; Meaney *et al*, 1985c). Corticosterone levels in the fetus remain high during parturition and then steadily decline over the next 2–3 days. At about this time the adrenal undergoes a period of apparent quiesence, which lasts until about the beginning of the second week of life. During this period basal corticosterone levels in plasma are generally only barely detectable with radioimmunoassay (~1 µg/dl), with little evidence for a diurnal rhythm. Beginning on about day 14 of life, basal corticosterone levels increase, and by the third week of life are clearly within the adult range. The sex difference in basal corticosterone levels emerges at the time of puberty, in response to the effects of gonadal hormones on CBG production and the metabolic clearance rate for corticosterone (Kitay, 1961; Gala and Westphal, 1965).

Despite the variations in plasma corticosterone levels over the first weeks after parturition, there is little change in basal ACTH levels. This discrepancy has led to considerable speculation concerning possible developmental changes in the sensitivity of the adrenal to ACTH and changes in the steroidogenic capacity of the adrenal. There is considerable controversy here, particularly with regard to adrenal sensitivity to ACTH. Note, however, that Walker *et al.* (1986a, b) reported that exogenous CRH results in a substantial increase in plasma corticosterone levels, suggesting that the adrenal is indeed capable of mobilizing a substantial corticosterone response. Moreover, Schroeder and Henning (1989) report little change in the *in vitro* response of adrenal tissue to either ACTH or dibutryl cAMP. Nevertheless, we are left with the finding that measurable, total corticosterone levels remain low during early postnatal life despite the presence of adult-like ACTH levels, as well as a number of reports of decreased adrenal sensitivity to ACTH.

The importance of age-related changes in the plasma CBG will now be considered. It has been shown that developmental changes in CBG concentrations are triggered by increased circulating levels of thyroid hormones (Henning *et al.*, 1975; Henning, 1978). As a consequence of the elevated CBG levels plasma corticosterone titers increase, since both the volume of distribution (V_d) and the steroid metabolic clearance rate are decreased. Thus, with the increase in CBG levels, there is a dramatic increase in the percentage of corticosterone bound to CBG and a decrease in tissue uptake of the steroid. Since CBG protects corticosterone from degradation, the $t_{1/2}$ for corticosterone in plasma increases (Schapiro, 1962; Schapiro *et al.*, 1962). Thus, Schroeder and Henning (1989) found that the increase in plasma total corticosterone occurred even in adrenalectomized pups implanted with fused pellets of corticosterone (i.e., a constant source of corticosterone). These authors also showed that the V_d for dexamethasone, which does not bind CBG, is unchanged over weeks 2–4 of life.

This raises a question about the level of free, biologically active corticosterone over this same period. Schroeder and Henning (1989) report that on day 12 about 50% of corticosterone is bound to plasma binders (CBG and weak binders such as albumin), whereas by day 22 about 90% is in the bound form. Thus, there is a disproportionately higher level of free corticosterone in the neonate. Since corticosterone that is not bound to CBG is biologically active, we would assume that the overall basal corticoid signal at the level of the intracellular receptor remains relatively constant over the neonatal period. In fact, at least for brain tissue, this appears to be the case. This question has been examined by measuring levels of receptor occupancy (hormone-bound receptors) in intact animals as a percentage of the total receptor population. As expected, this measure reliably reflects circulating corticosterone levels in the adult animal (Meaney *et al.*, 1988b; Reul and De Kloet, 1985). We (Meaney *et al.*, unpublished) recently found that levels of glucocorticoid receptor occupancy in both the hippocampus and the hypothalamus under basal corticosterone conditions remain fairly constant (i.e. 10–15% levels of receptor occupancy) over the first two weeks of life and do not differ from those in adults. This estimate of 10–15% occupancy is almost identical to that reported in earlier studies of glucocorticoid receptor occupancy in adult animals under basal corticosterone levels (Reul and De Kloet, 1985; Reul *et al.*, 1987a, b; Meaney *et al.*, 1988b). Thus, the percentage of the receptor pool that is hormone-bound does not vary greatly over development, despite the fluctuations in total corticosterone levels.

These findings suggest that it would be wrong to assume that the basal corticoid signal in the neonate is decreased, and in this regard measures of total corticosterone levels are misleading. At this point, we would conclude that the basal corticoid signal at the level of the intracellular glucocorticoid receptor is relatively unchanged over the course of postnatal development. What is different from the adult is the way in which this signal is maintained. Of course, at this point we must also consider the variation in receptor levels in target tissue. Thus, the hormone-receptor signal is a function of the percentage of receptors occupied and the total number of receptors. In the hippocampus and hypothalamus of the adult animal, where glucocorticoid receptor binding is about 2–3 times higher than in the neonate, we would assume that the hormone-receptor signal is greater. This difference occurs as a function of receptor density and reflects the importance of receptor regulation during development. In the case of the mineralocorticoid receptor, the results of receptor binding studies suggest that the hormone-receptor signal would be fairly consistent over development. However, the critical experiments that would address these predictions have not yet been performed and there are reports of age-related differences in the nuclear binding of heat- and ligand-activated glucocorticoid receptors (Sakly and Koch, 1981; Kalimi, 1984; Sharma and Timiras, 1987).

10.7 THE HPA STRESS RESPONSE DURING EARLY DEVELOPMENT

Schapiro (1962) was the first to propose that the HPA axis of the neonatal rat was comparatively unresponsive to stress. This hypothesis was formulated on the basis of a series of studies showing that the plasma total corticosterone response to stress in the neonate was either absent or greatly reduced in comparison with the adult rat. The temporal parameters of this so-called stress non-responsive period were intriguing. The fetal rat, as well as the newborn animal, has been found to produce a rather robust corticoid response to stress. However, the magnitude of the increase in plasma corticosterone levels falls precipitously after birth, and from about days 3 to 12 of life increases in plasma total corticosterone are indeed small. As an example, in adult rats resting levels of corticosterone in the AM phase of the cycle are about 1–3 µg/dl and rise to about 40–50 µg/dl following exposure to ether vapours. In the neonate, resting levels of corticosterone are about 0.5–1.0 µg/dl and increase to a maximal level of about 4 µg/dl.

There has been considerable debate as to whether these increases were statistically reliable and whether, in fact, there existed a non-responsive period in the neonatal rat. Sapolsky and Meaney (1986) concluded that although a stress-induced increase in plasma total corticosterone is often detectable, it is far smaller than that seen in the adult and proposed that the neonatal rat be considered as hyporesponsive to stress. This recommendation was also based on methodological grounds, since earlier reports were based on fluorometric analysis of plasma corticosterone levels which, given their greater variability and high blank values, are less sensitive to small, but statistically reliable differences.

Since this time far more thorough studies have been performed, with a greater focus on plasma ACTH levels. These studies (Walker *et al.*, 1991) have shown that the neonate does show a pituitary response to stress, but that (a) the magnitude of the response is generally muted during the first week of life and increases as a function of age, (b) the time course for the changes in ACTH are age-related, and (c) the pattern of ACTH secretion is highly variable depending upon the stressor. Thus, for example, plasma ACTH responses to ether are detectable in 1, 5, and 10-day-old pups with a peak response by 5 min after ether exposure. The ACTH response is considerably reduced at 10 min in contrast to adults, where a robust ACTH response is still apparent 20 min after ether exposure.

Walker *et al.* (1991) report that the magnitude of the response increased from 20–30% of controls in newborns to 300% on day 10. Moreover, the neonate was also capable of responding to two consecutive ether stressors on day 10 of life, with the second response greater than the first, underscoring the capacity of the pituitary to respond to stress. Likewise, Iny *et al.* (1987) found an adult-like increase in plasma β-endorphin levels following ether throughout the entire neonatal period. Interestingly, Meaney and Aitken (unpublished) found no evidence for an increase in plasma α-MSH-like material following stress in the neonate.

Widmaier (1989, 1990) has reported findings using insulin-induced hypoglycemia as a stressor. In this study 10-day-old pups produced significant increases in plasma ACTH, but only in the range of 125–135% of controls. In adults the same manipulation produced a 300% increase in plasma ACTH. Witek-Janusek (1988) found that whereas plasma ACTH responses to an endotoxin were moderately lower on days 1 and 2 of life, there was no difference in the response of animals 5 or 10 days of age and adult animals. This underscores the point derived from the studies by Walker *et al.* (1991) showing that variation in the magnitude and time course of the HPA response in neonates is dependent on the nature of the stressor. In all the above-mentioned studies, even in the presence of a robust ACTH response, corticosterone levels remained modest (< 10 µg/dl) compared to adults. These discrepancies between the magnitude in the pituitary and adrenal responses to stress are interesting and underscore the importance of any developmental changes in adrenal sensitivity to ACTH in the neonate.

In addition, more direct, physiological evidence for the importance of changes in adrenal sensitivity to ACTH (Stanton *et al.*, 1988) has shown that separation from the mother for a period of 12 h greatly potentiated the plasma corticosterone response of the pup to ether stress. The effect of maternal separation was, in part, mediated by post-separation changes in adrenal sensitivity to ACTH. Separated pups showed a greater *in vivo* corticosterone response to exogenously administered ACTH than pups left with their mothers. We (Meaney *et al.*, unpublished) found that maternal separation also results in increased ACTH in response to ether stress. Thus, the mother (and the thermal and nutritional needs provided for by the mother) serves as a regulator of the ontogeny of the HPA axis. Nevertheless, as Stanton *et al.* (1988) note, the plasma corticosterone response of separated pups is still considerably less than that observed in adults, indicating that changes in adrenal sensitivity alone cannot account for the dampened corticosterone response of the neonate.

There is also an interesting effect of the maternal separation manipulation on CBG in rat pups (Meaney *et al.*, unpublished). At day 6, CBG levels are extremely low, indeed barely detectable. By day 15 plasma CBG levels have increased to levels of about 3–4 pmol/mg protein, still well short of the ~20 pmol/mg protein seen in adults. However, when day 15 animals are separated from their mothers for 24 h, plasma CBG levels fall to about 1 pmol/mg; a substantial decrease. The decrease in plasma CBG is very likely to amplify the corticoid signal associated with stress in day 15 animals. This finding further emphasizes the importance of the mother as a regulator of HPA function in the neonatal rat.

Walker *et al.* (1991) have proposed an extension to the Henning hypothesis on corticosterone uptake and metabolism to account for the reduced stress-induced increase in levels of plasma corticosterone observed in the neonate. As observed by Schroeder and Henning (1989), the V_d for corticosterone is reduced in the neonate. According to Walker *et al.* (1991), the reduced CBG levels in the neonate suggest that corticosterone secreted by the adrenals would be more rapidly distributed into the body volume, since the reduced number of CBG binding sites would be rapidly occupied. The high levels of plasma CBG sites in the adult then serve to enhance the plasma signal by retaining corticosterone in the plasma compartment. This hypothesis states that the reduced corticosterone levels observed in the neonate are, in part, a methodological problem associated with differences in the compartmentalization of adrenal corticosterone in the neonate. The difference in distribution, together with the decrease in adrenal sensitivity to ACTH, provide a further explanation for the decreased corticosterone response in neonates.

The complexity of the interrelationships between various aspects of the HPA axis is also revealed by the effects of thyroid hormones on HPA function in the neonate. A complete discussion of the effects of thyroid hormones on HPA function is beyond the scope of this chapter, however, as mentioned earlier changes in plasma levels of corticosterone over the first three weeks of life emerge as a consequence of the effect of thyroid hormones on plasma CBG levels. Moreover, when rats are made hypothyroid prior to birth, plasma CBG levels are depressed, despite little change in plasma corticosterone levels compared to euthyroid pups. This suggests that the inhibitory free corticosterone signal over ACTH secretion is probably enhanced in the hypothyroid rats. As a consequence, ACTH responses to stress are suppressed until day 14 of life and reduced thereafter (Walker *et al.*, 1989). The important finding in these studies is that a single injection of thyroxine (T_4) 24h prior to stress could restore the ACTH response to stress in day 14 hypothyroid rats. The restoration of the normal pituitary ACTH response to stress was associated with a T_4-induced increase in plasma CBG, which decreased the free steroid available for feedback inhibition over ACTH release. Interestingly, corticosterone inhibits the conversion of T_4 into the biologically more potent, monodeiodinated, triiodothyronine (T_3).

The effects of plasma CBG on corticosteroid uptake and metabolism are also likely to affect the corticoid signal at the level of the receptor. If there is increased distribution of corticosterone in the body volume in neonates then this would imply that the stress-induced corticosterone signal at the level of the intracellular receptor might be disproportionately higher than would be expected on the basis of the plasma steroid levels. This is the same issue that emerged in considering the biological significance of the low basal corticosterone values in the neonate. Indeed, an examination of this question would also provide a more direct measure of the biological significance of the corticosterone signal following stress in the neonate. In a recent study (Meaney *et al.*, unpublished), using the maternal separation paradigm of Stanton and Levine, it was found that plasma corticosterone levels rose in response to ether stress to about 6–8 µg/dl in separated day 6 pups, compared to about 12–15 µg/dl in separated day 15 pups, and 40 µg/dl in adults. Nevertheless, the levels of receptor occupancy (as a percentage of the total receptor population) were comparable across all ages at about 70%. Thus, despite the reduced plasma total corticosterone, the percentage of receptors occupied by corticosterone following stress is comparable in the neonate and the adult. These findings, together with the recent studies measuring stress-induced changes in plasma ACTH seriously question the concept of a stress non- (or hypo) responsive period in the neonatal rat. There are unquestionably major differences in HPA function between the neonate and the adult, including tissue-specific differences in glucocorticoid receptor number, but the effect of the corticoid signal at the level of the glucocorticoid receptor, reflected in levels of receptor occupancy, is not appreciably different from that of the adult.

Of course, the density of brain glucocorticoid receptors is considerably lower in the neonate, such that the magnitude of the corticosterone effect at the level of the cell is likely to be reduced compared to the adult. However, this is not true for other tissues such as lung and pituitary where glucocorticoid receptor density is comparable to or even higher than in the adult. Thus, in the neonate, under maximally stressful conditions, the levels of corticosteroid receptors in target tissue can serve as a major rate-limiting step in determining the nature of the corticosterone signal at the cellular level. This factor underscores the importance of receptor regulation during tissue development.

10.8 GLUCOCORTICOID NEGATIVE-FEEDBACK DURING EARLY DEVELOPMENT

In the intact, adult animal glucocorticoids in circulation serve to inhibit HPA activity in a dose-dependent manner (Keller-Wood and Dallman, 1984; Dallman *et al.*, 1987). The major site for such glucocorticoid negative-feedback in the adult appears to be the brain, including both hypothalamic and extrahypothalamic regions (Dallman *et al.*, 1987; Levin *et al.*, 1987). Levin *et al.* (1987) found that lesions of the PVN of the hypothalamus block the increase in plasma ACTH that accompanies adrenalectomy, an effect that is probably due to the removal of CRH neurons. Levin *et al.* then administered exogenous CRH. If the pituitary is a major feedback target site in the adult, then the PVN lesion/adrenalectomy + CRH animals should have differed from PVN lesion/intact + CRH animals, since the presence of glucocorticoids would have attenuated the effects of exogenous CRH on ACTH release. In fact, when the lesioned animals were given exogenous CRH replacement there was an increase in plasma ACTH, but no difference between adrenalectomized and intact animals. This seminal experiment documented the importance of the brain as the major glucocorticoid target for the negative-feedback effects of circulating glucocorticoids in the adult animal.

In the adult animal the removal of the endogenous source of glucocorticoids, via adrenalectomy, removes negative-feedback inhibition and results in a profound increase in plasma ACTH levels which is reversed with corticosterone replacement (Dallman *et al.*, 1987). Similarly, in the neonate adrenalectomy results in a large increase (~400% 24 h after surgery) in plasma ACTH (Walker *et al.*, 1986b; Meaney *et al*, unpublished) that is reversed with corticosterone replacement (Walker *et al.*, 1991). Moreover, as in the adult, prior exogenous corticosterone administration decreases plasma ACTH responses to stress in neonates (Walker *et al.*, 1990, 1991). These findings suggest that glucocorticoid negative-feedback over HPA activity is functional in the neonate. Indeed, the results of a number of studies suggest that (a) there is exaggerated glucocorticoid negative-feedback in the neonate, (b) the relative importance of the pituitary versus the brain as critical sites for negative-feedback inhibition seems to differ between the neonate and the adult, and (c) that such differences are due to the changes in the levels of various corticosteroid receptors in brain, pituitary, and plasma over the course of early development.

Sakly and Koch (1981, 1983) found that pituitary intracellular CBG levels in neonatal rats are very low. At the same time, pituitary glucocorticoid receptor density is comparable to that in adults. Indeed, pituitary glucocorticoid receptor density is generally stable throughout development (Olpe and McEwen, 1976; Sakly and Koch, 1981, 1983). Thus, Sakly and Koch found that, with the decreased competition from intracellular CBG sites, nuclear uptake of a constant dose of corticosterone was greater in neonates than in adults. These studies provide a remarkable example of how changes in the extracellular and intracellular CBG levels can regulate the magnitude of the corticoid signal. In the presence of a stable glucocorticoid receptor population and a constant corticosterone level, a decrease in CBG sites results in an amplified corticoid signal at the level of the pituitary.

This increased corticoid signal appears to be associated with enhanced glucocorticoid negative feedback over pituitary ACTH release and may, in part, account for the apparent reduced ability of the neonate to secrete ACTH in response to certain forms of stress. The increased glucocorticoid negative feedback at the level of the pituitary has been demonstrated in a number of earlier studies. For instance, Walker *et al.* (1986b) demonstrated both *in vivo* and *in vitro* that corticosterone was more effective in suppressing ACTH release in the neonate than in older animals. Moreover, in day 5 pups corticosterone was equally effective in suppressing ACTH release as dexamethasone. In the adult, dexamethasone is normally more effective than corticosterone in suppressing ACTH release, since it does not bind with CBG. As the animal matures, CBG levels increase, compete with the glucocorticoid receptor for corticosterone, and corticosterone becomes increasingly less effective in suppressing ACTH release (i.e. an increase in the ID_{50}). In contrast, there is little change in the negative-feedback potency of dexamethasone over development (Walker *et al.*, 1986b). These findings support the hypothesis of Sakly and Koch (1981, 1983) who proposed that as extracellular and intracellular CBG levels rise there is a decrease in the nuclear uptake of corticosterone (due to the increasing intracellular competition between CBG and the glucocorticoid receptor) and thus a dampening of the negative-feedback effect of corticosterone on ACTH synthesis and release.

The situation at the level of the brain is somewhat less clear. Recall that mineralocorticoid receptor density is relatively stable over development, whereas glucocorticoid receptor density is much lower than that in the adult (Olpe and McEwen, 1976; Clayton *et al.*, 1977; Meaney *et al.*, 1985a, Rosenfeld *et al.*, 1988a; Sarrieau *et al.*, 1988a). Interestingly, Grino *et al.* (1989) found that adrenalectomy did not alter hypothalamic CRH mRNA levels in day 7 pups. The same manipulation results in a profound increase in CRH mRNA levels and CRH-ir in adults (Beyer *et al.*, 1988; Plotsky and Sawchenko, 1987). Not until day 14 did adrenalectomy increase CRH mRNA levels, suggesting that glucocorticoid negative feedback at the level of the hypothalamus is either absent or less effective in the neonate. Interestingly, glucocorticoid receptor density in hypothalamus and hippocampus increase to adult values by day 14 of life. However, Walker *et al.* (1990) has found that the IC_{50} for the inhibition of ACTH release in adrenalectomized neonates is about 1 nM, a value that would seem to imply a role for the central mineralocorticoid receptor, and this finding is consistent with the high levels of these receptors found in the hippocampus of the neonate.

Research on the role of corticosteroid receptors in the brain as mediators of glucocorticoid negative feedback over HPA function is really just beginning, and so what follows is somewhat speculative. However, the existing data suggest a pre-eminent role for the pituitary in mediating negative-feedback inhibition in the neonate. The findings of Sakly and Koch (1981, 1983) and of Walker *et al.* (1986b) suggest that the pituitary is the target of an enhanced corticoid signal, compared with the adult. As both intracellular and extracellular CBG levels rise, corticosterone becomes increasingly less effective at the level of the pituitary. Indeed, the findings of Levin *et al.* (1987) suggest that the pituitary is not a major negative-feedback site in the adult animal. With the increase in glucocorticoid receptors in the corticolimbic system, against a background of high levels of hippocampal mineralocorticoid receptors, the hypothalamus, hippocampus, and frontal cortex become more sensitive to corticosterone. As the animal grows, CNS drive over hypothalamic–pituitary function increases and the brain emerges as the major target for negative-feedback regulation of HPA function. Thus, as the animal ages, the relative importance of pituitary and brain target sites changes, and this switch is receptor mediated.

10.9 CONCLUSIONS: IMPLICATIONS FOR NEURAL DEVELOPMENT

The proposed stress non-responsive period in the neonatal rat has often found a very receptive audience among developmental neurobiologists. Elevated levels of glucocorticoids have frequently been shown to disrupt neuronal development. Such effects have been observed using indices of decreased neuronal proliferation, morphogenesis, synaptogenesis, and behavioral anomalies following exposure to elevated glucocorticoid levels in the early postnatal period (e.g., Schapiro, 1962; Balazs and Cotterrell, 1972; Howard and Benjamin, 1975; Bohn, 1980; Yehuda *et al.*, 1988, 1989). In all cases, even transiently elevated glucocorticoid levels often produce profound, long-term effects. These studies were often used as the basis for an understanding of the adaptive function for the so-called stress non-responsive period (Schapiro, 1962; Sapolsky and Meaney, 1986) and for the decreased glucocorticoid receptor density in brain over this period. Note, these findings are entirely consistent with the role of glucocorticoids in the cell cycle (Burnstein and Cidlowski, 1989). In general, an enhanced glucocorticoid signal is associated with the termination of cell proliferation.

At the same time, the integrity of at least one structure is seriously compromised by the complete absence of glucocorticoids. Slovitar *et al.* (1989), Gould *et al.* (1990) and McEwen and Gould (1990) have reported that adrenalectomy results in the loss of hippocampal granule cells. Interestingly, McEwen and Gould (1990) have shown that the effect of adrenalectomy is reversed by replacement with either low doses of corticosterone or aldosterone. Such low concentrations of corticosterone bind preferentially to the mineralocorticoid receptor, which has a higher affinity for corticosterone than does the glucocorticoid receptor. This interpretation is supported by the efficacy of the mineralocorticoid, aldosterone. Thus, it is interesting that over the course of neonatal development mineralocorticoid receptor density in the hippocampus is consistently at the same high levels as observed in adults. The evidence suggests that the damaging effects of elevated glucocorticoid levels are mediated by the interaction of the steroid with the glucocorticoid receptor (Sapolsky, 1990). It is certainly of interest that in the neonatal period glucocorticoid receptor density in the corticolimbic regions is low.

In general, the pituitary–adrenal unit appears to function in an analogous manner; maintaining a constant, tonic corticosterone signal but dampening the increases associated with stress (less so in the absence of the mother). The low levels of CBG in combination with high levels of pituitary glucocorticoid receptor levels result in increased glucocorticoid negative feedback inhibition over pituitary ACTH release. The enhanced pituitary negative feedback in combination with decreased adrenal sensitivity to ACTH serve to effectively limit corticosterone responses to stress. At the same time, the low CBG levels increase the fraction of circulating corticosterone in the free, biologically active form. This appears to result in almost the same resting-state level of hormone-receptor signal in the neonate as in the adult, despite the low levels of adrenal corticosterone secretion. At the same time the reduced adrenal signal associated with stress, in combination with the reduced glucocorticoid receptor density in the brain, would serve to protect the developing CNS of the neonate against the potentially damaging effects of elevated glucocorticoid exposure.

10.10 ENVIRONMENTAL REGULATION OF HPA DEVELOPMENT

There is clearly considerable fluctuation in HPA function over the course of early postnatal life in the rat. This period also represents a time of considerable plasticity in the HPA axis, such that the development of HPA function is profoundly altered by environmental events. Indeed, several years ago Levine (1957, 1962), Ader and Grota (1969) and others described the effects of postnatal 'handling' on the development of behavioral and endocrine responses to stress. As adults, handled rats exhibited attenuated fearfulness in novel environments and a less pronounced increase in the secretion of the adrenal glucocorticoids in response to a variety of stressors. These findings clearly demonstrated that the development of rudimentary, adaptive responses to stress could be modified by environmental events. We have followed these earlier studies, convinced that this paradigm provides an opportunity to examine how subtle variations in the early environment alter the development of specific neurochemical systems, leading to stable individual differences in biological responses to stimuli that threaten homeostasis. In this work we have shown how early handling influences neuronal differentiation in brain regions that regulate glucocorticoid negative-feedback inhibition over HPA activity. Specifically, handling increases glucocorticoid receptor density in the hippocampus and frontal cortex, enhancing the sensitivity of these structures to the negative-feedback effects of elevated circulating glucocorticoids, and increasing the efficacy of neural inhibition over ACTH secretion. These effects are reflected in the differential secretory pattern of ACTH and corticosterone in handled and non-handled animals under conditions of stress.

10.10.1   THE EFFECT OF HANDLING ON THE HPA RESPONSE TO STRESS

The handling procedure involves removing rat pups from their cage, placing the animals together in small containers, and after 15–20 min returning the animals to their cage and their mothers. The manipulation is performed daily for the first 21 days of life. In response to a wide variety of stressors handled (H) rats secrete less corticosterone and show a faster return to basal corticosterone levels following the termination of stress than do non-handled (NH) animals (Levine, 1957, 1962; Levine *et al.*, 1967; Ader and Grota, 1969; Hess *et al.*, 1969; Meaney *et al.*, 1985d, 1988a, 1989b). These differences are apparent as late as 24–26 months of age (Meaney *et al.*, 1988a, 1991), indicating that the handling effect persists over the entire life of the animal. The differences in HPA function are not due to changes in adrenal sensitivity to ACTH (Meaney *et al.*, 1989b, 1992) or in pituitary sensitivity to CRH (Meaney *et al.*, 1988b). Moreover, there are no differences between H and NH animals in the metabolic clearance rate for corticosterone (Grota, 1975; Meaney *et al.*, 1989b). Rather the difference lies in the fact that the NH animals show increased secretion of corticosterone both during and following stress.

Young adult H and NH animals do not differ in levels of corticosteroid-binding globulin (CBG), the principal plasma binder for corticosterone (Meaney *et al.*, 1985c, 1992) or in free corticosterone levels (Meaney *et al.*, 1992). This finding is of considerable importance since brain uptake of corticosterone appears to approximate the non-CBG bound (free + albumin-bound) portion of the steroid (Pardridge *et al.*, 1983). Thus, differences in total corticosterone are probably predictive of differences in brain uptake of the steroid. One further point of importance concerns the specificity of this difference. Young adult H and NH animals do not differ in basal corticosterone levels at any time point over the diurnal cycle (Meaney *et al.*, 1989b, 1992). Thus, the difference between the two groups of animals is specific to conditions of stress and is not seen under normal, resting conditions. This finding also indicates that differences in HPA activity observed during stress cannot be accounted for by differences in pre-stress, basal glucocorticoid levels.

H and NH animals also differ in plasma ACTH and ACTH secretagogue responses to stress. Levels of CRF-like bioactivity (Zarrow *et al.*, 1972) and plasma ACTH (Meaney *et al.*, 1989b; Viau *et al.*, in press) are higher both during and following stress in NH animals. These findings suggest that the mechanism(s) for differences between H and NH animals is located above the level of the pituitary and that the post-stress difference, at least, may be related to differential sensitivity of CNS negative-feedback processes.

10.10.2   THE EFFECT OF HANDLING ON HPA NEGATIVE-FEEDBACK PROCESSES

On the basis of the relative hypersecretion of ACTH and corticosterone by NH animals, we wondered whether H and NH animals might differ in negative-feedback sensitivity to circulating glucocorticoids. This idea was tested (Meaney *et al.*, 1989b) using a classical negative-feedback paradigm based on the finding that high levels of circulating glucocorticoids feedback onto the brain and/or pituitary to inhibit subsequent HPA activity (Keller-Wood and Dallman, 1984; Dallman *et al.*, 1987). Such delayed, negative feedback persists for hours after exposure to elevated glucocorticoid levels (Keller-Wood and Dallman, 1984). In these experiments, H and NH animals were injected with one of five doses of either corticosterone or dexamethasone 3 h before a 20-min immobilization stress. Both glucocorticoids were more effective in suppressing stress-induced HPA responses in the H animals (i.e., the ID_{50} for both corticosterone and dexamethasone was

5–10 times lower in the H animals). These data suggest that the H animals are indeed more sensitive to the negative-feedback effects of circulating glucocorticoids on HPA activity.

Since this delayed form of negative feedback is likely to be mediated by the binding of corticosterone to soluble intracellular receptors, we have measured both mineralocorticoid and glucocorticoid receptor sites in selected brain regions and pituitary of young adult H and NH animals (Meaney and Aitken, 1985b; Meaney *et al.*, 1985d, e, 1987, 1988a, 1989b, 1991; Mitchell *et al.*, 1990a; Sarrieau *et al.*, 1988a). The results of these studies have demonstrated that there are significant and regionally specific differences in glucocorticoid receptor binding capacity as a function of handling. H animals show increased glucocorticoid receptor binding capacity in the hippocampus, but not in septum, amygdala, hypothalamus, or pituitary. The difference in the receptor binding data is clearly related to the number of receptor sites, and not to the affinity of the receptor for [³H]radioligand, RU 28362. Moreover, the difference occurs in glucocorticoid receptors but not the mineralocorticoid receptor sites (measured using either radiolabeled aldosterone or corticosterone + 50-fold excess of cold RU 28362).

In one study Meaney *et al.* (1989b) demonstrated that this difference in hippocampal glucocorticoid receptor density is related to the more efficient suppression of post-stress HPA activity in the H animals. Chronic administration of corticosterone results in a 30–45% down-regulation of hippocampal glucocorticoid receptor binding sites (Olpe and McEwen, 1976; Sapolsky *et al.*, 1984; Spencer, 1990). The effect is highly specific to the hippocampus, such that receptor binding capacity in the hypothalamus and pituitary are unaffected. In this experiment H animals were treated for 5 days with corticosterone and allowed two days for steroid clearance. Hippocampal glucocorticoid receptor density was down-regulated in the H + corticosterone animals to levels that were indistinguishable from those of NH animals, and significantly less than that of the H + vehicle animals. There were no differences in glucocorticoid receptor density in the hypothalamus or pituitary. When the animals in these groups were exposed to a 20-min immobilization stress, it was found that the H+ corticosterone animals, like the NH animals, hypersecreted corticosterone 60 and 120 min post-stress in comparison to the H+ vehicle, control animals. These data suggest that the difference in negative-feedback efficiency between H and NH is related to the differences in hippocampal glucocorticoid receptor density. Thus, the chronic corticosterone treatment reversed both the handling-induced increase in hippocampal glucocorticoid receptor binding capacity and the difference in post-stress HPA activity. It appears, then, that the increase in glucocorticoid receptor sites in the hippocampus is a critical feature for the handling effect on HPA function. The increase in receptor density renders the hippocampus more sensitive to circulating glucocorticoids, enhancing the efficacy of negative-feedback inhibition over HPA activity, and serving to reduce post-stress secretion of ACTH and corticosterone in H animals.

The effect of postnatal handling on HPA negative-feedback probably involves glucocorticoid receptor differences in at least one other region. Handling also increases glucocorticoid receptor density in the frontal cortex (Meaney *et al.*, 1985e). Evidence has recently been provided for the role of the frontal cortex in the regulation of stress-induced HPA activity (Diorio *et al.*, unpublished). Ablations of the medial prefrontal cortex produce increased levels of both ACTH and corticosterone both during and following the termination of stress. Corticosterone implants directly into this region produce a 40–50% decrease in stress-induced ACTH and corticosterone levels. Interestingly, these effects are apparent only with more moderate (neurogenic?) stressors, in this case restraint. Neither ablations of the medial prefrontal cortex nor corticosterone implants into this region had any effect on ACTH or corticosterone levels observed using ether stress, a more severe stressor associated with 2–3 times higher levels of ACTH. Moreover, these effects were observed only during or following stress; neither treatment altered basal ACTH or corticosterone levels at any point over the diurnal cycle. These findings suggest that the handling effect on HPA function might well involve altered glucocorticoid receptor density in the frontal cortex.

10.10.3　THE EFFECT OF HANDLING ON HPA ACTIVITY DURING STRESS

We have focused largely on the mechanisms underlying the differences in post-stress levels of plasma ACTH and corticosterone. However, the majority of the earlier studies on handling were directed at the differences in corticosterone levels between H and NH rats achieved during stress. In general, H animals secreted lower levels of corticosterone during stress. However, this effect was often dependent on the gender of the animals (e.g., Weinberg *et al.*, 1978). Likewise, we have consistently found that H and NH female rats do not differ in corticosterone levels during stress (Meaney *et al.*, 1985c, 1991). Females, like males, differ in post-stress corticosterone levels and in hippocampal glucocorticoid receptor binding (Meaney *et al.*, 1985d, e, 1991). These findings suggest that differences in hippocampal corticosteroid receptor density are probably not related to differences in HPA activity during stress.

The results of an earlier study have implicated an entirely different receptor system in the effect of handling on HPA activity during stress. Handling results in

decreased levels of pituitary intracellular CBG levels in male, but not female rats. Recall that Sakly and Koch (1981, 1983) found that intracellular CBG levels in neonatal rats are very low and are associated with increased nuclear uptake of corticosterone. This increased corticoid signal appears to be associated with enhanced glucocorticoid negative feedback over pituitary ACTH release. Likewise, the reduced pituitary transcortin levels in the H male rats might allow for an increase in the nuclear uptake of corticosterone in the pituitary. Such an effect would be expected to reduce ACTH secretion during stress compared to NH rats, and this is indeed the common finding. It is, of course, interesting that H and NH female rats differ in neither HPA activity during stress nor in pituitary transcortin levels. We are currently examining the hypothesis that differences in stress-induced levels of ACTH release might be related to the effect of handling on pituitary transcortin levels. Regardless of the merit of this hypothesis, these findings remind us that the individual variation in HPA function is likely to involve a number of mechanisms situated at various levels of the axis.

The earlier studies on the effects of postnatal handling on HPA responses to stress were largely restricted to measures of plasma corticosterone. In a recent series of experiments Viau *et al.* (in press) have derived detailed measures of ACTH secretion during either restraint or ether stress. The results show that for both stressors, the integrated ACTH signal (pg/ml/min) is almost two-fold higher in the NH animals (thus far, these studies have only been performed with males). In the course of this work, we also found that basal median eminence levels of both CRH and AVP, but not oxytocin, were higher in the NH animals. The median eminence represents the projection site for the CRH and AVP neurons of the PVN of the hypothalamus. Therefore, these data most likely reflect differences in the axon terminal pool which is released in response to stress. Certainly such differences could explain the enhanced ACTH secretion in the NH animals in response to stress. These data are also consistent with the finding that hippocampal and cortical lesions are associated with increased CRH and AVP mRNA levels in the PVN (Herman *et al.*, 1989). Moreover, this finding would also explain why differences in HPA activity between H and NH animals are observed in response to a wide range of stressors, since any excitatory signal to the PVN neurons would probably result in greater release of CRH and/or AVP in the NH rats. Finally, this finding suggests that although plasma ACTH levels in H and NH animals do not differ under basal conditions, there is still evidence for differential negative-feedback activity in the resting state (at the level of the hypothalamus), and this difference becomes evident under conditions of stress.

10.11 MECHANISM OF ACTION OF HANDLING ON GLUCOCORTICOID RECEPTOR DEVELOPMENT

In our initial studies, animals were handled for the first 21 days of life. Based on earlier work (Levine, 1970), we wondered whether some portion of this 3-week period might represent a critical period for the handling effect on glucocorticoid receptors. In one study (Meaney and Aitken, 1985b), animals were handled daily and killed on either postnatal day 3, 7, 14, or 21 of life. We found that, in comparison to NH animals of the same age, H animals exhibited significantly increased hippocampal glucocorticoid receptor density as early as day 7 of life and that the magnitude of the effect did not increase thereafter. We also found that handling between days 1 and 7 of life was as effective in increasing hippocampal glucocorticoid receptor density as handling over the entire first 3 weeks. Handling over the second week of life (i.e. between days 8 and 14) was

somewhat less effective, whereas animals handled between days 15 and 21 did not differ from NH animals in glucocorticoid receptor binding. Thus, glucocorticoid receptor binding capacity appears to be especially sensitive to environmental regulation during the first week of life.

This temporal pattern corresponds to the normal developmental changes in glucocorticoid receptor density occurring over the early postnatal life (Meaney and Aitken, 1985a; Meaney *et al.*, 1985a; Olpe and McEwen, 1976; Pavlik and Buresova, 1984; Rosenfeld *et al.*, 1988a; Sarrieau *et al.*, 1988a). Glucocorticoid receptor density is low on postnatal day 3 (~30% of adult values) and increases steadily towards adult values which are achieved by about the third week of life. It is during this period of ontogenic variation that environmental events can influence the development of the receptor population. In contrast to the glucocorticoid receptor, mineralocorticoid receptor density does not vary significantly with age; hippocampal mineralocorticoid receptor density in early postnatal life is largely indistinguishable from that of adult rats (Rosenfeld *et al.*, 1988a; Sarrieau *et al.*, 1988a) and, as mentioned above, handling has no effect on hippocampal mineralocorticoid receptor binding (Meaney *et al.*, 1989b; Sarrieau *et al.*, 1988a). Thus, it is tempting to consider the relationship between the developmental pattern in hippocampal glucocorticoid receptor density and (a) the sensitivity of this receptor system to environmental stimuli, and (b) the timing of the critical period for these stimuli on glucocorticoid receptor development. However, the developmental pattern for glucocorticoid receptor binding in regions not affected by handling, such as the hypothalamus, amygdala and septum (Olpe and McEwen, 1976; Meaney *et al.*, 1985a), is identical to that of the hippocampus and the frontal cortex (Meaney and Aitken, 1985b). Thus, it is unlikely that the handling effect on glucocorticoid receptor density in the hippocampus and the frontal cortex can be explained simply by the status of the glucocorticoid receptor system during the first weeks of life.

Handling during the first week of life can involve a mild and transient drop in the pup's body temperature (although this feature of the manipulation is not critical for the effect on HPA activation during stress, see Levine, 1970). This is probably associated with the relatively modest thermoregulatory abilities of young rats over the first two weeks of life. Such changes in body temperature and/or some other sensory component of the handling manipulation activate the hypothalamic–pituitary–thyroid axis, leading to increased levels of circulating T_4 and increased intracellular levels of the biologically more potent T_4 metabolite, T_3. Handling results in a modest, but reliable (~30%) increase in plasma thyroid hormones (Meaney and Aitken, unpublished). Thus, Meaney *et al.* (1987) examined whether the effects of handling might be mediated by increased exposure to these thyroid hormones. Neonatal rat pups were exposed to elevated levels of either T_4, T_3, or the vehicle over the first week of life. Both T_4 and T_3 treatment resulted in significantly increased glucocorticoid receptor binding capacity in the hippocampus in animals examined as adults. Like the handling manipulation, neither T_4 nor T_3 treatment affected hypothalamic or pituitary glucocorticoid receptor density. Moreover, administration of the thyroid hormone synthesis inhibitor, propylthrouracil (PTU), to H pups for the first 2 weeks of life completely blocked the effects of handling on hippocampal glucocorticoid receptor binding capacity. These data are consistent with the idea that the thyroid hormones might mediate, in part at least, the effects of neonatal handling on the development of the hippocampal glucocorticoid receptor system.

Systemic injections of neonatal rat pups represent a fairly crude manipulation and although these data might implicate the thyroid hormones, there is no indication that the hippocampus is actually the critical site of action for thyroid hormone effect. In order to examine whether thyroid hormones might act directly on hippocampal cells we have turned to an *in vitro* system, using primary cultures of dissociated hippocampal cells (Banker and Cowan, 1977). The hippocampal cells are taken from embryonic rat pups (E20) and beginning on the fifth day after plating the cultures were exposed to 0, 1, 10, or 100 nM T_3. The cells were cultured in 10% fetal calf serum, stripped of thyroid hormones. Thus far, the results of several experiments have failed to detect any effect of thyroid hormones on glucocorticoid receptor density in cultured hippocampal cells (Meaney, unpublished). These *in vitro* data suggest that (a) the effects of the thyroid hormones on the glucocorticoid receptor binding capacity occur at some site distal to the hippocampal cells or (b) thyroid hormones interact at the level of the hippocampus with some other hormonal signal that is obligatory for the expression of the thyroid hormone effect.

Thyroid hormones have pervasive effects throughout the developing CNS and one such effect involves the regulation of central serotonergic neurons (e.g., Savard *et al.*, 1984). Thyroid hormones increase serotonin (5-HT) turnover in the hippocampus of the neonatal rat (Mitchell *et al.*, 1990a). Handling also increases hippocampal 5-HT turnover (Mitchell *et al.* 1990a; Smythe *et al.*, 1991), and thus both manipulations increase serotonergic stimulation of hippocampal neurons. There is also direct evidence for an effect of 5-HT on glucocorticoid receptor density in the neonatal rat. 5,7-Dihydroxytryptamine (5,7,-DHT) lesions of the raphe 5-HT neurons dramatically reduce the ascending serotonergic input into the hippocampus. Rat pups administered 5,7,-DHT on day 2 of life showed reduced hippocampal glucocorticoid receptor density as adults (Mitchell *et al.*, 1990a). Similar lesions of ascending catecholaminergic systems, using

6-hydroxydopamine, had no effect on the development of glucocorticoid receptor density in the hippocampus (Meaney, unpublished). Interestingly, this effect may also be specific to the neonatal period, since similar chemical lesions of the ascending 5-HT neurons in adult animals do not appear to affect hippocampal glucocorticoid receptor binding capacity (M. Lowy, personal communication). Finally, it was found that effects of postnatal handling on glucocorticoid receptor binding are blocked by concurrent administration of the 5-HT$_2$ receptor antagonist, ketanserin (Mitchell *et al.*, 1990a).

Recent studies (Smythe *et al.*, 1991) of 5-HT turnover have provided some insight into why the hippocampus and frontal cortex are selectively affected by handling. It was found that when rat pups were handled for the first seven days of life, and killed immediately after handling on day 7, that 5-HT turnover was significantly increased in the hippocampus and frontal cortex, but not in the hypothalamus or amygdala (regions where handling has no effect on glucocorticoid receptor density). These data suggest that handling selectively activates certain ascending 5-HT pathways and that it is this feature of the handling effect, together with the existence of a developing glucocorticoid receptor system, that underlies the sensitivity of this receptor system in specific brain regions to regulation by environmental events during the first weeks of life.

The next question then, is whether 5-HT might be mediating the effects of handling at the level of the hippocampal cells. We have recently found that 5-HT has a profound effect on glucocorticoid receptor density in cultured hippocampal cells (Mitchell *et al.*, 1990b, 1992). In hippocampal cells cultured in the presence of 10 nM 5-HT there was a two-fold increase in glucocorticoid receptor binding. The effect of 5-HT was dose-related, with an EC$_{50}$ of 4.3 nM, and required about 4 days for the maximal effect to occur. Shorter periods of exposure are ineffective, suggest-

ing that the effect of 5-HT may involve the increased synthesis of receptors. Moreover, the effect of 5-HT on glucocorticoid receptor binding appears to occur uniquely in the neuronal cell population. We found that there was no effect of 5-HT on glucocorticoid receptor binding in hippocampal glial-enriched cell cultures. This finding is not surprising, since our initial studies were performed with cultures composed largely (~75%) of neuron-like cells (Mitchell *et al.*, 1990b).

There are two very intriguing features of this effect that bear directly on the question of the hippocampal cell cultures as a model for neural differentiation (Meaney *et al.*, unpublished). First, the effects of 5-HT on glucocorticoid receptor binding in hippocampal cell cultures are restricted to the first 3 weeks in culture. Thus, cultures treated with 10 nM 5-HT for 7 days at any time during the first 3 weeks in culture show a significant increase in glucocorticoid receptor density; however, the effect is lost after this point. Cultures treated with 10 nM 5-HT for 7 days during weeks 3–4 after plating show no increase in glucocorticoid receptor binding. Second, and perhaps most exciting, the increase in glucocorticoid receptor binding capacity following exposure to 10 nM 5-HT persists as long as 40 days after removal of 5-HT from the medium. Thus, the effect of 5-HT on glucocorticoid receptor density observed in hippocampal culture cells mimics the long-term effects of neonatal handling. It will be of considerable interest to understand the cellular events underlying these features of the 5-HT effect on hippocampal development.

The effects of 10 nM 5-HT on glucocorticoid receptor density in cultured hippocampal cells was completely blocked by the 5-HT$_2$ receptor antagonists, ketanserin and mianserin (Mitchell *et al.*, 1990b). Moreover, the 5-HT$_2$ agonist 1-(2,5-dimethoxy-4-iodophenyl)-2-amino-propane (DOI), *m*-trifluromethyl-phenylpiperazine (TFMPP), and quipazine were also effective in increasing glucocorticoid receptor binding in hippocampal cultures. Selective agonists or antagonists of the 5-HT$_{1a}$ or 5-HT$_3$ receptors had no effect on glucocorticoid receptor binding. Using [^{125}I]-labeled 7-amino-8-iodoketanserin as radioligand, we have confirmed the presence of high affinity 5-HT$_2$ binding sites in our cultured hippocampal cells. Titeler *et al.* (1987) have provided evidence that the ketanserin-labeled 5-HT$_2$ site may exist in two states: an agonist state (5-HT$_{2H}$) with a high, nanomolar affinity for 5-HT and an antagonist state (5-HT$_{2L}$) with a low, micromolar affinity for 5-HT. In both states the receptor shows a high affinity for antagonists, such as ketanserin. The 5-HT$_{2H}$ site binds with high affinity to DOI, TFMPP, and quipazine, and Titeler *et al.* (1987) have reported a K_d of ~5 nM for 5-HT. This K_d is a close approximation of the EC$_{50}$ (4.3 nM) for the effect of 5-HT on glucocorticoid receptors in cultured hippocampal cells. Taken together, these findings suggest that this effect of 5-HT appears to be mediated via high-affinity, 5-HT$_2$ receptor.

We are currently examining the nature of the secondary messenger systems involved in this serotonergic effect on glucocorticoid receptor binding. In doing so, it has been found (Mitchell *et al.*, 1992) that low nanomolar concentrations of 5-HT (EC$_{50}$ = 7.2 nM) increase cAMP levels in cultured hippocampal cells, with no effect on cGMP levels. This increase in cAMP is blocked by ketanserin and mimicked by both quipazine and DOI. Moreover, treatment with the stable cAMP analog, 8-bromo-cAMP, or with 10 μM-forskolin produces a significant increase in glucocorticoid receptor density in cultured hippocampal neurons. The effect of 8-bromo-cAMP is concentration-related, and the maximal effect of 8-bromo-cAMP (~181%) is comparable to that for 5-HT (~195%). Interestingly, as with 5-HT, the effects of 8-bromo-cAMP on glucocorticoid receptor density were not apparent until at least 4 days of treatment. The similarity in the time courses

is of obvious interest, and the actual time period involved suggests that both treatments might exert their effects through an alteration in receptor synthesis. Indeed, the effect of 10 nM 5-HT on glucocorticoid receptor density in cultured hippocampal cells is blocked by either cycloheximide or actinomycin D (O'Donnell and Meaney, unpublished).

Taken together, these findings suggest that changes in cAMP concentrations may mediate the effects of 5-HT on glucocorticoid receptor synthesis in hippocampal cells. (Mitchell *et al.* (in press) have found that the cyclic nucleotide-dependent protein kinase inhibitor, H8 (Hidaka *et al.*, 1984) completely blocked the effects of 10 nM 5-HT on glucocorticoid receptor binding in hippocampal cell cultures. In contrast, the protein kinase C inhibitor, H7, had no such effect. These data suggest that activation of protein kinase A might, at some point, be involved in the serotonergic regulation of hippocampal glucocorticoid receptor development.

We must also consider the fact that we are working with intact cells over long incubation periods, and there is ample possibility for an interaction between second messenger systems. Indeed, we have preliminary data which suggest that treatment with a phorbol ester also increases glucocorticoid receptor binding in the hippocampal cell cultures (O'Donnell and Meaney, unpublished). The post-receptor events involved in the regulation of glucocorticoid receptor development are likely to be complex, but these studies should ultimately lead to an understanding of how environmental events can regulate glucocorticoid receptor gene expression and neuronal differentiation in the hippocampus.

10.12 CONCLUSIONS

Handling during the early postnatal period leads to increased glucocorticoid receptor binding in the hippocampus and is associated with enhanced negative-feedback control over HPA function. It is likely that this plasticity reflects a basic process, whereby the early environment is able to 'fine tune' the sensitivity and efficiency of certain neuro-endocrine systems that mediate the animal's response to stimuli which threaten home-ostasis. Such plasticity is likely to be of considerable importance to a species like the rat, which prospers in a tremendous range of ecological niches. In this sense, it is important to note that we do not consider H or NH animals to be superior to one another. They differ as a function of the variation in early stimulation afforded to the animals. What is of greatest interest is the possibility that the handling manipulation provides a paradigm for the study of the molecular processes through which the development of specific neuroendocrine systems is directed by the environment.

Over the course of this review it has been our intention to provide the reader with an appreciation of how changes in corticosteroid receptor systems alter HPA function, both in terms of regulating negative-feedback pro-cesses, and in determining target tissue sensitivity to circulating glucocorticoids. The changes in receptor density over develop-ment are tissue-specific. In some cases, such as the lung, intestine, and pituitary, there are high levels of glucocorticoid receptors in the neonate and substantial sensitivity to corticosteroids. In other regions, such as the corticolimbic systems of the brain, there are reduced levels of glucocorticoid receptors (although high levels of mineralocorticoid receptors). At one level, it is not hard to understand why such differences should exist. Glucocorticoids often subserve very different functions in these tissues. Even among regions where corticoid action is comparable, the consequences may be entirely different. Arresting cell proliferation in one tissue may have very different implications than in another structure. Thus, one of the central questions that emerges at this point concerns the processes underlying the tissue-specific regulation of corticosteroid receptor gene expression over development.

ACKNOWLEDGEMENTS

Supported by grants from the Medical Research Council of Canada (MRCC) to MJM. MJM was supported by a University Research Fellowship from the Natural Sciences and Engineering Research Council of Canada (NSERC) for much of the period when this research was completed, and currently holds an MRCC Scientist award. SB is a graduate fellow of the MRCC and the Canadian Heart and Stroke Foundation, AS is a postdoctoral fellow of the Fonds de la Recherche en Santé du Quebec and the Canadian Alzheimer's Society, JS and NS are NSERC post-doctoral fellows, and C.-D. W. is supported by a grant from the Swiss National Fund for Research. The authors would like to dedicate this chapter to the memory of Professor Shawn Schapiro, who was killed in an airplane accident shortly after beginning his pioneer-ing studies on the development of the HPA axis in the rat. Professor Schapiro was an eloquent and inspiring spokesperson for the wonders of developmental neuroendocrinol-ogy, and can justifiably be considered as a founder of this field of study.

REFERENCES

Adamson, W.T., Windh, R.T., Blackford, S. and Kuhn, C.M. (1991) Ontogeny of μ- and κ-opiate receptor control of the hypothalamic–pituitary–adrenal axis in rats. *Endocrinology*, **129**, 959–64.

Ader R. and Grota L.J. (1969) Effects of early experience on adrenocortical reactivity. *Physiol. Behav.*, **4**, 303–5.

Akana, S., Cascio, C.S., Du, J.-Z. *et al.* (1986) Reset in feedback in the adrenocortical system: an apparent shift in sensitivity of adrenocorti-cotropin to corticosterone between morning and evening. *Endocrinology*, **119**, 2325–32.

Almazan, G., Lefebvre, D.L. and Zingg, H.H. (1989) Ontogeny of hypothalamic vasopressin, oxytocin, and somatostatin gene expression. *Dev. Brain Res.*, **45**, 69–75.

Antoni, F.A. (1986) Hypothalamic control of ACTH secretion: advances since the discovery of 41-residue corticotropin-releasing factor. *Endocr. Rev.*, **7**, 351–70.

Arriza, J.L., Simerly, R.B., Swanson, L.W. and Evans, R. M. (1988) Neuronal mineralocorticoid receptor as a mediator of glucocorticoid response. *Neuron*, **1**, 887–900.

Balazs, R. and Cotterrell, M. (1972) Effect of hormonal state on cell number and functional maturation of the brain. *Nature*, **236**, 348–50.

Banker, G.A. and Cowan, W.M. (1977) Rat hippocampal neurons in dispersed cell culture. *Brain Res.*, **126**, 397–425.

Bartova, A. (1968) Functioning of the hypothalamo–pituitary–adrenal axis during postnatal development in rats. *Gen. Comp. Endocrinol.*, **10**, 235–9.

Baxter, J.D. and Tyrrell, J.B. (1987) The adrenal cortex, in *Endocrinology and Metabolism* (eds P. Felig, J.D. Baxter, A.E. Broadus and L.A. Frohman), McGraw-Hill, New York, pp. 385–511.

Beaumont, K. and Fanestil, D.D. (1983) Characterization of rat brain aldosterone receptors reveals high affinity for corticosterone. *Endocrinology*, **113**, 2043–51.

Beyer, H.S., Matta, S.G. and Sharp, B.M. (1988) Regulation of the messenger ribonucelic acid for corticotropin-releasing factor in the paraventricular nucleus and other brain sites of the rat. *Endocrinology*, **123**, 2117–23.

Bohn, M.C. (1980) Granule cell genesis in the hippocampus of rats treated neonatally with hydrocortisone, *Neuroscience*, **5**, 2003–12.

Bohn, M.C., McEwen, B.S., Luine, V.N. and Black, I.B. (1984) Development and characterization of glucocorticoid receptors in rat superior cervical ganglion. *Dev. Brain Res.*, **14**, 211–18.

Bradbury, M. and Dallman, M.F. (1989) Effects of type I and type II glucocorticoid receptor antagonists on ACTH levels in the PM. *Soc. Neurosci.*, **19**, 716.

Brindley, D.N. and Rolland, Y. (1989) Possible connections between stress, diabetes, obesity, hypertension and altered lipoprotein metabolism that may result in atherosclerosis. *Clin. Sci.*, **77**, 453–61.

Bugnon, C., Fellman, D., Gouget, A. *et al.* (1984) Corticoliberin neurons: cytophysiology, phylogeny, and ontogeny. *J. Steroid Biochem. Mol. Biol.*, **20**, 183–95.

Burnstein, K.L. and Cidlowski, J.A. (1989) Regulation of gene expression by glucocorticoids. *Annu. Rev. Physiol.*, **51**, 683–99.

Carson-Jurica, M.A., Schrader, W.T. and O'Malley, B.W. (1990) Steroid receptor family: structure and functions. *Endocr. Rev.*, **11**, 201–20.

Chatelain, A., Durand, P., Naaman, E. and Dupouy, J.P. (1989) Ontogeny of ACTH (1-24) receptors in rat adrenal glands during the perinatal period. *J. Endocrinol.*, **123**, 421–8.

Childs, G.V. (1987) Cytochemical studies of the regulation of ACTH secretion. *Ann. N Y Acad. Sci.*, **512**, 248–74.

Childs, G.V., Ellison, D.G. and Ramaley, J.A. (1982) Storage of anterior lobe adrenocorticotropin in corticotropes and a subpopulation of gonadotropes during the stress-nonresponsive period in the neonatal male rat. *Endocrinology*, **110**, 1676–92.

Clayton, C., Grosser, B. and Stevens, W. (1977) The ontogeny of corticosterone and dexamethasone receptors in rat brain. *Brain Res.*, **134**, 445–53.

Coté, T. and Yasumura, S. (1975) Effect of ACTH and histamine stress on serum corticosterone and adrenal cAMP levels in immature rats. *Endocrinology*, **96**, 1044–7.

D'Agostino, J. and Henning, S.J. (1981) Hormonal control of postnatal development of corticosteroid-binding globulin. *Am. J. Physiol.*, **240**, E402–406.

Dallman, M.F., Akana, S., Cascio, C.S. *et al.* (1987) Regulation of ACTH secretion: variations on a theme of B. *Recent Progr. Horm. Res.*, **43**, 113–73.

Dallman, M.F., Levin, N., Cascio, C.S. *et al.* (1989) Pharmacological evidence that the diurnal regulation of adrenocorticotropin secretion by corticosteroids is mediated by type I corticosterone-preferring receptors. *Endocrinology* **124**, 2844–50.

Danielson, M., Northrup, J.P. and Ringold, G.M. (1986) The mouse glucocorticoid receptor: mapping of functional domains by cloning, sequencing, and expression of wild-type and mutant receptor proteins. *EMBO J.*, **5**, 2513–22.

De Kloet, E.R. and McEwen, B.S. (1976) A putative glucocorticoid receptor and transcortin-like macromolecule in pituitary cytosol. *Biochim. Biophys. Acta*, **421**, 115–23.

Diamond, M.I., Miner, J.N., Yoshinaga, S.K. and Yamamoto, K.R. (1990) Transcription factor interactions: selectors of positive or negative regulation from a single DNA element. *Science*, **249**, 1266–72.

Dupouy, J.P., Coffigny, H. and Magre, S. (1975) Maternal and foetal corticosterone levels during late prenancy in rats. *J. Endocrinol.*, **65**, 347–52.

Edwards, C.R.W., Stewart, P.M., Burt, D. *et al.*

(1988) Localisation of 11β-hydroxysteroid dehydrogenase-tissue specific protector of the mineralcorticoid receptor. *Lancet,* **ii**, 986–9.

Emadian, S.M., Luttge, W.G. and Densmore, C.L. (1986) Chemical differentation of type I and type II receptors for adrenal steroids in brain cytosol. *J. Steroid Biochem. Mol. Biol.,* **24**, 953–61.

Evans, R.M. and Arriza, J.L. (1989) A molecular framework for the actions of glucocorticoid hormones in the nervous system. *Neuron,* **2**, 1105–12.

Feldman, S. and Conforti, N. (1976) Feedback effects of dexamethasone on adrenocortical responses in rats with fornix lesions. *Horm. Res.,* **7**, 56–60.

Feldman, S. and Conforti, N. (1980) Participation of the dorsal hippocampus in the glucocorticoid negative-feedback effect on adrenocortical activity. *Neuroendocrinology,* **30**, 52–5.

Fischette, C.T., Komisaruk, B. R., Edinger, H.M. *et al.* (1980) Differential fornix ablations and the circadian rhythmicity of adrenal corticosterone secretion. *Brain Res.,* **195**, 373–80.

Funder, J.W. and Sheppard, K. (1987) Adrenocortical steroids and the brain. *Annu. Rev. Physiol.,* **49**, 397–412.

Funder, J.W., Pearce, J.T., Smith, R. and Smith, A. I. (1988) Mineralocorticoid action: target tissue specificity is enzyme, not receptor, mediated. *Science,* **242**, 583–5.

Gaillard, R.C., Riondel, A., Muller, A.F. *et al.* (1984) RU 486: a steroid with antiglucocorticosteroid activity that only disinhibits the human pituitary–adrenal system at a specific time of day. *Proc. Natl. Acad. Sci. USA,* **81**, 3879–82.

Gala, R.R. and Westphal, U. (1965) Corticosteroid-binding globulin in the rat: studies on the sex differences. *Endocrinology,* **77**, 841–51.

Gibbs, D.M. (1986) Vasopressin and oxytocin: hypothalamic modulators of the stress response. *Psychoneuroendocrinology,* **11**, 131–40.

Goldman, L., Winget, C., Hollinshead, G. and Levine, S. (1978) Postweaning development of negative feedback in the pituitary–adrenal system of the rat. *Neuroendocrinology,* **12**, 199–211.

Gould, E., Woolley, C. and McEwen, B.S. (1990) Short-term glucocorticoid manipulations affect neuronal morphology and survival in the adult dentate gyrus. *Neuroscience,* **37**, 367–75.

Grino, M., Young, S.W. and Burgunder, J.M. (1989a) Ontogeny of expression of corticotropin-releasing factor gene in the hypothalamic paraventricular nucleus and of the proopiomelanocortin gene in rat pituitary. *Endocrinology,* **124**, 60–8.

Grino, M., Burgunder, J.M., Eskay, R.L. and Eiden, L.E. (1989b) Onset of glucocorticoid responsiveness of anterior pituitary corticotrophes during development is scheduled by corticotrophin-releasing factor. *Endocrinology,* **124**, 2686–92.

Grota, L.J. (1975) Effects of early experience on the metabolism and production of corticosterone in rats. *Dev. Psychobiol.,* **9**, 211–15.

Guillet, R. and Michaelson, S.M. (1978) Corticotropin responsiveness in the neonatal rat. *Neuroendocrinology,* **27**, 119–25.

Guillet, R., Saffran, M. and Michaelson, S.M. (1980) Pituitary–adrenal response in neonatal rats. *Endocrinology,* **106**, 991–4.

Hammond, G.L. (1990) Molecular properties of corticosteroid-binding globulin and sex-steroid binding proteins. *Endocr. Rev.,* **11**, 65–79.

Hary, L., Dupuoy, J.P. and Chatelain, A. (1984) Effect of norepinephrine on the pituitary adrenocorticotrophic activation by ether stress and on *in vitro* release of ACTH by the adrenohypophysis of male and female newborn rats. *Neuroendocrinology,* **39**, 105–13.

Henning, S. (1978) Plasma concentrations of total and free corticosterone during development in the rat. *Am. J. Physiol.,* **235**, E451–E456.

Henning, S.J., Ballard, P.L. and Kretchmer, N. (1975) A study of the cytoplasmic receptors for glucocorticoids in intestine of pre- and post-weaning rats. *J. Biol. Chem.,* **250**, 2073–9.

Herman, J.P., Schafer, M.K.-H., Young, E.A. *et al.* (1989) Evidence for hippocampal regulation of neuroendocrine neurons of the hypothalamo–pituitary–adrenocortical axis. *J. Neurosci.,* **9**, 3072–82.

Hess, J.L., Denenberg, V.H., Zarrow, M.X. and Pfeifer, W.D. (1969) Modification of the corticosterone response curve as a function of handling in infancy. *Physiol. Behav.,* **4**, 109–12.

Hidaka, H., Inagaki, M., Kawamato, S. and Sasaki Y. (1984) Isoquinolinesulfonamides, novel and potent inhibitors of cyclic nucleotide dependent protein kinase A and protein kinase *C*. *Biochemistry,* **23**, 5036–41.

Hollenberg, S.M., Weinberger, C., Ong, E.S. *et al.* (1985) Primary structure and expression of a functional human glucocorticoid receptor c DNA. *Nature,* **318**, 635–41.

Horner, H. C., Packan, D. R. and Sapolsky, R.M. (1990) Glucocorticoids inhibit glucose transport in cultured hippocampal neurons and glia. *Neuroendocrinology,* **52**, 57–64.

Hotta, M., Shibasaki, T., Masuda, A. *et al.* (1988) Ontogeny of pituitary responsiveness to corticotropin-releasing hormone in the rat. *Regul. Pept.*, **21**, 245–52.

Howard, E. and Benjamin, J.A. (1975) DNA, ganglioside and sulfatide in brains of rats given corticosterone in infancy, with an estimate of cell loss during development. *Brain Res.*, **92**, 73–87.

Hryb, D.J., Khan, M.S., Romas, N.A. and Rosner, W. (1986) Specific binding of human corticosteroid-binding globulin to cell membranes. *Proc. Natl. Acad. Sci. USA*, **83**, 3253–6.

Iny, L.J., Gianoulakis, C., Palmour, R. and Meaney, M.J. (1987) The β-endorphin response to stress during postnatal development in the rat. *Dev. Brain Res.*, **31**, 177–81.

Jacobson, L. and Sapolsky, R.M. (1991) The role of the hippocampus in feedback regulation of the hypothalamic–pituitary–adrenal axis. *Endocr. Rev.*, **12**, 118–34.

Johnson, J.T. and Levine, S. (1973) Influence of water deprivation on adrenocortical rhythms. *Neuroendocrinology*, **11**, 268–73.

Kalimi, M. (1984) Glucocorticoid receptors: from development to aging, a review. *Mech. Ageing Dev.*, **24**, 129–38.

Keller-Wood, M. and Dallman, M. (1984) Corticosteroid inhibition of ACTH secretion, *Endocr. Rev.*, **5**, 1–24.

Kido, H., Fukusen, N. and Katunuma, N. (1987) Inhibition by 1-(5-isoquinolonesulfonyl)-2-methylpiperazine, an inhibitor of protein kinase C, of enzyme induction by glucocorticoid and of nuclear translocation of glucocorticoid–receptor complexes. *Biochem. Biophys. Res. Commun.*, **144**, 152–9.

Kitay, J.I. (1961) Sex differences in adrenocortical secretion in the rat. *Endocrinology*, **68**, 818–24.

Koch, B. (1969) Fractions libre de la corticosterone plasmatique et la response hypophyso-surrenalieene à stress durant la periode postnatal chex le rat. *Horm. Metab. Res.*, **1**, 301–8.

Koch, B. and Lutz-Bucher, B. (1990) The vasopressin receptor system in the neonatal pituitary gland: evidence for reduced binding capacity and signal transmission. *Neuroendocrinology*, **51**, 592–8.

Koch, B., Sakly, M. and Lutz-Bucher, M. (1981) Modulation by transcortin-like binding sites of uptake and distribution of glucocorticoids by dispersed pituitary cells. *Mol. Cell. Endocrinol.*, **22**, 169–78.

Kovacs, K.J. and Makara, G.B. (1988) Corticosterone and dexamethasone act at different brain sites to inhibit adrenalectomy-induced adrenocorticotropin hypersecretion. *Brain Res.*, **474**, 205–10.

Krieger, D.T. (1974) Food and water restriction shifts corticosterone, temperature, activity, and brain amine periodicity. *Endocrinology*, **95**, 1195–201.

Krieger, D.T. and Hauser, H. (1978) Comparison of synchronisation of circadian corticosterone rhythms by photoperiod and food. *Proc. Natl. Acad. Sci. USA*, **75**, 1577–81.

Krozowski, Z.S. and Funder, J.W. (1983) Renal mineralocorticoid receptors and hippocampal corticosterone binding species have intrinsic steroid specificity. *Proc. Natl. Acad. Sci. USA*, **80**, 6056–60.

Kuhn, R.W., Green, A., Raymoure, W. and Siiteri, P. (1986) Immunocytochemical localization of corticosteroid-binding globulin in rat tissues. *J. Endocrinol.*, **108**, 31–43.

Lakshmi, V., Sakai, R.R., McEwen, B.S. and Monder, C. (1991) Regional distribution of 11β-hydroxysteroid dehydrogenase in rat brain. *Endocrinology*, **128**, 1741–8.

Lawson, A., Ahima, R., Krozowski, Z. and Harlan, R. (1991) Postnatal development of corticosteroid receptor immunoreactivity in rat hippocampus. *Dev. Brain Res.*, **62**, 69–79.

Levin, N., Shinsako, J. and Dallman, M.F. (1987) Corticosterone acts on the brain to inhibit adrenalectomy-induced adrenocorticotropin secretion. *Endocrinology*, **122**, 694–701.

Levine, S. (1957) Infantile experience and resistence to physiological stress. *Science*, **126**, 405–6.

Levine, S. (1962) Plasma-free corticosteroid response to electric shock in rats stimulated in infancy. *Science*, **135**, 795–6.

Levine, S. (1970) The pituitary–adrenal system and the developing brain. *Progr. Brain Res.*, **32**, 79–102.

Levine, S., Glick, D. and Nakane, P.K. (1972) Adrenal and plasma corticosterone and vitamin A in rat adrenal glands during postnatal development. *Endocrinology*, **80**, 910–14.

Levine, S., Haltmeyer, G.C., Karas, G.G. and Denenberg, V.H. (1967) Physiological and behavioral effects of infantile stimulation. *Physiol. Behav.*, **2**, 55–63.

Lian, J.B., Stein, G.S., Bortell, R. and Owen, T.A. (1991) Phenotype suppression: a postulated molecular mechanism for mediating the relationship of proliferation and differentiation by Fos/Jun interactions at AP-1 sites in steroid responsive promoter elements of tissue-specific genes. *J. Cell. Biochem.*, **45**, 9–14.

Luttge, W.G., Davida, M.M., Rupp, M.E. and Kang, C.G. (1989) High affinity binding and regulatory actions of dexamethasone-type I receptor complexes in mouse brain. *Endocrinology*, **125**, 1194–203.

Mains, R.E. and Eipper, B.A. (1990) The tissue-specific processing of pro-ACTH/endorphin. *Trends Endocrinol. Metab.*, **1**, 388–94.

Martin, C.E., Cake, M.H., Hartmann, P.E. and Cook, I.F. (1977) Relationship between foetal corticosteroids, maternal progesterone, and parturition in the rat. *Acta Endocrinol.*, **84**, 167–76.

McEwen, B.S. and Gould, E. (1990) Adrenal steroid influences on the survival of hippocampal neurons. *Biochem. Pharmacol.*, **40**, 2393–402.

McEwen, B.S., Lambdin, L.T., Rainbow, T.C. and de Nicola, A.F. (1986a) Aldosterone effects on salt appetite in adrenalectomized rats. *Neuroendocrinology*, **43**, 38–43.

McEwen, B.S., de Kloet, E.R. and Rostene, W.H. (1986b) Adrenal steroid receptors and actions in the nervous system. *Physiol. Rev.*, **66**, 1121–50.

Meaney, M.J. and Aitken, D.H. (1985a) [^{3}H]Dexamethasone binding in rat frontal cortex. *Brain Res.*, **328**, 176–80.

Meaney, M.J. and Aitken, D.H. (1985b) The effects of early postnatal handling on the development of hippocampal glucocorticoid receptors: temporal parameters. *Dev. Brain Res.*, **22**, 301–4.

Meaney, M.J., Sapolsky, R.M. and McEwen, B.S. (1985a) The development of the glucocorticoid receptor system in the rat limbic brain. I. Ontogeny and autoregulation. *Dev. Brain Res.*, **18**, 159–64.

Meaney, M.J., Sapolsky, R.M. and McEwen, B.S. (1985b) The development of the glucocorticoid receptor system in the rat limbic brain. II. An autoradiographic study. *Dev. Brain Res.*, **18**, 165–8.

Meaney, M.J., Sapolsky, R.M., Aitken, D.H. and McEwen, B.S. (1985c) [^{3}H]Dexamethasone binding in fetal rat limbic brain. *Dev. Brain Res.*, **23**, 297–300.

Meaney, M.J., Aitken, D.H., Bodnoff, S.R. *et al.* (1985d) The effects of postnatal handling on the development of the glucocorticoid receptor systems and stress recovery in the rat. *Progr. Neuropsychopharmacol. Biol. Psychiatr.*, **7**, 731–4.

Meaney, M.J., Aitken, D.H., Bodnoff, S.R. *et al.* (1985e) Early, postnatal handling alters glucocorticoid receptor concentrations in selected brain regions. *Behav. Neurosci.*, **99**, 760–5.

Meaney, M.J., Aitken, D.H. and Sapolsky, R.M. (1987) Thyroid hormones influence the development of hippocampal glucocorticoid receptors in the rat: a mechanism for the effects of postnatal handling on the development of the adrenocortical stress response. *Neuroendocrinology*, **45**, 278–83.

Meaney M.J., Aitken, D.H., Bhatnagar, S. *et al.* (1988a) Postnatal handling attenuates neuroendocrine, anatomical, and cognitive impairments related to the aged hippocampus. *Science*, **238**, 766–8.

Meaney, M.J., Viau, V., Bhatnagar, S. and Aitken, D.H. (1988b) Occupancy and translocation of hippocampal glucocorticoid receptors during and following stress. *Brain Res.*, **445**, 198–203.

Meaney, M.J., Viau, V., Aitken, D.H. and Bhatnagar, S. (1989a) Glucocorticoid receptors in brain and pituitary of the lactating rat. *Physiol. Behav.*, **43**, 209–12.

Meaney, M.J., Aitken, D.H., Sharma, S. *et al.* (1989b) Postnatal handling increases hippocampal type II, glucocorticoid receptors and enhances adrencocortical negative-feedback efficacy in the rat. *Neuroendocrinology*, **51**, 597–604.

Meaney, M.J., Aitken, D.H., Bhatnagav, S. and Sapolsky, R.M. (1991) Postnatal handling attenuates certain neuroendocrine, anatomical, and cognitive dysfunction associated with aging in female rats. *Neurobiol. Aging*, **12**, 31–8.

Meaney, M.J., Aitken, D.H., Sharma, S. and Viau, V. (1992) Basal ACTH, corticosterone, and corticosterone-binding globulin levels over the diurnal cycle, and hippocampal type I and type II corticosteroid receptors in young and old, handled and nonhandled rats. *Neuroendocrinology*, **55**, 204–13.

Miesfeld, R., Okref, S., Wikstrom, A.-C. *et al.* (1984) Characterization of a steroid hormone receptor gene and mRNA in wild-type and mutant cells. *Nature*, **312**, 779–81.

Meyer, J.S. (1986) Biochemical effects of corticosteroids on neural tissues. *Physiol. Rev.*, **65**, 970–1019.

Mitchell, J. B., Betito, K., Rowe, W. *et al.* (1992) Serotonergic regulation of type II corticosteroid receptor binding in cultured hippocampal cells: the role of serotonin-induced increases in cAMP levels. *Neuroscience*, **48**, 631–9.

Mitchell, J. B., Iny, L. J. and Meaney, M. J. (1990a) The role of serotonin in the development and environmental regulation of hippocampal type II corticosteroid receptors. *Dev. Brain Res.*, **55**, 231–5.

Mitchell, J. B., Rowe, W., Boksa, P. and Meaney, M. J. (1990b) Serotonin regulates type II corticosteroid receptor binding in hippocampal cell cultures. *J. Neurosci.*, **10**, 1745–52.

Moisan, M.-P., Seckl, J.R. and Edwards, C.R.W. (1990) 11β-hydroxysteroid dehydrogenase bioactivity and messenger RNA expression in rat forebrain: localization in hypothalamus, hippocampus, and cortex. *Endocrinology*, **127**, 1450–5.

Moisan, M.-P., Edwards, C.R.W. and Seckl, J.R. (1992) Ontogeny of 11β-hydroxysteroid dehydrogenase in rat brain and kidney. *Endocrinology*, **130**, 400–4.

Moisan, M.-P., Seckl, J.R. and Edwards, C.R.W. Effect of adrenalectomy and dexamethasone administration on 11β-hydroxysteroid dehydrogenase *in vitro* activity in rat brain and kidney. *J. Physiol.*, (in press).

Munck, A., Guyre, P.M. and Holbrook, N.J. (1984) Physiological functions of glucocorticoids in stress and their relations to pharmacological actions. *Endocr. Rev.*, **5**, 25–44.

O'Donnell, D. and Meaney, M.J. (1991) Steroidal regulation of type II corticosteroid receptors in primary cultures of dispersed hippocampal neurons. *Soc. Neurosci.*, **17**, 1554.

Olpe, H.R. and McEwen, B.S. (1976) Glucocorticoid binding to receptor-like proteins in rat brain and pituitary: ontogenetic and experimentally-induced changes. *Brain Res.*, **105**, 121–8.

Pardridge, W. M., Sakiyama, R. and Judd, H. L. (1983) Protein-bound corticosterone in human serum is selectively transported into rat brain and liver *in vivo*. *J. Clin. Endocrinol. Metab.*, **57**, 160–6.

Pavlik, A. and Buresova, M. (1984) The neonatal cerebellum: the highest level of glucocorticoid receptors in the brain. *Dev. Brain Res.*, **12**, 13–20.

Peiffer, A., Barden, N. and Meaney, M.J. (1991) Age-related changes in glucocorticoid receptor binding and mRNA levels in the rat brain and pituitary. *Neurobiol. Aging*, **12**, 475–9.

Perrot-Appelant, M., Racadot, O. and Milgrom, E. (1984) Specific localization of corticosteroid-binding globulin immunoreactivity in pituitary corticotrophs. *Endocrinology*, **115**, 559–69.

Plotsky, P.M. (1987) Regulation of hypophysiotropic factors mediating ACTH secretion. *Ann. NY Acad. Sci.*, **512**, 205–17.

Plotsky, P.M. and Vale, W.W. (1984) Hemorrhage-induced secretion of corticotropin-releasing factor-like immunoreactivity into the rat hypophysial portal circulation and its inhibition by glucocorticoids. *Endocrinology*, **114**, 164–9.

Plotsky, P.M. and Sawchenko, P.E. (1987) Hypophysial-portal plasma levels, median eminence content, and immunohistochemical staining of corticotropin-releasing factor, arginine vasopressin, and oxytocin after pharmacological adrenalectomy. *Endocrinology*, **120**, 1361–9.

Plotsky, P.M., Otto, S. and Sapolsky, R.M. (1986) Inhibition of immunoreactive corticotropin-releasing factor into the hypophysial-portal circulation by delayed glucocorticoid feedback. *Endocrinology*, **119**, 1126–30.

Pratt, W.B., Sanchez, E.R., Bresnick, E.H. *et al.* (1989) Interaction of the glucocorticoid receptor with the Mr 90,000 heat shock protein: an evolving model of ligand-mediated receptor transformation and translocation. *Cancer Res.* (Suppl.), **49**, 2222–9.

Purves, D. (1988) *Body and Brain: A Trophic Theory of Neural Connections.* Harvard University Press, Cambridge, MA.

Ratka, A., Sutanto, W., Bloemers, M. and De Kloet, E. R. (1989) On the role of brain mineralocorticoid (type I) and glucocorticoid (type II) receptors in neuroendocrine regulation. *Neuroendocrinology*, **50**, 117–23.

Reul, J.M.H.M. and De Kloet, E.R. (1985) Two receptor systems for corticosterone in rat brain: microdistribution and differential occupation. *Endocrinology*, **117**, 2505–11.

Reul, J.M.H.M. and De Kloet, E.R. (1986) Anatomical resolution of two types of corticosterone receptor sites in the rat brain with *in vitro* autoradiography and computerized image analysis. *J. Steroid Biochem. Mol. Biol.*, **24**, 269–72.

Reul, J.M.H.M., van den Bosch, F.R. and De Kloet, E.R. (1987a) Relative occupation of type-I and type-II corticosteroid receptors in rat brain following stress and dexamethasone treatment: functional implications. *J. Endocrinol.*, **115**, 459–67.

Reul, J., van den Bosch, F.R. and De Kloet, E.R. (1987b) Differential response of type I and type II corticosteroid receptors to changes in plasma steroid levels and circadian rhythmicity. *Neuroendocrinology*, **45**, 407–12.

Reul, J.M.H.M., Tonnaer, J. and De Kloet, E.R. (1988) Neurotrophic ACTH analogue promotes plasticity of type I corticosteroid receptors in brain of senescent male rats. *Neurobiol. Aging*, **9**, 253–7.

Ringold, G.M. (1985) Steroid hormone regulation of gene expression. *Annu. Rev. Pharmacol. Toxicol.*, **25**, 529–66.

Ritger, H., Veldhuis, H. and De Kloet, E.R. (1984) Spatial orientation and hippocampal corticosterone receptor systems of old rats: effects of ACTH4-9 analogue ORG2766. *Brain Res.*, **309**, 393–9.

Rivier, C. and Plotsky, P. M. (1986) Mediation by corticotropin-releasing factor of adrenohypophysial hormone secretion. *Annu. Rev. Physiol.*, **48**, 475–89.

Rosenfeld, P., Sutanto, W., Levine, S. and De Kloet, E. R. (1988a) Ontogeny of type I and type II corticosteroid receptors in the rat hippocampus. *Dev. Brain Res.*, **42**, 113–18.

Rosenfeld, P., Van Eekelen, J.A.M., Levine, S. and De Kloet, E.R. (1988b) Ontogeny of the type II glucocorticoid receptor in discrete rat brain regions: an immunocytochemical study. *Dev. Brain Res.*, **42**, 119–27.

Rosner, W. (1990) The functions of corticosteroid-binding globulin and sex hormone-binding globulin: recent advances. *Endocr. Rev.*, **11**, 80–91.

Rundle, S.E. and Funder, J.W. (1988) Ontogeny of corticotropin-releasing factor and vasopressin in the rat. *Neuroendocrinology*, **47**, 374–8.

Sakly, M. and Koch, B. (1981) Ontogenesis of glucocorticoid receptors in anterior pituitary gland: transient dissociation among cytoplasmic receptor density, nuclear uptake, and regulation of corticotropic activity. *Endocrinology*, **108**, 591–6.

Sakly, M. and Koch, B. (1983) Ontogenetical variations of transcortin modulate glucocorticoid receptor function and corticotropic activity in the pituitary gland. *Horm. Metab. Res.*, **15**, 92–6.

Sapolsky, R.M. (1990) Glucocorticoid hippocampal damage and the glutaminergic synapse. *Progr. Brain Res.*, **86**, 13–23.

Sapolsky, R.M. and Meaney, M.J. (1986) The maturation of the adrenocortical stress response in the rat. *Brain Res. Rev.*, **11**, 65–76.

Sapolsky, R.M., Krey, L.C. and McEwen, B.S. (1983a) The adrenocortical response in the aged rat: impairment of recovery from stress. *J. Exp. Gerontol.*, **18**, 55–63.

Sapolsky, R.M., Krey, L.C. and McEwen, B.S. (1983b) Corticosterone receptors decline in a site-specific manner in the aged rat. *Brain Res.*, **289**, 235–40.

Sapolsky, R.M., Krey, L.C. and McEwen, B.S. (1984a) Glucocorticoid-sensitive hippocampal neurons are involved in terminating the adrenocortical stress response. *Proc. Natl. Acad. Sci. USA*, **81**, 6174–7.

Sapolsky, R.M., Krey, L.C. and McEwen, B.S. (1984b) Stress down-regulated corticosterone receptors in a site-specific manner. *Endocrinology*, **114**, 287–92.

Sapolsky, R.M., Meaney, M.J. and McEwen, B.S. (1985) The development of the glucocorticoid receptor system in the rat limbic brain: negative-feedback control. *Dev. Brain Res.*, **18**, 176–81.

Sapolsky, R.M., Krey, L.C. and McEwen, B.S. (1986) The neuroendocrinology of stress and aging: the glucocorticoid cascade hypothesis. *Endocr. Rev.*, **7**, 284–301.

Sarrieau, A., Sharma, S. and Meaney, M. J. (1988a) Postnatal development and environmental regulation of hippocampal glucocorticoid and mineralocorticoid receptors in the rat. *Dev. Brain Res.*, **43**, 158–62.

Sarrieau, A., Dussaillant, M., Moguilewsky, M. *et al.* (1988b) Autoradiographic localization of glucocorticosteroid binding sites in rat brain after an *in vivo* injection of 3H-RU 28362. *Neurosci. Lett.*, **92**, 14–20.

Sato, S.M. and Mains, R.E. (1985) Post-translational processing of proadrenocorticotropin/endorphin-derived peptides during postnatal development in the rat pituitary. *Endocrinology*, **117**, 773–86.

Sato, S.M. and Mains, R.E. (1986) Regulation of adrenocorticotropin/endorphin-related peptide secretion in neonatal rat pituitary cultures. *Endocrinology*, **119**, 793–801.

Sato, S.M. and Mains, R.E. (1988) Plasticity in the ACTH-related peptides produced by primary cultures of neonatal rat pituitary. *Endocrinology*, **122**, 68–75.

Savard, P., Menard, Y., Di Paolo, T. and Dupont; A. (1984) Thyroid hormone regulation of serotonin metabolism in developing rat brain. *Brain Res.*, **292**, 99–108.

Schapiro, S. (1962) Pituitary ACTH and compensatory adrenal hypertrophy in stress non-responsive infant rats. *Endocrinology*, **71**, 986–9.

Schapiro, S., Geller, E. and Eidason, S. (1962) Corticoid response to stress in the steroid-inhibited rat. *Proc. Soc. Exp. Biol. Med.*, **109**, 935–7.

Schroeder, R.J. and Henning, S.J. (1989) Roles of plasma clearance and corticosteroid-binding globulin in the developmental increase of circulating corticosterone in infant rats. *Endocrinology*, **124**, 2612–18.

Seckl, J.R., Kelly, P.A. and Sharkey, J. (1991) Glycyrrhetinic acid, an inhibitor of 11β-hydroxysteroid dehydrogenase, alters local cerebral glucose utilization *in vivo*. *J. Steroid Biochem. Mol. Biol.*, **39**, 777–9.

Seizinger, B.R., Hollt, V. and Herz, A. (1984) Postnatal development of β-endorphin-related peptides in the rat anterior and intermediate lobes: evidence for contrasting development of proopiomelanocortin processing. *Endocrinology*, **115**, 136–42.

Selye, H. (1950) *The Physiology and Pathology of Exposure to Stress*. Acta, Montreal.

Sharma, R. and Timiras, P.S. (1987) Age-dependent activation of glucocorticoid receptors in the cerebral hemispheres of male rats. *Dev. Brain Res.*, **36**, 285–7.

Sheppard, K.E. and Funder, J.W. (1987) Equivalent affinity of aldosterone and corticosterone for type I receptors in kidney and hippocampus: direct binding studies. *J. Steroid Biochem. Mol. Biol.*, **28**, 737–42.

Siiteri, P.K., Murai, J.T., Hammong, G.L. *et al.* (1982) The serum transport of steroid hormones. *Recent Progr. Horm. Res.*, **38**, 457–505.

Singer, C.J., Khan, M.S. and Rosner, W. (1988) Characteristics of the binding of corticosteroid binding globulin to rat cell membranes. *Endocrinology*, **122**, 89–96.

Siperstein, E., Nichols, C.W., Griesbach, W.E. and Chaikoff, I.L. (1954) Cytological changes in rat anterior pituitary from birth to maturation. *Anat. Rec.*, **118**, 593–619.

Slovitar, R., Valiquette, G., Abrams, G. *et al.* (1989) Selective loss of hippocampal granule cells in the mature rat brain after adrenalectomy. *Science*, **243**, 535–8.

Smith, C.L. and Hammond, G.L. (1991) Ontogeny of corticosteroid-binding globulin biosynthesis in the rat. *Endocrinology*, **128**, 983–8.

Smythe, J.W., Rowe, W. and Meaney, M.J. (1991) The effects of neonatal handling on serotonin turnover and receptor binding. *Soc. Neurosci.*, **17**, 990.

Spencer, R.L., Young, E.A., Choo, P.H. and McEwen, B.S. (1990) Adrenal steroid type I and type II receptor binding: estimates of *in vivo* receptor number occupancy, and activation with varying level of steroid. *Brain Res.*, **514**, 37–48.

Stanton, M.E., Gutierrez, Y.R. and Levine, S. (1988) Maternal deprivation potentiates pituitary-adrenal stress responses in infant rats. *Behav. Neurosci.*, **102**, 692–700.

Stern, J. and Levine, S. (1974) Psychobiological aspects of lactation in rats. *Progr. Brain Res.*, **41**, 433–44.

Titeler, M., Lyon, R.A., Davis, K.H. and Clennon, R.A. (1987) Selectivity of serotoninergic drugs for muiltiple brain serotonin receptors. *Biochem. Pharmacol.*, **36**, 3265–71.

Van Dorp, A.W.V. and Deane, H.W. (1956) A morphological and cytological study of postnatal development of rat's adrenal cortex. *Anat. Rec.*, **107**, 265–81.

Van Eekelen, J.A.M., Rosenfield, P., Levine, S. *et al.* (1987) Postnatal disappearance of glucocorticoid receptor immunoreactivity in the super-chiasmatic nucleus of the rat. *Neurosci. Res. Commun.*, **1**, 129–33.

Van Eekelen, J.A.M., Bohn, M.C. and De Kloet, E.R. (1991) Postnatal ontogeny of mineralocorticoid and glucocorticoid receptor gene expression in regions of the rat tel- and diencephalon. *Dev. Brain Res.*, **61**, 33–43.

Veldhuis, H.D., van Koppen, C., van Ittersum, M. and de Kloet, E.R. (1982) Specificity of adrenal steroid receptor system in rat hippocampus. *Endocrinology*, **110**, 2044–51.

Viau, V. and Meaney, M.J. (1991) Variations in the hypothalamic–pituitary–adrenal response to stress during the estrus cycle in the rat. *Endocrinology*, **129**, 2503–11.

Viau, V., Sharma, S., Plotsky, P.M. and Meaney, M.J. Differences in plasma ACTH responses to stress between handled and nonhandled rats not dependent upon stress-induced secretion of corticosterone. *J. Neurosci.* (in press).

Walker, C.D., Perrin, M., Vale, W. and Rivier, C. (1986a) Ontogeny of the stress response in the rat: role of the pituitary and the hypothalamus. *Endocrinology*, **118**, 1445–51.

Walker, C.D., Sapolsky, R.M., Meaney, M.J. *et al.* (1986b) Increased pituitary sensitivity to glucocorticoid feedback during the stress non-responsive period in the rat. *Endocrinology*, **119**, 1816–21.

Walker, C.D., Sizonenko, P.C. and Auburt, M.L. (1989) Modulation of the neonatal pituitary and adrenocortical responses to stress by thyroid hormones in the rat: effects of hypothyroidism and hyperthyroidism. *Neuroendocrinology*, **50**, 265–73.

Walker, C.D., Akana, S.F., Cascio, C.S. and Dallman, M.F. (1990) Adrenalectomy in the neonate: adult-like adrenocortical system responses to both removal and replacement of corticosterone. *Endocrinology*, **127**, 832–42.

Walker, C.D., Scribner, K.A., Cascio, C.S. and Dallman, M.F. (1991) The pituitary–adreno-cortical system of neonatal rats is responsive to stress throughout development in a time-dependent and stressor-specific fashion. *Endocrinology*, **128**, 1385–95.

Wang, Y.-Q., Sato, M., Morita, Y. *et al.* (1989) Postnatal ontogeny of POMC gene expression in the rat pituitary: an analysis by *in situ* hybridization histochemistry. *Brain Res.*, **47**, 53–8.

Weinberg, J., Krahn, E. A. and Levine, S. (1978) Differential effects of handling on exploration in male and female rats. *Dev. Psychobiol.*, **11**, 251–9.

Widmaier, E.P. (1989) Development in rats of the brain–pituitary–adrenal response to hypoglycemia *in vivo* and *in vitro*. *Am. J. Physiol.*, **257**, E757–763.

Widmaier, E.P. (1990) Glucose homeostasis and hypothalamic–pituitary–adrenal axis during development in rats. *Am. J. Physiol.*, **258**, E601–613.

Widmaier, E.P., Lim, A.T. and Vale, W.W. (1990) Regulation of corticotropin-releasing factor *in vitro* by glucose. *Am. J. Physiol.*, **255**, E287–E292.

Wilkinson, C.W., Shinsako, J. and Dallman, M.F. (1979) Daily rhythms in adrenal responsiveness to adrenocorticotropin are determined primarily by the time of feeding in the rat. *Endocrinology*, **104**, 350–9.

Wilson, M., Greer, M. and Roberts, L. (1980) Hippocampal inhibition of pituitary– adrenocortical function in female rats. *Brain Res.*, **197**, 344–51.

Witek-Janusek, L. (1988) Pituitary–adrenal response to bacterial endotoxin in developing rats. *Am. J. Physiol.*, **255**, E525–E530.

Wolf, G., Armstrong, E.G. and Rosner, W. (1981) Synthesis *in vitro* of corticosteroid-binding globulin from rat liver messenger ribonucleic acid. *Endocrinology*, **108**, 805–12.

Yehuda, R., McDonald, D., Heller, H. J. and Meyer, J. (1988) Maze learning behavior in early adrenalectomized rats. *Physiol. Behav.*, **44**, 373–81.

Yehuda, R., Fairman, K. and Meyer, J. (1989) Enhanced brain cell proliferation following early adrenalectomy in the rat. *J. Neurochem.*, **53**, 241–8.

Zarrow, M.X., Campbel, P.S. and Denenberg, V.H. (1972) Handling in infancy: increased levels of the hypothalamic corticotropin releasing factor (CRF) following exposure to a novel situation. *Proc. Soc. Exp. Biol. Med.*, **356**, 141–3.

ESTROGEN RECEPTORS AND THE DEVELOPING NERVOUS SYSTEM

Samuel A. Sholl

11.1 INTRODUCTION

Estrogens such as estradiol-17β and estrone are secreted by the ovaries during adulthood. Estrogen secretion is crucial for the neuroendocrine regulation of reproductive events, including the positive and negative feedback control of pituitary hormone synthesis and secretion. Additionally, estrogens play an important role in the expression of proceptive behaviors. Like most steroids estrogens exert their effect on the CNS through binding to specific intracellular proteins. Estrogen-specific receptor protein(s) have been characterized in a variety of tissues including the CNS and can be found in the CNS throughout development into adulthood. The developmental appearance of receptors in specific regions of the CNS is undoubtedly important for the estrogen-mediated differentiation of CNS activity.

This chapter will focus on the estrogen receptor and its importance in the estrogen-mediated differentiation of specific regions of the CNS. After presenting background information on the general effects of estrogens on neural activity, the biochemistry of estrogens and the estrogen receptor will be reviewed briefly. Indirect and direct methods for detecting receptor activity will be included followed by a discussion of region-specific changes in receptor levels during the course of development. The possible significance of these changes with respect to the developmental impact of estrogens will be presented.

It should be noted that the importance of estrogens on the differentiation of CNS activity is unequivocal only in rodents, whereas in primates the role of estrogen remains controversial.

11.2 ORGANIZATIONAL EFFECTS OF STEROIDS ON THE CNS

In the rodent, androgen exposure during the fetal or neonatal period affects the later expression of both neuroendocrine and behavioral activities. These effects have been summarized (MacLusky and Naftolin, 1981) and include regulation of gonadotropin secretion, reproductive behaviors and non-reproductive behaviors such as play behavior, intraspecies aggression, scent marking and learning, which includes active avoidance and maze learning. Taste preferences, feeding and body weight, circadian rhythms, response to brain lesions in the septal area or globus pallidus, morphology of the orbital frontal cortex and ventromedial hypothalamus are also affected. The period when the brain is subject to the organization effects of androgens has become known as the 'critical period'. This period is most clearly defined in rats. Female rats exposed to androgens *in utero* or during the first five days after birth

Receptors in the Developing Nervous System Vol. 1: *Growth factors and hormones* Edited by Ian S. Zagon and Patricia J. McLaughlin. Published in 1993 by Chapman & Hall. ISBN 0 412 45240 5. Vols. 1 and 2 (set) ISBN 0 412 54520 9.

exhibit more masculine sexual behavior as adults than unexposed females (Harris, 1964; Gerall and Ward, 1966; Barraclough, 1967). The neuroendocrine system is also modified by androgen treatment. The efficacy of androgen action decreases progressively during the first five days after birth. It has been noted that the masculinizing effects of androgens in the rodent can be duplicated by the administration of fairly large doses of estrogens (Gorski 1963; Harris and Levine, 1965). This fact, together with the ability of neural tissue to convert androgens to estrogens (as discussed later) argues that it is really estrogens which are responsible for masculinizing the rodent brain.

Direct evidence of an estrogenic effect on neural differentiation was first described by Toran-Allerand (1976, 1980). Organotypic cultures of the newborn mouse hypothalamus–preoptic area responded to different concentrations of estrogen in the medium by variations in the extent of neurite outgrowth (Toran-Allerand, 1976, 1980, 1984). Moreover, responsivity of tissue to estrogen appears to be correlated topographically to its ability to concentrate radioactive estradiol (Toran-Allerand *et al.*, 1980) suggesting that estrogen action may be receptor dependent. Confirmational studies on the neurogenic action of estrogens have been conducted in the hypothalamus and preoptic area (Stumpf *et al.*, 1983; Uchibori and Kawashima, 1985), cerebral cortex (Uchibori and Kawashima, 1985) and mesencephalic dopaminergic neurons (A8–A10) (Reisert *et al.*, 1987).

Unlike the rodent the primate brain seems less susceptible to estrogen during fetal development, and evidence has been presented that androgens *per se* can have an organizational influence on primate brain differentiation. Specifically, modifications in the behavior of female rhesus monkeys have been produced by treatment *in utero* with dihydrotestosterone as well as testosterone (days 45–100 postconception) (Goy, 1981; Goy *et al.*, 1988). Since dihydrotestosterone is not aromatized, it has been suggested that androgens are directly involved in those neurogenic events which lead to masculine-like behaviors in primates. The day 45 to day 70 postconception period in monkeys is considered to be the 'critical period' in this species (Goy and McEwen, 1980), although unlike rats, this period may be less precisely defined depending upon the affected parameter. Several laboratories have detected androgen receptors in the fetal primate brain (Sholl and Pomerantz, 1986; Handa *et al.*, 1988), increasing the likelihood that androgens can influence fetal primate brain differentiation. Moreover in a recent report androgen receptor levels were examined in frontal and temporal lobes of the rhesus monkey on day 70 postconception (Sholl and Kim, 1990a). In this study, androgen receptors were found to be asymmetrically distributed between right and left hemispheres, and there was a distinct sex difference in the pattern of asymmetry. This asymmetry was not noted when the same tissue samples were analyzed for the enzymes aromatase and 5α-reductase. The latter converts testosterone into dihydrotestosterone. The lateralization of androgen receptor concentrations reinforces the argument that androgens can exert a major influence on brain organization in primates. However, preliminary work in my laboratory suggests that the estrogen receptor may also be asymmetrically distributed between the cerebral hemispheres but in a more random fashion. This finding, together with the reported presence of estrogen receptors in the primate brain during the critical period (Sholl and Kim, 1989) argues for an effect of estrogens on early brain organization in the primate. This possibility will be addressed more fully in section 11.3.

One of the most striking observations which demonstrates a sex-dependent organization of the brain was made originally by Gorski *et al.* (1978). An examination of the medial preoptic area of the rat indicated that it was sexually dimorphic. More specifically there was a 'sexually dimorphic nucleus' (SDN-POA) in this area which contained more neurons in males than females. The size of the SDN-POA also responded to perinatal hormone stimulation (Dohler *et al.*, 1982a). Since Gorski's original observation similar structures have been found in gerbils (Commins and Yahr, 1984), ferrets (Tobet *et al.*, 1986), guinea pigs (Hines *et al.*, 1985) and humans (Swaab and Fliers, 1985). In addition to the SDN-POA, another cell group comprising the SDN complex resides in the rostroventral periventricular gray of the POA and its adjacent area. This group is larger and more densely cellular in female rats than in males (Bleier *et al.*, 1982; Bleier and Byne, 1985) and is thought to correspond to the medial preoptic nucleus (MPN) which may be involved in the regulation of cyclic gonadotropin release in rats (Terasawa *et al.*, 1980). More recent evidence has revealed that the volume of the MPN depends on perinatal androgens (Ito *et al.*, 1986) in that the size of the MPN is significantly reduced in females exposed to testosterone propionate between days 17 and 21 of pregnancy. In addition to the MPN significant sex differences in morphology have also been found in the ventromedial nucleus (Matsumoto and Arai, 1983), bed nucleus of the stria terminalis (Hines *et al.*, 1985) and the medial amygdaloid nucleus (Mizukami *et al.*, 1983).

Sex differences in synaptic organization have also been noted. In the arcuate nucleus of the rat, the number of synapses on the dendritic spine is greater in females than males (Matsumoto and Arai, 1980). This number can be reduced in females by neonatal androgen treatment. Sex differences in the number of synapses on the dendritic shaft and number of synapses on the cell body have also been noted. In the ventrolateral part of the ventromedial nucleus of males, synapses on the dendritic shaft and spine are more abundant than in the dorsomedial area of this nucleus (Matsumoto and Arai, 1986). Interestingly, this difference in synaptic distribution between the two parts of the ventromedial nucleus is similar to that noted for the uptake of radioactive estrogen (Pfaff and Keiner, 1973) suggesting that the estrogen receptor might be involved in synapse formation. This possibility is supported by the finding that in the medial amygdaloid nucleus and arcuate nucleus of the female rat synaptogenesis is stimulated by long-term estrogen treatment between postnatal days 1 and 30 (Matsumoto and Arai, 1976; Arai and Matsumoto, 1978; Nishizuka and Arai, 1981).

It is unlikely that estrogens or androgens act independently to modify neural differentiation. Additional factors which have been implicated include insulin and the various growth factors. Toran-Allerand *et al.* (1988) found that insulin facilitated estradiol-induced neurite outgrowth from organotypic cultures of mouse olfactory bulb, hypothalamus and cerebral cortex. The high dose of insulin which was required suggested that a receptor for one of the insulin-like growth factors might be involved. More recently, Toran-Allerand and MacLusky (1989) noted in female rats on postnatal day 12 a co-localization of NGF receptors and estrogen binding in neurons of the medial septum and diagonal band of Broca. Komack *et al.* (1989) noted that basal forebrain NGF receptor mRNA levels are higher in female rats than males during the first two postnatal weeks. Finally, Yanase *et al.* (1988) reported that the administration of an antibody to NGF can inhibit the androgen-induced defeminization of female rats, whereas testosterone administration can elevate NGF levels in the mouse submaxillary gland (Ishii and Shooter, 1975). In primates there is only circumstantial evidence for an NGF-steroid interaction. Schatteman *et al.* (1988) found high NGF receptor levels in the rhesus monkey cerebellum on embryonic day 135 but not on either side of this period. A similar pattern has also been noted for androgen receptors in the monkey cerebellum (Pomerantz and Sholl, 1987).

11.3 SITES AND PATHWAYS OF ESTROGEN SYNTHESIS

In considering the role of estrogen receptors in neural development, it is appropriate first to establish the source of the estrogens which bind to those receptors. During fetal development estrogens may arise from the placenta, fetal ovaries or directly from the aromatization of androgens within neural tissue. The extent to which each of these sources contributes to the pool of available estrogen is species and stage specific. In many species a major source of estrogens is the aromatization of testicular androgens in CNS target tissues. In the rabbit George *et al.* (1978) first noted aromatase activity in the diencephalon on day 16 postconception. Activity subsequently increased to day 19 where it remained until approximately day 25. Similar observations have been reported for the rat (George and Ojeda, 1982) in which aromatase is first detectable in the hypothalamus and preoptic areas of the brain on day 16; there is an increase over the next 3–4 days followed by a decline at the end of gestation into the first few days of the postnatal period. Aromatization has also been detected in the human fetal brain. As reviewed by Naftolin *et al.* (1975), conversion of androstenedione into estrone and estradiol can be detected in the hypothalamus, limbic system, cerebral cortex and pituitary gland. More quantitative studies have been conducted in rhesus monkeys (*Macaca mulatta*), in which stage differences in aromatization within individual areas of the CNS have been noted (Sholl *et al.*, 1989; Sholl and Kim, 1990b). In the amygdala and medial basal hypothalamus aromatase activity increases between days 70 and 100 postconception followed by a decrease in activity by day 160. (The length of gestation in rhesus monkeys is approximately 168 days.) A similar late-term decrease has also been noted by Clark *et al.* (1988). In general, aromatase activity in estrogen target-regions of the primate brain and pituitary is higher than in areas which are relatively insensitive to estrogen. This generality has been confirmed in monkeys (Roselli and Resko, 1986, 1989; Roselli *et al.*, 1987b), rodents (Roselli *et al.*, 1984, 1985), birds and fish (Callard, 1984).

The concentration of aromatase activity in specific target regions of the CNS becomes important in those species in which circulating androgens can influence the course of neural development, specifically through their conversion into estrogens in those target areas. Factors which can moderate the concentration of estrogens in target tissue, and the efficacy of estrogen action, include: (a) changes in the availability of androgen to the aromatase enzyme, particularly from the fetal or neonatal testis; (b) co-concentration of aromatase and estrogen receptors, and (c) feedback regulation of androgens or estrogens on the aromatase enzyme. In rats, testosterone can be detected in the male fetus as early as day 17 postconception (Weisz and Ward, 1980). After a brief decline, testosterone concentrations rise through to postnatal day 5. This increase means that there is also a rise in the availability of substrate at a time when androgenization of the rat brain reaches a 'critical period' and depends on the extensive synthesis of estrogen from circulating androgen. The changes noted in fetal blood testosterone during the early postnatal period tend to parallel the changes in brain aromatase activity (MacLusky *et al.*, 1984) suggesting that testosterone may stimulate aromatase activity. In fact, evidence for this in the rat has been provided by Roselli *et al.* (1984, 1987a). In the fetal monkey, higher aromatase levels have been observed in the medial basal hypothalamus and amygdala than in the cerebral cortex or cerebellum (Roselli and Resko, 1986; Sholl and Kim, 1990b). Higher enzyme levels have also been observed in male versus female fetuses on both day 70 and 120 postconception (Roselli and Resko, 1986; Sholl and Kim, 1990b). These differences occur at a time when the fetal testes actively secrete testosterone (Resko *et al.*, 1980) suggesting that, as in the rat, androgens may stimulate aromatase activity. The importance of this pathway in the fetal primate is emphasized by the fact that aromatase activity in the hypothalamus and amygdala was considerably higher than in lung and heart tissue taken at the same stage (Sholl and Kim, 1990b). Brain aromatase activity in both sexes tends to decline beyond day 120 postconception (Clark *et al.*, 1988; Sholl and Kim, 1990b). The fact that female fetuses show developmental changes in aromatase activity argues that aromatase activity in the primate may be regulated by factors beyond the level of circulating androgen.

Estrogens in primates may also come from the placenta. During pregnancy, the placenta becomes a major source of estrogens in humans (Oakey, 1970), rhesus monkeys (Williams *et al.*, 1978; Ellinwood *et al.*, 1989) and baboons (Castracane and Goldzieher, 1986). In these primate species estrogens are synthesized in what has come to be termed the fetoplacental unit. This unit consists of the placenta and fetal adrenal, which together are responsible for the synthesis of estrogen. Specifically, C_{19}-steroids from the fetal adrenal are substrates for placental aromatase; the fetal adrenal lacks aromatase, whereas the placenta lacks some of the enzymes necessary for the synthesis of C_{19}-steroids. (Reviewed by Albrecht and Pepe, 1990.)

In the late-term fetal rhesus monkey the fetal ovaries also appear to secrete estrogen (Sholl and Kim, 1989) as evidenced by higher estrogen levels in the umbilical artery of female than male fetuses. The concentration of estrone was considerably greater than the more active estrogen, estradiol, suggesting that there may be significant oxidation of estradiol in the placenta or within the fetal ovary itself. The fetal primate CNS may therefore be protected from the potential effects of circulating estradiol by its rapid oxidation to estrone. In contrast to the primate, the neonatal rat exhibits elevated estradiol levels (Dohler and Wuttke, 1975), however, the concentration of

circulating α-fetoprotein is also elevated during the latter part of gestation and into the early postnatal period (Raynaud *et al.*, 1971). In the rat this protein appears to protect the CNS from the masculinizing effects of circulating estradiol (Mizejewski *et al.*, 1980), whereas in humans only a small percentage of circulating α-fetoprotein binds estrogen (Uriel *et al.*, 1975). Thus in primates, the rapid oxidation of estradiol, either by the placenta or in peripheral fetal tissues, may represent a more effective protection mechanism against the organizational influence of circulating estradiol.

The qualitative and quantitative patterns of circulating estrogen in both the primate and rodent and the general unavailability of potent estrogens to the CNS emphasize the importance of aromatization in specific target tissues. Further, one might predict that the potential effect of androgen exposure would be maximal if aromatase and estrogen receptors were co-localized in specific brain regions or neurons. Such co-localization remains to be demonstrated at the neuronal level, although general brain regions such as the amygdala, preoptic area and medial basal hypothalamus frequently exhibit both elevated aromatase and estrogen receptor levels.

11.4 BIOCHEMISTRY OF THE ESTROGEN RECEPTOR

The complete sequence of complementary DNA (cDNA) to the human estrogen receptor from MCF-7 cells was first determined in 1986 (Green *et al.*, 1986). When this cDNA was expressed in HeLa cells, the coded protein had the same apparent molecular mass and estrogen binding affinity as estrogen receptor from MCF-7 cells. The chromosomal gene was cloned and the genomic sequence was described more recently (Ponglikitmongkol *et al.*, 1988). Similarities have been noted between the estrogen receptor and receptors from the other steroid families. In general, steroid receptors have three characteristic regions. The first of these is the C1 region which is highly conserved between different steroid families (Carson-Jurica *et al.*, 1990). In fact, this conservancy has been used to screen for new kinds of steroid receptors. The C1 region forms two 'zinc fingers' (Green *et al.*, 1986; Greene *et al.*, 1986). At the base of each finger there are four cysteines which bind to a single zinc molecule. There are 12–13 amino acids in the loop, and an intermediate linkage region contains 15–17 amino acids. The C1 region has been identified through amino acid deletion and insertion studies as the region of the receptor which binds to DNA.

Steroid receptors modify the transcription of specific target genes through interaction with regulatory elements associated with those genes. These hormone regulatory elements tend to occur as dyads. As such it is not surprising that evidence has emerged that the estrogen receptor (as well as other steroid receptors) binds to DNA as a dimer (Gordon and Notides, 1986; Kumar and Chambon, 1988). Two additional conserved regions have been identified in steroid receptors, the C2 and C3 regions, consisting of approximately 42 and 22 amino acids, respectively, and lying nearer than C1 to the *C*-terminal end of the protein. In general, there is less homology between receptors in these regions than exists for the C1 region. Evidence has accumulated from insertional mutation and point mutation studies that hormone specificity resides in the 250 amino acids closest to the *C*-terminal end of the receptor (reviewed by Carson-Jurica *et al.*, 1990). Studies have been conducted in which the *C*-terminal end of the glucocorticoid and progesterone receptors have been deleted (Carson *et al.*, 1987; Godowski *et al.*, 1987; Hollenberg *et al.*, 1987). As a result of these deletions gene action was restored in the absence of the specific steroid. Results from such studies have argued that the hormone binding domain of the receptor acts as a repressor. This could be accomplished, for example, by maintaining the receptor in an inactive configuration.

Estrogen and other steroids cause the rapid 'transformation' of the inactive receptor to the active state (Grody *et al.*, 1982) (reviewed by Carson-Jurica *et al.*, 1990). This activational process is characterized by the transformation of the receptor from a large oligomeric complex (7–10S). A 90 kDa heat-shock protein (HSP 90) is found associated with various types of receptors before steroid activation, but not after the receptor binds to a steroid. The receptor oligomeric complex is unable to bind to DNA; however, on addition of a steroid and dissociation of the complex the receptor can be activated to bind to DNA. It is quite possible that HSP 90 serves both to maintain the receptor in an inactive state and also gives the receptor the confirmation necessary for steroid binding. In this regard there was a correlation between the loss of HSP 90 from oligomeric glucocorticoid receptor complexes and the capacity of those complexes to bind hormone (Bresnick *et al.*, 1989).

In addition to forming an oligomeric complex with HSP 90, glucocorticoid receptors also contain RNA which may indicate a role for this receptor in posttranscriptional processing (Yamamoto, 1985).

Some early studies of steroid receptors provided support for the theory that unbound receptors were primarily located in the cell cytoplasm. Further, the addition of steroids and the subsequent binding of these steroids to the receptor caused the rapid translocation of the steroid–receptor complex into the cell nucleus (Gorski *et al.*, 1968; Jensen *et al.*, 1968; Siiteri *et al.*, 1973). The subsequent use of immunocytochemical procedures, however, suggested that both the progesterone and estrogen receptor may be primarily nuclear even in the absence of the given steroid (King and Greene, 1984; Perrot-Applanat *et al.*, 1985). Despite this, one must question whether the antibodies which were used in the immunocytochemical studies recognized the antigenic determinate of the untransformed receptor, and also if fixation during the procedure affected the antigenicity of the unbound receptor.

Regardless of the location of the untransformed receptor, it is certain that following steroid-induced activation the receptor is in the cell nucleus, where it interacts with nuclear chromatin. If, in fact, the receptor is in the cytoplasm prior to activation, then it must somehow pass through the nuclear membrane. This could occur by means of either diffusion through the nuclear membrane or by interaction of the receptor with the nuclear pore. Diffusion does not appear to be a major route of passage, since the molecular weight exclusion limit is around 67 kDa (Paine *et al.*, 1975). This leaves an interaction with the nuclear pore as the more likely process. The interaction might be controlled by a 'translocation signal' (Carson-Jurica *et al.*, 1990) in the protein. In various receptors, including the estrogen receptor (Green *et al.*, 1986; Greene *et al.*, 1986) as well as the nuclear translocation signal of the SV40 T antigen, (Kalderon *et al.*, 1984; Lanford *et al.*, 1986) there are amino acid sequences which exhibit a high degree of homology. Moreover, in progesterone, androgen and glucocorticoid receptors these sequences are nearly identical. In the case of the progesterone and glucocorticoid receptor (Picard *et al.*, 1988; Guiochon-Mantel *et al.*, 1989), deletion mutations have provided evidence that specific amino acid sequences function as nuclear translocation signals. For the progesterone receptor two sequences appear to be involved, the primary sequence related to the T antigen as well as a smaller sequence which required activation of the DNA binding domain to function. In contrast nuclear localization of the rat glucocorticoid receptor appeared to depend on an amino acid sequence in the steroid binding domain.

Conclusive evidence for the functionality of specific domains of the receptor, and the specificity of those regions, has come more recently from domain-recombination studies. For example, a chimeric receptor was synthesized consisting of the human glucocorticoid receptor DNA binding region and the estrogen receptor steroid binding region (Green

and Chambon, 1987; Kumar *et al.*, 1987). The synthesized molecule activated mouse mammary tumor virus in an estrogen-responsive manner. Other cases have also been reported in which the steroid-binding domain of one receptor was combined with the DNA-binding domain of another with no loss in functionality.

It is likely that different regions of the receptor are involved in the activation of RNA transcription. Evidence for this comes from the analysis of glucocorticoid receptor mutations in the C1 region which resulted in a loss of activity (Severne *et al.*, 1988). Additionally, a mutation in the steroid binding domain of the estrogen receptor significantly affected overall activity (Kumar *et al.*, 1987). Mutations at the *N*-terminal end also appeared to have an effect on receptor activity.

There are several ways in which the activity of a receptor might be regulated, other than through its binding to a specific ligand. The first of these is through posttranslational phosphorylation. In this regard, after the synthesis of human estrogen receptor *in vitro* from cDNA the addition of tyrosine kinase was able to increase the binding capacity of the receptor with little effect upon its affinity (Migliaccio *et al.*, 1989). Mutation data suggested that the site of phosphorylation was at the *C*-terminal end. A second way in which receptor activity might be modulated is through an interaction of one receptor with another. Synergistic effects of different species of receptor may occur by interaction of receptors with different, closely associated DNA regulatory sites. Supporting this, synergistic effects of glucocorticoids and estrogen on the vitellogenin II gene have been reported (Ankenbauer *et al.*, 1988). Also, a heterodimer of retinoic acid and thyroid hormone receptors appears capable of positive and negative regulation of gene transcription (Glass *et al.*, 1989). Finally, it is known that steroid receptors can have affinities for multiple classes of steroids. For example, as evidenced in DNA-cellulose column effluent, the binding of dihydrotestosterone to the androgen receptor of fetal rhesus monkey brain can be competed with estradiol (Pomerantz *et al.*, 1985). Competitive effects of estradiol on testosterone binding have also been noted using cytosolic extracts from the rodent hypothalamus (Fox, 1975a). Tentative evidence (Sholl and Kim, 1989) has indicated that in the fetal monkey brain androgens may have an affinity for the estrogen receptor. Taken together, such synergism and competition studies indicate that the regulation of steroid receptor activity may be quite complex depending upon the hormonal mileu.

The effect of a specific steroid (say steroid A) through its receptor may be either accentuated or moderated by the presence of another steroid (say steroid B). Steroid B might heighten the efficacy of steroid A through an interaction of the two different species of receptors, one for steroid A and one for steroid B. Alternatively, steroid B might moderate the efficacy of steroid A through binding to steroid A's receptor. By so doing, the potential effect of steroid A might be significantly reduced, particularly if steroid A was present in high concentrations. In the latter case steroid B might have its own receptor or share a receptor with steroid A. In the primate evidence has accumulated that androgens might be more important than estrogens in organizing the developing fetal brain (see for example, Goy and McEwen, 1980; Goy *et al.*, 1988; Sholl and Kim, 1990a). However, in view of the fact that estrogens are present in the fetal circulation (Resko, 1974; Sholl and Kim, 1989), are synthesized in the CNS from androgens (Sholl *et al.*, 1989) and can compete with dihydrotestosterone binding to the androgen receptor (Pomerantz *et al.*, 1985), it is possible that the presence of estrogens in the fetal primate could modify the effects of androgen on brain differentiation. It remains to be determined how these different steroids might interact in the intricate orchestration of brain development.

11.5 ESTROGEN RECEPTOR DETECTION IN THE BRAIN

Estrogen receptors are generally present in femtomolar (10^{-15}M) concentrations in the developing brain. As such, it has been very difficult to characterize estrogen receptor levels, and many studies have taken indirect approaches toward the detection and localization of estrogen receptor activity. In the earliest studies investigators frequently infused or injected radioactive steroids. Following sacrifice of the animal, different brain regions were dissected, and radioactive estrogens were biochemically isolated and measured in those regions. Unfortunately, such studies yielded a composite picture of the dynamics of steroid uptake and localization in neural tissue. This composite included (a) the rate of steroid clearance from the site of injection and the body; (b) the degree of steroid metabolism in neural tissue; (c) the extent of both specific and nonspecific steroid binding in neural tissue; (d) the rate of dissociation from tissue binding sites and (e) possible regional differences in the accessibility of steroids as influenced by the vascular supply. Despite the absence of mechanistic detail, many important findings came from such investigations. It was generally established in various species that with the injection of either androgens or estrogens, one could obtain a localization of steroids in specific brain regions (McEwen *et al.*, 1970a, b; Gerlach *et al.*, 1976; Pfaff *et al.*, 1976; Sholl and Goy, 1981; Sholl *et al.*, 1982; Michael *et al.*, 1986, 1989). Moreover, it was frequently the case that these regions were important in the regulation of reproductive functions. The concentration of steroids in specific regions of the CNS suggested that there might be greater receptor concentrations in these areas. This was also used as evidence for the importance of various neural regions in reproductive and sexually dimorphic brain function. Sexual and developmental differences in the uptake of various steroids has also been reported periodically.

More precise information on the ability of the brain to concentrate steroids can be supplied by autoradiographic studies. In one of the first studies on the ability of the rat brain to concentrate estradiol (Pfaff and Keiner, 1973), adult ovariectomized rats were injected intraperitoneally with [³H]estradiol. Animals were killed 2h after injection and portions of the brain were frozen. Following freezing, cryostat sections were placed on slides containing photographic emulsion. Slides were developed after different periods of time and the concentrations of labeled cells were noted in various limbic and hypothalamic structures. Limbic structures included the medial and cortical nuclei of the amygdala, lateral septum, bed nucleus of the stria terminalis, diagonal band of Broca, olfactory tubercle, ventral hippocampus, and prepiriform and entorhinal cortex. Hypothalamic-preoptic areas included the medial preoptic area, medial anterior hypothalamus, ventromedial nucleus, arcuate nucleus and ventral premammillary nucleus. Other studies of a similar nature have also been performed by Stumpf and Sar (1976) and Sheridan *et al.* (1979). Although autoradiography may yield more precise information on the localization of estrogen in the brain, there are several inherent weaknesses in the methodology which might distort the results or the interpretation of the results. First, during the necessary long-term incubation of frozen sections on the photographic emulsion, radioactive steroids may diffuse away from their initial site. Second, before the animal is killed the estrogens could be potentially metabolized to other compounds (such has catecholestrogens (Reddy *et al.*, 1981)) which would still register on the photographic emulsion.

More recently, monoclonal antibodies to the estrogen receptor (Moncharmont *et al.*, 1982, 1984; Greene *et al.*, 1984) have been utilized to localize the receptor by immunocytochemistry. Receptors from either the human or calf can be used to produce monoclonal antibodies, which react with receptors from a wide range of species including humans (Garancis *et al.*,

1983), primates (McClellan *et al.*, 1986), rabbits (King and Greene, 1984) and rats (Sar and Parikh, 1986). Sar and Parikh (1986) used a monoclonal antibody to calf uterine cytosolic estrogen receptor to localize receptors in 24- and 50-day-old ovariectomized Sprague–Dawley rats. At 24h after ovariectomy, the rats were injected with estradiol-17β (1 μg/100g body weight). One hour after injection the rats were decapitated, and the brain, pituitary and uterus were removed and frozen in liquid propane. Frozen sections were thaw-mounted on glass slides and fixed in 4% paraformaldehyde in phosphate-buffered saline (PBS). After washing away the fixative with PBS, sections were treated with normal horse serum and incubated with an estrogen receptor monoclonal antibody. After incubation for 60 min, bound antibody was identified using the ABC technique (Sar, 1985). Localization of staining was similar in the brain of mature and immature rats. In the septal area stained neurons were noted in the region of the organum vasculosum lamina terminalis, in the nucleus septi lateralis and in the nucleus triangularis septi. In the preoptic area, localization of immunoreactivity was noted in the nucleus preopticus medialis, nucleus preopticus periventricularis, nucleus preopticus lateralis and nucleus interstitialis striae terminalis. In the anterior hypothalamus neuronal staining was evident in nucleus periventricularis hypothalami, nucleus paraventricularis and organum subfornicale. In the central hypothalamus localization of immunoreactivity was noted in the nucleus arcuatus hypothalami, the ventrolateral part of the nucleus ventromedialis hypothalami and in the nucleus premammillaris ventralis. Reactive neurons were also noted in the zona incerta, nucleus dorsomedialis and in the central, medial and cortical nuclei of the amygdala. Large differences were noted between regions in the degree of cell-labeling. For example, 60–80% of the neurons in the nucleus preopticus medialis stained positively whereas only 30–40% of cells in the cortical nucleus of the amygdala were stained.

As one might expect, a large number of cells in the anterior pituitary were also stained, as were luminal and glandular cells of the uterus. In all cases staining was noted in cell nuclei but not in cell cytoplasm, despite the fact that the antibody that was used recognized both occupied and unoccupied estrogen receptors (Moncharmont *et al.*, 1982). The use of monoclonal antibodies to identify cells which contain estrogen receptors has the advantage of being able to recognize the receptor even in the presence of its ligand. Additionally, the receptor can be precisely localized. The disadvantage of this technique is that it tends to be relatively insensitive and in some cases methods used in fixing tissue slices may have a deleterious effect on receptor antigenicity. Finally, this method will not provide a quantitative image of receptor changes.

The development of *in situ* hybridization (ISH) technology to detect mRNA along with the cloning of estrogen receptor cDNA has allowed the mapping of estrogen receptor mRNA in the brain (Pelletier *et al.*, 1988b) and pituitary (Pelletier *et al.*, 1988a) of the rat. This technique is very sensitive and can also provide an indication of the potential rate of receptor synthesis. In the studies by Pelletier *et al.* (1988a, b), adult female rats were fixed by perfusion of 4% paraformaldehyde in 0.1 M phosphate buffer. Subsequently, tissues were placed in 15% sucrose in phosphate buffer at 4° C overnight. Sections (10μm) were cut with a cryostat and stored at −80° C until used. The probe used consisted of a 72–mer synthetic oligonucleotide, corresponding to amino acids 1–24 of the human estradiol receptor (Green *et al.*, 1986). ISH was performed by first rinsing sections in 2 × SSC (SSC: 0.15M NaCI, 15 mM sodium citrate) followed by an incubation for 24h at 40° C in hybridization buffer containing 5 × SSC, yeast RNA, ATP, 8% dextran sulfate and 20% deionized formamide. The ^{35}S-labeled probe was applied to each section. After labeling, sections were rinsed, dehydrated, coated with Kodak NTB-2 emulsion and exposed to the emulsion for 3 days. After exposure, a high degree of labeling was noted in the cerebral cortex, hippocampus, amygdaloid nuclei and a few hypothalamic nuclei, including the supraoptic nucleus, paraventricular nucleus, ventromedial nucleus and anterior hypothalamic nucleus. With the exception of the hippocampus and cerebral cortex, high signal areas generally corresponded to those in which estradiol uptake experiments suggested that there were elevated estrogen receptor levels. In the case of the cerebral cortex and hippocampus, for example, previous studies had indicated little estrogen receptor activity in these regions in contrast to what was indicated by the ISH experiments of Pelletier *et al.* (1988b). Although the authors suggested that this discrepancy might be due to the higher sensitivity of the ISH technique, one cannot exclude the possibility that uptake experiments may not accurately measure receptor concentrations in the event that there is a rapid turnover in the receptor population in a particular neural region. Thus, ISH techniques might provide a better indication of the degree of receptor availability. Although currently ISH is capable of yielding only qualitative information on mRNA levels, one might expect that in the future 'solution hybridization' might be employed to provide a more quantitative assessment. (See for example, Blum and Roberts, 1989.)

At this time, quantitative information on estrogen receptors in neural tissue can only be obtained by using ligand-binding techniques on tissue extracts. Pioneering studies of rat and mouse neural tissue at different developmental stages have been conducted by several research groups using different methods (Fox, 1975b; McEwen *et al.*, 1975; Maclusky *et al.*, 1976; Westley and Salaman, 1976; Roy and McEwen, 1977). More recently, some of these techniques have been modified to measure estrogen receptor concentrations in primate neural tissue (Pomerantz *et al.*, 1985; Sholl and Kim, 1989). For the study of unoccupied receptor, brain tissue is frequently homogenized in a cold Tris–EDTA–glycerol–dithio-threitol buffer (MacLusky *et al.*, 1979). The buffer may also contain sodium molybdate (Sholl and Kim, 1989) to retard receptor transformation and help stabilize the receptor. The resultant homogenate is usually centri-fuged at a high speed to obtain an unoccupied, cytosolic receptor preparation. This fraction can then be incubated with an appropriate radioactive ligand for 2–24 h at 2–4° C. After the incubation, bound and free ligand can be separated by several techniques including gel filtration on Sephadex LH-20 mini columns (MacLusky *et al.*, 1979; Sholl and Kim, 1989) or filtration on DNA-cellulose columns (MacLusky *et al.*, 1979; Vito and Fox, 1982; Pomerantz *et al.*, 1985). On Sephadex LH-20 columns the ligand–receptor complex appears almost immediately in the eluate. In contrast, on DNA-cellulose columns ligand-bound receptor is retained until a high salt conc-entration (200–220mM) buffer is applied to the column. Early research also identified a 4S binding peak by sucrose gradient sedimen-tation (Fox, 1975a; Vito and Fox, 1982). This peak was noted in the presence of 150 mM NaCl; however, without salt, the peak was shifted to 5.0–7S. Such methods have been used to provide information on the binding characteristics of the estrogen receptor in neu-ral tissue. In general, the equilibrium disso-ciation constant (K_d) of the receptor is between 0.1 and 0.7 nM (MacLusky *et al.*, 1979, 1986; Sholl and Kim, 1989), whereas the concentra-tion of the unoccupied receptor can range from 1 to 15 fmol/mg protein depending on the tissue and the species. Binding constants can be determined by means of bound/unbound versus bound plots (Scatchard plots, Scatchard, 1949) or by computer analyzed bound/total versus total plots (Sholl and Kim, 1989). Binding characteristics may also depend on the ligand which is used. A popular choice has been tritiated Moxestrol (11β-methoxy-[³H]-R2858, DuPont NEN Research Products). Moxestrol is used, since it reportedly has a low affinity for circulating protein and a high affinity for the estrogen receptor (McEwen *et al.*, 1975; Raynaud *et al.*, 1978; MacLusky *et al.*, 1979; Sholl and Kim, 1989). Because of this, non-specific binding levels are substantially reduced when using crude receptor prep-arations. When estrogen receptor is analyzed on DNA-cellulose columns or by sucrose density gradient, estradiol may be employed as ligand. Early experiments using a crude extract from the hypothalamus and preoptic area of 4-week-old mice, and estradiol as ligand, revealed that estradiol-binding could be competed by testosterone (Fox, 1975a). The extension of these studies to *Tfm/Y* mutant mice, in which there is a deficiency in the androgen receptor, revealed that there was a significant reduction in the degree to which androgens could compete estrogen binding. This supported the conclusion that estrogens could also bind to the androgen receptor and that androgens do not have a significant affinity for the estrogen receptor. (In *Tfm/Y* mice androgen binding is significantly reduced, whereas estrogen binding is not markedly affected.) The ability of various steroids to compete estrogen binding to its receptor has also been studied in the rhesus monkey. In the neonatal and fetal rhesus monkey Moxestrol binding to a hypothal-amus–preoptic area (HPOA) receptor extract could be reduced by estradiol and diethyl-stilbestrol (DES) but not by progesterone, cortisol or testosterone (MacLusky *et al.*, 1986; Sholl and Kim, 1989). Also, when HPOA extracts from late pregnancy monkey fetuses were incubated with estradiol and applied to DNA–cellulose columns, estradiol binding in the 200–220 mM NaCl elution fraction was unaffected by the addition of a high concen-tration of dihydrotestosterone during the incubation (Pomerantz *et al.*, 1985). Such experiments indicate that as with the rodent the developing primate brain also contains a fairly specific estrogen receptor. However, the ability of estrogen to compete androgen binding to the androgen receptor (Fox, 1975a; Pomerantz *et al.*, 1985) has opened the possibility that neural tissue may be sensitive to a ratio of androgens to estrogens (Fox, 1975a) at the level of the androgen receptor. Thus for some effects of estrogen, the estrogen receptor need not be involved.

Ultimately, an indication of whether a steroid might be effective in regulating neural activity can be obtained by measuring the accumulation of that steroid in cell nuclei. Biochemical methods have been developed which permit the isolation of brain cell nuclei and the quantitation of occupied estrogen receptors in those nuclei. Roy and McEwen (1977) developed an exchange assay for nuclear estrogen receptors in the rat brain. More recently, this procedure was modified to estimate nuclear binding in smaller amounts of neural tissue (Brown *et al.*, 1990), specifically in rats at different developmental stages. An indirect approach to characterizing nuclear levels of occupied estrogen receptor has been developed by Michael *et al.* (1989) for the fetal primate brain. Briefly, this technique involves the injection of radioactive steroid into the umbilical vein. After 1h the fetus is killed and tissues are dissected. Nuclei are isolated by homogenizing tissues in a Triton X-100 buffer followed by centrifugation at different speeds and ultimately through sucrose. Radioactive steroids can be efficiently extracted from the purified nuclei and quantitated by high performance liquid chromatography (HPLC).

11.6 DEVELOPMENTAL CHANGES IN ESTROGEN RECEPTOR CONCENTRATIONS: DESCRIPTION AND SIGNIFICANCE

In the rat, unoccupied androgen and estrogen receptors are present in the HPOA of the brain seven days before birth (Vito and Fox, 1982). Both receptors are more abundant in the HPOA than in other areas of the brain. Estrogen receptor concentration (bound [^{3}H]estradiol/mg tissue) in the HPOA increases seven-fold through the last week of gestation and reaches a plateau during the first 2–4 days after birth. In contrast to the estrogen receptor, there is a more gradual increase in androgen receptor levels (measured as bound [^{3}H]dihydrotestosterone or [^{3}H]testosterone/ mg tissue), continuing through 20–25 days after birth. The elution pattern of binding peaks from DNA-cellulose columns established the identity of these receptors. The same authors also examined the development of androgen and estrogen receptors in the mouse and found a similar pattern, although levels of estrogen receptors in the mouse continued to increase beyond the neonatal period. MacLusky *et al.* (1979) specifically looked for sex differences in cytoplasmic estrogen receptor concentrations in the brain of perinatal rats. Using [^{3}H]Moxestrol as ligand and Sephadex LH-20 column chromatography to separate free and bound ligand, a significant increase in concentration was noted between day 21 of pregnancy and days 5–6 after parturition. This increase was observed primarily in limbic brain regions and the cerebral cortex but not in the brainstem. There did not appear to be any major difference between sexes. Nuclear receptors were also detected and were found more consistently and at higher levels in limbic areas of the male rat (both fetal day 21 and postnatal day 5) than in corresponding regions from the female. The significance of the sudden developmental increase in estrogen receptor concentrations in the rat has been related to the onset of the 'critical period' and the appearance of postmitotic neurons in this species (Vito and Fox, 1982).

More recently, estrogen nuclear binding capacity was examined in young (2.5 month), middle-aged (8–10 month) and old rats (19 month) (Brown *et al.*, 1990). Animals were gonadectomized and adrenalectomized prior to injecting a saturating dose of estradiol. In young females the periventricular preoptic area (PVP), preoptic area (POA) and ventromedial nucleus (VMN) of the brain exhibited higher nuclear estrogen receptor levels than comparable areas of the male. Between middle and old age there was a decrease in nuclear levels in the POA, arcuate-median eminence region and the VMN in female rats. These changes confirm and add to previous observations (Kanungo *et al.*, 1975; Jiang and Peng, 1981; Wise and Parsons, 1984; Rubin *et al.*, 1986). In addition, they suggest that the ovarian acyclicity of aged rodents may be linked to a change in brain estrogen receptor dynamics. In the same report cytosolic progesterone receptors were also examined in estrogen-primed animals; however, no age differences were noted, which emphasizes the potential importance of estrogen receptor dynamics in the aging process. Similar studies have not been conducted in primates.

In the monkey estrogen receptors were initially identified in brains from the late-term fetal (Pomerantz *et al.*, 1985) and neonatal periods (MacLusky *et al.*, 1986). The presence of these receptors suggested that they might be important for primate neural development. Unlike rodents, however, little importance had been attached to early estrogen exposure, at least in the monkey. This was primarily due to a series of observations that *in utero* exposure to dihydrotestosterone (a non-aromatizable androgen) can lead to the masculinization of postnatal behavioral patterns (Goy and McEwen, 1980; Goy, 1981; Goy *et al.*, 1988). However, these findings did not exclude the possibility that estrogens might exert an additional influence on primate brain differentiation, either directly or via the local aromatization of circulating androgens to estrogens. To provide more information on unoccupied estrogen receptor levels in the differentiating primate brain, Sholl and Kim (1989) examined brain estrogen receptor concentrations in the rhesus monkey fetus on days 70, 100 and 160 postconception. For this study, [^{3}H]Moxestrol was used as a ligand and free and bound ligand were separated on Sephadex LH-20 columns. Before incubation with the ligand, cytosol was pretreated with Lipidex 1000 to improve the sensitivity and

reproducibility of the assay. No sex difference was noted at any stage or in any of the regions which were examined. A significant change in receptor levels occurred during the course of development in both the medial basal hypothalamus (MBH) and cerebellum (CB) but not in the amygdala (AMG) or cerebral cortex. In the MBH the significance of this change could be attributed to a progressive increase in receptor concentration between days 70 and 160 postconception. Based on one sample from a 14-year-old female monkey, this increase continued into adulthood. In contrast to the MBH, CB levels exhibited a progressive decline with fetal age. A previous report found that in monkeys the positive feedback effect of estrogens on luteinizing hormone release becomes apparent at puberty (Terasawa, 1985). If estrogen receptors are involved in this feedback phenomenon then the increasing MBH levels which were noted would offer an explanation for the onset of this feedback mechanism. Additionally, the possible correlation between estrogen receptor feedback and receptor levels provides a clue regarding the minimal level of estrogen receptor needed for the initiation of a given neural response. In this regard, it was also noted that AMG receptor levels throughout development were only one-third of those in the adult MBH. Thus it is tempting to speculate that estrogens may be less effective in modulating the differentiation of limbic brain regions. An interesting finding which emerged from this study was that describing receptor changes in the CB, with values being highest on day 70 (considered to be around the 'critical stage' of development in monkeys (Goy and McEwen, 1980). This early elevated receptor concentration may be significant in terms of the ontogeny of sex differences in motor coordination and activity (for example, 'rough and tumble' play behavior.) This may be true particularly if the CB contains significant levels of aromatase activity through which androgens from the fetal testes could be converted into estrogens. This appears to be the case, since monkey cerebellar aromatase activity is relatively higher on days 70 and 100 postconception than on day 160 (Sholl and Kim, 1990b).

Sex differences have been described in humans for both verbal and spatial abilities (Wittig and Petersen, 1979). This difference suggests that early brain development, including the lateralization of brain activity, may be susceptible to hormonal modification. With the knowledge that prenatal androgen exposure in monkeys can modify postnatal behavioral patterns, a study was conducted to determine whether there might be a lateralization of androgen receptor activity in the cerebral cortex of the fetal monkey (Sholl and Kim, 1990a). Results of this study are summarized in Fig. 11.1. Briefly, rhesus monkey fetuses were studied on day 70 postconception. Analysis of unoccupied androgen receptor levels in frontal and temporal cortical areas revealed a striking asymmetry. In contrast, aromatase activity in the same tissue samples was symmetrically distributed (data not shown). Androgen receptor levels in the right frontal lobe of male subjects were higher than in the left frontal lobe for every subject, whereas in female fetuses there was no consistent pattern in the distribution of receptor levels between the two sides of the frontal lobe. A sex-dependent asymmetry was also noted in the temporal lobe, but in this case receptor concentrations on the left side of male subjects were consistently higher than on the right side. In the report of these findings hypotheses were presented which relate to the potential importance of asymmetrical receptor levels with respect to sexual differences in brain lateralization (Sholl and Kim, 1990a).

With the possibility that estrogens are also involved in the differentiation of the primate brain, it is important to determine whether cortical estrogen receptors in monkeys are asymmetrically distributed along with androgen receptors. Preliminary evidence of asymmetrical estrogen receptor levels is presented in Fig. 11.2. Summarizing these findings, estrogen receptor levels in a majority of cases were asymmetrically distributed in both male and female fetuses (postconception day 70.) However, unlike androgen receptors, several subjects exhibited undetectable levels, and in others levels between sides were similar. In neither sex was one side predominantly favored over the other in those cases where values were asymmetrical. The asymmetry in estrogen receptors is not without precedence, since Sandhu *et al.*, (1986) demonstrated a significant sex-dependent right–left side asymmetry in estrogen receptor concentrations in the rat cerebral cortex on postnatal day 2. Moreover, in line with these findings, it has been demonstrated that when estradiol is implanted unilaterally into the rat hypothalamus there are differential behavioral and neuroendocrine effects depending on the side of exposure (Nordeen and Yahr, 1982; Roy and Lynn, 1987). In the monkey it is possible that a consistent pattern of estrogen receptor asymmetry does not develop until after day 70 postconception. If that is the case, it would suggest that any estrogens which might be derived locally from androgens via aromatization could be important at a later stage in the organization or activation of sex-dependent asymmetrical brain activity.

11.7 *IN UTERO* DIETHYLSTILBESTROL (DES) EXPOSURE IN HUMANS

In order to demonstrate the importance of a specific mechanism with respect to a particular developmental process, one can either try to block that mechanism or perturb it in some fashion. In the case of rodents, the importance of estrogen receptors in the sexual differentiation of the brain has been demonstrated by estrogen receptor blocking studies and experiments in which estrogens were injected during the neonatal period. In the case of primates, such studies are more difficult and frequently one must rely on

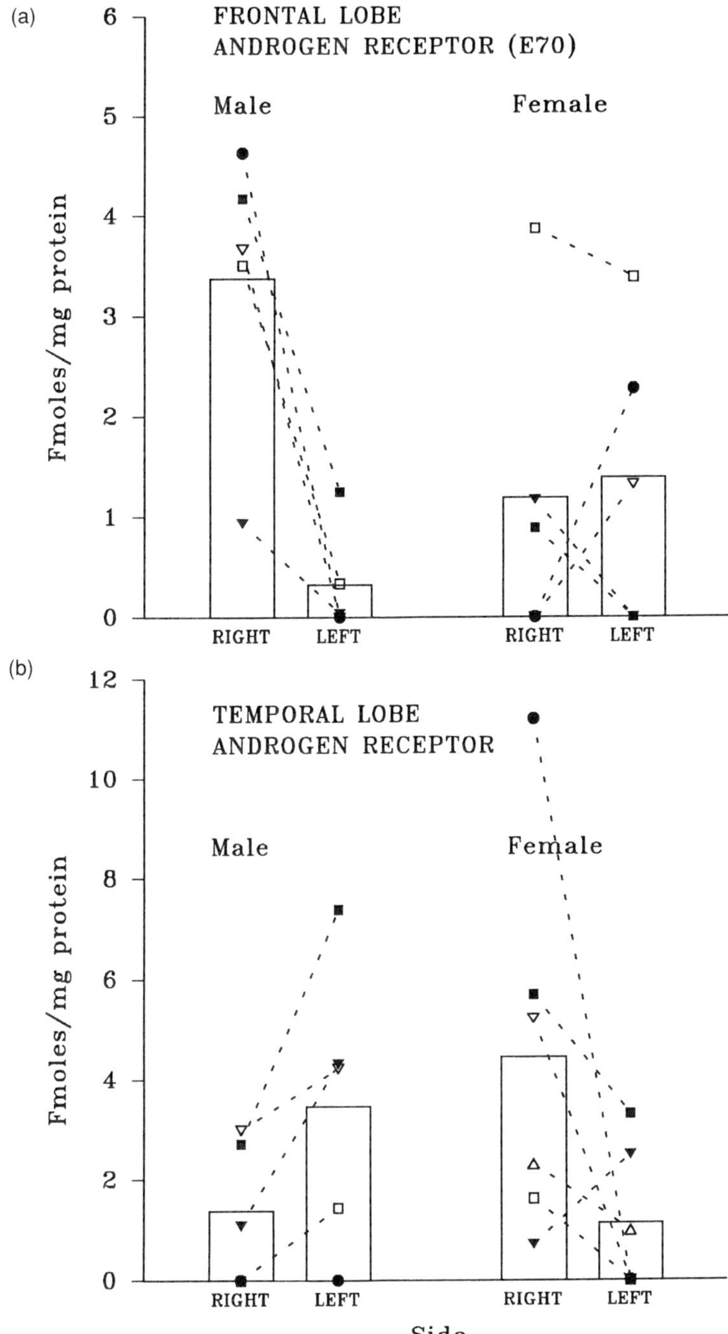

Fig. 11.1.   Unoccupied androgen receptor levels in (a) the frontal and (b) the temporal areas of the fetal monkey (*Macaca mulatta*) cerebral cortex on embryonic day 70. Dashed lines connect right and left side values in individual male and female subjects. Open bars represent the means of the respective groups. These data have been described previously (Sholl and Kim, 1990a).

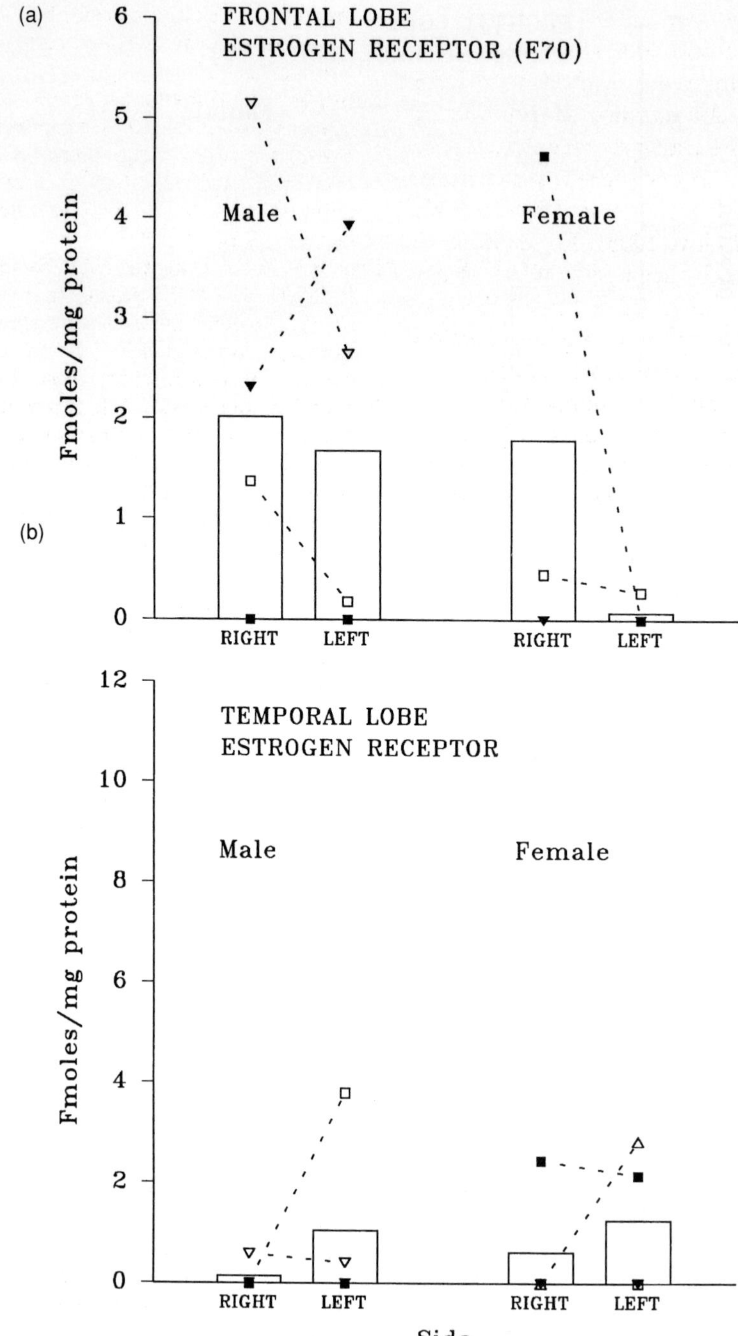

Fig. 11.2. Unoccupied estrogen receptor levels in (a) the frontal and (b) the temporal areas of the fetal monkey (*Macaca mulatta*) cerebral cortex on embryonic day 70. Dashed lines connect right and left side values in individual male and female subjects. Open bars represent the means of the respective groups.

indirect evidence. For example, from the 1940s to 1971 between 0.4 and 2.8 million human pregnancies were treated with DES (Edelman, 1986). As mentioned previously DES can bind to the estrogen receptor. It also has effects similar to androgen on the brain and behavior of lower mammals (Dohler *et al.*, 1982b; Hines and Goy, 1985). In addition to this, Michael and Bonsall (1990) have demonstrated in fetal monkeys that DES can accumulate in brain cell nuclei when injected into the umbilical circulation. There have been several attempts to demonstrate that *in utero* exposure of human females to DES has an effect on their behavior. Unfortunately, results from these studies are not in agreement (Ehrhardt *et al.*, 1989; Lish *et al.*, 1991). Moreover, a recent study emphasized the inconsistency of behaviors in women exposed to DES (Lish *et al.*, 1991). Whether this represents the final word in the DES controversy is not known. However, it is unlikely that any further study of these subjects will provide evidence that fetal estrogens are capable of influencing primate brain differentiation.

REFERENCES

Albrecht, E.D. and Pepe, G.J. (1990) Placental steroid hormone biosynthesis in primate pregnancy. *Endocr. Rev.* **11**, 124–50.

Ankenbauer, W., Strahle, U. and Schutz, G. (1988) Synergistic action of glucocorticoid and estradiol responsive elements. *Proc. Natl. Acad. Sci. USA*, **85**, 7526–30.

Arai, Y. and Matsumoto, A. (1978) Synapse formation of the hypothalamic arcuate nucleus during post-natal development in the female rat and its modification by neonatal estrogen treatment. *Psychoneuroendocrinology*, **3**, 31–45.

Barraclough, C.A. (1967) Modifications in reproductive function after exposure to hormones during prenatal and early postnatal period, in *Neuroendocrinology,* Vol. II (ed L. Martini), Academic Press, New York, pp. 62–100.

Bleier, R. and Byne, W. (1985) Septum and hypothalamus, in *The Rat Nervous System*, Vol. I (ed G. Paxinos), Academic Press, Sydney, pp. 87–118.

Bleier, R., Byne, W. and Siggelkow, I. (1982) Cytoarchitectonic sexual dimorphism of the medial preoptic and anterior hypothalamic areas in guinea pig, rat, hamster, and mouse. *J. Comp. Neurol.*, **212**, 118–30.

Blum, M. and Roberts, J.L. (1989) Quantitation of nuclear low-level gene expression in central nervous system using solution hybridization and *in situ* hybridization, in *Methods in Neurosciences*, Vol. 1 (ed P.M. Conn), Academic Press, New York, pp. 293–303.

Bresnick, E.H., Dalman, F.C., Sanchez, E.R. and Pratt, W. B. (1989) Evidence that the 90-kDa heat shock protein is necessary for the steroid binding conformation of the L cell glucocorticoid receptor. *J. Biol. Chem.*, **264**, 4992–7.

Brown, T.J., MacLusky, N.J., Shanabrough, M. and Naftolin, F. (1990) Comparison of age- and sex-related changes in cell nuclear estrogen-binding capacity and progestin receptor induction in the rat brain. *Endocrinology*, 126, 2965–72.

Callard, G.V. (1984) Aromatization in brain and pituitary: an evolutionary perspective, in *Metabolism of Hormonal Steroids in the Neuroendocrine Structures* (eds F. Celotti, F. Naftolin and L. Martini), Raven Press, New York, pp. 79–102.

Carson, M.A., Tsai, M.J., Conneely, O.M. *et al.* (1987) Structure–function properties of the chicken progesterone receptor A synthesized from complementary deoxyribonucleic acid. *Mol. Endocrinol.* **1**, 791–801.

Carson-Jurica, M.A., Schrader, E.T. and O'Malley, W. (1990) Steroid receptor family: structure and functions. *Endocr. Rev.* **11**, 201–20

Castracane, V.D. and Goldzieher, J.W. (1986) Timing of the luteal–placental shift in the baboon (*Papio cynocephalus*). *Endocrinology,* **118**, 506–12.

Clark, A.S., MacLusky, N.J. and Goldman-Rakic, P.S. (1988) Androgen binding and metabolism in the cerebral cortex of the developing rhesus monkey. *Endocrinology*, **123**, 932–40.

Commins, D. and Yahr, P. (1984) Adult testosterone levels influence the morphology of a sexually dimorphic area in the mongolian gerbil brain. *J. Comp. Neurol.*, **224**, 132–40.

Dohler, K.D. and Wuttke, W. (1975) Changes with age in levels of serum gonadotropins, prolactin, gonadal steroids in prepubertal male and female rats. *Endocrinology*, **97**, 898–907.

Dohler, K.D., Coquelin, A., Davis. F. *et al.* (1982a) Differentiation of the sexually dimorphic nucleus in the preoptic area of the rat brain is determined by the perinatal hormone environment. *Neurosci. Lett.* **33**, 295–8

Dohler, K.D., Hines, K., Coquelin, A. *et al.* (1982b) Pre- and postnatal influence of diethystilbestrol on differentiation of the sexually dimorphic nucleus in the preoptic area of the female rat brain. *Neuroendocrinol. Lett.*, 4, 361–5.

Edelman, D.A. (1986) *DES/Diethylstilbestrol: New Perspectives*, MTP Press, Boston.

Ehrhardt, A.A., Meyer-Bahlburg, H.F., Rosen, L.R. *et al.* (1989) The development of gender-related behaviour in females following prenatal exposure to diethylstilbestrol (DES). *Horm. Behav.*, 23, 526–41.

Ellinwood, W.E., Stanczyk, F.Z., Lazur, J.J. and Novy, M.J. (1989) Dynamics of steroid biosynthesis during luteal–placental shift in rhesus monkeys. *J. Clin. Endocrinol. Metab.*, 69, 348–55.

Fox, T.O. (1975a) Androgen- and estrogen-binding macromolecules in developing mouse brain: biochemical and genetic evidence. *Proc. Natl. Acad. Sci. USA*, 72, 4303–7.

Fox, T.O. (1975b) Oestradiol receptor of neonatal mouse brain. *Nature*, 258, 441–4.

Garancis, J.C., Miller, L.S., Tomita, J.T. *et al.* (1983) Immunoperoxidase localization of estrogen receptors in human breast carcinoma. *Cancer Detect. Prev.*, 6, 235–9.

George, F.W. and Ojeda, S.R. (1982) Changes in aromatase activity in the rat brain during embryonic, neonatal, and infantile development. *Endocrinology*, 111, 522–9.

George, F.W., Tobleman, W.T., Milewich, L. and Wilson, J.D. (1978) Aromatase activity in the developing rabbit brain. *Endocrinology*, 102, 86–91.

Gerall, A.A. and Ward, J.L. (1966) Effects of prenatal exogenous androgen on the sexual behaviour of the female albino rat. *J. Comp. Physiol. Psychol.*, 62, 370–5.

Gerlach, J.L., McEwen, B.S., Pfaff, D.W. *et al.* (1976) Cells in regions of rhesus monkey brain and pituitary retain radioactive estradiol, corticosterone and cortisol differently. *Brain Res.* 103, 603–12.

Glass, C.K., Lipkin, S.M., Devary, O.V. and Rosenfeld, M.G. (1989) Positive and negative regulation of gene transcription by a retinoic acid–thyroid hormone receptor heterodimer. *Cell*, 59, 697–708.

Godowski, P.J., Rusconi, S., Miesfeld, R. and Yamamoto, K.R. (1987) Glucocorticoid receptor mutants that are constitutive activators of transcriptional enhancement. *Nature*, 325, 365–8.

Gordon, M.S. and Notides, A.V. (1986) Computer modeling of estradiol interactions with the estrogen receptor. *J. Steroid Biochem.*, 25, 177–81.

Gorski, J., Toft, D., Shyamala, G. *et al.* (1968) Hormone receptors: studies on the interaction of estrogen with uterus. *Recent Progr. Horm. Res.*, 24, 45–80.

Gorski, R.A. (1963) Modification of ovulatory mechanisms by postnatal administration of estrogen to the rat. *Am. J. Physiol.*, 205, 842–4.

Gorski, R.A., Gordon, J.H., Shryne, J.E. and South, A.M. (1978) Evidence for a morphological sex difference within the medial preoptic area of the rat brain. *Brain Res.*, 148, 333–46.

Goy, R.W. (1981) Differentiation of male social traits in female rhesus macaques by prenatal treatment with androgens: variation in type of androgen, duration, and timing of treatment, in *Fetal Endocrinology* (eds M.J. Novy and J.A. Resko), Academic Press, New York, pp. 319–39.

Goy, R.W. and McEwen, B.S. (1980) *Sexual Differentiation of the Brain*, MIT Press, Cambridge, MA.

Goy, R.W., Uno, H. and Sholl, S.A. (1988) Psychological and anatomical consequences of prenatal exposure to androgens in female rhesus, in *Toxicity of Hormones in Perinatal Life* (eds T. Mori and H. Nagasawa), CRC Press, Boca Raton, FL, pp. 127–42.

Green, S. and Chambon, P. (1987) Oestradiol induction of a glucocorticoid-responsive gene by a chimeric receptor. *Nature*, 325, 75–8.

Green, S., Walter, P., Kumar, V. *et al.* (1986) Human oestrogen receptor cDNA: sequence expression and homology to v-erb-A. *Nature*, 320, 134–9.

Greene, G.L., Sobel, N.B., King, W.J. and Jensen, E.V. (1984) Immunocytochemical studies of estrogen receptors. *J. Steroid Biochem.*, 20, 51–6.

Greene, G.L., Gilna, P., Waterfield, M. *et al.* (1986) Sequence and expression of a functional human glucocorticoid receptor cDNA. *Science*, 231, 1150–4.

Grody, W.W., Schrader, W.T. and O'Malley, B.W. (1982) Activation, transformation, and subunit structure of steroid hormone receptors. *Endoc. Rev.*, 3, 141–63.

Guiochon-Mantel, A., Loosfelt, H., Lescop, P. *et al.* (1989) Mechanisms of nuclear localization of the progesterone receptor: evidence for interaction between monomers. *Cell*, 57, 1147–54.

Handa, R.J., Connolly, P.B. and Resko, J.A. (1988) Ontogeny of cytosolic androgen receptors in the brain of the fetal rhesus monkey. *Endocrinology*, 122, 1890–6.

Harris, G.W. (1964) Sex hormones, brain development, and brain function. *Endocrinology*, 75, 627–48.

Harris, G.W. and Levine, S. (1965) Sexual differentiation of the brain and its experimental control. *J. Physiol.*, 181, 379–400.

Hines, M. and Goy, R.W. (1985) Estrogens before birth and development of sex-related reproductive traits in the female guinea pig. *Horm. Behav.*, **19**, 331–47.

Hines, M., Davies, F.C., Coquelin, A. *et al.* (1985) Sexually dimorphic regions in the medial preoptic area and the bed nucleus of the stria terminalis of the guinea pig brain: a description and an investigation of their relationship to gonadal steroids in adulthood. *J. Neurosci.*, **5,** 40–7.

Hollenberg, S.M., Giguere, V., Segui, P. and Evans, R.M. (1987) Co-localization of DNA-binding and transcriptional activation functions in the human glucocorticoid receptor. *Cell,* **49,** 39–46.

Ishii, D.M. and Shooter, E.M. (1975) Regulation of nerve growth factor synthesis in mouse submaxillary glands by testosterone. *J. Neurochem.*, **25**, 843–51.

Ito, S., Murakami, S., Yamanouchi, K. and Arai, Y. (1986) Perinatal androgen exposure decreases the size of the sexually dimorphic medial preoptic nucleus in the rat. *Proc. Jpn. Acad.*, **62,** 408.

Jensen, E.V., Suzuki, T., Kawashima, T. *et al.* (1968) A two-step mechanism for the interaction of estradiol with rat uterus. *Proc. Natl. Acad. Sci. USA* **59**, 632–8

Jiang, M.J. and Peng, M.T. (1981) Cytoplasmic and nuclear binding of estradiol in the brain and pituitary of old female rats. *Gerontology*, **27**, 51–7.

Kalderon, D., Roberts, B.L., Richardson, W.D. and Smith, A.E. (1984) A short amino acid sequence able to specify nuclear location. *Cell,* **39,** 499–509.

Kanungo, M.S., Patnaik, S.K. and Koul, O. (1975) Decrease in 17β-estradiol receptor in brain of aging rats. *Nature,* **253,** 366–7.

King, W.J. and Greene, G.L. (1984) Monoclonal antibodies localize estrogen receptor in the nuclei of target cells. *Nature,* **307,** 745–7.

Komack, D.R., Lu, B. and Black, I.B. (1989) Sexually dimorphic expression of the NGF receptor gene in rat brain. *Neurosci. Abst.*, **19,** 954.

Kumar, V. and Chambon, P. (1988) The estrogen receptor binds tightly to its responsive element as a ligand-induced homodimer. *Cell,* **55,** 145–56.

Kumar, V., Green, S., Stack, G. *et al.* (1987) Functional domains of the human estrogen receptor. *Cell,* **51,** 941–51.

Lanford, R.E., Kanda, P. and Kennedy, R.C. (1986) Induction of nuclear transport with synthetic peptide homologous to the SV40 antigen transport signal. *Cell,* **46,** 575–82.

Lish, J.D., Ehrhardt, A.A., Meyer-Bahlburg, H.F. *et al.* (1991) Gender-related behaviour development in females exposed to diethylstilbestrol (DES) *in utero*: an attempted replication. *J. Am. Acad.Child Adolesc. Psychiatry,* **30,** 29–37.

MacLusky, N.J., and Naftolin, F. (1981) Sexual differentiation of the central nervous system. *Science,* **211,** 1294–302.

MacLusky, N.J., Chaptal, C., Lieberburg, I. and McEwen, B.S. (1976) Properties and subcellular inter-relationships of presumptive estrogen receptor macromolecules in the brains of neonatal and prepubertal female rats. *Brain Res.,* **114,** 158–65.

MacLusky, N.J., Chaptal, C., Lieberburg, I. and McEwen, B.S. (1979) The development of estrogen receptor systems in the rat brain: perinatal development. *Brain Res.* **178,** 129–42.

MacLusky, N.J., Philip, A., Hurlburt, C. and Naftolin, F. (1984) Estrogen metabolism in neuroendocrine structures, in *Metabolism of Hormonal Steroids in the Neuroendocrine Structures* (eds F. Celotti, F. Naftolin and L. Martini), Raven Press, New York, pp. 103–16.

MacLusky, N.J., Naftolin, F. and Goldman-Rakic, P.S. (1986) Estrogen formation and binding in the cerebral cortex of the developing rhesus monkey. *Proc. Natl. Acad. Sci. USA,* **83,** 513–6.

Matsumoto, A. and Arai, Y. (1976) Effect of estrogen on early postnatal development of synaptic formation in the hypothalamic arcuate nucleus of female rats. *Neurosci. Lett.* **2,** 79–82.

Matsumoto A. and Arai, Y. (1980) Sexual dimorphism in 'wiring pattern' in the hypothalamic arcuate nucleus and its modification by neonatal hormonal environment. *Brain Res.,* **190,** 238–42.

Matsumoto, A. and Arai, Y. (1983) Sex difference in volume of the ventromedial nucleus of the hypothalamus in the rat. *Endocrinol. J.,* **30,** 277–80.

Matsumoto, A. and Arai, Y (1986) Male–female difference in synaptic organization of the ventromedial nucleus of the hypothalamus in the rat. *Neuroendocrinology,* **42,** 232–6.

McClellan, M., West, N.B. and Brenner, R.M. (1986) Immunocytochemical localization of estrogen receptors in the macaque endometrium during the luteal-follicular transition. *Endocrinology,* **119,** 2467–75.

McEwen, B.S., Pfaff, D.W. and Zigmond, R.E. (1970a) Factors influencing sex hormone uptake by rat brain regions. II. Effects of neonatal treatment and hypophysectomy on testosterone uptake. *Brain Res.,* **21,** 17–28.

McEwen, B.S., Pfaff, D.W. and Zigmond, R.E. (1970b) Factors influencing sex hormone uptake by rat brain regions. III. Effects of competing steroids on testosterone uptake. *Brain Res.*, **21**, 29–38.

McEwen, B.S., Plapinger, L., Chaptal, C. *et al.* (1975) Role of fetoneonatal estrogen binding proteins in the associations of estrogen with neonatal brain cell nuclear receptors. *Brain Res.*, **96**, 400–6.

Michael, R.P. and Bonsall, R.W. (1990) The uptake of tritiated diethylstilbesterol by the brain, pituitary gland, and genital tract of the fetal macaque: a combined chromatographic and autoradiographic study. *J. Clin. Endocrinol. Metab.*, **71**, 868–74.

Michael, R.P., Bonsall, R.W. and Rees, H.D. (1986) The nuclear accumulation of [^{3}H]testosterone and [^{3}H]estradiol in the brain of the female primate: evidence for the aromatization hypothesis. *Endocrinology*, **118**, 1935–44.

Michael, R.P., Bonsall, R.W. and Rees, H.D. (1989) The uptake of [^{3}H]testosterone and its metabolites by the brain and pituitary gland of the fetal macaque. *Endocrinology*, **124**, 1319–26.

Migliaccio, A., Di Domenico, M., Green, S. *et al.* (1989) Phosphorylation on tyrosine of *in vitro* synthesized human estrogen receptor activates its hormone binding. *Mol. Endocrinol.* **3**, 1061–9.

Mizejewski, G.J., Vonnegut, M. and Simon, R. (1980) Neonatal androgenization using antibodies to alpha–fetoprotein. *Brain Res.*, **188**, 273–7.

Mizukami, S., Nishizuka, M. and Arai, Y. (1983) Sexual difference in nuclear volume and its ontogeny in the rat amygdala. *Exp. Neurol.*, **79**, 569–75.

Moncharmont, B., Su, J.L. and Parikh, I. (1982) Monoclonal antibodies against estrogen receptor: interaction with different molecular forms and functions of the receptor. *Biochemistry*, **21**, 6916–21.

Moncharmont, B., Anderson, W.L., Rosenberg, B. and Parikh, I. (1984) Interaction of estrogen receptors of calf uterus with a monoclonal antibody: probing of various molecular forms. *Biochemistry*, **23**, 3907–12.

Naftolin, F., Ryan, K.J., Davies, I.J. *et al.* (1975) The formation of estrogens by central neuroendocrine tissues, in *Recent Progress in Hormone Research*, Vol. 31 (ed R.O. Greep), Academic Press, New York, pp. 295–319.

Nishizuka, M. and Arai, Y. (1981) Organizational action of estrogen on synaptic pattern in the amygdala: implications for sexual differentiation of the brain. *Brain Res.*, **213**, 422–6.

Nordeen, E.J. and Yahr, P. (1982) Hemispheric asymmetries in the behavioural and hormonal effects of sexually differentiating mammalian brain. *Science*, **218**, 391–4.

Oakey, R.E. (1970) The progressive increase in oestrogen production in human pregnancy: an appraisal of the factors responsible. *Vitam. Horm.*, **28**, 1–36.

Paine, P.L., Moore, L.C. and Horowitz, S.B. (1975) Nuclear envelope permeability. *Nature*, **254**, 109–14.

Pelletier, G., Liao, N., Follea, N. and Gorindan, M.V. (1988a) Distribution of estrogen receptors in the rat pituitary as studies by *in situ* hybridization. *Mol. Cell. Endocrinol.*, **56**, 29–33.

Pelletier, G., Liao, N., Follea, N. and Govindan, M.V. (1988b) Mapping of estrogen receptor-producing cells in the rat brain by *in situ* hybridization. *Neurosci. Lett.*, **94**, 23–8.

Perrot-Applanat, M., Logeat, F., Groyer-Picard, M.T. and Milgrom, E. (1985) Immunocytochemical study with monoclonal antibodies to progesterone receptor in human breast tumors. *Cancer Res.*, **47**, 2652–61.

Pfaff, D.W. and Keiner, M. (1973) Atlas of estradiol-concentrating cells in the central nervous system of the female rat. *J. Comp. Neurol.*, **151**, 121–58.

Pfaff, D.W., Gerlach, J.L., McEwen, B.S. *et al.* (1976) Autoradiographic localization of hormone-concentrating cells in the brain of the female rhesus monkey. *J. Comp. Neurol.*, **170**, 279–94.

Picard, D., Salser, S.J. and Yamamoto, K.R (1988) A movable and regulatable inactivation function within the steroid binding domain of the glucocorticoid receptor. *Cell*, **54**, 1073–80.

Pomerantz, S.M. and Sholl, S.A. (1987) Analysis of sex and regional differences in androgen receptors in fetal rhesus monkey brain. *Dev. Brain Res.*, **36**, 151–4.

Pomerantz, S.M., Fox, T.O., Sholl, S.A. *et al.* (1985) Androgen and estrogen receptors in fetal rhesus monkey brain and anterior pituitary. *Endocrinology*, **116**, 83–9.

Ponglikitmongkol, M., Green, S. and Chambon, P. (1988) Genomic organization of the human oestrogen receptor gene. *EMBO J.*, **7**, 3385–8.

Raynaud, J., Mercier-Bodard, C. and Baulieu, E. (1971) Rat estradiol binding plasma protein (EBP). *Steroids*, **18**, 767–88.

Raynaud, J., Martin, P.M., Bouton, M. and Ojasoo, T. (1978) 11β-methoxy-17-ethynyl-1,3,5(10)-estratriene-3,17β-diol (moxestrol), a tag for estrogen receptor binding sites in human tissue. *Cancer Res.*, **38**, 3044–50.

Reddy, V.V., Hanjani, P. and Rajan, R. (1981) Synthesis of catechol estrogens by human uterus and leiomyoma. *Steroids*, **37**, 195–203.

Reisert, I., Han, V., Lieth, E. *et al.* (1987) Sex steroids promote neurite growth in mesencephalic tyrosine hydroxylase immunoreactive neurons *in vitro. Int. J. Dev. Neurosci.*, **5**, 91–8.

Resko, J.A. (1974) Sex steroids in the circulation of the fetal and neonatal rhesus monkey: a comparison between male and female fetuses, in *International Symposium on Sexual Endocrinology*, Vol. 32, Inserm, Lyon, France, pp. 195–204.

Resko, J.A., Ellinwood, W.E., Pasztor, L.M. and Buhl, A.E. (1980) Sex steroids in the umbilical circulation of fetal rhesus monkeys from the time of gonadal differentiation. *J. Clin. Endocrinol. Metab.*, **50**, 900–5.

Roselli, C.E. and Resko, J.A. (1986) Effects of gonadectomy and androgen treatment on aromatase activity in the fetal monkey brain. *Biol. Reprod.*, **35**, 106–12.

Roselli, C.E. and Resko, J.A. (1989) Testosterone regulates aromatase activity in discrete brain areas of male rhesus monkeys. *Biol. Reprod.*, **40**, 929–34.

Roselli, C.E., Ellinwood, W.E. and Resko, J.A. (1984) Regulation of brain aromatase activity in rats. *Endocrinology*, **114**, 192–200.

Roselli, C.E., Horton, L.E. and Resko, J.A. (1985) Distribution and regulation of aromatase activity in the rat hypothalamus and limbic system. *Endocrinology*, **117**, 2471–7.

Roselli, C.E., Salisbury, R.L. and Resko, J.A. (1987a) Genetic evidence for androgen-dependent and independent control of aromatase activity in the rat brain. *Endocrinology*, **121**, 2205–10.

Roselli, C.E., Stadelman, H., Horton, L.E. and Resko, J.A. (1987b) Regulation of androgen metabolism and luteinizing hormone-releasing hormone content in discrete hypothalamic and limbic areas of male rhesus macaques. *Endocrinology*, **120**, 97–106.

Roy, E.J. and Lynn, D.M. (1987) Asymmetry in responsiveness of the hypothalamus of the female rat to estradiol. *Physiol. Behav.*, **40**, 267–9.

Roy, E.J. and McEwen, B.S. (1977) An exchange assay for estrogen receptors in cell nuclei of the adult rat brain. *Steroids,* **30**, 657–69.

Rubin, B.S., Fox, T.O. and Bridges, R.S. (1986) Estrogen binding in nuclear and cytosolic extracts from brain and pituitary of middle-aged female rats. *Brain Res.*, **383**, 60–7.

Sandhu, S., Cook, P. Diamond, M.C. (1986) Rat cerebral cortical estrogen receptors: male–female, right–left. *Exp. Neurol.*, **92**, 186–96.

Sar, M. (1985) Application of avidin–biotin complex technique for the localization of estradiol receptor in target tissues using monoclonal antibodies, in *Techniques in Immunocytochemistry* (eds G.R. Bullock and P. Petrusz), Academic Press, New York, pp. 43–54.

Sar, M. and Parikh, I. (1986) Immunohistochemical localization of estrogen receptor in rat brain, pituitary and uterus with monoclonal antibodies. *J. Steroid Biochem.*, **24**, 497–503.

Scatchard, G. (1949) The attraction of protein for small molecules and ions. *Ann. N. Y. Acad. Sci.*, **51**, 660–72.

Schatteman, G.C., Gibbs, L., Lanahan, A.A. *et al.* (1988) Expression of NGF receptor in the developing and adult primate central nervous system. *J. Neurosci.*, **8**, 860–73.

Severne, Y., Wieland, S., Schaffner, W. and Rusconi, S. (1988) Metal binding 'finger' structures in the glucocorticoid receptor defined by site-directed mutagenesis. *EMBO J.*, **7**, 2503–8.

Sheridan, P.J., Buchanan, J.M., Anselmo, V.C. and Martin, P.M. (1979) Equilibrium: the intracellular distribution of steroid receptors. *Nature*, **282**, 579–82.

Sholl, S.A. and Goy, R.W. (1981) Dynamics of testosterone, dihydrotestosterone and estradiol-17β uptake and metabolism in the brain of the male guinea pig. *Psychoneuroendocrinology*, **6**, 105–11.

Sholl, S.A. and Kim, K.L. (1989) Estrogen receptors in the rhesus monkey brain during fetal development. *Dev. Brain Res.*, **50**, 189–96.

Sholl, S.A. and Kim, K.L. (1990a) Androgen receptors are differentially distributed between right and left cerebral hemispheres of the fetal male rhesus monkey. *Brain Res.*, **516**, 122–6.

Sholl, S.A. and Kim, K.L. (1990b) Aromatase, 5α-reductase, and androgen receptor levels in the fetal monkey brain during early development. *Neuroendocrinology*, **52**, 94–8.

Sholl, S.A. and Pomerantz, S.M. (1986) Androgen receptors in the cerebral cortex of fetal female monkeys. *Endocrinology*, **119**, 1625–31.

Sholl, S.A., Goy, R.W. and Uno, H. (1982) Differences in brain uptake and metabolism of testosterone in gonadectomized, adrenalectomized male and female rhesus monkeys. *Endocrinology*, **111**, 806–13.

Sholl, S.A., Goy, R.W. and Kim, K.L. (1989) 5α-reductase, aromatase, and androgen receptor levels in the monkey brain during fetal development. *Endocrinology*, **124**, 627–34.

Siiteri, P.K., Schwarz, B.E., Moriyama, I. *et al.* (1973)

Estrogen binding in the rat and human, in *Receptors for Reproductive Hormones*, (eds B.W. O'Malley and A. R. Means), Plenum Press, New York, pp. 97–112.

Stumpf, W.E. and Sar, M. (1976) Autoradiographic localization of estrogen, androgen, progestin and glucocorticosteroid in target tissues and non-target tissues, in *Receptors and Mechanism of Action of Steroid Hormones*, Vol. 8, (ed J.R. Pasqualini), Dekker, New York, pp. 41–84.

Stumpf, W.E., Sar, M., Reisert, I. and Pilgrim, C. (1983) Estrogen receptor sites in the developing central nervous system and their relationships to catecholamine systems. *Monogr. Neural Sci.*, **9**, 205–12.

Swaab, D.F. and Fliers, E. (1985) A sexually dimorphic nucleus in the human brain. *Science,* **228**, 1112–15.

Terasawa, E. (1985) Developmental changes in the positive feedback effect of estrogen on luteinizing hormone release in ovariectomized female rhesus monkeys. *Endocrinology*, **117**, 2490–7.

Terasawa, E., Wiegand, S.J. and Bridson, W.E. (1980) A role for the medial preoptic nucleus on afternoon of proestrus in female rats. *Am. J. Physiol.*, **238**, E533–9

Tobet, S.A., Zahniser, D.J. and Baum, M.J. (1986) Sexual dimorphism in the preoptic/anterior hypothalamic area of adult ferrets: effects of adult exposure to sex steroids. *Brain Res.*, **364**, 240–57.

Toran-Allerand, C.D. (1976) Sex steroids and the development of the newborn mouse hypothalamus and preoptic area *in vitro*: implication for sexual differentiation. *Brain Res.*, **106**, 407–12.

Toran-Allerand, C.D. (1980) Sex steroids and the development of the newborn mouse hypothalamus and preoptic area *in vitro*. II. Morphological correlates and hormonal specificity. *Brain Res.*, **189**, 413–27.

Toran-Allerand, C.D. (1984) On the genesis of sexual differentiation of the central nervous system: morphogenetic consequences of steroidal exposure and possible role of alpha–fetoprotein. *Progr. Brain Res.*, **61**, 63–98.

Toran-Allerand, C.D. and MacLusky, N.J. (1989) Co-localization of NGF and estrogen receptors: implications for the basal forebrain. *Neurosci. Abstr.*, **15**, 954.

Toran-Allerand, C.D., Gerlach, J.L. and McEwen, B.S. (1980) Autoradiographic localization of [^{3}H]estradiol related to steroid responsiveness in cultures of the newborn mouse hypothalamus and preoptic area. *Brain Res.*, **184**, 517–22.

Toran-Allerand, C.D., Ellis, L. and Pfenninger, K.H. (1988) Estrogen and insulin synergism in neurite growth enhancement *in vitro*: mediation of steroid effects by interactions with growth factors. *Dev. Brain Res.*, **41**, 87–100.

Uchibori, M. and Kawashima, S. (1985) Effect of sex steroids on the growth of neuronal processes in neonatal rat hypothalamus–preoptic area and cerebral cortex in primary culture. *Int. J. Dev. Neurosci.*, **3**, 169–76.

Uriel, J., Souillon, D. and Dupiers, M. (1975) Affinity chromatography of human, rat and mouse alpha-fetoprotein on estradiol-sepharose adsorbents. *FEBS Lett.*, **53**, 305–8

Vito, C.C. and Fox, T.O. (1982) Androgen and estrogen receptors in embryonic and neonatal rat brain. *Dev. Brain Res.*, **2**, 97–110.

Weisz, J. and Ward, I.L. (1980) Plasma testosterone and progesterone titers of pregnant rats, their male and female fetuses and neonatal offspring. *Endocrinology*, **106**, 306–16.

Westley, B.R. and Salaman, D.F. (1976) Role of estrogen receptor in androgen-induced sexual differentiation of the brain. *Nature*, **262**, 407–8.

Williams, R.F., Johnson, D.K. and Hodgen, G.D. (1978) Ovarian estradiol secretion during early pregnancy in monkeys: luteal versus extra-luteal secretion and effect of chorionic gonadotropin. *Steroids*, **32**, 539–45.

Wise, P.M. and Parsons, B. (1984) Nuclear estradiol and cytosol progestin receptor concentrations in the brain and the pituitary gland and sexual behaviour in ovariectomized estradiol-treated middle-aged rats. *Endocrinology,* **115**, 810–16.

Wittig, M.A. and Petersen, A.C. (1979) *Sex-Related Differences in Cognitive Functioning*, Academic Press, New York.

Yamamoto, K. (1985) Steroid receptor regulated transcription of specific genes and gene networks. *Annu. Rev. Genet.*, **19**, 209–52.

Yanase, M., Honmura, A., Akaishi, T. and Sakuma, Y. (1988) Nerve growth factor-mediated sexual differentiation of the rat hypothalamus. *Neurosci. Res.*, **6**, 181–5.

VASOPRESSIN AND OXYTOCIN RECEPTORS AND THE DEVELOPING BRAIN

Gerard J. Boer

12.1 INTRODUCTION

The nonapeptides vasopressin (VP) and oxytocin (OT) were originally recognized as neurohormones, synthesized in hypothalamic magnocellular neurons of the supraoptic (SON) and paraventricular (PVN) nuclei and axonally transported to the neurohypophysis for secretion into the general circulation. The two nonapeptides differ in amino acid sequence at positions 3 and 8, but both contain a disulfide bridge formed by the cysteines at positions 1 and 6 (Fig. 12.1). Investigations into VP and OT receptors have long been focused on the peripheral target organs of these neuropeptides, i.e., kidney, liver, blood vessels and blood platelets for VP, and uterus and mammary glands for OT. At least two types of

VP receptors and one type of OT receptor have been characterized in this way. VP receptor subtypes were termed V1 and V2 receptors, originally on the basis of different second messenger systems (i.e. phosphoinositol turnover and activation of adenylate cyclase) (Michell *et al.*, 1979). Later, these subtypes were further characterized on a pharmacological basis (see review by Jard *et al.*, 1987).

Both neuropeptides have also been recognized as neurotransmitters in the nervous system. Immunocytochemical and *in situ* hybridization studies have shown parvocellular neurons of the PVN to synthesize VP or OT, as well as neurons in several other brain nuclei within and outside the hypothalamus. Most of these cells have synaptic projections to other brain areas (see review, by Buijs, 1987).

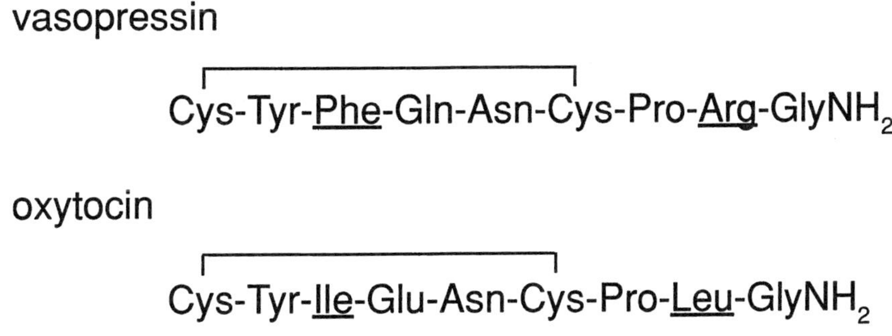

Fig. 12.1. Amino acid sequences of the neuropeptides vasopressin and oxytocin.

Receptors in the Developing Nervous System Vol. 1: *Growth factors and hormones* Edited by Ian S. Zagon and Patricia J. McLaughlin. Published in 1993 by Chapman & Hall. ISBN 0 412 45240 5. Vols. 1 and 2 (set) ISBN 0 412 54520 9.

Apart from these anatomical studies, De Wied (1983) had shown in behavioral studies that central effects of VP and OT occur (for reviews see Ganten and Pfaff, 1986; Gash and Boer, 1987). The first indications of the presence of central binding sites for VP and OT soon followed (Van Leeuwen and Wolters, 1983; Yamamura *et al.*, 1983; Brinton *et al.*, 1984; Audigier and Barberis, 1985). The VP binding could be pharmacologically characterized as of the V1 type (Dorsa *et al.*, 1984; see also reviews by Van Leeuwen, 1987; Tribollet, 1992).

Studies on the early appearance of brain VP and OT and the brain growth impairment of the VP-deficient Brattleboro rat (Boer *et al.*, 1982; Boer, 1985) stimulated interest in the possible role of VP and its receptors in neural maturational processes in the brain. The time course of development for the various VP and OT cell groups was found to be different, whereas there was not always a relationship with the appearance of stainable VP/OT innervation and receptors in target areas. This is of particular interest since early ligand–receptor interactions may play a developmental role (Csaba, 1986)

This chapter will describe briefly the brain VP and OT systems and the pharmacological properties of the VP and OT receptors. Next, an overview of the ontogeny of these receptors will be given, and an attempt will be made to discuss the functional role of these receptors in brain development. We will mainly describe rat studies.

12.2 VASOPRESSIN AND OXYTOCIN SYSTEMS

In the adult rat, immunocytochemistry reveals the PVN, SON and suprachiasmatic nucleus (SCN) as obvious sites of VP synthesis, as well as various other nuclei (De Vries and Buijs, 1983; Sofroniew, 1985; Buijs, 1987; Caffé *et al.*, 1987). Two neuroendocrine systems were recognized: (a) the magnocellular neurons of the PVN and SON which are part of the hypothalamo-neurohypophysial system, and (b) the parvocellular VP neurons of the PVN which comprise a hypophysiotropic pathway (PVN–hypophysial portal system). However, other VP neurons do not release VP into the bloodstream but have central projections. The VP cells of the SCN innervate the periventricular and mediodorsal hypothalamus, the periventricular thalamic area and the organum vasculosum laminae terminalis and are involved in circadian rhythmicity (see review Buijs, 1987). Furthermore, other VP cell groups, though in smaller numbers, exist in the bed nucleus of the stria terminalis (BNST) and medial amygdala. These VP systems are under control of gonadal steroids. The VP neurons of the BNST especially provide an important network of projections to the lateral septum and lateral habenula and – to a lesser extent – to the cingulate gyrus, dorsal Raphe nucleus and locus coeruleus. Finally, VP cell groups are localized in the dorsomedial hypothalamus and locus coeruleus (co-localized with norepinephrine in the latter nucleus; Caffé *et al.*, 1985).

Any of these VP cell systems has its own timetable of ontogeny (see review Boer, 1987). The magnocellular VP neurons are among the first macroneurons appearing in the brain. Two days later, on embryonic day 14 (ED14), these cells start to express the VP precursor. Except for those in the locus coeruleus, all other VP neurons are probably generated later and their maturation generally occurs much more slowly and sometimes even takes place postnatally in the rat (Boer, 1991). This would mean that, if an early hormonal VP signal in the absence of a functional blood–brain barrier (Johanson, 1980) could reach particular fetal brain areas, it would be well before any synaptic contact by VP fibers had been established.

OT is also synthesized in magnocellular neurons of the SON and PVN, which, together with the large VP-producing neurons, constitute the hypothalamo-neurohypophysial neuroendocrine system. These cells also mature rapidly, although processing of the OT polypeptide precursor to form free OT develops slightly later than is seen for the similar process for VP in VP neurons (Whitnall *et al.*, 1985). To date, the PVN is the only site in the brain where small non-secretory OT neurons have been found. Therefore, these neurons are probably the sole source of the OT fiber pathways in the brain (Buijs *et al.*, 1985). There is still insufficient information on the early ontogeny of the extrahypothalamic OT fibers to be specific about their early appearance and maturation.

12.3 VASOPRESSIN AND OXYTOCIN RECEPTOR SUBTYPES

Originally, tritiated VP and OT were used in both membrane and autoradiographic binding studies. These ligands were appropriate for describing the V1 and V2 VP and the OT receptors outside the brain, since in many peripheral target organs only one of these sites is specifically present. In the brain, however, intermingling of [3H]VP and [3H]OT binding sites is seen on autoradiograms of tissue

sections (e.g. Snijdewint *et al.*, 1989). Moreover, the observation that [3H]OT labels at a high and a low affinity site in rat hippocampal synaptic plasma membranes, both sites revealing high affinity for VP, led Audigier and Barberis (1985) to conclude that the supposed OT receptor (high affinity site) had almost equipotent affinity to VP. Thus, [3H]VP labels OT receptors nearly as efficiently as VP receptors (Table 12.1). Displacement studies for [3H]OT and [3H]VP binding supported the presence of different specificities: (a) OH[Thr[4], Gly[7]]OT, recognized as a highly selective OT agonist on the rat uterus and mammary gland, exhibited high affinity on the OT high affinity binding site and very low affinity on the VP site, (b) [Phe[2],Orn[8]]vasotocin, a selective VP agonist and (c) the VP antagonist d(CH$_2$)$_5$[Tyr(Et)[2],Val[4]]VP both had a much lower affinity for OT- than for VP-binding sites (cf. Table 12.1). Thus, two separate binding sites exist in the brain (Audigier and Barberis, 1985), and this created the need for more selectively labeled ligands than [3H]VP and -OT.

VP binding sites in septum, amygdala and dorsal hindbrain exhibit pharmacological

Table 12.1. Survey of affinities reported for VP and OT and their commonly used ligands in assays of the various subtypes of their receptors

	Type	Brain V1a	Liver V1a	Anterior pituitary V1b	Kidney V1/V2	Uterus OT	Brain OT
VP	V1/V2 agonist	0.4–2.9	0.15–18	0.4–8.0	1.5–1.9	0.8–1.6	1.0
dDAVP	V2 agonist	760–>1000	190–>1000	34–110	0.8–1.0	320	
d(CH$_2$)$_5$[Tyr(Me)[2]]VP	V1 antagonist	0.3–1.0	0.01–1.5	310[a]	9.0–16	13	[a]
dP[Tyr(Me)[2]]VP	V1 antagonist	1.7	8	3	16		
d(CH$_2$)$_5$[Sar[7]]VP	V1 antagonist	1.6	2.0–3.0		0.5[b]		
OT	OT agonist	30–2000	70–360	100–110	1.2–>100	1.9	1.8–6.0
HO[Thr[4], Gly[7]]OT	OT agonist		2500		2.5	0.6–6.3	1.0–20
d(CH$_2$)$_5$[Tyr(Me)[2], Thr[4],Orn[8],Tyr (NH$_2$)[9]] vasotocin	OT antagonist	6	14–26		10–15	0.03–3.3	0.15

Data (nM) are compiled from the literature cited in this paper and are based on saturation or displacement analysis on membrane fractions of the organs indicated (see also Jard *et al.*, 1987).
[a] No labeling of the radioactive labeled compound in autoradiograms.
[b] The ^{125}I-compound labels V1 sites of the inner renal medulla in autoradiograms.

properties similar to the peripheral V1-type of VP receptors (Dorsa *et al.*, 1983, 1984; Cornett and Dorsa, 1985; Table 12.1), including its effects on inositol phospholipid metabolism (Stephens and Logan, 1986; Shewey and Dorsa, 1988), though different G proteins mediate activation of phospholipase C than seen for V1 liver receptors (Swank and Dorsa, 1990). OT is not able (Stephens and Logan, 1986) or not potent enough (Shewey and Dorsa, 1988) to stimulate brain inositol phospholipid metabolism. The latter observation may indicate that the central OT receptor might differ from the peripheral one for which coupling to phospholipid metabolism is well established (Flint *et al.*, 1986).

Improved ligands have only recently been developed (Table 12.1). For brain V1 receptors the tritiated antagonist [^{3}H]d(CH$_2$)$_5$[Tyr-(Me)^2]VP (Van Leeuwen *et al.*, 1987; Dorsa *et al.*, 1988), the iodinated VP antagonist [^{125}I]d[CH$_2$)$_5$ [Tyr-(Me)^2,Tyr(NH$_2$)^9]VP (Elands *et al.*, 1988a), the iodinated linear VP antagonist [^{125}I]phenylacetyl-D-Tyr(Me)-Phe-Glu-Asn-Arg-Pro-Arg-Tyr(NH$_2$) (Schmidt *et al.*, 1991), and the iodinated antagonist [^{125}I]d(CH$_2$)$_5$-[Sar^7]VP became available (Phillips *et al.*, 1988; Kelly *et al.*, 1989; Gerstberger and Fahrenholz, 1989). The latter two in particular appeared to be powerful new and sensitive tools for specific binding studies that, moreover, like [^{3}H]d(CH$_2$)$_5$[Tyr(Me)$_2$]VP and [^{3}H]dDAVP can fully distinguish the V1 sites from the V2 receptors (V2 sites, so far, have never been shown to be present in the brain). [^{125}I]d(CH$_2$)$_5$[Sar^7]VP competition by unlabeled VP analogs gave the following order of potencies in the brain: d(CH$_2$)$_5$[D-Tyr^2(Et), Val^4,dGly^9]VP $\approx$ d(CH$_2$)$_5$-[Tyr(Me)^2]VP $\approx$ VP $\approx$ d(CH$_2$)$_5$[Sar^7]VP > d[D-Arg^8]VP (dDAVP) $\approx$ OT (Phillips *et al.*, 1988, 1990). This is consistent with V1 subtyping in the liver with the same ligand (Kelly *et al.*, 1989) and in the brain with both [^{3}H]VP and [^{3}H]d(CH$_2$)$_5$ [Tyr(Me)^2]VP as ligands (Dorsa *et al.*, 1988) for which the sequence is d(CH$_2$)$_5$[Tyr(Me)^2]VP $\approx$ VP > dDAVP $\approx$ OT.

The V1 ligands [^{125}I]d[CH$_2$)$_5$[Tyr(Me)^2, Tyr(NH$_2$)^9]VP and ^{125}I-linear VP antagonist have, until now, hardly been used in brain studies. One should also mention here that V1 receptors can be subdivided in V1a and V1b subclasses, the latter so far only being recognized distinctly in the anterior pituitary, the target in the PVN–hypophysial portal pathway (Antoni, 1984; Baertchi and Friedli, 1985; Jard *et al.*, 1986). The main difference with the V1a receptor in brain and liver is that the affinity of the VP antagonist d(CH$_2$)$_5$[Tyr(Me)^2]VP is much lower in the anterior pituitary (Table 12.1). Specific binding sites for the metabolically derived VP fragment [pGlu^4,Cyt^6]VP$_{4-9}$ also exist in the brain (De Kloet *et al.*, 1985a) but these have hardly been investigated further.

Highly selective ligands of the OT receptor have also been developed (Table 12.1). The tritiated agonist [^{3}H][Thr^4,Gly^7]OT (Elands *et al.*, 1988b) and the iodinated antagonist [^{125}I]d[CH$_2$)$_5$[Tyr(Me)^2,Thr^4,Orn^8,Tyr(NH$_2$)^9] vasotocin ([^{125}I]OTA; Elands *et al.*, 1987) appeared powerful ligands for both autoradiographic and membrane binding assays. These ligands possess even less affinity for the VP receptor than OT (Table 12.1), so that proper discrimination between VP and OT receptors in the brain could be reached (Elands *et al.*, 1988c; Tribollet, 1992).

At least part of the V1- and OT-type receptors is localized on neurons, since application of exogenous VP, OT or their agonists alters the rate of firing of single neurons in regions where binding sites were detectable (Joëls, 1987; Raggenbass *et al.*, 1988, 1989). However, the occurrence of VP and OT binding sites has also been reported on astroglia (Cholewinski *et al.*, 1988), the significance of which is unknown.

Approaches to molecular identification of the receptors generally develop rapidly. However, although strategies are described towards cloning and characterization of VP and OT receptors (Fahrenholz *et al.*, 1988; Richter, 1988; Trinder *et al.*, 1991) no conclusive data are available as yet. The first preliminary report on VP receptor mRNA claims a V1 type to be present in rat brain (Ostrowski *et al.*, 1991), but a further search for other possible subtypes of VP/OT is still necessary.

12.4 APPEARANCE OF VP RECEPTOR

The ontogeny of brain VP binding sites has, so far, mainly been studied by autoradiography using [³H]VP as ligand (Petracca *et al.*, 1986; Snijdewint *et al.*, 1989; Tribollet *et al.*, 1991a, b). Better techniques and blockade of [³H]VP binding to OT sites by incubation in the presence of the OT agonist HO[Thr⁴,Gly⁷]-OT permitted Tribollet *et al.* (1991a) to describe the most detailed pattern of appearance. However, all authors reported a similar sequence of appearance in the various brain areas, as well as the transient presence of VP binding sites in some neonatal rat brain regions.

The earliest detection of VP binding sites is, at ED16, in the ventral pontine reticular formation in a group of neurons tentatively identified as the pericentral part of the pontine reticular tegmental nucleus (Tribollet *et al.*, 1991a). Two days later binding is detectable in the suprachiasmatic nucleus and the facial nucleus. Tribollet *et al.* (1991a) identified the latter small nucleus by electrophysiology: antidromically identified by stimulation of the genu of the facial nerve, these neurons were VP sensitive. The early, very dense [³H]VP labeling in the lateral reticular nucleus of the brainstem, described by Snijdewint *et al.* (1989), should therefore be the same facial nucleus. At ED20 the septal area, choroid plexuses, locus coeruleus, nucleus of solitary tract and the spinal trigeminal nucleus could be labeled, whereas binding was only seen after birth in anterior olfactory nucleus, nucleus accumbens, central amygdala, in some thalamic nuclei, hippocampus, dorsal Raphe nucleus, area postrema, hypoglossal nucleus and inferior olive. Finally, at postnatal day 5 (PD5), the superficial gray

layer of the colliculus, the hypothalamic sigmoid nucleus and zona incerta became labeled (Fig. 12.2). Petracca *et al.* (1986) and Snijdewint *et al.* (1989) also showed binding in the cingulate cortex, and Snijdewint *et al.* (1989) additionally showed binding in the dorsal part of the caudate putamen and bed nucleus of the stria terminalis. However, Tribollet *et al.* (1991a) provided evidence that this reflects binding to the OT-type receptor (see below).

In many of these areas the density of receptor labeling increases with age, first rapidly, then more slowly. At PD10 all areas that express VP binding sites become visible (Fig. 12.2). However, in various areas the labeling was more intense in the infant that in the adult state. Zona incerta, dorsal Raphe nucleus, locus coeruleus, spinal trigeminal and facial nucleus all lose their VP receptor content in the second and third week postnatally. The adult distribution of VP receptor sites in the brain is reached at the time of weaning (Tribollet *et al.*, 1991a). The changes in the facial nucleus in particular are remarkable, since VP receptor density is among the highest seen in rat brain (Snijdewint *et al.*, 1989; Tribollet *et al.*, 1991a) and is totally absent in adulthood (Fig. 12.3 and 12.4). This transiently present VP receptor is of the V1 type in extracellular electrophysiological characterization studies (Tribollet *et al.*, 1991a).

In the first paper on the ontogeny of VP receptors, Petracca *et al.*, (1986) recognized the transient presence of [³H]VP binding in the cingulate gyrus, which was later confirmed by Snijdewint *et al.* (1989). Szot *et al.* (1989) characterized this binding in one of the few studies that used a saturation analysis on a crude membrane preparation from rat neonatal brain. The density of [³H]VP binding in the cingulate gyrus at PD8 was about 60% of the adult value, but the apparent affinity was much higher ($K_d = 0.9$ nM) in comparison to the value in adulthood (5.7 nM). This could explain why the autoradiographical signal nearly disappeared

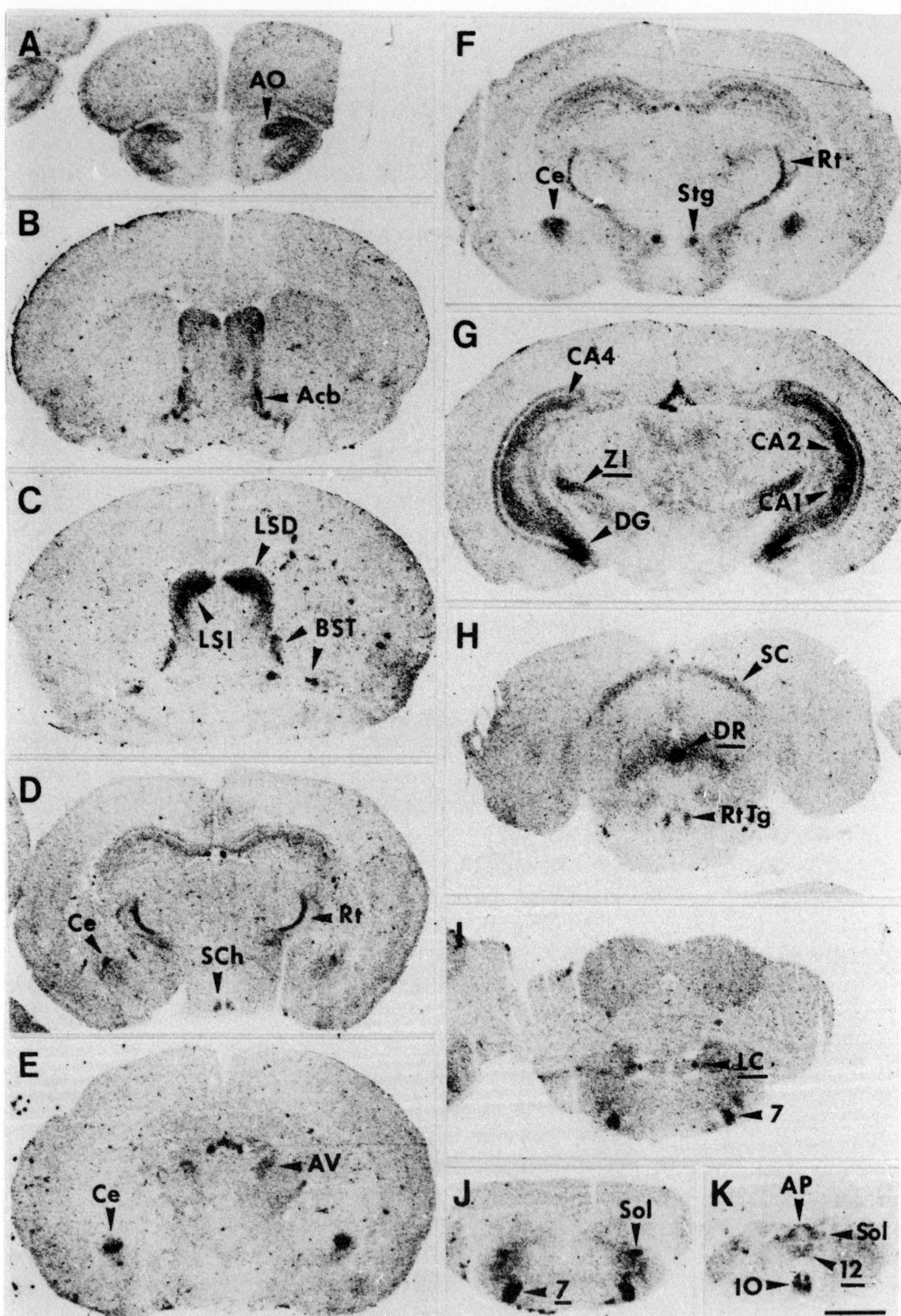

Fig. 12.2. Autoradiographic map of [³H]VP binding sites in coronal sections of the brain taken from frontal to caudal (A through K) of a PD10 rat. Incubation with [³H]VP was performed in the presence of a specific OT antagonist to block the equal potency of VP to bind OT receptors.

Fig. 12.2 *(Contd)*   At PD10 all those brain regions are labeled that express the VP receptor only transiently in the perinatal period (areas indicated by underlining the abbreviation). Arrowheads indicate the nuclei in which specific bindings were found. Abbreviations: Acb, nucleus accumbens; AO, anterior olfactory nucleus; AP, area postrema; AV, anteroventral thalamic nucleus; BST, bed nucleus of the stria terminalis; CA1/CA2, hippocampal fields; Ce, central amygdaloid nucleus; DG, dentate gyris; DR, dorsal Raphe nucleus; LC, locus coeruleus; LSD, dorsal part lateral septum; LSI, intermediate part lateral septum; Rt, reticular thalamic nucleus; RtTg, reticulotegmental nucleus of the pons; SC, superior colliculus; SCh, suprachiasmatic nucleus; Sol, nucleus of the solitary tract; Stg, stigmoid hypothalamic nucleus; ZI, zona incerta; 7, facial nucleus; 10, dorsal motor nucleus of the vagus nerve; 12, hypoglossal nucleus. Bar 2 mm. Slightly adapted from Tribollet *et al.* (1991a).

at a later age. The pharmacological profile of the [^{3}H]VP binding was similar at both ages, but different from a V1a subtyping and with no or little displacement of ligand by OT or its agonist OH[Thr^4,Gly^7]OT (Szot *et al.*, 1989). Due to the low affinity for the V1 antagonist d(CH$_2$)$_5$[Tyr(Me)^2]VP, a V1b subtyping was proposed (the VP receptor type of

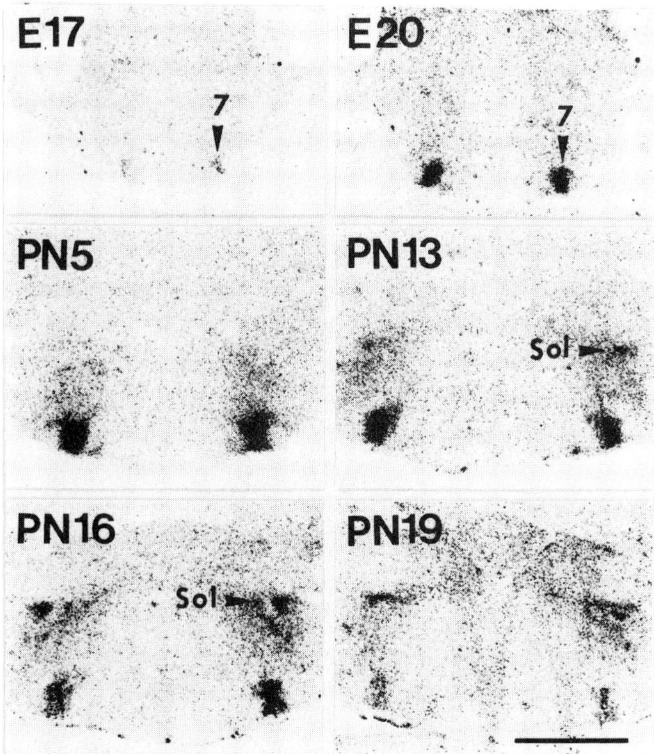

Fig. 12.3.   Evanescent presence of specific [^{3}H]VP binding in the facial nucleus (7) of developing rat. Receptor autoradiography on coronal brain stem sections (cf. Fig. 12.2J) from animals at various embryonic (E) and postnatal (PN) ages. See legend to Fig. 12.2 for details. Sol, nucleus of the solitary tract. Bar 2 mm. From Tribollet *et al.* (1991b), reproduced by permission of the publisher.

the anterior pituitary). However, in autoradiographic studies [³H]OT and [¹²⁵I]OTA also label the neonatal cingulate gyrus (Snijdewint *et al.*, 1989; Shapiro and Insel, 1989; Tribollet *et al.*, 1989). Since, moreover, incubation with the OT agonist OH[Thr⁴, Gly⁷]OT prevents binding of [³H]VP in this area (Tribollet *et al.*, 1991a), it seems as if receptors of the OT-type are also present. However, information as to whether the labeled OT ligands have affinity for the V1b receptor is, surprisingly, not yet known, and this information is needed to solve these apparent discrepancies.

12.5 APPEARANCE OF OT RECEPTOR

The development of OT receptors in rat brain has been studied autoradiographically, using [³H]OT (Snijdewint *et al.*, 1989) and, with higher sensitivity, using [¹²⁵I]OTA (Shapiro and Insel, 1989; Tribollet *et al.*, 1989). Specific binding for [¹²⁵I]OTA was detected first, at ED14, in the posterior dorsal neural tube, the presence of which suggests location on a still migrating group of cells that later, at ED18, forms the dorsal motor nucleus of the vagus nerve (Tribollet *et al.*, 1989). [³H]OT binding was not visible prenatally in hippocampal areas, caudate putamen and parietal cortex but only appeared at PD1 (Snijdewint *et al.*, 1989). The ontogenic pattern of OT receptors in both studies, and that of the VP receptors (see above), have a number of characteristics in common: in some areas binding site density steadily increases up to adult levels, whereas in other areas a transient presence of binding is seen (Fig. 12.4).

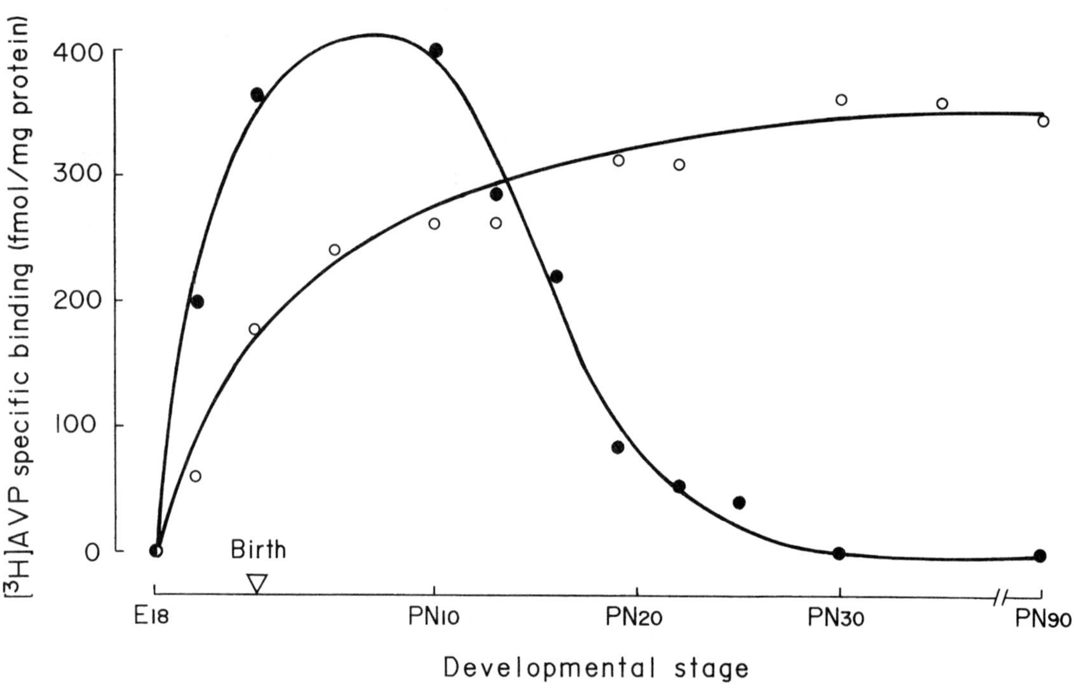

Fig. 12.4. Quantification of the ontogeny of [³H]VP binding in rat septum (open circles) and facial nucleus (closed circles) exemplifying the two patterns of development. Septal VP receptors have an initial progressive increase in density followed by a slower increase to adult values. Receptors in the facial nucleus are only present transiently peaking approximately in the first postnatal week (cf. Figs 12.3 and 12.7). From Tribollet *et al.* (1991b), reproduced by permission of the publisher.

At ED20 [^{125}I]OTA binding becomes visible in the anterior olfactory nucleus, caudate putamen, nucleus accumbens, at PD1 in the dorsal and lateral hippocampus and at PD3 in the bed nucleus of the stria terminalis, various septal, thalamic and amygdaloid nuclei and some cortical areas. Finally, at PD5, it becomes visible, albeit diffusely, throughout the hypothalamus and amygdala. Areas like the cingulate gyrus, nucleus accumbens, caudate putamen complex, the lateral area of the bed nucleus of the stria terminalis, dorsal hippocampus, anterior and paraventricular thalamic nuclei and the mammillary complex all lose

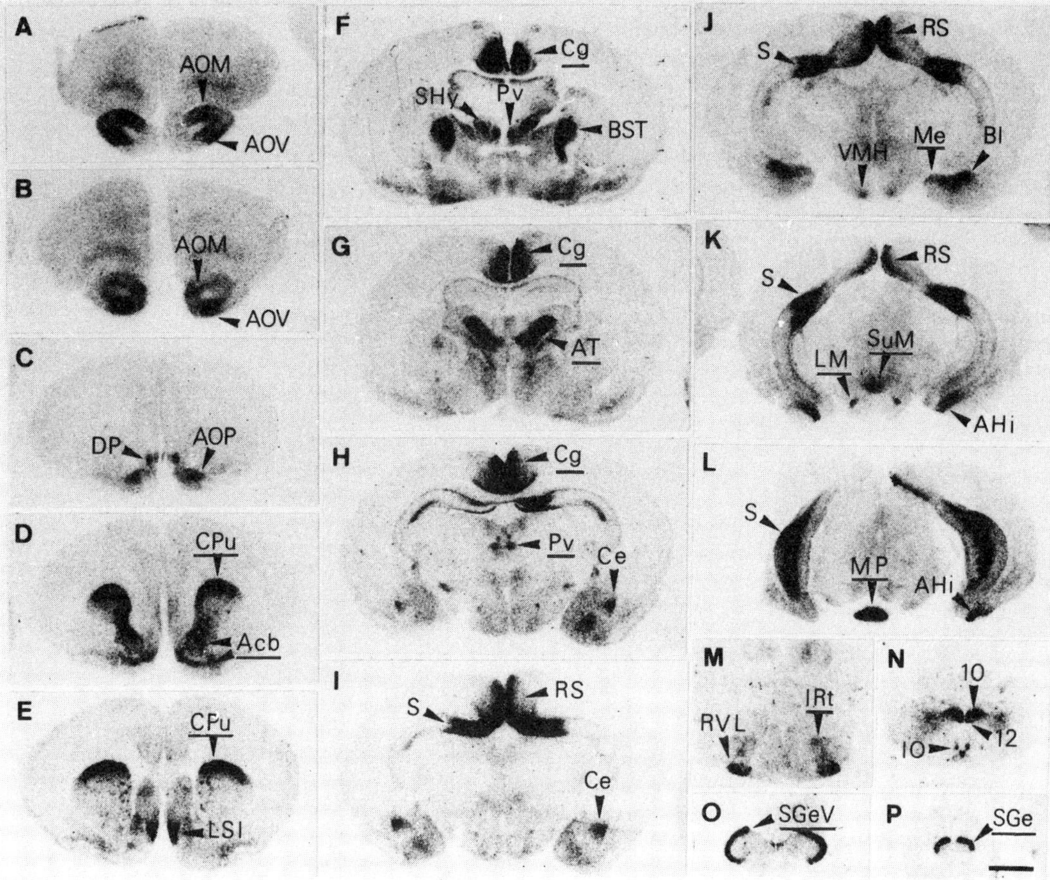

Fig. 12.5. Autoradiographic map of iodinated oxytocin antagonist ([^{125}I]OTA) binding sites in rostrocaudal coronal sections of the brain of a PD10 rat. As for the [^{3}H]VP sites in Fig. 12.2, brain regions of transient expression of OT receptor during development are indicated by underlining their abbreviation. For abbreviations see the legend to Fig. 12.2 and the following: AHi, amygdalohippocampal area; AOM/AOV/AOP, medial, ventral and posterior parts of the anterior olfactory nucleus; AT, anterior thalamus; B1, basolateral amygdaloid nucleus; Cg, cingulate cortex; CPu, caudate putamen; DP, dorsal peduncular nucleus; IRt, intermediate reticular nucleus; Me, medial amygdaloid nucleus; MP, posterior part mammillary nucleus; Pv, paraventricular thalamic nucleus; IM, lateral mammillary nucleus; RS, retrosplenial cortex; RVL, rostroventral reticular nucleus; S, subiculum; SGe/SGeV, substantia gelatinosa; SHy, septohypothalamic nucleus; SuM, supramammillary nucleus; VMH, ventromedial hypothalamic nucleus. Bar 2 mm. Slightly adapted from Tribollet *et al.* (1989).

labeling intensity after PD10–PD13 (Fig. 12.5) when compared with adult brain slices (Tribollet *et al.*, 1989). Also, the faint labeling of OT-sites in the hypothalamus and amygdaloid disappeared around PD19–PD22. The sharp peak indicating presence of [^{125}I]OTA binding in the cingulate gyrus, thalamus and dorsal hippocampus are very remarkable (Shapiro and Insel, 1989; Tribollet *et al.*, 1989). In the hippocampus, the disappearance of [^{125}I]OTA binding in the dorsal part is accompanied by a concomitant appearance in the ventral part. The pattern of appearance and changes is roughly the same when [^{3}H]OT is used as a ligand, albeit that the first signs of labeling were always found at a later stage (Snijdewint *et al.*, 1989), possibly due to the lower sensitivity of the ^{3}H- as compared to the ^{125}I-autoradiography.

Shapiro and Insel (1989) showed that analysis of [^{125}I]OTA binding on crude membranes of the cingulate gyrus at PD10 revealed an affinity (K_d = 99 pM) similar to that found in adult hippocampus (K_d =73 pM; Elands *et al.*, 1987). Also, the pharmacological profile of displacement being HO[Thr^4,-Gly^7] OT $\approx$ OT > VP>d(CH$_2$)$_5$[Tyr(Me)^2]VP is similar for various areas, such as the cingulate gyrus, septum and dorsal hippocampus (Shapiro and Insel, 1989). Thus, transient changes in density are not likely to be explained by ontogenetic changes in OT receptor affinity.

Unlike VP sites, OT sites experience a second stage of appearance after PD35, a time

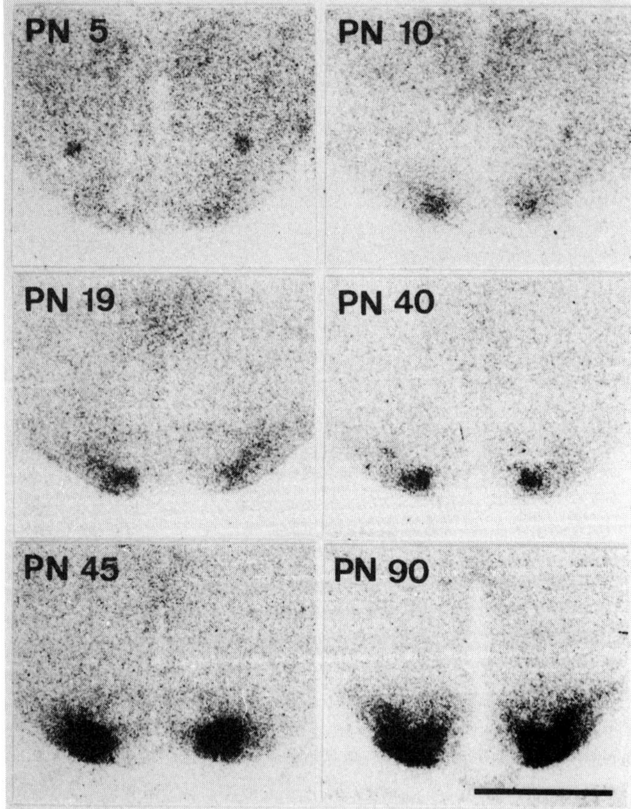

Fig. 12.6.　Late appearance of OT receptors in the hypothalamic ventromedial nucleus of rat. Receptor autoradiography using [^{125}I]OTA on coronal sections (cf. Fig. 12.5J) at various postnatal (PN) ages. Bar 2 mm. From Tribollet *et al.* (1989), reproduced by permission of the publisher.

corresponding to the onset of puberty (Shapiro and Insel, 1989; Tribollet *et al.*, 1989). The ventromedial hypothalamic nucleus, only slightly labeled at PD10 and PD15 by [^{125}I]OTA and [^{3}H]OT respectively (Snijdewint *et al.*, 1989; Tribollet *et al.*, 1989) greatly enhanced binding capacity for [^{125}I]OTA after P35 (Fig. 12.6). A similar pattern is seen for the Calleja islands, ventral pallidum and, to a lesser extent, for the bed nucleus of the stria terminalis. Thus, the adult distribution pattern of OT binding sites is reached rather late in ontogeny (around PD60) and is only established in the post-weaning period towards puberty.

12.6 RELATIONSHIP WITH APPEARANCE OF VP AND OT INNERVATION

Generally, one might say that [^{3}H]OT binding sites appear later than [^{3}H]VP binding sites of the brain (Snijdewint *et al.*, 1989). Since [^{125}I]OTA allows the detection of much lower densities of OT receptors than [^{3}H]OT, it is not surprising that [^{125}I]OTA autoradiography revealed an appearance of the OT receptor on ED14, possibly even on migrating neurons (see above). Future binding experiments with the use of iodinated VP antagonists will probably also show labeling of VP sites earlier than has been seen, so far, in the developing rat brain.

The fact that in the brain VP sites mature earlier than OT sites coincides with the observation of the difference in appearance of the immunostaining of the nonapeptides in the VP and OT networks. VP is synthesized from its precursor as early as ED16, whereas full processing of the OT precursor develops much more slowly and OT cannot be visualized until ED20 (Whitnall *et al.*, 1985; Altstein and Gainer, 1988). However, the production of bioactive forms of both peptides from their respective precursors seems not to be needed for initiating expression of the VP and OT receptor in rat brain. In the brain of the Brattleboro di/di rat, which is VP-deficient due to a recessive mutation in the ability to synthesize VP, [^{3}H]-VP binding is seen to appear as early as in age-matched heterozygous +/di control rats (Petracca *et al.*, 1986; Snijdewint *et al.*, 1989) (Table 12.2).

Table 12.2. Concentration of [^{3}H]-VP binding sites (fmol/mg brain grey matter) in sections of brain areas subjected to receptor autoradiography, comparison of developmental pattern in VP-deficient Brattleboro di/di rat and its heterozygous +/di control

		ED20	PD1	PD5	PD15
Lateral septum	+/di	2.1±0.7(9)	4.5±0.5(15)	9.5±0.7(15)	14.7±1.6(16)
	di/di	2.1±0.9(11)	6.1±0.7(17)	12.7±10(14)[*]	13.9±1.5(22)
Suprachiasmatic nucleus	+/di	–	8.5±0.8(2)	–	4.8±0.6(2)
	di/di	–	14.9±3.3(6)	–	15.8±1.9(9)[*]
Central amygdala	+/di	–	5.9±0.5(28)	12.0±1.3(15)	15.2±0.5(41)
	di/di	–	7.0±0.7(21)	11.2±0.8(32)	12.6±0.5(55)[*]
Facial nucleus	+/di	7.6±1.4(9)	15.6±1.4(12)	17.1±2.3(9)	24.2±2.3(5)
	di/di	10.8±0.7(14)[*]	19.0±0.7(11)[*]	8.8±1.3(10)[*]	n.d.[*]

Data are means ± SEM (half of the number of 10 µm sections from two rat pups used for incubation with [^{3}H]VP. [*]denotes statistically significant difference as compared to +/di values.
Note that [^{3}H] VP binding appeared in both genotypes at the same age. For the septal area of the brain no major differences were observed, for the suprachiasmatic nucleus a three-fold enhanced density appeared at PD15 in the di/di mutant, for the amygdala a slight decrease is found. For the facial nucleus a transient present of [^{3}H]VP binding is shown for normal rat, i.e. no labeling in adults (Snijdewint *et al.*, 1989; Tribollet *et al.*, 1991a), in di/di rats fading out of labeling is apparently reached as early as P15 (n.d. not detectable). In this study no binding was seen at E18. Data from Snijdewint *et al.* (1989).

Moreover, detailed comparison for various brain areas of first appearance of VP fiber innervation and VP receptor respectively showed that, even though [³H]-VP autoradiography might not be too sensitive, first receptor labeling can precede staining of incoming VP fibers. Labeling of the septal area, for instance, is present prenatally, whereas the presence of VP input is not immunocytochemically visible until PD10 (De Vries *et al.*, 1981; Boer 1987, 1991). The abovementioned data, taken together with the very early appearance of OT receptors at ED14 – about 6 days before the actual availability of OT itself – lead to the conclusion that the peptides are not priming the onset of receptor gene expression (Petracca *et al.*, 1986; Freund-Mercier *et al.*, 1988; Snijdewint *et al.*, 1989).

12.7 LATE APPEARANCE OF OT RECEPTORS

The late major increase of OT binding sites in the hypothalamic ventromedial nucleus, cell groups of the olfactory tubercle (Calleja islands and ventral pallidum) and bed nucleus of the stria terminalis seems to point to an involvement of gonadal steroids. These hormones are increased in the sexual maturation period between PD25 and PD60 (Farris, 1949) whereas in adulthood these regions contain a high density of estrogen receptors (Pfaff and Keiner, 1973; Axelson and Van Leeuwen, 1990). Moreover, OT receptors in adult brain are subject to regulation by the gonadal system. Ovariectomy or castration decreased OT receptor density in the ventromedial nucleus and the islands of Calleja (Tribollet *et al.*, 1990) whereas estrogen treatment increased OT receptor content (De Kloet *et al.*, 1985b; Coirini *et al.*, 1989) as did testosterone (Johnson *et al.*, 1991). It thus looks as if the above areas become sensitized to central OT after the onset of puberty. The influence of centrally applied OT on sexual (Caldwell *et al.*, 1986; Elands *et al.*, 1991) and maternal behavior (Pedersen *et al.*, 1982) and of OT applied in the ventromedial nucleus of

the hypothalamus on female lordosis behavior (Schumacher *et al.*, 1990), further indicates the functional importance of the late stage OT receptor development.

12.8 POSSIBLE SIGNIFICANCE OF TRANSIENT RECEPTOR EXPRESSION

The transient expression of VP and OT receptors in the developing brain is an intriguing observation, although it is not unique. Peak levels of other peptide receptors during ontogeny have also been described: opiates (Kent *et al.*, 1982; McLaughlin and Zagon, 1991), neurotensin (Kiyama *et al.*, 1987), CRH (Insel *et al.*, 1988), substance P (Quirion and Dam, 1986) and somatostatin receptors (Gonzalez *et al.*, 1989). In some instances, for example, somatostatin binding at the extragranular germinal zones of the rat cerebellum, the disappearance seems directly related to the loss of neurogenesis. However, the evanescent expression of VP and OT receptors is not seen at the germinal zones. It does, therefore, not reflect any possible direct control on cell generation and differentiation, but is more an indication of a particular ligand/receptor interaction or of a disappearance of cells or neurites that express the receptors. The VP and OT receptors in the areas of transient expression may therefore subserve functions in development that are distinct from their role in synaptic transmission in the other areas that remained labeled into adulthood. None of the areas that lose the VP- and OT-binding are recognized as receiving main VP or OT input. Thus, a process of downregulation of receptor expression, due to locally enhanced receptor occupation by release of VP or OT from newly arriving fibers, cannot easily explain the results. However, the knowledge about the precise ontogenic profile of the outgrowth of VP- and OT- fiber pathways is still limited, because early immunocytochemical staining for the peptide fibers is not intense. On the other hand, with VP absent, i.e. in the Brattleboro di/di rat, brain transient expression of [³H]VP

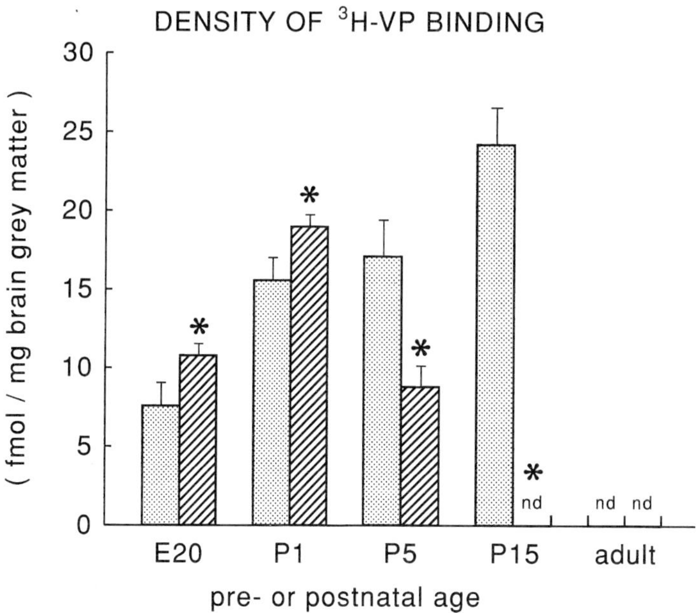

Fig. 12.7.   Density of [³H]VP binding in the facial nucleus of heterozygous +/di (dotted bars) and VP-deficient di/di Brattleboro rats throughout postnatal life (hatched bars). Data are derived from Table 12.2 (perinatal period) and from unpublished results (adulthood). Note the different developmental profile between the two genotypes. Bars indicate average ± SEM. Asterisk denotes statistically significant difference at *P* < 0.5.

and [³H]OT binding is also seen (Petracca *et al.*, 1986; Snijdewint *et al.*, 1989; Table 12.2, Fig. 12.7), supporting the idea that ontogenic changes in the expression of VP are indeed not necessary for this phenomenon. Since in the Brattleboro di/di rat the release of OT is enhanced (Boer *et al.*, 1988a) a role of (hormonal) OT in the control of transient expression of VP/OT receptors in the brain cannot be excluded.

12.9 (EARLY) RECEPTOR REGULATION

There are still only a limited number of studies on brain VP and OT receptor regulation. With respect to autoregulation in adulthood, centrally applied VP failed to change [³H]VP binding in rat septum and amygdala in one study (Swank and Dorsa, 1991) but decreased binding in another study (Shewey

et al., 1989). Centrally increased release of VP following osmotic stress had no effects (Landgraf *et al.*, 1991), nor had the opposite, i.e. depletion of VP following castration (Poulin and Pittman, 1991). However, centrally applied VP (and OT) sensitized the *in vitro* V1-mediated septal phosphoinositide metabolism (Lebrun *et al.*, 1990; Poulin, 1991; Poulin and Pittman, 1991). Chronic intracerebroventricular treatment with the V1 antagonist d(CH₂)₅[Tyr(Me)²]VP did not have an effect on [³H]VP maximal binding either, though the affinity constant (K_d) increased (Swank and Dorsa, 1991). Chronic activation of VP receptors therefore alters the receptor effector mechanisms rather than total receptor content in the brain. The possibility that VP and/or OT may alter the expression of proteins involved in the intracellular messenger process will have to be considered.

In the single study that described the effects on [³H]VP binding sites of neonatal exposure to VP (given as repeated subcutaneous injection), no change was found in total density or affinity in the septum, neither acutely (Fig. 12.8) nor 50 days after the treatment (Szot *et al.*, 1992). Similar treatments, however, showed an up-regulation for the V1 receptors in the liver (Szot *et al.*, 1992) and a down-regulation for the V2 receptors in the kidney (Handelmann and Sayson, 1984; but not seen by Boer, 1991; Szot *et al.*, 1992) whereas neonatal treatment with the V1 antagonists d(CH₂)₅[Tyr(Me)²]VP and dP[Tyr-(Me)²]VP showed a decrease of liver (Myers *et al.*, 1992) and of kidney VP binding sites respectively (Boer, 1991). So peripheral V1 and V2 receptors might be more sensitive to autoregulation than the central V1 receptor. However, V1 receptor density in the brain is very low compared to that in the liver, so that technical difficulties may have prevented changes from being seen. On the other hand, as in adult brain, the autoregulation of receptor efficacy might be better found at the level of postreceptor signal transduction. This has yet to be investigated.

The above-mentioned data become confusing when, in addition, observations in the VP-deficient Brattleboro di/di rat are considered. Adult septum of this mutant revealed increased [³H]VP binding in crude membrane preparations and a higher K_d value as compared to +/di controls (Shewey and Dorsa, 1986). Central supplementation of VP resulted in normal receptor density and affinity as well as normalizing the level of *in vitro* VP-stimulated phosphoinositide turnover (Shewey *et al.*, 1989). Thus regulation via receptor occupation by VP may exist but could in the Brattleboro di/di rat be put at a different setpoint, perhaps due to the absence of VP in the developing period. Receptor autoradiography revealed no differences in septal [³H]VP binding in both adult (Freund-Mercier *et al.*, 1988) and young Brattleboro rat as compared to their controls (Snijdewint *et al.*, 1989; Table 12.2). But the differences seen in [³H]VP binding in the course of Brattleboro rat development might point to adaptational effects (Snijdewint *et al.*, 1989). Higher levels of neonatal [³H]VP binding were found in suprachiasmatic nucleus and central amygdala, as well as in the transiently labeled

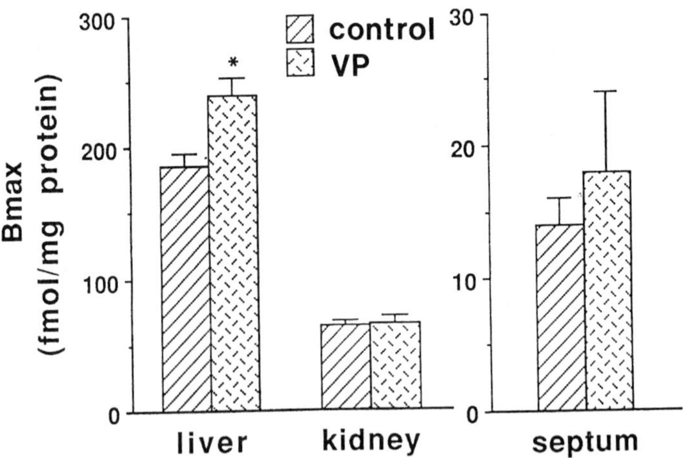

Fig. 12.8. [³H]VP binding density in liver, kidney and septum of the brain of PD60 rats chronically treated with VP (40 µg/100 g body wt./day) during the first week of life. Bars indicate mean ± SEM and the asterisk denotes a significant difference at $P < 0.05$. Slightly adapted from Szot *et al.* (1992), reproduced by permission of the publisher.

dorsal hippocampus, cingulate cortex and facial nucleus. Homozygous di/di rats, in which the latter area was described as lateral reticular nucleus by Snijdewint *et al.* (1989), lost the [³H]VP labeling even earlier than heterozygous +/di rats, which clearly indicates altered maturational processes (Table 12.2; Fig. 12.7). The suprachiasmatic nucleus keeps its high density of [³H]VP sites throughout life (Freund-Mercier *et al.*, 1988).

No studies have been performed on possible changes in brain OT receptors on exposure to OT (or VP). As stated above, OT receptors in the ventromedial hypothalamus, cell groups of the olfactory tubercle and bed nucleus of the stria terminalis develop after the period of neurogenesis and during the onset of puberty. As discussed above, the [³H]OT and [¹²⁵I]OTA binding in these areas is, in adulthood, under the control of gonadal hormones. There are no major sex differences for the response of OT receptors on gonadal hormones (Tribollet *et al.*, 1990), nor is the density of [³H]VP affected in either sex (Poulin and Pittman, 1991; Tribollet *et al.*, 1990). It is, however, not yet known whether the expression of OT sites in these late maturing areas would respond to neonatal manipulation of gonadal hormones, and whether this could have lasting effects on OT receptor expression. Only the observation by Freund-Mercier *et al.* (1988) possibly points to developmental setting of OT sites. They describe a two-fold decrease of [³H]VP/[³H]OT binding in the ventromedial nucleus of the hypothalamus in Brattleboro di/di rats. This decrease in OT-type receptors may be caused by the enhanced OT release which is seen throughout the postnatal period in the Brattleboro rat (Boer *et al.*, 1988a).

12.10   POSSIBLE TROPHIC EFFECTS OF VP

VP production and VP receptors are present in the brain from early developmental stages onwards. In the areas of transient appearance of VP receptors, a VP action distinct from that seen in the adult might be expected. This may be a role in growth and development of the functional organization of a particular cell system or of areas whose development depends on the maturation of efferents of such cell groups. Since in some regions VP and OT receptors appear before actual VP or OT synapse formation took place, these sites may receive the earlier available hormonal VP and, slightly later, OT signal. The blood–brain barrier in the rat does not 'close' until between the second and third postnatal week (Johanson, 1980) whereas pituitary release becomes functional at ED20/21 (Pitzel *et al.*, 1982).

Indications that VP may have a function in brain development first evolved from the stunted neurogenesis in the VP-deficient Brattleboro di/di rat (Boer *et al.*, 1982; Boer and Patel, 1983). This disturbance was particularly present in the cerebellum, an area that has never been shown to contain VP receptors either transiently in the developing period, or in adulthood. Still, VP is recognized as a mitogenic factor in a variety of *in vivo* and *in vitro* peripheral, i.e. non-neuronal, cell systems (see reviews by Boer, 1985, 1987). This effect appeared to be V1 receptor-mediated in 3T3 fibroblast cultures (Collins and Rozengurt, 1983), the subtype of VP receptor which is also present in the brain. Attempts to correct the stunted brain growth of the Brattleboro rat by neonatal systemic supplementation of VP failed (Boer *et al.*, 1984; Boer, 1985), though slight positive effects were seen in the offspring of pregnant Brattleboro females chronically treated with [Lys⁸]VP (Snijdewint *et al.*, 1985, 1988). Chronic VP treatment of normal fetal (Oosterbaan *et al.*, 1985) and neonatal rat (Boer, 1991), however, impaired brain growth (Table 12.3), also seen upon treatment with its congener vasotocin (Goldstein and Rodica, 1986). Recently we observed that cerebellar cell acquisition at PD10 is indeed severely affected, dose-dependently, by subcutaneously injected VP (Fig. 12.9). Thus, subcutaneous VP supplementation

Table 12.3. Decreased brain weights and body weight of PD35–40 rat pups neonatally treated with VP

	Control	VP
Treatment PD1–9 (5µg daily s.c.), measurement on PD40		
Body (g)	141±4	146±6
Brain (g)	1.74±0.03	1.65±0.01*
Cerebellum (mg)	229±4	219±5*
Treatment PD1–21 (2 × 2.5 µg daily s.c.), measurement on PD35		
Body (g)	117±5	115±3
Brain (g)	1.69±0.02	1.60±0.02*
Cerebellum (mg)	214±4	196±3*

Data are average ± SEM ($n = 8$)
* statistically significant difference from control at $P<0.05$.
From Boer (1991).

in the Brattleboro rat might have given positive as well as negative, i. e. opposite effects on brain development.

VP has also been shown to promote neurite growth in cultured embryonic neurons, at least those from amphibians (Brinton and Gruener, 1987; Fig. 12.10) and to potentiate the effects of 3,3′,5-triiodo-thyronine (T_3) on survival and maturation of cultured rat hippocampal neurons dissociated postnatally at PD5 (Clos and Gabrion, 1989). The neurotrophism of neurite outgrowth is probably also V1-mediated and shows an optimal effect at 1 µMVP, indicating a narrow dose window of stimulatory effect (Fig. 12.10). The potentiating effects of VP on T3-mediated maturation of hippocampal neurons specifically reflects the acetylcholinergic neurons. One may wonder whether this effect, seen at PD5, is related to activation of VP on the transiently present OT receptor in the dorsal hippocampus up to PD19 (Tribollet *et al.*, 1989). The T3/VP interaction might also be a V1 receptor-mediated phenomenon, since in rat neonates a hypothyroid state greatly enhanced total binding of the V1 antagonist $[^3H]d(CH_2)_5[Tyr(Me)^2]VP$ (Rami *et al.*, 1989).

Whether exogenous VP is involved in making the brain VP receptors play a role in

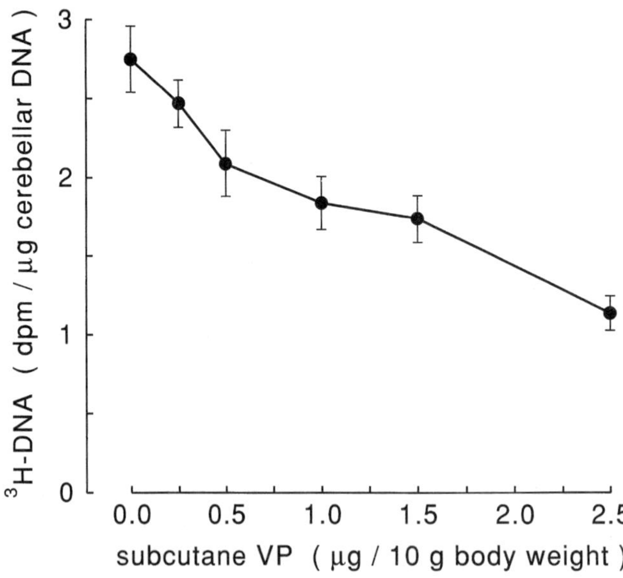

Fig. 12.9. Dose-dependent reduction of neurogenesis of PD10 rats upon subcutaneous injections of VP. Cell acquisition was measured (according to Boer and Patel, 1983) 10 min after the VP bolus as a 30 min period of [³H]thymidine incorporation into cerebellar DNA. Points indicate average ± SEM ($n = 6$).

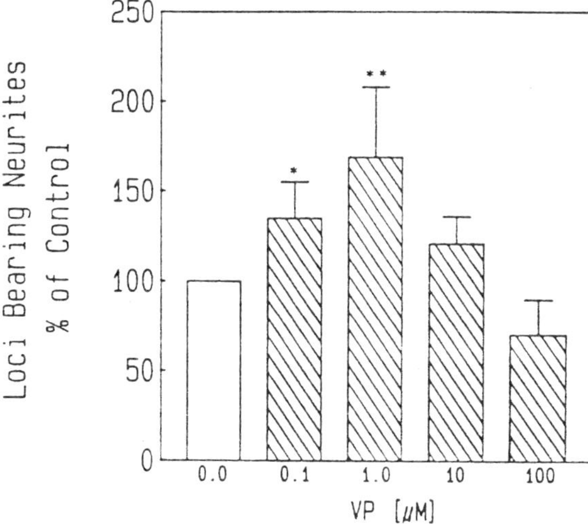

Fig. 12.10. Effect of 24h post-plating exposure to various levels of VP on neurite outgrowth in a dissociated cell culture of embryonic neurons prepared from the dorsomedial part of the spinal cord of *Xenopus laevis*. Stimulation of VP on neurite outgrowth, measured as number of loci (number of cell bodies) bearing neurites, has a peak at around a 1 µM concentration. Bars indicate mean ± SEM (*n* = 3–5) and asterisks denote statistically significant differences from control. From Brinton and Gruener (1987), reproduced by permission of the publisher.

neurogenesis and/or cell survival and maturation still remains to be established since its metabolic and cardiovascular action might also be involved (cf. Boer, 1987). For OT, no data on neurotrophism are available, except that enhanced maternal OT, which advances labor, gave an impaired brain and cerebellar growth as measured at PD70 (Boer, 1991). However, maternal circulating OT (and VP) is not supposed to enter the fetal compartment (Oosterbaan *et al.*, 1985) so that a direct effect on fetal brain is unlikely to occur.

12.11   ORGANIZATIONAL EFFECTS OF VP

Abundant evidence that VP can alter the development of functional organization of the brain is obtained from behavioral studies.

This is in line with the general concept of behavioral teratology that drugs interfering with neurotransmission in the developing brain can alter the adult functional capacities (Boer and Swaab, 1985; Boer *et al.*, 1988b). Activation of the brain V1 receptor can, moreover, increase the mRNA for the proto-oncogene *c-fos* in septum and hippocampus of the adult mouse (Giri *et al.*, 1990), indicating a possible role of VP in neuroadaptation (it can, however, be indirect via presynaptic modulation of neurotransmitter release).

Chronic subcutaneous infusion of VP in pregnant rat (1 µg/24 h) has been reported to affect memory retrieval in the offspring at PD60 (Tinius *et al.*, 1987), whereas repeated injection (1–2 µg daily) impaired learning in female rats in one study (Swenson *et al.*, 1990) and facilitates relearning in another study (Ermisch *et al.*, 1986). Similar injections with OT did not have similar effects (Swenson *et al.*, 1990). As stated above, one may question whether these behavioral changes are caused by direct effects of VP on the fetal brain because of the placental barrier. However, direct treatment of rat neonates alters later behavior as well. Improved performance of relearning was observed in rat neonatally treated with 1 µg VP daily between birth and PD30 (Ermisch *et al.*, 1986), facilitated learning and memory after 0.2 µg dDAVP treatment daily up to PD4 or PD14 (Chen *et al.*, 1988) and increased grooming behavior after VP 5 µg/day up to PD9 (Boer, 1991). Again OT failed to give lasting effects on the parameters tested (Chen *et al.*, 1988). Reduction of the effects of endogenous VP by treatment with the V1 antagonist dP[Tyr(Me)^2]VP (1 µg/day for the first 7 days of life) enhanced lordosis response in female rats measured after PD75 (Meyerson *et al.*, 1988), whereas a single injection with VP antiserum on PD2 resulted in deficits in learning (Fig. 12.11) and memory, and in hyperanalgesia 3 months later (Moratella *et al.*, 1986, 1987).

All behaviors that have been found to be long-lastingly affected by pre- or postnatal

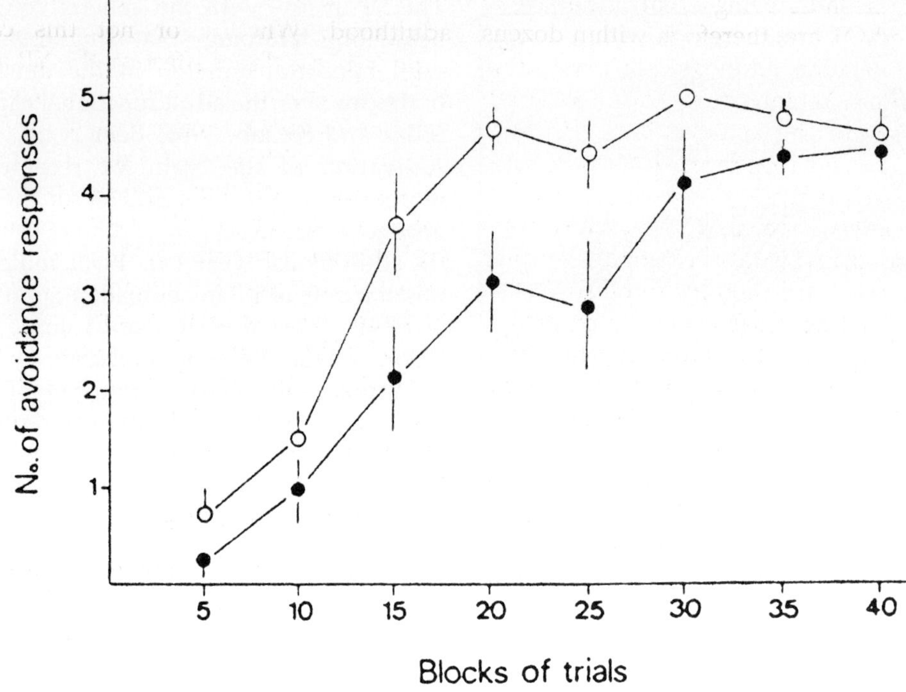

Fig. 12.11.   Effect of a single subcutaneous neonatal injection of VP antiserum (100 µl; closed circles) on the acquisition by adult rat in active avoidance behavior in the shuttle box. Control rats (open circles) received a neonatal injection with immunoglobulin from normal rabbit serum. Points indicate average avoidance response per block of five trials (n = 7–10). From Moratella *et al.* (1987), reproduced by permission of the publisher.

manipulation of VP activity are VP-sensitive in normal untreated adult rat. One might, therefore, postulate that functional alterations in VP and interrelated neuronal pathways have been induced by the changed VP levels in early development. So far, hardly any research has been directed this way. Meyerson *et al.* (1988) describe an increase in the hypothalamic VP content on early V1 antagonist treatment and Szot *et al.* (1992) see no change in [³H]VP binding at PD60 after early neonatal VP application, but further neurochemical data are presently not available. Studies on the basal or challenged activity of VP (or OT) brain systems in animals that were exposed early on to the peptides are therefore still needed. Despite a lack of knowledge about any under-

lying mechanism of the lasting behavioral effects the studies suggest that perinatally applied VP has an organizing role in the development of the brain. Of course these effects may still be indirect, since VP has its targets on organs like liver, kidney and adenohypophysis, the proper function of which in the mother also determines general development.

12.12   CONCLUDING REMARKS

Many VP and OT cell systems exist in the mammalian brain. They comprise a variety of neurosecretory and neurotransmitter pathways, each with its own time course of functional development and targeted specifically on the V1 subclass of VP receptors or on OT receptors, the latter being equally sensitive to VP. VP and OT are, therefore, within dozens of communication pathways. In many developing brain regions these sites were expressed before the actual VP or OT fiber ingrowth, i.e. synapse formation, took place. In other areas receptor expression peaked in the neonatal period and sometimes completely faded away in adulthood. Receptor ontogeny has been thought to be subject to ligand imprinting (Csaba, 1986), a process of receptor adaptation for the establishment of the final homeostatic setpoint of receptor number. For VP and OT brain receptors, however, density is only affected in a minor way, by altered levels of the peptides. Perinatal treatments of rat with VP have not been shown to down- or up-regulate the VP receptor level in the brain, although peripheral target areas like liver (Szot *et al.*, 1992), kidney (Handelmann and Sayson, 1984) and arteries (Csaba *et al.*, 1980) developed lasting changes. Moreover, in the VP-deficient Brattleboro di/di rat onset of regional receptor expression is similar to strain mates of the +/di (VP-producing) genotype, whereas only in a few brain areas are VP and OT density altered (increased or decreased; cf. Table 12.2).

Neonatal imbalance of VP activity, however, affects the organization of the brain in such a way that behavior that normally comes under modulation of VP is adapted in a long-lasting or permanent way. The underlying mechanisms of these so-called functional neuroteratological effects (Boer *et al.*, 1988b; Boer, 1991) are still barely understood. They could reside on the level of VP or OT neurotransmission, in a way similar to sensitization of adult brain to VP; no changes in VP receptor density but altered post-receptor signal transduction efficacy (Poulin and Pittman, 1992). Such tests have not yet been performed with brain regions of rat neonatally exposed to VP, OT or their congeners. Additionally, VP may play a neurotrophic role and influence the establishment of neuronal pathways, i.e. centrally regulated physiological and behavioral performance in adulthood. Whether or not this can be regarded as a direct effect on the VP and/or OT receptor of the developing nervous system remains an open question. Neonatal treatment of the rat with the V2 agonist dDAVP, a compound with low affinity for the V1 receptor of the brain, still has long-term effects on behavior performances (Chen *et al.*, 1988). However, as long as the underlying structural or neurochemical changes are not exposed, central VP and OT receptor-mediated modulation of growth and organizational processes are still the most obvious explanations.

ACKNOWLEDGEMENTS

The authors would like to express their gratitude to Dr D.F. Swaab, Dr M.G.P. Feenstra, and Dr F.W. van Leeuwen (Netherlands Institute for Brain Research, Amsterdam) for their critical remarks, and to Mrs W. Verweij for her editorial efforts.

REFERENCES

Altstein, M. and Gainer, H. (1988) Differential biosynthesis and posttranslational processing of vasopressin and oxytocin in rat brain during embryonic and postnatal development. *J. Neurosci.*, **8**, 3967–77.

Antoni, F.A. (1984) Novel ligand specificity of pituitary vasopressin receptors in the rat. *Neuroendocrinology*, **39**, 186–8.

Audigier, S. and Barberis, C. (1985) Pharmacological characterization of two specific binding sites for neurohypophyseal hormones in hippocampal synaptic plasma membranes of the rat. *EMBO J.*, **4**, 1407–12.

Axelson, J.F. and Van Leeuwen, F.W. (1990) Differential localization of estrogen receptors in various vasopressin synthesizing nuclei of the rat brain. *J. Endocrinol.*, **2**, 209–16.

Baertschi, A.J. and Friedli, M. (1985) A novel type of vasopressin receptor on anterior pituitary corticotrophs? *Endocrinology*, **116**, 499–502.

Boer, G.J. (1985) Vasopressin and brain development: studies using the Brattleboro rat. *Peptides*, **6** (Suppl.1), 49–62.

Boer, G.J. (1987) Development of vasopressin systems and their functions, in *Vasopressin; Principles and Properties* (eds D.M. Gash and G.J. Boer), Plenum Press, New York, pp. 117–74.

Boer, G.J. (1991) Neuropeptides: a new class of transmitters, a new class of functional teratogens, in *Functional Neuroteratology of Short-Term Exposure to Drugs* (eds T. Fujii and G.J. Boer), Teikyo University Press, Tokyo, pp. 73–85.

Boer, G.J. and Patel, A.J. (1983) Disorders of cell acquisition in the brain of rats deficient in vasopressin (Brattleboro mutant). *Neurochem. Int.*, **5**, 463–9.

Boer, G.J. and Swaab, D.F. (1985) Neuropeptide effects on brain development to be expected from behavioral teratology. *Peptides*, **6** (Suppl. 2), 21–8.

Boer, G.J., Van Rheenen-Verberg, C.M.H. and Uylings, H.B.M. (1982) Impaired brain development of the diabetes insipidus Brattleboro rat. *Dev. Brain Res.*, **3**, 557–75.

Boer, G.J., Kragten, R., Kruisbrink, J. *et al.* (1984) Vasopressin fails to restore postnatally the stunted brain development in the Brattleboro rat, but affects water metabolism permanently. *Neurobehav. Toxicol. Teratol.*, **6**, 103–9.

Boer, G.J., Van Heerikhuize, J.J. and Van der Woude, T.P. (1988a) Elevated serum oxytocin of the vasopressin-deficient Brattleboro rat is present throughout life and is not sensitive to treatment with vasopressin. *Acta Endocrinol.*, **1117**, 442–450.

Boer, G.J., Snijdewint, F.G.M. and Swaab, D.F. (1988b) Neuropeptides and functional neuroteratology. *Progr. Brain Res.*, **73**, 245–64.

Brinton, R.E. and Gruener, R. (1987) Vasopressin promotes neurite growth in cultured embryonic neurons. *Synapse*, **1**, 329–34.

Brinton, R.E., Wamsley, J.K., Gee, K.W. *et al.* (1984) [^{3}H]Oxytocin binding sites in the rat brain demonstrated by quantitative light microscopic autoradiography. *Eur. J. Pharmacol.*, **103**, 365–7.

Buijs, R.M. (1987) Vasopressin localization and putative functions in the brain, in *Vasopressin, Principles and Properties* (eds J.M. Gash and G.J. Boer), Plenum Press, New York, pp. 91–115.

Buijs, R.M., De Vries, G.J. and Van Leeuwen, F.W. (1985) Anatomic distribution and synaptic rebase of oxytocin in the central nervous system, in *Oxytocin: Clinical and Laboratory Aspects* (eds J.A. Amico and A.G. Robinson), Excerpta Medica, Amsterdam, pp. 77–86.

Caffé, A.R., Van Leeuwen, F.W., Buijs, R.M. *et al.* (1985) Co-existence of vasopressin, neurophysin and noradrenaline immunoreactivity in medium-sized cells of the locus coeruleus and subcoeruleus in the rat brain. *Brain Res.*, **338**, 160–4.

Caffé, A.R., Van Leeuwen, F.W. and Luiten, P.G.M. (1987) Vasopressin cells in the medial amygdala of the rat project to the lateral septum and ventral hippocampus. *J. Comp. Neurol.*, **261**, 237–52.

Caldwell, J.D., Prange, A.J. and Pedersen, C.A. (1986) Oxytocin facilitates the sexual receptivity of estrogen-treated female rats. *Neuropeptides*, **7**, 175–89.

Chen, X.F., Chen, Z.F., Liu, R.Y. *et al.* (1988) Neonatal administrations of a vasopressin analog (DDAVP) and hypertonic saline enhance learning behavior in rats. *Peptides*, **9**, 717–21.

Cholewinski, A.J., Hanley, M.R. and Wilkin, G.P. (1988) A phosphoinositide-linked peptide response in astrocytes: evidence for regional heterogeneity. *Neurochem. Res.*, **13**, 389–94.

Collins, M.K.L. and Rozengurt, E. (1983) Vasopressin induces selective desensitization of its mitogenic response in Swiss 3T3 cells. *Proc. Natl. Acad. Sci. USA*, **80**, 1924–7.

Clos, J. and Gabrion, J. (1989) A thyroid hormone–vasopressin interaction promotes survival and maturation of hippocampal neurons dissociated postnatally. *Neurochem. Res.*, **14**, 919–25.

Coirini, H., Johnson, A.E. and McEwen, B.S. (1989) Estradiol modulation of oxytocin binding in the ventromedial hypothalamic nucleus of male and female rats. *Neuroendocrinology*, **50**, 193–8.

Cornett, L.E. and Dorsa, D.M. (1985) Vasopressin receptor subtypes in dorsal hindbrain and renal medulla. *Peptides*, **6**, 85–9.

Csaba, G. (1986) Receptor ontogeny and hormonal imprinting. *Experientia*, 42, 750–9.

Csaba, G., Ronai, A., Laszlo, V. *et al.* (1980) Amplification of hormone receptors by neonatal oxytocin and vasopressin treatment. *Horm. Metab. Res.*, **12**, 28–31.

De Kloet, E.R., Voorhuis, T.A.M., Burbach, J.P.H. *et al.* (1985a) Autoradiographic localization of binding sites for the arginine–vasopressin (VP) metabolite, VP$_{4-9}$, in rat brain. *Neurosci. Lett.*, **56**, 7–11.

De Kloet, E.R., Voorhuis, T.A.M. and Elands, J. (1985b) Estradiol induces oxytocin binding sites in rat hypothalamus ventromedial nucleus. *Eur. J. Pharmacol.*, **118**, 185–6.

De Vries, G.J. and Buijs, R.M. (1983) The origin of the vasopressinergic and oxytocinergic innervation of the rat brain with special reference to the lateral septum. *Brain Res.*, **273**, 301–17.

De Vries, G.J., Buijs, R.M. and Swaab, D.F. (1981) Ontogeny of the vasopressinergic neurons of the suprachiasmatic nucleus and their extrahypothalamic projections in the rat brain – presence of a sex difference in the lateral septum. *Brain Res.*, **218**, 67–78.

De Wied, D. (1983) Central actions of neurohypophyseal hormones. *Progr. Brain Res.*, **60**, 155–67.

Dorsa, D.M., Majumbar, L.A., Petracca, F.M. *et al.* (1983) Characterization and localization of ^{3}H-arginine vasopressin binding to rat kidney and brain. *Peptides*, **4**, 699–706.

Dorsa, D.M. Petracca, F.M., Baskin, D.G. *et al.* (1984) Localization and characterization of vasopressin binding sites in the amygdala of the rat brain. *J. Neurosci.*, **4**, 1764–70.

Dorsa, D.M., Brot, M.D., Shewey, L.M. *et al.* (1988) Interaction of a vasopressin antagonist with vasopressin receptors in the septum of the rat brain. *Synapse*, **2**, 205–11.

Elands, J., Barberis, C., Jard, S. *et al.* (1987) ^{125}I-labelled d(CH$_2$)$_5$[Tyr(Me)^2,Thr^4,Tyr-NH$_2$^{9}]OVT: a selective oxytocin receptor ligand. *Eur. J. Pharmacol.*, **147**, 197–207.

Elands, J., Barberis, C., Jard, S. *et al.* (1988a) ^{125}I-d(CH$_2$)$_5$[Tyr(Me)^2,Tyr(NH$_2$)^9]AVP iodination and binding characteristics of a vasopressin receptor ligand. *FEBS Lett.*, **229**, 251–5.

Elands, J., Barberis, C. and Jard, S. (1988b) [^{3}H][Thr^4,Gly^7]OT: a highly selective ligand for central and peripheral OT receptors. *Am. Physiol. Soc.*, **254**, 31–8.

Elands, J., Beetsma, A., Barberis, C. *et al.* (1988c) Topography of the oxytocin receptor system in rat brain: an autoradiographical study with a selective radioiodinated oxytocin antagonist. *J. Chem. Neuroanat.*, **1**, 293–302.

Elands, J., Van Doremalen, E., Spruyt, B. *et al.* (1991) Oxytocin receptors in the rat hypothalamic ventromedial nucleus: a study of possible mediators of female sexual behaviour, in *Vasopressin* (eds S. Jard and R. Jaminson), Colloque INSERM/John Sibbey Eurotest Ltd., Paris, pp. 311–19.

Ermisch, A., Koch, M. and Barth, T. (1986) Learning performance of rats after pre- and postnatal application of arginine–vasopressin. *Monogr. Neural Sci.*, **12**, 142–7.

Fahrenholz, F., Korjo, E., Plage, G. *et al.* (1988) Renal V$_2$ vasopressin receptor proteins: identification and enrichment. *J. Recept. Res.*, **8**, 283–94.

Farris, E.J. (1949) Breeding of the rat, in *The Rat in Laboratory Investigation* (eds E.J. Farris and J.Q. Griffith), Lippincott, Philadelphia, pp. 1–18.

Flint, A.P.F., Leat, W.M.F., Sheldrick, E.L. *et al.* (1986) Stimulation of phosphoinositide hydrolysis by oxytocin and the mechanism by which oxytocin controls prostaglandin synthesis in the ovine endometrium. *Biochem. J.*, **237**, 797–806.

Freund-Mercier, M.J., Dietl, M.M. and Stoeckel, M.E. (1988) Quantitative autoradiographic mapping of neurohypophysial hormone binding sites in the rat forebrain and pituitary gland-II. Comparative study on the Long–Evans and Brattleboro strains. *Neuroscience*, **26**, 273–81.

Ganten, D. and Pfaff, D. (eds) (1986) Neurobiology of oxytocin. *Current Topics in Neuroendocrinology*, 6, Springer-Verlag, Berlin.

Gash, D.M. and Boer, G.J. (eds) (1987) *Vasopressin, Principles and Properties*, Plenum Press, New York.

Gerstberger, R. and Fahrenholz, F. (1989) Autoradiographic localization of V1 vasopressin binding sites in rat brain and kidney. *Eur. J. Pharmacol.*, **167**, 105–16.

Giri, P.R., Dave, J.R., Tabakoff, B. *et al.* (1990) Arginine vasopressin induces the expression of c-fos in the mouse septum and hippocampus. *Mol. Brain Res.*, **7**, 131–7.

Goldstein, R. and Rodica, B. (1986) Pharmacological evidence that effects of vasotocin on brain maturation are mediated by GABA mechanisms. *Int. J. Dev. Neurosci.*, **4**, 305–9.

Gonzalez, B.J., Leroux, P., Bodenant, C. *et al.* (1989) Ontogeny of somatostatin receptors in the rat brain: biochemical and autoradiographic study. *Neuroscience*, **29**, 629–44.

Handelmann, G.E. and Sayson, S.C. (1984) Neonatal exposure to vasopressin decreases vasopressin binding sites in the adult kidney. *Peptides*, **5**, 1217–19.

Insel, T.R., Battaglia, G., Fairbanks, D.W. *et al.* (1988) The ontogeny of brain receptors for corticotropin-releasing factor and the development of their functional association with adenylate cyclase. *J. Neurosci.*, **216**, 4151–8.

Jard, S., Gaillard, R.C., Guillon, G. *et al.* (1986) Vasopressin antagonists allow demonstration of a novel type of vasopressin receptor in the rat adenohypophysis. *Mol. Pharmacol.*, **30**, 171–7.

Jard, S., Barberis, C., Audigier, S. *et al.* (1987) Neurohypophysial hormone receptor systems in brain and periphery. *Progr. Brain Res.*, **72**, 173–87.

Joëls, M. (1987) Electrophysical actions of vasopressin in extrahypothalamic regions of the central nervous system, in *Vasopressin, Principles and Properties* (eds J.M. Gash and G.J. Boer), Plenum Press, New York, pp. 257–74.

Johanson, C.E. (1980) Permeability and vascularity of the developing brain: cerebellum vs cerebral cortex. *Brain Res.*, **190**, 3–16.

Johnson, A.E., Coirini, H., Insel, T.R. *et al* (1991) The regulation of oxytocin receptor binding in the ventromedial hypothalamic nucleus by testosterone and its metabolites. *Endocrinology*, **128**, 891–6.

Kelly, J.M., Abrahams, J.M., Phillips, P.A. *et al.* (1989) [^{125}I][d(CH$_2$)$_5$,Sar^7]AVP: a selective radioligand for V$_1$ vasopressin receptors. *J. Recept. Res.*, **9**, 27–41.

Kent, J.L., Pert, C.B. and Herkenham, M. (1982) Ontogeny of opiate receptors in rat forebrain: visualization by *in vitro* autoradiography. *Dev. Brain Res.*, **2**, 487–502.

Kiyama, H., Nagaki, S.I., Kito, S. *et al.* (1987) Ontogeny of [^{3}H]neurotensin binding sites in the rat cerebral cortex: autoradiography study. *Dev. Brain Res.*, **31**, 303–06.

Landgraf, R., Szot, P. and Dorsa, D.M. (1991) Vasopressin receptors in the brain, liver and kidney of rats following osmotic stimulation. *Brain Res.*, **544**, 287–90.

Lebrun, C.J., Gruber, M.G., Meister, M. *et al.* (1990) Central vasopressin pretreatment sensitizes phosphoinositol hydrolysis in the rat septum. *Brain Res.*, **531**, 167–72.

McLaughlin, P.J. and Zagon, I.S. (1991) Opioids, receptors, and the ontogeny and dysgenesis of the mammalian nervous system, in *Functional Neuroteratology of Short-Term Exposure to Drugs* (eds T. Fujii and G.J. Boer), Teikyo University Press, Tokyo, pp. 23–35.

Meyerson, B.J., Höglund, U., Johansson, C. *et al.* (1988) Neonatal vasopressin antagonist treatment facilitates adult copulatory behaviour in female rats and increases hypothalamic vasopressin content. *Brain Res.*, **473**, 344–51.

Michell, R.H., Kirk, C.J. and Billah, M.M. (1979) Hormonal stimulation of phosphatidyl inositol breakdown with particular reference to the hepatic effects of vasopressin. *Biochem. Soc. Trans.*, **7**, 861–5.

Moratella, R., Sanchez-Franco, F. and Del Rio, J. (1986) Long-term hyperanalgesia in rats induced by neonatal administration of vasopressin antiserum. *Life Sci.*, **38**, 109–15.

Moratella, R., Borrell, J., Sanchez-Franco, F. *et al.* (1987) Neonatal administration of vasopressin antiserum induces long-term deficits on active and passive avoidance behaviour in rats. *Behav. Brain Res.*, **23**, 231–7.

Myers, K.M., Szot, P. and Dorsa, D.M. (1992) Postnatal treatment with a V1-antagonist reduces ^{3}H-arginine–vasopressin binding sites in the liver of 8 day Long–Evans rats. *Neurosci. Lett.* (in press).

Oosterbaan, H.P., Swaab, D.F. and Boer, G.J. (1985) Oxytocin and vasopressin in the rat do not readily pass from the mother to the amniotic fluid in late pregnancy. *J. Dev. Physiol.*, **7**, 55–62.

Ostrowski, N.L., Young, W.S., O'Carroll, A.M. *et al.* (1991) The distribution of vasopressin V$_1$ receptor mRNA in rat brain. *Soc. Neurosci. Abstr.*, **17**, 801.

Pedersen, C.A., Ascher, J.A., Monroe, Y.L. *et al.* (1982) Oxytocin induces maternal behavior in virgin female rats. *Science*, **216**, 648–50.

Petracca, F.M., Baskin, D.G., Diaz, J. *et al.* (1986) Ontogenetic changes in vasopressin binding site distribution in rat brain: an autoradiographic study. *Dev. Brain Res.*, **28**, 63–8.

Pfaff, D.W. and Keiner, M. (1973) Atlas of estradiol-concentrating cells in the central nervous system of the female rat. *J. Comp. Neurol.*, **151**, 121–58.

Phillips, P.A., Abrahams, J.M., Kelly, J.M. *et al.* (1988) Localization of vasopressin binding sites in rat brain by *in vitro* autoradiography using a radioiodinated V1 receptor antagonist. *Neuroscience*, **27**, 749–61.

Phillips, P.A., Abrahams, J.M., Kelly, J.M. *et al.* (1990) Localization of vasopressin binding sites in rat tissues using specific V1 and V2 selective ligands. *Endocrinology*, **126**, 1478–4.

Pitzel, L., Lein, B. Scheffels, E. *et al.* (1982) Neurohypophyseal hormone content and release in fetal and newborn rats. *Neuroendocrinol. Lett.*, **4**, 349–54.

Poulin, P. (1991) Regulation of vasopressin receptors, Thesis, University of Calgary, Calgary, Canada.

Poulin, P. and Pittman, Q.J. (1991) Septal arginine vasopressin (AVP) receptor regulation in rats depleted of septal AVP following long term castration. *J. Neurosci.*, **11**, 1531–9.

Poulin, P. and Pittman, Q.J. (1992) Arginine vasopressin (AVP)-induced sensitization in brain: facilitated inositol phosphate production without changes in receptor number. *J. Neuroendocrinol.* (in press).

Quirion, R. and Dam, T.V. (1986) Ontogeny of substance P receptor binding sites in the rat brain. *J. Neurosci.*, **6**, 2187–99.

Raggenbass, M., Dubois-Dauphin, M., Tribollet, E. *et al.* (1988) Direct excitatory action of vasopressin in the lateral septum of the rat brain. *Brain Res.*, **459**, 60–9.

Raggenbass, M., Tribollet, E., Dubois-Dauphin, M. *et al.* (1989) Correlation between oxytocin neuronal sensitivity and oxytocin receptor binding: an electrophysiological and autoradiographical study comparing rat and guinea pig hippocampus. *Proc. Natl. Acad. Sci. USA*, **86**, 750–4.

Rami, A., Barberis, C. and Clos, J. (1989) Effects of hypothyroidism on high-affinity vasopressin binding sites in developing hippocampal synaptosomes. *Synapse,* **3**, 200–4.

Richter, D (1988) Molecular events in expression of vasopressin and oxytocin and their cognate receptors. *Am. J. Physiol.*, **255**, F207–F219.

Schmidt, A., Audigier, S., Barberis, C. *et al.* (1991) A radioiodinated linear vasopressin antagonist: a ligand with high affinity and specificity for V_{1a} receptors. *FEBS*, **282**, 77–81.

Schumacher, M., Coirini, H., Pfaff, D.W. *et al.* (1990) Behavioral effects of progesterone associated with rapid modulation of oxytocin receptors. *Science,* **250**, 691–4.

Shapiro, L.E. and Insel, T.R. (1989) Ontogeny of oxytocin receptors in rat forebrain: a quantitative study. *Synapse,* **4**, 259–66.

Shewey, L.M. and Dorsa, D.M. (1986) Enhanced binding of ^{3}H-arginine^{8}-vasopressin in the Brattleboro rat. *Peptides,* **7**, 701–4.

Shewey, L.M. and Dorsa, D.M. (1988) V_1-type vasopressin receptors in rat brain septum: binding characteristics and effects in inositol phospholipid metabolism. *J. Neurosci.*, **8**, 1671–7.

Shewey, L.M., Boer, G.J., Szot, P. *et al.* (1989) Regulation of vasopressin receptors and phosphoinositide hydrolysis in the septum of heterozygous and homozygous Brattleboro rats. *Neuroendocrinology,* **50**, 292–8.

Snijdewint, F.G.M., Boer, G.J. and Swaab, D.F. (1985) Body and brain growth following continuous perinatal administration of arginine- and lysine-vasopressin to the homozygous Brattleboro rat. *Dev. Brain Res.*, **22**, 269–77.

Snijdewint, F.G.M., Boer, G.J. and Swaab, D.F. (1988) Prenatal development of the Brattleboro rat is influenced by genotype and lysine vasopressin treatment of the mother. *Biol. Neonate,* **53**, 295–304.

Snijdewint, F.G.M., Van Leeuwen, F.W. and Boer, G.J. (1989) Ontogeny of vasopressin and oxytocin binding sites in the brain of Wistar and Brattleboro rats as demonstrated by light-microscopical autoradiography. *J. Chem. Neuroanat.*, **2**, 3–17.

Sofroniew, M.V. (1985) Vasopressin, oxytocin and their related neurophysins, in *Handbook of Chemical Neuroanatomy, Vol. 4, GABA and Neuropeptide in the CNS* (eds A. Björklund, T. Hökfelt), Elsevier, Amsterdam, pp. 93–146.

Stephens, L.R. and Logan, S.D. (1986) Arginine–vasopressin stimulates inositol phospholipid metabolism in rat hippocampus. *J. Neurochem.*, **46**, 649–51.

Swank, M.W. and Dorsa, D.M. (1990) Guanine nucleotides and pertussis toxin alter agonist binding to rat septal V_1-vasopressin receptors. *Mol. Cell. Neurosci.*, **1**, 117–20.

Swank, M.W. and Dorsa, D.M. (1991) Chronic treatment with vasopressin analogues alters affinity of vasopressin receptors in the septum and amygdala of the rat brain. *Brain Res.*, **544**, 342–4.

Swenson, R.R., Beckwith, B.E., Lamberty, K.J. *et al.* (1990) Prenatal exposure to AVP or caffeine but not oxytocin alters learning in female rats. *Peptides,* **11**, 927–32.

Szot, P., Myers, K.M., Swank, M. *et al.* (1989) Characterization of a ^{3}H-arginine^{8}-vasopressin binding site in the cingulate gyrus of the rat pup. *Peptides,* **10**, 1231–7.

Szot, P., Myers, K.M. and Dorsa, D.M. (1992) Effect of vasopressin administration and deficiency upon ^{3}H-AVP binding sites in the CNS and periphery during development. *Peptides,* **13**, 389–94.

Tinius, T.P., Beckwith, B.E., Preussler, D.W. *et al.* (1987) Prenatal administration of arginine vasopressin impairs memory retrieval in adult rats. *Peptides,* **8**, 493–9.

Tribollet, E. (1992) Vasopressin and oxytocin receptors in the rat brain, in *Handbook of Chemical Neuroanatomy, Vol. 11, Neuropeptide Receptors in the CNS* (eds A. Björklund, T. Hökfelt and M.J. Kuhar), Elsevier, Amsterdam, pp. 289–320.

Tribollet, E., Charpak, S., Schmidt, A. *et al.* (1989) Appearance and transient expression of oxytocin receptors in fetal, infant, and peripubertal rat brain studied by autoradiography and electrophysiology. *J. Neurosci.*, **9**, 1764–73.

Tribollet, E., Audigier, S., Dubois-Dauphin, M. *et al.* (1990) Gonadal steroids regulate oxytocin receptors but not vasopressin receptors in the brain of male and female rats. An autoradiographical study. *Brain Res.*, **511**, 129–40.

Tribollet, E., Goumaz, M., Raggenbass, M. *et al.* (1991a) Early appearance and transient expression of vasopressin receptors in the brain of rat fetus and infant. An autoradiographical and electrophysiological study. *Dev. Brain Res.*, **58**, 13–24.

Tribollet, E., Goumaz, M., Raggenbass, M. *et al.* (1991b) Appearance and transient expression of vasopressin and oxytocin receptors in the rat brain. *J. Recept. Res.*, **11**, 333–46.

Trinder, D., Mooser, V., Kelly, J.M. *et al.* (1991) Characterization of monoclonal antibodies to a rat liver vasopressin receptor. *Clin. Exp. Pharmacol. Physiol.*, **18**, 345–8.

Van Leeuwen, F.W. (1987) Vasopressin receptors in the brain and pituitary, in *Vasopressin, Principles and Properties* (eds D.M. Gash and G.J. Boer), Plenum Press, New York, pp. 477–96.

Van Leeuwen, F.W. and Wolters, P. (1983) Light microscopic autoradiographic localization of [³H]-arginine-vasopressin binding sites in the rat brain and kidney. *Neurosci. Lett.*, **41**, 61–6.

Van Leeuwen, F.W., Van der Beek, E.M., Van Heerikhuize, J.J. *et al.* (1987) Quantitative light microscopic autoradiographic localization of binding sites labelled with [³H]vasopressin antagonist d(CH₂)₅Tyr(Me)VP in the rat brain, pituitary and kidney. *Neurosci. Lett*, **80**, 121–6.

Whitnall, M.H., Key, S., Ben-Barak, Y. *et al.* (1985) Neurophysin in the hypothalamo-neurohypophysial system. II. Immunocytochemical studies of the ontogeny of oxytocinergic and vasopressinergic neurons. *J. Neurosci.*, **5**, 98–109.

Yamamura, H.I., Gee, K.W., Brinton, R.E. *et al.* (1983) Light microscopic autoradiographic visualization of [³H]-arginine vasopressin binding sites in rat brain. *Life Sci.*, **32**, 1919–24.

ADDENDUM

Shortly after finalizing this review there was a breakthrough in the cloning and sequencing of the VP and OT receptor cDNAs. Both V1a, V2 and OT receptors have been cloned and characterized from rat liver (Morel *et al.*, 1992), rat and human kidney (Lolait *et al.*, 1992; Birnbaumer *et al.*, 1992) and human myometrium (Kimura *et al.*, 1992). All receptors appeared prototypic G-protein-coupled receptors with seven putative transmembrane domains. Sequence similarity between these receptors is considerable, which may mirror the similarity of the peptide hormones themselves (Sharif and Hanley, 1992). Since screening of rat and human genomic DNA digests, using a probe for the V1a receptor, revealed five to nine bands (Morel *et al.*, 1992), sub- or isotypes of VP (and/or OT) receptors are to be expected.

In situ hybridization of the V1a and V2 ³⁵S-labeled cRNA probes showed V1a receptor mRNA in all brain areas where receptor autoradiography already showed receptor expression, whereas V2 receptor expression was completely absent (Ostrowski *et al.*, 1992). Other areas, however, such as the cerebellum, the inferior olive and the arcuate nucleus, were also labeled for V1a receptor expression.

REFERENCES

Birnbaumer, M., Seibold, A., Gilbert, S. *et al.* (1992) Molecular cloning of the receptor for human antidiuretic hormone. *Nature*, **357**, 333-5.

Kimura, T., Tanizawa, O., Mori, K. *et al.* (1992) Structure and expression of a human oxytocin receptor. *Nature*, **356**, 526-9.

Lolait, S.J., O'Carroll, A-M., McBride. O.W. *et al.* (1992) Cloning and characterization of a vasopressin V2 receptor and possible link nephrogenic diabetes insipidus. *Nature*, **357**, 336-9.

Morel, A., O'Carroll, A-M., Brownstein, M.J. *et al.* (1992) Molecular rat cloning and expression of a rat V1a arginine vasopressin receptor. *Nature*, **356**, 523-6.

Ostrowski, N.L., Lolait, S.L., Bradley, D.J *et al.* (1992) Distribution of V1a and V2 vasopressin receptor messenger ribonucleic acids in rat liver, kidney, pituitary and brain. *Endocrinology*, **131**, 533-5.

Shariff, M. and Hanley, M.R. (1992) Stepping up the pressure. *Nature*, **357**, 279-80.

INDEX

RECEPTORS IN THE DEVELOPING NERVOUS SYSTEM

VOLUME 2

RECEPTORS IN THE DEVELOPING NERVOUS SYSTEM

VOLUME 2
Neurotransmitters

Edited by

Ian S. Zagon and
Patricia J. McLaughlin

Department of Neuroscience and Anatomy
The Milton S. Hershey Medical Center
The Pennsylvania State University
Hershey, Pennsylvania, USA

CHAPMAN & HALL
London · Glasgow · New York · Tokyo · Melbourne · Madras

Published by Chapman & Hall, 2–6 Boundary Row, London SE1 8HN

Chapman & Hall, 2–6 Boundary Row, London SE1 8HN, UK

Blackie Academic & Professional, Wester Cleddens Road, Bishopbriggs, Glasgow G64 2NZ, UK

Chapman & Hall Inc., 29 West 35th Street, New York, NY10001, USA

Chapman & Hall Japan, Thomson Publishing Japan, Hirakawacho Nemoto Building, 6F, 1–7–11 Hirakawa-cho, Chiyoda-ku, Tokyo 102, Japan

Chapman & Hall Australia, Thomas Nelson Australia, 102 Dodds Street, South Melbourne, Victoria 3205, Australia

Chapman & Hall India, R. Seshadri, 32 Second Main Road, CIT East, Madras 600 035, India

First edition 1993

© 1993 Chapman & Hall

Typeset in 10/12 Pt Palatino by Expo Holdings, Malaysia
Printed in Great Britain at the University Press, Cambridge

ISBN 0 412 49400 0   Vols. 1 and 2 (set) 0 412 54520 9

A catalogue record for this book is available from the British Library

Library of Congress Cataloging-in-Publication data

Receptors in the developing nervous system / edited by Ian S. Zagon and
    Patricia J. McLaughlin.—1st ed.
        p. cm.
    Includes bibliographical references and index.
      Contents: v. 1. Receptors related to growth factors and hormones
    —v. 2. Neurotransmitters.
      ISBN 0-412-54520-9 (set: alk. paper). — ISBN 0-412-49400-0 (hb: v. 2:
    alk. paper)
      1. Developmental neurology. 2. Neurotransmitter receptors. I. Zagon,
    Ian S. II. McLaughlin, Patricia J.
    [DNLM: 1. Growth Substances. 2. Nervous System—embryology.
    3. Neuroregulators. 4. Receptors, Endogenous Substances.
    5. Receptors, Sensory, WL 101 R295]
    QP363. 5.R43 1993
    599' .0333—dc20
    DNLM/DLC   92-49100
    for Library of Congress   CIP

For Eileen
And my grandparents,
 Mary and Abraham Shaffer
I. S. Z.

For my parents,
 Katharine and Charles
P. J. M.

CONTENTS

CONTRIBUTORS

JOHN D. ALVARO
Laboratory of Molecular Psychiatry,
    Department of Psychiatry and Program in
    Neuroscience, Abraham Ribicoff Research
    Facility, Yale University School of
    Medicine, 34 Park St, New Haven, CT
    06508, USA.

LUCIO G. COSTA
Department of Environmental Health, SC-34,
    University of Washington, Seattle, WA
    98195, USA.

THAN-VINH DAM
Douglas Hospital Research Centre and
    Department of Psychiatry, Faculty of
    Medicine, McGill University, Verdun,
    Quebec, Canada H4H IR3.

ANGEL LUIS DE BLAS
Division of Molecular Biology and
    Biochemistry, School of Biological Sciences,
    University of Missouri-Kansas City,
    Kansas City, MO 64110-2499, USA.

RONALD S. DUMAN
Laboratory of Molecular Psychiatry,
    Department of Psychiatry and Program in
    Neuroscience, Abraham Ribicoff Research
    Facility, Yale University School of
    Medicine, 34 Park St, New Haven, CT
    06508, USA.

F. JAVIER GARCIA-LADONA
Institute of Pathology, Department of
    Neuropathology, University of Basel,
    Switzerland.

GAIL E. HANDELMANN
Department of Pharmacology and
    Toxicology, University of Utah, Salt Lake
    City, UT 84112, USA.

FRANCES M. LESLIE
Department of Pharmacology, University of
    California at Irvine, Irvine, CA 92717, USA.

EDYTHE D. LONDON
Addiction Research Center, National
    Institute on Drug Abuse, Baltimore, MD
    21224, USA.

SANDRA E. LOUGHLIN
Department of Anatomy and Neurobiology,
    University of California at Irvine, Irvine,
    CA 92717, USA.

GUADALUPE MENGOD
Department of Neurochemistry, Centro
    Investigacion y Desarrollo, Consejo
    Superior de Investigaciones Cientificas
    (CSIC), Jordi Girona, 18–26, Barcelona,
    Spain.

TIMOTHY H. MORAN
Department of Psychiatry, Johns Hopkins
    University School of Medicine, Baltimore,
    MD 21205, USA.

JOSÉ M. PALACIOS
Department of Neurochemistry, Centro
    Investigacion y Desarrollo, Consejo
    Superior de Investigaciones Cientificas
    (CSIC), Jordi Girona, 18–26, Barcelona,
    Spain, and Research Institute, Laboratorios
    Almirall, Barcelona, Spain.

REBECCA M. PRUSS
Marion Merrell Dow Research Institute, 2110 E.
  Galbraith Road, Cincinnati, OH 45215, USA.

RÉMI QUIRION
Douglas Hospital Research Centre and
  Department of Psychiatry, Faculty of
  Medicine, McGill University, Verdun,
  Quebec, Canada H4H 1R3.

PAUL H. ROBINSON
Westminster Hospital, London SW1, UK.

THOMAS ROTHE
University of Leipzig, Paul Flechsig Institute
  for Brain Research, Department of
  Neurochemistry, Leipzig, Germany.

REINHARD SCHLIEBS
University of Leipzig, Paul Flechsig Institute
  for Brain Research, Department of
  Neurochemistry, Leipzig, Germany.

JOHN D. STEPHENSON
Institute of Psychiatry, London SE5, UK.

ANN TEMPEL
Hillside Hospital, Division of Long Island
  Jewish Medical Center, The Long Island
  Campus for the Albert Einstein College of
  Medicine, Glen Oaks, New York 11004,
  USA.

PATRICIA M. WHITAKER-AZMITIA
Department of Psychiatry, State University
  of New York, Stony Brook, NY 11794,
  USA.

STEPHEN R. ZUKIN
Albert Einstein College of Medicine, Yeshiva
  University, Bronx, New York, NY 10461,
  USA.

CONTENTS TO VOLUME ONE

CONTRIBUTORS TO VOLUME ONE

MARTIN ADAMO
Section on Molecular and Cellular
    Physiology, Diabetes Branch, National
    Institute of Diabetes and Digestive and
    Kidney Diseases, National Institutes of
    Health, Bethesda, MD 20892, USA

SEEMA BHATNAGAR
Developmental Neuroendocrinology
    Laboratory, Douglas Hospital Research
    Centre, Department of Psychiatry and
    Neurology and Neurosurgery, McGill
    University, Montreal, Canada H4H 1R3

GERARD J. BOER
Netherlands Institute for Brain Research,
    Meibergdreef 33, 1105AZ Amsterdam Z0,
    The Netherlands

CAROLYN A. BONDY
Developmental Endocrinology Branch,
    National Institute of Child Health and
    Human Development, National Institutes
    of Health, Bethesda, MD 20892, USA

JEAN-GUY CHABOT
Douglas Hospital Research Centre and
    Department of Psychiatry, Faculty of
    Medicine, McGill University, Verdun,
    Québec, Canada H4H 1R3

KWEN-JEN CHANG
Department of Anesthesiology and
    Pharmacology, Duke University Medical
    Center, and Division of Cell Biology,
    Wellcome Research Laboratories,
    Burroughs Wellcome Co.
    Research Triangle Park, NC 27709, USA

MOSES V. CHAO
Department of Cell Biology and Anatomy,
    Hematology/Oncology Division, Cornell
    University Medical College, 1300 York
    Avenue, New York, NY 10021, USA

ERROL B. DE SOUZA
Central Nervous System Diseases Research,
    The DuPont Merck Pharmaceutical Co.,
    Wilmington, DE 19880, USA

JEAN DE VELLIS
Departments of Anatomy and Psychiatry,
    Mental Retardation Research Center, Brain
    Research Institute, Laboratory of
    Biomedical and Environmental Sciences,
    UCLA School of Medicine, University of
    California, Los Angeles, CA 90024, USA

ARACELI ESPINOSA DE LOS MONTEROS
Departments of Anatomy and Psychiatry,
    Mental Retardation Research Center, Brain
    Research Institute, Laboratory of
    Biomedical and Environmental Sciences,
    UCLA School of Medicine, University of
    California, Los Angeles, CA 90024, USA

DIMITRI E. GRIGORIADIS
Central Nervous System Diseases Research,
    The DuPont Merck Pharmaceutical Co.,
    Wilmington, DE 19880, USA

JEFFREY A. HEROUX
Central Nervous System Diseases Research,
    The DuPont Merck Pharmaceutical Co.,
    Wilmington, DE 19880, USA

THOMAS R. INSEL
Laboratory of Neurophysiology NIMH
   Poolesville, MD 20837, USA

SATYABRATA KAR
Douglas Hospital Research Centre and
   Department of Psychiatry, Faculty of Medi-
   cine, McGill University, Verdun, Québec,
   Canada H4H 1R3

STEPHEN L. KINSMAN
The Kennedy Krieger Institute, 707 North
   Broadway, Baltimore, MD 21205, USA

DEREK LEROITH
Section of Molecular and Cellular
   Physiology, Diabetes Branch, National
   Institute of Diabetes and Digestive and
   Kidney Diseases, National Institutes of
   Health, Bethesda, MD 20892, USA

PATRICIA J. MCLAUGHLIN
Department of Neuroscience and Anatomy,
   The Pennsylvania State University, The
   M.S. Hershey Medical Center, Hershey, PA
   17033, USA

MICHAEL J. MEANEY
Developmental Neuroendocrinology
   Laboratory, Douglas Hospital Research
   Centre, Departments of Psychiatry and
   Neurology and Neurosurgery, McGill
   University, Montreal, Canada H4H 1R3

DAJAN O'DONNELL
Developmental Neuroendocrinology
   Laboratory, Douglas Hospital Research
   Centre, Departments of Psychiatry and
   Neurology and Neurosurgery, McGill
   University, Montreal, Canada H4H 1R3

LUIS F. PARADA
Molecular Embryology Group, ABL-Basic
   Research Program, NCI-Frederick Cancer
   Research, P O Box B, Frederick, MD 21701,
   USA

DANIEL H. POLK
UCLA School of Medicine, Perinatal
   Laboratories, Harbor-UCLA Medical
   Center, Torrance, CA 90509, USA

RÉMI QUIRION
Douglas Hospital Research Centre and
   Department of Psychiatry, Faculty of
   Medicine, McGill University, Verdun,
   Québec, Canada H4H 1R3

MOHAN K. RAIZADA
Department of Physiology, University of
   Florida, Gainesville, FL 32610, USA

CHARLES T. ROBERTS
Section on Molecular and Cellular
   Physiology, Diabetes Branch, National
   Institute of Diabetes and Digestive and
   Kidney Diseases, National Institutes of
   Health, Bethesda, MD 20892, USA

ALAIN SARRIEAU
Developmental Neuroendocrinology
   Laboratory, Douglas Hospital Research
   Centre, Departments of Psychiatry and
   Neurology and Neurosurgery, McGill
   University, Montreal, Canada H4H 1R3

NOLA SHANKS
Developmental Neuroendocrinology
   Laboratory, Douglas Hospital Research
   Centre, Departments of Psychiatry and
   Neurology and Neurosurgery, McGill
   University, Montreal, Canada H4H 1R3

SAMUEL A. SHOLL
Wisconsin Regional Primate Research Center,
   University of Wisconsin, Madison,
   WI53715–1299, USA

JAMES SMYTHE
Developmental Neuroendocrinology
   Laboratory, Douglas Hospital Research
   Centre, Departments of Psychiatry and
   Neurology and Neurosurgery, McGill
   University, Montreal, Canada H4H 1R3

YING-FU SU
Department of Anesthesiology and Pharmacology, Duke University Medical Center, Durham, NC 27710, USA

VICTOR VIAU
Developmental Neuroendocrinology Laboratory, Douglas Hospital Research Centre, Departments of Psychiatry and Neurology and Neurosurgery, McGill University, Montreal, Canada H4H 1R3

CLAIRE-DOMINIQUE WALKER
Department of Physiology, University of California at San Francisco, San Francisco, CA 94143, USA

HAIM WERNER
Section on Molecular and Cellular Physiology, Diabetes Branch, National Institute of Diabetes and Digestive and Kidney Diseases, National Institutes of Health, Bethesda, MD 20892, USA

IAN S. ZAGON
Department of Neuroscience and Anatomy, The Pennsylvania State University, The M.S. Hershey Medical Center, Hershey, PA 17033, USA

PREFACE

Receptors for cell hormones, growth factors, and neurotransmitters are involved in the control and modulation of an enormous array of biological processes. The development of these receptors has distinct spatial and temporal arrangements, and alterations in this pattern during embryogenesis can have significant consequences for the well-being of the fetus, infant, child and adult. The developing nervous system is particularly dependent on receptors because its period of structural and functional organization extends through both prenatal and postnatal phases. Moreover, receptors are a key element in neural communication in both the developing and adult organism, so that the ontogeny of receptors is crucial in determining the myriad connections forming the circuitry of the nervous system.

The purpose of these two volumes is to provide a comprehensive review of receptors in the developing nervous system, placing basic and clinical information into perspective in order to formulate future scientific inquiry. A number of themes are maintained throughout the books. First, the receptors discussed in these books are part of a unit, and both the receptor and neurotransmitter, hormone, or growth factor must be considered. Therefore, the authors have spent some time in each chapter introducing the compound(s) that interacts with each receptor. Second, some receptors appear transiently and are responsible for participating in a biological process related to ontogeny, disappearing by maturation. Third, a receptor in the process of evolving into a structure important in the mature animal may also be vital in establishing the framework of particular pathways or the architecture of the brain.

Fourth, alterations in the development of neural receptors may have profound implications for the structure and function of the organism. As much as possible, the repercussions of disrupting the orchestration of receptor development in the nervous system are discussed. In many instances, however, we are just beginning to learn about some receptors and the authors may not be in a position to discuss the consequences of receptor dysfunction.

In designing these two volumes, we have asked major figures in each field to review the literature, to apprise the audience of their latest findings, and to provide a perspective on the role of receptors in the developing nervous system. These books are intended to summarize not only where we have been, but also to map directions for future research efforts, with the ultimate goals of understanding processes of normal development and the prevention or treatment of receptor dysfunction. The book is intended to be part of a continuing dialog about an important and emerging field. Given the broad scope of the subject matter, it is our expectation that the information provided by these experts will be of interest to basic and clinical researchers and graduate students in the field of developmental neurobiology, as well as to individuals in psychology, cellular and molecular biology, endocrinology, embryology, neuroscience, pharmacology, and in clinical professions such as pediatrics, neonatology, and neurology. The style adopted by the contributors should also ensure that the contents of the two volumes will be accessible to undergraduate students enrolled in advanced courses concerned with neuroscience and cell and molecular biology.

The subject matter with respect to receptors and the developing nervous system has been divided into two volumes. The editors have attempted to group central themes and topics within three major areas: receptors related to growth factors and hormones (Volume I) and to neurotransmitters (Volume II). The reader should be aware that these are rather arbitrary divisions, and that a particular compound may have multiple functions. For example, opioid receptors known to be involved in neurotransmission may also have a trophic influence on development, and could also be considered a growth factor. An introductory chapter on the biology of receptors has been included in Volume I. For convenience, an index of the subject matter in each volume is included. Finally, since Volumes I and II should be viewed as a single, integral entity, we have included the chapter titles and authors of both volumes in each book.

We thank all the authors for their thoughtful and stimulating chapters, and for the prompt attention shown to editorial requests. We have learned a great deal from reviewing these chapters and trust that the reader will have an equally enjoyable experience.

Ian S. Zagon
Patricia J. McLaughlin

DEVELOPMENTAL EXPRESSION OF ADRENERGIC RECEPTORS IN THE CENTRAL NERVOUS SYSTEM

Ronald S. Duman and John D. Alvaro

1.1 INTRODUCTION

The norepinephrine (NE) neurotransmitter receptor system is involved in numerous processes in the nervous system, including vigilance, attention, arousal, memory, neuroendocrine function, drug addiction, anxiety, depression, mania and stress-related disorders. The receptors for NE and epinephrine have been divided into α and β adrenergic receptors (αAR and βAR, respectively) based on functional and ligand binding studies. The development of more selective pharmacological agents has resulted in the characterization of α_1AR, α_2AR, β_1AR, and β_2AR subtypes, and molecular cloning has isolated further subtypes of each pharmacological type. Adrenergic receptors belong to the G protein-coupled receptor superfamily, a group of proteins which share a common feature of having seven hydrophobic transmembrane domains and which include dopaminergic, serotonergic, neuropeptide and other receptors. Members of this family couple with specific G proteins which then interact with effector systems such as enzymes or ion channels. Adrenergic receptor coupled effector systems regulate intracellular pathways that result in immediate or short-term changes in neuronal function as well as more long-term adaptive changes which may be dependent on regulation of gene expression.

The formation and function of adrenergic receptor systems during development ultimately influence a number of vital neuronal and behavioral processes. A clear understanding of the ontogeny of adrenergic receptors and the factors which regulate their development will help define the normal physiological processes regulated by these receptors. Moreover, the mechanisms which underlie psychiatric abnormalities associated with adrenergic receptors will be elucidated. The focus of this chapter is to review studies of the development of adrenergic receptors and their coupling factors/effector systems in the nervous system. The studies discussed in this review will focus on the development of adrenergic receptors in rat since the majority of the work has been conducted with this species.

1.2 DEVELOPMENT OF THE NOREPINEPHRINE NEUROTRANSMITTER SYSTEM IN THE NERVOUS SYSTEM

Both NE and epinephrine are effective neurotransmitter ligands at αAR and βAR, although NE is more widely distributed and more abundant than epinephrine in the central nervous system. For this reason the majority of

Receptors in the Developing Nervous System Vol. 2: Neurotransmitters. Edited by Ian S. Zagon and Patricia J. McLaughlin. Published in 1993 by Chapman & Hall. ISBN 0 412 49400 0. Vols. 1 and 2 (set) ISBN 0 412 54520 9.

studies have focused on the NE system. The distribution of NE- and epinephrine-containing cells and their projections has been reviewed by Moore and Bloom (1979) and Foote *et al.* (1983) and is summarized below. The majority of NE-containing cells are located in the pons and medulla of the brainstem. The nucleus locus coeruleus is a nearly homogeneous population of cells which contains approximately 40% of the NE cell bodies of the brain. The noradrenergic neurons located in the lateral tegmentum are more diffusely organized than those of the locus coeruleus and are spread throughout the lateral tegmental fields from the most rostral aspect of the medulla through the pons. Epinephrine-containing cell bodies are often intermingled with NE-containing cells in the lateral tegmentum and dorsal medulla.

NE projections throughout the brain are extensive and in many brain regions follow a very uniform plexus pattern of innervation. The locus coeruleus noradrenergic neurons project to many areas of the central nervous system including the cerebral cortex, thalamus, amygdala, striatum, hippocampus, hypothalamus, olfactory bulb, brainstem, cerebellum, and spinal cord; locus coeruleus is the primary source of NE to most of these brain regions. The lateral tegmental NE neurons project to many of the same brain areas and provide the primary NE input to some of these regions, specifically amygdala, septum and some hypothalamic nuclei such as the paraventricular nucleus and the supraoptic nucleus. Epinephrine-containing cells send projections primarily to the hypothalamus and midbrain regions.

Norepinephrine neurons differentiate and begin to express catecholamine during embryogenesis, and levels of NE continue to increase after birth (for reviews see Coyle, 1977; Foote *et al.*, 1983). In the rat, locus coeruleus neurons appear between embryonic day (ED) 10 and 13. The presence of catecholamine histofluorescence and the immuno-

chemical staining of NE biosynthetic enzymes (tyrosine hydroxylase and dopamine β-hydroxylase) in these neurons appear as early as ED14. Catecholamine projections to the thalamus and hypothalamus appear around ED14, and ascending NE fibers to the cerebral cortex and subcortical regions appear around ED16. Neurons in the hippocampus appear around ED13–ED18 and in the cerebellar Purkinje neurons around ED14–ED15. This time course for development of LC neurons precedes the development of neurons in terminal field brain regions of this neurotransmitter. During the last week of gestation in the rat (ED15–ED22) there is a linear increase in levels of NE biosynthetic enzymes, and by ED18 high affinity uptake of NE can be demonstrated in synaptosomes.

At birth the levels of NE and its biosynthetic enzymes are between 10 and 20% of adult levels (Fig. 1.1). The innervation of cerebral cortex progresses to the adult pattern, determined by NE histofluorescence and tyrosine hydroxylase immunocytochemistry, by the end of postnatal day (PD) 7 in the rat. Although the adult pattern of NE innervation is present throughout the brain by the first week after birth, the arborization of NE terminals continues; adult levels of NE and tyrosine hydroxylase are not attained until 4–5 weeks after birth.

1.3   DEVELOPMENT OF β_1 AND β_2 ADRENERGIC RECEPTORS IN THE NERVOUS SYSTEM

The physiology, pharmacology and function of βAR subtypes have been widely studied. These receptors play a role in the central control of cardiovascular function, are involved in memory, and have been implicated in psychiatric illnesses, particularly depression (Minneman, 1981; Heninger and Charney, 1987). β_1AR and β_2AR subtypes have been defined based on physiological and pharmacological studies, and individual cDNA clones

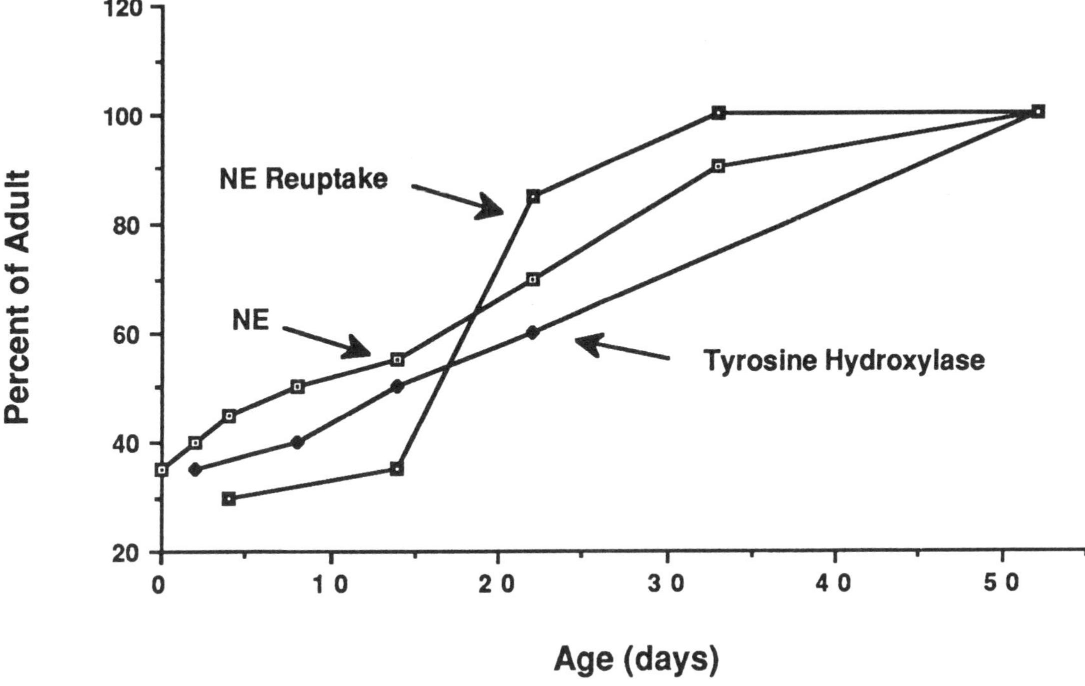

Fig. 1.1. Development of NE, tyrosine hydroxylase and NE re-uptake in rat brain. Levels of endogenous NE, tyrosine hydroxylase enzyme activity, and uptake of [³H]NE were determined at different times after birth in lateral cerebral cortex. The results are expressed as percentages of adult values and are derived from Johnston and Coyle (1980).

have been isolated for each receptor subtype (Dixon *et al.*, 1986; Frielle *et al.*, 1987). The predicted amino acid sequence for each receptor contains seven putative hydrophobic transmembrane domains and the human β_1AR and β_2AR are 54% homologous at the amino acid level. Both βAR subtypes are positively coupled to adenylate cyclase via the stimulatory G protein, $G_{s\alpha}$, and activation of the cyclic AMP system represents the primary second messenger pathway through which βARs regulate cellular function.

A number of studies have examined the development of βAR in fetal and postnatal rat brain. Many of these studies utilized the non-selective ligands ^{125}I-labeled hydroxyben-zylpindolol ([^{125}I]HYP) and [³H]dihydroalpre-nolol (DHA) which, under the assay con-ditions used, label both β_1AR and β_2AR sub-types. In fetal forebrain [³H]DHA binding was observed as early as ED16 (Bruinink and Lichtensteiger, 1984). On PD1 levels of [^{125}I]HYP binding in cerebral cortex were reported to be approximately 10–20% of adult levels (Fig. 1.2), increased slightly during the first week, and then increased four- to five-fold to adult levels by PD16–PD21 (Harden *et al.*, 1977). The increase was shown to be due to an increase in the total number of binding sites (B_{max}) with no change in receptor affinity (K_d). Other studies have reported a similar pattern for the development of total βAR binding but with a time course which was either slightly faster (Keshles and Levitski, 1984) or slower (Waddington and Banks, 1981; Bruinink and Lichtensteiger, 1984). Levels of NE, re-uptake

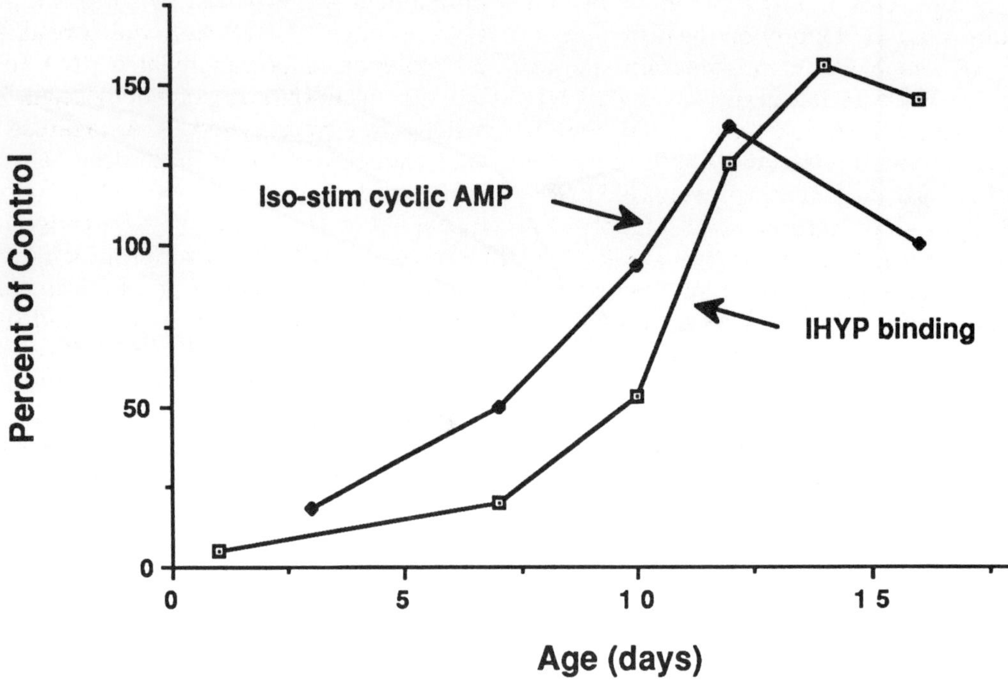

Fig. 1.2.   Development of [^{125}I]HYP binding and isoproterenol-stimulated cyclic AMP in rat cerebral cortex. Levels of [^{125}I] HYP binding were determined in homogenates of cerebral cortex, and isoproterenol-stimulated cyclic AMP accumulation (Iso-stim cyclic AMP) was determined in slices of cerebral cortex. Results are expressed as percentages of control values which were either adult or P16 levels of binding or cyclic AMP, respectively. The data are derived from Harden *et al.* (1977).

sites, and catecholamine synthetic enzymes develop more slowly and do not reach adult levels until 5–6 weeks after birth. Because βAR reach adult levels much sooner than does NE, it is believed that the developmental expression of βAR is independent of expression of neurotransmitter. The latter point is supported by a later study demonstrating the presence of βAR in rats which had been treated at birth with 6-hydroxydopamine (6-OHDA), a neurotoxin which destroys catecholamine terminals (Minneman *et al.*, 1979). However, levels of NE are not completely eliminated by this treatment and it is conceivable that exposure to NE during gestation is sufficient to stimulate expression and continued development of βAR.

The development of β_1AR and β_2AR subtypes has also been measured by utilizing subtype-selective antagonists. Pittman *et al.* (1980) examined the development of β_1AR and β_2AR in cerebral cortex and cerebellum, two brain regions which express the subtypes at measurably different levels in the adult. In adult cerebral cortex approximately 65% of βAR ligand binding is to β_1ARs whereas in adult cerebellum about 80% of βAR ligand binding is to β_2ARs. In both cerebral cortex and cerebellum Pittman *et al.* (1980) found that levels of β_1AR increased four- to fivefold by PD21, a time course similar to that reported previously for total βAR binding. In contrast, the development of β_2AR followed a much slower pattern of development in both brain regions, reaching adult levels by PD28 in cerebral cortex and by PD42 in cerebellum. Results of a recent study on the development of β_1AR and β_2AR in rat forebrain support these findings (Erdtsieck-Ernste *et al.*, 1991). In this study β_1AR and β_2AR ligand binding are also reported to be present in forebrain as early as ED13 and increase severalfold before birth. This study also demonstrates that the amount of β_2AR in forebrain is greater than 50% of total βAR ligand binding before PD4 (including gestation) but less than 30% of total βAR ligand binding after PD4.

The slower development of β_2AR binding in brain together with the reports that levels of β_2AR are not influenced by treatments which alter synaptic levels of NE (i.e. 6-OHDA or desipramine treatment) lead to the hypothesis that β_2ARs are located on glia which do not have direct noradrenergic synaptic connections (Minneman *et al.*, 1979). However, autoradiographic studies have demonstrated that β_2AR ligand binding is regulated by 6-OHDA and desipramine treatments in certain brain regions (Ordway *et al.*, 1988; Johnson *et al.*, 1989). These findings suggest that β_2ARs may be located on either neurons or glia. In contrast, levels of β_1AR ligand binding are clearly regulated by treatments which alter synaptic levels of catecholamine and therefore are thought to be located on neurons receiving direct noradrenergic synaptic input.

In addition to the use of binding assays, another approach for studying the development of βAR subtypes is to measure levels of β_1AR and β_2AR mRNA by either hybridization blot analysis (Northern blot) or solution hybridization. The advantage of this approach is that the DNA and RNA probes used for these studies are extremely specific and under stringent conditions hybridize only with mRNA of the appropriate receptor subtype. We have used hybridization blot analysis to study the development of β_1AR and β_2AR mRNA (Duman *et al.*, 1989). For these studies an enriched mRNA fraction was used since β_1AR and β_2AR mRNAs are expressed at very

low levels in brain as well as in other tissues. Enrichment was achieved by subjecting total RNA to oligo(dT) selection which results in the isolation of polyadenylated RNA (most but not all mRNA is polyadenylated). The mRNA was fractionated on an agarose gel and transferred to nitrocellulose. Human β_1AR and hamster β_2AR cDNA clones (provided by Dr Lefkowitz, Duke University) were labeled with ^{32}P by nick translation and hybridized with the nitrocellulose filters containing mRNA. At birth, β_1AR and β_2AR mRNA were present at 20% and 50% of adult levels respectively, and both then gradually increased to adult levels by PD25. This time course was parallel to the development of β_1AR and β_2AR ligand binding determined in the same study and is approximately the same as reported by others. These results indicate that development of both βAR subtypes is dependent on activation of gene transcription, expression of mRNA and the resultant *de novo* receptor synthesis.

A number of studies have examined the development of βAR-stimulated cyclic AMP formation as a measure of the ontogeny of functional βAR. βAR-stimulated cyclic AMP formation in brain slices can be observed as early as ED17 (Walton *et al.*, 1979). After birth the development of βAR-stimulated cyclic AMP formation in slices of cerebral cortex (Fig. 1.2) corresponds to the expression of βAR binding sites (Perkins and Moore, 1973; Harden *et al.*, 1977). These studies also examined both the development of cyclic AMP accumulation in brain slices and the ontogeny of levels of adenylate cyclase activity in brain homogenates. Basal levels of cyclic AMP formation and adenylate cyclase activity on PD1 are present at approximately 30–40% of adult levels and reach adult levels by PD21. Stimulation of adenylate cyclase activity by Mn^{2+}, which directly activates the catalytic unit, follows a similar time course of development (Keshles and Levitski, 1984). Stimulation of adenylate cyclase by fluoride or Gpp(NH)p, which activate $G_{s\alpha}$, also follows a similar time course. These results indicate that the level of functional $G_{s\alpha}$ and adenylate cyclase present at birth are sufficient to support βAR-stimulated cyclic AMP formation and therefore suggest that the development of functional βAR is dependent on the expression of the receptors and not on the development of the effector system.

1.4   DEVELOPMENT OF α_1AR IN THE NERVOUS SYSTEM

α_1ARs appear to mediate the actions of NE on locomotor activity and behavioral excitation (Bylund and U'Prichard, 1983; Berridge and Dunn, 1989). α_{1A} and α_{1B} subtypes have been defined on the basis of physiological and pharmacological studies (Morrow and Creese, 1986; Minneman *et al.*, 1988), and molecular cloning studies have confirmed these subtypes by isolation of α_{1A} and α_{1B} cDNA clones (Cotecchia *et al.*, 1988; Lomasney *et al.*, 1991). Like the other adrenergic receptor subtypes α_{1A} and α_{1B} receptor proteins have the characteristic seven transmembrane spanning domains.

Both subtypes appear to be localized on postsynaptic elements in brain and display a regional heterogeneity in the nervous system (Table 1.1; Cotecchia *et al.*, 1988; Wilson and Minneman, 1989; Lomasney *et al.*, 1991).

Activation of α_1ARs stimulates phospholipase C and the phosphatidylinositol second messenger pathway which has two components: (1) formation of inositol triphosphate which stimulates the release of intracellular calcium (Berridge and Irvine, 1989), and (2) formation of diacylglycerol which together with calcium stimulates protein kinase C (Nishizuka, 1988). In addition, α_1ARs appear indirectly to regulate the cyclic AMP system in brain by enhancing the response to other stimulatory ligands (Duman *et al.*, 1986). Finally, α_1ARs have also been reported to regulate voltage-dependent calcium channels independent of an effect on phospholipase C (Han *et al.*, 1987). The α_{1B} subtype has been proposed to couple to phospholipase C whereas the α_{1A} may be coupled to Ca^{2+} influx. G_q appears to mediate receptor stimulation of the phosphatidylinositol pathway, although the specific G protein subtypes which mediate the effects of α_1ARs on this second messenger

Table 1.1.   Adrenergic receptor subtypes

Receptor subtype	G protein	Effector system	Regional distribution
α_1 *subtypes*			
α_{1A}	G_i/G_o	Ca^{2+} influx	CTX>HP>BSM>CB
α_{1B}	G_q	IP_3/DG ↑	CTX>BSM>>>CB>>HP
α_{1C}	G_q	IP_3/DG ↑	Bovine tissues only
α_2 *subtypes*			
α_{2A}(C10)[a]	G_i/G_o	Cyclic AMP ↓ K^+ channel ↑	CTX=BSM>MID>NS>CB>HP
α_{2B}(C2)	G_i/G_o	Cyclic AMP ↓	Very low levels in brain
α_{2C}(C4)	G_i/G_o	Cyclic AMP ↓	NS>CTX>CB>HP>MID>BSM
β *subtypes*			
β_1	G_S	Cyclic AMP ↑	CTX>HP>NS>CB
β_2	G_S	Cyclic AMP ↑	CB>NS>CTX>HP
β_3	G_S	Cyclic AMP ↑	Fat tissue

[a] C2, C4 and C10 refer to the cDNA clones which appear to encode for each receptor subtype; these cDNA clones or their rat homologues were also used for studies of the regional distribution of mRNA. Brain regions examined include: CTX, cerebral cortex; HP, hippocampus; BSM, brainstem; CB, cerebellum; MID, midbrain; NS, neostriatum.

system and on ion channels have not been identified.

The development of α_1AR ligand binding and NE-stimulated inositol phosphate accumulation have been examined in rat brain. Using [³H]prazosin, a selective α_1AR ligand, the levels of α_1AR ligand binding in rat whole brain (minus cerebellum) are reported to be approximately 10–20% of adult levels during the first week after birth; by PD14 the levels of ligand binding are 40–50% of adult levels and then increase to adult levels by approximately PD21 (Morris *et al.*, 1980). Similar results have been observed using [³H]WB4101 as a ligand although some regional differences were noted; in hippocampus, frontal cortex, and hypothalamus levels of [³H]WB4101 ligand binding do not reach adult levels until PD30 (Hartley and Seeman, 1983; Bylund and U'Prichard, 1983).

The appearance of α_1AR ligand binding appears to correlate qualitatively with the development of NE-stimulated inositol phosphate formation in rat cerebral cortex in that there is a general increase to adult levels in both measures during the first 3 weeks of postnatal development (Fig. 1.3). However, there are some differences (Fig. 1.3). Between PD1 and PD14 NE-stimulated inositol phosphate formation in cerebral cortex is approximately 20% of adult levels (whereas ligand binding has reached 40–50%) and then increases to adult levels by PD21–PD24 (Schoepp and Rutledge, 1985). Similar results were observed in hippocampal slices in a separate study (Nicolleti *et al.*, 1986). One other study also

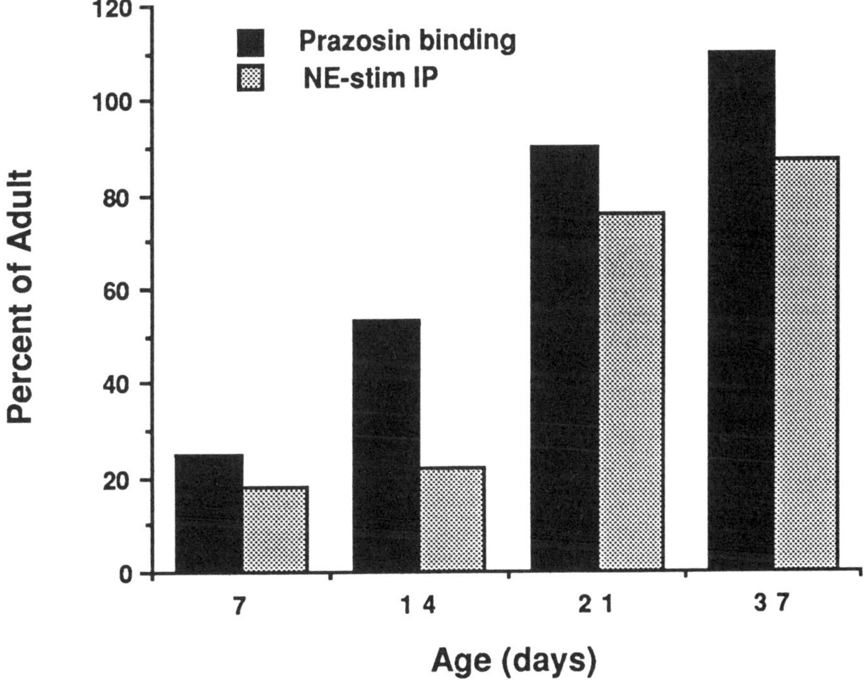

Fig. 1.3.   Development of [³H]prazosin binding and NE-stimulated inositol phosphate production in rat cerebral cortex. Levels of [³H]prazosin ligand binding were determined in homogenates of cerebral cortex; K_d = 0.15 nM; B_{max} = 125 fmol/mg protein. NE-stimulated inositol phosphate accumulation was determined in slices of cerebral cortex. Results are expressed as percentage of adult values. The data are derived from Schoepp and Rutledge (1985).

reported a complete lack of correlation between levels of ligand binding and NE-stimulated inositol phosphate accumulation (Rooney and Nahorski, 1987). These discrepancies cannot be explained by the time course for development of the effector enzyme as basal and glutamate- or acetylcholine-stimulation of inositol phosphate accumulation are expressed at adult levels or higher at birth. The development of the α subunit of G_q has not been examined.

Taken together the results suggest that the ontogeny of α_1AR coupling to and stimulation of inositol phosphate formation lags behind the developmental expression of the receptors themselves. However, it should be noted that the ligand binding studies used to measure the receptor levels were carried out either with [^{3}H]prazosin, which does not distinguish between α_{1A} and the α_{1B}, or [^{3}H]WB4101, which only has a 10–20-fold selectivity for these subtypes. By using these ligands, a combination of both α_{1A} and α_{1B} adrenergic receptor subtypes was measured. Because it is conceivable and likely that the α_{1A} and α_{1B} subtypes display individual developmental profiles, the discrepancy between the expression of α_1AR ligand binding and NE-stimulated inositol phosphate production may be due to the measurement of ligand binding to both subtypes when only α_{1B} appears to stimulate this second messenger system. Under the appropriate conditions, some ligands, such as [^{3}H]WB4101, can be used to distinguish one subtype from the other (Morrow and Creese, 1986; Minneman *et al.*, 1988). Using these conditions future studies should be able to differentiate the developmental time courses for the expression of α_{1A} and α_{1B} ligand binding sites.

As with the βAR subtypes, an alternative approach to studying the development of α_1AR subtypes is to examine mRNA levels using subtype-specific cDNA clones. Using an α_{1B} cDNA probe, McCune and Voigt (1991) examined the developmental expression of α_{1B} mRNA. Levels of α_{1B} mRNA increase severalfold during the first week of postnatal development and reach adult levels by about PD21 (McCune and Voigt, 1991). This time course corresponds to that of total α_1AR ligand binding. The development of α_{1A} mRNA in brain remains to be examined. By combining studies of α_1AR subtype mRNA levels with binding assays that use more selective ligands, the time course for the developmental expression of the α_{1A} and α_{1B} subtypes may be revealed. Such studies should elucidate the apparent discrepancies between α_1AR ligand binding and NE-stimulation of the phosphatidylinositol system in brain.

1.5 DEVELOPMENT OF α_2AR IN THE NERVOUS SYSTEM

The α_2ARs influence a variety of physiological functions including regulation of blood pressure, locomotor activity, antinociception, anxiety, memory and drug addiction (Aghajanian, 1978; Bylund and U'Prichard, 1983; Zeng and Lynch, 1991). α_2ARs were initially thought to be localized to presynaptic adrenergic terminals where they are known to mediate negative feedback inhibition of neurotransmitter release. Subsequently, α_2ARs have also been found on postsynaptic sites throughout the nervous system. Multiple α_2AR subtypes have been defined based on relative affinities of receptors for oxymetazoline and the α_1AR antagonist prazosin (Bylund *et al.*, 1988). The α_{2A} subtype has higher affinity for oxymetazoline relative to prazosin whereas the α_{2B} receptor has higher affinity for prazosin relative to oxymetazoline. An α_{2C} receptor subtype has also been defined which is closely related to α_{2B} based on relative affinities for these two compounds (Murphy and Bylund, 1988).

Molecular cloning studies have confirmed in part this subdivision of α_2AR subtypes (Table 1.1). An α_{2A} subtype has been isolated and is also referred to as C10 because it is located on chromosome number 10 (Kobilka *et al.*, 1987). Additional α_2AR subtypes have been cloned but their relationship to pharmacologically defined receptor subtypes is not clear. One subtype referred to as α_2-C4 (located on chromosome number 4) was isolated from human kidney and appears to have ligand binding characteristics most similar to that of an α_{2B} receptor (Regan *et al.*, 1988). However, based on its regional distribution in peripheral tissues (it is not present in neonatal rat lung) and that it is glycosylated, this receptor appears to be more closely related to the α_{2C} subtype (Lorenz *et al.*, 1990; Zeng and Lynch, 1991). A third subtype was subsequently isolated and is referred to as either RNGα_2 (Zeng *et al.*, 1990) or α_2-C2 (Lomasney *et al.*, 1990). This receptor also has ligand binding properties characteristic of an α_{2B} subtype, and this assignment is supported by its presence in neonatal rat lung and by the fact that this receptor is not glycosylated.

Activation of $\alpha_2 ARs$ has been shown to regulate cellular function in a number of ways depending on the neuronal system (Table 1.1). Activation of α_{2A} receptors inhibits the adenylate cyclase second messenger system in brain (Duman and Enna, 1986). In addition, the indirect effect of αAR activation to enhance the neurotransmitter receptor-stimulation of cyclic AMP in brain appears to be mediated in part by an $\alpha_2 AR$ (Duman *et al.*, 1986). Finally, stimulation of $\alpha_2 ARs$ activates K^+ channels in locus coeruleus neurons (Williams *et al.*, 1985) and inhibits Ca^{2+} channels (Bean, 1989). In most cases $\alpha_2 ARs$ couple with pertussis toxin-sensitive G proteins (G_i/G_o) to influence both ion channels and adenylate cyclase.

A number of studies have described the developmental expression of $\alpha_2 AR$ binding but in all cases [^{3}H]clonidine was used. [^{3}H]clonidine is a non-selective ligand for $\alpha_2 AR$ subtypes and also labels nonadrenergic imidazoline binding sites (for discussion of imidazoline binding sites see Michel *et al.*, 1988 and Michel and Insel, 1989). In addition, clonidine is an agonist and therefore its affinity is influenced by the interaction of the receptor

with G proteins. A more selective ligand would be [^{3}H]rauwolscine which only labels $\alpha_2 AR$ and displays some selectivity for the subtypes.

The multiple limitations of [^{3}H]clonidine ligand binding complicate the interpretation of developmental studies using this ligand. However, with these problems in mind it is useful to examine the results of these studies. In whole rat brain minus cerebellum, levels of [^{3}H]clonidine binding at birth are approximately 35% of adult levels, increase to about 50% of the adult level by PD10, and reach adult levels by PD20 (Morris *et al.*, 1980). This study reports that [^{3}H]clonidine labels a single class of binding sites with a K_d of 2.3 nM and that the developmental increase reflects an increase in the total number of binding sites. The same study reports that levels of NE do not reach adult levels until 5–6 weeks after birth. The results indicate that the expression of [^{3}H]clonidine binding sites precedes the expression of adult levels of neurotransmitter, a trend that has also been observed for the other adrenergic receptor subtypes.

Subsequent studies reported a similar pattern of development for [^{3}H]clonidine binding in rat cerebral cortex (Nomura *et al.*, 1982) and more specifically in mesolimbic and hypothalamic regions (Hartley and Seeman, 1983). In hippocampus, levels of [^{3}H]clonidine binding increase to adult levels by PD10, and in frontal cortex, adult levels of ligand binding are reached 5–6 weeks after birth (Fig. 1.4). The latter study also reported that [^{3}H] clonidine labels a single class of sites with a K_d in the nanomolar range (1.5 nM). However, Nomura *et al.* (1982) reported that in neonates [^{3}H]clonidine binding displays a subnanomolar K_d (0.27 nM) whereas on PD7 and in adults the K_d is 1.6 nM.

In subsequent studies Nomura *et al.* (1984) investigated the nature of the high and low affinity [^{3}H]clonidine binding sites observed at different times of development. The presence of a single class of low affinity site in neonatal rat cortex was confirmed, but after the first week of development a low as well as the high

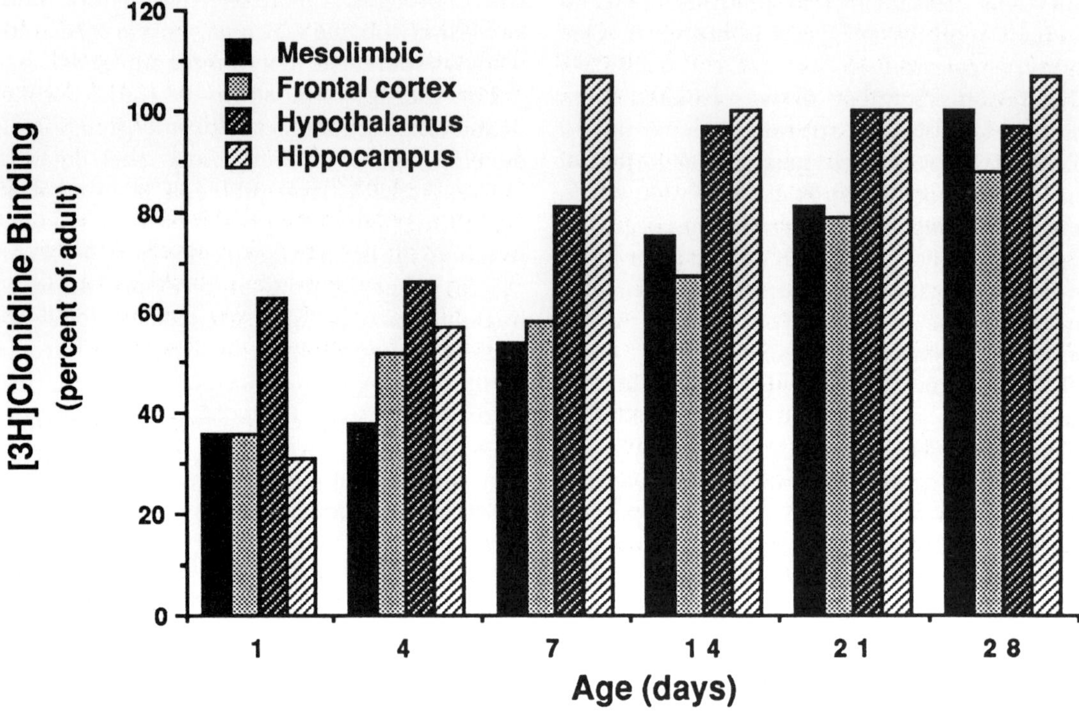

Fig. 1.4.   Development of [³H]clonidine binding in different rat brain regions. The K_d was approximately 1.5 nM in all brain regions. The B_{max} for each brain region was (in fmol/mg protein): mesolimbic, 160; frontal cortex, 120; hypothalamus, 115; hippocampus, 75. The results are expressed as percentages of adult values and are derived from Hartley and Seamon (1983).

affinity [³H]clonidine binding site (0.7 and 7.0 nM, respectively) was seen. Both sites are present at approximately 60% of adult levels on PD7 and reach adult levels by PD30. In a later study (Kitamura *et al.*, 1989) it was shown that in the presence of GTP, which converts the majority of receptors into low affinity binding sites, there is little or no change in levels of [³H]clonidine binding during development. In contrast, in the presence of Mn^{2+}, which increases the level of high affinity binding, there is an increase in the expression of [³H] clonidine binding during development. From these results the authors concluded that expression of the high affinity ligand–receptor–G protein ternary complex is the rate limiting step in the development of functional clonidine binding sites. This pattern of expression may be related to the expression of G

protein subunits which couple with clonidine binding sites although the G protein subunits have been reported to be present at 35–50% of adult levels at birth (section 1.6).

One study has examined the development of functional clonidine binding sites by measuring the expression of K^+-induced [³H]NE release from cerebral cortical slices and inhibition of neurotransmitter release by clonidine (Nomura *et al.*, 1982). High K^+ induced release of [³H]NE is observed on ED18 and increases to near adult levels by PD7. Clonidine inhibition of neurotransmitter release is not observed on PD1 but by PD7 clonidine inhibition of [³H]NE release is observed and is approximately 70% of that seen in adult brain slices. These results indicate that clonidine receptor binding sites are functionally expressed at this time during development which corresponds to the appearance of low affinity [³H]clonidine binding sites. Moreover, these results also demonstrate the developmental expression of the presynaptic α_2AR which regulates the release of NE.

While measurement of α_2AR subtypes by ligand binding poses some difficulty due to the lack of selective ligands for these receptors, expression of receptor mRNA, specifically α_2-C10 and α_2-C4 receptor mRNA, has been examined using specific cDNA probes (McCune and Voigt, 1991). The developmental time course of α_2-C4 and α_2-C10 has been confirmed and extended in our laboratory (Fig. 1.5). Levels of α_2-C4 are present, but are very low on ED18, increase several fold by PD8, and increase to adult levels by PD16. These results are similar to the time course for development of [³H]clonidine ligand binding described above. In contrast, levels of α_2-C10

are found to be highest at birth and then gradually decrease to adult levels by PD16. The high levels of α_2-C10 mRNA suggest that levels of α_{2A} receptor protein and ligand binding are present at higher levels at birth than in adults. However, it is possible that turnover of mRNA and/or receptor protein is higher in neonates and that levels of functional receptor protein are at low levels. Analysis of α_{2A} ligand binding or immunoblot analysis of α_{2A} receptor protein will be required to determine the nature of this discrepancy.

1.6 DEVELOPMENT OF G PROTEINS AND EFFECTOR SYSTEMS IN THE NERVOUS SYSTEM

The regulation of cellular function by adrenergic receptors and other members of this receptor family occurs through coupling of

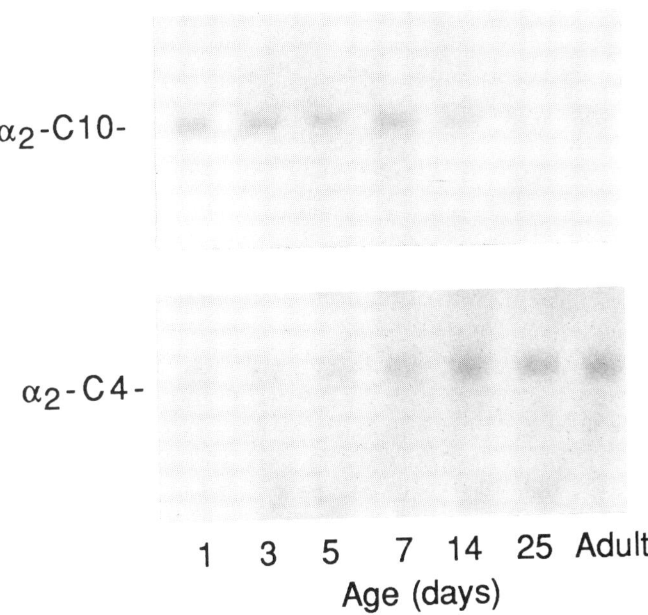

Fig. 1.5.   Development of α_2-C4 and α_2-C10 mRNA in rat whole brain. Total RNA was isolated from rat whole brain at different times of development and subjected to hybridization blot analysis. The RNA (20 µg) was fractionated on an agarose gel, transferred to nitrocellulose, and then hybridized with ³²P-labeled α_2-C4 or α_2-C10 cDNA (probes were kindly provided by S. Lanier). The filters were then exposed for 3 days without an intensifying screen.

receptors to transmembrane signalling proteins known as GTP binding proteins or G proteins (Stryer and Bourne, 1986; Gilman, 1987; Simon *et al.*, 1991). The interaction of ligand bound receptors with G proteins results in the hydrolysis of GTP and then coupling of G proteins with effector systems. The effector systems regulated by receptor-coupled G proteins include second messenger enzymes such as adenylate cyclase, phospholipase C and phospholipase A_2, and K^+ and Ca^{2+} channels (Table 1.1). Activation of these second messengers results in regulation of intracellular pathways, such as protein kinases, which have acute or immediate effects on cellular function as well as long-term effects including regulation of gene transcription. Developmental expression of receptors which are functional depends on the expression of the G proteins and effector systems which couple these receptors to intracellular second messenger pathways. For this reason it is useful to review studies of the development of specific G proteins and effector systems.

G proteins are heterotrimers composed of α, β and γ subunits (Stryer and Bourne, 1986; Gilman, 1987; Simon *et al.*, 1991). Specific α subunits appear to exist for each effector system whereas the $\beta\gamma$ subunits appear to subserve a number of α subunits. For example, receptor stimulation and inhibition of adenylate cyclase is mediated by $G_{S\alpha}$ and $G_{i\alpha}$, respectively, while regulation of phospholipase C is mediated by $G_{o\alpha}$. Receptor regulation of ion channels is also thought to be mediated by subtypes of $G_{i\alpha}$ and/or $G_{o\alpha}$. Studies of the structure and function of G proteins have been aided by the use of cholera and pertussis toxins which catalyze the ADP-ribosylation of the α subunits of G_s and G_i/G_o, respectively. Molecular cloning studies have identified at least 16 G protein α subunits, at least three β subunits, and four types of γ subunits, all encoded by separate genes (Simon *et al.*, 1991). However, with the exception of $G_{S\alpha}$, the receptor/effector system regulated by each of these subtypes has not yet been determined.

The development of G protein subunits in brain has been examined using a variety of approaches. Levels of $G_{o\alpha}$ and G_β immunoreactivity in cerebral cortex were reported to increase approximately two-fold between ED16 and birth, at which time levels of these G proteins subunits were equal to the levels observed in adults (Kitamura *et al.*, 1989). A somewhat different time course was reported in forebrain (Milligan *et al.*, 1987); levels of $G_{o\alpha}$, G_β and $G_{i\alpha}$ immunoreactivity were reported to be present at 25, 35 and 50%, respectively, of adult levels and then increased to adult levels by PD30. A third study reported a similar developmental pattern for $G_{o\alpha}$ immunoreactivity in whole brain (Chang *et al.*, 1988). Levels of pertussis toxin-catalyzed ADP-ribosylation are reported to be approximately 35% of adult levels at birth and increase to adult levels by PD16, a somewhat slower pattern of development compared with levels of $G_{i\alpha}$ and $G_{o\alpha}$ immunoreactivity.

The developmental expression of mRNA for these G protein subunits has also been studied using subunit specific cDNA probes (Fig. 1.6, Duman *et al.*, 1989). Levels of $G_{o\alpha}$ and G_β in whole brain are present at or above adult levels at birth, increase to 150–200% of adult levels by PD3–PD7, and then decrease to adult levels by PD14. Levels of $G_{i1\alpha}$ and $G_{i2\alpha}$ mRNA follow a similar pattern of development although the magnitude of increase above adult levels is not as great. The differences in development of G protein immunoreactivity and mRNA may result from the different brain regions being examined, although levels of $G_{o\alpha}$ mRNA and immunoreactivity were both determined in rat whole brain. Alternatively, the stability or rate of turnover of mRNA and protein may change during development. One other possibility is that the antibodies used recognize more than one G protein subtype whereas the cDNA probes are relatively more selective. Thus, although sufficient levels of some G protein subtypes are present during early stages it is possible that some of the

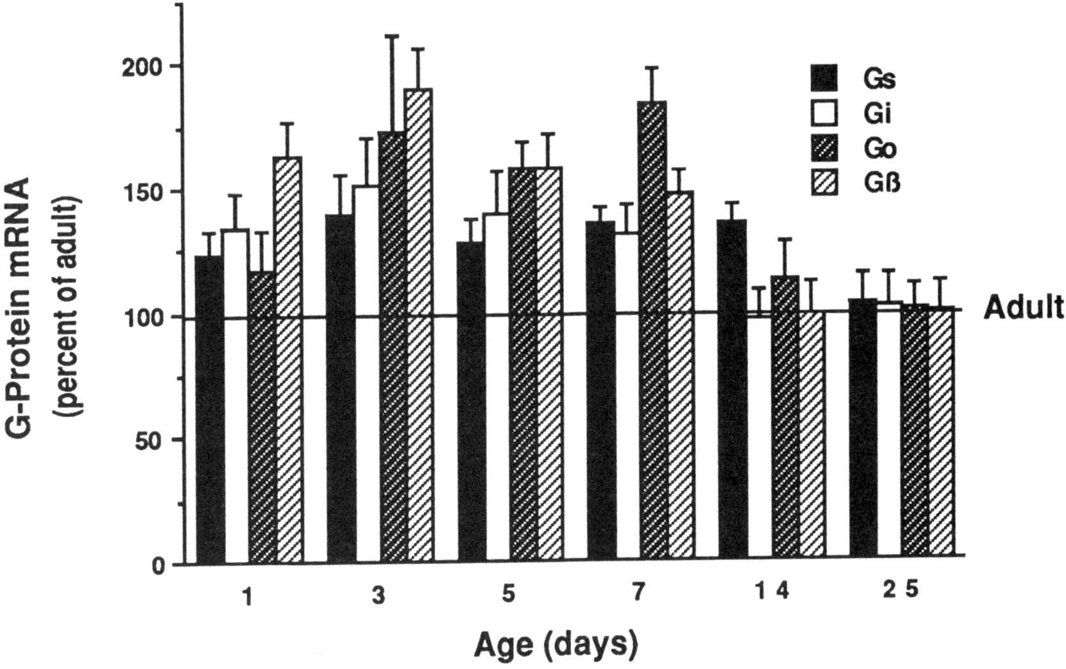

Fig. 1.6. Development of G protein subunit mRNA in rat brain. Total RNA was isolated from whole brain at different times of development and submitted to hybridization blot analysis for levels of $G_{s\alpha}$, $G_{i2\alpha}$, $G_{o\alpha}$, and G_β. The results are expressed as percentages of adult values and are derived from Duman *et al.* (1989).

many other subtypes are not expressed until later periods of development.

The measurement of guanine nucleotide-stimulated adenylate cyclase has been used as a measure of the development of functional $G_{s\alpha}$. These studies have demonstrated that levels of Gpp(NH)p-stimulated adenylate cyclase are approximately 35% of adult levels at birth and then gradually increase to adult levels by PD21 (section 1.3). The development of $G_{s\alpha}$ has been further studied by measurement of cholera toxin catalyzed ADP-ribosylation of $G_{s\alpha}$ in homogenates of cerebral cortex and by measuring levels of mRNA for $G_{s\alpha}$. Levels of cholera toxin-stimulated ADP-ribosylation of $G_{s\alpha}$ at birth are reported to be present at adult levels, increase about two-fold by day 12 and then gradually decrease to adult levels (Kitamura *et al.*, 1989). Similarly, levels of $G_{s\alpha}$ mRNA in whole brain are found

to be expressed at adult levels on PD1, increase slightly, and then return to adult levels by PD21 (Fig. 1.6, Duman *et al.*, 1989). These results indicate that adult levels of $G_{s\alpha}$ are present at birth and support the conclusion that development of functional βAR is dependent on expression of receptors.

Although little information is available on the development of most of the effector systems, the development of adenylate cyclase has been studied. At least two forms of the enzyme have been characterized in brain, calmodulin-sensitive (type 1) and calmodulin-insensitive (type 2) adenylate cyclase, and cDNA clones for each have been identified (Krupinski *et al.*, 1989; Feinstein *et al.*, 1991). A number of studies have examined the development of adenylate cyclase and under the conditions used, the level of enzyme activity measured would have been a combination of both enzyme types. Basal and Mn^{2+}-stimulated enzyme activity are reported to be present at 30–40% of adult levels at birth and increase to adult levels by PD21 (Harden *et al.*, 1977; Keshles and Levitski, 1984). The development of type 2 adenylate cyclase mRNA in whole brain has been studied in our laboratory. Expression of levels of type 2 adenylate cyclase mRNA follows a time course similar to that of enzyme activity (Fig. 1.7). Additional studies will be needed to examine the development of type 1 adenylate cyclase mRNA.

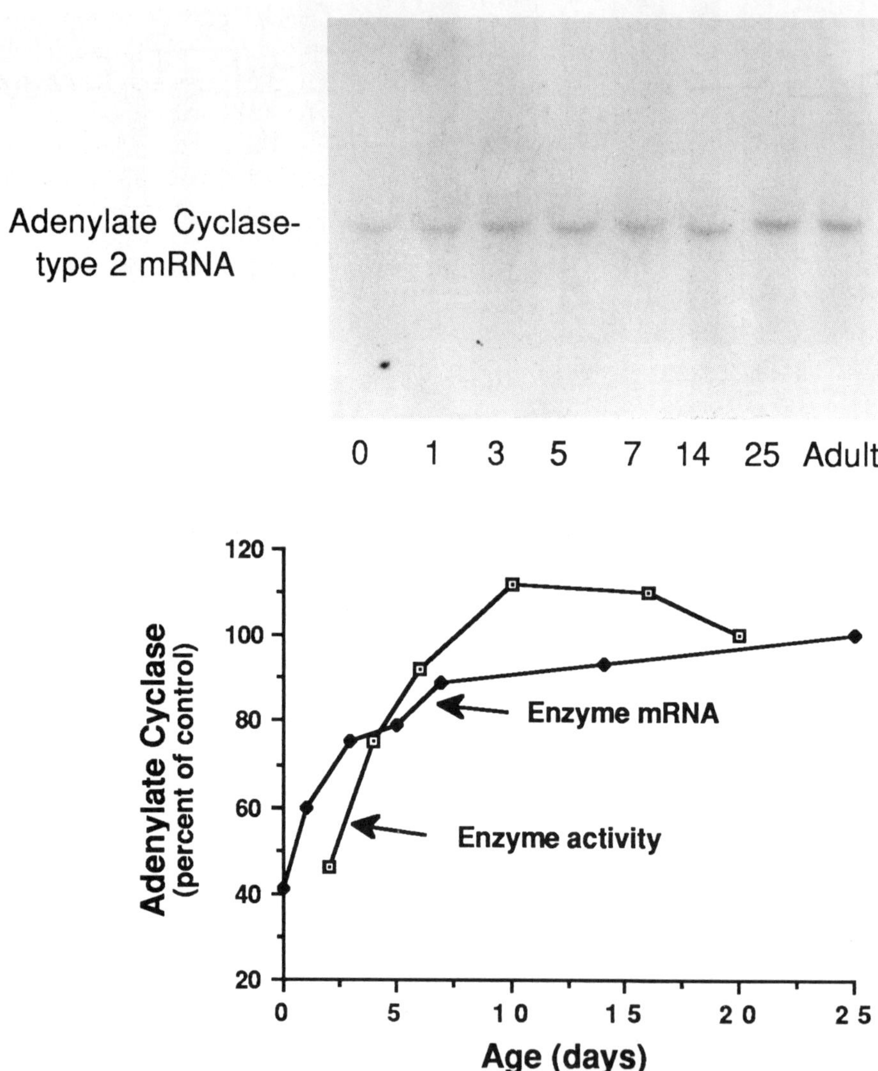

Fig. 1.7.   Development of adenylate cyclase enzyme activity and type 2 mRNA in rat brain. Levels of Mn^{2+} plus forskolin-stimulated adenylate cyclase activity were determined in homogenates of cerebral cortex and are expressed as percentages of levels at PD21 (derived from Keshles and Levitski, 1984). Levels of adenylate cyclase type 2 mRNA were determined in rat whole brain and are expressed as percentages of adult levels. A 20 µg sample of total RNA was subjected to hybridization blot analysis using ^{32}P-labeled adenylate cyclase type 2 cDNA (kindly provided by R. Reed, Johns Hopkins).

Although the development of receptor-stimulated inositol phosphate accumulation in brain has been examined (section 1.4), the ontogeny of phospholipase C activity and mRNA has not been studied. Similarly, the expression of other phospholipases as well as K^+ and Ca^{2+} channels have not been examined. In addition, it is conceivable that adrenergic receptors as well as other neurotransmitter receptors regulate additional, as yet unidentified, effectors, some of which are expressed only during early stages of development. Examination of these possibilities must await future studies.

1.7 ADRENERGIC RECEPTORS AND IMPLICATIONS FOR DEVELOPMENT

The early appearance of NE during gestation and the extensive innervation of the NE neurotransmitter system throughout the nervous system have led to the hypothesis that NE may act as a neurotrophic factor. Consistent with this hypothesis are reports that NE enhances the growth of neurons and glia in cerebellar explants (Vernadakis and Gibson, 1974). Destruction of noradrenergic nerve terminals in neonatal rats leads to significant morphological changes in both the cerebral cortex and the cerebellum, and the extent of these changes correlates with the reduction in levels of norepinephrine (Konkol *et al.*, 1978; Felten *et al.*, 1982; Lovell, 1982). Such changes may underlie the influence of neonatal 6-OHDA on locomotor activity (Pappas *et al.*, 1975) and attention deficits (Mason and Iverson, 1979). Additional studies will be needed to test this hypothesis further and to elucidate the role of NE in central nervous system development.

Although these findings suggest that the noradrenergic system may have neurotrophic actions, the mechanisms by which NE could regulate neuronal growth have not been identified. Ornithine decarboxylase (ODC), which has been shown to be obligatory for neuronal development, is under βAR regulation during development (Morris *et al.*, 1983;

Morris and Slotkin, 1985). ODC catalyzes the initial reaction in the synthesis of polyamines which have been shown to play a significant role in the development and maturation of cells. βAR-stimulation of ODC expression appears to be mediated by $β_2AR$ and activation of the cyclic AMP second messenger system. Maximal βAR-stimulated ODC activity correlates with major periods of replication, differentiation and outgrowth during development. Although these results support the hypothesis that ODC activity is critical to neuronal development and maturation, that βAR-stimulated ODC is required for such development is unclear. Additional studies will be required to further elucidate the role of βAR-stimulated ODC in neuronal development.

1.8 SUMMARY AND FUTURE PROSPECTS

The developmental expression of most adrenergic receptor subtypes matures to adult levels by the end of the third week after birth, whereas adult levels of NE are not attained until approximately 6 weeks after birth. This has led investigators to conclude that the expression of adrenergic receptors precedes that of the NE neurotransmitter system. However, it has also been noted that although adult levels of NE, its biosynthetic enzymes, and re-uptake sites do not attain adult levels until a later postnatal time, the adult pattern of noradrenergic innervation is achieved much earlier, by the first to second week after birth. The development of adrenergic receptors appears to correspond more closely with this latter time course. According to this interpretation, the expression of neurotransmitter and receptors may develop in parallel although it is difficult to identify whether NE or adrenergic receptors appears first during development. One possibility is that expression of NE and adrenergic receptors are independent events and that continued expression of receptors is dependent on the presence of neurotransmitter. Alternatively, expression of adrenergic receptors may be turned on earlier so that the cells expressing these receptors are fully functional when NE innervation occurs.

The embryonic expression and further development of both adrenergic receptors and the norepinephrine neurotransmitter system are controlled by regulation of gene expression. The development of brain and other tissues is controlled by turning on and off sets of genes encoding proteins needed for each period of development. Thus, to understand fully the development of adrenergic receptors, the gene transcription factors which control expression of these receptors must be identified. Regulation of β_1AR and β_2AR gene transcription has been examined in cultured cells. The genes for both receptors have been shown to be under the control of thyroid hormone (Bahouth, 1991), and β_2AR gene expression is also regulated by glucocorticoid hormone (Hadcock and Malbon, 1988). It is possible that expression of fetal hormones plays a role in the ontogeny of these receptors. Alternatively, maternal levels of these hormones may regulate fetal gene transcription since these hormones are lipophilic and therefore can penetrate the placental barrier.

In addition, both the β_1AR and β_2AR genes contain a cyclic AMP response element (CRE), and β_2AR gene expression has been shown to be regulated by the cyclic AMP system, presumably through activation of a CRE binding protein or CREB (Collins *et al.*, 1990). The regulation and function of these gene transcription factors and a variety of other classes of transcription factors are being studied in cultured cells although the role of cyclic AMP in the developmental expression of adrenergic receptors has not been examined. In addition, rapid advances are being made in studies of the expression and function of a variety of nerve growth factors which may also participate in the development of the central nervous system. Identification of the transcription factors, growth factors, and other determinants of gene expression will contribute to a more comprehensive understanding of both the development of adrenergic recept-

ors and regulation of these receptors in adults. These studies will also help elucidate the mechanisms which underlie perturbations of adrenergic receptor function that occur during development, some of which could underlie abnormalities of receptor function in adults.

REFERENCES

Aghajanian, G.K. (1978) Tolerance of locus coeruleus neurons to morphine and suppression of withdrawal response by clonidine. *Nature*, **276**, 186–8.

Bahouth, S. (1991) Thyroid hormones transcriptionally regulate the β1-adrenergic receptor gene in cultured ventricular myocytes. *J. Biol. Chem.*, **266**, 15863–9.

Bean, B.P. (1989) Neurotransmitter inhibition of neuronal calcium currents by changes in channel voltage dependence. *Nature*, **340**, 153–6.

Berridge, C.W. and Dunn, A.J. (1989) Restraint-stress-induced changes in exploratory behavior appear to be mediated by norepinephrine-stimulated release of CRF. *J. Neurosci.*, **9**, 3513–21.

Berridge, M.J. and Irvine, R.F. (1989) Inositol phosphates and cell signalling. *Nature*, **341**, 197–205.

Bruinink, A. and Lichtensteiger, W. (1984) β-Adrenergic binding sites in fetal rat brain. *J. Neurochem.*, **43**, 578–81.

Bylund, D.B. and U'Prichard, D.C. (1983) Characterization of α_1- and α_2-adrenergic receptors. *Int. Rev. Neurobiol.*, **24**, 343–431.

Bylund, D.B., Ray-Prenger, C. and Murphy, T.J. (1988) Alpha-2A and alpha-2B adrenergic receptor subtypes: antagonist binding in tissues and cell lines containing only one subtype. *J. Pharmacol. Exp. Ther.*, **245**, 600–7.

Chang, K.-J., Puch, W., Blanchard, S.G. *et al.* (1988) Antibody specific to the α subunit of the guanine nucleotide-binding regulatory protein G_o: developmental appearance and immunocytochemical localization in brain. *Proc. Natl. Acad. Sci. USA*, **85**, 4929–33.

Collins, S., Altschmied, J., Herbsman, O. *et al.* (1990) A cAMP response element in the β_2-adrenergic receptor gene confers transcriptional autoregulation by cAMP. *J. Biol. Chem.*, **265**, 19330–5.

Cotecchia, S., Schwinn, D.A., Randall, R.R. *et al.* (1988) Molecular cloning and expression of the cDNA for the hamster α_1-adrenergic receptor. *Proc. Natl. Acad. Sci. USA*, **85**, 7159–63.

Coyle, J.T. (1977) Biochemical aspects of neurotransmission in the developing brain. *Int. Rev. Neurobiol.*, **20**, 65–103.

Dixon, R.A.F., Kobilka, B.K., Strader, D.J. *et al.* (1986) Cloning of the gene and cDNA for mammalian β-adrenergic receptor and homology with rhodopsin. *Nature*, **321**, 75–9.

Duman, R.S. and Enna, S.J. (1986) A procedure for measuring α_2-adrenergic receptor-mediated inhibition of cyclic AMP accumulation in rat brain slices. *Brain Res.*, **384**, 391–4.

Duman, R.S., Karbon, E.W., Harrington, C. and Enna, S.J. (1986) An examination of the involvement of phospholipases A_2 and C in the α-adrenergic and γ-aminobutyric acid receptor modulation of cyclic AMP accumulation in rat brain slices. *J. Neurochem.*, **47**, 800–10.

Duman, R.S., Saito, N. and Tallman, J.F. (1989) Development of β-adrenergic receptor and G protein messenger RNA in rat brain. *Mol. Brain Res.*, **5**, 289–96.

Erdtsieck-Ernste, B.H.W., Feenstra, M.G.P. and Boer, G.J. (1991) Pre- and postnatal developmental changes of adrenoceptor subtypes in rat brain. *J. Neurochem.*, **57**, 897–903.

Feinstein, P.G., Schrader, K.A., Bakalyar, H.A. *et al.* (1991) Molecular cloning and characterization of a Ca^{2+}/calmodulin insensitive adenylyl cyclase from rat brain. *Proc. Natl. Acad. Sci. USA*, **88**, 10173–7.

Felten, D.L., Hallman, K. and Johsson, G. (1982) Evidence for a neurotrophic role of noradrenaline neurons in the postnatal development of rat cerebral cortex. *J. Neurocytol.*, **11**, 119–35.

Foote, S.L., Bloom, F.E. and Aston-Jones, G. (1983) Nucleus locus ceruleus: new evidence of anatomical and physiological specificity. *Physiol. Rev.*, **63**, 844–914.

Frielle, T., Collins, S., Daniel, K.W. *et al.* (1987) Cloning of the cDNA for the human β_1-adrenergic receptor. *Proc. Natl. Acad. Sci. USA*, **84**, 7920–4.

Gilman, A.G. (1987) G-proteins: transducers of receptor-generated signals. *Annu. Rev. Biochem.*, **56**, 615–49.

Hadcock, J.R. and Malbon, C.C. (1988) Regulation of β-adrenergic receptors by 'permissive' hormones: glucocorticoids increase steady-state levels of receptor mRNA. *Proc. Natl. Acad. Sci. USA*, **85**, 8415–19.

Han, C., Abel, P.W. and Minneman, K.P. (1987) α_1-adrenoceptor subtypes linked to different mechanisms for increasing intracellular Ca^{2+} in smooth muscle. *Nature*, **329**, 333–5.

Harden, T.K., Wolfe, B.B., Sporn, J.R. *et al.* (1977) Ontogeny of β-adrenergic receptors in rat cerebral cortex. *Brain Res.*, **125**, 99–108.

Hartley, E. and Seeman, P. (1983) Development of receptors for dopamine and noradrenaline in rat brain. *Eur. J. Pharmacol.*, **91**, 391–7.

Heninger, G.R. and Charney, D.S. (1987) Mechanism of action of antidepressant treatments: implications for the etiology and treatment of depressive disorders, in *Psychopharmacology, the Third Generation of Progress* (ed. H.Y. Meltzer), Raven Press, New York, pp. 535–44.

Johnson, E.W., Wolfe, B.B. and Molinoff, P.B. (1989) Regulation of subtypes of β-adrenergic receptors in rat brain following treatment with 6-hydroxydopamine. *J. Neurosci.*, **9**, 2297–305.

Johnston, M.V. and Coyle, J.T. (1980) Ontogeny of neurochemical markers for noradrenergic, GABAergic, and cholinergic neurons in neocortex lesioned with methylazoxymethanol acetate. *J. Neurochem.*, **34**, 1429–41.

Keshles, O. and Levitzki, A. (1984) The ontogenesis of β-adrenergic receptors and of adenylate cyclase in the developing rat brain. *Biochem. Pharmacol.*, **33**, 3231–3.

Kitamura, Y., Mochii, M., Kodama, R. *et al.* (1989) Ontogenesis of α_2-adrenoceptor coupling with GTP-binding proteins in the rat telencephalon. *J. Neurochem.*, **53**, 249–57.

Kobilka, B.K., Matsui, H., Kobilka, T.S. *et al.* (1987) Cloning, sequencing, and expression of the gene coding for the human platelet α_2-adrenergic receptor. *Science*, **238**, 650–6.

Konkol, R.J., Bendeich, E.G. and Breese, G.R. (1978) A biochemical and morphological study of the altered growth pattern of central catecholamine neurons following 6-hydroxydopamine. *Brain Res.*, **140**, 125–35.

Krupinski, J., Coussen, F., Bakalyar, H.A. *et al.* (1989) Adenylyl cyclase amino acid sequence: possible channel- or transporter-like structure. *Science*, **244**, 1558–64.

Lomasney, J.W., Lorenz, W., Allen, L.F. *et al.* (1990) Expansion of the α_2-adrenergic receptor family: cloning and characterization of a human α_2-adrenergic receptor subtype, the gene for which is located on chromosome 2. *Proc. Natl. Acad. Sci. USA*, **87**, 5094–8.

Lomasney, J.W., Cotecchia, S., Lorenz, W. *et al.* (1991) Molecular cloning and expression of the cDNA for the α_{1A}-adrenergic receptor. *J. Biol. Chem.* **266**, 6365–9.

Lorenz, W., Lomasney, J.W., Collins, S. *et al.* (1990) Expression of three α_2-adrenergic receptor subtypes in rat tissues: implications for α_2 receptor classification. *Mol. Pharmacol.*, **38**, 599–603.

Lovell, J. (1982) Effects of 6-hydroxydopamine-induced norepinephrine depletion on cerebellar development. *Dev. Neurosci.*, **5**, 359–68.

Mason, S.T. and Iversen, S.D. (1979) Theories of the dorsal bundle extension effect. *Brain Res. Rev.*, **1**, 107–37.

McCune, S.K. and Voigt, M.M. (1991) Regional brain distribution and tissue ontogenic expression of a family of alpha-adrenergic receptor mRNAs in the rat. *J. Mol. Neurosci.*, **3**, 29–37.

Michel, M.C. and Insel, P.A. (1989) Are there multiple imidazoline binding sites? *Trends Pharmacol. Sci.*, **10**, 342–5.

Michel, M.C., Brodde, O.-E., Schnepel, B. *et al.* (1988) [^{3}H]Idazoxan and some other α_2-adrenergic drugs also bind with high affinity to a nonadrenergic site. *Mol. Pharmacol.*, **35**, 324–30.

Milligan, G., Streaty, R.A., Gierschik, P. *et al.* (1987) Development of opiate receptors and GTP-binding regulatory proteins in neonatal rat brain. *J. Biol. Chem.*, **262**, 8626–30.

Minneman, K.P. (1981) Adrenergic receptor molecules, in *Neurotransmitter Receptors, Part 2 (Receptors and Recognition. Series B, Vol. 10)* (eds H.I. Yamamura and S.J. Enna), Chapman & Hall, London.

Minneman, K.P., Dibner, M.D., Wolfe, B.B. and Molinoff, P.B. (1979) β_1- and β_2-adrenergic receptors in rat cerebral cortex are independently regulated. *Science*, **204**, 866–8.

Minneman, K.P., Han, C. and Abel, P.W. (1988) Comparison of α_1-adrenergic receptor subtypes distinguished by chlorethylclonidine and WB 4101. *Mol. Pharmacol.*, **33**, 509–14.

Moore, R.Y. and Bloom, R.E. (1979) Central catecholamine neuron systems: anatomy and physiology of the norepinephrine and epinephrine systems. *Annu. Rev. Neurosci.*, **2**, 113–68.

Morris, G. and Slotkin, T.A. (1985) Beta-2 adrenergic control of ornithine decarboyxlase activity in brain regions of the developing rat. *J. Pharmacol. Exp. Ther.*, **233**, 141–7.

Morris, G., Seidler, F.J. and Slotkin, T.A. (1983) Stimulation of ornithine decarboyxlase by histamine or norepinephrine in brain regions of the developing rat: evidence for biogenic amines as trophic agents in neonatal brain development. *Life Sci.*, **32**, 1565–71.

Morris, M.J., Dausse, J.-P., Devynck, M.-A. and Meyer, P. (1980) Ontogeny of α_1 and α_2-adrenoceptors in rat brain. *Brain Res.*, **190**, 268–71.

Morrow, A.L. and Creese, I. (1986) Characterization of α_1-adrenergic receptor subtypes in rat brain: a reevaluation of [^{3}H]WB4104 and [^{3}H]prazosin binding. *Mol. Pharmacol.*, **29**, 321–30.

Murphy, T.J. and Bylund, D.B. (1988) Characterization of α_2-adrenergic receptors in the OK cell, an opossum kidney cell line. *J. Pharmacol. Exp. Ther.*, **244**, 571–8.

Nicoletti, F., Iadarola, M.J., Wroblewski, J.T. and Costa, E. (1986) Excitatory amino acid recognition sites coupled with inositol phospholipid metabolism: developmental changes and interaction with α_1-adrenoceptors. *Proc. Natl. Acad. Sci. USA*, **83**, 1931–5.

Nishizuka, Y. (1988) The molecular heterogeneity of protein kinase C and its implications for cellular regulation. *Nature*, **334**, 661–5.

Nomura, Y., Yotsumoto, I. and Nishimoto, Y. (1982) Ontogeny of influence of clonidine on high potassium-induced release of noradrenaline and specific [^{3}H]clonidine binding in the rat brain cortex. *Dev. Neurosci.*, **5**, 198–204.

Nomura, Y., Kawai, M., Mita, K. and Segawa, T. (1984) Developmental changes of cerebral cortical [^{3}H]clonidine binding in rats: influences of guanine nucleotide and cations. *J. Neurochem.*, **42**, 1240–5.

Ordway, G.A., Gambarana, C. and Frazer, A. (1988) Quantitative autoradiography of central beta adrenoceptor subtypes: comparison of the effects of chronic treatment with desipramine or centrally administered l-isoproterenol. *J. Pharmacol. Exp. Ther.*, **247**, 379–89.

Pappas, B.A., Peters, D.A.V., Sobrian, S.K. *et al.* (1975) Early behavioural and catecholaminergic effects of 6-hydroxydopamine and guanethidine in the neonatal rat. *Pharmacol. Biochem. Behav.*, **3**, 683–5.

Perkins, J.P. and Moore, M.M. (1973) Characterization of the adrenergic receptors mediating the rise in cyclic 3',5'-adenosine monophosphate in rat cerebral cortex. *J. Pharmacol. Exp. Ther.*, **185**, 371–8.

Pittman, R.N., Minneman, K.P. and Molinoff, P.B. (1980) Ontogeny of β_1- and β_2-adrenergic receptors in rat cerebellum and cerebral cortex. *Brain Res.*, **188**, 357–68.

Regan, J.W., Kobilka, T.S., Yang-Feng, T.L. *et al.* (1988) Cloning and expression of a human kidney cDNA for an α_2-adrenergic receptor subtype. *Proc. Natl. Acad. Sci. USA*, **85**, 6301–5.

Rooney, T.A. and Nahorski, S.R. (1987) Postnatal ontogeny of agonist and depolarization-induced phosphoinositide hydrolysis in rat cerebral cortex. *J. Pharmacol. Exp. Ther.*, **243**, 333–41.

Schoepp, D.D. and Rutledge, C.O. (1985) Comparison of postnatal changes in alpha1-adrenoceptor binding and adrenergic stimulation of phosphoinositide hydrolysis in rat cerebral cortex. *Biochem. Pharmacol.*, **34**, 2705–11.

Simon, M.I., Strathmann, M.P. and Gautam, N. (1991) Diversity of G proteins in signal transduction. *Science*, **252**, 802–8.

Stryer, L. and Bourne, H.R. (1986) G proteins: a family of signal transducers. *Annu. Rev. Cell Biol.*, **2**, 391–419.

Vernadakis, A. and Gibson, D.A. (1974) Role of neurotransmitter substances in neural growth, in *Perinatal Pharmacology: Problems and Priorities* (eds J. Dancis and J.C. Hwang), Raven Press, New York, pp. 65–76.

Waddington, G. and Banks, P. (1981) The development of pre- and postsynaptic components of the noradrenergic system in the rat cerebellum. *J. Neurochem.*, **37**, 576–81.

Walton, K.G., Miller, E. and Baldessarini, R.J. (1979) Prenatal and early postnatal β-adrenergic receptor-mediated increase of cyclic AMP in slices of rat brain. *Brain Res.*, **177**, 515–22.

Williams, J.T., Henderson, G. and North, R.A. (1985) Characterization of alpha 2-adrenoceptors which increase potassium conductance in rat locus coeruleus neurons. *Neuroscience*, **14**, 95–101.

Wilson, K.M. and Minneman, K.P. (1989) Regional variations in α_1-adrenergic receptor subtypes in rat brain. *J. Neurochem.*, **53**, 1782–6.

Zeng, D. and Lynch, K.R. (1991) Distribution of α_2-adrenergic receptor mRNAs in the rat CNS. *Mol. Brain Res.*, **10**, 219–25.

Zeng, D., Harrison, J.K., D'Angelo, D.D. *et al.* (1990) Molecular characterization of a rat α_{2B}-adrenergic receptor. *Proc. Natl. Acad. Sci. USA*, **87**, 3102–6.

MUSCARINIC RECEPTORS AND THE DEVELOPING NERVOUS SYSTEM

2

Lucio G. Costa

2.1 INTRODUCTION

The cholinergic system plays a primary role in a number of important behaviors, and cholinergic dysfunctions may result in several neurological disorders (e.g. dementias, affective disorders; Singh *et al.*, 1985). In the central nervous system, muscarinic receptors constitute the majority of receptors of the cholinergic type. Though the muscarinic action of acetylcholine has been known since the beginning of the century, a significant advancement in our knowledge of the biochemical and molecular characteristics of muscarinic receptors has occurred only in the last decade. A number of excellent reviews on muscarinic receptors give a comprehensive discussion of the subject (Nathanson, 1987; Mitchelson, 1988; Brown, 1989; Goyal, 1989; Mei *et al.*, 1989; van Delft *et al.*, 1989; Hulme *et al.*, 1990). The most striking findings of the last decade are probably the discovery of the existence of several muscarinic receptor subtypes and the characterization of their second messenger responses. This chapter will briefly review some basic concepts of muscarinic receptor structures and functional responses and will then discuss in more detail our current knowledge of their ontogeny and their potential roles in brain development. A discussion of the development of muscarinic receptors in other tissues, such as the heart, can be found in Nathanson (1989).

2.2 MUSCARINIC RECEPTOR SUBTYPES

The first strong evidence of muscarinic receptor heterogeneity came from studies with

Table 2.1. Subtypes of muscarinic receptors

Molecular sequence	m_1	m_2	m_3	m_4	m_5
Amino acid residues					
Human	460	466	590	479	532
Rat	460	466	589	478	531
Pharmacological subtype	M_1	M_2	M_3	—	—
Selective antagonists	Pirenzepine	AFDX-384	*p*-Fluorohexa-	—	—
	(+) Telenzepine	Methoctramine	hydrosiladiferidol		

Adapted from Mei *et al.* (1989) and Levine and Birdsall (1989).

Receptors in the Developing Nervous System Vol. 2: Neurotransmitters. Edited by Ian S. Zagon and Patricia J. McLaughlin. Published in 1993 by Chapman & Hall. ISBN 0 412 49400 0. Vols. 1 and 2 (set) ISBN 0 412 54520 9.

the muscarinic antagonist pirenzepine, which revealed a unique regional distribution of binding, different from that of other antagonists, such as quinuclidinyl benzylate (QNB; Hammer *et al.*, 1980). This finding led to the classification of muscarinic receptors into two subtypes: a high affinity pirenzepine site, found primarily in the CNS, which was named M_1, and a site with low affinity for pirenzepine, present in heart and smooth muscle, which was named M_2. A third type of receptor, named M_3, was soon identified pharmacologically in exocrine glands and smooth muscle. New tools in molecular biology have since led to the cloning of five distinct rat and human muscarinic receptors (Table 2.1). Muscarinic receptors from mouse, *Drosophila* and chick have also been cloned (Shapiro *et al.*, 1988; Tietje *et al.*, 1990). The rat genes are designated m_1 to m_5 and their products correspond to the pharmacologically defined subtypes (M_1, M_2, etc.; Kubo *et al.*, 1986; Peralta *et al.*, 1987; Buckley *et al.*, 1988; Bonner, 1989).

Sequence analysis of cloned muscarinic receptors has indicated that the receptor protein is composed of seven transmembrane domains (α helical segments), which are highly conserved among subtypes. However, the third inner cytoplasmatic loop differs among subtypes, and this portion of the protein is believed to infer different functional characteristics in the receptor. Each muscarinic receptor subtype, unlike other receptors coupled to ion channels (e.g. $GABA_A$ or cholinergic nicotinic receptors) which consist of several subunits, but similar to other G protein-coupled receptors, appears to be the product of a single gene. The distribution of muscarinic receptors in the CNS has been determined by measuring receptor recognition sites by radioligand binding assays, as well as subtype mRNAs by *in situ* hybridization (Potter *et al.*, 1983; Buckley *et al.*, 1988). These studies have shown that regions such as the cerebral cortex and the hippocampus contain mostly M_1, and to a minor extent, M_2 and M_3 receptors; the striatum has a preponderance of M_4, M_1 and possibly M_2 subtypes, whereas cerebellum and medulla pons contain almost exclusively M_2 receptors. Low levels of the M_5 subtype have been found in the hippocampus and substantia nigra (Vilaró *et al.*, 1990). The pre- and postsynaptic location of receptor subtypes has also received much attention. For example, in the cerebral cortex and the hippocampus M_1 and M_3 receptors are believed to be located postsynaptically, whereas M_2 receptors might be located presynaptically in cholinergic terminals, and appear to be involved in the modulation of acetylcholine release (Kilbinger, 1987). In the striatum, on the other hand, an M_3 muscarinic autoreceptor appears to regulate the release of acetylcholine (De Boer *et al.*, 1990).

2.3 SECOND MESSENGER SYSTEMS

Activation of muscarinic receptors in the nervous system can lead to several biochemical responses, the most important or, at least, the most studied of which, are an increase in the levels of cyclic GMP, a decrease in the level of cyclic AMP and an increased turnover of membrane phospholipids, particularly phosphoinositides (Nathanson, 1987).

2.3.1 REGULATION OF CYCLIC AMP METABOLISM

In several tissues, activation of muscarinic receptors leads to an inhibition of adenylate cyclase and to a decrease in the rate of formation of cyclic AMP (adenosine 3',5'-monophosphate) (Olianas *et al.*, 1983; Harden, 1989). Similar to activation of adenylate cyclase by beta adrenergic agonists, muscarinic receptor-mediated inhibition of this enzyme involves a G protein, Gi. Inhibition of adenylate cyclase leads to a decrease in the intracellular levels of cyclic AMP. Although a direct association of cyclic AMP with muscarinic receptor-mediated responses has been unambiguously demonstrated in only a few cases, in several tissues, particularly those rich in the M_2 receptor subtype, physiological effects of acetylcholine involve alterations of cyclic AMP metabolism (Harden, 1989). In addition to causing inhibition of adenylate cyclase through an interaction with Gi, there is limited evidence that activation of muscarinic receptors may lead to activation of phosphodiesterase. This also will lead to a decrease in intracellular levels of cyclic AMP (Harden, 1989). The observation that muscarinic receptors can stimulate phosphodiesterase, together with the observation that activation of alpha$_1$-adrenoceptors can produce a similar event, has led to a proposed mechanism for such stimulation. Since both alpha$_1$-adrenoceptors and muscarinic receptors are coupled to phospholipase C, this model considers the interaction between the phosphoinositide and the cyclic AMP pathway (Harden, 1989). Elevation of cytoplasmatic Ca^{2+} levels occurs as a consequence of the increased inositol 1,4,5-trisphosphate (Ins-1,4,5-P_3) levels, and an activation of a Ca^{2+} calmodulin-regulated phosphodiesterase ensues. Thus, whereas inhibition of cyclic AMP accumulation through interaction with Gi can be seen as a direct mechanism, the muscarinic receptor-mediated attenuation of cyclic AMP accumulation probably should be considered more as a modulatory mechanism that follows receptor-stimulated phosphoinositide hydrolysis (Harden, 1989). Although detailed pharmacological studies on muscarinic receptor-mediated activation of phosphodiesterase have not been carried out, it would appear that all subtypes of muscarinic receptor are capable of affecting cyclic AMP levels, the M_2 subtype, directly, via inhibition of adenylate cyclase, and the M_1 and M_3 subtypes indirectly, as a consequence of an increase of intracellular calcium which follows phosphoinositide hydrolysis. It is becoming apparent, however, that several muscarinic receptor subtypes couple to each of the second-messenger response systems with varying degrees of efficiency.

2.3.2 REGULATION OF CYCLIC GMP METABOLISM

Activation of muscarinic receptors also causes the elevation of intracellular cyclic GMP (guanosine 3',5'-monophosphate) levels. This effect has generally been considered to be indirect, involving an additional second messenger, since muscarinic agonists cannot activate guanylate cyclase (the enzyme which converts GTP into cyclic GMP) in broken cell preparations (McKinney and Richelson, 1989). The most likely candidates for this role are calcium ions or a metabolite of arachidonic acid. Evidence for a role of calcium comes from experiments showing that extracellular calcium ions are necessary for the muscarinic receptor stimulation of cyclic GMP (Schultz *et al.*, 1973), and that calcium channel antagonists block this effect (El-Fakahany and Richelson, 1983). The role of a metabolite of arachidonic acid is suggested by the finding that inhibitors of lipoxygenase (but not cyclo-oxygenase) block the cyclic GMP response (Snider *et al.*, 1984; McKinney and Richelson, 1986). The muscarinic receptor involved in the cyclic GMP response appears to be the M_1 subtype (McKinney and Richelson, 1989), with a pharmacological profile similar to that of stimulation of phosphoinositide metabolism. Some evidence has suggested a link between these two biochemical responses (e.g. Ohsako and Deguchi, 1981), but this is not always the case (e.g. Kendall, 1986).

Increases in intracellular levels of cyclic GMP contribute to some of the effects of muscarinic receptor stimulations in various tissues such as smooth muscle, heart and brain (McKinney and Richelson, 1989). Most of these effects are probably mediated by protein phosphorylation regulated by cyclic GMP-dependent protein kinases. The cyclic GMP pathway has recently been accorded much attention, since it has been discovered that nitric oxide (NO), synthesized from L-arginine by NO synthase, binds to a heme moiety which is attached to guanylate cyclase and stimulates the accumulation of cyclic GMP (Ignarro, 1991; Snyder and Bredt, 1991). It has been shown that NO mediates gluta-mate-stimulation of cyclic GMP formation in the cerebellum (Garthwaite *et al.*, 1989), as well as norepinephrine-induced increases in cyclic GMP in astrocytes (Agullo and Garcia, 1991). It has been recently reported that the accumulation of cyclic GMP stimulated by activation of muscarinic receptors is mediated by the NO pathway (Castoldi *et al.*, 1993).

2.3.3 ACTIVATION OF THE PHOSPHOINOSITIDE/PROTEIN KINASE C PATHWAY

Interaction of acetylcholine with the appro-priate subtype of muscarinic receptor acti-vates a phosphoinositidase (phosphoinositi-dase C) which hydrolyzes phosphatidyli-nositol 4,5-bisphosphate (PtdIns 4,5-P_2) to Ins 1,4,5-P_3 and 1,2-diacylglycerol (DG) (Gonzales and Crews, 1984; Berridge, 1987; Costa, 1990). A GTP-binding protein is believed to couple the receptor to phosphoinositidase C (Chiu *et al.*, 1988; Harden, 1990; Fain, 1990; Candura *et al.*, 1992b). The exact nature of such protein (possibly Gp or Gq) has not been fully elucidated, although the current view is that there might be more than one protein (Lo and Hughes, 1987; Smrcka *et al.*, 1991).

Ins 1,4,5-P_3 binds to specific and saturable receptor sites in, or near, the endoplasmic reticulum (Worley *et al.*, 1989) and causes the mobilization of calcium ions in the cytosol (Stundermann *et al.*, 1988). Ins 1,4,5-P_3 is then dephosphorylated by phosphatases to gener-ate Ins 1,4-P_2, Ins 1-P and inositol. This latter reaction, catalyzed by an Ins 1-P phosphatase, is inhibited by lithium ions (Drummond *et al.*, 1987). Ins 1,4,5-P_3 can also be phosphorylated by a 3-kinase to form inositol 1,3,4,5-tetrakisphosphate (Ins 1,3,4,5-P_4), which in turn is dephosphorylated to Ins 1,3,4-P_3 (Irvine *et al.*, 1988). It has been suggested that Ins 1,3,4-P_3 might be involved in long-term cellular effects (e.g. regulation of gene transcription or cell division), but evidence for this is still lacking. Ins 1,3,4,5-P_4 might play a role in modulating the mobilization of intracellular Ca^{2+} by Ins 1,4,5-P_3 and/or in regulating Ca^{2+} entry from outside the cell (Irvine *et al.*, 1988). All these inositol phos-phates are formed, in different amounts and over different time-courses, following activa-tion of muscarinic receptors by agonists (Stephens and Logan, 1989). Ins 1,3,4,5,6-P_5 and Ins-P_6 may also be formed, and it has been suggested that they serve as a storage for phosphates (the same role they have in higher plants). Both have been shown to induce dose-dependent changes in heart rate and blood pressure in the rat (Vallejo *et al.*, 1987) and specific binding sites for Ins-P_6 have been identified in mammalian brain (Hawkins *et al.*, 1990). The cyclic inositol (1:2 cyclic),4,5-trisphosphate is also capable of mobilizing Ca^{2+} and is formed and metabolized at a slower rate; it could therefore act as a long-term messenger, although it is still of doubt-ful physiological significance. Thus, a major consequence of the activation of the phospho-inositide system by muscarinic agonists is a change in the intracellular concentration of Ca^{2+} which plays a pivotal role in synaptic transmission and in cell functions (Llinas, 1982; Kudo *et al.*, 1988; Boess *et al.*, 1990).

The other product of phosphoinositide hydrolysis, DG, is capable of activating a novel protein kinase, PKC (Nishizuka, 1988). In most tissues, PKC is present in its inactive form in the cytosol and is translocated to membranes when cells are stimulated; activation of PKC requires Ca^{2+} and phospho-lipids, particularly phosphatidylserine. It is now evident that at least seven subspecies of PKC exist in nerve tissue, and their struc-tures, deduced by analysis of their DNA sequences, have been elucidated (Nishizuka, 1988; Kikkawa *et al.*, 1988). These different PKCs, one of which, the gamma, is expressed only in the brain and spinal cord, but not in other tissues, including peripheral nerves, differ in their specific activities in different brain areas, in their developmental profile, and in their activation requirements. In addition to DG, the gamma form can also be activated by arachidonic acid and by other eicosanoids (Kikkawa *et al.*, 1988; Rana and Hokin, 1990), suggesting that production of these compounds upon receptor stimulation might have differential intracellular effects depending on the forms of PKC present in a certain cell, and that PKC might be regulated independently of DG and Ca^{2+}, i.e. in a manner not necessarily linked to phosphoinositide turnover. In recent years, it has been shown that activation of muscarinic receptors can also cause the hydrolysis of membrane phosphatidylcholine; this hydrolysis appears to be mediated both by phospholipase C and by phospholipase D (Martinson *et al.*, 1989; Qian and Drewes, 1989; Diaz-Meco *et al.*, 1989; Horwitz, 1990; Sandmann and Wurtman, 1991). Hydrolysis by phospholipase C generates phosphorylcholine and DG, whereas hydrolysis by phospholipase D leads to the formation of choline and phosphatidic acid. The latter can be metabolized to DG by the action of a phosphohydrolase. Thus, significant amounts of DG, available for activation of PKC, can be formed by hydrolysis of phosphoinositides and phosphatidylcholine on activation of muscarinic receptors.

PKC is the receptor for a class of tumor promoters, the phorbol esters (Ashendel, 1985) and the use of these compounds as direct activators of PKC is proving extremely useful for investigating the role of PKC in cellular functions. A large number of proteins have been shown to be substrates for PKC; these include such diverse molecules as receptors, including the muscarinic receptors (Haga *et al.*, 1990), ion channels, cytoskeletal proteins and enzymes (Nishizuka, 1988). PKC has been shown to provide both a 'positive forward action' leading to a synergistic interaction with the Ca^{2+} pathway (e.g. in control of gene expression), and a 'negative feedback

control' over various steps of the cellsignalling process (Berridge, 1987; Nishizuka, 1988). Muscarinic receptor-induced phosphoinositide hydrolysis has been shown to be inhibited by phorbol esters in brain slices and neurons in cultures (Labarca *et al.*, 1984; Gonzales *et al.*, 1987). Furthermore, activation of PKC is also believed to be involved in muscarinic receptor desensitization and down regulation (Liles *et al.*, 1986; Jia *et al.*, 1989; however, see also Lai *et al.*, 1990; Hoover and Toews, 1990).

It is now well established that phorbol esters enhance the release of various neurotransmitters in a number of neuronal preparations (Kikkawa and Nishizuka, 1986), an effect believed to be due to activation of PKC, since it is not induced by inactive phorbol esters and is inhibited by PKC antagonists, such as H-7. The involvement of PKC in neurotransmitter release might also be linked to its role in the maintenance of long-term potentiation (LTP), a rapidly induced, persistent increase in synaptic efficacy (Brown *et al.*, 1988). Since LTP is a leading candidate for a synaptic mechanism of rapid learning in mammals, a role of PKC in memory formation has also been suggested (Chiarugi *et al.*, 1989). There is also increasing evidence that phosphoinositol metabolism plays a relevant role in the control of cell proliferation (Vicentini and Villareal, 1986). For example, in the nervous system, proliferation of astrocytes has been shown to be associated with activation of the phosphoinositide/PKC system by acetylcholine (Ashkenazi *et al.*, 1989). Stimulation of phosphoinositide metabolism differs in different brain areas, the highest effect being found in cerebral cortex, hippocampus and striatum (Gonzales and Crews, 1984; Balduini *et al.*, 1990a). The m_1, m_3 and m_5 receptor subtypes induce a strong phosphoinositide response, whereas m_2 and m_4 receptors weakly activate phosphoinositide hydrolysis (Peralta *et al.*, 1988; Forray and El-Fakahany, 1990). Ongoing studies with chimeric receptors are attempting to define the specific structural domain(s) conferring selectivity of coupling to phosphoinositide hydrolysis (Wess *et al.*, 1990).

2.4 REGULATION OF MUSCARINIC RECEPTORS

The sensitivity of muscarinic receptors is regulated by the prolonged presence of an agonist or an antagonist at the receptor site (Nathanson, 1989; El-Fakahany and Cioffi, 1990). Activation of muscarinic receptors by an agonist leads, in a very short time, to receptor desensitization; persistent activation results in a down-regulation of muscarinic receptors. Prolonged exposure to muscarinic antagonists, on the other hand, causes a supersensitivity of muscarinic receptors, often accompanied by receptor up-regulation, a phenomenon similar to that observed after denervation. A large number of studies have investigated the modulation of muscarinic receptors in several tissues and model systems, from cell cultures to intact animals. These findings have been summarized in several recent reviews (Nathanson, 1987; 1989; Costa, 1988; El-Fakahany and Cioffi, 1990). Prolonged exposure to direct muscarinic agonists such as oxotremorine, and indirectly to agonists, such as cholinesterase inhibitors, has been shown to cause a decrease in the sensitivity of muscarinic receptors. This alteration of muscarinic receptor sensitivity is manifested in decreases in receptor density measured by radioligand binding, in receptor mRNA levels (Wang *et al.*, 1990), in muscarinic receptor-mediated cellular responses and in various behavioral effects elicited by cholinergic agonists. Conversely, repeated treatments with muscarinic antagonists cause an increase in the density of muscarinic receptors and a super-sensitivity to the behavioral effects of muscarinic agonists.

In addition to muscarinic agonists (including cholinesterase inhibitors) and antagonists, a number of other agents can modify muscarinic receptor density, affinity or sensitivity following *in vivo* administration. These include drugs, such as barbiturates (Nordberg *et al.*, 1980), hormones, such as estrogens (Dohanick *et al.*, 1982), environmental neurotoxicants, such as mercury (Aronstam and Eldefrawi, 1979), or physical agents, such as repeated noise (Lai *et al.*, 1989). A loss of muscarinic receptors, particularly of the M_2 subtype, has been found to occur with aging (Araujo *et al.*, 1990); a severe loss of M_2 autoreceptors has also been shown in brain of patients with Alzheimer's disease (Mash *et al.*, 1985).

2.5 MUSCARINIC RECEPTORS IN THE DEVELOPING NERVOUS SYSTEM

2.5.1 RECEPTOR BINDING

A large number of studies have examined the ontogenesis of muscarinic receptors in the CNS. Most of these studies have been carried out in rodents, particularly in rats, although there is some limited information for other species, including human; data also exist for the development of muscarinic receptors in neurons in primary cultures. One early study (Coyle and Yamamura, 1976) measured the binding of the specific but non-selective antagonist [³H]quinuclidinyl benzilate (QNB) to rat brain membranes. At 15 days of gestation, the whole brain exhibited low, but detectable, levels of binding, approximately 1% of adult levels. Whole brain [³H]QNB binding increased in a relatively linear fashion during the last week of gestation and first month *post partum*, when the concentration of specific binding sites for [³H]QNB reached adult levels. An analysis of muscarinic receptors in brain regions revealed a similar developmental pattern, with adult levels of [³H]QNB binding being reached at 28 days of age. The rate of muscarinic receptor development was, however, different in different regions; for example, at two weeks after birth, the medulla pons and hypothalamus had nearly 80% concentration of the adult muscarinic receptors, whereas the parietal cortex and striatum had less than 40% of adult levels (Coyle and Yamamura, 1976). The region-specific development of muscarinic receptors was later confirmed by Kuhar *et al.* (1980), who found a faster development of [³H]methyl scopolamine binding in medulla pons than in cortex or diencephalon. A differential development of high and low affinity agonist binding sites was also found, with the high affinity site appearing only 6–7 days after the low affinity sites (Kuhar *et al.*, 1980).

Several additional studies have examined the ontogenesis of muscarinic receptors, labeled by non-selective antagonists such as [³H]QNB, in rat or mouse brain or brain areas, by means of receptor binding or receptor autoradiography, with similar results (Aronstam *et al.*, 1979; Rotter *et al.*, 1979; Egozi *et al.*, 1980; Dudai *et al.*, 1980; East and Dutton, 1980; Mallol *et al.*, 1984; Hohman *et al.*, 1985; Large *et al.*, 1986; Balduini *et al.*, 1987; Represa *et al.*, 1989). Similar developmental profiles were also observed in rat cerebral and hippocampal cells in primary culture (Dudai and Yavin, 1978; Fernandez-Tomè and Segal, 1987), and in rabbit or chick brain (Yavin and Harel, 1979; Enna *et al.*, 1976).

A number of studies have also investigated the development of muscarinic receptors in the human fetal and neonatal brain (Brooksbank *et al.*, 1978). Egozi *et al.* (1986) measured the binding of *N*-[³H]methyl-4-piperidyl benzilate in several regions of human fetal brain from week 14 to week 24 of gestation, and found a gradual increase in binding. Longer gestational ages (up to 40 weeks) were studied by Ravikumar and Sastry (1985) using [³H]QNB. These investigators found an increase in receptor density from week 16 to 20, a lag period between 20 and 24 weeks, and a rapid increase of receptors during the third trimester reaching a maximum at birth and declining thereafter. Some of these observations were also made by Nordberg and Winblad (1981). On the other hand, Gremo *et al.* (1987) reported a decrease in receptor density between prenatal weeks 28–32 and birth in

most brain areas. It should be noted that in all these animal and human studies the affinity of the antagonist ligands did not change significantly during development.

Recent attention has focused on the development of muscarinic receptor subtypes in brain tissue. In human fetal frontal cortex the density of M_1 and M_2 sites increased from gestational week 8 to week 22 (Perry *et al.*, 1986). The ratio of M_1/M_2 sites also increased with age, as might be expected if presynaptic elements, those containing M_2 receptors, are established prior to functional neurotransmission, including postsynaptic activity. On the other hand, an autoradiographic study in the rat indicated that M_1 receptors appear early (gestational day 18 in striatum and 20 in hippocampus), whereas M_2 receptors develop more slowly and increase significantly only during the second and third weeks of development (Aubert *et al.*, 1990). Evans *et al.* (1985) measured the postnatal development of muscarinic receptors labeled with either [³H]QNB or [³H]pirenzepine in mouse cerebral cortex. Binding of both ligands increased with age and reached adult levels at about postnatal day 21. Interestingly, in this study the density of binding sites was found to be almost identical for both ligands, suggesting that the murine cerebral cortex contains predominantly M_1 sites. These findings were confirmed by an *in vitro* autoradiographic study (Miyoshi *et al.*, 1987). No data are available, to date, on the development of mRNA levels for the five muscarinic receptor subtypes.

In summary, most receptor binding studies have shown that muscarinic receptor binding sites increase with development in all species and, in the rat, they reach maximal density at approximately 30 days of age.

2.5.2 DEVELOPMENT OF MUSCARINIC RECEPTOR-COUPLED SECOND MESSENGER SYSTEMS

In recent years, the development of muscarinic receptor-coupled second messenger systems, particularly the phosphoinositide/ protein kinase C pathway, has begun to receive some attention. An interesting and surprising finding was that the accumulation of inositol phosphates stimulated by muscarinic agonists had a different developmental profile from that of muscarinic receptor binding sites. Although muscarinic receptors in rat cerebral cortex (labeled by [^{3}H]QNB) increase during development, as previously discussed, the phosphoinositide response increases from birth to postnatal days 7–10 and then gradually declines (Fig. 2.1; Balduini *et al.*, 1987, 1990a,b; Heacock *et al.*, 1987; Rooney and Nahorski, 1987). Lee *et al.* (1990) attributed this developmental pattern to a decrease in the incorporation of [^{3}H]inositol into tissues of older rats. However, this explanation is not fully convincing because (1) data were usually corrected to account for differ-ences in incorporation; (2) incorporation was actually greater on postnatal days 1 and 3, when the effect of muscarinic agonists was less than that observed on days 7 or 10; (3) such a developmental pattern could not be observed for other neurotransmitters. Indeed, the developmental pattern of receptor-stimulated phosphoinositide metabolism has been shown to be neurotransmitter receptor- and brain region-specific (Balduini *et al.*, 1991b).

In addition to rat cortical slices, similar findings have been also reported in a synaptoneurosomal preparation (Dudek *et al.*, 1989). This developmental pattern has also been observed in other tissues, such as the rat cochlea (Bartolami *et al.*, 1990). In brain regions other than the cerebral cortex, the developmental profile of muscarinic receptor-stimulated phosphoinositide metabolism is

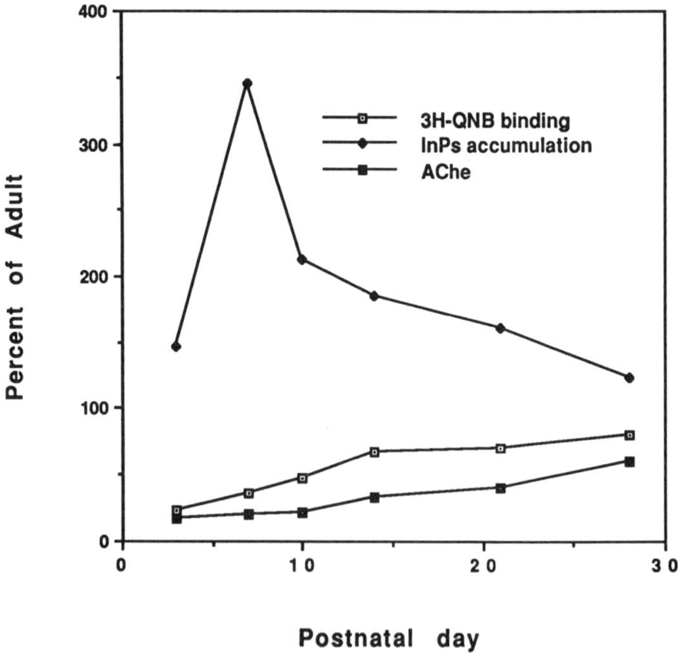

Fig. 2.1.   Postnatal development of [^{3}H]QNB binding, carbachol-stimulated inositol phosphates (InsPs) accumulation and acetylcholinesterase (AChE) activity in rat cerebral cortex, expressed as percentages of adult (75-day-old) values. Adult values were: [^{3}H]QNB binding, 1907 ± 128 fmol/mg of protein; carbachol (1 mM)-stimulated InsPs accumulation, 428 ± 24% of basal; AChE activity, 136 ± 21.3 μmol acetylthiocholine hydrolyzed/min/mg of protein. Adapted from Balduini *et al.* (1987).

different. Indeed, while the hippocampus resembles the cerebral cortex, with a maximal stimulation on day 7, the effect of carbachol in brainstem and cerebellum declines from gestational day 19 to adulthood (Balduini *et al.*, 1991b). A similar decrease from day 4 to adulthood has also been observed in the rat retina (Osborne, 1988).

In order to understand the possible mechanism(s) underlying the enhanced stimulation of phosphoinositide metabolism seen in immature rats, relative to adults, a series of pharmacological and biochemical characteristics of the muscarinic receptor-stimulated inositol phosphate metabolism were compared in brain from 7-day-old and adult rats (Balduini *et al.*, 1990a). No significant age differences were found with respect to the effects of agonists, partial agonists and antagonists, calcium and sodium ion dependency, sensitivity to phorbol esters and regional distribution. It was found, however, that a high concentration of potassium ions (at least 12 mM) was capable of potentiating the effect of carbachol on phosphoinositide metabolism in cortices from adult rats, but not in 7-day-old animals (Balduini *et al.*, 1990b). Eva and Costa (1986) suggested that in adult animals K^+ might change the coupling efficiency of the muscarinic recognition site with its transducer. One might speculate that in the 7-day-old animal this 'positive coupler' would already be operating and, therefore, the potentiating effect of K^+ would be minimal. This hypothesis would agree with that of Heacock *et al.* (1987), who suggested the presence of an enhanced receptor–effector coupling as the mechanism underlying the enhanced sensitivity of phosphoinositide metabolism in 7-day-old rats.

In rat cortical membranes, the GTP analog GTP[S] (guanosine 5'-O-(3-thiotrisphosphate)) stimulates phosphoinositide metabolism to a higher degree in adult, than in 7-day-old rats (Candura *et al.*, 1992b). This is not surprising since this nucleotide, as well as AlF_4^- which shows a similar pattern of stimulation, is expected to interact with all G proteins, not solely those coupled to the muscarinic receptors. However, in membranes from neonatal rats, the stimulatory effect of carbachol, which is seen only in the presence of micromolar concentrations of GTP (or GTP[S]) (Wallace and Claro, 1990), is similar to that observed in adult animals (Candura *et al.*, 1992b). Thus, although the enhanced response to muscarinic agonist in the neonatal rat, as observed in slices or synaptoneurosomes, is not seen in this membrane preparation, the effect of muscarinic agonists in 7-day-old animals is still remarkable, since only a fraction (about 30%) of adult muscarinic receptors are present at this developmental stage (Balduini *et al.*, 1987). A somewhat related similar observation was made in electrophysiological recordings from neonatal rat hippocampal slices, which showed that the responses elicited by muscarinic agonists (depolarization of the membrane potential and/or increase in firing frequency) were similar to those observed in mature animals (Reece and Schwartzkroin, 1991).

A few studies have investigated the development of muscarinic receptor-induced inositol phosphate formation in neurons in primary culture. In striatal cultures from mouse brain (prepared from 14–15-day-old fetuses) the effects of carbachol increased up to 14 days *in vitro*, roughly equivalent to day 7–8 *in vivo* (Weiss *et al.*, 1988). Longer times were not investigated. On the other hand, the effect of glutamate peaked earlier and then declined, a finding that led these investigators to suggest that the excitatory amino acid response precedes synaptogenesis, whereas maximal response to carbachol requires complete neuronal differentiation (Weiss *et al.*, 1988). Similar results were obtained in the same system by Schmidt *et al.* (1991), who found a doubling of the carbachol response after 13 days in culture, i.e. after synaptogenesis. Other investigators used primary cultures to study muscarinic receptor-mediated inositol responses (e.g. Ambrosini and Meldolesi, 1989; Akins *et al.*, 1990), but, no time-course studies of developmental maturation were conducted.

One aspect that has also been addressed by studying brain tissue in culture is that of the neuronal and/or glial localization of the muscarinic response. Whereas initial findings suggested that muscarinic receptor-mediated inositide hydrolysis is primarily a neuronal response (Gonzales *et al.*, 1985), subsequent studies showed that a robust phosphoinositide response is also present in astrocytes (Pearce and Murphy, 1988). In slice preparations, the observed response is probably the sum of the effects in both neurons and glial cells, and the contribution of astrocytes to events observed in slices cannot be ignored, especially as these glial cells can outnumber neurons by an order of magnitude (Pearce and Murphy, 1988).

Lee *et al.* (1990) provide the only information on the development of muscarinic receptor-mediated inhibition of cyclic AMP accumulation in rat cerebral cortex. They found that this action of muscarinic agonists does not occur until postnatal week 3, when it is already maximal (about 20% inhibition of forskolin-stimulated cyclic AMP accumulation). This is surprising, since muscarinic receptors of the M_2 subtype are present even in the first two weeks after birth. Since the subunits of G proteins are also present, it has been proposed that during the early stages of development, the interactions between muscarinic receptors and G proteins may be less efficient, thus leading to a lack of muscarinic receptor-mediated inhibition of cyclic AMP accumulation (Lee *et al.*, 1990). The mechanism(s) underlying such diminished coupling remains, however, obscure. Muscarinic receptor-mediated inhibition of cyclic AMP accumulation has also been studied in primary cultures of striatal and brainstem neurons from 14–15-day-old mouse embryos (Ellis *et al.*, 1990). Cells were maintained *in vitro* for seven to eight divisions, up to a developmental stage when there is little synapse formation; no time-course studies on developmental changes were conducted. No information is available on the development of muscarinic receptor-mediated cyclic GMP accumulation in nerve tissue.

2.5.3 DEVELOPMENT OF OTHER PARAMETERS OF THE CHOLINERGIC SYSTEM

In order to have a better understanding of the possible role(s) of muscarinic receptors in brain development, it is useful to compare their ontogeny with that of other parameters of the cholinergic system. The development of activities of cholineacetyltransferase (ChAT) and acetylcholinesterase, of high affinity sodium-dependent choline uptake, and of levels of choline and acetylcholine in different brain areas, have been studied by several investigators, and the results have been summarized in various reviews (Bajgar *et al.*, 1979; Muller *et al.*, 1985; Hohman and Ebner, 1985; Vaca, 1988); only those aspects which may be relevant to a discussion of the development of muscarinic receptors will be briefly summarized here. Activity of ChAT in rat brain is usually low at birth and increases with age; its development, however, is rather slow. For example, only 19% of adult ChAT activity is present after one week (Fig. 2.2; Ladinsky *et al.*, 1972). Represa *et al.* (1989) similarly reported that ChAT activity in rat striatum was approximately 25% of adult levels at postnatal day 6; in other brain areas, the development was somewhat different. For example, in the septum, a spike in activity was seen on day 9 (up to 40% of adult levels), followed by a decline and by a slow increase (Represa *et al.*, 1989). A study by Large *et al.* (1986) showed that in the olfactory bulb the development of ChAT parallels that of muscarinic receptors.

The development of ChAT and several markers of the cholinergic system was also investigated by Coyle and Yamamura (1976). In whole brain levels of ChAT and high affinity choline uptake were almost undetect-

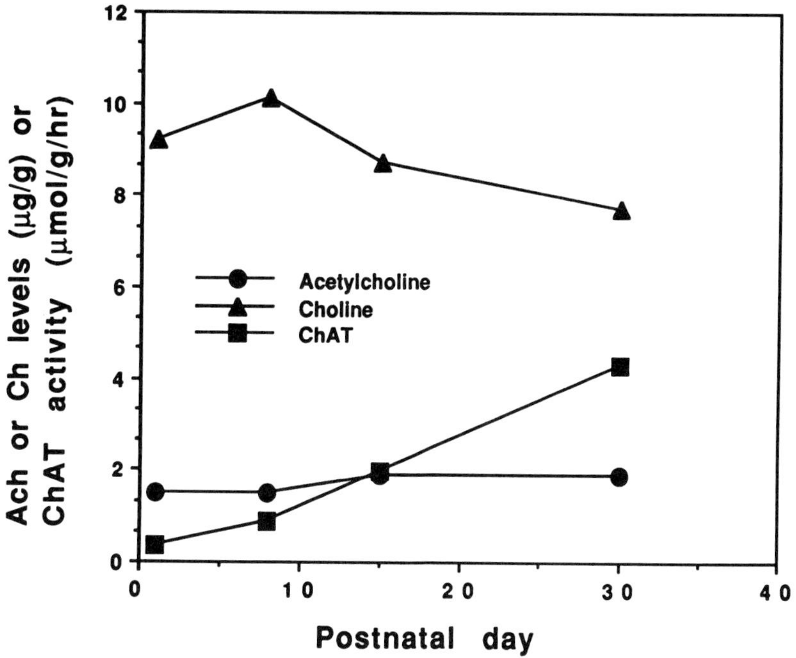

Fig. 2.2. Levels of acetylcholine and choline (μg/g wet wt) and activity of cholineacetyltransferase (ChAT; μmol/g wet wt per h) in brain of developing rat. Data adapted from Ladinsky *et al.* (1972).

able at birth and increased rapidly after the first week. On the other hand, levels of acetylcholine were already 30–40% of adult levels in the prenatal brain and during the first postnatal week. Particularly in the medulla-pons, the levels of acetylcholine in newborn rats were 87% of adult values (Coyle and Yamamura, 1976). Ladinsky *et al.* (1972) similarly reported levels of acetylcholine to be as high as 73% of adult levels by postnatal day 1 (Fig. 2.2). High levels of acetylcholine in the neonatal brain have also been reported by other investigators (Kasa *et al.*, 1982; Dörner *et al.*, 1982).

Choline levels, which are considered to be the rate-limiting factor in the biosynthesis of acetylcholine, are actually higher in the neonatal than in the adult rat brain (Fig. 2.2; Ladinsky *et al.*, 1972). A study by Kotas and Prince (1987) confirmed these findings. Levels of acetylcholine in the hippocampus of the 7-day-old rat were already 40% of adult levels, whereas choline levels were three-fold higher than in adults. However, a significantly higher percentage of the choline is incorporated into phosphorylcholine at one week of age, than in adult animals. The rate of hemocholinium-3-sensitive uptake increased with age, and reached 37% of adult values at one week of age. High affinity choline uptake sites were also measured by Aubert *et al.* (1990), and found to mature more slowly than M_1 receptors, but similarly to M_2 (presynaptic) receptors. On the other hand, acetylcholinesterase activity is very low at birth (20% of adult values) and increases steadily to reach adult levels at 35 days (Fig. 2.1).

A noteworthy feature of the development of these parameters is that there is a continual apparent lag in the development of ChAT activity, compared to acetylcholine levels. Only by postnatal day 20 does the level of ChAT activity become commensurate with the endogenous levels of acetylcholine (Ladinsky *et al.*, 1972). Two factors may, therefore, play a role in the high concentration of acetylcholine in the neonatal brain: the activity of the metabolizing enzyme, acetylcholinesterase, which is quite low, and the availability of high levels of choline. Studies in *Xenopus* embryos have shown that acetylcholine can be released from the growth cone upon stimulation at the soma (Sun and Pao, 1987). The rate of release of acetylcholine by slices of immature brain, whether resting or electrically stimulated, can greatly exceed those evident in mature synapses (Pedata *et al.*, 1983). Furthermore, the turnover of acetylcholine is high in the adult brain, whereas it is lower in immature neurons (Purpura, 1972). High levels of acetylcholine would, therefore, be available for interaction with muscarinic receptors and generation of second messenger signals.

The regional developmental studies also indicate that the maturation of ChAT is more rapid in the cerebral regions of the brain, probably reflecting the fact that cell division ceases, and synaptogenesis commences in these regions much in advance of the subcortical and cortical areas (Coyle and Yamamura, 1976). Navarro *et al.* (1989) measured ChAT and choline uptake in the developing rat cerebral cortex: both parameters increased with age. However, the uptake/ChAT ratio peaked on postnatal day 9, just at the beginning of a major phase in synaptogenesis, and then declined. Only the cortex displayed a postnatal spike of uptake/ChAT; in the midbrain–brainstem, for example, this ratio decreased monotonically after birth. This developmental pattern resembles those observed for muscarinic receptor-stimulated phosphoinositide metabolism in these two areas (Balduini *et al.*, 1987, 1991b). Although the interrelationship between those observations is still unclear, it appears that in certain brain regions, such as the cerebral cortex, the unique development of cholinergic activity may render this system vulnerable to developmental insult (Hohman *et al.*, 1988; Navarro *et al.*, 1989).

In human brain, Perry *et al.* (1986) found high activity of cholineacetyltransferase in fetal cortex, even within the first three weeks of gestation. During months 2–6 of gestation levels of cholineacetyltransferase declined by approximately 50%, while the proportion of G4/G1 molecular forms of AChE (as shown also by others; Muller *et al.*, 1985), and M_1 and M_2 muscarinic receptors increased three- to five-fold (Perry *et al.*, 1986). This finding suggests that M_2 receptors are not solely located presynaptically, or that they are not confined to cholinergic axons.

2.6 POSSIBLE ROLE OF MUSCARINIC RECEPTORS IN THE DEVELOPMENT OF THE NERVOUS SYSTEM

Though a large number of studies have investigated the possible role of neurotransmitters and their receptors in brain development (see reviews by Lanier *et al.*, 1976; Coyle, 1977; Lauder, 1988; Mattson, 1988; Meier *et al.*, 1991; Mattson and Hauser, 1991), there is only limited information on the cholinergic system, particularly muscarinic receptors. Most findings on the development of various parameters of the cholinergic system (see previous sections) indicate that muscarinic receptor formation precedes the development of the presynaptic cholinergic marker ChAT (Brooksbank *et al.*, 1978; Rotter *et al.*, 1979). Thus, muscarinic receptors may have an effect on synaptogenesis, though, in turn, the latter may influence the final distribution of receptors (Rotter *et al.*, 1979).

Recent findings have suggested that an important role in brain development may be carried out by the second messenger system cascade of events which follows activation of muscarinic receptors. In particular, the peculiar developmental profile of muscarinic receptor-stimulated phosphoinositide metabolism, which peaks between postnatal days 6–8, in correspondence with the brain growth spurt, has led to the hypothesis that this system may have a fundamental role in the regulation of neurocytomorphogenesis and glial cell proliferation (Balduini *et al.*, 1987). Recent observations have lent support to the hypothesis that this effect of acetylcholine may have functional relevance during brain development. Hohman *et al.* (1988) have shown that neonatal lesions of the basal forebrain cholinergic neurons result in abnormal cortical development. The main features of this disruption include a delay in the developmental expression of critical neuronal components in neocortex, such as the large pyramidal cells in layer V, and a more prolonged disruption of the laminar boundaries that characterize the cortex (Hohman *et al.*, 1988). Though the mechanisms that mediate this disruption in cortical morphogenesis linked to transient loss of forebrain cholinergic neurons remain to be defined, it has been suggested that the ontogenetic expression of muscarinic receptor second messenger coupling in neocortex supports a role of this receptor in modulating neocortical differentiation (Hohman *et al.*, 1988).

A second study has shown that activation of muscarinic receptors in nerve growth cones isolated from neonatal rat brain can stimulate the phosphorylation of the nerve-specific protein B-50 (or GAP43), a major substrate for PKC (Van Hoof *et al.*, 1989). Nerve growth cones are the highly motile tips of axons and dendrites that guide the trailing neurite to the correct target (Smalheiser, 1990; Kater and Mills, 1991). Neurotransmitters are thought to modulate neurite outgrowth and maturation of a growth cone into a synaptic terminal (Mattson, 1988) and the phosphoprotein B-50 is presumed to mediate signal transduction in growth cone membranes (Van Hoof *et al.*, 1989). The finding that a muscarinic agonist can stimulate phosphorylation of B-50 in nerve growth cones adds evidence to the hypothesis that it may play a relevant role in synaptogenesis.

Finally, studies by Ashkenazi *et al.* (1989) have shown that activation of those subtypes of muscarinic receptors which elicit a strong phosphoinositide response, stimulates DNA synthesis in primary astrocytes derived from prenatal rat brain, in an age-dependent fashion. The strong correlation between phosphoinositide response and DNA synthesis and its age-dependence led the authors to conclude that activation of phosphoinositide hydrolysis by acetylcholine is involved in its mitogenic action and is an important factor in astroglial cell growth during brain development (Ashkenazi *et al.*, 1989). Although the same mechanism could also be operating earlier for neurons in their dividing phase, another hypothesis is that for non-dividing cells, neurotransmitter-driven growth processes could involve an increase in cytoplasmatic surface area mediated by an enhancement of process formation and branching (Hanley, 1989). Though higher than ChAT activity, muscarinic receptor density is low in the neonatal brain (Balduini *et al.*, 1987); however, as discussed earlier, acetylcholine is present in significant amounts. It is conceivable, therefore, that acetylcholine could have profound influences on brain development through its 'amplified' phosphoinositide response during the brain growth spurt. An interesting, possibly related, observation (Mattson, 1989) was that in cultured hippocampal neurons acetylcholine potentiates glutamate-induced neurodegeneration, by activating muscarinic receptors. This effect of acetylcholine (which was devoid of neurotoxicity when present alone) may be mediated by increases in intracellular calcium (Mattson, 1989). Clearly, the potential role(s) of muscarinic receptors in brain development remain mostly unexplored. Additional roles of muscarinic receptors in early embryogenesis would also be worth exploring. Filogamo and Marchisio (1971) suggested that acetylcholine may serve a function other than as a neurotransmitter in early embryogenesis and may not be localized exclusively in cholinergic neurons.

Muscarinic receptors have been identified electrophysiologically in human ovarian oocytes, and sperm-carried acetylcholine has been suggested to be involved in activation processes triggered by sperm–egg interaction (Eusebi *et al.*, 1984). Thus, muscarinic receptors may play a significant role at different stages of development.

2.7   MUSCARINIC RECEPTORS AS TARGETS FOR DEVELOPMENTAL NEUROTOXICITY

Drugs, environmental toxicants and disease states have been shown to alter the development of brain muscarinic receptors. Whether these changes are a direct consequence of the action of the exogenous agents, or represent an adaptive response to other lesions, remains, in most cases, to be determined. Neonatal treatment of mice with L-thyroxine or betamethasone caused an accelerated accumulation of muscarinic receptors in the mouse cortex; however, at 30 days after birth, there was a reduction in the quantity of receptors in several brain areas (Ben-Baruch *et al.*, 1981). Undernutrition induced by restriction of the blood supply to the fetus by ligation of the uterine vessels five days before birth (a model known as intrauterine growth retardation or IUGR), has been shown to delay the development of muscarinic receptors in the hippocampus (Represa *et al.*, 1989). Alterations in the development of cerebellar muscarinic receptors have been reported in X-ray irradiated rats, homozygous Gunn rats (jj) and staggerer mice (Soreq *et al.*, 1982).

Exposure to various environmental neurotoxicants (e.g. pesticides, metals) during brain development has been shown to affect the ontogenesis of muscarinic receptors. Two pyrethroid insecticides, bioallethrin and deltamethrin, and the organochlorine DDT, have been shown to alter the development of muscarinic receptors (Eriksson and Nordberg, 1986, 1990). Exposure to the organophosphorus compound, cholinesterase inhibitor, diiso-propylfluorophosphate from postnatal days 8 to 14, resulted in a gradual decrease in the number of muscarinic binding sites in mouse brain (Levy, 1981). However, when the treatment was stopped the density of sites approached the control levels within a week. Similar observations were made by Ben-Barak *et al.* (1981) in Tetram (*O,O*-diethyl *S*-(β-diethyl-amino) ethyl phosphorothiolate)-treated rats. Prenatal exposure to diiso-propylfluorophosphate also caused a delay in the development of brain muscarinic receptors, however, full recovery was observed by day 20 (Michalek *et al.*, 1985). Stamper *et al.* (1988) exposed developing rats to the organophosphorus insecticide parathion from postnatal day 5 to day 20, and observed a decreased receptor density in cerebral cortex on days 21 and 28, which was paralleled by small, but significant, deficits in tests of spatial memory in both the T-maze and the radial arm maze. Developmental exposure of rats to lead (postnatal days 0–21, via the mother's milk) was shown to cause a selective decrease in muscarinic receptor density in the visual cortex (Costa and Fox, 1983), which could be related to observed visual acuity deficits (Fox *et al.*, 1982). Ethanol was reported to decrease muscarinic receptors in the hippocampus of 21-day-old rats following exposure from birth (Serbus and Light, 1990). Similar changes were, however, not observed in the hippocampus of rats exposed to ethanol prenatally (Wigal *et al.*, 1990), or in cerebral cortex of rats exposed between postnatal days 4 and 10 (Balduini and Costa, 1989).

The peculiar developmental profile of the muscarinic receptor-stimulated phosphoinositide metabolism and its possible functional implications (see previous sections) led to the formulation of the hypothesis that this system could represent a target for the developmental neurotoxicity of ethanol. Administration of ethanol to rats during the brain growth spurt (i.e. during the first two postnatal weeks, equivalent to the late second and third trimester of pregnancy in humans) has been shown to cause microencephaly, an effect which is seen in 80% of Fetal Alcohol Syndrome cases (West, 1986). Ethanol (given at the daily dose of 4 g/kg, by gavage, from day 4 to day 9) also caused a significant decrease of muscarinic receptor-stimulated phosphoinositide response in cerebral cortex and hippocampus of 7- and 10-day-old rats, i.e. at the peak of the brain growth spurt, but not at other ages (Balduini and Costa, 1989). The finding that exposure of adult rats to dosages of ethanol resulting in nearly identical blood alcohol levels (approx. 250 mg/dl) had no effects on muscarinic receptor-stimulated phosphoinositide metabolism suggested that this system may represent a molecular target specifically relevant for the developmental neurotoxicity of ethanol. A series of *in vitro* studies (Balduini and Costa, 1990; Balduini *et al.*, 1991a; Candura *et al.*, 1991, 1992a) have addressed this hypothesis. Ethanol was found to inhibit the effect of muscarinic receptor agonists on phosphoinositide metabolism at concentrations equivalent to those present following exposure *in vivo*. The effect was region-dependent, in accordance with the reported differential developmental effects of ethanol in brain areas (West, 1986). For example, in cerebral cortex, hippocampus and cerebellum, the effect of ethanol was more pronounced than in brainstem, which is not affected by ethanol following *in vivo* administration. The effect of ethanol was neurotransmitter-specific: no effect on norepinephrine-, serotonin-, or histamine-stimulated phosphoinositide metabolism was observed. The effect of ethanol was also strongly age-dependent: inhibition of phosphoinositide metabolism was more pronounced in the 7-day-old rat and less so in younger or older animals. Ethanol also inhibited carbachol-stimulated inositol metabolism in cortical membrane preparations from neonatal rats. Inhibition of muscarinic receptor-stimulated phosphoinositide metabolism was also produced by *n*-propanol and *t*-butanol, two aliphatic alcohols which cause

microencephaly when administered during the brain growth spurt in the rat (Grant and Samson, 1982, 1984). Altogether, these data provide strong, though still circumstantial, evidence that the metabolism of phosphoinositides stimulated by activation of muscarinic receptors might represent a relevant and specific target for the developmental neurotoxicity of ethanol. Further studies should investigate the exact molecular mechanism of this effect of ethanol and the consequence of such inhibition on intracellular calcium mobilization and activation of protein kinase C.

2.8 CONCLUSIONS

Over the last several years there has been considerable growth in the knowledge of muscarinic receptors. Five subtypes of muscarinic receptors have been cloned and the role of each structural domain in agonist and antagonist binding and in coupling to second messenger systems is being investigated. While molecular biologists lead the way, research in chemistry and pharmacology lags behind in discovering new chemicals that possess receptor subtype specificity. Particularly for the m_4 and m_5 subtype, no selective agent is yet available. Application of the novel tools for studying muscarinic receptors will find its ways also in investigations on their development in the nervous system. Our knowledge on the development of receptor subtypes and on their second messenger system is still limited. The interaction of acetylcholine with other neurotransmitter systems during brain development is also worthy of further investigation. Progress in developmental neurobiology would certainly be necessary to place all these findings in a functional physiological perspective.

ACKNOWLEDGEMENTS

Work by the author was supported in part by grants from the National Institute of

Environmental Health Sciences (ES-04696), the Alcohol and Drug Abuse Institute, University of Washington, and the Fondazione Clinica del Lavoro, Pavia. Thanks are due to Drs A.F. Castoldi and S.M. Candura for their useful comments and to Ms Chris Sievanen for secretarial assistance.

REFERENCES

Agullo, L. and Garcia, A. (1991) Norepinephrine increases cyclic GMP in astrocytes by a mechanism dependent on nitric oxide synthesis. *Eur. J. Pharmacol.*, **206**, 343–6.

Akins, P.T., Surmeier, D.J. and Kitai, S.T. (1990) M_1 muscarinic acetylcholine receptor in cultured rat neostriatum regulates phosphoinositide hydrolysis. *J. Neurochem.*, **54**, 266–73.

Ambrosini, A. and Meldolesi, J. (1989) Muscarinic and quisqualate receptor-induced phosphoinositide hydrolysis in primary cultures of striatal and hippocampal neurons. Evidence for differential mechanisms of activation. *J. Neurochem.*, **53**, 825–33.

Araujo, D.M., Lapchak, P.A., Meaney, M.J. *et al.* (1990) Effects of aging on nicotinic and muscarinic autoreceptor function in the rat brain: relationship to presynaptic cholinergic markers and binding sites. *J. Neurosci.*, **10**, 3069–78.

Aronstam, R. and Eldefrawi, M.E. (1979) Transition and heavy metal inhibition of ligand binding to muscarinic acetylcholine receptors from rat brain. *Toxicol. Appl. Pharmacol.*, **48**, 489–96.

Aronstam, R.S., Kellog, C. and Abood, L.G. (1979) Development of muscarinic cholinergic receptors in inbred strains of mice: identification of receptor heterogeneity and relation to audiogenic seizure susceptibility. *Brain Res.*, **162**, 231–41.

Ashendel, C.L. (1985) The phorbol ester receptor: a phospholipid-regulated protein kinase. *Biochim. Biophys. Acta*, **822**, 219–42.

Ashkenazi, A., Ramachandran, J. and Capon, D.J. (1989) Acetylcholine analogue stimulates DNA synthesis in brain-derived cells via specific muscarinic receptor subtypes. *Nature*, **340**, 146–50.

Aubert, I., Cecyre, D., Araujo, D.M. *et al.* (1990) Comparative expression of cholinergic markers during brain development. *Soc. Neurosci. Abst.*, **16**, 536.

Bajgar, J., Hrdina, V., Petr, R. and Golda, V. (1979) Development of cholinergic nervous system in the brain of normal and hypertensive rats. *Dev. Neurosci.*, **2**, 94–100.

Balduini, W. and Costa, L.G. (1989) Effects of ethanol on muscarinic receptor-stimulated phosphoinositide metabolism during brain development. *J. Pharmacol. Exp. Ther.*, **250**, 541–7.

Balduini, W. and Costa, L.G. (1990) Developmental neurotoxicity of ethanol: *in vitro* inhibition of muscarinic receptor-stimulated phosphoinositide metabolism in brain from neonatal but not adult rats. *Brain Res.*, **512**, 248–52.

Balduini, W., Murphy, S.D. and Costa, L.G. (1987) Developmental changes in muscarinic receptor-stimulated phosphoinositide metabolism in rat brain. *J. Pharmacol. Exp. Ther.*, **241**, 421–7.

Balduini, W., Murphy, S.D. and Costa, L.G. (1990a) Characterization of cholinergic muscarinic receptor-stimulated phosphoinositide metabolism in brain from immature rats. *J. Pharmacol. Exp. Ther.*, **253**, 573–9

Balduini, W., Murphy, S.D. and Costa, L.G. (1990b) Potassium ions potentiate the muscarinic receptor-stimulated phosphoinositide metabolism in cerebral cortex slices: a comparision of neonatal and adult rats. *Neurochem. Res.*, **15**, 33–9.

Balduini W., Candura, S.M., Manzo, L. *et al.* (1991a) Time-concentration-, and age-dependent inhibition of muscarinic receptor-stimulated phosphoinositide metabolism by ethanol in the developing rat brain. *Neurochem. Res.*, **16**, 1235–40.

Balduini, W., Candura, S.M. and Costa, L.G. (1991b) Regional development of carbachol-, glutamate-, norephinephrine- and serotonin-stimulated phosphoinositide metabolism in rat brain. *Dev. Brain Res.*, **62**, 115–20.

Bartolami, S., Guiramund, J., Lenoir, M. *et al.* (1990) Carbachol-induced inositol phosphate formation during rat cochlea development. *Hear. Res.*, **47**, 229–34.

Ben-Barak, J., Gazit, H., Silman, I. and Dudai, Y. (1981) In vivo modulation of the number of muscarinic receptors in rat brain by cholinergic ligands. *Eur. J. Pharmacol.*, **74**, 73–81.

Ben-Baruch, G., Egozi, Y., Kloog, Y. *et al.* (1981) Altered ontogenesis of muscarinic cholinergic receptor in mouse brain: effect of L-thyroxine and betamethasone. *Endocrinology*, **109**, 235–9.

Berridge, M.J. (1987) Inositol triphosphate and diacylglycerol: two interacting second messengers. *Annu. Rev. Biochem.*, **56**, 159–93.

Boess, F.G., Balasubramian, M.K., Brammer, M.J. and Campbell, I.C. (1990) Stimulation of muscarinic acetylcholine receptors increases synaptosomal free calcium concentration by protein kinase-dependent opening of L-type calcium channels. *J. Neurochem.*, **55**, 230–6.

Bonner, T.I. (1989) The molecular basis of muscarinic receptor diversity. *Trends Neurosci.*, **12**, 148–51.

Brooksbank, B.W.L., Martinez, M., Atkinson, D.J. and Balazs, R., (1978) Biochemical development of the human brain. I. Some parameters of the cholinergic system. *Dev. Neurosci.*, **1**, 267–84.

Brown, J.H. (ed.) (1989) *The Muscarinic Receptors*. Humana Press, Clifton, NJ.

Brown, T.H., Chapman, P.F., Kairiss, E.W. and Keenan, C.L. (1988) Long-term synaptic potentiation. *Science*, **242**, 724–8.

Buckley, N.J., Bonner, T.I. and Brann, M.R. (1988) Localization of a family of muscarinic receptors mRNAs in rat brain. *J. Neurosci.*, **8**, 4646–52.

Candura, S.M., Balduini, W. and Costa, L.G. (1991) Interaction of short chain aliphatic alcohols with muscarinic receptor-stimulated phosphoinositide metabolism in cerebral cortex from neonatal and adult rats. *Neurotoxicology*, **12**, 23–32.

Candura, S.M., Manzo, L. and Costa, L.G. (1992a) Inhibition of muscarinic receptor- and G-protein-dependent phosphoinositide metabolism in cerebrocortical membranes from neonatal rats by ethanol. *Neurotoxicology*, **13**, 281–8.

Candura, S.M., Manzo, L. and Costa, L.G. (1992b) Guanine nucleotide- and muscarinic agonist-dependent phosphoinositide metabolism in synaptoneurosomes from cerebral cortex of immature rats. *Neurochem. Res.*, **17**, 1133–41.

Castoldi, A.F., Manzo, L. and Costa, L.G. (1993) Cyclic GMP formation induced by muscarinic receptors is mediated by nitric oxide synthesis in rat primary cultures. *Brain Res.* (in press).

Chiarugi, V.P., Ruggiero, M. and Corradetti, R. (1989) Oncogenes, protein kinase C, neuronal differentiation and memory. *Neurochem. Int.*, **14**, 1–9.

Chiu, A.S., Li, P.P. and Warsh, J.J. (1988) G-protein involvement in central nervous system muscarinic receptor-coupled polyphosphoinositide hydrolysis. *Biochem. J.*, **256**, 995–9.

Costa, L.G. (1988) Organophosphorus compounds, in *Recent Advances in Nervous System Toxicology* (eds C.L. Galli, L. Manzo and P.S. Spencer), Plenum Press, New York, pp. 203–46.

Costa, L.G. (1990) The phosphoinositide/protein kinase C system as a potential target for neurotoxicity. *Pharmacol. Res.*, **22**, 393–408.

Costa, L.G. and Fox, D.A. (1983) A selective decrease of cholinergic muscarinic receptors in the visual cortex of adult rats following developmental lead exposure. *Brain Res.*, **276**, 259–66.

Coyle, J.T. (1977) Biochemical aspects of neurotransmission in the developing brain. *Int. Rev. Neurobiol.*, **20**, 65–103.

Coyle, J.T. and Yamamura, H.I. (1976) Neurochemical aspects of the ontogenesis of cholinergic neurons in the rat brain. *Brain Res.*, **118**, 429–40.

De Boer, P., Westerink, B.H.C., Rollema, H. *et al.* (1990) An M_3-like muscarinic autoreceptor regulates the *in vivo* release of acetylcholine in rat striatum. *Eur. J. Pharmacol.*, **179**, 167–72.

Diaz-Meco, M.T., Larrodera, P., Lopez-Barahona, M. *et al.* (1989) Phospholipase C-mediated hydrolysis of phosphatidylcholine is activated by muscarinic agonists. *Biochem. J.*, **263**, 115–20.

Dohanick, G.P., Witcher, J.A., Weaver, D.R. and Clemens, L.G. (1982) Alteration of muscarinic binding in specific brain areas following estrogen treatment. *Brain Res.*, **241**, 347–50.

Dörner, G., Bluth, R. and Tönjer, R. (1982) Acetylcholine concentrations in the developing brain appear to affect emotionality and mental capacity in later life. *Acta Biol. Med. Germ.*, **41**, 721–3.

Drummond, A.H., Joels, L.A. and Hughes, P.J. (1987) The interaction of lithium ions with inositol lipid signalling systems. *Biochem. Soc. Trans.*, **15**, 32–5.

Dudai, Y. and Yavin, E. (1978) Ontogenesis of muscarinic receptors and acetylcholinesterase in differentiating rat cerebral cortex in culture. *Brain Res.*, **155**, 368–73.

Dudai, Y., Ben-Barak, J., Silman, I. and Gazik, H. (1980) Ontogenesis and modulation of cholinergic receptors in rat brain, in *Neurotransmitters and Their Receptors* (eds U.Z. Littauer, Y. Dudai, I. Silman *et al.*), Wiley, New York, pp. 217–39.

Dudek, S.M., Bowen, W.D. and Bear, M.F. (1989) Postnatal changes in glutamate-stimulated phosphoinositide turnover in rat neocortical synaptoneurosomes. *Dev. Brain. Res.*, **47**, 123–8.

East, J.M. and Dutton, G.R. (1980) Muscarinic binding sites in developing normal and mutant mouse cerebellum. *J. Neurochem.*, **34**, 657–61.

Egozi, Y., Kloog, Y. and Sokolovsky, M. (1980) Studies on postnatal changes of muscarinic receptors in mouse brain, in *Neurotransmitters and Their Receptors* (eds U.Z. Littauer, Y. Dudai, I. Silman *et al.*), Wiley, New York, pp. 201–15.

Egozi, Y., Sokolovsky, M., Scheijter, E. *et al.* (1986) Divergent regulation of muscarinic binding sites and acetylcholinesterase in discrete regions of the developing human fetal brain. *Cell. Mol. Neurobiol.*, **6**, 55–70.

El-Fakahany, E.E. and Cioffi, C.L. (1990) Molecular mechanisms of regulation of neuronal muscarinic receptor sensitivity. *Membr. Biochem.*, **9**, 9–27.

El-Fakahany, E.E. and Richelson, E. (1983) Effect of some calcium antagonists on muscarinic receptor-mediated cyclic GMP formation. *J. Neurochem.*, **40**, 705–10.

Ellis, J., Huyler, J.H., Kemp, D.E. and Weiss, S. (1990) Muscarinic receptors and second-messenger responses of neurons in primary culture. *Brain Res.*, **511**, 234–40.

Enna, S.J., Yamamura, H.I. and Snyder, S.H. (1976) Development of muscarinic cholinergic and GABA receptor binding in chick embryo brain. *Brain Res.*, **101**, 177–83.

Eriksson, P. and Nordberg, A. (1986) The effects of DDT, DDOH-palmitic acid and a chlorinated paraffin on muscarinic receptors and the sodium-dependent choline uptake in the central nervous system of immature mice. *Toxicol. Appl. Pharmacol.*, **85**, 121–7.

Eriksson, P. and Nordberg, A. (1990) Effects of two pyrethroids, Bioallethrin and Deltamethrin, on subpopulations of muscarinic and nicotinic receptors in the neonatal mouse brain. *Toxicol. Appl. Pharmacol.*, **102**, 456–63.

Eusebi, F., Pasetto, N. and Siracusa, G. (1984) Acetylcholine receptors in human oocytes. *J. Physiol.*, **346**, 321–30.

Eva, C. and Costa, E. (1986) Potassium ions facilitation of phosphoinositide turnover activation by muscarinic receptor agonists in rat brain. *J. Neurochem.*, **46**, 1429–35.

Evans, R.A., Watson, M., Yamamura, H.I. and Roeske, W.R. (1985) Differential ontogeny of putative M_1 and M_2 muscarinic receptor binding sites in the murine cerebral cortex and heart. *J. Pharmacol. Exp. Ther.*, **235**, 612–18.

Fain, J.N. (1990) Regulation of phosphoinositide-specific phospholipase C. *Biochim. Biophys. Acta*, **1053**, 81–8.

Fernandez-Tomè, P. and Segal, M. (1987) Ontogenesis of muscarinic receptors in cultured rat hippocampal cells. *Dev. Brain Res.*, **35**, 158–60.

Filogamo, G. and Marchisio, P.C. (1971) Acetylcholine system and neural development. *Neurosci. Res.*, **4**, 29–64.

Forray, C. and El-Fakahany, E.E. (1990) On the involvement of multiple muscarinic receptor subtypes in the activation of phosphoinositide metabolism in rat cerebral cortex. *Mol. Pharmacol.*, **37**, 893–902.

Fox, D.A., Wright, A.A. and Costa, L.G. (1982) Visual acuity deficits following neonatal lead exposure: cholinergic interactions. *Neurobehav. Toxicol. Teratol.*, **4**, 689–93.

Garthwaite, J., Garthwaite, G., Palmer, R.M.J. and Moncada, S. (1989) NMDA receptor activation induces nitric oxide synthesis from arginine in rat brain slices. *Eur. J. Pharmacol.*, **172**, 413–16.

Gonzales, R.A. and Crews, F.T. (1984) Characterization of the cholinergic stimulation of phosphoinositide hydrolysis in rat brain slices. *J. Neurosci.*, **4**, 3120–7.

Gonzales, R.A., Feldstein, J.B., Crews, F.T. and Raizada, M.K. (1985) Receptor-mediated inositide hydrolysis is a neuronal response: comparison of primary neuronal and glial cultures. *Brain Res.*, **345**, 350–5.

Gonzales, R.A., Greger, P.H., Baker, S.P. *et al.* (1987) Phorbol esters inhibit agonist-stimulated phosphoinositide hydrolysis in neuronal primary cultures. *Dev. Brain Res.*, **37**, 759–66.

Goyal, R.K. (1989) Muscarinic receptor subtypes. Physiology and clinical implications. *N. Engl. J. Med.*, **321**, 1022–9.

Grant, K.A. and Samson, H.H. (1982) Ethanol and tertiary butanol-induced microencephaly in the neonatal rat: comparison of brain growth parameters. *Neurobehav. Toxicol. Teratol.*, **4**, 315–21.

Grant, K.A. and Samson, H.H. (1984) *n*-Propanol-induced microencephaly in the neonatal rat. *Neurobehav. Toxicol. Teratol.*, **6**, 165–9.

Gremo, F., Palomba, M., Marchisio, A.M. *et al.* (1987) Heterogeneity of muscarinic cholinergic receptors in the developing human fetal brain: regional distribution and characterization. *Early Hum. Dev.*, **15**, 165–77.

Haga, K., Haga, T. and Ichiyama, A. (1990) Phosphorylation by protein kinase C of the muscarinic acetylcholine receptor. *J. Neurochem.*, **54**, 1639–44.

Hammer, R., Berrie, C.P., Birdsall, N.J.M. *et al.* (1980) Pirenzepine distinguishes between different subclasses of muscarinic receptors. *Nature*, **283**, 90–2.

Hanley, M.R. (1989) Mitogenic neurotransmitters. *Nature*, **340**, 97.

Harden, T.K. (1989) Muscarinic cholinergic receptor-mediated regulation of cyclic AMP metabolism, in *The Muscarinic Receptors* (ed. J.H. Brown), Humana Press, Clifton, NJ, pp. 221–58.

Harden, T.K. (1990) G Protein-dependent regulation of phospholipase C by cell surface receptors. *Am. Rev. Respir. Dis.*, **141**, 5119–22.

Hawkins, P.T., Reynolds, D.J.M., Poyner, D.R. and Hanley, M.R. (1990) Identification of a novel inositol phosphate recognition site: specific ³H-inositol hexakisphosphate binding to brain regions and cerebellar membranes. *Biochem. Biophys. Res. Commun.*, **167**, 819–27.

Heacock, A.M., Fisher, S.K. and Agranoff, B.W. (1987) Enhanced coupling of neonatal muscarinic receptors in rat brain to phosphoinositide turnover. *J. Neurochem.*, **48**, 1904–11.

Hohman, C.F. and Ebner, F.F. (1985) Development of cholinergic markers in mouse forebrain. I. Choline acetyltransferase enzyme activity and acetylcholinesterase histochemistry. *Dev. Brain Res.*, **23**, 225–41.

Hohman, C.F., Pert, C.C. and Ebner, F.F. (1985) Development of cholinergic markers in mouse forebrain. II. Muscarinic receptor binding in cortex. *Dev. Brain Res.*, **23**, 243–53.

Hohman, C.F., Brooks, A.C. and Coyle, J.T. (1988) Neonatal lesions of the basal forebrain cholinergic neurons result in abnormal cortical development. *Dev. Brain Res.*, **42**, 253–64.

Hoover, R.K. and Toews, M.L. (1990) Activation of protein kinase C inhibits internalization and down regulation of muscarinic receptors in 1321N1 human astrocytoma cells. *J. Pharmacol. Exp. Ther.*, **253**, 185–91.

Horwitz, J. (1990) Carbachol and bradykinin increase the production of diacylglycerol from sources other than inositol-containing phospholipids in PC-12 cells. *J. Neurochem.*, **54**, 983–91.

Hulme, E.C., Birdsall, N.J.M. and Buckley, N.J. (1990) Muscarinic receptor subtypes. *Annu. Rev. Pharmacol. Toxicol.*, **30**, 633–73.

Ignarro, L.J. (1991) Signal transduction mechanisms involving nitric oxide. *Biochem. Pharmacol.*, **41**, 485–90.

Irvine, R.F., Moor, R.M., Pollock, W.K. *et al.* (1988) Inositol phosphates: proliferation, metabolism and function. *Philos. Trans. R. Soc. Lond. [Biol.]*, **320**, 281–98.

Jia, W.G., Shaw, C., van Huizen, F. and Cynader, M.S. (1989) Phorbol 12,13-dibutyrate regulates muscarinic receptors in rat cerebral cortical slices by activating protein kinase C. *Mol. Brain Res.*, **5**, 311–15.

Kasa, P., Bansaghy, K., Rakonczay, Z. and Guyla, K. (1982) Postnatal development of the acetylcholine system in different parts of the rat cerebellum. *J. Neurochem.*, **39**, 1726–32.

Kater, S.B. and Mills, L.R. (1991) Regulation of growth cone behavior by calcium. *J. Neurosci.*, **11**, 891–9.

Kendall, D.A. (1986) Cyclic GMP and inositol phosphate accumulation do not share common origins in rat brain slices. *J. Neurochem.*, **47**, 1483–9.

Kikkawa, U., and Nishizuka, Y. (1986) The role of protein kinase C in transmembrane signalling. *Annu. Rev. Cell Biol.*, **2**, 149–78.

Kikkawa, U., Ogita, K., Shearman, M.S. *et al.* (1988) The heterogeneity and differential expression of protein kinase C in nervous tissue. *Philos. Trans. R. Soc. Lond. [Biol.]*, **320**, 313–24.

Kilbinger, H. (1987) Control of acetylcholine release by muscarinic autoreceptors, in *International Symposium on Muscarinic Cholinergic Mechanisms* (eds S. Cohen and M. Sokolovsky), Freund, London, pp. 219–28.

Kotas, A.M. and Prince, A.K. (1987) High-affinity uptake of choline, a marker for cholinergic nerve terminals, is not specific in developing rat brain. *Dev. Brain Res.*, **35**, 175–81.

Kubo, T., Fukuda, R., Mikami, A. *et al.* (1986) Cloning, sequencing and expression of complementary DNA encoding the muscarinic acetylcholine receptor. *Nature*, **323**, 411–16.

Kudo, Y., Ogura, A. and Iijima, T. (1988) Stimulation of muscarinic receptor in hippocampal neuron induces characteristic increase in cytosolic free Ca²⁺ concentration. *Neurosci. Lett.*, **85**, 345–50.

Kuhar, M.J., Birdsall, N.J.M., Burgen, A.S.V. and Hulme, E.C. (1980) Ontogeny of muscarinic receptors in rat brain. *Brain Res.*, **184**, 375–83.

Labarca, R., Janowsky, A., Patel, J. and Paul, S.M. (1984) Phorbol esters inhibit agonist-induced [³H] inositol-1-phosphate accumulation in rat hippocampal slices. *Biochem. Biophys. Res. Commun.*, **123**, 703–9.

Ladinsky, H., Consolo, S., Peri, G. and Garattini, S. (1972) Acetylcholine, choline and choline acetyltransferase activity in the developing brain of normal and hypothyroid rats. *J. Neurochem.*, **19**, 1947–52.

Lai, H., Carino, M.A. and Wen, Y.F. (1989) Repeated noise exposure affects muscarinic cholinergic receptors in the rat brain. *Brain Res.*, **488**, 361–4.

Lai, W.S., Rogers, T.B. and El-Fakahany, E.E. (1990) Protein kinase C is involved in desensitization of muscarinic receptors induced by phorbol esters but not by receptor agonists. *Biochem. J.*, **267**, 23–9.

Lanier, L.P., Dunn, A.J. and van Hartesveldt, C.

(1976) Development of neurotransmitters and their function in brain. *Rev. Neurosci.*, **2**, 195–257.

Large, T.H., Lambert, M.P., Gremillion, M.A. and Klein, W.L. (1986) Parallel postnatal development of choline acetyltransferase activity and muscarinic acetylcholine receptors in the rat olfactory bulb. *J. Neurochem.*, **46**, 671–80.

Lauder, J.M. (1988) Neurotransmitters as morphogens. *Progr. Brain Res.*, **73**, 365–87.

Lee, W., Nichlaus, K.J., Manning, D.C. and Wolfe, B.B. (1990) Ontogeny of cortical muscarinic receptor subtypes and muscarinic receptor-mediated responses in rat. *J. Pharmacol. Exp. Ther.*, **252**, 482–90.

Levine, R.R. and Birdsall, N.J.M. (1989) Nomenclature for muscarinic receptor subtypes recommended by symposium. *Trends Pharmacol. Sci.* Suppl., *Subtypes of Muscarinic Receptors*, p. vii.

Levy, A. (1981) The effect of cholinesterase inhibition on the ontogenesis of central muscarinic receptors. *Life Sci.*, **29**, 1065–70.

Liles, W.C., Hunter, D.D., Meier, K.E. and Nathanson, N.M. (1986) Activation of protein kinase C induces rapid internalization and subsequent degradation of muscarinic acetylcholine receptors in neuroblastoma cells. *J. Biol. Chem.*, **261**, 5307–16.

Llinas, R.R. (1982) Calcium in synaptic transmission. *Sci. Am.*, **247**, 56–65.

Lo, W.W.Y. and Hughes, J. (1987) Receptor-phosphoinositidase C coupling. Multiple G proteins? *FEBS Lett.*, **244**, 1–3.

Mallol, J., Sarraga, M.C., Bartolomè, M. *et al.* (1984) Muscarinic receptor during postnatal development of rat cerebellum: an index of cholinergic synapse formation? *J. Neurochem.*, **42**, 1641–9.

Martinson, E.A., Goldstein, D. and Brown, J.H. (1989) Muscarinic receptor activation of phosphatidylcholine hydrolysis. *J. Biol. Chem.*, **264**, 14748–54.

Mash, D.C., Flynn, D.D. and Potter, L.T. (1985) Loss of M_2 muscarinic receptors in the cerebral cortex in Alzheimer's disease and experimental cholinergic denervation. *Science*, **228**, 1115–17.

Mattson, M.P. (1988) Neurotransmitters in the regulation of neuronal cytoarchitecture. *Brain Res. Rev.*, **13**, 179–212.

Mattson, M.P. (1989) Acetylcholine potentiates glutamate-induced neurodegeneration in cultured hippocampal neurons. *Brain Res.*, **497**, 402–6.

Mattson, M.P. and Hauser, K.F. (1991) Spatial and temporal integration of neurotransmitter signals in the development of neural circuitry. *Neurochem. Int.*, **19**, 17–24.

McKinney, M. and Richelson, E. (1986) Blockade of N1E-115 murine neuroblastoma muscarinic receptor function by agents that affect the metabolism of arachidonic acid. *Biochem. Pharmacol.*, **35**, 2389–97.

McKinney, M. and Richelson, E. (1989) Muscarinic receptor regulation of cyclic GMP and eicosanoid production, in *The Muscarinic Receptors* (ed J.H. Brown), Humana Press, Clifton, NJ, pp. 309–40.

Mei, L., Roeske, W.R. and Yamamura, H.I. (1989) Molecular pharmacology of muscarinic receptor heterogeneity. *Life Sci.*, **45**, 1831–51.

Meier, E., Hertz, L. and Schousboe, A. (1991) Neurotransmitters as developmental signals. *Neurochem. Int.*, **19**, 1–15.

Michalek, H., Pintor, A., Fortuna, S. and Bisso, G.M. (1985) Effects of diisopropylfluorophosphate on brain cholinergic systems of rats at early developmental stages. *Fundam. Appl. Toxicol.*, **5**, S204–12.

Mitchelson, F. (1988) Muscarinic receptor differentiation. *Pharmacol. Ther.*, **37**, 357–423.

Miyoshi, R., Kito, S., Shimizu, M. and Matsubayashi, H. (1987) Ontogeny of muscarinic receptors in the rat brain with emphasis on the differentiation of M_1- and M_2-subtypes – semiquantitative *in vitro* autoradiography. *Brain Res.*, **420**, 302–12.

Muller, F., Dumez, Y. and Massoulie, J. (1985) Molecular forms and solubility of acetylcholinesterase during the embryonic development of rat and human brain. *Brain Res.*, **331**, 295–302.

Nathanson, N.M. (1987) Molecular properties of the muscarinic acetylcholine receptor. *Annu. Rev. Neurosci.*, **10**, 195–236.

Nathanson, N.M. (1989) Regulation and development of muscarinic receptor number and function, in *The Muscarinic Receptors* (ed. J.H. Brown), Humana Press, Clifton, NJ, pp. 419–54.

Navarro, H.A., Seidler, F.J., Eylers, J. *et al.* (1989) Effects of prenatal nicotine exposure on development of central and peripheral cholinergic neurotransmitter systems. Evidence for cholinergic trophic influences in developing brain. *J. Pharmacol. Exp. Ther.*, **251**, 894–900.

Nishizuka, Y. (1988) The molecular heterogeneity of protein kinase C and its implication for cellular regulation. *Nature*, **334**, 661–5.

Nordberg, A. and Winblad, B. (1981) Cholinergic receptors in human hippocampus. Regional distribution and variance with age. *Life Sci.*, **29**, 1937–44.

Nordberg, A., Wahlström, G. and Larsson, C. (1980) Increased number of muscarinic binding sites in brain following chronic barbiturate treatment to rat. *Life Sci.*, **26**, 231–7.

Ohsako, S. and Deguchi, T. (1981) Stimulation by phosphatidic acid of calcium influx and cyclic GMP synthesis in neuroblastoma cells. *J. Biol. Chem.*, **256**, 10945–8.

Olianas, M.C., Onali, P., Neff, N.H. and Costa, E. (1983) Adenylate cyclase activity of synaptic membranes from rat striatum. Inhibition by muscarinic receptor agonists. *Mol. Pharmacol.*, **23**, 393–8.

Osborne, N.N. (1988) Muscarinic stimulation of inositol phosphate formation in rat retina: developmental changes. *Vision Res.*, **29**, 871–81.

Pearce, B. and Murphy, S. (1988) Neurotransmitter receptors coupled to inositol phospholipid turnover and Ca^{2+} flux: consequences for astrocyte function, in *Glial Cell Receptors* (ed. H.K. Kimelberg), Raven Press, New York, pp. 197–221.

Pedata, F., Slavikova, J., Kotas, A. and Pepeu, G. (1983) Acetylcholine release from rat cortical slices during postnatal development and aging. *Neurobiol. Aging*, **4**, 31–5.

Peralta, E.G., Ashkenazi, A., Winslow, J.W. *et al.* (1988) Differential regulation of PI hydrolysis and adenylyl cyclase by muscarinic receptor subtypes. *Nature*, **334**, 434–7.

Peralta, E.G., Ashkenazi, A., Winslow, J.W. *et al.* (1987) Distinct primary structures, ligand-binding properties and tissue-specific expression of four human muscarinic acetylcholine receptors. *EMBO J.*, **6**, 3923–9.

Perry, E.K., Smith, C.J., Atach, J.R. *et al.* (1986) Neocortical cholinergic enzyme and receptor activities in the human fetal brain. *J. Neurochem.*, **47**, 1262–9.

Potter, L.T., Flynn, D.D., Hanchett, H.E. *et al.* (1983) Independent M_1 and M_2 receptors: ligands, autoradiography and functions. *Trends Pharmacol. Sci.* (Suppl), *Subtypes of Muscarinic Receptors*, pp. 22–31.

Purpura, D.P. (1972) Intracellular studies of synaptic organization in the mammalian brain, in *Structure and Function of Synapses* (eds G.D. Pappas and D.P. Purpura), Raven Press, New York, pp. 257–302.

Qian, Z. and Drewes, L.R. (1989) Muscarinic acetylcholine receptor regulates phosphatidyl-choline phospholipase D in canine brain. *J. Biol. Chem.*, **264**, 21720–4.

Rana, R.S. and Hokin, L.E. (1990) Role of phosphoinositides in transmembrane signaling. *Physiol. Rev.*, **70**, 115–64.

Ravikumar, B.V. and Sastry, P.S. (1985) Muscarinic cholinergic receptors in human fetal brain: characterization and ontogeny of ^{3}H-quiniclidinyl benzilate binding sites in frontal cortex. *J. Neurochem.*, **44**, 240–6.

Reece, L.J. and Schwartzkroin, P.A. (1991) Effects of cholinergic agonists on immature rat hippocampal neurons. *Dev. Brain Res.*, **60**, 29–42.

Represa, A., Chanez, C., Flexor, M.A. and Ben-Ari, V. (1989) Development of the cholinergic system in control and intra-uterine growth retarded rat brain. *Dev. Brain Res.*, **47**, 71–9.

Rooney, T.A. and Nahorski, S.R. (1987) Postnatal ontogeny of agonist and depolarization-induced phosphoinositide hydrolysis in rat cerebral cortex. *J. Pharmacol. Exp. Ther.*, **243**, 333–41.

Rotter, A., Field, P.M. and Raisman, G. (1979) Muscarinic receptors in the central nervous system of the rat. III. Postnatal development of binding of ^{3}H-propylbenzilylcholine mustard. *Brain Res. Rev.*, **1**, 185–205.

Sandmann, J. and Wurtman, R.J. (1991) Stimulation of phospholipase D activity in human neuroblastoma (LA-N-2) cells by activation of muscarinic acetylcholine receptors or by phorbol esters: relationship to phosphoinositide turnover. *J. Neurochem.*, **56**, 1312–19.

Schmidt, B.H., Manzoni, O.J.J., Royer, M. *et al.* (1991) Cholinergic inositol phosphate formation in striatal neurons is mediated by distinct mechanisms. *Eur. J. Pharmacol.*, **206**, 87–94.

Schultz, G., Hardman, J.G., Schultz, K. *et al.* (1973) The importance of calcium ions for the regulation of guanosine 3′,5′-monophosphate levels. *Proc. Natl. Acad. Sci. USA*, **70**, 3889–93.

Serbus, D.C. and Light, K.E. (1990) Cholinergic alterations of hippocampus and cerebellum at postnatal days 21 and 25 following twice daily ethanol exposure of rats throughout the suckling period. *Alcoholism*, **14**, 336.

Shapiro, R.A., Scherer, N.M., Habecher, B.A. *et al.* (1988) Isolation, sequence and functional expression of the mouse M_1 muscarinic acetylcholine receptor gene. *J. Biol. Chem.*, **263**, 18397–403.

Singh, M.M., Warburton, D.M. and Lal, H. (eds) (1985) *Central Cholinergic Mechanisms and Adaptive Dysfunctions*. Plenum Press, New York.

Smalheiser, N.R. (1990) Neuronal growth cones: an extended view. *Neuroscience*, **38**, 1–11.

Smrcka, A.V., Hepler, J.R., Brown, K.D. and Sternweis, P.C. (1991). Regulation of phosphoinositide-specific phospholipase C activity by purified Gq. *Science*, **251**, 804–7.

Snider, R.M., McKinney, M., Forray, C. and Richelson, E. (1984) Neurotransmitter receptors mediate cyclic GMP formation by involvement of arachidonic acid and lipoxygenase. *Proc. Natl. Acad. Sci. USA*, **81**, 3905–9.

Snyder, S.H. and Bredt, D.S. (1991) Nitric oxide as a neuronal messenger. *Trends Pharmacol. Sci.*, **12**, 125–7.

Soreq, H., Gurwitz, D., Eliyahu, D. and Sokolovsky, M. (1982) Altered ontogenesis of muscarinic receptors in agranular cerebellar cortex. *J. Neurochem.*, **39**, 756–63.

Stamper, C.R., Balduini, W., Murphy, S.D. and Costa, L.G. (1988) Behavioral and biochemical effects of postnatal parathion exposure in the rat. *Neurotoxicol. Teratol.*, **10**, 261–6.

Stephens, L.R. and Logan, S.D. (1989) Formation of [^{3}H] inositol metabolites in rat hippocampal formation slices prelabeled with [^{3}H] inositol and stimulated with carbachol. *J. Neurochem.*, **52**, 713–21.

Stundermann, K.A., Harris, G.D. and Lovenberg, W. (1988) Characterization of inositol 1,4,5-triphosphate-stimulated calcium release from rat cerebellum microsomal fractions. *Biochem. J.*, **255**, 667–83.

Sun, Y.A. and Pao, M.M. (1987) Evoked release of acetylcholine from the growing embryonic neuron. *Proc. Natl. Acad. Sci. USA*, **84**, 2540–4.

Tietje, K.M., Goldman, P.S. and Nathanson, N.M. (1990) Cloning and functional analysis of a gene encoding a novel muscarinic acetylcholine receptor expressed in chick heart and brain. *J. Biol. Chem.*, **265**, 2828–34.

Vaca, K. (1988) The development of cholinergic neurons. *Brain Res. Rev.*, **13**, 262–86.

Vallejo, M., Jackson, T., Lightman, S. and Hanley, M.R. (1987) Occurrence and extracellular actions of inositol pentakis- and hexakis-phosphate in mammalian brain. *Nature*, **330**, 656–8.

Van Hoof, C.O.M., De Graan, P.N.E., Oestreicher, A.B. and Gispen, W.H. (1989) Muscarinic receptor activation stimulates B50/GAP43 phosphorylation in isolated nerve growth cones. *J. Neurosci.*, **9**, 3753–9.

van Delft, A.M.L., Hagan, J.J. and Tonnaer, J.A.D.M. (1989) Muscarinic receptors in the central nervous system. *Progr. Pharmacol. Clin. Pharmacol.*, **7**, 93–117.

Vicentini, L.M. and Villareal, M.L. (1986) Inositol phosphates turnover, cytosolic Ca^{2+} and pH: putative signals for the control of cell growth. *Life Sci.*, **38**, 2269–76.

Vilarò, M.T., Palacios, J.M. and Mengod, G. (1990) Localization of m$_5$ muscarinic receptor mRNA in rat brain examined by *in situ* hybridization histochemistry. *Neurosci. Lett.*, **114**, 154–9.

Wallace, M.A. and Claro, E. (1990) Comparison of serotoninergic to muscarinic cholinergic stimulation of phosphoinositide-specific phospholipase C in rat brain cortical membranes. *J. Pharmacol. Exp. Ther.*, **255**, 1296–300.

Wang, S.Z., Hu, J., Long, R.M. *et al.* (1990) Agonist-induced down regulation of m$_1$ muscarinic receptors and reduction of their mRNA level in a transfected cell line. *FEBS Lett.*, **276**, 185–8.

Weiss, S., Schmidt, B.H., Sebben, M. *et al.* (1988) Neurotransmitter-induced inositol phosphate formation in neurons in primary culture. *J. Neurochem.*, **50**, 1425–33.

Wess, J., Bonner, T.J. and Brann, M.R. (1990) Chimeric m$_2$/m$_3$ muscarinic receptors: role of carboxyl terminal receptor domains in selectivity of ligand binding and coupling to phosphoinositide hydrolysis. *Mol. Pharmacol.*, **38**, 872–7.

West, J.R. (1986) *Alcohol and Brain Development.* Oxford University Press, New York.

Wigal, S.B.E., Amsel, A. and Wilcox, R.E. (1990) Cholinergic alterations of hippocampus and cerebellum at postnatal days 21 and 25 following twice daily ethanol exposure of rats throughout the suckling period. *Alcoholism*, **4**, 336.

Worley, P.F., Baraban, J.M. and Snyder, S.H. (1989) Inositol 1,4,5-triphosphate receptor binding: autoradiographic localization in rat brain. *J. Neurosci.*, **9**, 339–46.

Yavin, E. and Harel, S. (1979) Muscarinic binding sites in the developing rabbit brain. *FEBS Lett.*, **97**, 151–4.

3

THE ROLE OF SEROTONIN AND SEROTONIN RECEPTORS IN DEVELOPMENT OF THE MAMMALIAN NERVOUS SYSTEM

Patricia M. Whitaker-Azmitia

The neurons which produce the neurotransmitter serotonin, make up one of the most widely distributed neuronal systems in the mammalian brain. This neuronal network is also one of the earliest developing systems. The turnover rate of serotonin (i.e. the ratio of metabolite to neurotransmitter) is higher in the immature mammalian brain than at any other time in life (Hamon and Bourgoin, 1981). This distribution, developmental pattern and high level of activity are the key elements in the role which serotonin plays in the immature brain – a role as a growth factor directing both proliferation and maturation (Lauder and Krebs, 1978; Buznikov, 1984; Lauder, 1983; Chubakov *et al.*, 1986; Lauder *et al.*, 1988; Lauder and Zimmerman, 1988; Lauder, 1990).

This chapter will first discuss what is known about the development of the serotonin system itself, since development of this system is so important to overall development of the brain. Then there will be discussion of which serotonin receptors mediate the developmental role of serotonin, since for any neurotransmitter to elicit a response, it must have a receptor.

3.1 DEVELOPMENT OF THE SEROTONIN NEURONAL SYSTEM

The anatomical development of the serotonin neuronal system has been most extensively studied in the rat, although the development has been described in several species, including chick (Wallace, 1985; Wallace *et al.*, 1986), leech (Stuart *et al.*, 1987), sheep (Tillet, 1988), xenopus (Messenger and Warner, 1989), fruitfly (Valles and White, 1988), cat (Gu *et al.*, 1990), mouse (Fujimiya *et al.*, 1986; Ni and Jonakait, 1989) and primate (see references in Jacobs and Azmitia, 1992). However, the following sections refer to the rat, unless noted otherwise.

The cell bodies of serotonin neurons develop as bilateral groups in two distinct regions of the brainstem of rat. First, a rostral collection of cells (caudal to the mesencephalic flexure) on embryonic day (ED) 12 and then in more dorsal regions near the pontine flexure on ED14 (Aitken and Tork, 1988). Through varying degrees of migration, these clusters develop into the adult pattern of nine groups (B 1–9) by ED18 (Lidov and Molliver, 1982a). The rostral group eventually will

Receptors in the Developing Nervous System Vol. 2: Neurotransmitters. Edited by Ian S. Zagon and Patricia J. McLaughlin. Published in 1993 by Chapman & Hall. ISBN 0 412 49400 0. Vols. 1 and 2 (set) ISBN 0 412 54520 9.

become B 4–9 and give rise to ascending fibers. The rostral group develops into B 1–3 and projects mainly into the spinal cord (Wallace and Lauder, 1983; Aitken and Tork, 1988).

As the serotonin cells migrate into nuclei, they continue to divide, with mitosis continuing up to ED18 (Aitken and Tork, 1988). After completion of mitosis, the cells begin to migrate in a new direction – towards the midline. Eventually, six of the nine nuclei become fused nuclei (the midline or raphe nuclei). Fusion takes place in a rostral to caudal direction, beginning at ED18 and ending at postnatal day (PD) 6.

Axonal projections from the nuclei develop very rapidly once the cells are present. In the rostral group, visualized by ED12, fibers are identified within 24 h entering into the diencephalon. In general the fibers appear to migrate along preformed non-serotonergic pathways, although there are some exceptions (Lidov and Molliver, 1982b; Wallace and Lauder, 1983). By ED17, these fibers have reached the frontal neocortical pole (Wallace and Lauder, 1983). Although the nuclei have become fused, they maintain their original projections, that is the projections are mostly ipsilateral (Levitt and Moore, 1978; Goto and Sano, 1984). Complete, adult-like innervation of the cerebral cortex is extremely heterogeneous, continuing up to PD21, since any one particular region is only innervated once it has reached a relatively mature architecture (Lidov and Molliver, 1982b).

Descending pathways also begin to develop almost immediately after the cells are differentiated, with the caudal-most regions of the spinal cord being reached by ED17. However, as with the ascending projections, there is a period of redistribution of terminals, with the final adult pattern being reached at PD21 (Rajaofetra *et al.*, 1989).

Dendritic arborization begins later than axonal growth, with a rapid increase from ED 19 until PD7. At the same time, the serotonin nuclei become less densely packed, as other cell types begin to develop in the vicinity (Lidov and Molliver, 1982b).

As the serotonin neurons are developing, the amount of serotonin produced, released and metabolized also increases rapidly. The enzyme responsible for the synthesis of serotonin, tryptophan hydroxylase, is virtually saturated by the very high levels of tryptophan. Only during this perinatal period are levels of tryptophan so high that this enzyme is saturated. The increased levels of tryptophan are caused by lack of tryptophan binding and an increased carrier system compared to that in the adult (Hamon and Bourgoin, 1979).

Like all neurotransmitter systems, a variety of factors are involved in the regulation of normal serotonin development. If any of these factors are altered, in concentration or time of exposure, the serotonin system may never recover fully. The remainder of this section describes some of these factors, although of course there are likely to be many more yet to be studied. Since serotonin itself is a growth factor, any of the factors which influence serotonin development may then indirectly influence the development of many other systems.

3.1.1. GROWTH AND ATTACHMENT FACTORS

Although many growth factors have been described in brain, few have actually been studied for a direct effect on specific neurotransmitter systems. For the serotonin system, nerve growth factor (NGF), epidermal growth factor (EGF) and insulin have all been shown to be without effect (Azmitia *et al.*, 1990a). However, for some time the presence of a soluble, proteinaceous growth factor for serotonin neurons has been suggested from lesioning studies and transplant studies. Moreover, this factor was shown to be selective for serotonin neurons, having no effect on noradrenergic neurons (Zhou *et al.*, 1987). In our work with astroglial cells, we have identified a substance which may be identical to this factor – the astroglial specific protein, S-100β; this is discussed in more detail later.

In addition to S-100, other astroglial-derived proteins appear to be important. Developing serotonin neurons show selective adhesion to astrocytes and display a greater rate and amount of neurite outgrowth, as compared to their growth on fibroblasts (Lieth *et al.*, 1990). At least some of this attachment and guidance may be due to the presence of laminin (Zhou and Azmitia, 1988).

3.1.2   NEUROTRANSMITTER EFFECTS

The normal course of serotonin development is also influenced by the levels of several neurotransmitters and neuromodulators. This includes an influence of serotonin itself on its own neurons – a circumstance referred to as autoregulation of development. Since this influence takes place through serotonin receptors, this work will be discussed later.

Dopamine is also very important in the regulation of serotonin neuronal development. Since the earliest studies by Breese *et al.* (1984) and more recently by Jackson *et al.* (1988), removal of dopamine by selective lesioning of neonates has been shown to lead to overgrowth of serotonin terminals into the caudate nucleus. Since dopamine had been shown to inhibit neurite outgrowth through a dopamine D_1 receptor (Lankford *et al.*, 1988), we proposed that dopamine may have a tonically inhibitory effect on serotonin development through this receptor. Removal of dopamine would therefore cause overgrowth into regions where D_1 receptors could be localized on serotonin terminals, such as the caudate. To test this hypothesis, animals were treated prenatally with a selective D_1 receptor agonist, SKF 38393. At 90 days of age these animals showed significant loss of serotonin terminals, plus altered behavioral sensitivity to both serotonin and dopamine agonists (Whitaker-Azmitia *et al.*, 1990b).

Substance P also has an influence on developing serotonin neurons, but this influence appears to be important only if the endogenous serotonin influence is lacking (Jonsson and Hallman, 1983). Also, ACTH and related peptides are stimulatory to growth, in the absence of trophic signals from a target region (Azmitia and De Kloet, 1987). Finally, enkephalin is inhibitory to the growth of serotonin neurons (Davila-Garcia and Azmitia, 1989).

3.1.3   EXOGENOUS FACTORS

Changes in serotonin levels and the subsequent development of brain have been shown to be induced by diet and malnutrition (Resnick and Morgane, 1984; Spear and Scalzo, 1985). This is presumably due to the high demand for tryptophan for normal development. Tryptophan, as an essential amino acid, is often lacking in poor diets.

Work by Peters has focused on the role of stress in altering the development of the serotonin system. This work has shown changes in serotonin behaviors and serotonin receptors by stress at various pre- and postnatal timepoints (Peters, 1982, 1988a, b, 1990). Some of these changes may be due to the influence of corticosteroids on the activity of tryptophan hydroxylase.

Finally, drugs of abuse may seriously damage the development of the serotonin system. For example, cocaine, 3,4-methylenedioxy amphetamine (MDMA) (Azmitia *et al.*, 1990b) and alcohol (Goodrich *et al.*, 1986; Schambra *et al.*, 1990) have all been shown to cause permanent changes.

3.2   SEROTONIN RECEPTORS INVOLVED IN DEVELOPMENT

In order for a neurotransmitter to function as a growth factor, there must be receptors present for that neurotransmitter. Characterizing these receptors, in terms of their pharmacology, transduction systems and ontogeny, is thus very important in understanding that neurotransmitter's developmental role.

Our initial studies, begun before all the currently known subtypes of adult serotonin receptors had been described, used 5-methoxytryptamine (5-MT), which is useful in pharmacological studies for its biphasic effects on serotonin receptors. At low doses, this agent stimulates release-regulating autoreceptors, whereas at high doses, it stimulates a postsynaptic receptor.

High affinity serotonin receptors are present prenatally and are, in a sense, functional, as they show appropriate compensatory changes in density in response to chronic stimulation or blockage (Whitaker-Azmitia *et al.*, 1987). To test directly for a role of these receptors in influencing development of serotonin neurons, ED14 rat serotonergic neurons were grown in culture for four days and exposed to varying doses of 5-MT. The growth responses of the serotonin neurons were biphasic, leading to the conclusion that two serotonin receptors were involved in regulation of development. The highest affinity receptor, presumably the presynaptic release-regulating autoreceptor, was inhibitory to growth of terminals. The lower affinity site was stimulatory (Whitaker-Azmitia and Azmitia, 1986a).

In a whole-animal model, similar results were obtained by treating animals prenatally from gestational day 12 until birth, with three different doses of 5-MT. A low dose of 5-MT (1 mg/kg) caused a decrease in terminal density in forebrain, up to PD30. Conversely, 3.0 mg/kg 5-MT caused significant increases up to PD30.

In addition to the neurochemical studies, the animals were tested in several behavioral paradigms – the neonatal serotonin syndrome as described by Spear and Ristene (1982) at PD5, spontaneous alternation and open field activity at PD15, and a passive avoidance paradigm (lick suppression) at PD30. In the neonatal animals, changes in response to quipazine were found in those pups which had been treated prenatally with 1.0 or 3.0 mg/kg, but there was no effect in those animals treated with the lowest dose, 0.1 mg/kg. By PD15, there were still significant behavioral changes in the animals treated prenatally with 1.0 or 3.0 mg/kg, but for the first time there were also significant changes in the lowest dose animals, those treated with 0.1 mg/kg. Finally, by PD30, the animals treated with the highest dose, 3.0 mg/kg were not impaired in the lick suppression paradigm. Interestingly, the animals which were the most impaired were those which had been treated with the lowest dose prenatally, 0.1 mg/kg. In addition to testing unchallenged behaviors, the animals treated prenatally with 1.0 mg/kg 5-MT were also tested at 30 days of age for open field activity, in response to challenges of 5-MT or apomorphine. These animals were found to have a blunted response to both receptor active drugs (Shemer *et al.*, 1991).

The classification of the serotonin receptors involved in development has not been a straightforward task. Although in searching for these receptors it has been assumed that they will be identical to those found in the adult brain, this may not necessarily be the case (Connell and Wallis, 1989). It is possible that the pharmacological profile of the receptor is different, or that the signal transduction mechanism could change. In these cases, strategies other than radioligand binding may be necessary, such as mRNA analysis or the use of specific antibodies. It is also possible that the receptor will be a transiently expressed receptor, which is lacking altogether in the adult brain. With these limitations in mind, we have tested selective drugs, and drawn some conclusions on how they may relate to brain development.

At present, there are thought to be seven different types of serotonin receptor in the adult brain. These are described in the following sections.

3.2.1   5-HT$_{1a}$ RECEPTORS

In adult brain, the 5-HT$_{1a}$ receptor can be both negatively and positively coupled to the production of cAMP. It is often referred to as the limbic serotonin receptor, as it is highly localized in these regions, particularly the hippocampus. It is virtually absent from the basal ganglia but can be found in the raphe nuclei, where it may act as a cell body autoreceptor. Drugs acting at this receptor are clinically useful in the treatment of depression and anxiety. The most commonly used radiolabel for this site is 8-[^{3}H]hydroxy-DPAT. This compound is also commonly used as an antagonist in behavioral and functional studies, as are ipsaperone, buspirone, geperone and tandospirone.

The 5-HT$_{1a}$ receptor is one of a group of neurotransmitter receptors which can be termed a 'transiently expressed receptor'; that is, at specific times in development very high amounts are expressed which then decrease as the animal ages. This peak in receptor number has been shown in rat (Daval *et al.*, 1987) and human fetal tissue (Bar-Peled *et al.*, 1991). A loss of this receptor has been reported in both Down's syndrome and Alzheimer's disease (Middlemiss *et al.*, 1986).

The 5-HT$_{1a}$ receptor appears to be the receptor which is stimulatory to growth of serotonin neurons, as well as to neurons in target regions.

Astroglial cells possess a number of different neurotransmitter receptors, including serotonin receptors, which decrease in number as the astrocyte matures (Whitaker-Azmitia and Azmitia, 1986b). Stimulating this receptor with selective receptor agonists causes production of growth-promoting media, principally by agonists selective for the 5-HT$_{1a}$ receptor (Whitaker-Azmitia and Azmitia, 1989). At the same time as the astrocytes release this growth factor, they attain a mature morphology. To characterize the growth-promoting substance(s), the astroglial-conditioned media (GCM) were preincubated with various antibodies, to see which would eliminate the growth-promoting properties. One such antibody tested was against S-100β. This protein was considered a candidate for several reasons: it has recently been shown to be on chromosome 21 (Allore *et al.*, 1988) (trisomy of which is responsible for Down's syndrome, which has long been considered to have a serotonergic involvement); to be overexpressed in Alzheimer's disease (Griffin *et al.*, 1989); and to be transiently expressed in the fetal rat brain at the time and in the place that serotonin neurons are developing (Van Hartesveldt *et al.*, 1986). Moreover, S-100β is a neurite extension factor (Kligman and Marshak, 1985). These studies indicated that S-100β is indeed at least one of the substances released by stimulation of astroglial 5-HT$_{1a}$ receptors (Whitaker-Azmitia *et al.*, 1990a). This protein can then act as a trophic factor for both serotonin neurons (Azmitia *et al.*, 1990) and cortical neurons (Kligman and Marshak, 1985).

In our most recent studies, we have examined the role of 5-HT$_{1a}$ receptors in the development of whole animals and found that the timing of stimulation of the 5-HT$_{1a}$ receptor is crucial. Selective agonists have limited effect prenatally, but at early postnatal timepoints the effects are profound.

In whole animal studies, pregnant rats were treated from gestational day 12 until birth with 1.0 mg/kg 8-OH-DPAT (a 5-HT$_{1a}$ receptor agonist) or 1.0 mg/kg spiroxatrine (a 5-HT$_{1a}$ receptor antagonist). Both of these drugs were toxic to the fetus, causing a significant amount of fetal reabsorption (based on low litter sizes) as well as high neonatal death rates. However, in the animals that survived, no changes were seen in behavior. In addition, there were no changes in serotonin terminal density, up until 60 days of age, when significant increases were observed in both hippocampus and brainstem.

When given at critical time periods postnatally, either from PD5–9 or PD11–15, 8-OH-DPAT has more significant effects than after prenatal administration. Interestingly, the results on behavioral and neurochemical development of the animals appears to be, in many cases, completely opposite. Treatment with the agonist at the earlier timepoint accelerates development – the animals opened their eyes sooner and gained weight more rapidly. At the later timepoint for treatment, the animals showed an increase in anxiety-related behaviors and development was delayed – the animals gained weight more slowly and showed delayed olfactory responses. Other workers have shown that depletion of serotonin by *p*-chlorophenyl-alanine on PD8–16 but not PD1–7, leads to decreased anxiety and more environmental reactivity in the rats when they reached adulthood (Farabollini *et al.*, 1988).

In tissue culture studies, 8-OH-DPAT induces tryptophan hydroxylase activity in embryonic mouse hypothalamus cells (de Vitry *et al.*, 1986).

3.2.2   5-HT$_{1b/d}$ RECEPTORS

The 5-HT$_{1b}$ receptor occurs in rat brain largely in the basal ganglia, where it functions predominantly as a release-regulating auto-receptor. There are also reports that this receptor is negatively coupled to adenylate cyclase. The receptor has been difficult to study pharmacologically, since the drugs which are potent at this site also affect other serotonin receptors and other neuro-transmitter systems. The most commonly used agonists are mCPP (*m*-chloro-phenylpiperazine) and TFMPP (trifluoro-methphenylpiperazine).

The 5-HT$_{1b}$ receptor has not been charac-terized for developmental density changes, in terms of a radiolabel binding assay. However, the receptor has been studied for behavioral effects. By four days of age, rat pups respond to mCPP by changes in mouthing and probing behaviors, as well as showing a

general behavioral activation (Kirstein and Spear, 1988). There are few data yet to show a role for this receptor in mediating the role of serotonin as a growth factor. The 5-HT$_{1b}$ receptor has been shown to increase the synthesis of DNA in fibroblasts in culture, through inhibition of the production of cAMP (Seuwen *et al.*, 1988).

Drugs selective for the 5-HT$_{1b}$ receptor were found to have no effect on direct application to serotonin neurons growing in culture or on the production of growth factors by astrocytes. In studies using whole animals, rats treated with TFMPP (1 mg/kg) from gestational day 12 until birth had normal physical appearance but their weight gain was significantly greater than controls. In general, the animals were slow, slept excessively and displayed a gait abnormality. However, there were no neurochemical changes at postnatal days 15, 30 or 60. Although this receptor may be the most likely receptor responsible for the developmentally inhibitory effects of 5-MT, since it is a release-regulating autoreceptor, the agonists current-ly available may not be selective enough to adequately test this hypothesis.

3.2.3   5-HT$_{1c}$

The 5-HT$_{1c}$ receptor is highly concentrated in the choroid plexus, although it can be found in other regions at much lower concentra-tions. The receptor is linked to the inositol phosphate second messenger system and is thus in may ways related to the 5-HT$_2$ receptor. In radioligand binding assays, a number of receptor antagonists are useful, including [^{3}H]mesulergine and iodo-[^{3}H]LSD, however, there are no selective agonists. This has limited the ability of researchers to test for a role of this receptor in development. In one report, based on transfection of the gene for this receptor into fibroblasts, there is evidence that the 5-HT$_{1c}$ receptor may act as a protooncogene (Julius *et al.*, 1989). There are no studies on behavioral activation of the 5-

HT$_{1c}$ receptor in either young or adult animals. However, the one study on number of receptors shows a linear increase in binding sites up to the fourth week, when adult levels are reached. Thus, there appears to be no transient peak in these receptors during development (Zilles *et al.*, 1986).

3.2.4 5-HT$_2$ RECEPTORS

The 5-HT$_2$ receptor is coupled to the production of inositol phosphates and is widely distributed through the brain, but principally in cortex. The receptor has also been localized to astroglial cells. There are a number of selective agonists, including DOI, as well as antagonists, specifically ketanserin. Both of these drugs are also used in radioligand binding studies.

The 5-HT$_2$ receptor apparently does not show a transiently increased level of expression during development. Rather, the receptors are not detectable at birth, and increase slowly thereafter to adult levels (Murrin *et al.*, 1985). Interestingly, however, the second messenger system to which this receptor is linked, phosphatidylinositol hydrolysis, does show a developmental peak, with the production of inositol phosphates by serotonin approximately ten-fold greater in the immature brain than in the adult brain (Claustre *et al.*, 1988). Moreover, 3-day-old rat pups respond to DOI by behavioral activation (Kirstein and Spear, 1988) and the effect of LSD on inhibiting firing of dorsal raphe serotonergic neurons is much greater in neonatal animals than in adult animals (Smith and Gallager, 1989). This suggests that the coupling efficiency of the receptor may change with time.

In our studies, using prenatal treatment with DOI, we observed no developmental changes in the offspring, in terms of serotonergic behaviors or neurochemistry.

Although 5-HT$_2$ receptors may not influence development, they may be what is referred to as programmable receptors, that is

events during development may affect the number, affinity or function of these receptors in the adult brain (Whitaker-Azmitia, 1991). In the case of 5-HT$_2$ receptors, both prenatal and postnatal stress to the mother significantly increases the number of 5-HT$_2$ receptors in the offspring, even after they have become adults (Peters, 1988a, b). Since the 5-HT$_2$ receptor is the predominant serotonin receptor in cortex, postulated to be involved in a variety of animal behaviors and human disease states, it is important that more research is done on examining the factors which program this receptor.

3.2.5 5-HT$_3$ RECEPTORS

The 5-HT$_3$ receptor is a ligand-gated ion channel principally transporting potassium. The distribution is quite limited, being concentrated in caudate, spinal cord, area postrema and hippocampus. Selective agonists include phenylbiguanide and 2-methyl-serotonin. Antagonists include MDL 72222, odansetron and zacopride. 5-HT$_3$ antagonists have recently become available for use as antiemetics.

In tissue culture studies of 5-HT$_3$ receptor active drugs, 5-HT$_3$ agonists are inhibitory to development of ED14 serotonin neurons and inhibit proliferation of the neuronal/glial cell line NG 108-15. Although these tissue culture studies suggest that the 5-HT$_3$ receptor is inhibitory to the development of serotonin neurons, the results are more divergent in whole animals.

Animals were treated prenatally from gestational day 12 until birth, with 5.0 mg/kg phenylbiguanide, or 5.0 mg/kg MDL 72222 or the equivalent volume of saline. Animals were tested at PD16 for spontaneous alternation (a measure of ascending serotonin terminal function) and at days 10, 18 and 30 for a tail flick response, that is latency to remove their tails from a heated light (a measure of descending serotonin terminal function). Interestingly, the tail flick responses were significantly altered. This is the first case in which we have found serotonergic drugs to have an effect on the development of descending serotonin neurons. No changes were observed in spontaneous alternation, but immunocytochemical analysis of serotonin terminals in hippocampus showed a significant decrease at 30 and 60 days of age. This confirms the tissue culture observations. At the same time, serotonin terminal density in spinal cord was increased (Bell *et al.*, 1991).

The data from these studies, therefore, suggest that the 5-HT$_3$ receptor is not simply related to development in a positive or negative manner. The message may be more complicated, and may result in a positive influence into spinal cord and a negative influence into brain.

3.2.6   5-HT$_4$ RECEPTORS

One of the most intriguing questions regarding the role of serotonin receptors in development has been the observation that serotonin stimulates adenylate cyclase activity in immature brain, but that the activity of this system declines with age (Nelson *et al.*, 1980). Since very little positively coupled cyclase is present in adult brain, it has been difficult pharmacologically to classify this receptor. Recently, however, the 5-HT$_4$ receptor has been described in immature rat colliculi. This receptor has a unique pharmacology from other serotonin receptors and is positively coupled to cyclase. A receptor which stimulates cAMP is of course particularly interesting in development, given the role that this second messenger system plays in cellular differentiation (McMahon, 1974).

3.3   CONCLUSION

Serotonin has clearly been shown to play a very important role in regulating the development of the mammalian brain. In order for serotonin to function in this capacity, there must be specific receptors present in the developing brain, which can respond to serotonin by activating specific transduction mechanisms. There are at least two receptors involved. One receptor is active at low concentrations of 5-MT, is related to the release of serotonin, and is inhibitory to the growth of serotonin neurons. The pharmacology of this receptor may not be identical to any receptor which occurs in the adult brain. The other receptor is stimulatory to growth of serotonin neurons, and is likely to be related to the 5-HT$_{1a}$ receptor. Since this receptor also releases the growth factor S-100β, this may be the mechanism by which serotonin directs development in target tissues. Other receptors have not been as well characterized for a role in development. However, the 5-HT$_3$ receptor and possibly the 5-HT$_4$ receptor may also be involved. Finally, the 5-HT$_2$ receptor, although not involved in regulating development, is greatly influenced by events during development, and can thus be termed the programmable serotonin receptor.

ACKNOWLEDGEMENTS

The author gratefully acknowledges the contributions made by collaborators in both the work described and in the formulation of hypotheses. These collaborators include Efrain Azmitia, Ann Shemer and Jean Lauder.

This work has been supported by grants from the National Institute for Child Health and Human Development and from the National Institute on Neurological Disease and Stroke.

REFERENCES

Aitken, A.R. and Törk, I. (1988) Early development of serotonin-containing neurons and pathways as seen in wholemount preparations of the fetal rat brain. *J. Comp. Neurol.*, **274**, 32–47.

Allore, R., O'Hanlon, D., Price, R. *et al.* (1988) Gene encoding the beta subunit of S-100 protein is on chromosome 21: implications for Down Syndrome. *Science*, **239**, 1311–13.

Azmitia, E.C. and De Kloet, E.R. (1987) ACTH

neuropeptide stimulation of serotonergic neuronal maturation in tissue culture: modulation by hippocampal cells. *Progr. Brain Res.*, **72**, 311–17.

Azmitia, E.C., Dolan, K. and Whitaker-Azmitia, P.M. (1990a) S-100 beta, but not NGF, EGF, insulin or calmodulin is a CNS serotonergic growth factor. *Brain Res.*, **576**, 354–6.

Azmitia, E.C., Murphy, R.B., and Whitaker-Azmitia, P.M. (1990b) MDMA (ecstasy) effects on cultured serotonergic neurons: evidence for Ca2(+)-dependent toxicity linked to release. *Brain Res.*, **510**, 97–103.

Bar-Peled, O., Gross-Isseroff, R., Ben-Hur, H. *et al.* (1991) Fetal human brain exhibits a prenatal peak in the density of serotonin 5HT$_{1A}$ receptors. *Neurosci. Lett.*, **127**, 173–6.

Bell, J., Zhang, X.N. and Whitaker-Azmitia, P.M. (1991) 5-HT$_3$ receptor-active drugs alter development of spinal serotonergic innervation: lack of effect of other serotonergic agents. *Brain Res.*, **571**, 293–7.

Breese, G.R., Baumeister, A.A., McCown, T.J. *et al.* (1984) Behavioral differences between neonatal and adult 6-hydroxydopamine-treated rats to dopamine agonists: relevance to neurological symptoms in clinical syndromes with reduced brain dopamine. *J. Pharmacol. Exp. Ther.*, **231**, 343–54.

Buznikov, G.A. (1984) The action of neurotransmitters and related substances on early embryogenesis. *Pharmacol. Ther.*, **25**, 23–59.

Chubakov, A.R., Gromova, E.A., Konovalov, G.V. *et al.* (1986) The effects of serotonin on the morpho-functional development of rat cerebral neocortex in tissue culture. *Brain Res.*, **369**, 285–97.

Claustre, Y., Rouquier, L. and Scatton, B. (1988) Pharmacological characterization of serotonin-stimulated phosphoinositide turnover in brain regions of the immature rat. *J. Pharmacol. Exp. Ther.*, **244**, 1051–6.

Connell, L.A. and Wallis, D.I. (1989) 5-Hydroxytryptamine depolarizes neonatal rat motorneurones through a receptor unrelated to an identified binding site. *Neuropharmacology*, **28**, 625–34.

Daval, G., Verge, D., Becerril, A. *et al.* (1987) Transient expression of 5-HT$_{1A}$ receptor binding sites in some areas of the rat CNS during postnatal development. *Int. J. Dev. Neurosci.*, **5**, 171–89.

Davila-Garcia, M.I. and Azmitia, E.C. (1989) Effects of acute and chronic administration of leu-enkephalin on cultured serotonergic neurons. *Dev. Brain Res.*, **49**, 97–103.

DeVitry, F., Hamon, M., Catelon, J. *et al.* (1986) Serotonin initiates and autoamplifies its own synthesis during mouse central nervous system development. *Proc. Natl. Acad. Sci. USA*, **83**, 8629–33.

Farabollini, F., Hole, D.R. and Wilson, C.A. (1988) Behavioral effects in adulthood of serotonin depletion by *p*-chlorophenylalanine given neonatally to male rats. *Int. J. Neurosci.*, **41**, 187–99.

Fujimiya, M., Kimura, H. and Maeda, T. (1986) Postnatal development of serotonin nerve fibers in the somatosensory cortex of mice studied by immunohistochemistry. *J. Comp. Neurol.*, **246**, 191–201.

Goodrich, C.A., Baker, P.C. and Bauman, G.P. (1986) Biochemical and functional effects of fenfluramine in maturing mice. *Gen. Pharmacol.*, **17**, 456–60.

Goto, M. and Sano, Y. (1984) Ontogenesis of the central serotonin neuron system of the rat – an immunohistochemical study. *Neurosci. Res.*, **1**, 3–18.

Griffin, W.S.T., Stanley, L.C., Ling, C. *et al.* (1989) Brain interleukin 1 and S-100 immunoreactivity are elevated in Down Syndrome and Alzheimer Disease. *Proc. Natl. Acad. Sci. USA*, **86**, 7611–15.

Gu, Q., Patel, B. and Singer, W. (1990) The laminar distribution and postnatal development of serotonin-immunoreactive axons in the cat primary visual cortex. *Exp. Brain Res.*, **81**, 256–66.

Hamon, M. and Bourgoin, S. (1979) Ontogenesis of tryptophan transport in the rat brain. *J. Neural Transm.*, Suppl., **15**, 93–105.

Hamon, M. and Bourgoin, S. (1981) Possible role of serotonin and other monoamines as growth factors during brain development, in *Physiological and Biochemical Basis for Perinatal Medicine* (eds M. Morset-Couchard and A. Minkowski), Karger, Basel, pp. 286–95.

Jackson, D., Bruno, J.P., Stachowiak, M.K. and Zigmond, M.J. (1988) Inhibition of striatal acetylcholine release by serotonin and dopamine after the intracerebral administration of 6-hydroxydopamine to neonatal rats. *Brain Res.*, **457**, 267–73.

Jacobs, B.L. and Azmitia, E.C. (1992) Structure and function of the brain serotonin system. *Physiol. Rev.*, **72**, 165–229.

Jonsson, G. and Hallman, H. (1983) Effect of substance P on the 5, 7-dihydroxytryptamine induced alteration of the postnatal development of central serotonin neurons. *Med. Biol.*, **61**, 105–12.

Julius, D., MacDermott, A.B., Axel, R. and Jessell, T.M. (1989) Molecular characterization of a functional cDNA encoding the serotonin 1c receptor. *Science*, **241**, 558–64.

Kligman, D. and Marshak, D. (1985) Purification and characterization of a neurite extension factor from bovine brain. *Proc. Natl. Acad. Sci. USA.*, **82**, 7136-9.

Kirstein, C.L. and Spear, L.P. (1988) 5-HT$_{1A}$, 5-HT$_2$ receptor agonists induce differential behavioral responses in neonatal rat pups. *Eur. J. Pharmacol.*, **150**, 339–45.

Lankford, K.L., Fernando, F.G. and Klein, W.L. (1988) D$_1$ type dopamine receptors inhibit growth cone motility in cultured retina neurons: evidence that neurotransmitters act as morphogenic growth regulators in the developing central nervous system. *Proc. Natl. Acad. Sci. USA*, **85**, 4567–71.

Lauder, J.M. (1983) Hormonal and humoral influences on brain development. *Psychoneuroendocrinology*, **8**, 121–55.

Lauder, J.M. (1990) Ontogeny of the serotonergic system in the rat: serotonin as a developmental signal. *Ann. NY Acad. Sci.*, **600**, 297–313.

Lauder, J.M. and Krebs, H. (1978) Serotonin as a differentiation signal in early neurogenesis. *Dev. Neurosci.*, **1**, 15–30.

Lauder, J.M. and Zimmerman, E.F. (1988) Sites of serotonin uptake in epithelia of the developing mouse palate, oral cavity, and face: possible role in morphogenesis. *J. Craniofacial Genet. Dev. Biol.*, **8**, 265–76.

Lauder, J.M., Tamir, H. and Sadler, T.W. (1988) Serotonin and morphogenesis. I. Sites of serotonin uptake and binding protein immunoreactivity in the midgestation mouse embryo. *J. Craniofac. Genet. Dev. Biol.*, **102**, 709–20.

Levitt, P. and Moore, R.Y. (1978) Developmental organization of raphe serotonin neuron groups in the rat. *Anat. Embryol. (Berl.)*, **154**, 241–51.

Lidov, H.G. and Molliver, M.E. (1982a) Immunohistochemical study of the development of serotonergic neurons in the rat CNS. *Brain Res. Bull.*, **9**, 559–604.

Lidov, H.G. and Molliver, M.E. (1982b) An immunohistochemical study of serotonin neuron development in the rat: ascending pathways and terminal fields. *Brain Res. Bull.*, **8**, 389–430.

Lieth, E., McClay, D.R. and Lauder, J.M. (1990) Neuronal-glial interactions: complexity of neurite outgrowth correlates with substrate adhesivity of serotonergic neurons. *Glia*, **3**, 169–79.

McMahon, D. (1974) Chemical messengers in development: a hypothesis. *Science*, **185**, 1012–21.

Messenger, N.J. and Warner, A.E. (1989) The appearance of neural and glial cell markers during early development of the nervous system in the amphibian embryo. *Development*, **107**, 43–54.

Middlemiss, D.N., Palmer, A.M., Edel, N. and Bowen, D.M. (1986) Binding of the novel serotonin agonists 8-hydroxy-2-(dipropylamino) tetralin in normal and Alzheimer brain. *J. Neurochem.*, **46**, 993–6.

Murrin, C.L., Gibbens, D.L. and Ferrer, J.R. (1985) Ontogeny of dopamine, serotonin and spirodecanone receptors in rat forebrain – an autoradiographic study. *Dev. Brain Res.*, **23**, 91–109.

Nelson, D.L., Herbert, A., Adrien, J. *et al.* (1980) Serotonin-sensitive adenylate cyclase and ^{3}H-serotonin binding sites in the CNS of the rat – II. Respective regional and subcellular distributions and ontogenic developments. *Biochem. Pharmacol.*, **29**, 2455–63.

Ni, L. and Jonakait, G.M. (1989) Ontogeny of substance P-containing neurons in relation to serotonin-containing neurons in the central nervous system of the mouse. *Neuroscience*, **30**, 257–69.

Peters, D.A. (1982) Prenatal stress: effects on brain biogenic amine and plasma corticosterone levels. *Pharmacol. Biochem. Behav.*, **17**, 721–5.

Peters, D.A. (1988a) Both prenatal and postnatal factors contribute to the effects of maternal stress on offspring behavior and central 5-hydroxytryptamine receptors in the rat. *Pharmacol. Biochem. Behav.*, **30**, 669–73.

Peters, D.A. (1988b) Effects of maternal stress during different gestational periods on the serotonergic system in adult rat offspring. *Pharmacol. Biochem. Behav.*, **31**, 829–43.

Peters, D.A. (1990) Maternal stress increases fetal brain and neonatal cerebral cortex 5-hydroxytryptamine synthesis in rats: a possible mechanism by which stress influences brain development. *Pharmacol. Biochem. Behav.*, **35**, 943–7.

Rajaofetra, N., Sandillon, F., Geffard, M. and Privat, A. (1989) Pre- and postnatal ontogeny of serotonergic projections to the rat spinal cord. *J. Neurosci. Res.*, **22**, 305–21.

Resnick, O. and Morgane, P.J. (1984) Ontogeny of the levels of serotonin in various parts of the brain in severely protein malnourished rats. *Brain Res.*, **303**, 163–70.

Schambra, U.B., Lauder, J.M., Petrusz, P. and Sulik, K.K. (1990) Development of neurotransmitter systems in the mouse embryo following acute ethanol exposure: a histological and immunocytochemical study. *Int. J. Dev. Neurosci.*, **8**, 507–22.

Seuwen, K., Magnaldo, I. and Pouysségur, J. (1988) Serotonin stimulates DNA synthesis in fibroblasts acting through 5-HT$_{1B}$ receptors coupled to a Gi-protein. *Nature*, **335**, 254–6.

Shemer, A.V., Azmitia, E.C. and Whitaker-Azmitia, P.M. (1991) Dose-related effects of prenatal 5-methoxytryptamine (5-MT) on development of serotonin terminal density and behaviour. *Dev. Brain Res.*, **59**, 59–63.

Smith, D.A. and Gallager, D.W. (1989) Electrophysiological and biochemical characterization of the development of alpha 1-adrenergic and 5-HT1 receptors associated with dorsal raphe neurons. *Dev. Brain Res.*, **46**, 173–86.

Spear, L.P. and Ristine, L.A. (1982) Suckling behavior in neonatal rats: psychopharmacological investigations. *J. Comp. Phys. Behav.*, **96**, 244–55.

Spear, L.P. and Scalzo, F.M. (1985) Ontogenic alterations in the effects of food and/or maternal deprivation on 5-HT/5-HIAA and 5-HIAA/5-HT ratios. *Brain Res.*, **350**, 143–57.

Stuart, D.K., Blair, S.S. and Weisblat, D.A. (1987) Cell lineage, cell death, and the developmental origin of identified serotonin- and dopamine-containing neurons in the leech. *J. Neurosci.*, **7**, 1107–22.

Tillet, Y. (1988) Early ontogeny of serotonin-immunoreactivity in the sheep brain. An immunohistochemical study. *Anat. Embryol.*, **178**, 429–40.

Vallés, A.M. and White, K. (1988) Serotonin-containing neurons in *Drosophila melanogaster*: development and distribution. *J. Comp. Neurol.*, **268**, 414–28.

Van Hartesveldt, C., Moore, B. and Hartman, B.K. (1986) Transient midline raphe glial structure in the developing rat. *J. Comp. Neurol.*, **253**, 175–84.

Wallace, J.A. (1985) An immunocytochemical study of the development of central serotoninergic neurons in the chicken embryo. *J. Comp. Neurol.*, **236**, 443–53.

Wallace, J.A. and Lauder, J.M. (1983) Development of the serotonergic system in the rat embryo: an immunocytochemical study. *Brain Res. Bull.*, **10**, 459–79.

Wallace, J.A., Allgood, P.C., Hoffman, T.J. *et al.* (1986) Analysis of the change in number of serotonergic neurons in the chick spinal cord during embryonic development. *Brain Res. Bull.*, **17**, 297–305.

Whitaker-Azmitia, P.M. (1991) Role of serotonin and other neurotransmitter receptors in brain development: basis for developmental pharmacology. *Pharmacol. Rev.*, **43**, 553–61.

Whitaker-Azmitia, P.M. and Azmitia, E. (1986a) Autoregulation of fetal serotonergic neuronal development: role of high affinity serotonin receptors. *Neurosci. Lett.*, **67**, 307–12.

Whitaker-Azmitia, P.M. and Azmitia, E.C. (1986b) ^{3}H-5-Hydroxytryptamine binding to brain astroglial cells: differences between intact and homogenized preparations and mature and immature cultures. *J. Neurochem.*, **46**, 1186–90.

Whitaker-Azmitia, P.M. and Azmitia, E.C. (1989) Stimulation of astroglial serotonin receptors produces media which regulates development of serotonergic neurons. *Brain Res.*, **497**, 80–5.

Whitaker-Azmitia, P.M., Lauder, J.M., Shemer, A. and Azmitia, E.C. (1987) Prenatal plasticity of high affinity serotonin receptors: further evidence for a role of receptors in development. *Dev. Brain Res.*, **33**, 285–90.

Whitaker-Azmitia, P.M., Murphy, R. and Azmitia, E.C. (1990a) Stimulation of astroglial 5-HT$_{1a}$ receptors releases the serotonergic growth factor, protein S-100, and alters astroglial morphology. *Brain Res.*, **528**, 155–8.

Whitaker-Azmitia, P.M., Quartermain, D. and Shemer, A.V. (1990b) Prenatal treatment with SKF 38393, a selective D-1 receptor agonist: longterm consequences on ^{3}H-paroxetine binding and on dopamine and serotonin receptor sensitivity. *Dev. Brain Res.*, **57**, 181–5.

Zhou, F.C. and Azmitia, E.C. (1988) Laminin facilitates and guides fiber growth of transplanted neurons in adult brain. *J. Chem. Neuroanat.*, **1**, 133–46.

Zhou, F.C., Auerbach, S. and Azmitia, E. (1987) Denervation of serotonergic fibers in the hippocampus induced a trophic factor which enhances the maturation of transplanted serotonergic neurons but not norepinephrine neurons. *J. Neurosci. Res.*, **17**, 235–46.

Zilles, K., Rath, M., Schleicher, A. *et al.* (1986) Ontogenesis of serotonin (5-HT) binding sites in the choroid plexus of the rat brain. *Brain Res.*, **380**, 201–3.

Than-Vinh Dam, Gail E. Handelmann and Rémi Quirion

4.1 INTRODUCTION

It is now established that mammalian neurokinins (NK) are derived from two different but related genes namely preprotachykinin (PPT) I and II (Nawa *et al.*, 1983, 1984; Kawaguchi *et al.*, 1986; Kotani *et al.*, 1986; Krause *et al.*, 1987, 1989; Nakanishi, 1987). The RNA transcribed from the PPT-I gene is alternatively spliced to yield different mRNAs encoding the α-, β- and γ-preprotachykinin (α-, β- or γ-PPT, respectively) precursors. Among the various NKs, it has been shown that substance P (SP) may arise from the post-translational processing of α-, β- or γ-PPT whereas neurokinin A (NKA) is generated by the processing of β- or γ-PPT (Nawa *et al.*, 1984; Kawaguchi *et al.*, 1986; Kotani *et al.*, 1986; Krause *et al.*, 1987, 1989; Nakanishi, 1987). Neurokinin B (NKB) is exclusively derived from the PPT-II gene (Kotani *et al.*, 1986; Nakanishi, 1987; Krause *et al.*, 1989).

The ratio between these various NKs varies according to the differential splicing of the PPT in RNAs in brain and peripheral tissues (Nawa *et al.*, 1984; Krause *et al.*, 1987). For example, in the adult rat brain, γ-PPT mRNA represents up to 80% of total PPT-I derived mRNAs whereas α-PPT mRNA is almost absent (Krause *et al.*, 1987, 1988). Moreover, it was recently shown that the three PPT-I mRNAs can potentially generate multiple biologically active peptides, in addition to the three better known NKs (Dam *et al.*, 1990c; Helke *et al.*, 1990). Extended forms of NKA such as neuropeptide K and neuropeptide γ (γ-PTT-(72–92)-peptide amide), as well as NKA(3–10) are known to exist in various brain regions and peripheral tissues (Tatemoto *et al.*, 1985; MacDonald *et al.*, 1988). It is likely that similar extended forms of either SP and NKB could exist in some tissues (MacDonald *et al.*, 1988).

The diversity of the biological effects induced by these various NKs (subsequent sections and Watson *et al.*, 1983; Buck *et al.*, 1984, 1986; Quirion, 1985; Maggio, 1988; Quirion and Dam, 1988; Regoli *et al.*, 1988, 1989; Krause *et al.*, 1989; Rovero *et al.*, 1989; Takeda and Krause, 1989b; Helke *et al.*, 1990) provides indirect evidence for the existence of multiple classes of NK receptors. This hypothesis has been supported by receptor binding (Cascieri and Liang, 1983; Viger *et al.*, 1983; Buck and Burcher, 1986; Lee *et al.*, 1986; Quirion and Dam, 1986; Iversen *et al.*, 1987; Maggio, 1988) and autoradiographic data (Mantyh *et al.*, 1984a; Quirion and Dam, 1985, 1988). Thus far, three major classes of NK receptors (NK-1, NK-2 and NK-3) have been found and have been recently cloned (Masu *et al.*, 1987; Yokota *et al.*, 1989; Hershey and Krause, 1990; Shigemoto *et al.*, 1990). All three

Receptors in the Developing Nervous System Vol. 2: Neurotransmitters. Edited by Ian S. Zagon and Patricia J. McLaughlin.
Published in 1993 by Chapman & Hall. ISBN 0 412 49400 0. Vols. 1 and 2 (set) ISBN 0 412 54520 9.

show important sequence homologies and are members of the seven transmembrane domain G-protein-coupled receptor family, which also includes catecholamine, dopamine and serotonin receptors, the *mas* oncogene, the rhodopsin receptors and several receptors of undefined specificity (O'Dowd *et al.*, 1989). SP, NKA (and its extended forms) and NKB preferentially, but not exclusively, bind to the NK-1, NK-2 and NK-3 receptor class, respectively (Quirion and Dam, 1988; Dam *et al.*, 1990c; Helke *et al.*, 1990).

We have previously reported on the differential distribution of these various classes of NK receptors in mammalian brain (Quirion *et al.*, 1983; Quirion and Dam, 1985, 1988; Dam and Quirion, 1986; Dam *et al.*, 1988) and on the ontogenic profile of NK-1 (Quirion and Dam, 1986) and other NK receptors, especially in the cortical areas (Dam *et al.*, 1988). These results revealed, for example, that the density of NK-1 binding sites in the brainstem is extremely high during the first postnatal week but is very low in the same area in the adult. Moreover, the respective distribution of NK-2 and NK-3 receptor binding sites undergoes modifications during postnatal brain maturation. This suggests the possible involvement of NKs and NK receptors in the developmental functional organization of the CNS. This chapter reviews some of the possible functional roles of NKs in the CNS and focuses on the ontogenic profile of NKs and NK receptor subtypes in the rat brain. The possible regulation of NKs with regard to brain maturation is also discussed.

4.2 BIOLOGICAL FUNCTIONS OF NEUROKININS

4.2.1 NOCICEPTION AND OTHER SENSORY PROCESSES

NKs play an important role in the transmission of sensory information from peripheral receptors to the central nervous system. In the peripheral nervous system, SP and NKA are contained in primary sensory gan-

glion neurons which give rise to the small calibre, unmyelinated C fibers (Holzer *et al.*, 1982; Ogawa *et al.*, 1985; Takano *et al.*, 1986). Primary sensory neurons are contained in the dorsal root ganglia, which project to the spinal cord. The peripheral terminals of the dorsal root ganglia are distributed in the skin and other organs and are found in close apposition to blood vessels, sweat glands, hair follicles, and in autonomic ganglia such as the colonic and hypogastric ganglia. These neurons are activated by painful stimulation, particularly that caused by heat or chemical irritants. Activation of the dorsal root ganglion cells induces the release of SP in the superficial laminae (I and II) of the dorsal horn of the spinal cord (Otsuka and Konishi, 1976). NK-1, NK-2 and NK-3 receptors are concentrated in the dorsal horn, particularly in laminae I and II, where the C fibers terminate (Charlton and Helke, 1985; Buck *et al.*, 1986; Yashpal *et al.*, 1990, 1991a). Several lines of evidence support the role of SP in nociception. SP depolarizes those dorsal horn neurons which are known to be activated by painful stimuli (Randic and Miletic, 1977; Sastry, 1979). Intrathecal administration of SP causes a specific scratching syndrome, as though the animals were responding to a perceived noxious stimulus (Lembeck and Gamse, 1982). Intrathecal SP also temporarily decreases the latency to respond to noxious cutaneous stimulation, suggesting sensitization to this stimulation (Lembeck and Donnerer, 1981). Additional support comes from experiments using the neurotoxin capsaicin. Capsaicin administered to neonatal animals selectively destroys SP-containing neurons in sensory ganglia, but has no effect on SP-containing neurons in the CNS (Gamse *et al.*, 1980). Capsaicin administered to adults does not destroy sensory neurons, but temporarily depletes their supply of SP. Treatment with capsaicin in both neonates and adults causes a dramatic decrease in sensitivity to painful stimulation of the skin, particularly when the pain is induced by heat or chemical irritants. Finally, SP immunoreactivity is absent in the dorsal horn of patients having diminished sensitivity to pain as a result of familial dysautonomia (Pearson *et al.*, 1982). Together, these data indicate a role for NKs in nociception at the level of the spinal cord.

NKs may also play a role in sensory information processing in higher central nervous system structures. SP-containing neurons are found in the trigeminal ganglia and visceral sensory ganglia, which send primary sensory afferent fibers directly to the brain (Cuello *et al.*, 1982). The peripheral terminals of the trigeminal ganglia are distributed in various tissues of the head, including glands, dental pulp, the nasal mucosa and the tongue. The dental C fibers, which contain SP are especially reactive to heat stimuli (Byers, 1984). At the level of the brainstem, there is abundant SP innervation of the general somatic sensory nuclei, such as the substantia gelatinosa rolandi, the nucleus of the tractus spinalis trigemini and the nucleus principalis trigemini. The general visceral sensory nuclei of the brainstem are also innervated by SP fibers (Cuello and Kanazawa, 1978). The origin of these fibers is most likely the respective sensory ganglia outside the brain. Sensory fibers from other organ systems, such as the cardiovascular, gastrointestinal and respiratory systems, terminate in the nucleus of the tractus solitarius, where integration of central peripheral afferent information occurs. This nucleus contains a high density of NK receptors and SP is excitatory to neurons in the nucleus (Morin-Surun *et al.*, 1984). SP binding sites are also found in sensory brain regions such as the central gray, superior and inferior colliculi and cerebral cortex. In cortical areas, SP is excitatory to neurons in the somatosensory cortex, particularly in laminae Vb and VIb (Lamour *et al.*, 1983). Although all three NK receptors are present in cortex, NK-3 receptors are especially dense in these deep cortical layers (Dam *et al.*, 1988, 1990a, b, c; Mantyh *et al.*, 1989a; Stoessl and Hill, 1990).

In addition to their role in the conduction of pain sensation, NKs may also play a role in the processing of olfactory and visual information. NK-1 receptors are contained in a number of components of the olfactory system. NK-1 and NK-2 receptors are present in high concentration in the olfactory bulb, olfactory tubercule, primary olfactory cortex and septum (Shults *et al.*, 1984; Dam *et al.*, 1990b, c). Surprisingly, only a few scattered SP fibers are present in these structures suggesting that another NK may be the endogenous ligand for the receptors present in the olfactory system. SP sensory fibers are also contained in the nasal mucosa, but their function there appears to be related to the regulation of local blood supply (Holzer *et al.*, 1982; St-Jarne *et al.*, 1989) as will be discussed below.

SP is also highly represented in visual pathways. SP is contained in most layers of the eye and in the optic nerve and SP-containing terminals are present in the superior colliculi and lateral geniculate (Cuello and Kanazawa, 1978). In most parts of the eye, the SP fibers are of sensory origin (Miller *et al.*, 1981). However, sensory denervation does not reduce SP contents in the retina (Tervo *et al.*, 1982). The rat retina contains both SP immunoreactivity, primarily in amacrine cells, and also NK-1 receptors suggesting a local action of SP within the retina as well as a role in the transmission of information to visual centers in the brain. Neonatal treatment with monosodium glutamate, which destroys retinal ganglion cells and the inner nuclear layers of the retina, severely reduced NK-1 receptor content in the retina. This suggests that NK-1 are located on the ganglion cells and/or on cells of the inner nuclear layer of the retina (Lee and Cheng, 1988). SP applied locally enhances light-evoked excitation of retinal ganglion cells, at least in non-mammalian species (Dick *et al.*, 1980; Glickman *et al.*, 1980). This suggests a possible role for SP as a positive modulator of the transmission of visual information.

The possibility that SP might also be involved in the conduction of gustatory information is suggested by the finding that SP-containing nerve terminals are present in the taste buds of the tongue (Lundberg *et al.*, 1980; Nishimoto *et al.*, 1982), presumably derived from the glossopharyngeal nerve (Pernow, 1983). The SP terminals, however, do not make synaptic contact with underlying cells in the taste buds (Yamasaki *et al.*, 1985). Therefore, although SP may be released in the vicinity of the taste buds, the SP fibers are probably not involved in gustatory neurotransmission.

4.2.2 REGULATION OF CARDIOVASCULAR ACTIVITY

Another well-characterized activity of NKs is in the regulation of blood pressure and local blood supply. A unique feature of the primary sensory neurons is their symmetrical nature: both their central and peripheral branches contain and release SP. SP is supplied to sensory nerve branches by fast axonal transport and much of the SP produced in sensory ganglion cells is directed toward the periphery. SP nerve networks derived from these sensory neurons are present in vascular beds in virtually every tissue. Peripheral administration of SP causes vasodilation leading to hypotension (Hassessian *et al.*, 1988; Helke *et al.*, 1990). In addition, SP acts centrally to alter blood pressure. In contrast to its action in the periphery, central administration of SP or NK-3 agonists causes hypertension (Unger *et al.*, 1981; Nagashima *et al.*, 1989), apparently through the activation of the sympathetic nervous system (Unger *et al.*, 1981).

SP released from peripheral sensory nerve endings regulates local blood flow (Lembeck and Holzer, 1979). Antidromic stimulation of sensory C fibers, which contain and release SP, causes vasodilation (Hinsey and Gasser, 1930). Treatment with capsaicin to deplete SP in C fibers abolishes the antidromic vasodilation (Lembeck and Holzer, 1979). Local administration of SP mimicks the effects of antidromic nerve stimulation on vasodilation in dental pulp and the eye (Olgart *et al.*, 1977;

Bill *et al.*, 1979). SP is active only in tissues with intact endothelium. The smooth muscle relaxation appears to be due to the stimulation of an endothelial NK-1 receptor, whose activation brings about the release of a vasodilator factor (D'Orléans-Juste *et al.*, 1985). This endothelium-derived relaxing factor has recently been identified as nitric oxide (Schini *et al.*, 1990). NK-2 receptors also appear to be localized in smooth muscle membrane (D'Orléans-Juste *et al.*, 1986; Mastrangelo *et al.*, 1986) and may be similarly involved in vasodilation. In general, however, NK-1 agonists are more potent in their effects on blood pressure than are NK-2 and NK-3 agonists (Hassessian *et al.*, 1988).

SP is present in cerebrovascular nerves and can be depleted by capsaicin, suggesting that primary sensory afferents innervate cerebral blood vessels (Duckles and Buck, 1982). Bundles of SP fibers run in the pia matter of the spinal cord; they appear to arrive there via both the dorsal and ventral roots, again suggesting that they are of sensory origin (Dalsgaard *et al.*, 1982). SP fibers have also been demonstrated around pial arteries in the medulla oblongata (Edvinsson *et al.*, 1981; Edvinsson and Uddman, 1982). As in other tissues, the release of SP at these nerve terminals causes vasodilation and thereby increases sensory nerves; on the other hand, they may represent a pathway for vascular headache (Mayberg *et al.*, 1981).

NKs administered centrally affect blood pressure. NK receptors are located in several brain areas implicated in central control of blood pressure, including the nucleus of the tractus solitarius. Increased numbers of SP receptors were found in several brain regions in spontaneously hypertensive rats (Shigematsu *et al.*, 1987), which are supersensitive in terms of their cardiovascular response to centrally administered SP (Unger *et al.*, 1981).

In the spinal cord, NK-1 binding sites are located postsynaptically on preganglionic sympathetic neurons in the intermediolateral cell column. This location indicates a role for SP in regulating sympathetic outflow from these neurons which influence the cardiovascular system. SP terminals in this region are derived from projections of the ventral medulla (Helke *et al.*, 1982) and SP is excitatory to the preganglionic neurons (Gilbey *et al.*, 1983; Backman and Henry, 1984). Intrathecal administration of a stable SP analog increased blood pressure and heart rate, and these effects were blocked by a sympathetic ganglionic blocker (Keeler *et al.*, 1985). Therefore, SP neurons in the ventral medulla, which send axons to the intermediolateral cell column, may be capable of elevating sympathetic activity and thereby play a role in the maintenance of vasomotor tone.

4.2.3 NON-VASCULAR SMOOTH MUSCLE MOTILITY

In addition to its effects on vascular smooth muscle, SP also influences motility of non-vascular smooth muscle (Lundberg *et al.*, 1980). A major site of this activity is the gastrointestinal tract. In the gastrointestinal tract, the majority of SP is contained in intrinsic neurons (Costa *et al.*, 1980; Holzer *et al.*, 1980). The intrinsic neurons, including those which contain SP, are embedded in the intestinal wall in small ganglia connected by small nerve bundles to form two ganglionated plexuses, the myenteric and submucosal plexuses. From these plexuses, nerve bundles emerge to supply the muscle layers, blood vessels, glands and villi. Although SP is found in only a small proportion of myenteric and submucous neurons, these neurons produce a profuse innervation of muscle, mucosa and the submucous ganglia (Costa *et al.*, 1980). SP is also contained in some extrinsic sensory nerves which supply the submucosal blood vessels and the submucous ganglia (Furness *et al.*, 1982). SP induces contraction of intestinal muscle and it is more potent than acetylcholine in this regard. Both longitudinal and circular muscle layers are contracted directly by SP

(Bury and Mashford, 1977). In addition, SP has been shown, *in vitro*, to stimulate cholinergic neurons, which in turn contract intestinal muscle (Holzer and Lembeck, 1980).

Distinct NK-1 and NK-3 receptors have been characterized in guinea-pig ileum. The NK-1 receptors are present on smooth muscle while the NK-3 receptors are located on cholinergic neurons (Laufer *et al.*, 1985; Cascieri *et al.*, 1986). Despite the presence of NK-3 receptors, NKB is not detectable by radioimmunoassay in the guinea-pig ileum (Too *et al.*, 1989) suggesting that other NKs may act as endogenous ligands at this site. NKA, on the other hand, is found in high concentrations in the ileum (Takano *et al.*, 1986), although NK-2 receptors are not detectable. In general, NK-2 and NK-3 agonists are more potent than NK-1 agonists in stimulating smooth muscle contraction (Erspamer *et al.*, 1980; Regoli *et al.*, 1989; Jacques and Couture, 1990).

4.2.4 SALIVATION

SP innervation of the salivary glands is derived from the otic and trigeminal ganglia parasympathetic nerves (Sharkey and Templeton, 1984) and SP stimulates salivation by a direct action on secretory cells in the salivary glands. The effect is dose-dependent and mediated by NK-1 receptors (Buck and Burcher, 1985; Murray *et al.*, 1988). NKA, neuropeptide K and neuropeptide γ are also present in salivary glands (Takano *et al.*, 1986; Takeda *et al.*, 1990). NKA stimulates salivation although less potently than SP (Murray *et al.*, 1988). Neuropeptide K, on the other hand, is even more potent than SP in producing salivation; it also potentiates the actions of SP on the salivary glands (Takeda and Krause, 1989a). Neuropeptide γ can potentiate the effect of SP on salivation although it has little sialogogic activity alone (Takeda and Krause, 1989b). These results may be consistent with the presence of more than one NK receptor functioning in the salivary glands.

4.2.5   PROINFLAMMATORY ACTIONS AND WOUND HEALING

In addition to vasodilation, release of SP from peripheral terminals of sensory nerves produces several proinflammatory responses. One of these is plasma extravasation or edema (Lembeck and Holzer, 1979). In the rat, cutaneous plasma extravasation can be induced by sensory nerve stimulation or by injection of SP either into an artery or into the spinal cord (Jacques and Couture, 1990). In the rat paw, plasma extravasation, induced either by sensory nerve stimulation or by intra-arterial SP, is blocked by histamine blockers (Lembeck and Holzer, 1979). Direct release of histamine from mast cells by SP has been demonstrated (Erjavec *et al.*, 1981). Therefore, SP appears to mediate edema by activation of mast cells. The action of SP on mast cells is not mediated by NK-2 or NK-3 receptors, because NKA and NKB also stimulate local edema but have no effect on mast cells (Brain and Williams, 1989). No selective membrane receptors for SP have been identified in those preparations and recent evidence suggests that SP may act on mast cells via a receptor-independent mechanism. Mousli *et al.* (1990a, b) have found that SP directly activates G proteins *in vitro*. They propose that SP crosses the mast cell membrane to interact directly with the intracellular G protein subunits. This implies a specific interaction of SP with the G protein that other NKs do not share (by virtue of basic charged *N*-terminal amino acid residues).

Release of SP in the joints also appears to induce proinflammatory activity. The joints are innervated by small unmyelinated sensory nerve fibers containing SP (Wyke, 1981; Pernow, 1983) and SP has been detected in synovial fluids (Inman *et al.*, 1986). SP stimulates rheumatoid synoviocytes to produce prostaglandin E_2 and collagenase and also causes them to proliferate (Lotz *et al.*, 1987). These actions may explain the ability of SP to increase the severity of adjuvant-induced arthritis in rats (Levine *et al.*, 1984).

Studies of various populations of immune cells indicate that NKs induce a variety of responses in these cells. NKA and physalaemin stimulate mouse thymocyte proliferation in tissue culture, although SP had no effect (Soder and Hellstrom, 1989). SP also stimulates proliferation of human T-lymphocytes and enhances the mitogenic response of these cells to phytohemaglutinin (Payan *et al.*, 1983), and there are specific SP receptors on human IM-9 lymphoblasts (Payan *et al.*, 1984). SP also stimulates phagocytosis by macrophages and polymorphonuclear leukocytes (Bar-Shavit *et al.*, 1980). SP stimulates IgA synthesis by organs of the secretory immune system of the mucosal tissues and IgM synthesis by the spleen (Stanisz *et al.*, 1986). SP promotes chemotaxis of monocytes and neutrophils (Ruff *et al.*, 1985), although the chemotaxis may be mediated by a receptor-independent mechanism (Kroegel *et al.*, 1990). SP stimulates lysosomal enzyme release from human, rat and rabbit neutrophils (Serra *et al.*, 1988). Finally, SP induces generation of oxygen radicals and thromboxane A_2 in macrophages and neutrophils (Hartung *et al.*, 1986). Local release of NKs may therefore act as immunoregulatory factors through a variety of mechanisms. In turn, immune system cell products may regulate SP synthesis. The cytokine interleukin 1β increases SP contents of superior cervical ganglion explants (Freidin and Kessler, 1990). In dissociated cultures, this action required the presence of non-neuronal cells, suggesting an indirect effect on the neurons via glia or fibroblasts. The immunosuppressant dexamethasone prevented the interleukin effect. This suggests that an increase in SP synthesis is a response to local injury or inflammation.

Within the central nervous system, certain immune functions are carried out by glial cells. Recent evidence suggests that NK receptors are expressed by glial cells. *In vitro* studies indicate that cortical astrocytes from the mouse express SP receptors in culture (Torrens *et al.*, 1986). NK-1 receptors were also found on cells of a human astrocytoma cell line (Lee *et *al.*, 1989). Binding of SP to these cells was inhibited by guanyl-5'-imidodiphosphate, which is a characteristic of G-protein-coupled receptors (Snyder, 1979), indicating that the glial receptors share some of the characteristics of the neuronal NK-1 receptors. SP added to these cultures increased RNA synthesis. *In vivo*, NK-1 receptors are also present on Schwann cells of the giant squid. SP causes a long-lasting hyperpolarization of these Schwann cell membranes (Evans *et al.*, 1990). In a mammalian system, NK-1 receptors were found to be expressed by reactive astrocytes forming a glial scar following section of the optic nerve in adult rabbits (Mantyh *et al.*, 1989b).

SP release in response to irritation or tissue damage can therefore regulate inflammatory and immune reactions. In addition, SP has actions on cutaneous tissue which would promote wound healing. For example, NKs are mitogens for several cell types. SP and NKA stimulate DNA synthesis in cultured arterial smooth muscle cells and human skin fibroblasts. This effect is inhibited by the SP antagonist spantide (Nilsson *et al.*, 1985). However, because NKA was more potent than SP, the effect is probably mediated via NK-2 or NK-3 receptors. SP but not NKA stimulated DNA synthesis in cultured mouse epidermal cells, although the effect required the presence of serum, indicating the requirement for a synergistic growth factor (Tanaka *et al.*, 1988). SP also stimulates neovascularization in the rabbit cornea (Ziche *et al.*, 1990), which may result from the proliferation of endothelial cells, as SP stimulates endothelial cell proliferation in tissue culture (Dalsgaard *et al.*, 1989; Ziche *et al.*, 1990). The effect of SP on cultured endothelial cells appears to be mediated via NK-1 receptors, as NK-2 or NK-3 agonists had no effect. SP also stimulates endothelial cell migration via an NK-1 receptor-mediated mechanism, which also plays a major role in the process of neovascularization (Ziche *et al.*, 1991). These results suggest that the SP innervation of

blood vessels has two functions, vasodilation and stimulation of growth. Therefore NK released during inflammation may participate in wound healing processes.

4.2.6 GROWTH AND TROPHIC EFFECTS

In addition to its possible functions in promoting cell proliferation in the immune systems and during tissue repair, SP may have other roles as a trophic factor or growth-promoting factor in various tissues. For example, the atrophy of the salivary glands which occurs after denervation or feeding a liquid diet can be prevented by daily infusions of SP (Mansson *et al.*, 1990). In addition, administration of physalaemin but not eledoisin increases salivary gland weight in intact rats (Bertaccini *et al.*, 1966; Cantalamessa *et al.*, 1975). Therefore, SP may have a trophic effect on salivary glands, probably via NK-1 receptors. The fact that other parasympathetic agonists did not have this effect on salivary glands (Mansson *et al.*, 1990) suggests that this is a function which is distinct from increasing nerve reflex activity or nervous tone.

High doses of SP also stimulate neurite outgrowth in embryonic dorsal root ganglia in tissue culture (Narumi and Fujita, 1978) and in cultured neuroblastoma cells (Narumi and Maki, 1978). The mechanism appears to involve the production of cAMP. The doses of SP required to produce this effect are six orders of magnitude greater than those required for nerve growth factor. Although this comparison would suggest that SP is not a particularly potent stimulator of neurite outgrowth, experiments *in vivo* suggest that SP may stimulate regrowth of neurites following damage. SP counteracts the toxic effects of the specific neurotoxin 6-hydroxydopamine in the neonate (Jonsson and Hallman, 1982b). Treatment of newborn rats with 6-hydroxydopamine causes permanent degeneration of the distant nerve terminals of noradrenergic neurons and excessive growth of those nerve terminals which innervate nearby targets. SP injected intracisternally blocked these effects. SP may act by stimulating neurite regrowth following the toxic damage. In kittens, SP infusion into the fourth ventricle increased the rate of reinnervation of a cortical area denervated by local application of 6-hydroxydopamine (Nakai and Kasamatsu, 1984). Because SP did not have this effect when infused at the site of the lesion, SP is probably acting on the noradrenergic cell bodies in the locus coeruleus. However, the effect on reinnervation is probably not simply due to excitation of the noradrenergic neurons because other compounds excitatory to these cells, such as bethanechol chloride, did not stimulate reinnervation.

Wall *et al.* (1982) have suggested that SP may play a role in the maintenance of synaptic connections between sensory afferents and central neurons. They found that neonatal capsaicin treatment of mice resulted in a loss of specificity of the connections between individual whiskers and their receptive cells in the somatosensory cortex. In addition, they found that capsaicin treatment of sensory nerves in the adult expanded the receptive fields of dorsal horn neurons (Wall *et al.*, 1982). These data suggest that disorganization of somatotrophic maps occurs in the absence of SP. Alternatively, however, such disorganization may result from sprouting from intact nerve terminals in the vicinity of the denervated neurons (Shortland *et al.*, 1990).

Activity of NKs in stimulating growth is also suggested by their second messenger. All three NK receptors have been shown to be associated with inositol phospholipid hydrolysis in different tissues (Mantyh *et al.*, 1984b; Dam *et al.*, 1986; Bristow *et al.*, 1987; Nakanishi, 1987; Guard *et al.*, 1988). Receptors acting on the inositol phospholipid pathway may ultimately stimulate cell division. The agonist-induced hydrolysis of inositol phospholipid into inositol triphosphate and diacylglycerol is an early cellular response to mitogenic stimulation. Inositol triphosphate and diacylglycerol are believed to function as

second messengers through their ability to mobilize calcium from intracellular stores. The ability of SP to stimulate DNA synthesis, at least in a lymphoblast cell line, is closely related to its ability to mobilize intracellular calcium (Payan *et al.*, 1986). Finally, the NK-2 receptor shares sequence homology to the receptor-like protein coded by the *mas* oncogene (44% homology). The *mas* oncogene has been predicted to encode a membrane receptor which activates a critical component in a growth regulatory pathway (Young *et al.*, 1986). Therefore, the NK-2 receptor and *mas* could define a subfamily of receptors with growth-control activities (Hanley and Jackson, 1987). Additionally, the fact that the distribution of NK-1 and NK-2 receptors undergoes major reorganization during brain ontogeny, suggests neurotrophic roles for NKs in the CNS (Quirion and Dam, 1986; Dam *et al.*, 1988).

4.2.7   MOTOR FUNCTION

A role for NKs in motor function is suggested by the location of NKs in various motor systems of the brain and spinal cord. SP and its receptors are found in many of the structures comprising the extrapyramidal motor system. Some of these structures are the striatum, substantia nigra, globus pallidus, inferior olivary complex and cerebellum. In fact, one of the heaviest SP innervations in the midbrain is found in the substantia nigra (Cuello *et al.*, 1982). The SP network of the substantia nigra originates in the striatum and SP terminals have been seen in association with dendrites of the substantia nigra (Somogyi *et al.*, 1982). The substantia nigra apparently does not contain NK-1 receptors which is surprising in light of its dense SP innervation. Low levels of NK-2 and NK-3 receptors, however, are found in both the substantia nigra and striatum (Saffroy *et al.*, 1988; Mantyh *et al.*, 1989a; Dam *et al.*, 1990a,c). SP- and dopamine-containing neurons in these structures appear to participate in a mutual regulation which is important in the control of movement (Bannon *et al.*, 1987) and sensory motor integration (Iversen, 1982). SP increases the firing rate of nigral dopamine neurons (Davies and Dray, 1976) and infusion of SP into the substantia nigra produces a grooming response in rats which is characteristic of activation of the striatonigral system (Iversen, 1982). The potential clinical importance of SP in the substantia nigra is underscored by the fact that injection of a SP antagonist into this brain region reduced muscle tone in spastic rats (Turski *et al.*, 1990).

In the ventral horn of the spinal cord, SP terminals of ventral medulla projections are located in close apposition to motor neurons (Ljungdahl *et al.*, 1978; Helke *et al.*, 1982). NK-1 receptors are present on motor neurons (Helke *et al.*, 1985, 1986; Yashpal *et al.*, 1990), which are depolarized by SP (Otsuka and Yanagisawa, 1980). SP fibers and NK-1 receptors are also present in the phrenic motor nucleus of the spinal cord, which controls the diaphragm (Charlton and Helke, 1985). These findings indicate a direct influence of SP on spinal motor mechanisms. NK-2 and NK-3 receptor sites are more discretely distributed in the cord, mostly present in the superficial laminae of the dorsal horn (Yashpal *et al.*, 1990, 1991a).

SP may also influence motor behavior through the mesolimbic dopamine system, serotonergic pathways and the corticospinal tract. When infused into the A10 dopamine cell body region, SP stimulated locomotor and exploratory behavior (Kelly *et al.*, 1979) which suggest that SP increases the functional activity of these dopaminergic neurons. Evidence suggests that NK-3 receptor activation may influence serotonergic neurons. In support of this hypothesis, the NK-3 agonist, senktide, has been shown to elicit 5-HT-mediated motor behaviors following intracisternal or subcutaneous administration in the mouse and rat (Stoessl *et al.*, 1988). In the corticospinal tract, SP is excitatory to the giant pyramidal cells of Betz in motor cortex

which give rise to the descending fibers of the tract (Phillis and Limacher, 1974).

4.2.8. MEMORY

In the limbic system, SP cell bodies are found in the hippocampus, septum and amygdala suggesting that NKs might influence learning and memory. In fact, the heaviest SP fiber network in the forebrain is the medial amygdaloid nucleus (Cuello and Kanazawa, 1978) suggesting that SP may play an important role in the function of this structure. In addition, high concentrations of NK-1 receptors are found in the hippocampus and amygdala (Saffroy *et al.*, 1988). Several reports indicate that SP, NKA and neuropeptide K can each modulate memory retention after central or peripheral administration (Houston and Staubli, 1979; Hecht *et al.*, 1979; Staubli and Houston, 1980; Kafetzopoulos *et al.*, 1986; Flood *et al.*, 1990). Whether this modulation occurs through direct effects on mechanisms of memory storage or through indirect effects such as increased sensitivity to negative reinforcement, remains to be determined.

4.2.9 ENDOCRINE AND EXOCRINE REGULATION

The presence of NKs and their receptors in the hypothalamus, pituitary and adrenal glands suggests a possible role for these peptides in endocrine regulation. Numerous reports have indicated that SP alters the release of many hormones, including gonadotropins, prolactin, growth hormone and TRH (for references see O'Donohue *et al.*, 1990). In contrast, SP inhibits insulin release from the pancreas (Brown and Vale, 1976). Moreover, SP stimulates exocrine pancreatic secretion *in vitro* (Konturek *et al.*, 1981) and inhibits hepatic bile output (Holm *et al.*, 1978). Finally, the central inhibition of gastric acid output appears to be mediated entirely by NK-2 and NK-3 receptors. Central administration of NK-2 and NK-3 agonists caused this inhibition, while selective NK-1 agonists had no effect (Improta and Broccarde, 1990).

4.2.10 ONTOGENY OF NK-LIKE IMMUNOREACTIVITY IN THE BRAIN

Both NKs and their receptors (see below) are present early in fetal development. SP immunoreactivity is detectable in the mouse brain at embryonic day (ED) 12 (Ni and Jonakait, 1981), in the rat brain at ED14 (Johansson *et al.*, 1981; Inagaki *et al.*, 1982a) and in the human brain after week 11 of gestation (Yew *et al.*, 1990). SP appears in the spinal cord and dorsal root ganglia soon after (Gilbert and Emson, 1979; Senba *et al.*, 1982; Charnay *et al.*, 1983). SP innervation of some brainstem nuclei and peripheral structures, such as the taste buds of the tongue, occurs postnatally (Sakanaka *et al.*, 1982; Yamasaki *et al.*, 1985). In the cerebellum, there is a transient SP innervation in the newborn, which can be traced to the lower brainstem, but which disappears within two weeks (Inagaki *et al.*, 1982b). In general, the distribution of SP and PPT mRNA in the nervous system is comparable to that of the adult at an early stage of development, but the concentrations of SP and the prohormone mRNA increase steadily after birth (Brene *et al.*, 1990; Walker *et al.*, 1991). Adult concentrations are reached between 7 and 60 days after birth, depending on the tissue. NKA, NKB and neuropeptide K are also detectable in the rat brain at birth. Their levels increase during the second week of birth, then decline to adult levels after postnatal day (PD) 15 (Diez-Guerra *et al.*, 1989).

4.3   ONTOGENIC PROFILE OF NK RECEPTOR SUBTYPES

4.3.1   ONTOGENY OF NK-1 RECEPTORS

The distribution of NK-1 binding sites undergoes major modifications during ontogeny (Figs 4.1–4.5) of the rat brain. For example,

although high densities of NK-1 binding sites are present very early in most brainstem nuclei, very low concentrations of sites are seen in the same region at PD21 as in the adult CNS (Quirion and Dam, 1988; Dam *et al.*, 1988). A detailed examination of the profile is presented in the following sections.

PD1

At one day postnatally, NK-1 binding sites are concentrated in most nuclei of the brainstem (Fig. 4.1E). High densities of NK-1 binding sites are also present in the striatum (Fig. 4.1A–C), olfactory tubercle (Fig. 4.1A,B), dentate gyrus of the hippocampus (Fig. 4.1C,D), various hypothalamic and amygdaloid nuclei (Fig. 4.1C,D), the habenula (Fig. 4.1D), amygdalohippocampal area (Fig. 4.1D), the colliculi (Fig. 4.1E) and the entorhinal cortex (Fig. 4.1E). Moderate densities of sites are seen in the lateral septum (Fig. 4.1A,B) and various thalamus nuclei (Fig. 4.1C). Low densities are found in most cortical areas (Fig. 4.1). The corpus callosum is devoid of specific labeling (Fig. 4.1).

PD4

The distribution of NK-1 binding sites is relatively similar at one and four days after birth (Figs 4.1 and 4.2). Very high densities of sites are found in various brainstem nuclei (Fig. 4.2E), the locus coeruleus (not shown) and inferior olive (not shown). High densities of sites are present in the striatum (Fig. 4.2A, B), habenula (Fig. 4.2C), dentate gyrus (Fig. 4.2C,D), certain thalamic nuclei (Fig. 4.2C) and the superior colliculus (Fig. 4.2D). Moderate densities are found in the lateral septum (Fig. 4.2A,B) and various hypothalamic nuclei (Fig. 4.2C). Low to moderate densities are present in most cortical areas (Fig. 4.2). Very low densities of sites are present in the substantia nigra throughout postnatal ontogeny (Figs 4.1–4.5).

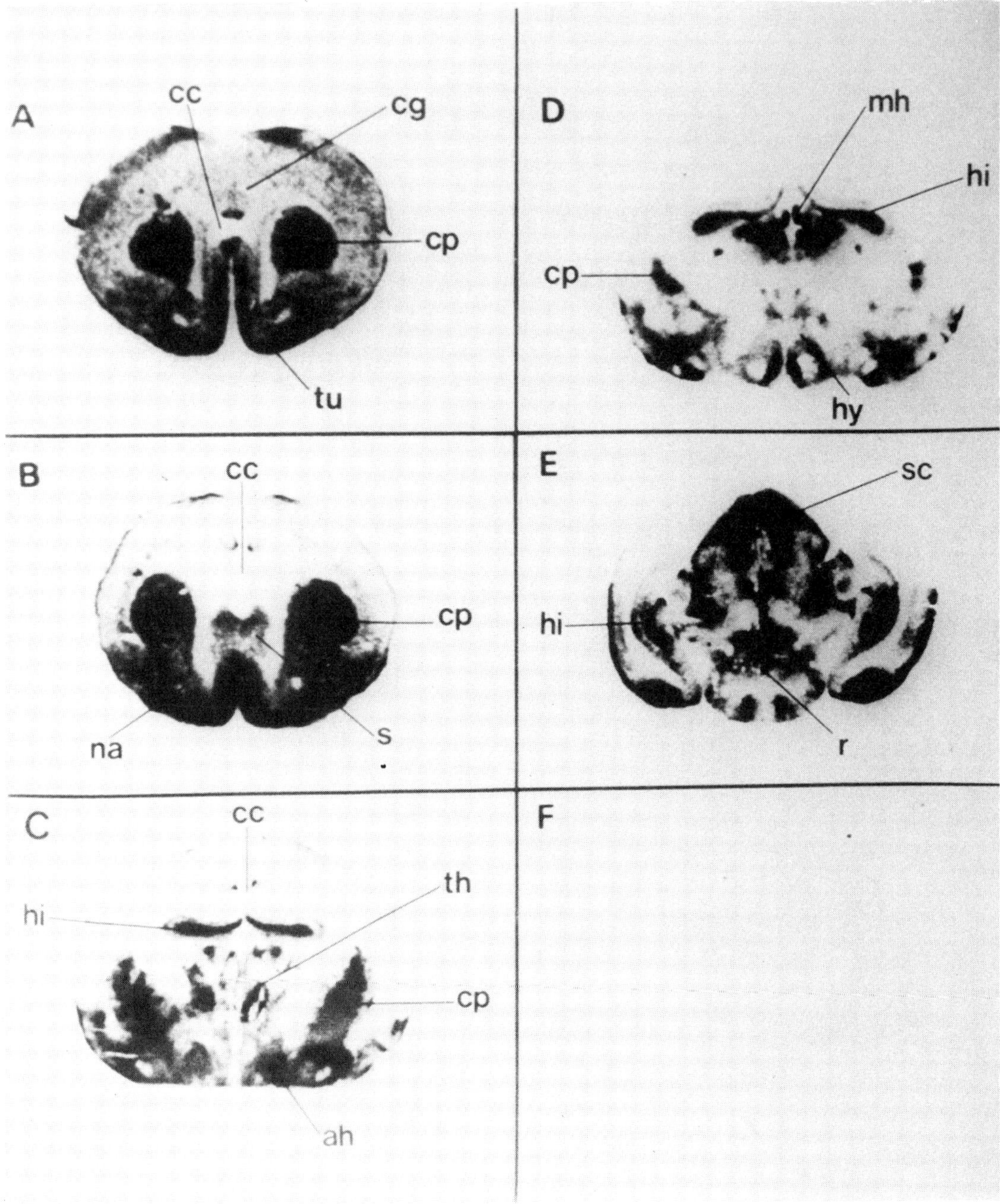

Fig. 4.1.   Photomicrographs of the distribution of NK-1 binding sites in coronal brain sections from one day old rats (PD1). High densities of sites are present in the striatum (A,B), olfactory tubercle (A,B), dentate gyrus (C,D), amygdala (C,D) and colliculus (E). Moderate densities are found in the lateral septum (A,B) and hypothalamus (C,D). Section incubated in presence of 1.0 µM SP (F). Abbreviations: ah, anterior hypothalamus; cc, corpus callosum; cg, cingulate cortex; cp, caudate putamen; hi, hippocampus; hy, hypothalamus; mh, medial habenula; na, nucleus accumbens; r, raphe; s, septum; sc, superior colliculus; th, thalamus; tu, olfactory tubercle.

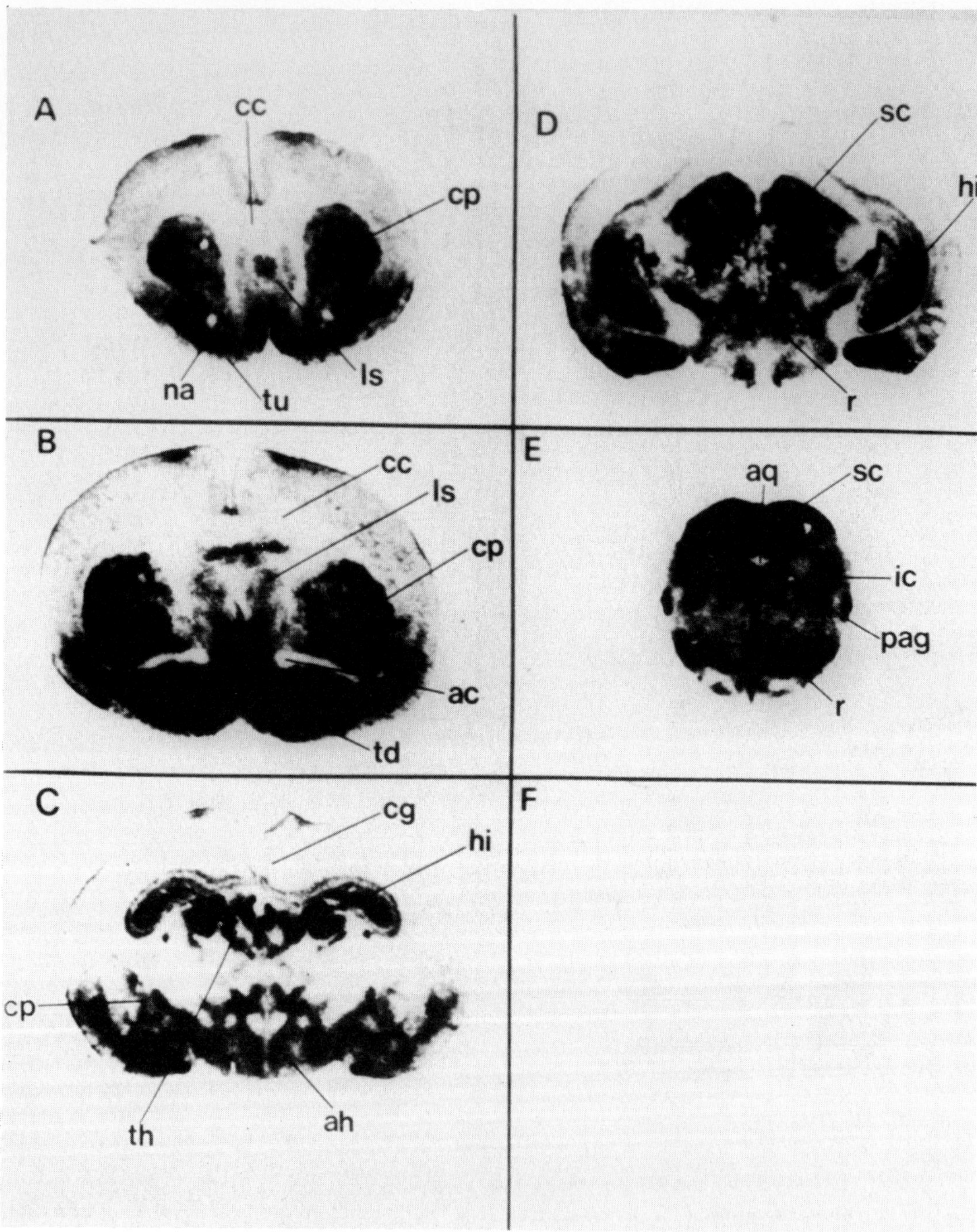

Fig. 4.2.   Photomicrographs of the distribution of NK-1 binding sites in coronal brain sections from four-day-old rats (PD4). High densities of sites are present in the striatum (A,B), olfactory tubercle (A,B), dentate gyrus (C,D), certain thalamic nuclei (C), superior colliculus (D,E) and most brainstem nuclei (E). Moderate densities of sites are found in the lateral septum (A,B) and hypothalamus (C). Section incubated in the presence of 1.0 µM SP (F). Abbreviations: ac, anterior commissura; ah, anterior hypothalamus; aq, cerebral aqueduct; cc, corpus callosum; cg, cingulate cortex; cp, caudate putamen; hi, hippocampus; ic, inferior colliculus; ls, lateral septum; na, nucleus accumbens; pag, periaqueductal gray matter (central gray); r, raphe; sc, superior colliculus; td, tractus diagonalis; th, thalamus; tu, olfactory tubercle.

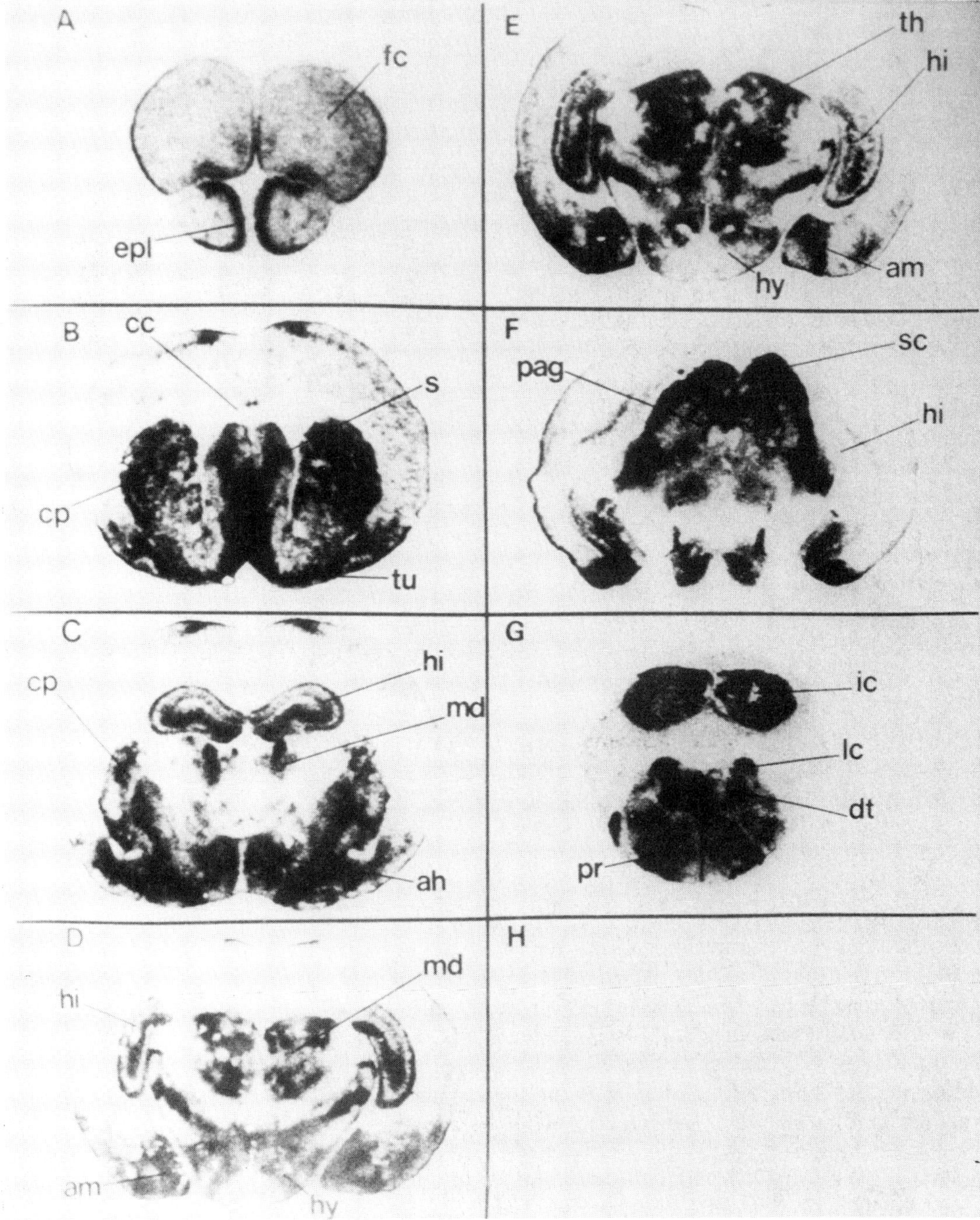

Fig. 4.3. Photomicrographs of the distribution of NK-1 binding sites in coronal brain sections from seven-day-old rats (PD7). High densities of sites are seen in the external plexiform layer of the olfactory bulb (A), striatum (B,C), septum (B), dentate gyrus (C–E), certain thalamic nuclei (C–E), superior colliculus (F), medial geniculate nuclei (F), inferior colliculus (G) and most brainstem nuclei (G) present at this level. Section incubated in the presence of 1.0 μM SP (H). Abbreviations: ah, anterior hypothalamus; am, amygdala; cc, corpus callosum; cp, caudate putamen; dt, dorsal tegmentum; epl, external plexiform layer of the olfactory bulb; fc, frontal cortex; hi, hippocampus; hy, hypothalamus; ic, inferior colliculus; lc, locus coeruleus; md, mediodorsal thalamic nuclei; s, septum; sc, superior colliculus; th, thalamus; tu, olfactory tubercle.

PD7

Very high densities of NK-1 sites are still present over most upper and lower brainstem nuclei (Fig. 4.3). High densities of SP binding sites are found in the external plexiform layer of the olfactory bulb (Fig. 4.3A), striatum (Fig. 4.3B,C), lateral septum (Fig. 4.3B), olfactory tubercle (Fig. 4.3B), dentate gyrus of the hippocampus (Fig. 4.3C,D), amygdala (Fig. 4.3C,D) habenula (Fig. 4.3C), anterior hypothalamic nucleus (Fig. 4.3C), certain thalamic nuclei (Fig. 4.3C,D), the colliculi (Fig. 4.3E,F,G), central gray matter (Fig. 4.3F), entorhinal cortex (Fig. 4.3E,F), medial geniculate nuclei (Fig. 4.3F), locus coeruleus (Fig. 4.3G), olive nuclei and nucleus of the trigeminal nerve. Low densities of sites are present in cortex (Fig. 4.3), and only background levels are seen in the cerebellum (Fig. 4.3G). White matter areas are devoid of NK-1 binding sites (Fig. 4.3).

PD14

At 14 days after birth, the autoradiographic distribution of NK-1 binding sites in rat brain is relatively similar to that seen in adult brain (Quirion *et al.*, 1983; Shults *et al.*, 1984), except for the higher densities in the brainstem (Fig. 4.4F,H) and the somewhat higher density in the hypothalamus (Fig. 4.4D,E). In the forebrain, NK-1 sites are found in various areas including the external plexiform layer of the olfactory bulb (Fig. 4.4A), striatum (Fig. 4.4B–D), olfactory tubercle (Fig. 4.4B,C), lateral septum (Fig. 4.4C), habenula (Fig. 4.4D,E), certain thalamic nuclei (Fig. 4.4D,E), amygdalohippocampal area (Fig. 4.4D,E), amygdala (Fig. 4.4D,E), zona incerta (Fig. 4.4E) and dentate gyrus (Fig. 4.4D,E). At this age, the laminar distribution of NK-1 binding sites in the hippocampus is highly apparent (Fig. 4.4D–F). Low to moderate densities of sites are also present, in a laminated fashion, in various cortical areas (Fig. 4.4). More caudally, high densities of NK-1 sites are found in the superior colliculus (Fig. 4.4F,G), central gray matter (Fig. 4.4F,G), medial geniculate nuclei (Fig. 4.4F,G), pre- and parasubiculum (Fig. 4.4G), locus coeruleus (Fig. 4.4H) and certain brainstem nuclei (Fig. 4.4H).

PD21

At 21 days after birth, low densities of NK-1 sites are seen in most brainstem nuclei (Fig. 4.5) as in the adult rat. However, as shown in Fig. 4.5, high densities of sites are seen in the striatum (Fig. 4.5B–F), olfactory tubercle (Fig. 4.5B–D), septum (Fig. 4.5C,D), amygdala (Fig. 4.5F,G), amygdalohippocampal area (Fig. 4.5G,H), habenula (Fig. 4.5F,G), anterior hypothalamus (Fig. 4.5E,F), dentate gyrus (Fig. 4.5F–H), superior colliculus (Fig. 4.5I,J), central gray matter (Fig. 4.5J) and locus coeruleus (Fig. 4.5K). Moreover, NK-1 binding sites are distributed in a laminar fashion in the cortex and hippocampus (Fig. 4.5). Very low densities of sites are seen in the substantia nigra (Fig. 4.5I,J) and white matter areas such as the corpus callosum are devoid of specific labeling (Fig. 4.5).

4.3.2   ONTOGENY OF NK-2 RECEPTORS

The distribution of NK-2 binding sites in rat brain is markedly altered during ontogeny (Figs 4.6–4.11). As in the case of NK-1 sites (Quirion and Dam, 1986), NK-2 sites are present in high amounts in different brainstem nuclei from PD1 to PD7 (Figs 4.6–4.8). However, the density of these receptors diminishes thereafter to reach adult levels rather rapidly (Figs 4.9–4.10).

PD1

One day after birth, very high densities of NK-2 sites are observed in the amygdalohippocampal area (Fig. 4.6C), inferior colliculus (Fig.

4.6E), locus coeruleus (Fig. 4.6E), raphe nucleus (Fig. 4.6D–G) and periaqueductal gray matter (Fig. 4.6F). Regions moderately enriched with NK-2 sites include the striatum (Fig. 4.6A,B), layers III and IV of the prefrontal cortex (Fig. 4.6A,B), the lateral septum (Fig. 4.6A), certain thalamic nuclei (Fig. 4.6B,C), diagonal band of Broca (Fig. 4.6B), hippocampus (Fig. 4.6B,C), medial habenula (Fig. 4.6C) and zona incerta (Fig. 4.6C).

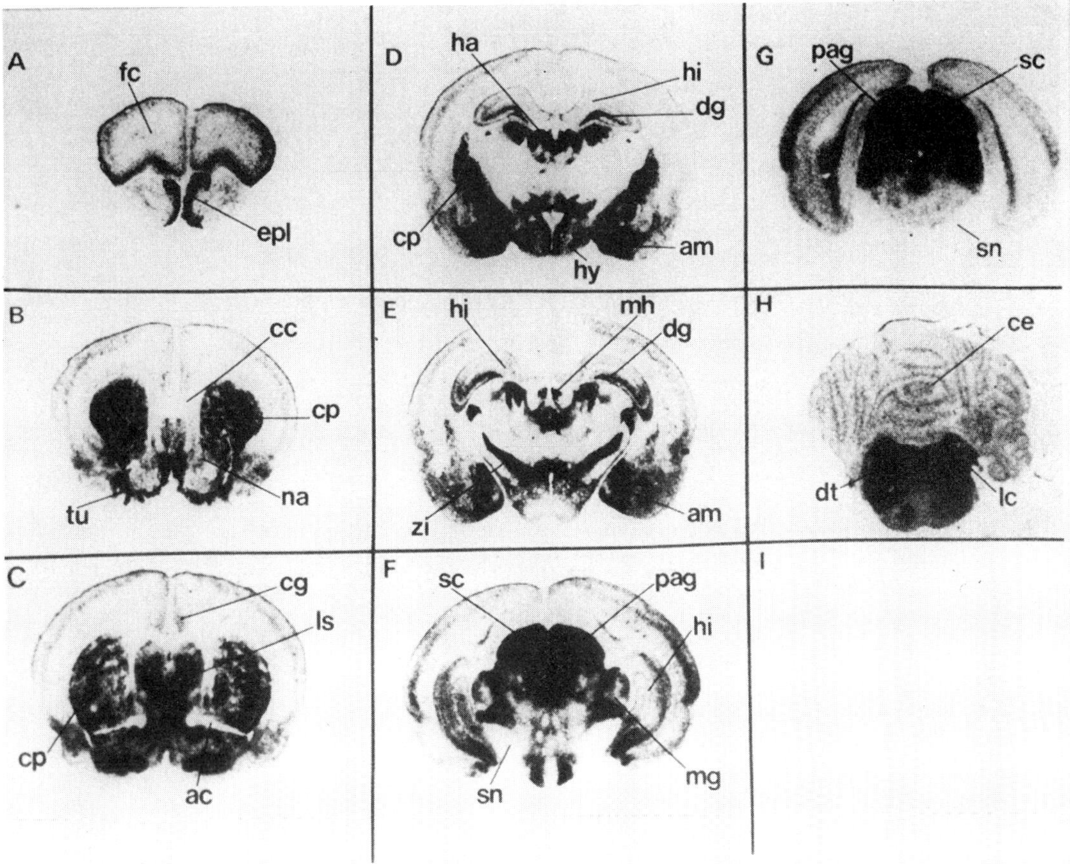

Fig. 4.4. Photomicrographs of the distribution of NK-1 binding sites in coronal brain sections from 14-day-old rats (PD14). High densities of sites are found in the external plexiform layer of the olfactory bulb (A), olfactory tubercle (B,C), striatum (B–D), septum (C), habenula nuclei (D,E), dentate gyrus (D,E), amygdalohippocampal area (D,E), certain hypothalamic nuclei (D,E), superior colliculus (F,G), central gray matter (F,G) and most brainstem nuclei including the locus coeruleus (H). Moderate densities are seen in the hippocampus (D–F) and pre- and parasubiculum (G). Very low densities are present in the substantia nigra (F,G). Section incubated in the presence of 1.0 μM SP (I). Abbreviations: ac, anterior commissura; am, amygdala; cc, corpus callosum; ce, cerebellum; cg, cingulate cortex; cp, caudate putamen; dg, dentate gyrus; dt, dorsal tegmentum; epl, external plexiform layer of the olfactory bulb; fc, frontal cortex; ha, habenula; hi, hippocampus; hy, hypothalamus; lc, locus coeruleus; ls, lateral septum; mh, medial habenula; mg, medial geniculate nucleus; na, nucleus accumbens; pag, periaqueductal gray matter (central gray); sc, superior colliculus; sn, substantia nigra; tu, olfactory tubercle; zi, zona incerta.

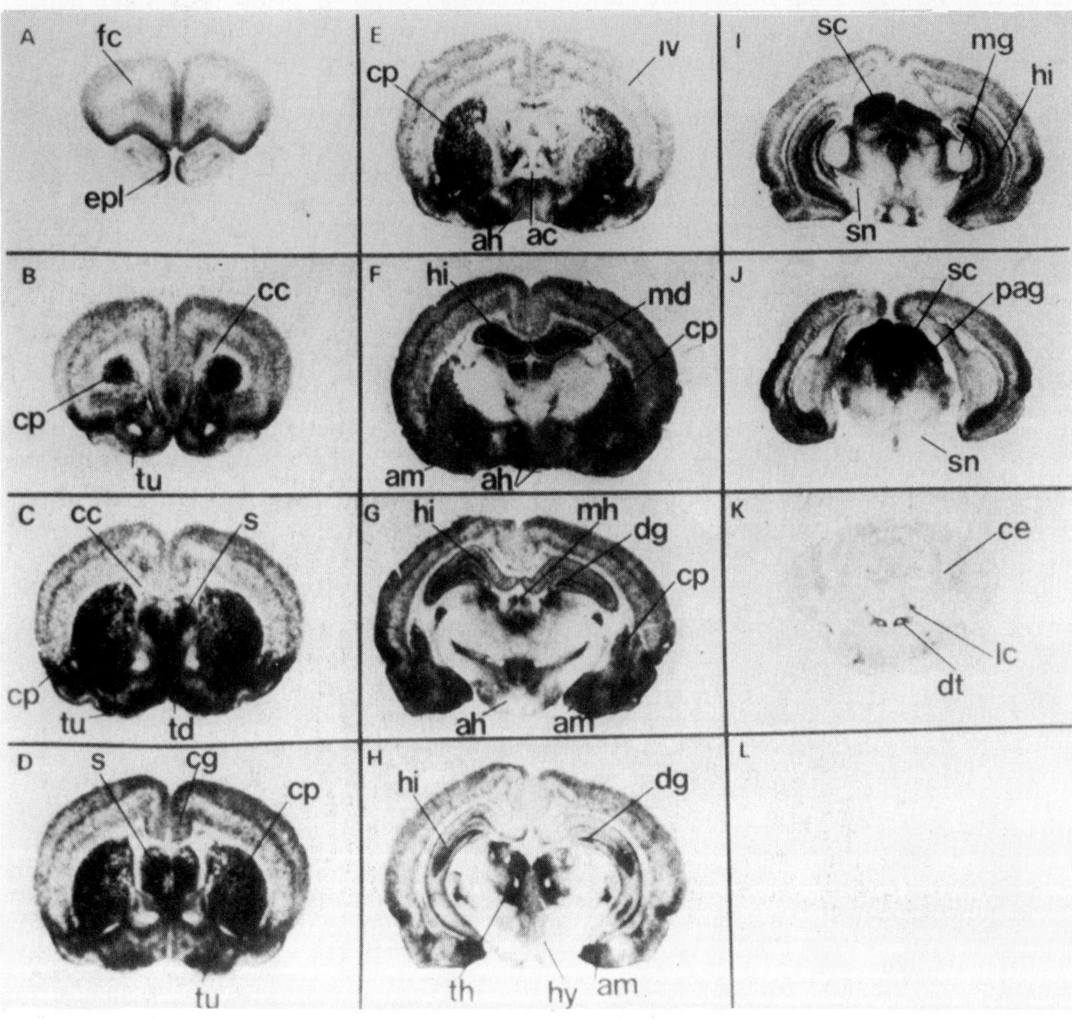

Fig. 4.5.   Photomicrographs of the distribution of NK-1 binding sites in coronal brain sections from 21-day-old rats (PD21). High densities of sites are found in various regions including the external plexiform layer of the olfactory bulb (A), striatum (B,F), lateral septum (C,D), olfactory tubercle (C,D), habenula (F,G), hypothalamus (E–G), dentate gyrus (F,I), certain thalamic nuclei (G,H), amygdalohippocampal area (G,H), superior colliculus (I,J), central gray matter (J) and locus coeruleus (K). At this age, very few brainstem nuclei contain high densities of SP binding sites (K). The substantia nigra is virtually devoid of NK-1 binding sites (I,J). Section incubated in the presence of 1.0 µM SP (L). Abbreviations: ac, anterior commissura; ah, anterior hypothalamus; am, amygdala; cc, corpus callosum; ce, cerebellum; cg, cingulate cortex; cp, caudate putamen; dg, dentate gyrus; dt, dorsal tegmentum; epl, external plexiform layer of the olfactory bulb; fc, frontal cortex; hi, hippocampus; hy, hypothalamus; lc, locus coeruleus; md, medial thalamic nuclei; mg, medial geniculate nucleus; mh, medial habenula; pag, periaqueductal gray matter (central gray); s, septum; sc, superior colliculus; sn, substantia nigra; td, tractus diagonalis; th, thalamus; tu, olfactory tubercle; IV, fourth layer of the cortex.

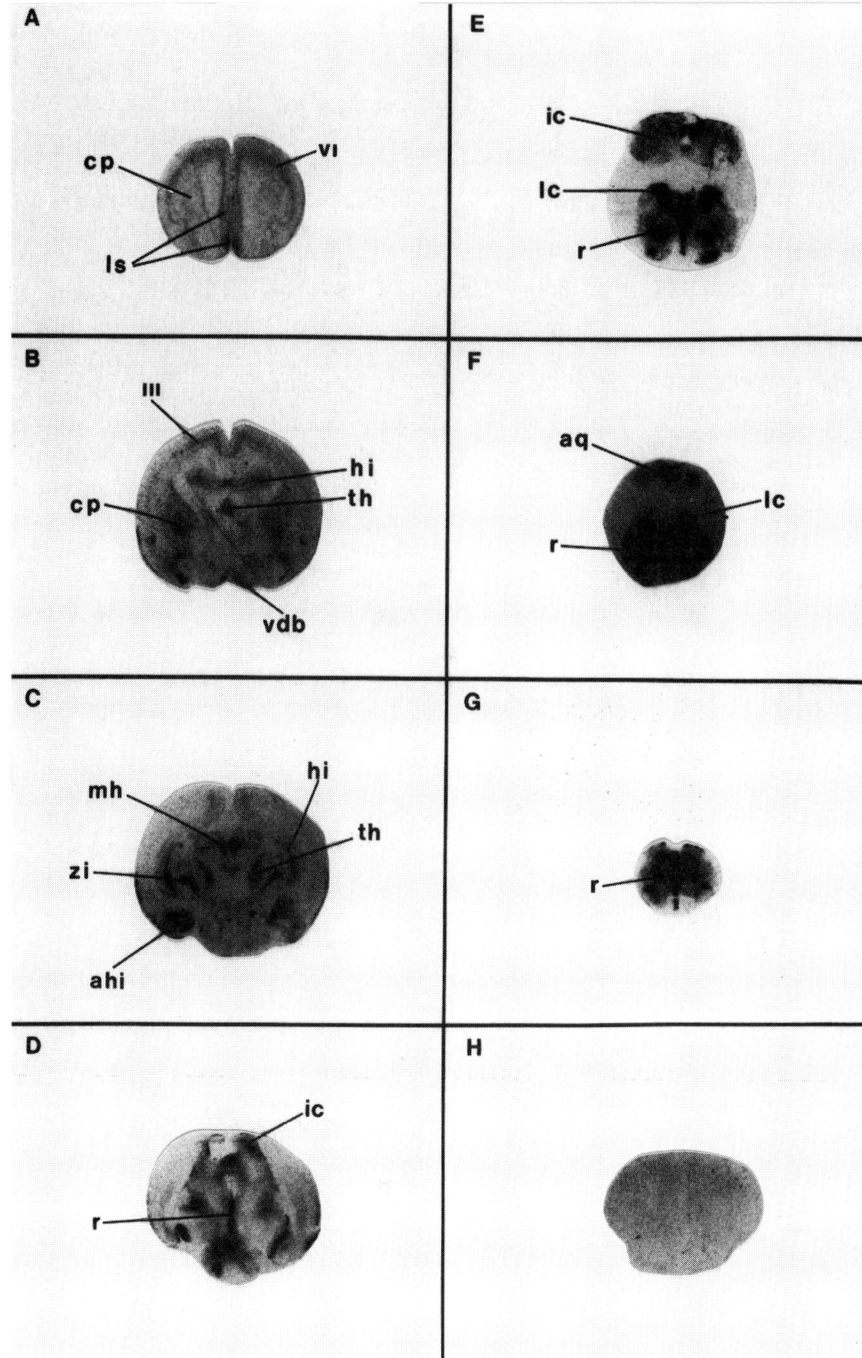

Fig. 4.6. Photomicrographs of the distribution of NK-2 receptor binding sites in coronal brain sections from 1-day-old rats (PD1). Sections were incubated with 50 pM of ^{125}I-labeled NKA. Non-specific (NS) labeling seen in the presence of 1.0 μM unlabeled NKA is shown in H. Abbreviations: ahi, amygdalo-hippocampal area; aq, cerebral aqueduct; cp, caudate putamen; hi, hippocampus; ic, inferior colliculus; lc, locus coeruleus; ls, lateral septum; mh, medial habenula; r, raphe; th, thalamus; vdb, diagonal band of Broca; zi, zona incerta; III, VI, layer III, VI of the frontal cortex.

PD4

Four days after birth, the density of NK-2 sites in the cortex (laminae III and IV) is markedly increased. However, its distribution remains similar to that seen at PD1 (Figs 4.6 and 4.7). Very high densities of NK-2 sites are seen in the frontal cortex (Fig. 4.7A–C) and the external plexiform layer of the olfactory bulb (Fig. 4.7A). Other regions such as the nucleus of diagonal band of Broca (Fig. 4.7B), amygdalohippocampal area (Fig. 4.7C–E), mediodorsal thalamic nucleus (Fig. 4.7D), superior and inferior colliculi (Fig. 4.7E–G), raphe nucleus (Fig. 4.7F,G) and medial habenula (Fig. 4.7D) are also enriched with NK-2 sites at this age. Lower but still significant quantities of sites are found in other regions including striatum, hippocampus (Fig. 4.7B–D) and periaqueductal gray area (Fig. 4.7C,D).

PD7

Seven days after birth, the distribution of NK-2 sites is rather similar to that observed at early ages (Figs 4.6, 4.7 and 4.8) with the exception of a certain scattering of NK-2 sites in laminae III and IV of the occipital cortex (Fig. 4.8E,F). High densities of sites are still located in laminae VI of the frontal cortex (Fig. 4.8B–D), external plexiform layer of the olfactory bulb (Fig. 4.8A,B), diagonal band of Broca (Fig. 4.8C), various amygdaloid nuclei (Fig. 4.8D), the amygdalohippocampal area (Fig. 4.8E,F), medial geniculate nucleus (Fig. 4.8G), superior and inferior colliculi (Fig. 4.8G–I), mediodorsal thalamic nucleus (Fig. 4.8D,E), periaqueductal gray matter (Fig. 4.8G), ventral tegmental area (Fig. 4.8I), locus coeruleus (Fig. 4.8H) and raphe (Fig. 4.8H). Lower densities of NK-2 sites are also found in the striatum (Fig. 4.8B–D), hippocampal formation (Fig. 4.8D–F) and pontine nucleus (Fig. 4.8G).

PD21

Major modifications are observed in the distribution of NK-2 sites in 21-day-old rats

(Fig. 4.9). For example, the density of specific NK-2 labeling in the cerebral cortex is markedly diminished as compared to that observed in younger animals. Moreover, the laminar distribution of NK-2 sites in the cortex is altered, being mostly concentrated in mid-layers (Fig. 4.9D–F). Marked differences are also seen in most brainstem nuclei which are mostly devoid of specific labeling at this age (Fig. 4.9H,I). However, very high densities of NK-2 sites are still observed in the external plexiform layer of olfactory bulb (Fig. 4.9A–C), diagonal band of Broca (Fig. 4.9D), preoptic, suprachiasmatic and periventricular nuclei of the hypothalamus (Fig. 4.9E,F), amygdalohippocampal area (Fig. 4.9H), entorhinal cortex (Fig. 4.9E,G,H), medial habenula (Fig. 4.9H), inferior colliculus (Fig. 4.9I) and locus coeruleus (Fig. 4.9I). Moderate amounts of labeling are also present in the hippocampal formation (Fig. 4.9G,H) and the striatum (Fig. 4.9D–G). The cerebellum is usually devoid of specific labeling (Fig. 4.9I).

PD35

The distribution of NK-2 sites in 35-day-old rats is very similar to that observed in adults (Fig. 4.10) (Dam *et al.*, 1990c). By this age, the density of NK-2 sites in the cortex is very low (Fig. 4.10A–E). Only the entorhinal cortex seems to retain specific NK-2 labeling (Fig. 4.10F,G). In subcortical regions, substantial quantities of NK-2 sites are seen in the diagonal band of Broca (Fig. 4.10A), septal nuclei (Fig. 4.10B), amygdalohippocampal area (Fig. 4.10D), ventral hippocampal region (Fig. 4.10E), parasubiculum (Fig. 4.10F,G), presubiculum (Fig. 4.10F,G), superior colliculus (Fig. 4.10E,F) and periaqueductal gray (Fig. 4.10G). Moderate densities of labeling are also detected in the striatum (Fig. 4.10A–C), certain thalamic nuclei (Fig. 4.10C), medial habenula (Fig. 4.10D) and zona incerta (Fig. 4.10C). Other regions are apparently devoid of specific NK-2 sites.

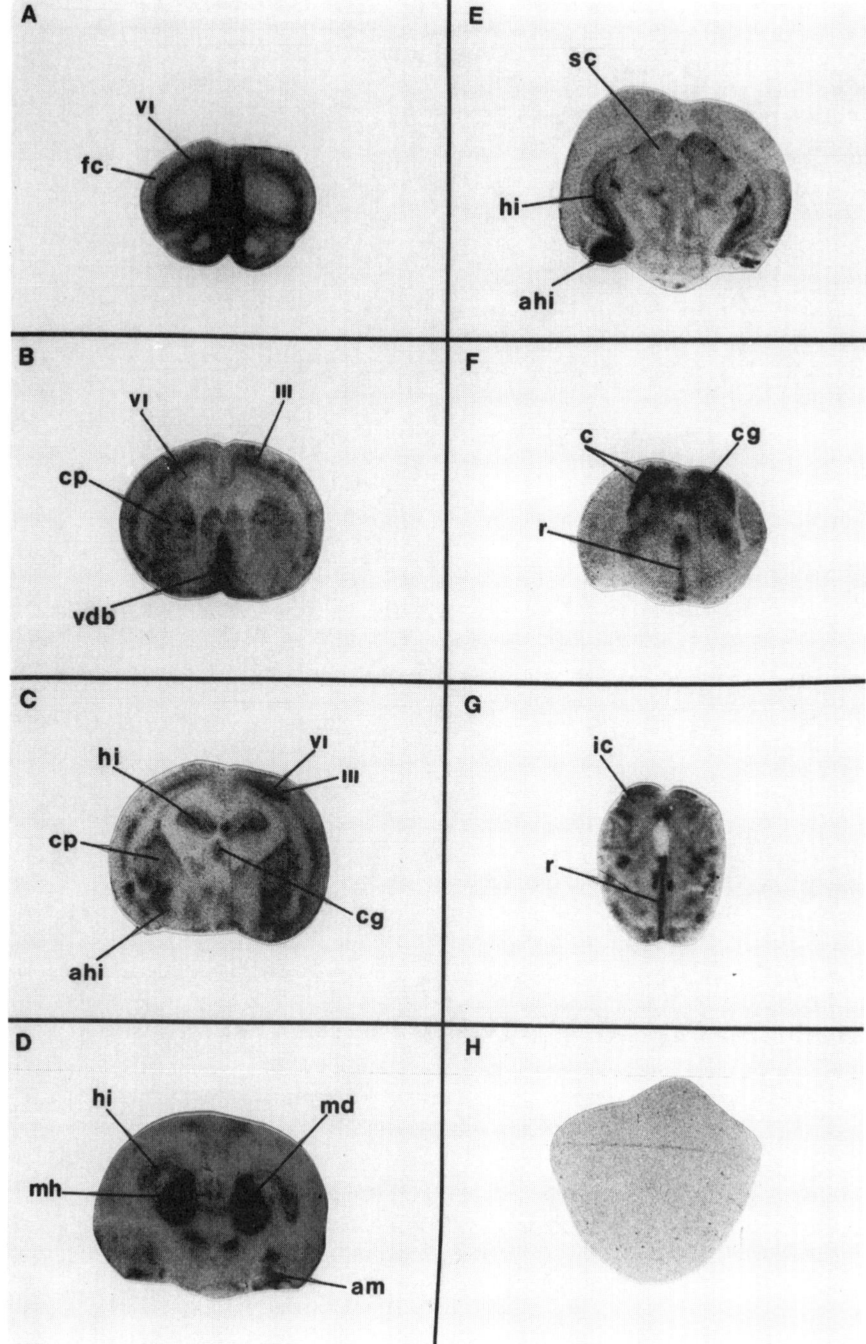

Fig. 4.7.   Photomicrographs of the distribution of NK-2 receptor binding sites in coronal brain sections from 4-day-old rats (PD4). Sections were incubated with 50 pM ^{125}I-labeled NKA. Non-specific (NS) labeling seen in presence of 1.0 μM unlabeled NKA is shown in section H. Abbreviations: ahi, amygdalohippocampal area; am, amygdala; c, colliculus; cg, central gray; cp, caudate putamen; fc, frontal cortex; hi, hippocampus; ic, inferior colliculus; md, mediodorsal thalamic nuclei; mh, medial habenula; r, raphe; sc, superior colliculus; vdb, diagonal band of Broca; III, VI, layer III, VI of the frontal cortex.

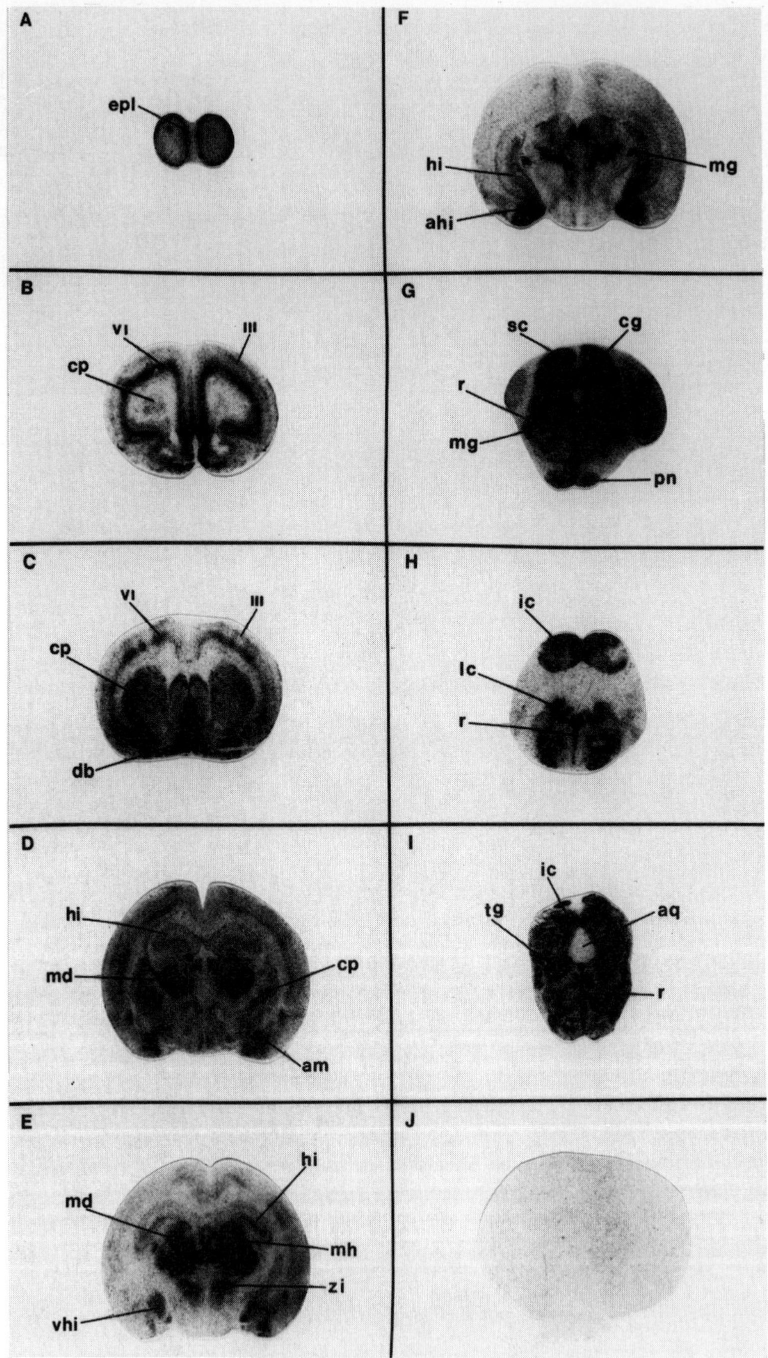

Fig. 4.8.   Photomicrographs of the distribution of NK-2 receptor binding sites in coronal brain sections from 7-day-old rats (PD7). Sections were incubated with 50 pM of ^{125}I-labeled NKA. Non-specific (NS) labeling seen in the presence of 1.0 µM unlabeled NKA is shown in section J. Abbreviations: ahi, amygdalohippocampal area; am, amygdala; aq, cerebral aqueduct; cg, central gray; cp, caudate putamen; db, diagonal band of Broca; epl, external plexiform layer of the olfactory bulb; hi, hippocampus; ic, inferior colliculus; lc, locus coeruleus; md, mediodorsal thalamic nuclei; mg, medial geniculate nucleus; mh, medial habenula; pn, pontine nuclei; r, raphe; sc, superior colliculus; tg, tegmental nuclei; vhi, ventral hippocampal area; zi, zona incerta; III, VI, layer III, VI of the frontal cortex.

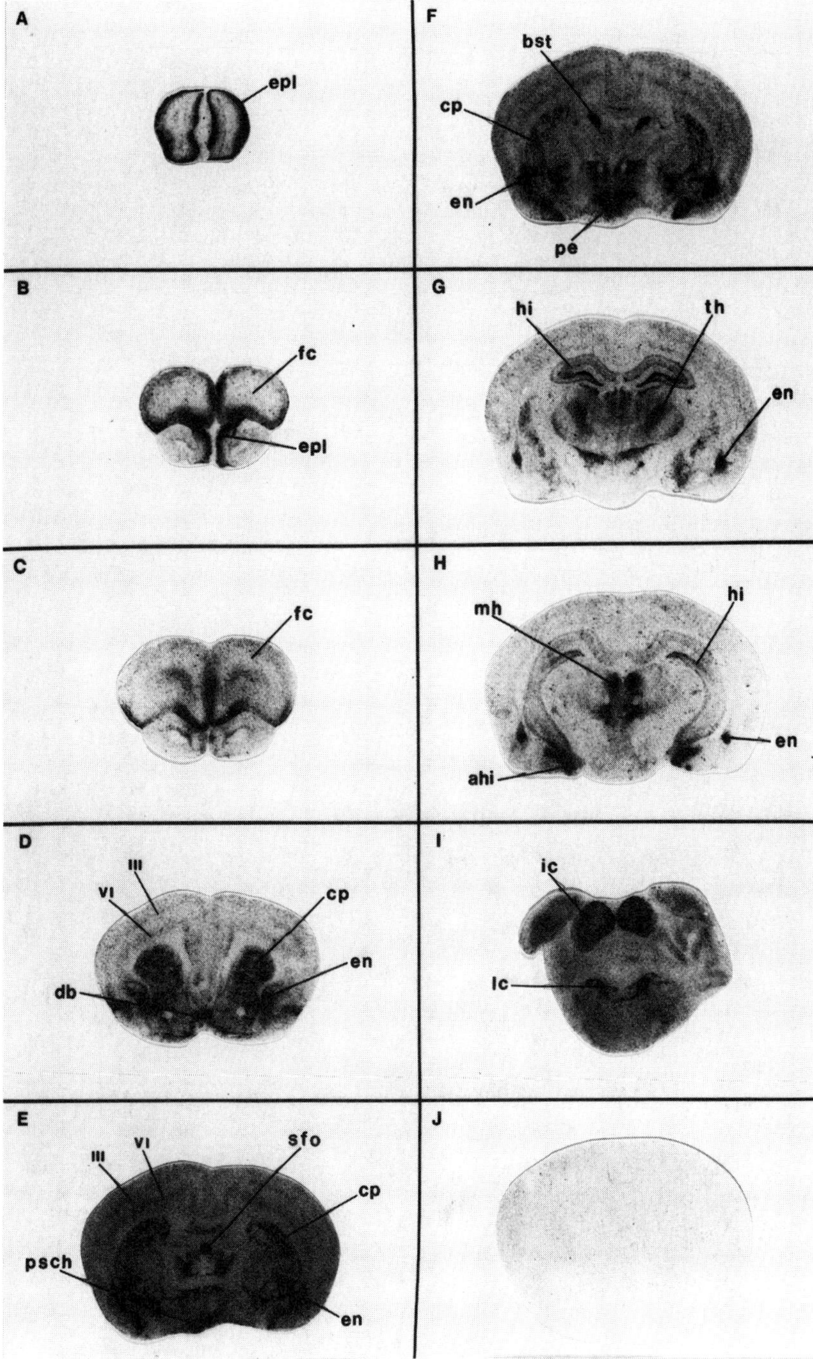

Fig. 4.9. Photomicrographs of the distribution of NK-2 receptor binding sites in coronal brain sections from 21-day-old rats (PD21). Sections were incubated with 50 pM ^{125}I-labeled NKA. Non-specific (NS) labeling seen in the presence of 1.0 μM unlabeled NKA is shown in section J. Abbreviations: ahi, amygdalohippocampal area; bst, stria terminal bed nuclei; cp, caudate putamen; db, diagonal band of Broca; en, entorhinal cortex; epl, external plexiform layer of the olfactory bulb; fc, frontal cortex; hi, hippocampus; ic, inferior colliculus; lc, locus coeruleus; mh, medial habenula; pe, periventricular hypothalamic nuclei; psch, preoptic suprachiasmatic nuclei; sfo, subfornical organ; th, thalamus; III,VI, layer III, VI of the frontal cortex.

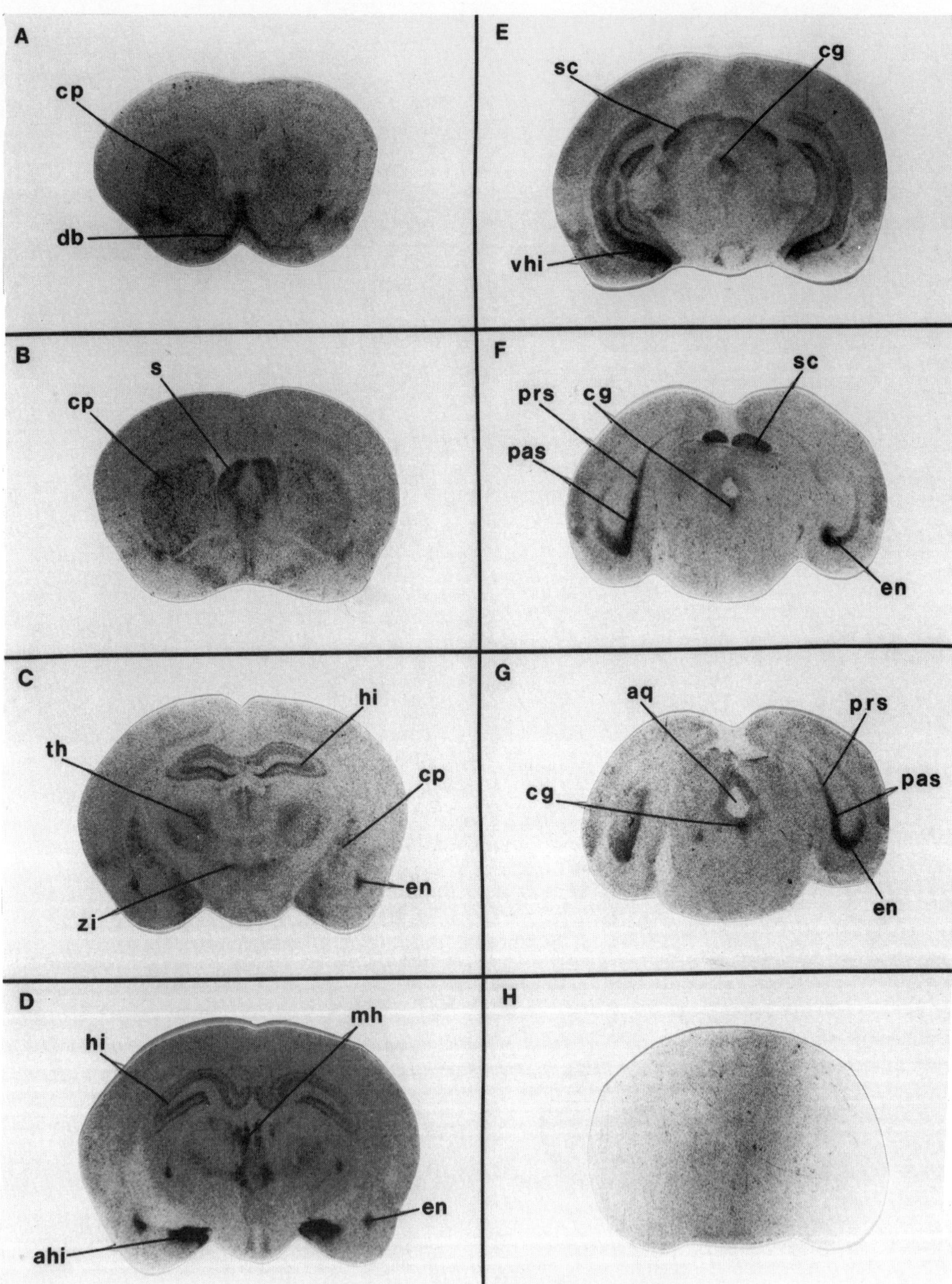

Fig. 4.10. Photomicrographs of the distribution of NK-2 receptor binding sites in coronal brain sections from 35-day-old rats (PD35). Sections were incubated with 50 pM ^{125}I-labeled NKA. Non-specific (NS) labeling seen in the presence of 1.0 µM unlabeled NKA is shown in section H. Abbreviations: ahi, amygdalohippocampal area; aq, cerebral aqueduct; cg, central gray; cp, caudate putamen; db, diagonal band of Broca; en, entorhinal cortex; hi, hippocampus; mh, medial habenula; pas, parasubiculum; prs, presubiculum; s, septum; sc, superior colliculus; th, thalamus; vhi, ventral hippocampus; zi, zona incerta.

4.3.2   ONTOGENY OF NK-3 RECEPTORS

The distribution of NK-3 binding sites in the rat brain does not undergo major modification during the first postnatal week (Figs 4.11–4.13). However, minor modifications are observed thereafter (Figs 4.14–4.16).

ED20

One day before birth, high densities of NK-3 sites are present in the external plexiform layer of the olfactory bulb (Fig. 4.11A), striatum (Fig. 4.11B), septal area (Fig. 4.11B), hippocampal formation (Fig. 4.11D) as well as various thalamic, hypothalamic and amygdaloid nuclei (Fig. 4.11C,D). At this age, a low density of labeling is seen in most cortical layers (Fig. 4.11A–D).

PD1

One day after birth, NK-3 sites are concentrated in the external plexiform layer of the olfactory bulb (Fig. 4.12A), lateral septum (Fig. 4.12B), CA_2 and CA_3 subfields of the hippocampus (Fig. 4.12C,D), various nuclei of the thalamus, hypothalamus and amygdala (Fig. 4.12C, D) and the zona incerta (Fig. 4.12C). As at ED20, low densities of NK-3 binding sites are present in cortical areas.

PD6

As shown in Fig. 4.13, the distribution of NK-3 sites at this age is rather similar to that observed at PD1. High densities of NK-3 binding sites are found in the external plexiform layer of the olfactory bulb (Fig. 4.13A), striatum (Fig. 4.13B,C), diagonal band of Broca (Fig. 4.13B), CA_2 and CA_3 subfields of

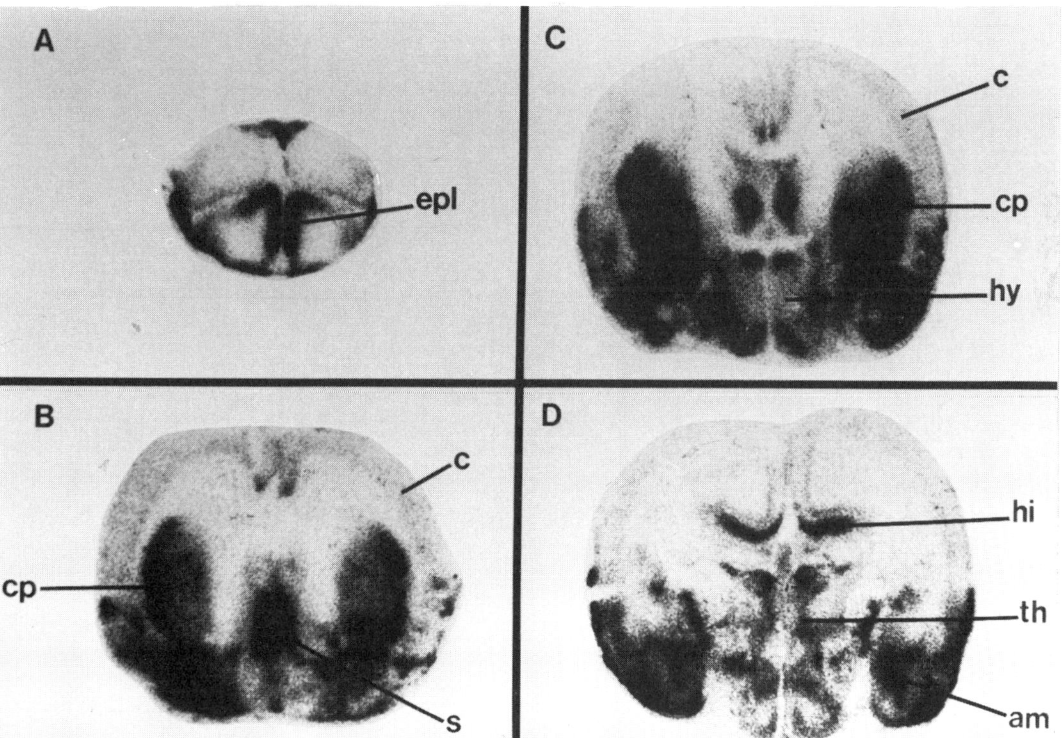

Fig. 4.11.   Photomicrographs of the distribution of NK-3 receptor binding sites in coronal brain sections from embryonic 20-day-old rats (ED20). Abbreviations: am, amygdala; c, colliculus; cp, caudate putamen; epl, external plexiform layer of the olfactory bulb; hi, hippocampus; hy, hypothalamus; s, septum; th, thalamus.

the hippocampus (Fig. 4.13C,D), zona incerta (Fig. 4.13D), various thalamic, hypothalamic and amygdaloid nuclei (Fig. 4.13C,D), medial habenula (Fig. 4.13D), superior colliculus and central gray (Fig. 4.13E). However, it is interesting to note that in the frontal cortex, moderate densities of sites are now detected in laminae III and IV (Fig. 4.13B) as well as in the retrosplenial cortex (Fig. 4.13B).

PD14

As observed at younger ages, at PD14, high densities of NK-3 sites are seen in the external and internal plexiform layers of the olfactory bulb (Fig. 4.14A), nucleus of the horizontal limb of the diagonal band of Broca (Fig. 4.14C,D), olfactory tubercle (Fig. 4.14C,D), medial habenula (Fig. 4.14E), ventral hippo-

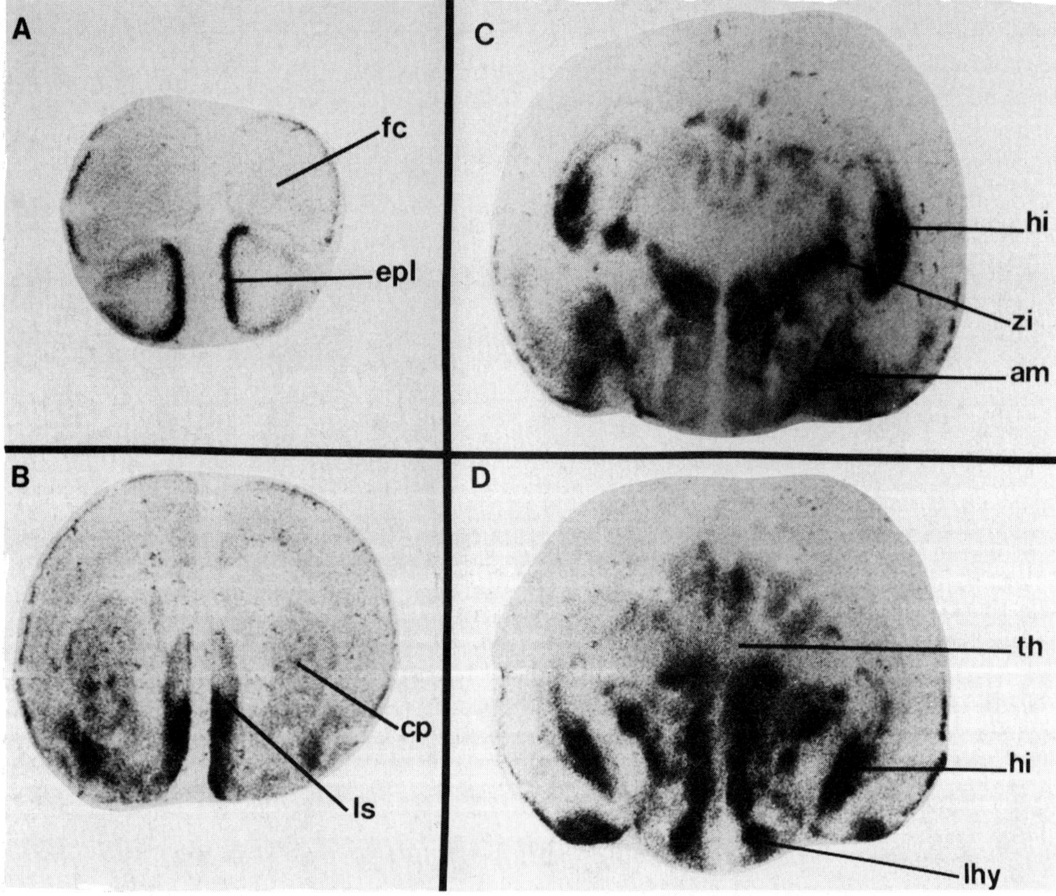

Fig. 4.12.   Photomicrographs of the distribution of NK-3 receptor binding sites in coronal brain sections from 1-day-old rats (PD1). Abbreviations: am, amygdala; cp, caudate putamen; epl, external plexiform layer of the olfactory bulb; fc, frontal cortex; hi, hippocampus; lhy, lateral hypothalamic area; ls, lateral septum; th, thalamus; zi, zona incerta.

campal commissura (Fig. 4.14F), supraoptic and paraventricular hypothalamic nuclei (Fig. 4.14E), certain nuclei of the amygdalohippocampal area (Fig. 4.14E,F), superior and inferior colliculi (Fig. 4.14G,H), central gray (Fig. 4.14G), ventral tegmental area (Fig. 4.14G) and the interpeduncularis nucleus (Fig. 4.14G). However, in the frontal cortex, very high densities of NK-3 binding sites are now clearly evident in laminae IV, even

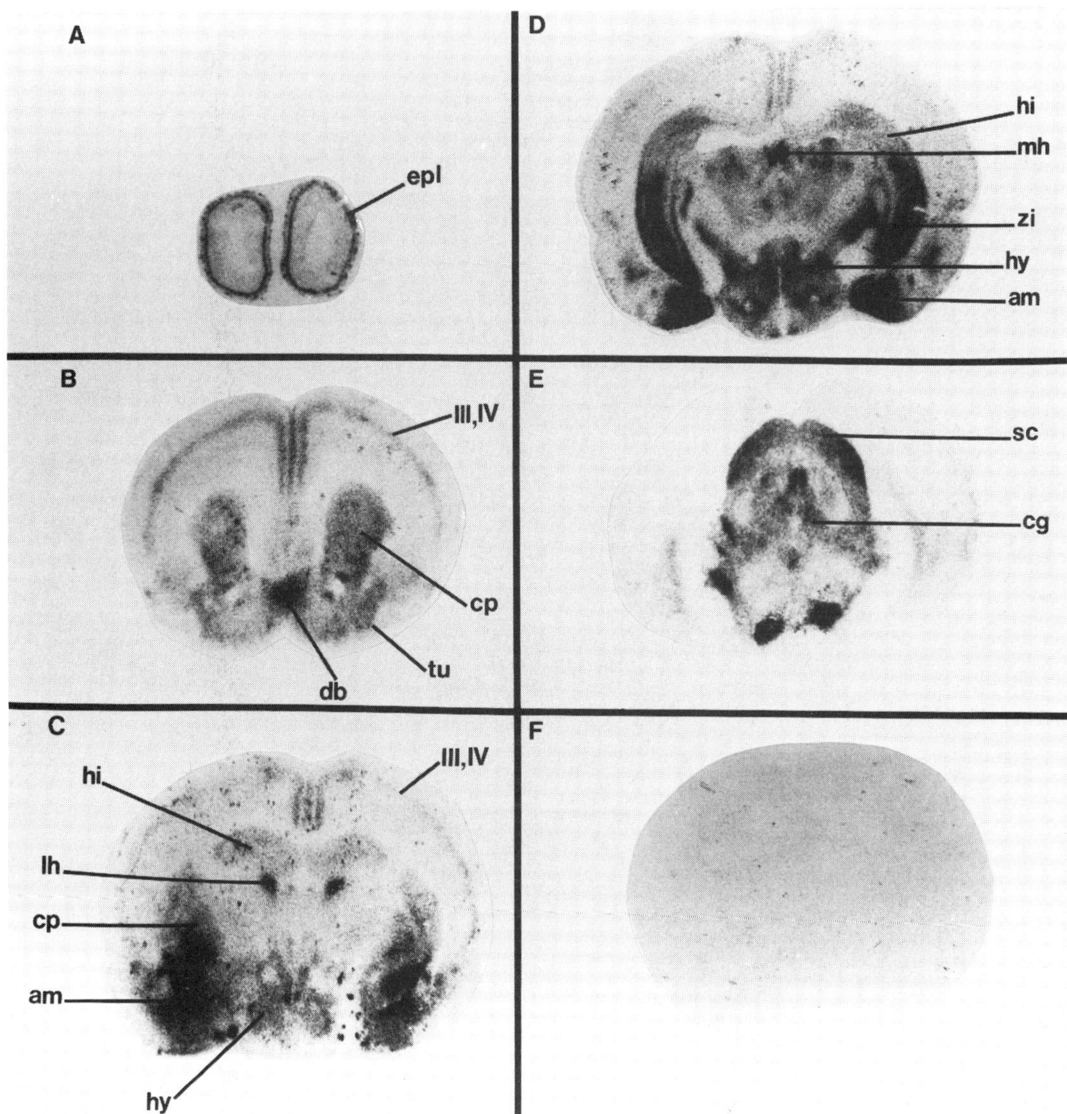

Fig. 4.13. Photomicrographs of the distribution of NK-3 receptor binding sites in coronal brain sections from 6-day-old rats (PD6). Sections were incubated with 50 pM ^{125}I-labeled BH-eledoisin. Non-specific (NS) labeling seen in the presence of 1.0 μM unlabeled eledoisin is shown in section F. Abbreviations: am, amygdala; cg, central gray; cp, caudate putamen; db, diagonal band of Broca; epl, external plexiform layer of the olfactory bulb; hi, hippocampus; hy, hypothalamus; lh, lateral hypothalamic area; mh, medial habenula; sc, superior colliculus; tu, olfactory tubercle; zi, zona incerta; III, IV, layer III, IV of the frontal cortex.

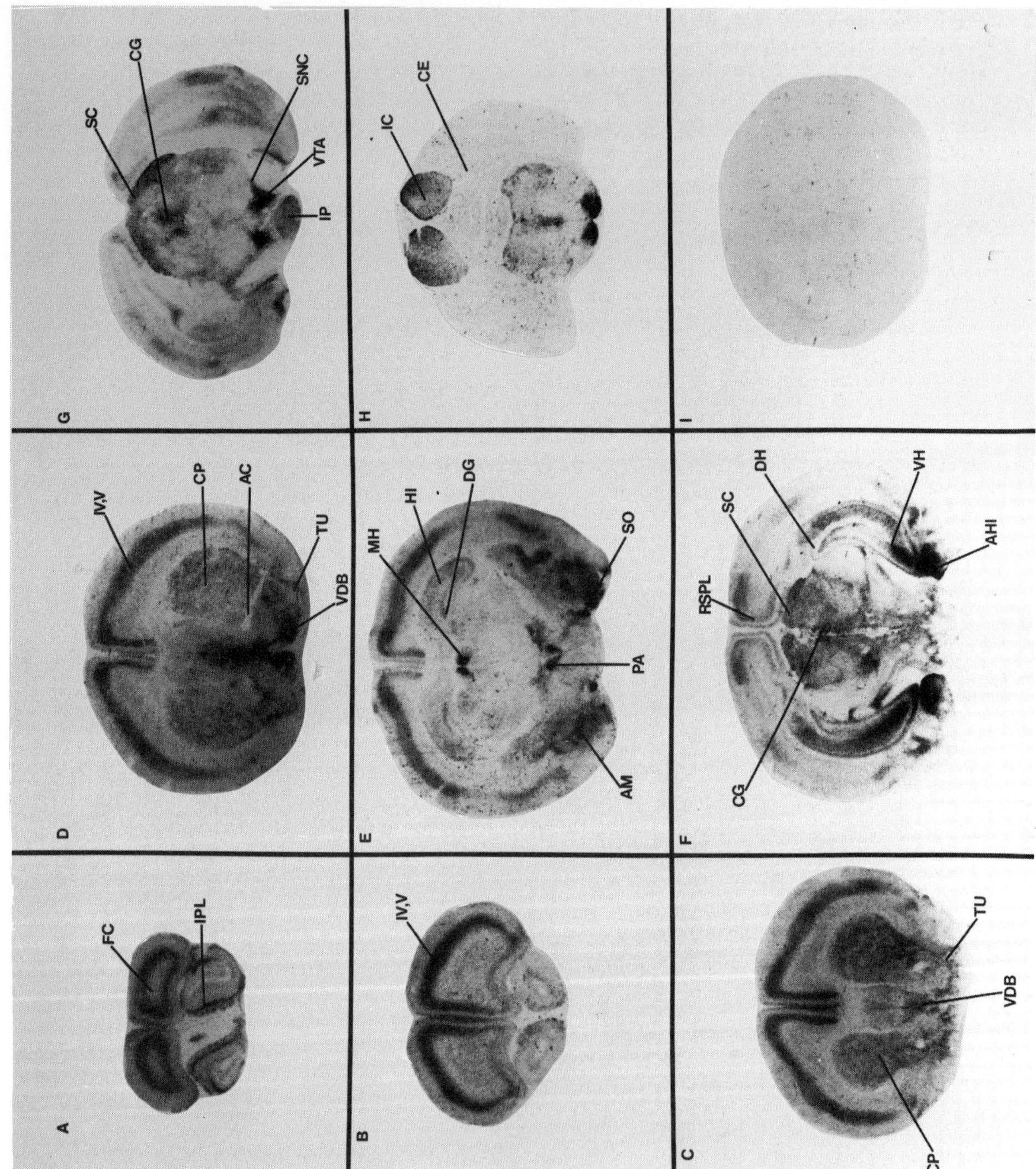

Fig 4.14 (*caption on next page*)

expanding to laminae V (Fig. 4.14A–E). In the occipital cortex, only the retrosplenial portion is enriched with NK-3 sites (Fig. 4.14F,G). The nucleus accumbens and the cerebellum are clearly devoid of NK-3 binding sites (Fig. 4.14D,H).

PD28

At this age (Fig. 4.15), the distribution of NK-3 sites is rather similar to that observed at PD14 (Fig. 4.14) except for the superficial layers of the cortex which are now relatively enriched with NK-3 sites (Fig. 4.15) in contrast to the apparent absence of labeling detected at earlier ages (Fig. 4.12–4.14). Lobules 9 and 10 of the cerebellum also contain NK-3 sites at PD28 (Fig. 4.15G).

PD35

The CNS distribution of NK-3 binding sites of 35-day-old rats is shown in Fig. 4.16. As at PD28 high densities of sites are seen in laminae IV and V of the cortex (Fig. 4.16A–H), in the supraoptic nucleus (Fig. 4.16B), amygdalohippocampal area (Fig. 4.16C,D), zona incerta (Fig. 4.16D), interpeduncular nucleus (Fig. 4.16G), superior colliculus (Fig. 4.16H), locus coeruleus (Fig. 4.16I) and nucleus tractus solitarius (Fig. 4.16J,K). Moderate densities are found in the striatum (Fig. 4.16A–D), septum (Fig. 4.16A,B), dorsal hippocampus (Fig. 4.16D,E), substantia nigra pars compacta (Fig. 4.16F), inferior colliculus (Fig. 4.16I) as well as lobules 9 and

10 of the cerebellum (Fig. 4.16J). Other regions of the cerebellum are devoid of specific labeling NK-3 (Fig. 4.16J). Thus, NK-3 sites are similarly distributed at PD28 and PD35 in the rat brain suggesting their complete profile of maturation during the second and third postnatal week. This is further supported by the discrete localization of NK-3 sites observed in adult (3-month-old) brain tissue.

4.4 POSSIBLE SIGNIFICANCE OF ONTOGENIC RECEPTOR MODIFICATIONS

4.4.1 NK-1 RECEPTORS

NK-1 binding sites appear very early during embryonic development since highly significant amounts of NK-1 labeling are present three days before birth in the rat brain (Quirion and Dam, 1986). Moreover, the density of NK-1 sites increases markedly 1 day before birth suggesting that SP might play a very important role in the early maturation process and organization of the CNS (Quirion and Dam, 1986).

It is evident that the distribution of NK-1 sites undergoes major modifications during postnatal ontogeny. This is especially striking in the brainstem since very high densities of sites are seen in this region up to 14 days after birth but not thereafter. This suggests that SP, by activating NK-1 receptors, is associated with the ontogenic development and maturation of this brain region. This is of special interest since it has been shown that SP can act as a trophic factor for brainstem

Fig. 4.14. Photomicrographs of the distribution of NK-3 receptor binding sites in coronal brain sections from 14-day-old rats (PD14). Sections were incubated with 50 pM ^{125}I-labeled BH-eledoisin. Non-specific (NS) labeling seen in the presence of 1.0 µM unlabeled eledoisin is shown in section I. Abbreviations: AC, anterior commissura; AHI, amygdalohippocampal area; AM, amygdala; CE, cerebellum; CG, central gray; CP, caudate putamen; DG, dentate gyrus; DH, dorsal hippocampus; FC, frontal cortex; HI, hippocampus; IC, inferior colliculus; IP, interpeduncular nuclei; IPL, internal plexiform layer of the olfactory bulb; MH, medial habenula; PA, paraventricular hypothalamic nuclei; RSPL, retrosplenial cortex; SC, superior colliculus; SNC, substantia nigra pars compacta; SO, supraoptic nuclei; TU, olfactory tubercle; VDB, diagonal band of Broca; VH, ventral hippocampus; VTA, ventral tegmental area; IV, V, layer IV, V of the frontal cortex.

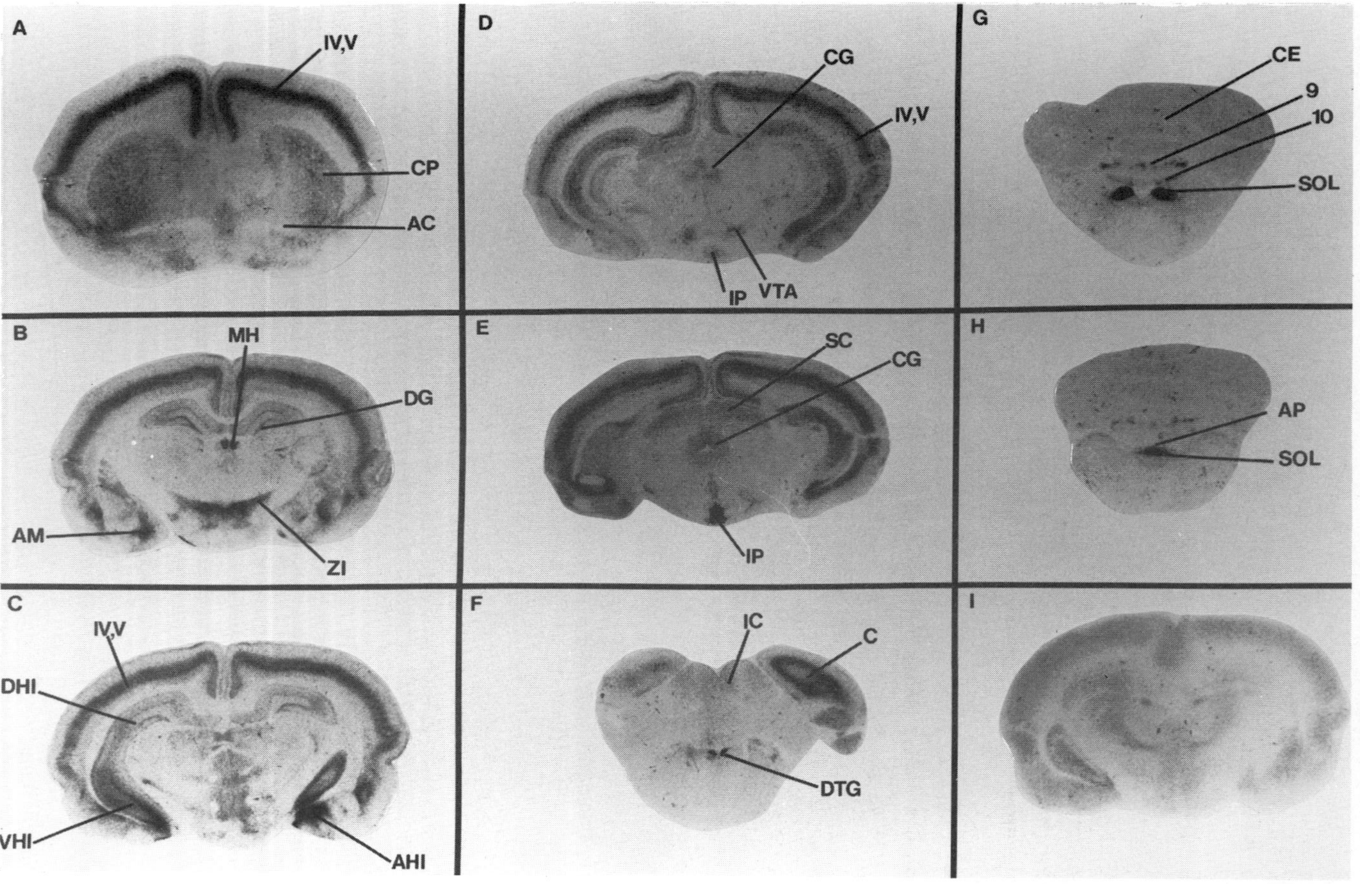

Fig. 4.15. Photomicrographs of the distribution of NK-3 receptor binding sites in coronal brain sections from 28-day-old rats (PD28). Sections were incubated with 50 pM ^{125}I-labeled BH-eledoisin. Non-specific (NS) labeling seen in the presence of 1.0 μM unlabeled eledoisin is shown in section I. Abbreviations: AC, anterior commissura; AHI, amygdalohippocampal area; AM, amygdala; C, cortex; CE, cerebellum; CG, central gray; CP, caudate putamen; DG, dentate gyrus; DHI, dorsal hippocampus; DTG, dorsal tegmentum; IC, inferior colliculus; IP, interpeduncular nuclei; MH, medial habenula; SC, superior colliculus; SOL, tractus solitarius nuclei; VHI, ventral hippocampus; VTA, ventral tegmental area; ZI, zona incerta; IV, V, layer IV, V of the frontal cortex; 9, 10, lobules 9 and 10 of the cerebellum.

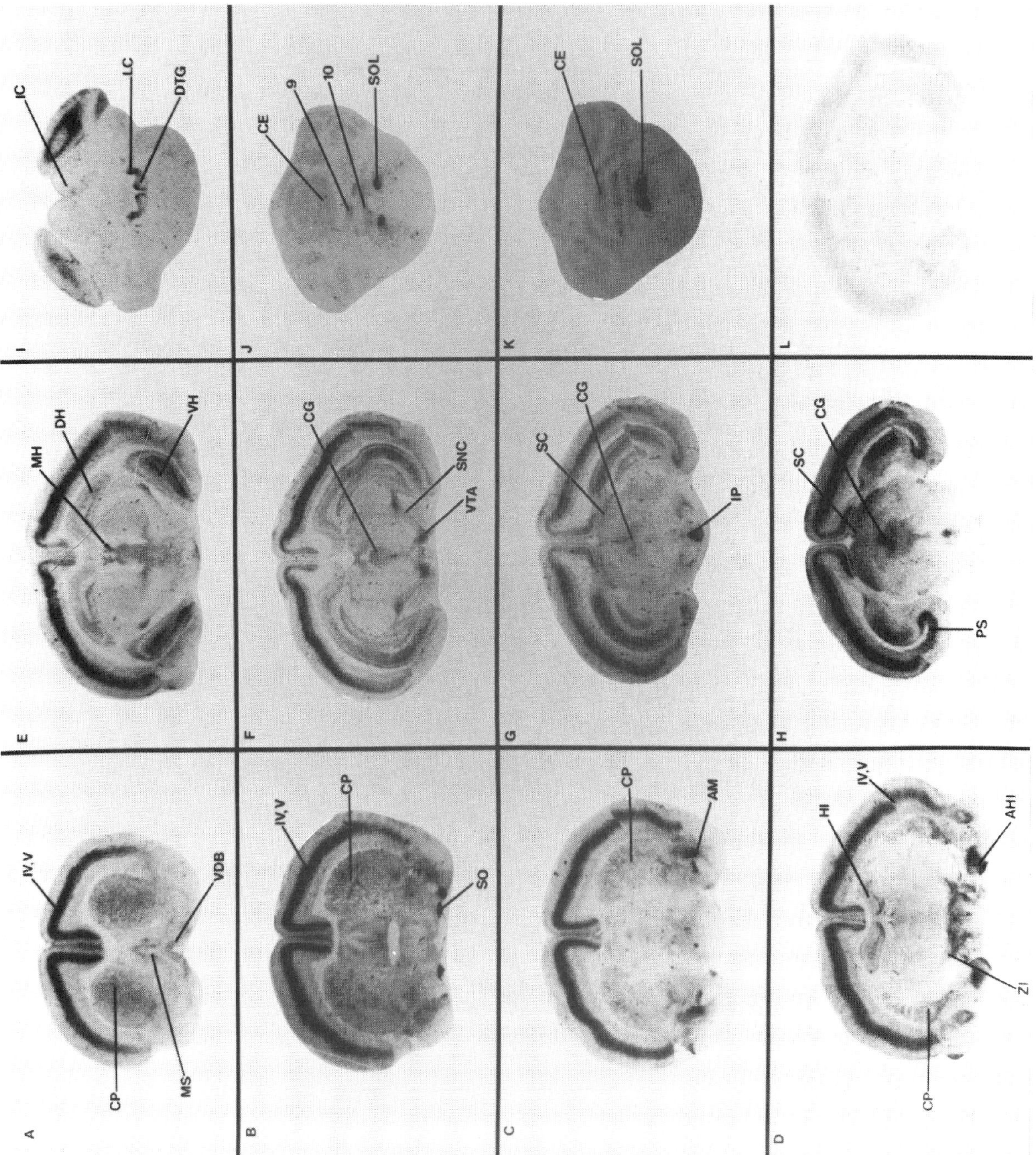

Fig. 4.16.   Photomicrographs of the distribution of NK-3 receptor binding sites in coronal brain sections from 35-day-old rats (PD35). Sections were incubated with 50 pM ^{125}I-labeled BH-eledoisin. Non-specific (NS) labeling seen in the presence of 1.0 μM unlabeled eledoisin is shown in section L. Abbreviations: AHI, amygdalohippocampal area; AM, amygdala; CE, cerebellum; CG, central gray; CP, caudate putamen; DH, dorsal hippocampus; DTG, dentate gyrus; HI, hippocampus; IC, inferior colliculus; IP, interpeduncular nuclei; LC, locus coeruleus; MS, median septum; MH, medial habenula; PS, presubiculum; SC, superior colliculus; SNC, substantia nigra pars compacta; SO, supraoptic nuclei; SOL, tractus solitarius nuclei; VDB, diagonal band of Broca; VH, ventral hippocampus; VTA, ventral tegmental area; ZI, zona incerta; IV, V, layer IV, V of the frontal cortex; 9, 10, lobule 9, 10 of the cerebellum.

serotonergic neurons during ontogeny following neurotoxin treatment (Jonsson and Hallman, 1983), can prevent the degeneration of damaged noradrenergic neurons (Jonsson and Hallman, 1982a,b), can stimulate neurite outgrowth in embryonic chick dorsal root ganglia (Narumi and Fujita, 1978) and neuroblastoma cells (Naruma and Maki, 1978) and can accelerate the growth and maturation of catecholaminergic cells of the locus coeruleus (Nakai and Kasamatsu, 1984). Moreover, it has recently been shown that growth cones are enriched with NK-1 receptors suggesting that SP and NK-1 receptors may be directly involved in growth of certain neuronal propagation by modulating growth cones (Lockerbie *et al.*, 1988).

SP could also be an important factor involved in the maturation of other brain pathways on the basis of its early appearance during embryonic development. For example, it has been shown that SP is present in rat brain at gestation day 14, reaching adult level between days 5 and 15 after birth (Inagaki *et al.*, 1982a; Sakanaka *et al.*, 1982). Similar data have been obtained in human fetal brain with high densities of SP-like fibers identified, especially throughout the lower brainstem during embryogenesis (Namura *et al.*, 1982; Charnay *et al.*, 1983; Del Fiacco *et al.*, 1984). This correlates well with the high density of NK-1 binding sites visualized here in the brainstem of neonatal rats.

4.4.2   NK-2 SITES

As for NK-1 sites, the distribution of NK-2 receptors undergoes major modifications during postnatal ontogeny, with only low densities of specific labeling detected in adult animals. Very high densities of NK-2 are present very early (PD1) in the superior and inferior areas of brainstem. Subsequently, the concentrations of NK-2 sites in these regions gradually decrease to reach the adult level by PD21. This suggests that, in addition to SP, NKA could play an important role during the

development and organization of this brain region. The ontogeny profile of NKA-like immunoreactivity remains to be established. However, it appears that SP and NKA are distributed in a similar manner in the adult rat brain (Maggio and Hunter, 1984; Deutch *et al.*, 1985; Dalsgaard *et al.*, 1985; Linderfors *et al.*, 1985; Lee *et al.*, 1986) and are derived from the same precursor (Nakanishi, 1987; Krause *et al.*, 1989). Since SP appears very early during brain embryogenesis (Inagaki *et al.*, 1982a,b; Sakanaka *et al.*, 1982; Del Fiacco *et al.*, 1984; Paulin *et al.*, 1986), it is probable that NKs are also present during the early phase of CNS development. It is also noteworthy that NKA is able to stimulate cell growth in connective tissues (Nilsson *et al.*, 1985) and that the NK-2 receptor is related to the *mas* oncogene (Young *et al.*, 1986). It is likely that NKs, by activating NK-2 receptors, would modulate the growth and maturation of certain brain pathways.

In the basal ganglia, moderate levels of NK-2 sites are present during the various phases of brain development. Moreover, their distribution remains homogeneous and does not reflect the high amount of NKA present in this region in adult rat brain (Minamino *et al.*, 1984). This is rather different from the heterogeneous redistribution of NK-1 sites observed in the striatum during brain maturation (Quirion and Dam, 1986) demonstrating that NK-2 receptors are distinct from NK-1 sites.

This is supported further by the redistribution of NK-1 and NK-2 sites observed in cortical areas during postnatal ontogeny (Dam *et al.*, 1988). NK-2 sites appear very early after birth (PD1); their densities markedly increasing by PD4 in frontal cortex and PD7 in remaining areas. However, by the end of the second week, the density of cortical NK-2 sites begins to decrease and totally disappears by PD35. This is clearly different from NK-1 receptors which are present in various cortical areas in neonatal and adult brain (Quirion and Dam, 1986; Dam *et al.*, 1988).

4.4.3   NK-3 SITES

As for other NK receptors, NK-3 sites appear very early during embryonic development. For example, very high densities of NK-3 sites are observed in the external plexiform layer of the olfactory bulb and in the hippocampus at ED20. Thereafter, the amounts of NK-3 sites remain relatively constant during the first postnatal week. However, the distribution of NK-3 sites undergoes major modification during the second week of ontogenic development to reach the adult level by the end of the third week. As for NK-1 (Quirion and Dam, 1986) and NK-2 sites, various brainstem nuclei are enriched with NK-3 sites early postnatally but not during adulthood. Overall, the distribution of NK-3 sites is not as modified as that of NK-1 and NK-2 receptors during ontogeny suggesting that NK-3 receptors are less likely to mediate trophic effects of NKs.

The ontogenic profile of cortical NK-3 sites is unique. They appear later (PD6) than NK-1 and NK-2 sites and concentrate mostly in the deeper laminae (IV and V) of the frontal cortex. Moreover, NK-3 sites present in the cortex do not disappear (as for NK-2 sites) suggesting that they could be involved in the maintenance of normal cortical functions. It is thus of interest that Lamour *et al.* (1983) have shown that the excitatory responses elicited by NKs in deeper cortical laminae could be related to the activation of NK-3 sites. The distribution of NKB, the putative endogenous ligand of the NK-3 subtype, remains to be fully established either in adult tissues or during brain development.

4.5   ONTOGENIC PROFILE OF NK FUNCTIONS

The development of NK-mediated functions has not been explored as extensively as the development of the NK immunoreactivity and receptors. It appears, however, at least from studies in the spinal cord, that the presence of NK fibers and receptors alone does not predict the onset of function. SP is present in dorsal root ganglia, sensory nerves, spinal cord and skin at birth in the rat (Fitzgerald and Gibson, 1984) indicating that the distribution of SP in the primary sensory afferent system occurs early in development. SP-containing C fibers enter the dorsal horn on ED19–20 in the rat (Senba *et al.*, 1982; Fitzgerald, 1987). On PD 1, SP fibers are concentrated in laminae 1 but on following days they become concentrated in laminae 2. By PD8, the staining pattern is comparable to that in the adult, but the adult density is reached on PD15. Foot withdrawal responses to pinching and heating of the hindfoot skin are present at birth and are exaggerated in amplitude and duration compared to the adult (Fitzgerald and Gibson, 1984). It is possible that the high density of NK-1 receptors in the spinal cord at birth may play a role in this exaggerated response (Charlton and Helke, 1986; Yashpal *et al.*, 1991b). Withdrawal responses to irritant chemicals and neurogenic extravasation, however, did not occur until PD 10. The reason for this developmental difference in functions of the C fibers is not apparent, although it suggests that there may be different synaptic pathways for the responses to these different stimuli and that certain elements in these different pathways may develop at different times.

In comparison, in the peripheral nervous system, the responses to SP in the porcine trachea and rabbit bladder are in place at birth (Haxhiu *et al.*, 1990; Zderic *et al.*, 1990). Similar to the spinal cord, the response of the bladder to SP was greater in the neonate than in the adult. This may also suggest a greater receptor number in the developing animal.

4.6   FACTORS WHICH CAN INFLUENCE THE DEVELOPMENT OF SP AND NK-1 RECEPTORS

4.6.1   SP CONTENT AND APPEARANCE

Several factors have been identified which influence the development of SP-containing neurons. For example, postnatal administration of nerve growth factor increases the content of SP in the rat dorsal root ganglia (Kessler and Black, 1980). Rats exposed during development to anti-nerve growth factor antibodies had a significant decrease in SP levels in the dorsal root ganglia and spinal cord and also showed increased sensitivity to pain (Otten *et al.*, 1982).

Recent evidence suggests that the presence of dopamine may regulate the postnatal development of the SP content of the striatum. As described previously, there is a close interrelationship between dopamine- and SP - containing neurons in the adult striatum. Depletion of striatal dopamine by 6-hydroxydopamine administration on PD3 depressed the expression of PPT mRNA and SP content in the striatum at least through PD35 (Sivam *et al.*, 1991). The distribution of SP neurons in the striatum was unaffected, however, by neonatal dopamine depletion (Snyder-Keller, 1991), suggesting that early developmental abnormalities in the dopamine system could have a long-term influence on the SP system as well.

The development of SP neurons in the dorsal horn of the spinal cord appears to be influenced by thyroid hormone and serotonin. Rats made hypothyroid on the first day after birth had twice the content of SP in the dorsal horn at maturity than normal rats. Similarly, inhibition of serotonin synthesis by PCPA treatment during the neonatal period increased SP in the spinal cord (Savard *et al.*, 1983).

Contact with the appropriate target organ may also play a role in regulating SP in developing neurons. In dissociated culture of superior cervical ganglion neurons, the content of SP increased dramatically when the cells were grown on pineal or salivary gland cells (normal targets) compared to when they were grown on non-target cells such as heart or intestine (Kessler *et al.*, 1984). This effect was not mimicked by nerve growth factor.

4.6.2   SP RECEPTOR SITES

The presence of SP itself may be an important factor in regulating the development of SP receptors. SP receptors were up-regulated in adult rats which were administered SP (1 μg/day) during the first week after birth. The increased receptor number was apparent in the salivary gland and various brain regions, including the hypoglossal nucleus, dorsal tegmental nucleus, locus coeruleus, septofimbrial nucleus, dorsal raphe and central gray (Handelmann *et al.*, 1987). SP concentrations in microdissected brain regions were not affected by the neonatal treatment. The increase in receptor expression was sufficient to alter some SP-mediated behaviors in the adult rats. For example, the sensitivity of the rat to painful stimulation of the paws was increased. In addition, the rats salivated in response to lower doses of SP injected intravenously and they produced a greater volume of saliva (Handelmann *et al.*, 1984). There were no changes, however, in cardiovascular responses to SP. These data indicate that neonatal exposure to increased levels of SP increased the expression of SP receptors in several tissues and also increased the sensitivity of the tissues to SP. In contrast, adult rats which had been treated with a specific substance P antiserum on PD2 had an attenuated cardiovascular response to SP (De Felipe *et al.*, 1989). Treated animals also failed to demonstrate an increased nociceptive threshold in response to SP administered intracerebroventricularly, in contrast to control rats. This suggests that neonatal exposure to a SP antiserum, which might be expected to decrease the amount of SP available synaptically, caused decreased sensitivity to SP in the adults. These investigators, however, failed to observe differences in SP binding to membranes derived from the spinal cord and the central gray matter. This may suggest alterations in transduction mechanisms instead of receptor density and/ or affinity.

The amount of SP present during development can therefore have a long-lasting impact on SP receptors. SP can also regulate SP receptors in the adult but the change in receptor number is in the opposite direction. In the adult, the number of postsynaptic SP receptors increases following removal of the SP innervation (Helke *et al.*, 1986). The mechanism of action by which SP influences the development of its receptors is unknown. The presence of SP may up-regulate the density of SP receptors expressed on a given cell. Alternatively, given the possibility that SP has trophic influences on cells during development, an excess of SP could promote the survival of cells containing SP receptors. In either event, it would be expected that factors or events such as those described above could have long-lasting impact on behaviors mediated by SP.

Although there is little other information concerning the developmental regulation of SP receptors, it has been shown that SP receptor gene transcription is blocked by glucocorticoids in adult brain (Ihara and Nakanishi, 1990). If a similar action occurs during development, the ontogeny of SP receptors could be influenced by increased circulating glucocorticoids resulting from maternal or fetal stress.

4.7 SUMMARY

NKs are neuropeptides with a very broad distribution in the central and peripheral nervous systems. Thus, factors which influence the development of NK pathways would also affect the functioning of numerous organ systems.

The early appearance of NKs and their receptors in the fetal nervous system suggests that the NKs themselves might play an important role in the maturation of nervous and other tissues. In fact, a number of the functions of the NKs already characterized would be important in development. For example, the effects of these peptides on cell proliferation and migration would be of obvious importance during ontogeny. The appearance of SP receptors on growth cones (Lockerbie *et al.*, 1988) also suggests a role in synaptic development of the nervous system and it has been proposed that SP is important in maintaining synaptic connections once they are formed (Wall *et al.*, 1982). The regulation of local blood flow by NKs could also be important for the supply of nutrients and blood-borne trophic factors necessary for cell growth and survival. Finally, the regulation of hormones such as growth hormone and insulin by NKs would have a broad influence on developmental processes. Although the significance of the NKs in neural development is still only poorly understood, these neuropeptides are likely to have a far-reaching impact on processes of growth and development throughout the body, as their receptors are differentially expressed by many cell types during various developmental periods.

ACKNOWLEDGEMENTS

This work was supported by grants from the Canadian Parkinson Foundation and the Scottish Rite Foundation for Schizophrenia. TVD is a holder of a studentship from the Fonds de Chercheur et d'aide à la recherche, RQ is a Chercheur-Boursier of the Fonds de la Recherche en Santé du Québec.

REFERENCES

Backman, S.B. and Henry, J.L. (1984) Effects of substance P and thyrotropin-releasing hormone on sympathetic preganglionic neurones in the upper thoracic intermediolateral nucleus of the cat. *Can. J. Physiol. Pharmacol.*, **62**, 248–51.

Bannon, M.J., Freeman, A.S., Chiodo, L.A. *et al.* (1987). The electrophysiological and biochemical pharmacology of the mesolimbic and mesocortical dopamine neurons, in *Handbook of Psychopharmacology* (eds L.L. Iversen, S.D. Iversen and S.H. Snyder), Plenum, New York, Vol. 19, pp 329–74.

Bar-Shavit, Z., Goldman, R., Stabinsky, Y. *et al.* (1980) Enhancement of phagocytosis: a newly found activity for substance P residing in its N-terminal tetrapeptide sequence. *Biochem. Biophys. Res. Commun.*, **94**, 1445–51.

Bertaccini, G., De Caro, G. and Cheli, R. (1966) Enlargement of salivary glands in rats after chronic administration of physalaemin or iso-prenaline. *J. Pharm. Pharmacol.*, **18**, 312–16.

Bill, A., Stjernschantz, J., Mandahl, A. *et al.* (1979) Substance P: release on trigeminal nerve stimula-tion, effects in the eye. *Acta Physiol. Scand.*, **106**, 371–3.

Brain, S.D. and Williams, T.J. (1989) Interactions between the tachykinins and calcitonin gene-related peptide lead to the modulation of oedema formation and blood flow in rat skin. *Br. J. Pharmacol.*, **97**, 77–82.

Brene, S., Linderfors, N., Friedman, W.J. and Persson, H. (1990) Preprotachykinin A mRNA expression in the rat brain during development. *Dev. Brain Res.*, **57**, 151–62.

Bristow, D.R., Curtis, N.R., Suman-Chauhan, N. *et al.* (1987) Effects of tachykinins on inositol phos-pholipid hydrolysis in slices of hamster urinary bladder. *Br. J. Pharmacol.*, **90**, 211–17.

Brown, M. and Vale, W. (1976) Effect of neurotensin and substance P on plasma insulin, glucagon and glucose levels. *Endocrinology*, **98**, 819–22.

Buck, S.H. and Burcher, E. (1985) The rat submaxillary gland contains predominantly P-type tachykinin binding sites. *Peptides*, **6**, 1079–84.

Buck, S.H. and Burcher, E. (1986) The tachykinins: a family of peptides with a brood of receptors. *Trends Pharmacol. Sci.*, **7**, 65–8.

Buck, S.H., Burcher, E., Shults, C.W. *et al.* (1984) Novel pharmacology of substance K-binding sites: a third type of tachykinin receptor. *Science*, **226**, 987–9.

Buck, S.H., Helke, C.J., Burcher, E. *et al.* (1986) Pharmacologic characterization and autoradio-graphic distribution of binding sites for iodinated tachykinins in the rat central nervous system. *Peptides*, **7**, 1109–20.

Bury, R.W. and Mashford, M.L. (1977) A pharmacological investigation of synthetic sub-stance P on the isolated guinea-pig ileum. *Clin. Exp. Pharmacol. Physiol.*, **4**, 453–61.

Byers, M.R. (1984) Dental sensory receptors. *Int. Rev. Neurobiol.*, **25**, 39–94.

Cantalamessa, F., De Caro, G. and Perfumi, M. (1975) Effects of chronic administration of ele-doisin or physalaemin on the rat salivary glands. *Pharmacol. Rev. Commun.*, **7**, 259–71.

Cascieri, M.A. and Liang, T. (1983) Characterization of the substance P receptor and the inhibition of radioligand binding by guanine nucleotides. *J. Biol. Chem.*, **258**, 5158–64.

Cascieri, M.A., Chicchi, G.G., Friedinger, R.M. *et al.* (1986) Conformationally constrained tachykinin analogs which are selective ligands for the eledoisin binding site. *Mol. Pharmacol.*, **29**, 34–8.

Charlton, C.G. and Helke, C.J. (1985) Autoradio-graphic localization and chracterization of spinal cord substance P binding sites: high densities in sensory, autonomic, phrenic and Onuf's motor nuclei. *J. Neurosci.*, **5**, 1653–61.

Charlton, C.G. and Helke, C.J. (1986) Ontogeny of substance P receptors in rat spinal cord: quan-titative changes in receptor number and differ-ential expression in specific loci. *Dev. Brain Res.*, **29**, 81–91.

Charnay, Y., Paulin, C., Chayvialle, J.A. and Dubois, P.M. (1983) Distribution of substance P-like immunoreactivity in the spinal cord and dorsal root ganglia of the human foetus and infant. *Neuroscience*, **10**, 41–55.

Costa, M., Cuello, A.C., Furness, J.B. and Frano, R. (1980) Distribution of enteric neurons showing immunoreactivity for substance P in the guinea pig ileum. *Neuroscience*, **5**, 323–31.

Cuello, A.C. and Kanazawa, I. (1978) The distribu-tion of substance P immunoreactive fibers in the rat central nervous system. *J. Comp. Neurol.*, **178**, 129–56.

Cuello, A.C., Priestley, J.V. and Matthews, M.R. (1982) Localization of substance P in neuronal pathways. *Ciba Found. Symp.*, **9**, 55–83.

Dalsgaard, C.-J., Risling, M. and Cuello, C. (1982) Immunohistochemical localization of substance P in the lumbrosacral spinal pia matter and ventral roots of the cat. *Brain Res.*, **246**, 168–71.

Dalsgaard, C.-J., Haegerstrand, A., Theodorsson-Norheim, E. *et al.* (1985) Neurokinin A-like immunoreactivity in rat primary sensory neur-ons: coexistence with substance P. *Histochemistry*, **83**, 37–39.

Dalsgaard, C.-J., Hultgardh-Nilsson, A., Haeger-strand, A. and Nilsson, J. (1989) Neuropeptides as growth factors. Possible roles in human diseases. *Regul. Pept.*, **25**, 1–9.

Dam, T.V. and Quirion, R. (1986) Pharmacological characterization and autoradiographic localiza-tion of substance P receptors in guinea pig brain. *Peptides*, **7**, 855–64.

Dam, T.V., Escher, E. and Quirion, R. (1986) Tachykinins and phosphatidyl inositol turnover in rat brain, in *Substance P and Neurokinins* (eds J.

Henry, R. Couture, A.C. Cuello *et al.*) Springer-Verlag, New York, pp. 75–7.

Dam, T.V., Escher, E. and Quirion, R. (1988) Evidence for the existence of three classes of neurokinin receptors in brain. Differential ontogeny of neurokinin-1, neurokinin-2 and neurokinin-3 binding sites in rat cerebral cortex. *Brain Res.*, **453**, 372–6.

Dam, T.V., Escher, E. and Quirion, R. (1990a) Visualization of neurokinin-3 receptor sites in rat brain using the highly selective ligand [^{3}H] senktide. *Brain Res.*, **506**, 175–9.

Dam, T.V., Martinelli, B. and Quirion, R. (1990b) Autoradiographic distribution of brain neurokinin-1/substance P receptors using a highly selective ligand [^{3}H]-Sar^9,Met(O$_2$)-substance P. *Brain Res.*, **531**, 333–7.

Dam, T.V., Takeda, Y., Krause, J.E. *et al.* (1990c) Gamma-preprotachykinin (72–92)-peptide amide: an endogenous preprotachykin-I gene-derived peptide that preferentially binds to neurokinin-2 receptors. *Proc. Natl. Acad. Sci. USA*, **87**, 246–50.

Davies, J. and Dray, A. (1976) Substance P in the substantia nigra. *Brain Res.*, **107**, 623–7.

De Felipe, M.C., Molinero, M.T. and Del Rio, J. (1989) Long-lasting neurochemical and functional changes in rats induced by neonatal administration of substance P antiserum. *Brain Res.*, **485**, 301–8.

Del Fiacco, M., Dessi, M.L. and Leranti, M.C. (1984) Topographical localization of substance P in the human postmortem brainstem. An immunohistochemical study in the newborn and adult tissue. *Neuroscience*, **12**, 591–611.

Deutch, A.Y., Maggio, J.E., Bannon, M.J. *et al.* (1985) Substance K and substance P differentially modulate mesolimbic and mesocortical systems. *Peptides*, **6** (Suppl.2), 113–22.

Dick, E., Miller, R.F. and Behbehani, M.M. (1980) Opioids and substance P influence ganglion cells in amphibian retina. *Invest. Opthalmol. Visual Sci.*, **19** (ARVO Suppl.) 132.

Diez-Guerra, F.J., Veira, J.A., Augood, S. and Emson, P.C. (1989) Ontogeny of the novel tachykinins neurokinin A, neurokinin B and neuropeptide K in the rat central nervous system. *Regul. Pept.*, **25**, 87–97.

D'Orléans-Juste, P., Dion, S., Mizrahi, J. and Regoli, D. (1985) Effects of peptides and non-peptides on isolated arterial smooth muscles: role of endothelium. *Eur. J. Pharmacol.*, **114**, 9–21.

D'Orléans-Juste, P., Dion, S., Drapeau, G. and Regoli, D.(1986) Different receptors are involved in the endothelium-mediated relaxation and the smooth muscle contraction of the rabbit pulmonary artery in response to substance P and related neurokinins. *Eur. J. Pharmacol.*, **125**, 37–44.

Duckles, S.P. and Buck, S.H. (1982) Substance P in the cerebral vasculature: depletion by capsaicin suggests a sensory role. *Brain Res.*, **245**, 171–4.

Edvinsson, L. and Uddman, R. (1982) Immunohistochemical localization and dilatory effect of substance P on human cerebral vessels. *Brain Res.*, **232**, 466–71.

Edvinsson, L., McCulloch, J. and Uddman, R. (1981) Substance P: immunohistochemical localization and effect upon cat pial arteries *in vitro* and *in situ*. *J. Physiol. (Lond.)*, **318**, 251–8.

Erjavec, F., Lembeck, F. and Florjanc-Irman, T. (1981) Release of histamine by substance P. *N.-S. Arch. Pharmacol.*, **317**, 67–70.

Erspamer, G.F., Erspamer, V. and Piccinnelli, D. (1980) Parallel bioassay of physalaemin and kassinin, a tachykinin dodecapeptide from the skin of the African frog *Kassina senegalensis*. *N.-S. Arch. Pharmacol.*, **311**, 61–5.

Evans, P.D., Reale, V., Merzan, R.M. and Villegas, J. (1990) Substance P modulation of the membrane potential of the Schwann cell of the squid giant nerve fibre. *Glia*, **3**, 393–404.

Fitzgerald, M. (1987) Prenatal growth of fine diameter primary afferents into the rat spinal cord: a transganglionic tracer sensory study. *J. Comp. Neurol*, **261**, 98–104.

Fitzgerald, M. and Gibson, S. (1984) The postnatal physiological and neurochemical development of peripheral sensory C fibres. *Neuroscience*, **13**, 933–44.

Flood, J.F., Baker, M.L., Hernandez, E.N. and Morley, J.E. (1990) Modulation of memory retention by neuropeptide K. *Brain Res.*, **520**, 284–90.

Freidin, M. and Kessler, J.A. (1990) Cytokine regulation of substance P expression in sympathetic neurons. *Proc. Natl. Acad. Sci. USA*, **88**, 3200–3.

Furness, J.B., Papka, R.F., Della, N.G. *et al.* (1982) Substance P-like immunoreactivity in nerves associated with the vascular system in guinea pigs. *Neuroscience*, **7**, 447–59.

Gamse, R., Holzer, P. and Lembeck, F. (1980) Decrease of substance P in primary afferent neurons and impairment of neurogenic plasma extravasation by capsaicin. *Br. J. Pharmacol.*, **68**, 207–13.

Gilbert, R.F.T. and Emson, P.C. (1979) Substance P in rat CNS and duodenum during development. *Brain Res.*, **171**, 166–70.

Gilbey, M.P., McKenna, K.E. and Schramm, L.P. (1983) Effects of substance P on sympathetic preganglionic neurons. *Neurosci. Lett.*, **41**, 157–9.

Glickman, R.D., Adolph, A.R. and Dowling, J.E. (1980) Does substance P have a physiological role in the carp retina? *Invest. Opthalmol. Visual Sci.*, **19** (ARVO Suppl.), 281.

Guard, S., Watling, K.J. and Watson, S.P. (1988) Neurokinin-3 receptors are linked to inositol phospholipid hydrolysis in the guinea pig ileum longitudinal muscle-myenteric plexus preparation. *Br. J. Pharmcol.*, **94**, 148–54.

Handelmann, G.E., Selsky, J.H. and Helke, C.J. (1984) Substance P administration to neonatal rats increases adult sensitivity to substance P. *Physiol. Behav.*, **33**, 297–300.

Handelmann, G.E., Shults, C. and O'Donohue, T.L. (1987) A developmental influence of substance P on its receptors. *Int. J. Dev. Neurosci.*, **5**, 11–15.

Hanley M.R. and Jackson, T. (1987) Substance K receptor: return of the magnificent seven. *Nature*, **329**, 766–7.

Hartung, H.-P., Wolters, K. and Toyka, K.V. (1986) Substance P: binding properties and studies on cellular responses in guinea pig macrophages. *J. Immunol.*, **136**, 3856–63.

Hassessian, H., Drapeau, G. and Couture, R. (1988) Spinal action of neurokinins producing cardiovascular responses in the conscious freely moving rat: evidence for a NK-1 receptor mechanism. *N.-S. Arch. Pharmacol.*, **338**, 649–54.

Haxhiu, M.A., Haxhiu-Poskurica, B., Moracic, V. *et al.* (1990) Reflex and chemical responses of tracheal submucosal glands in piglets. *Respir. Physiol.*, **82**, 267–77.

Hecht, K., Oehme, P., Poppei, M. and Hect, T. (1979) Conditioned-reflex learning of normal juvenile and adults rats exposed to action of substance P and of an SP analogue. *Pharmazie*, **34**, 419–23.

Helke, C.J., Neil, J.J., Massari, V.J. and Loewy, A.D. (1982) Substance P neurons project from the ventral medulla to the intermediolateral cell column and ventral horn in the rat. *Brain Res.*, **243**, 147–52.

Helke, C.J. , Charlton, C.G. and Wiley, R.G. (1985) Suicide transport of ricin demonstrates the presence of substance P receptors on medullary somatic and autonomic motor neurons. *Brain Res.*, **328**, 190–5.

Helke, C.J., Charlton, C.G. and Wiley, R.G. (1986) Studies on the cellular localization of spinal cord substance P receptors. *Neuroscience*, **19**, 523–33.

Helke, C.J., Krause, J.E., Mantyh, P.W. *et al.* (1990) Diversity in mammalian tachykinin peptidergic neurons: multiple peptides, receptors and regulatory mechanisms. *FASEB J.*, **4**, 1606–15.

Hershey, A.D. and Krause, J.E. (1990) Molecular characterization of a functional cDNA encoding the rat substance P receptor. *Science*, **247**, 958–62.

Hinsey, J.C. and Gasser, H.S. (1930) The component of the dorsal root mediating vasodilation and the Sherrington contraction. *Am. J. Physiol.*, **92**, 679–789.

Holm, I., Thulin, L. and Hellgren, M (1978) Anticholeretic effect of substance P in anesthetized dogs. *Acta Physiol. Scand.*, **102**, 274–80.

Holzer, P. Lembeck, F. (1980) Neurally mediated contraction of ileal longitudinal muscle by substance P. *Neurosci. Lett.*, **17**, 101–5.

Holzer, P., Gamse, R. and Lembeck, F. (1980) Distribution of SP in the gastrointestinal tract: lack of effect of capsaicin pretreatment. *Eur. J. Pharmacol.*, **61**, 303–7.

Holzer, P., Bucsics, A. and Lembeck, F. (1982) Distribution of capsaicin-sensitive nerve fibers containing immunoreactive substance P in cutaneous and visceral tissues of the rat. *Neurosci. Lett.*, **31**, 253–7.

Huston, J.P. and Staubli, U. (1979) Post-trial injection of substance P into lateral hypothalamus and amygdala, respectively, facilitates and impairs learning. *Behav. Neural Biol.*, **27**, 244–8.

Ihara, H. and Nakanishi, S. (1990) Selective inhibition of expression of the substance P receptor mRNA in pancreatic acinar AR42J cells by glucocorticoids. *J. Biol. Chem.*, **265**, 22441–5.

Improta, G. and Broccarde, M. (1990) Tachykinin: effects on gastric secretion and emptying in rats. *Pharmacol. Res.*, **22**, 605–10.

Inagaki, S., Sakanaka, M., Shiosaka, S. *et al.* (1982a) Ontogeny of substance P-containing neuron system of the rat: immunohistochemical analysis. I. Forebrain and upper brainstem. *Neuroscience* , **7**, 251–77

Inagaki, S., Sakanaka, M., Shiosaka, S. *et al.* (1982b) Experimental and immunohistochemical studies of the cerebellar substance P of the rat: localization, postnatal ontogeny and ways of entry to the cerebellum. *Neuroscience*, **7**, 639–45.

Inman, R.D., Chiu, B. and Marshall, K.W. (1986) Substance P and arthritis analysis of plasma and synovial fluid levels. *Arthritis Rheum.*, **29**, Suppl. 1, S9.

Iversen, S.D. (1982) Behavioral effects of substance

P through dopaminergic pathways in the brain. *Ciba Found. Symp.*, **91**, 307–24.

Iversen, L.L., Foster, A.C., Watling, K.J. *et al.* (1987) Multiple receptors and binding sites for tachykinins, in *Substance P and Neurokinins* (eds J.L. Henry, R. Couture, A.C. Cuello *et al.*), Springer-Verlag, New York, pp. 40–3.

Jacques, L. and Couture, R. (1990) Studies on the vascular permeability induced by intrathecal substance P and bradykinin in the rat. *Eur. J. Pharmacol.*, **184**, 9–20.

Johansson, O., Hökfelt, T., Pernow, B. *et al.* (1981) Immunohistochemical support for three putative transmitters in one neuron: coexistence of 5-hydroxytryptamine-, substance P- and thyrotropin releasing hormone-like immunoreactivity in medullary neurons projecting to the spinal cord. *Neuroscience*, **6**, 1857–81.

Jonsson, G. and Hallman, H. (1982a) Substance P counteracts neurotoxin damage on norepinephrine neurons in rat brain during ontogeny. *Science*, **215**, 75–7.

Jonsson, G. and Hallman, H. (1982b) Substance P modifies the 6-hydroxydopamine induced alteration of postnatal development of central noradrenaline neurons. *Neuroscience*, **7**, 2909–18.

Jonsson, G. and Hallman, H. (1983) Effect of substance P on the 5,7-dihydroxytryptamine induced alteration of postnatal development of central serotonin neurons. *Med. Biol.*, **61**, 105–12.

Kafetzopoulos, E., Holzhauer, M.S. and Houston, J.P. (1986) Substance P injected into the region of the nucleus basalis magnocellularis facilitates performance of an inhibitory avoidance task. *Psychopharmacology*, **90**, 281–3.

Kawaguchi, Y., Hoshimaru, M., Nawa, H. and Nakanishi, S. (1986) Sequence analysis of cloned cDNA for rat substance P precursor: existence of a third substance P precursor. *Biochem. Biophys. Res. Commun.*, **139**, 1040–6.

Keeler, J.R., Charlton, C.G. and Helke, C.J. (1985) Cardiovascular effects of spinal cord substance P: studies with a stable receptor agonist. *J. Pharmacol. Exp. Ther.*, **233**, 755–60.

Kelley, A.E., Stinus, L. and Iversen, S.D. (1979) Behavioral activation induced in the rat by substance P infusion into ventral tegmental area: implication of dopaminergic A10 neurones. *Neurosci. Lett.*, **11**, 335–9.

Kessler, J.A. and Black, I.B. (1980) Nerve growth factor stimulates the development of substance P in sensory ganglia. *Proc. Natl. Acad. Sci. USA*, **77**, 649–52.

Kessler, J.A., Adler, J.E., Jonakait, G.M. and Black, I.B. (1984) Target organ regulation of substance P in sympathetic neurons in culture. *Dev. Biol.*, **103**, 71–9.

Konturek, S.J., Jaworek, J., Tasler, J. *et al.* (1981) Effect of substance P and its C-terminal hexapeptide on gastric and pancreatic secretion in the dog. *Am. J. Physiol.*, **21**, G74–G81.

Kotani, H., Hoshimaru, M., Nawa, H. and Nakanishi, S. (1986) Structure and gene organization of bovine neuromedin K precursor. *Proc. Natl. Acad. Sci. USA*, **83**, 7074–8.

Krause, J.E., Chirgwin, J.M., Carter, M.S. *et al.* (1987) Three rat preprotachykinin mRNAs encode the neuropeptides substance P and neurokinin A. *Proc. Natl. Acad. Sci. USA*, **84**, 881–5.

Krause, J.E., Cremins, J.D., Carter, M.S. *et al.* (1988) Solution hybridization-nuclease protection assays for sensitive detection of differentially spliced substance P- and neurokinin A-encoding messenger ribonucleic acids. *Methods Enzymol.*, **168**, 634–52.

Krause, J.E., MacDonald, M.R. and Takeda, Y. (1989) The polyprotein nature of substance P precursors. *BioEssays*, **10**, 62–9.

Kroegel, C., Giembycz, M.A. and Barnes, P.J. (1990) Characterization of eosinophil cell activation by peptides. Differential effects of substance P, melittin and FMET-Leu-Phe. *J. Immunol.*, **145**, 2581–7.

Lamour, Y., Dutar, P. and Jobert, A. (1983) Effects of neuropeptides on rat cortical neurons: laminar distribution and interaction with the effect of acetylcholine. *Neuroscience*, **10**, 107–17.

Laufer, R., Wormser, U., Friedman, Z.Y. *et al.* (1985) Neurokinin B is a preferred agonist for a neuronal substance P receptor and its action is antagonized by enkephalin. *Proc. Natl. Acad. Sci. USA*, **82**, 7444–8.

Lee, C.-M. and Cheng, W.T. (1988) Effects of neonatal monosodium glutamate treatment on substance P binding sites in the rat retina. *Neurosci. Lett.*, **92**, 310–14.

Lee, C.-M., Campbell, N.J., Williams, B.J. and Iversen, L.L. (1986) Multiple tachykinin binding sites in peripheral tissues and in brain. *Eur. J. Pharmacol.*, **130**, 209–17.

Lee, C.-M., Kum, W., Cockram, C.S. *et al.* (1989) Functional substance P receptors on a human astrocytoma cell line (U-373 MG). *Brain Res.*, **488**, 328–31.

Lembeck, F. and Donnerer, J. (1981) Time course of capsaicin-induced functional impairments in comparison with changes in neuronal substance P content. *N.-S. Arch. Pharmacol.*, **316**, 240–3.

Lembeck, F. and Gamse, R. (1982) Substance P in peripheral sensory processes. *Ciba Found. Symp.* **91**, 35–54.

Lembeck, F. and Holzer, P. (1979) Substance P as a neurogenic mediator of antidromic vasodilation and neurogenic plasma extravasation. *N.-S. Arch. Pharmacol.*, **310**, 175–83.

Levine, J.D., Clark, R., Devor, M. *et al.* (1984) Intraneuronal substance P contributes to the severity of experimental arthritis. *Science*, **226**, 547–9.

Lindefors, N.E., Brodin, E., Theodorsson-Norheim, E. and Ungerstedt, U. (1985) Calcium-dependent potassium-stimulated release of neurokinin A and neurokinin B from rat brain regions *in vitro*. *Neuropeptides*, **6**, 453–61.

Ljungdahl, A., Hökfelt, T. and Nilsson, G. (1978) Distribution of substance P like immunoreactivity in the central nervous system of the rat. I. Cell bodies and nerve terminals. *Neuroscience*, **3**, 861–943.

Lockerbie, R.O., Beaujouan, J.-C., Saffroy, M. and Glowinski, J. (1988) An isolated growth cone-enriched fraction from developing rat brain has substance P binding sites. *Dev. Brain Res.*, **40**, 1–9.

Lotz, M., Carson, D.A. and Vaughan, J.H. (1987) Substance P activation of rheumatoid synovio-cytes: neural pathway in pathogenesis of arthritis. *Science*, **235**, 893–5.

Lundberg, J.M., Hökfelt, T., Arggård, A. *et al.* (1980) Peripheral peptide neurons: distribution, axonal transport and some aspects on possible function, in *Neural Peptides and Neuronal Communication* (eds E. Costa and M. Trabucchi), Raven Press, New York, pp. 25–36.

MacDonald, M.R., McCourt, D.W. and Krause, J.E. (1988) Posttranslational processing α, β and γ-preprotachykinins: cell-free translation and early posttranslational processing events. *J. Biol. Chem.*, **263**, 15176–83.

Maggio, J.E. (1988) Tachykinins. *Annu. Rev. Neurosci.*, **11**, 13–28.

Maggio, J.E. and Hunter, J.C. (1984) Regional distribution of kassinin-like immunoreactivity in rat central and peripheral tissues and the effect of capsaicin. *Brain Res.*, **307**, 370–3.

Mansson, B., Nilsson, B.-O. and Ekstrom, J. (1990) Effects of repeated infusions of substance P and vasoactive intestinal peptide on the weights of salivary glands subjected to atrophying influences in rats. *Br. J. Pharmacol.*, **101**, 853–8.

Mantyh, P.W., Hunt, S.P. and Maggio, J.E. (1984a) Substance P receptors; localization by light microscopic autoradiography in rat brain using [^{3}H] SP as the radioligand. *Brain Res.*, **307**, 147–65.

Mantyh, P.W., Pinnock, R.D., Downes, C.P. *et al.* (1984b) Correlation between inositol phospholipid hydrolysis and substance P receptors in rat CNS. *Nature*, **309**, 795–7.

Mantyh, P.W., Gates, T., Mantyh, C.R. and Maggio, M.D. (1989a) Autoradiographic localization and characterization of tachykinin receptor binding sites in the rat brain and peripheral tissues. *J. Neurosci.*, **9**, 258–79.

Mantyh, P.W., Johnson, D.J., Boehmer, C.G. *et al.* (1989b) Substance P receptor binding sites are expressed by glia *in vivo* after neuronal injury. *Proc. Natl. Acad. Sci. USA*, **86**, 5193–7.

Mastrangelo, D., Mathison, R., Huggel, H.J. *et al.* (1987) The rat isolated portal vein: a preparation sensitive to neurokinins, particularly neurokinin B. *Eur. J. Pharmacol.*, **134**, 321–6.

Masu, Y., Nakayama, K., Tamaki, H. *et al.* (1987) cDNA cloning of bovine substance K through oocyte expression system. *Nature*, **329**, 836–8.

Mayberg, M., Langer, R.S., Zervas, N.T. and Moskowitz, M.A. (1981) Perivascular meningeal projections from cat trigeminal ganglia: possible pathway for vascular headaches in man. *Science*, **213**, 228–30.

Miller, A., Costa, M., Furness, J.B. and Chubb, I.W. (1981) Substance P immunoreactive sensory nerves supply the rat iris and cornea. *Neurosci. Lett.*, **23**, 243–9.

Minamino, N., Kangawa, K., Fukuda, A. and Matsuo, H. (1984) Neuromedin L: a novel mammalian tachykinin identified in porcine spinal cord. *Neuropeptides*, **4**, 157–66.

Morin-Surun, M.P., Jordan, D., Champagnat, J. *et al.* (1984) Excitatory effects of iontophoretically applied substance P on neurons in the nucleus tractus solitarius of the cat: lack of interaction with opiates and opioids. *Brain Res.*, **307**, 388–92.

Mousli, M., Bronner, C., Landry, Y. *et al.* (1990a) Direct activation of GTP-binding regulatory protein (G proteins) by substance P and compound 48/80. *FEBS Lett.*, **259**, 260–2.

Mousli, M., Bueb, J.L., Bronner, C. *et al.* (1990b) G-protein activation: a receptor-independent mode of action for cationic amphiphilic neuropeptides and venom peptides. *Trends Pharmacol. Sci.*, **11**, 358–62.

Murray, M., Saffroy, M., Torrens, Y. *et al.* (1988) Tachykinin binding sites in the interpeduncular nucleus of the rat: normal distribution, postnatal development and the effects of lesions. *Brain Res.*, **459**, 76–92.

Nagashima, A., Takano, Y., Tateisha, K. *et al.* (1989) Central pressor actions of neurokinin B: increases in neurokinin B contents in discrete nuclei in spontaneously hypertensive rats. *Brain Res.*, **499**, 198–203.

Nakai, K. and Kasamatsu, T. (1984) Accelerated regeneration of central catecholamine fibers in cat occipital cortex: effects of substance P. *Brain Res.*, **323**, 374–9.

Nakanishi, S. (1987) Substance P precursor and kininogen: their structures, gene organizations and regulation. *Physiol. Rev.*, **67**, 1117–42.

Namura, H., Shiosaka, S., Inagaki, S. *et al.* (1982) Distribution of substance P-like immunoreactivity in the lower brainstem of the human fetus: an immunohistochemical study. *Brain Res.*, **252**, 315–25.

Narumi, S. and Fujita, T. (1978) Stimulatory effects of substance P and nerve growth factor (NGF) on neurite outgrowth in embryonic chick dorsal root ganglia. *Neuropharmacology*, **17**, 73–6.

Narumi, S. and Maki, Y. (1978) Stimulatory effects of substance P on neurite extension and cyclic AMP levels in cultured neuroblastoma cells. *J. Neurochem.*, **30**, 1321–6.

Nawa, H., Hirose, T., Takashima, H. *et al.* (1983) Nucleotide sequences of cloned cDNA for two types of bovine brain substance P precursor. *Nature*, **306**, 32–6.

Nawa, H., Kotani, H. and Nakanishi, S. (1984) Tissue specific generation of two preprotachykinin mRNAs from one gene by alternative RNA splicing. *Nature*, **312**, 729–34.

Ni, L. and Jonakait, G.M. (1988) Development of substance P-containing neurons in the central nervous system in mice: an immunocytochemical study. *J. Comp. Neurol.*, **275**, 493–510.

Nilsson, J., von Euler, A.M. and Dalsgaard, C.J. (1985) Stimulation of connective tissue cell growth by substance P and substance K. *Nature*, **315**, 61–3.

Nishimoto, T., Akai, M., Inagaki, S. *et al.* (1982) On the distribution and origins of substance P in the papillae of the rat tongue: an experimental and immunohistochemical study. *J. Comp. Neurol.*, **207**, 85–92.

Nomura, H., Shiosaka, S., Inagaki, S. *et al.* (1982) Distribution of substance P-like immunoreactivity in the lower brainstem of the human fetus: an immunohistochemical study. *Brain Res.*, **252**, 315–25.

O'Donohue, T.L., Helke, C.J., Shults, C.W. *et al.* (1990) Tachykinin receptors, in *Handbook of Chemical Neuroanatomy* Vol. **9**. *Neuropeptides in the CNS*, Part II (eds T. Hökfelt and M.J. Kuhar), Elsevier, Amsterdam, pp. 395–442.

O'Dowd, B.F., Lefkowitz, R.J. and Caron, M.G. (1989) Structure of the adrenergic and related receptors. *Annu. Rev. Neurosci.*, **12**, 67–83.

Ogawa, T., Kanazawa, I. and Kimura, S. (1985) Regional distribution of substance P, neurokinin A and neurokinin B in rat spinal cord, nerve roots and dorsal root section or spinal transection. *Brain Res.*, **359**, 152–7.

Olgart, L., Gazelius, B., Brodin, E. and Nilsson, G. (1977) Release of substance P-like immunoreactivity from the dental pulp. *Acta Physiol. Scand.*, **101**, 510–12.

Otsuka, M. and Konishi, S. (1976) Release of substance P-like immunoreactivity from isolated spinal cord of newborn rat. *Nature*, **264**, 83–4.

Otsuka, M. and Yanagisawa, M. (1980) The effects of substance P and baclofen on motorneurons of isolated spinal cord of the newborn rat. *J. Exp. Biol.*, **89**, 201–14.

Otten, U., Rüegg, U.T., Hill, R.C. *et al.* (1982) Correlation between substance P content of primary sensory neurones and pain sensitivity in rats exposed to antibodies to nerve growth factor. *Eur. J. Pharmacol.*, **85**, 351–3.

Paulin, C., Charnay, Y., Dubois, P.M. and Chayvialle, J.A. (1986) Localisation de substance P dans le système nerveux du foetus humain: résultats préliminaires. *C.R. Acad. Sci. (Paris)*, **29**, 253–60.

Payan, D.G., Brewster, D.R. and Goetzl, E.J. (1983) Specific stimulation of human T lymphocytes by substance P. *J. Immunol.*, **131**, 1613–15.

Payan, D.G., Brewster, D.R. and Goetzl, E.J. (1984) Stereospecific receptors for substance P on cultured human IM-9 lymphoblasts. *J. Immunol.*, **133**, 3260–5.

Payan, D.G., McGillis, J.P. and Organist, M.L. (1986) Binding characteristics and affinity labelling of protein constituents of the human IM-9 lymphoblast receptor for substance P. *J. Biol. Chem.*, **261**, 14321–9.

Pearson, J., Brandeis, L. and Cuello, A.C (1982) Depletion of substance P-containing axons in the substantia gelatinosa of patients with diminished pain sensitivity. *Nature*, **295**, 61–3.

Pernow, B. (1983) Substance P. *Pharmacol. Rev.*, **35**, 85–141.

Phillis, J.W. and Limacher, J.J. (1974) Substance P

excitation of cerebral cortical Betz cells. *Brain Res.*, **69**, 158–63.

Quirion, R. (1985) Multiple tachykinin receptors. *Trends Neurosci.*, **8.**, 183–5.

Quirion, R. and Dam, T.V. (1985) Multiple tachykinin receptors in guinea pig brain. High densities of substance K (neurokinin A) binding sites in the substantia nigra. *Neuropeptides*, **6**, 191–204.

Quirion, R. and Dam, T.V. (1986) Ontogeny of substance P receptor binding sites in rat brain. *J. Neurosci.*, **6**, 2187–99.

Quirion, R. and Dam, T.V. (1988) Multiple neurokinin receptors: recent developments. *Regul. Pept.*, **22**, 18–25.

Quirion, R., Shults, C., Moody, T. *et al.* (1983) Autoradiographic distribution of substance P receptors in rat central nervous system. *Nature*, **303**, 714–16.

Randic, M. and Miletic, V. (1977) Effects of substance P in cat dorsal horn neurons activated by noxious stimuli. *Brain Res.*, **128**, 164–9

Regoli, D., Drapeau, G., Dion, S. and Couture, R. (1988) New selective agonists for neurokinin receptors: pharmacological tools for receptors characterization. *Trends Pharmacol. Sci.*, **9**, 290–5.

Regoli, D., Dion, S., Rhaleb, N.E. *et al.* (1989) Selective agonists for receptors of substance P and related neurokinins. *Biopolymers*, **28**, 81–90.

Rovero, P., Pestellini, V., Rhaleb, N.E. *et al.* (1989) Structure–activity studies of neurokinin A. *Neuropeptides*, **13**, 263–70.

Ruff, M.R., Wahl, S.M. and Pert, C.B. (1985) Substance P receptor-mediated chemotaxis of human monocytes. *Peptides*, **6**, 107–11.

Saffroy, M., Beaujouan, J.C., Torrens, Y. *et al.* (1988) Localization of tachykinin binding sites (NK1, NK2, NK3 ligands) in the rat brain. *Peptides*, **9(2)**, 227–41.

Sakanaka, M., Inagaki, S., Shiosaka, S. *et al.* (1982) Ontogeny of substance P-containing neuron system of the rat: immunohistochemical analysis. II Lower brainstem. *Neuroscience*, **7**, 1097–126.

Sastry, B.R. (1979) Substance P effects on spinal nociceptive neurons. *Life Sci.*, **24**, 2169–77.

Savard, P., Mërand, Y., Bëdard, P. *et al.* (1983) Comparative effects of neonatal hypothyroidism and euthyroidism on TRH and substance P content of lumbar spinal cord in saline and PCPA-treated rats. *Brain Res.*, **277**, 263–8.

Schini, V.B., Katusic, Z.S. and Vanhoutte, P.M. (1990) Neurohypophyseal peptides and tachy-kinins stimulate the production of cyclic GMP in cultured porcine aortic endothelial cells. *J. Pharmacol. Exp. Ther.*, **255**, 994–1000.

Senba, E., Shiosaka, S., Hara, Y. *et al.* (1982) Ontogeny of the peptidergic system in the rat spinal cord: immunohistochemical analysis. *J. Comp. Neurol.*, **208**, 54–66.

Serra, M.C., Bazzoni, F., Della Bianca, V. *et al.* (1988) Activation of human neutrophils by substance P. Effect on oxidative metabolism, exocytosis, cytosolic Ca^{2+} concentration and inositol phosphate formation. *J. Immunol.*, **141**, 2118–24.

Sharkey, K.A. and Templeton, D. (1984) Substance P in the rat parotid gland: evidence for a dual origin from the otic and trigeminal ganglia. *Brain Res.*, **304**, 392–6.

Shigematsu, K., Niwa, M., Kurihara, M. *et al.* (1987) Alterations in substance P binding in brain nuclei of spontaneously hypertensive rats. *Am. J. Physiol.*, **252**, H301–6.

Shigemoto, R., Yokota, Y., Tsuchida, K. and Nakanishi, S. (1990) Cloning and expression of a rat neuromedin K receptor cDNA. *J. Biol. Chem.*, **265**, 623–8.

Shortland, P., Molander, C., Woolf, C.J. and Fitzgerald, M. (1990) Neonatal capsaicin treatment induces invasion of the substantia gelatinosa by the terminal arborizations of hair follicle afferents in the rat dorsal horn. *J. Comp. Neurol.*, **296**, 23–31.

Shults, C.W., Quirion, R., Chronwall, B. *et al.* (1984) A comparison of the anatomical distribution of substance P and substance P receptors in the rat central nervous system. *Peptides*, **5**, 1097–128.

Sivam, S.P., Krause, J.E., Breese, G.R. and Hong, J.S. (1991) Dopamine-dependent postnatal development of enkephalin and tachykinin neurons of rat basal ganglia. *J. Neurochem.*, **56**, 1499–508.

Snyder, S.H. (1979) Peptide and neurotransmitter receptors in the brain: regulation by ions and guanyl nucleotides, in *Central Regulations of the Endocrine System* (eds K. Fuxe, T. Hökfelt and R. Luft), Plenum, New York, pp. 109–17.

Snyder-Keller, A.M. (1991) Development of striatal compartmentalization following pre- or postnatal dopamine depletion. *J. Neurosci.*, **11**, 810–21.

Soder, O. and Hellstrom, P.M. (1989) The tachykinins neurokinin A and physalaemin stimulate murine thymocyte proliferation. *Int. Arch. Allergy Appl. Immunol.*, **90**, 91–6.

Somogyi, P., Priestley, J.V., Cuello, A.C. et al. (1982) Synaptic connections of substance P immunoreactive nerve terminals in the substantia nigra of the rat: a correlated light and electron microscopic study. *Cell Tissue Res.*, **223**, 469–86.

Stanisz, A.M., Befus, D. and Bienenstock, J. (1986) Differential effects of vasoactive intestinal peptide, substance P and somatostatin on immunoglobulin synthesis and proliferations by lymphocytes from Peyer's patches, mesenteric lymph nodes and spleen. *J. Immunol.*, **136**, 152–6.

Staubli, U. and Huston, J.P. (1980) Facilitation of learning by post-trial injection of substance P into the medial septal nucleus. *Behav. Brain Res.*, **1**, 245–55.

Stjarne, P., Lundblad, L., Anggård, A. *et al.* (1989) Tachykinins and calcitonin gene-related peptide: co-existence in sensory nerves of the nasal mucosa and effects on blood flow. *Cell Tissue Res.*, **256**, 439–46.

Stoessl, A.J. and Hill, D.R. (1990) Autoradiographic visualization of NK-3 tachykinin binding sites in the rat brain, utilizing [^{3}H] senktide. *Brain Res.*, **534**, 1–7.

Stoessl, A.J., Dourish, C.T. and Iversen, S.D. (1988) The NK-3 tachykinin agonist senktide elicits 5-HT mediated behaviour following central or peripheral administration in mice and rats. *Br. J. Pharmacol.*, **94**, 285–7.

Takano, Y., Nagashima, A., Masui, H. *et al.* (1986) Distribution of substance K (neurokinin A) in the brain and peripheral tissues of rat. *Brain Res.*, **369**, 400–4.

Takeda, Y. and Krause, J.E. (1989a) Neuropeptide K potently stimulates salivary gland secretion and potentiates substance P-induced salivation. *Proc. Natl. Acad. Sci. USA*, **86**, 392–6.

Takeda, Y. and Krause, J.E. (1989b) γ-preprotachykin-(72-92)-peptide amide potentiates substance P-induced salivation. *Eur. J. Pharmacol.*, **161**, 267–71.

Takeda, Y., Takeda, J., Smart, B. and Krause, J.E. (1990) Regional distribution of neuropeptide γ, and other tachykinin peptides derived from the substance P gene in the rat. *Regul. Pept.*, **28**, 323–33.

Tanaka, T., Ikai, K. and Imamura, S. (1988) Effects of substance P and substance K on the growth of cultured keratinocytes. *J. Invest. Dermatol.*, **90**, 399–401.

Tatemoto, K., Lundberg, J.M., Jörnvall, H. and Mutt, V. (1985) Neuropeptide K: isolation, structure and biological activities of a novel brain tachykinin. *Biochem. Biophys. Res. Commun.*, **128**, 947–53.

Tervo, K., Tervo, T., Erenko, L. *et al.* (1982) Effect of sensory and sympathetic denervation on substance P immunoreactivity in nerve fibers of the rabbit eye. *Exp. Eye Res.*, **34**, 577–85.

Too, H.-P., Cordova, J.L. and Maggio, J.E. (1989) A novel radioimmunoassay for neuromedin K. I. Absence of neuromedin K-like immunoreactivity in guinea pig ileum and urinary bladder. II. Heterogeneity of tachykinins in guinea pig tissues. *Regul. Pept.*, **26**, 93–105.

Torrens, Y., Beaujouan, J. L., Saffroy, M. *et al.* (1986) Substance P receptors in primary cultures of cortical astrocytes from the mouse. *Proc. Natl. Acad. Sci. USA*, **83**, 9216–20.

Turski, L., Klockgether, T., Schwarz, M. *et al.* (1990) Substantia nigra: a site of action of muscle relaxant drugs. *Ann. Neurol.*, **28**, 341–8.

Unger, T., Rascher, W., Schuster, C. *et al.* (1981) Central blood pressure effects of substance P and angiotensin II: role of sympathetic nervous system and vasopressin. *Eur. J. Pharmacol.*, **71**, 33–42.

Viger, A., Beaujouan, J.C., Torrens, Y. and Glowinski, J. (1983) Specific binding of a ^{125}I-substance P derivative to rat brain synaptosomes. *J. Neurochem.*, **40**, 1030–9.

Walker, P.D., Green, T.L., Jonakait, G.M. and Hart, R.P. (1991) A comparison of substance P peptide and preprotachykinin mRNA levels during development of rat medullary raphe and neostriatum. *Int. J. Dev. Neurosci.*, **9**, 47–55.

Wall, P.D., Fitzgerald, M., Nussbaumer, J.C. *et al.* (1982) Somatotrophic maps are disorganized in adult rodents treated neonatally with capsaicin. *Nature*, **295**, 691–3.

Watson, S.P., Sandberg, B.E.B., Hanley, M.R. and Iversen, L.L. (1983) Tissue selectivity of substance P alkyl esters: suggesting multiple receptors. *Eur. J. Pharmacol*, **87**, 77–84.

Wyke, B. (1981) The neurology of joints: a review of general principles. *Clin. Rheum. Dis.*, **7**, 233–39.

Yamasaki, H., Kubota, Y. and Tohyama, M. (1985) Ontogeny of substance P-containing fibers in the taste buds and the surrounding epithelium. I Light microscopic analysis. *Dev. Brain Res.*, **18**, 301–5.

Yashpal, K., Dam, T.V. and Quirion, R. (1990) Quantitative autoradiographic distribution of multiple neurokinin binding sites in rat spinal cord. *Brain Res.*, **506**, 259–66.

Yashpal, K., Dam, T.V. and Quirion, R. (1991a) Effects of dorsal rhizotomy on neurokinin receptor sub-types in the rat spinal cord: a quantitative autoradiographic study. *Brain Res.*, **552**, 240–7.

Yashpal, K., Kar, S., Quirion, R. and Henry, J.L. (1991b) Noxious cutaneous stimulation produces analgesia in the tail-flick test and decreases binding of substance P in the dorsal horn. *Soc. Neurosci. (Abst.)*, **17**, 1006.

Yew, D.T., Luo, C.B., Zheng, D.R. *et al.* (1990) Development and localization of enkephalin and substance P in the nucleus of tractus solitarius in the medulla oblongata of human fetuses. *Neuroscience*, **34**, 491–8.

Yokota, Y., Sasai, Y., Tanaka, K. *et al.* (1989) Molecular characterization of a functional cDNA for rat substance P receptor. *J. Biol. Chem.*, **264**, 17649–52.

Young, D., Waitches, G., Birchmeier, C. *et al.* (1986) Isolation and characterization of a new cellular oncogene encoding a protein with multiple potential transmembrane domains. *Cell*, **15**, 711–19.

Zderic, S.A., Duckett, J.W., Wein A.J. *et al.* (1990) Development factors in the contractile response of the rabbit bladder to both autonomic and non-autonomic agents. *Pharmacology*, **41**, 119–23.

Ziche, M., Morbidelli, C., Pacini, M. *et al.* (1990) Substance P stimulates neovascularization *in vivo* and proliferation of cultured endothelial cells. *Microvasc. Res.*, **40**, 264–78.

Ziche, M., Morbidelli, L., Geppetti, P. *et al.* (1991) Substance P induces migration of capillary endothelial cells: a novel NK-1 selective receptor mediated activity. *Life Sci.*, **48**, PL7–11.

Paul H. Robinson, John D. Stephenson and Timothy H. Moran

5.1 INTRODUCTION

The term cholecystokinin (CCK) was first
used to describe a hormonal factor identified
in extracts from the intestine which caused
gall bladder contraction (Ivy and Oldberg,
1928). Later, Harper and Raper (1943) demonstrated that intestinal extracts could also
stimulate pancreatic secretion and termed the
responsible factor, pancreozymin. Further
experiments (Jorpes and Mutt, 1973) showed
that pancreozymin and CCK were identical.
CCK was first characterized as a 33-amino
acid peptide in which the C-terminal pentapeptide was identical to that found in gastrin.
Subsequently, several molecular forms of
CCK·have been identified including CCK-39,
CCK-33, CCK-8 and CCK-4 (Dockray, 1981).
These have been shown to have a number of
pharmacological actions in the gut, the
physiological significance of which remains
to be determined.

CCK was originally identified in the central
nervous systems of rats (Vanderhaegen *et al.*,
1975) and non-human primates (Strauss and
Yalow, 1978) as gastrin-like immunoreactivity
and it was subsequently shown that the majority of brain CCK was in the form of the
octapeptide (Dockray *et al.*, 1985). Little is
known about the specific functions of CCK in
the brain although it has been proposed to

function both as a neurotransmitter or a
neuromodulator in peripheral and central
neurons (Larsson and Rehfeld, 1979).

Neuronal CCK can be detected in both cell
bodies and terminals but occurs mainly in
synaptic vesicles (Emson *et al.*, 1980). Its biosynthetic pathway, in both cortical and subcortical rat brain regions, was described by
Goltermann (1982) and Goltermann *et al.*
(1980, 1981). CCK has been shown to be
released from brain slices in response to a
depolarizing stimulus (Emson *et al.*, 1980) and
to be inactivated by deamination (Williams *et
al.*, 1981). Receptors for CCK have been
identified in brain (Innis and Snyder, 1980;
Saito *et al.*, 1980) and their differential distribution mapped in autoradiographic studies
(Zarbin *et al.*, 1983; Moran and McHugh,
1990). Finally, iontophoretically applied CCK
increases neuronal firing in a variety of brain
regions (Dodd and Kelly, 1981; Bunney *et al.*,
1985). Together, these findings support the
proposed function of CCK as a transmitter or
modulator within the CNS.

5.2 CCK RECEPTOR SUBTYPES

There are two types of CCK receptors (Moran
et al., 1986). Type A receptors are present in
the pancreas (Innis and Snyder, 1980), pylo-
rus (Smith *et al.*, 1984) and vagal afferents

Receptors in the Developing Nervous System Vol. 2: Neurotransmitters. Edited by Ian S. Zagon and Patricia J. McLaughlin.
Published in 1993 by Chapman & Hall. ISBN 0 412 49400 0. Vols. 1 and 2 (set) ISBN 0 412 54520 9.

(Moran *et al.*, 1990). In the rat CNS, type A receptors are restricted to the nucleus tractus solitarius, area postrema, interpeduncular nucleus and other discrete brain structures (Moran *et al.*, 1986; Hill *et al.*, 1987). In contrast, the type B CCK receptor is widely distributed in the brain (Moran *et al.*, 1986; Moran and McHugh, 1990). The basis for the distinction between the subtypes is pharmacological and was originally based on differences in the relative affinities of various CCK agonists for the two receptor subtypes. Thus, sulfated forms of CCK bind with higher affinity to the CCK-A receptors than unsulfated CCK, pentagastrin, CCK-4 and gastrin. In contrast, sulfated and non-sulfated forms of CCK and gastrin show smaller differences in their binding affinities to the CCK-B receptor (Innis and Snyder, 1980; Moran *et al.*, 1986). The development of selective CCK receptor antagonists has confirmed this receptor differentiation (Hill *et al.*, 1987; Dourish and Hill, 1989; Chang and Lotti, 1986; Lotti and Chang, 1989). Antagonists at central CCK-B receptors also have a high affinity for peripheral gastrin receptors suggesting that CCK-B and gastrin receptors may be structurally related (Yu *et al.*, 1990).

5.3 AUTORADIOGRAPHIC RECEPTOR DISTRIBUTION

The distribution of CCK receptors in the adult rat brain has been mapped using autoradiographic techniques by a variety of investigators (Zarbin *et al.*, 1983; Moran and McHugh, 1990; Van Dijk *et al.*, 1984). We have already discussed the central distribution of type A CCK receptors. Although they have been identified autoradiographically in only a restricted number of sites, electrophysiological and functional data suggest that they may be more widely distributed although at such a low density that their presence is difficult to detect by standard receptor techniques (Crawley, 1988; Wang *et al.*, 1988). Type B CCK receptors are widely distributed

in the rat brain. Within the olfactory bulb, high densities are evident in the internal granular layer, the inferior olfactory nucleus and caudate putamen. CCK receptors are also found in a variety of limbic structures including the hippocampal formation, the hippocampus, subiculum, dentate gyrus, inferior amygdala and nucleus accumbens (especially in its medial aspects). Binding is found throughout the depth of the cortex, but is concentrated in cortical laminae, II to IV. The highest densities of cortical CCK receptors are found throughout the full rostral-caudal extent of the cingulate cortex.

Only a few areas within the diencephalon contain significant densities of CCK binding. These include the reticular thalamic nuclei, paraventricular thalamic nucleus, ventral medial hypothalamus and the supraoptic nucleus.

Within the midbrain, low levels of binding are found in the dorsal raphe and the several components of the substantia nigra. Type B CCK receptors were only found at a few sites in the brainstem. Low levels of binding are evident in the superior spinal and medial vestibular nucleus, suprageniculate nuclei and cuneate nuclei. Lateral aspects of the nucleus tractus solitarius also contain type B CCK receptors. Low densities of CCK binding have also been demonstrated in the spinal tract of the trigeminal nerve. CCK receptors are present in the cerebellum of the guinea pig and primate, but not in the rat cerebellum.

5.4 DEVELOPMENT OF CENTRAL CCK AND ITS RECEPTORS

Studies of the development of levels of CCK suggest that increase of the peptide is primarily postnatal in altricial subprimate species such as the rat, whereas in precocial species such as the guinea pig, there is significant prenatal development of CCK levels (Goldman *et al.*, 1985).

Although the distribution of CCK receptors and the classification of receptor subtypes

Ontogeny of CCK Binding to Forebrain Membrane Preparations

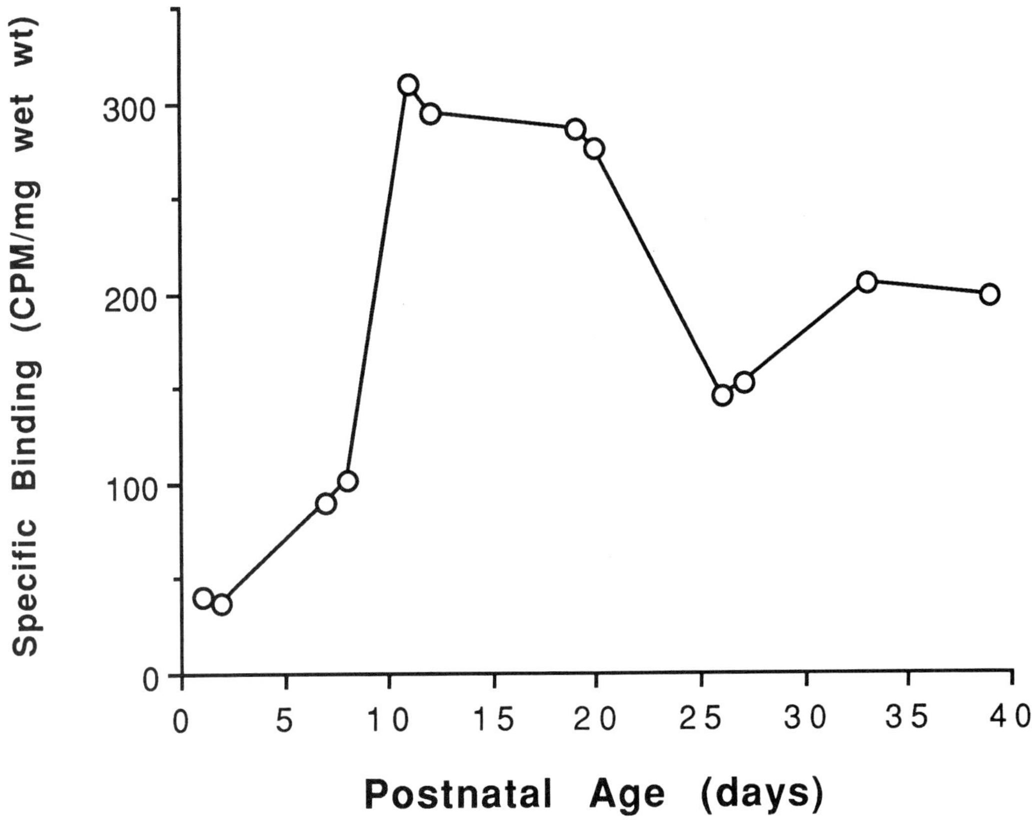

Fig. 5.1.   Age-associated changes in CCK receptor numbers in developing rat brain homogenates. Reproduced from Hays *et al.* (1981).

have been made in the adult, the ontogeny of CCK receptors in the CNS has been examined only rarely. Hays *et al.* (1981), examining binding to rat brain homogenates, demonstrated that the ontogeny of CCK receptors was primarily postnatal. Figure 5.1 (adapted from Hays *et al.* (1981)) shows that specific binding of CCK to membrane preparations of the rat forebrain increased from birth until it reached a plateau between postnatal day (PD) 12 and PD20. After this time, levels fell and then gradually approached adult levels. Autoradiographic studies have confirmed that the primary development of CCK receptors in the rat brain occurs between birth and PD20.

Pelaprat *et al.* (1988) demonstrated that, at birth, CCK receptors were absent from most regions of the rat brain and were confined to selected areas, namely ventromedial thalamic nucleus, endopyriform cortex and medial nucleus of amygdala. Labeling gradually increased during the first three weeks of life to above levels found in the adult and thereafter declined to adult levels over the fourth week.

The distribution of CCK receptors in the forebrain of a 1-day-old rat is shown in Fig. 5.2A. This shows that binding of ^{125}I-labeled Bolton Hunter CCK-8 (Garcha *et al.*, 1987) is limited to the central cingulate and pyriform

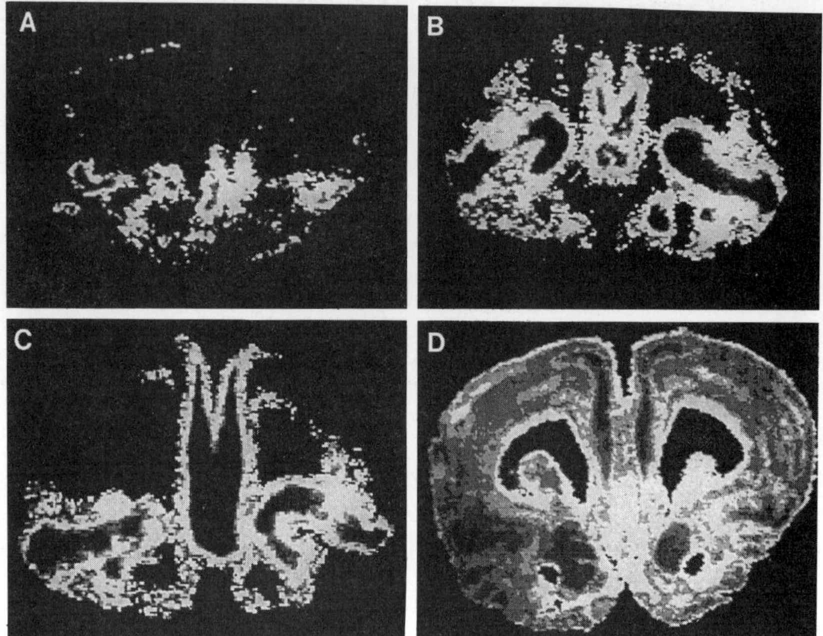

Fig. 5.2.   Computerized microdensitometry images from autoradiographs made using ^{125}I-labeled Bolton Hunter-CCK-8, to demonstrate the ontogeny of specific CCK binding in the rat brain. A: PD1 rat forebrain, showing specific binding in cingulate and pyriform cortex. B: PD6 rat forebrain: cingulate and pyriform cortex binding has increased in extent and density. C: PD10 rat forebrain: extensive binding, still confined to cingulate cortex and pyriform cortex. D: adult rat brain: binding is present throughout the cortex and in subcortical structures.

cortices. In contrast to the results reported by Pelaprat *et al.* (1988), binding was not observed in all pups. This difference may be attributed to small variations in the maturity of one-day-old pups during a period of rapid receptor development. By PD6 binding was consistently observed and had increased in both of these areas (Fig. 5.2B).

As rat pups mature, cortical binding becomes more extensive but is still concentrated in limbic areas. Up to PD10, CCK receptors remained confined to paleocortical areas, in which they increased in density (Fig. 5.2C). Between PD10 and PD15 there was rapid development of receptors in the neocortex which, by PD15, adopted approximately the distribution observed in the adult (Fig. 5.2D).

To date, no studies specifically examining the relative ontogeny of types A and B CCK receptors have been conducted.

5.5   DEVELOPMENT OF CCK-CONTAINING NEURONS IN RAT CNS

CCK neuron systems develop contemporaneously with CCK receptors (Noyer *et al.*, 1980; Cho *et al.*, 1983; Kiyama *et al.*, 1983; Duchemin *et al.*, 1987). CCK neurons first appear in the developing ventral tegmental area and in the primordium of the medial forebrain bundle around gestational age 15. Thereafter, CCK cell numbers increase in the forebrain and the upper brainstem until birth and continue to rise until around PD10. In the ventral tegmental area, however, maximum levels of CCK neurons are found in the late gestational period and then fall in the adult, suggesting that CCK processing occurs largely in terminal regions. Moreover, in the hypothalamic dorsal medial nucleus high levels are found up to PD5 and disappear by PD10. This temporal pattern of development of CCK containing neurons is similar to that of CCK receptors in the rodent brain.

5.6 OTHER CCK RECEPTOR SYSTEMS

The ontogeny of CCK receptors in the brain can be contrasted with their development in the upper gastrointestinal tract. Previous studies have shown that the locus of binding sites of CCK in the circular muscle of the rat pyloric sphincter is evident at 17 days of gestation and persists as a dense area of binding through development into adulthood (Robinson *et al.*, 1988). CCK binding sites are also transiently found in other areas of the developing rat stomach. The gastric mucosa has extensive binding in the fetus that persists until PD 10–PD 15. The antral circular muscle is also densely packed with CCK binding sites in fetal and early postnatal life. In both these cases binding declines in both density and area with increasing age. Thus, in contrast to the brain, CCK receptors in the GI tract may play a role in the functional ontogeny of the GI tract.

5.7 CONCLUSIONS

CCK is one of a number of gastrointestinal peptides which is found in the rat CNS. We have shown that at birth, receptors for CCK are confined to the pyriform and cingulate cortical regions. The major increase in receptor density and distribution occurred during the first two weeks of postnatal development, with the 15-day-old rat showing a pattern of CCK distribution similar to that seen in the adult. This developmental pattern accords with those described by Hays *et al.* (1981) and Pelaprat *et al.* (1988) and parallel changes in CCK concentration (Cho *et al.*, 1983). The last

authors showed that CCK immunoreactivity appeared first in the ventral tegmental area and pyriform cortex one week after birth and rapidly spread to other forebrain areas. In contrast to CCK, other neuropeptides, e.g. somatostatin (Shiosaka *et al.*, 1982), are present in high concentration in the fetal rat nervous system. However, the late appearance of CCK in the rat nervous system contrasts with its early prenatal development in non-primate precocial species, such as the guinea pig and chicken (Goldman *et al.*, 1985) and in non-human primate, the macaque monkey (Hayashi *et al.*, 1989). The lack of CCK binding in the developing rat brain also contrasts with its early appearance in the stomach, which showed extensive binding four days before birth (Robinson *et al.*, 1988). This suggests that in the rat, CCK may be involved in gut, but not brain, development.

Little is known about the CCK fiber system, but the observations that CCK first appears in the ventral tegmental area and pyriform cortex (Cho *et al.*, 1983; Kiyama *et al.*, 1983) and coexists with dopamine in several brain regions, suggest that the CCK tract from the mesencephalon to the limbic cortex (Hokfelt *et al.*, 1980) may be identical to the dopaminergic mesolimbic pathway. This co-localization provides the rationale for investigating CCK function in psychiatric states such as schizophrenia in which abnormal dopamine function has been postulated.

CCK has been proposed to be an endogenous regulator of satiety, although its site of action remains the subject of debate. The demonstration that CCK induces satiety in rats at PD1, when CCK receptors are abundant in the stomach but not in the CNS (Robinson *et al.*, 1987, 1988), suggests that either gastric mechanisms are most important for regulating CCK-induced satiety in the neonate or that the effect is mediated through highly localized brain regions in which CCK receptors are present at birth but are not easily visualized.

REFERENCES

Bunney, B.S., Chiodo, L.A. and Freeman, A.S. (1985) Further studies on the specificity of proglumide as a selective CCK antagonist in the central nervous system. *Ann. NY Acad. Sci.*, **448**, 345–51.

Chang, R.S.L. and Lotti, V.J. (1986) Biochemical and pharmacological characterization of an extremely potent and selective non-peptide cholecystokinin antagonist. *Proc. Natl. Acad. Sci. USA*, **83**, 4923–6.

Cho, H.J., Shiotani, Y., Shiosaka, S. *et al.* (1983) Ontogeny of cholecystokinin-8 containing neuron system of the rat: an immunohistochemical analysis. I. Forebrain and upper brainstem. *J. Comp. Neurol.*, **218**, 25–41.

Crawley, J.N. (1988) Behavioral analysis of antagonists of the peripheral and central effects of CCK, in *Cholecystokinin Antagonists* (eds R. Y. Wang and R. Schoenfeld), Alan R. Liss, New York, pp. 243–62.

Dockray, G.J. (1981) Cholecystokinin, in *Gut Hormones*, 2nd edn (eds S. R. Bloom and J. M. Polak), Churchill Livingstone, Edinburgh, pp. 228–39.

Dockray, G.J., Desmond, H., Gayston, R.J. *et al.* (1985) Cholecystokinin and gastrin forms in the central nervous system. *Ann. NY Acad. Sci.*, **448**, 32–43.

Dodd, J. and Kelly, J.S. (1981) The actions of cholecystokinin and related peptides in pyramidal neurons of the mammalian hippocampus. *Brain Res.*, **205**, 337–50.

Dourish, C.T. and Hill, D.R. (1987) Classification and function of CCK receptors. *Trends Pharmacol. Sci.*, **8**, 207–8.

Duchemin, A.M., Quach, T.T., Iadarola, M.J. *et al.* (1987) Expression of the cholecystokinin gene in rat brain during development. *Dev. Neurosci.*, **9**, 61.

Emson, P.C., Lee, C.M. and Rehfeld, J.F. (1980) Cholecystokinin octapeptide: vesticular localization and calcium dependent release from rat brain *in vitro*. *Life Sci.*, **26**, 2157–63.

Garcha, G.S., McHugh, P.R., Moran, T.H. *et al.* (1987) Development of cholecystokinin receptors and suppression of ingestion and gastric emptying in the rat. *J. Physiol. (Lond.)*, **387**, 93.

Goldman, S.A., Monahan, J.W. and Schneider, B. S. (1985). The regional and subcellular development of cholecystokinin immunoreactivity in vertebrate brain. *Brain Res.*, **354**, 237–46.

Goltermann, N.R. (1982) *In vivo* synthesis of cholecystokinin in rat cerebral cortex: identification of COOH-terminal peptides with labelled amino acids. *Peptides*, **3**, 733–7.

Goltermann, N.R., Rehfeld, J.F. and Roigaard-Petersen, H. (1980) In vivo biosynthesis of cholecystokinin in rat cerebral cortex. *J. Biol. Chem.*, **255**, 6181–5.

Goltermann, N.R., Stengaart-Petersen, V., Rehfeld, J.F. *et al.* (1981) Newly synthesized cholecystokinin in subcellular fractions of the rat brain. *J. Neurochem.*, **36**, 959–65.

Harper, A.A. and Raper, H.S. (1943) Pancreozymin a stimulant of the secretion of pancreatic enzymes in extracts of the small intestine. *J. Physiol. (Lond.)*, **102**, 115–25.

Hayashi, M., Yamashita, A., Shimizu, K. and Oshima, K. (1989) Ontogeny of cholecystokinin-8 and glutamic acid decarboxylase in cerebral neocortex of the macaque monkey. *Exp. Brain Res.*, **74**, 249–55.

Hays, S.E., Goodwin, F.K. and Paul, S.M. (1981) Cholecystokinin receptors in brain: effect of obesity, drug treatment and lesions. *Peptides*, **2**, 21–6.

Hill, D.R., Campbell, N.J., Shaw, T.M. *et al.* (1987) Autoradiographic localization and biochemical characterization of peripheral type CCK receptors in rat CNS using highly selective nonpeptide CCK antagonists. *J. Neurosci.*, **7**, 2967–76.

Hökfelt, T.L., Skirboll, L., Rehfeld, J.F. *et al.* (1980) A subpopulation of mesencephalic dopamine neurons projecting to limbic areas contains a cholecystokinin-like peptide: evidence from immunohistochemistry combined with retrograde tracing. *Neuroscience*, **5**, 2093–124.

Innis, R.B. and Snyder, S.-H. (1980) Distinct cholecystokinin receptors in brain and pancreas. *Proc. Natl. Acad. Sci. USA*, **77**, 6919–21.

Ivy, A.C. and Olberg, E. (1982) A hormone mechanism for gall-bladder contraction and evacuation. *Am. J. Physiol.*, **86**, 599–613.

Jorpes, J.E. and Mutt, V. (1973) Secretin and cholecystokinin (CCK), in *Secretin, Cholecystokinin, Pancreozymin and Gastrin. Handbook of Experimented Pharmacology*, Vol. **34** (eds J. E. Jorpes and V. Mutt), Springer-Verlag, Berlin, pp. 1–179.

Kiyama, H., Shiosaka, S., Kubota, Y. *et al.* (1983) Ontogeny of cholecystokinin-8 containing neuron system of the rat: an immunohistochemical analysis. II. Lower brainstem. *Neuroscience*, **10**, 1341–59.

Larsson, L.I. and Rehfeld, J.F. (1979) Localization

and molecular heterogeneity of cholecystokinin in the central and peripheral nervous system. *Brain Res.*, **165**, 201–18.

Lotti, V.J. and Chang, R.S.L. (1989) A new potent and selective non-peptide gastrin antagonist and brain cholecystokinin receptor (CCK-B) ligand: L-365,260. *Eur. J. Pharmacol.*, **162**, 273–80.

Moran, T.H. and McHugh, P.R. (1990) Cholecystokinin receptors, in *Handbook of Chemical Neuroanatomy*: Vol. **9**, *Neuropeptides in the CNS*, Part II, (eds A. Bjorklund, T. Hokfeld and M.J. Kuhar), Elsevier, Amsterdam, pp. 455–76.

Moran, T.H., Robinson, P.H., Goldrich, M.S. *et al.* (1986) Two brain CCK receptors: implications for behavioral actions. *Brain Res.*, **362**, 175–9.

Moran, T.H., Norgren, R., Crosby, R.J. and McHugh, P.R. (1990) Central and peripheral vagal transport of cholecystokinin binding sites occurs in afferent fibers. *Brain Res.*, **526**, 95–102.

Noyer, M., Diem Bui, N., Deschodt-Lanckman, M. *et al.* (1980) Postnatal development of the cholecystokinin–gastrin family of peptides in the brain and gut of rat. *Life Sci.*, **27**, 2197–203.

Pelaprat, D., Dusart, I. and Peschanski, M. (1988) Postnatal development of cholecystokinin binding sites in the rat forebrain and midbrain: an autoradiographic study. *Dev. Brain Res.*, **44**, 119–32.

Robinson, P.H., Moran, T.H., Goldrich, M. and McHugh, P.R. (1987) Development of cholecystokinin receptors in rat upper gastrointestinal tract. *Am. J. Physiol.*, **252**, G529–G534.

Robinson, P.H., Moran, T.H. and McHugh, P.R. (1988) Cholecystokinin inhibits independent ingestion in neonatal rats. *Am. J. Physiol.*, **255**, R14–R20.

Saito, A., Sankaran, H., Goldfine, I.D. *et al.* (1980) Cholecystokinin receptors in brain: characterization and distribution. *Science*, **208**, 1155–6.

Shiosaka, S., Takatsuki, K., Sakanaka, M. *et al.* (1982) Ontogeny of somatostatin containing neuron system in the rat. Immunohistochemical observations. II. Forebrain and diencephalon. *J. Comp. Neurol.*, **204**, 211–24.

Smith, G.T., Moran, T.H., Coyle, J.T. *et al.* (1984) Anatomical localization of cholecystokinin receptors to the pyloric sphincter. *Am. J. Physiol.*, **246**, R127–130.

Strauss, E. and Yalow, R.S. (1978) Species specificity of cholecystokinin in gut and brain of several mammalian species. *Proc. Natl. Acad. Sci. USA*, **75**, 486–9.

Vanderhaeghen, J.J., Signeau, J.C. and Gepts, W. (1975) New peptide in vertebrate CNS reacting with antigastrin antibodies. *Nature*, **251**, 604–5.

Van Dijk, A., Richards, J.G., Trzeciak, A. *et al.* (1984) Cholecystokinin receptors: biochemical demonstration and autoradiographical localization in rat brain and pancreas using [^{3}H]cholecystokinin-8 as a radioligand. *J. Neurosci.*, **4**, 1021–33.

Wang, R.Y., Kasser, R.J. and Hu, X.-T. (1988) Cholecystokinin receptor subtypes in the rat nucleus accumbens, in *Cholecystokinin Antagonists* (eds R. Y. Wang and R. Schoenfeld), Alan R. Liss, New York, pp. 199–216.

Williams, R.G., Goyku, R.J., Zoo, N.Y. *et al.* (1981) Changes in brain cholecystokinin octapeptide following lesions of the media forebrain bundle. *Brain Res.*, **212**, 221–30.

Yu, D.H., Huang, S.C., Wank, S.A. *et al.* (1990) Pancreatic receptors for cholecystokinin: evidence for three receptor classes. *Am. J. Physiol.*, **258**, G86–G95.

Zarbin, M.A., Innis, R.B., Wamsley, J.K. *et al.* (1983) Autoradiographic localization of cholecystokinin receptors in rodent brain. *J. Neurosci.*, **3**, 877–906.

GABA$_A$/BENZODIAZEPINE RECEPTORS IN THE DEVELOPING MAMMALIAN BRAIN

6

Angel Luis de Blas

6.1 THE GABA$_A$ BENZODIAZEPINE RECEPTORS

GABA (γ-aminobutyric acid) is the main inhibitory transmitter in the brain. Up to 30–40% of the brain synapses use GABA as a neurotransmitter (Bloom and Iversen, 1971). When GABA binds to the GABA$_A$ receptor (GABAR) it opens the associated Cl$^-$ channel allowing the flow of this anion inside the neuron, therefore hyperpolarizing the neuronal membrane and making the cell less reactive to excitatory neurotransmitters. Benzodiazepines (BZDs) are the most widely prescribed drugs. They are potent anxiolytic, antiepileptic and muscle relaxing agents. The psychotropic effects produced by the BZDs (i.e. librium and valium) result from their binding to brain benzodiazepine receptors (BZDRs) that are present in the membranes of many brain neurons (Bräestrup and Squires, 1977; Möhler and Okada, 1977). The BZDR is part of a protein complex that also includes the GABAR and a chloride channel. Barbiturates and some steroids also bind to these receptor-channel protein complexes affecting the efficacy of GABA in opening the Cl$^-$ channel. The BZDs potentiate the inhibitory effect of GABA by increasing the frequency of channel opening events induced by GABA. In this way, the total Cl$^-$ current flowing through the channel is increased. For reviews of the pharmacological, biochemical and molecular biology of the GABAR/BZDR see Olsen and Tobin (1990) and Schwartz (1988).

There are other drugs that specifically bind to the various elements of the GABAR complex. Thus, muscimol (MUS) is an agonist and bicuculline is an antagonist of GABA. Picrotoxinin and TBPS (butylbicyclophosphorothionate) bind to the Cl$^-$channel and block it. There are also specific drugs for the BZDR which are agonists (e.g. diazepam), antagonists (e.g. Ro15-1788) and inverse-agonist (e.g. β-carboline carboxylate ethyl ester, β-CCE). The inverse agonists decrease the effectiveness of GABA in opening the Cl$^-$ channel. Therefore they are proconvulsant and anxiogenic substances with effects opposite to diazepam and other agonist benzodiazepines. The antagonists block the effects of both agonists and inverse agonists but they do not affect in one way or the other the GABA-induced opening of the Cl$^-$ channel.

In addition to the GABA$_A$ receptors, the brain also has GABA$_B$ receptors with different pharmacological and molecular properties. Thus (–)baclofen is a specific agonist for the GABA$_B$ receptors whereas bicuculline is not a ligand for this receptor. In addition, the GABA$_B$ receptor is not a Cl$^-$ channel. It belongs to the family of receptors coupled to G proteins. The

Receptors in the Developing Nervous System Vol. 2: Neurotransmitters. Edited by Ian S. Zagon and Patricia J. McLaughlin. Published in 1993 by Chapman & Hall. ISBN 0 412 49400 0. Vols. 1 and 2 (set) ISBN 0 412 54520 9.

activation of this receptor regulates the activity of K$^+$ and Ca^{2+} channels. There is little information available on the developmental aspects of the GABA$_B$ receptors. This chapter will discuss exclusively the development of the GABA$_A$/benzodiazepine receptor complex.

The GABAR/BZDR complex has been solubilized with various detergents and purified in several laboratories by using ligand

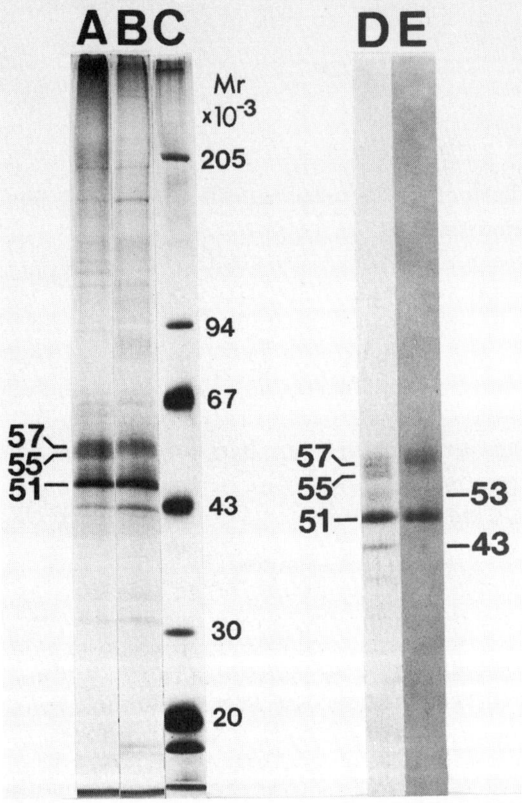

Fig. 6.1.   SDS-PAGE of the GABAR/BZDR complex. The bovine brain receptor was purified by immunoaffinity chromatography (IAC) on mAb 62-3G1 (lanes A and D) or by Ro 7-1986/1 affinity chromatography (AC) (lanes B and E). Each lane contained 1 µg of the purified receptor. SDS-PAGE was in a 5–20% polyacrylamide gradient (lanes A–C) or 10% polyacrylamide (lanes D and E). The gels were stained by a silver method. Each lane shows a different receptor preparation. Lane C has M_r markers. Reprinted from Park *et al.* (1991), with permission.

affinity chromatography (AC) on immobiliz-ed BZDs (Sigel and Barnard, 1984; Taguchi and Kuriyama, 1984; Vitorica *et al.*, 1988). In addition, there has been recent development of a novel immunoaffinity chromatography (IAC) purification method of the receptor (Park *et al.*, 1991; Park and De Blas, 1991a,b) by using one of the monoclonal antibodies (62-3G1) to the GABAR/BZDR complex that was developed in our laboratory (Vitorica *et al.*, 1988). The receptor preparations purified by either method (AC or IAC) showed several peptides in the range 50 000–60 000 daltons (Fig. 6.1; Sigel and Barnard, 1984; Vitorica *et al.*, 1988; Park *et al.*, 1991; Park and De Blas, 1991a,b). In addition to the M_r, the peptides can be identified by: (1) photoaffinity labeling (PAL) with [^{3}H]flunitrazepam (FNZ) or [^{3}H]muscimol (Fig. 6.2); (2) reactivity with subunit specific monoclonal and polyclonal antibodies; (3) peptide mapping and (4) extent of glycosylation (Park *et al.*, 1991; Park and De Blas, 1991a,b). Recent cloning studies and receptor subunit expression after mRNA injection in *Xenopus laevis* oocytes or transfec-tion into mammalian cells indicate that the neuronal receptor can be functionally recon-stituted with a combination of α, β and γ subunits (Pritchett *et al.*, 1989a; Ymer *et al.*, 1989; Verdoorn *et al.*, 1990). To date, six variants of the α subunit (α$_1$–α$_6$), three of the β (β$_1$–β$_3$), two of the γ (γ$_1$–γ$_2$), one δ and one ρ have been cloned from mammalian brain (Schofield *et al.*, 1987; Olsen and Tobin, 1990; Cutting *et al.*, 1991). Combinations of four or five of these subunits form a receptor–chan-nel complex. These membrane-spanning sub-units are arranged such that they form a trans-membrane Cl$^-$ channel in a fashion similar to the nicotinic acetylcholine and glycine recep-tors. All the subunits are similarly organized having a long extracellular amino-terminal peptide and four hydrophobic transmem-brane α-helix domains (Schofield *et al.*, 1987). All the subunits of the complex participate in the formation of the transmembrane Cl$^-$ chan-nel. The reconstitution of the receptor func-

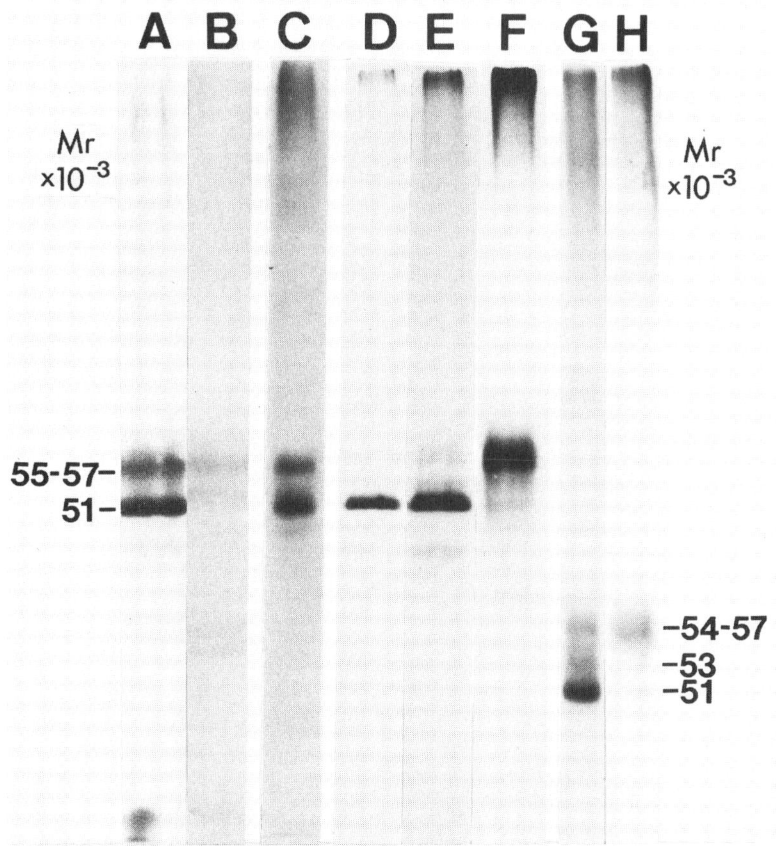

Fig. 6.2.   PAL of the purified GABAR/BZDR complex from bovine cerebral cortex (fluorographs). Lanes A, C, D, E, and G show [³H]FNZ labeling, whereas lanes B, F, and H show [³H]MUS labeling. Lanes A and B contain membranes and lanes C–H purified receptors. In lane C, the membrane receptor was PAL, solubilized, and purified by IAC on mAb 62-3G1. In lanes D–F, the receptor complex was purified by Ro 7-1986/1 AC, followed by PAL. In lanes G and H, the receptor was immunoadsorbed on the mAb 62-3G1 gel beads, washed, and then PAL. The x-ray films were exposed for 26 days (lanes A and B), 20 days (lane C), 10 days (lanes D, F, and G), or 55 days (lane H). Lane E is a 60-day exposed fluorograph identical to lane D. SDS-PAGE was in 5–20% polyacrylamide gradient (lanes A and B) or 10% polyacrylamide (lanes C–H). Lanes G and H were from SDS-PAGE in which samples were run for a longer time in order to resolve the PAL peptides better. Reprinted from Park *et al.* (1991), with permission.

tion by expression studies together with the biochemical studies on the purified receptor suggest that the α_1 subunits bind BZDs whereas the β subunits bind muscimol and GABA. The γ_1 and γ_2 subunits seem to be necessary for the functional coupling of the GABAR and BZDR as well as for the binding of BZDs to the α subunits (Pritchett *et al.*, 1989a; Verdoorn *et al.*, 1990; Ymer *et al.*, 1990).

The cloning studies have also shown a high degree of heterogeneity in the subunit composition of the GABAR/BZDR complex. Earlier studies had also suggested receptor heterogeneity. Thus photoaffinity labeling (PAL) with [³H]FNZ had revealed differences in the peptide composition of the BZDRs in brain, cerebellum and hippocampus (Fig. 6.3; Sieghart and Karobath, 1980; Sieghart and

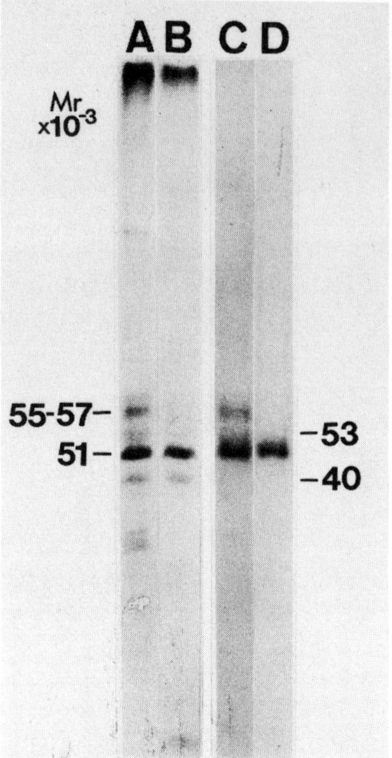

Fig. 6.3. IAC-purification of the [³H]FNZ PAL membrane receptors (fluorographs). Lanes A and B show the PAL membranes and lanes C and D show the mAb 62-3G1 IAC purified receptors after being PAL in membranes. Lanes A and C represent receptors from bovine cerebral cortex, and lanes B and D from bovine cerebellum. Reprinted from Park and de Blas (1991b), with permission.

Drexler, 1983; Park and De Blas, 1991b). Ligand binding studies had previously shown that CL218-872, β-CCE and zolpidem could distinguish two BZDR types (named type I and type II). The cerebellar BZDR is mostly type I whereas in the cerebral cortex both types I and II BZDRs co-exist (Klepner *et al.*, 1979; Sieghart *et al.*, 1983). The cloning and expression of the receptors in mammalian cells have shown that the alpha subunits dictate whether the BZDR is type I or II. The type I BZDR is associated with the α_1 subunit whereas α_2, α_3 and α_5 are associated with type II. In addition, α_5 has lower affinities for zolpidem than α_2 and α_3 indicating heterogeneity of

the type II BZDR (Pritchett *et al.* 1989b; Pritchett and Seeburg, 1990). The recombinant receptors composed of α_6 (α_6, β_2 and γ_2) bind both [³H]MUS and the benzodiazepine [³H]R015– 4513. However, it neither binds other benzodiazepines nor β-carbolines (Lüddens *et al.*, 1990). In addition, Ymer *et al.* (1990) have shown that the γ subunits also affect the pharmacology of the receptors. The recombinant receptors with γ_1 subunit (α_1, β_1 and γ_1) have much lower affinity for the antagonists and inverse agonists than the recombinants with the γ_2 subunit (α_1, β_1 and γ_2).

The significance of this subunit heterogeneity is not yet known but functional implications have been suggested from the aforementioned pharmacological specificities as well as from localization studies using *in situ* hybridization techniques (Shivers *et al.*, 1989) and from functional studies of the receptors expressed either in *Xenopus laevis* oocytes or transfected mammalian cells (Levitan *et al.*, 1988; Verdoorn *et al.*, 1990; Sigel *et al.*, 1990). It seems that the subunit heterogeneity allows the existence of a large combinatorial variety of GABAR and BZDR types showing differential sensitivities to GABA, benzodiazepines and other ligands.

The localization of the GABAR/BZDR in the brain was first revealed by radioligand autoradiography such as [³H]FNZ, [³H]MUS and [³⁵S]TBPS which are specific for the BZDR, GABAR and Cl⁻ channel respectively. In this way complete brain maps for each ligand have been generated (Palacios *et al.*, 1981; McCabe and Wamsley, 1986). These studies have shown that there is no total co-localization of these three binding sites (McCabe and Wamsley, 1986). These results have been interpreted by others as an indication that not all the GABAR and BZDR are physically coupled to one another (Unnerstall *et al.*, 1981). This interpretation is supported by the recent cloning and functional reconstitution studies which indicate the necessity of the γ_2 subunits (in addition to α and β subunits) for: (1) the binding of BZDs to the

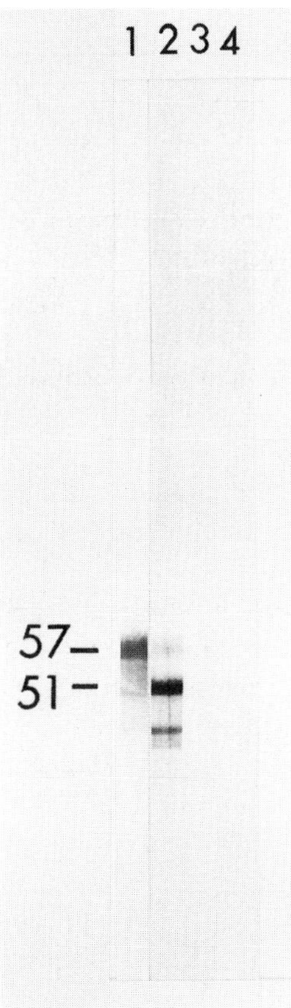

Fig. 6.4. Immunoblots with mAbs of AC-purified GABAR/BZDR complex. Lanes 1–4 are the mAb 62-3G1, a mouse antiserum, the mAb 62-5F6, and the mAb 62-2G4 respectively. Reprinted from Vitorica *et al.* (1988), with permission.

α_1 subunit of the receptor complex and (2) for the BZD-dependent stimulation of the GABA-induced opening of the Cl⁻ channel. In the absence of the γ subunits, the α and β subunits form GABA-gated Cl⁻ channels but are insensitive to BZDs (Pritchett *et al.*, 1989a; Verdoorn *et al.*, 1990; Ymer *et al.*, 1990). Therefore, the ligand-binding mismatch as well as the heterogeneity of the receptors regarding the ligand-binding properties indicate that the radioligand binding maps cannot discriminate all the receptor types defined at the mRNA level.

Monoclonal (mAb) and polyclonal antibodies raised in our laboratory (Fig. 6.4; Vitorica *et al.*, 1987, 1988) have been used to study the receptor localization at the cellular and subcellular level by both light and electron microscopy immunocytochemistry (De Blas *et al.*, 1988; Juiz *et al.*, 1989; Yazulla *et al.*, 1989). Similar localization has been found by using mAbs 62-3G1 (De Blas *et al.*, 1988) and bd-17 (Richards *et al.*, 1987) which specifically recognize both the β_2 and β_3 subunits of the GABAR (Ewert *et al.*, 1990, 1992) or the mAb bd-24 (Richards *et al.*, 1987) which is specific for α_1 subunit (Ewert *et al.*, 1990). These studies complement and expand the earlier radioligand autoradiography mapping and have allowed the subcellular localization of the GABAR/BZDR complex in the brain to be revealed (Figs. 6.5–6.7). Given the subunit heterogeneity of the receptors and the selectivity of the available antibodies for few (although the most abundant) subunits, such as α_1, β_2, and β_3, the immunocytochemical maps of the various receptor types are still incomplete. New antibodies specific for all the other subunits need to be developed for completing the mapping.

The mapping of the various receptor subunit mRNAs has also been done with *in situ* hybridization techniques by using specific cDNA or cRNA probes (Siegel, 1988; Wisden *et al.*, 1988, 1989; Khrestchatisky *et al.*, 1989; Pritchett *et al.*, 1989a; Shivers *et al.*, 1989; Lüddens *et al.*, 1990; Malherbe *et al.*, 1990; Olsen and Tobin, 1990). These studies complement the radioligand autoradiography and immunocytochemistry mapping studies. They have revealed which subunits co-exist in the same cells, which presumably indicates which subunits are made and assembled into mature receptors by a particular cell. The *in situ* hybridization studies have some limitations, such as the almost exclusive labeling of the cell bodies, not showing where in the neuron the receptors are localized. It is conceivable that the same cell might assemble different combinations of receptor subunits in different synapses or cell areas (i.e. dendrites vs axons). It is also possible that an expressed

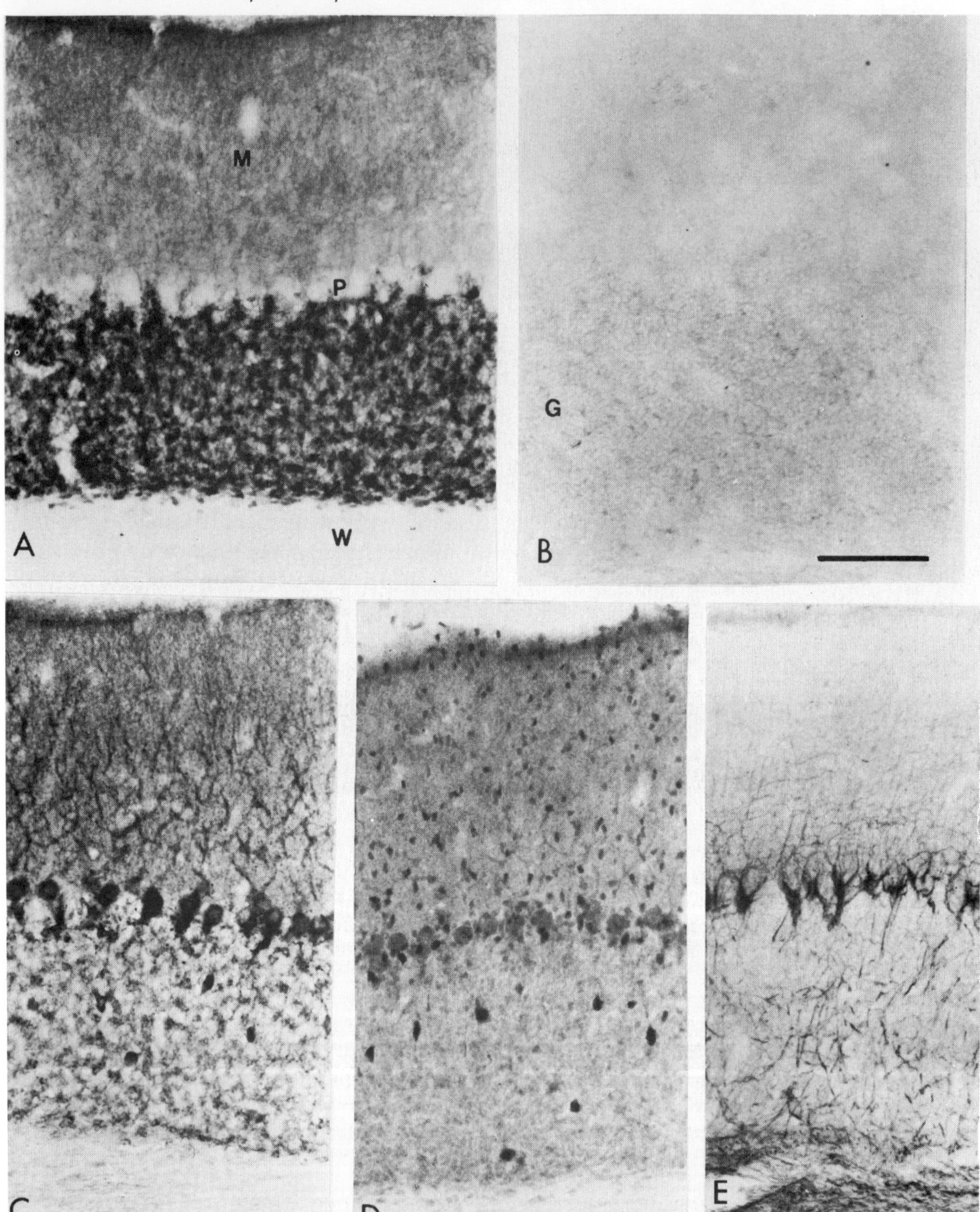

Fig. 6.5.   Rat cerebellum immunocytochemistry with mAb 62-3G1 (A, B), anti-GAD (C), anti-GABA (D), and anti-NFP (E). B, The mAb 62-3G1 (1 ml) was incubated with 9 µg of the AC-purified GABAR/BZDR complex previous to the incubation with the brain tissue. M, molecular layer; P, Purkinje cell layer; G, granule layer; W, white matter. Notice the large GABA-containing Golgi II cells in the granule layer (C, D). Bar, 100 µm. Reprinted from de Blas *et al*. (1988), with permission.

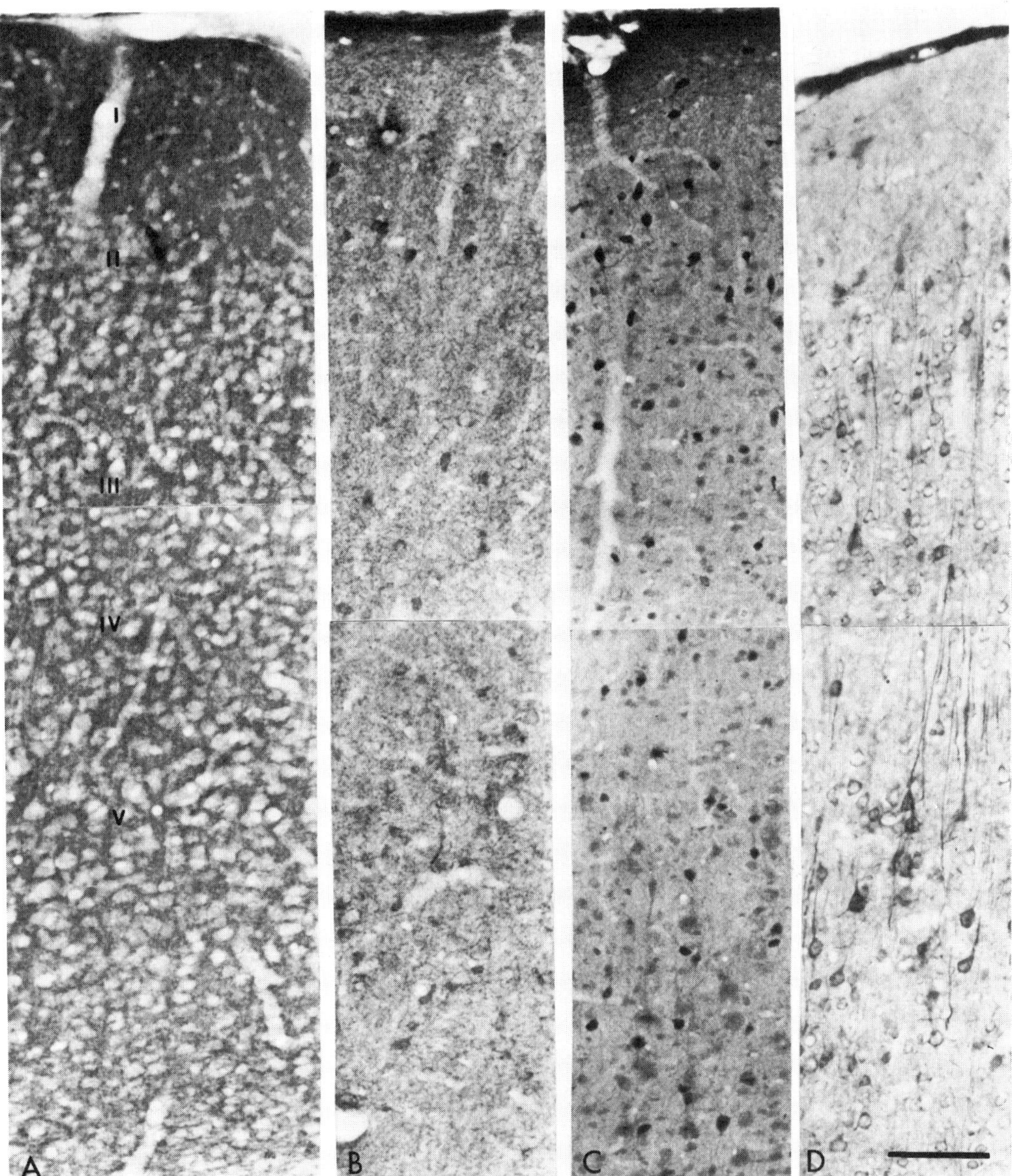

Fig. 6.6. Rat cerebral cortex immunocytochemistry with 62-3G1 (A), anti-GAD (B), anti-GABA (C), and the mAb 8-6A2 (D). Notice that the GABAR, GAD, and GABA immunoreactivities are distributed through all layers. The mAb 8-6A2 was used as a marker for neurons and for the identification of the pyramidal cells (De Blas *et al.*, 1984). Bar, 100 μm. Reprinted from De Blas *et al.* (1988), with permission.

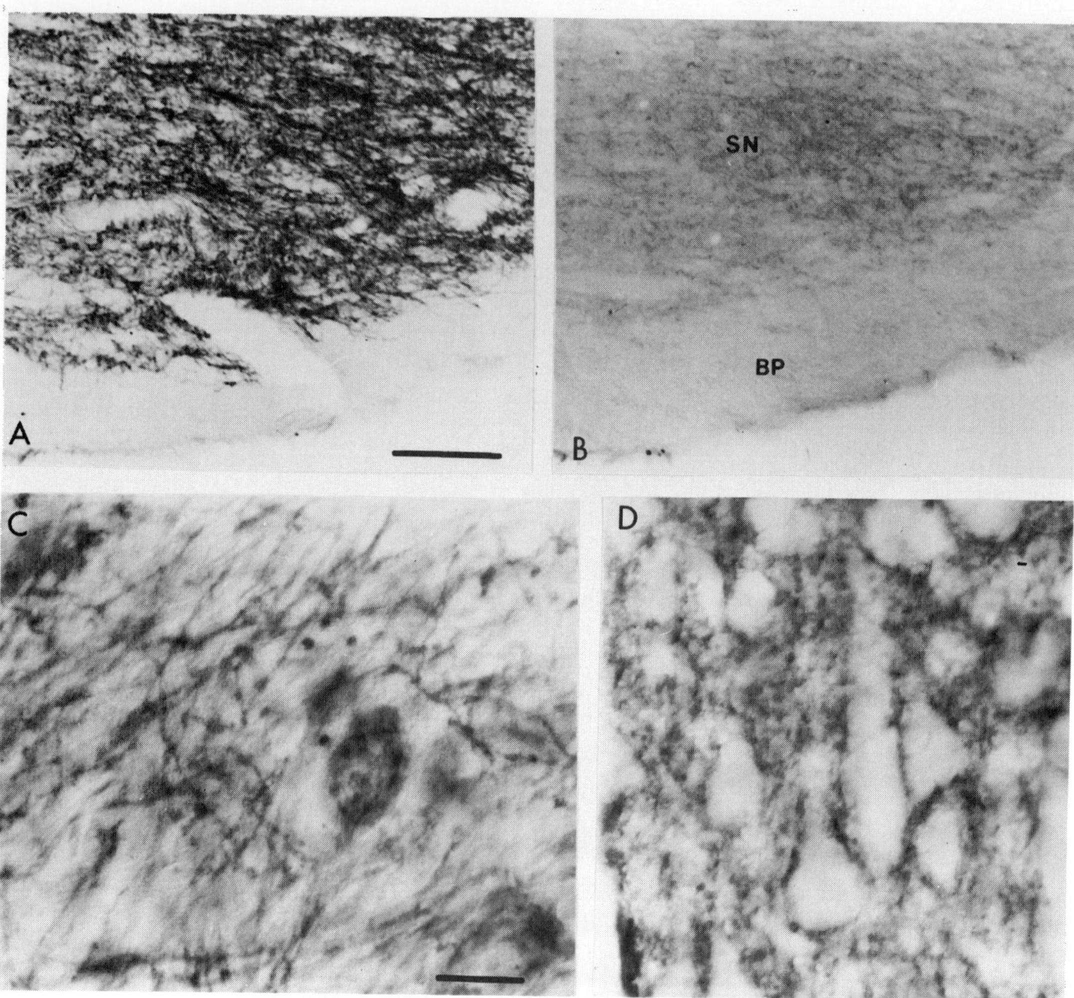

Fig. 6.7. Rat substantia nigra (A–C) and cerebral cortex (D) immunocytochemistry with 62-3G1. B, Control in which mAb was incubated with 9 μg of purified GABAR/BZDR complex previous to the incubation with the tissue. SN, substantia nigra; BP, basis pedunculus. Notice in (C) the presence of receptor clusters on dendrites and cell bodies, and in (D) the association of the reaction product with the neuronal surface of the cortical neurons. Bar, 100 μm (A, B); 20 μm (C, D). Reprinted from De Blas *et al.* (1988), with permission.

mRNA for a specific subunit might not be assembled into a mature receptor in the absence of other subunits. In addition, the different relative abundance of the subunit mRNAs and the limited sensitivity of the assays also make it difficult to interpret correctly the qualitative *in situ* hybridization results.

6.2 THE GABAR/BZDR DURING MAMMALIAN BRAIN DEVELOPMENT

6.2.1 RADIOLIGAND BINDING STUDIES

The first studies on the ontogenesis of the GABAR in the rat brain were done by Coyle and Enna (1976), who reported that GABARs were already present in the embryo at embryonic day (ED)15 although they remained at low levels during birth (25% of the adult values) and up to postnatal day (PD)8, when GABAR levels increased dramatically. The time courses for the GABA synthesizing enzyme, glutamate decarboxylase (GAD), and for the GABAR were similar. However, the neurotransmitter GABA developed earlier than both GABAR and GAD. At birth, the levels of GABA were 50% of the adult values reaching maximum levels around ten days earlier than GABAR or GAD. These results, which have been observed by other authors (see below) suggest that either the GABA turnover in embryos is slower than in adults or that there are alternative synthetic pathways for GABA. This early development of the GABA system suggests that tonic inhibition predominates over excitation during the early development of the brain. The results also suggest a role for the GABA system in the trophic and synaptic interactions that occur during development.

The development of the BZDR in the rat brain has been studied by several groups (Bräestrup and Nielsen, 1978; Palacios *et al.*, 1979; Candy and Martin, 1979; Lippa *et al.*, 1981; Chisholm *et al.*, 1983; Vitorica *et al.*, 1990a). The development of the BZDR in the spinal cord (Bruning *et al.*, 1990) and in the mouse brain (Garrett and Tabakoff, 1985) has also been studied. In the rat, maximum [^{3}H] FNZ binding was observed by the third postnatal week whereas maximum [^{3}H]MUS binding was reached by the fourth postnatal week. The [^{3}H]MUS binding developed more slowly than the [^{3}H]FNZ binding. This developmental mismatch (as well as the distribution mismatch discussed above) has been interpreted by Palacios *et al.* (1979) as evidence that not all the BZDR and GABAR co-exist. The developmental studies have also shown that at birth most of the BZDRs are type II which rapidly increase during the first postnatal week reaching the adult levels by the third or fourth week. The type I receptors however, appear later, increasing after the first week and reaching adult levels by the second postnatal week. These observations probably reflect the different time courses of expression of the various receptor subunits associated with types I and II BZDRs as discussed above. The time course of GABAR expression during development determined by the radioligand binding assays coincides with the results obtained with a different assay based on the expression of GABAR in *Xenopus laevis* oocytes after the injection of polyA$^+$ mRNA obtained from rats of various ages (Carpenter *et al.*, 1988).

6.2.2 PHOTOAFFINITY LABELING, IMMUNOPRECIPITATION AND NORTHERN BLOT STUDIES

In new born animals, the photoaffinity labeling of the membrane bound GABAR/BZDR with [^{3}H]FNZ reveals two peptides of 59 and 55 kDa, whereas in adults, the labeled peptides are of 55 and 51 kDa (Eichinger and Sieghart, 1986). We have investigated the [^{3}H]FNZ photolabeled peptides during the development of the rat brain (Vitorica *et al.*, 1990a) by immunoprecipitation of the solubilized and photolabeled receptor using subunit specific antibodies: the monoclonal antibody (62-3G1) which is specific for both β_2 and β_3 subunits (Ewert *et al.*, 1992) and an antiserum to the α_1 subunit. Both antibodies to the purified GABAR/BZDR were raised and characterized in our laboratory (Vitorica *et al.*, 1987, 1988). Age-dependent heterogeneity of the [^{3}H]FNZ photolabeled subunits was revealed. Peptides of 59, 57, 53 and 51 kDa were immunoprecipitated from the brain of newborn rats, the higher M_r peptides being the most abundant. During early development there is a progressive decrease of the higher M_r peptides and an increase of the 51 kDa peptide. By PD20 the pattern of immunoprecipitated photolabeled peptides had reached the characteristics of the adult animal in which the main photolabeled peptide is the 51 kDa peptide (Fig. 6.8). During the different stages of development, the photolabeling of

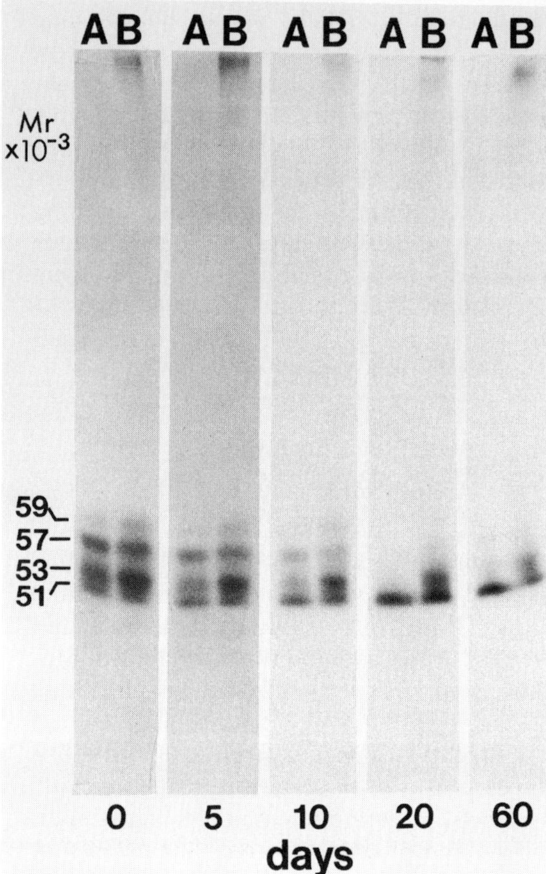

Fig. 6.8.   Fluorographs of the [³H]FNZ photoaffinity-labeled GABAR/BZDR during the postnatal development of the rat cerebral cortex. The immunoprecipitates of the triton X-100 (TX)-SDS-solubilized GABAR/BZDR by mAb 62-3G1 and by rabbit antiserum A are shown in lanes A and B, respectively. Reprinted from Vitorica *et al.* (1990a), with permission.

the 51 kDa peptide, but not the others, could be inhibited by Cl218-872 indicating that this peptide was associated with type I BZDR (Fig. 6.9). In support of the results obtained with binding assays already discussed above, the results with photolabeling and the immunoprecipitation also indicated that at birth only type II BZDR were present in the brain and that type I receptors appeared later. These results also showed that the antibodies recognize both types I and II BZDRs.

The differential expression of receptor subunits during the development of the rat brain has also been observed by Garrett *et al.* (1990) using Northern blot hybridization with ³²P-random primed α_1 and β_1 cDNA probes. The β_1 mRNA was highest at birth, then decreased to adult levels by PD5–PD7. In contrast, the α_1 mRNA was low at birth but increased thereafter reaching the adult levels by PD14– PD25. Levitan *et al.* (1988) have shown that the amount of α_3 mRNA is prominent in the cerebellum of the 12-day-old calf, then declines with age.

The various studies using subunit specific probes (photoaffinity labeling, antibodies or DNA) have shown that during development there is differential regulation of the expression of the various GABAR/BZDR subunits. The coordinated expression of some subunits and their assembly into functional receptors are areas that will be explored in the near future as the appropriate tools become available.

6.2.3   RADIOLIGAND AUTORADIOGRAPHY AND *IN SITU* HYBRIDIZATION

The prenatal development of the BZDR in the rat CNS was described by Schlumpf *et al.* (1983) using [³H]FNZ and [³H]Ro15-1788 autoradiography. The BZDR was first detected at ED14 in spinal cord and lower brainstem progressing in a caudorostral gradient. At ED14 and ED15 the expression was extended to the brainstem, mesencephalon and parts of the diencephalon. At ED16, the BZDR was also expressed in the corpus striatum, olfactory bulb and the frontoventral part of the neocortex. By ED21 the receptor was expressed in the remaining neocortex. In the neocortex, the BZDR appeared first in the superficial layer. Then between ED16 and ED18 the receptor was also expressed in the subcortical plate and at ED21 the BZDR appeared in the cortical plate with an inside–out density gradient. Based on these results the authors postulate that the expression of the BZDR is related to

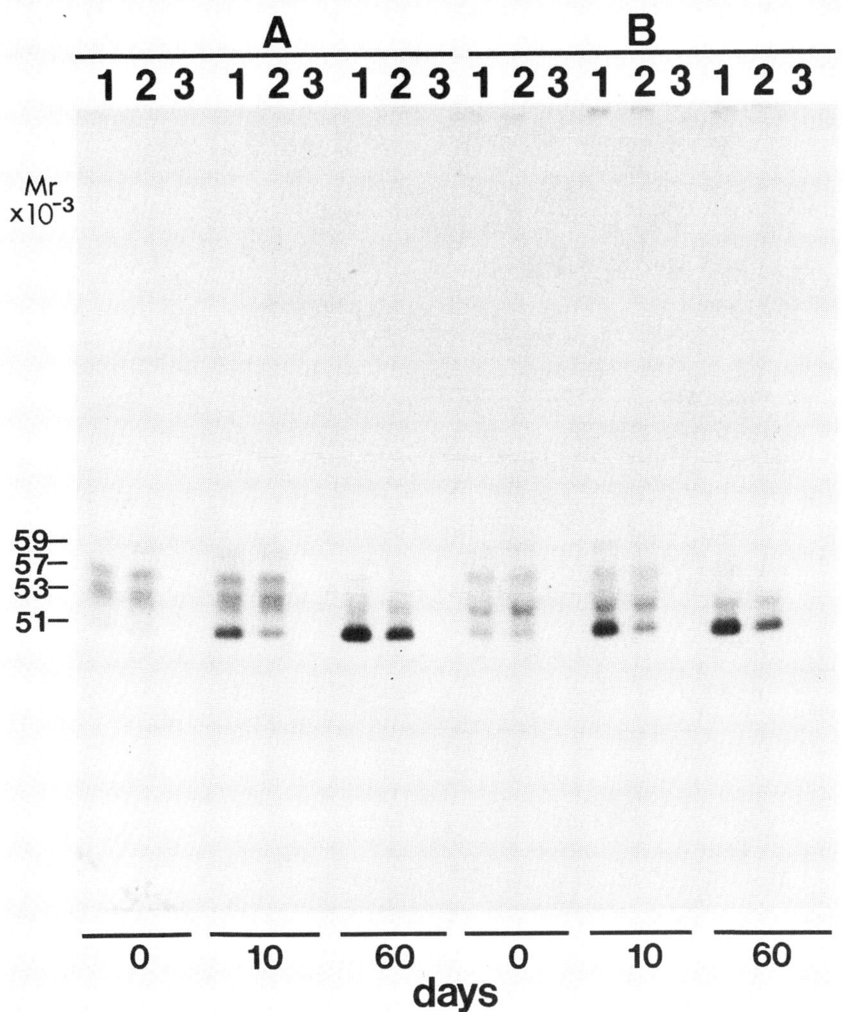

Fig. 6.9.   Fluorographs of [³H]FNZ photoaffinity-labeled and immunoprecipitated GABAR/BZDR from 0-, 10- and 60-day-old rats. The membranes were photoaffinity-labeled with [³H]FNZ, solubilized, and immunoprecipitated with either the mAb 62-3G1 (A) or rabbit antiserum A (B). Lanes 1, 2, and 3 show photoaffinity labeling in controls and in the presence of 300 nM Cl218-872 and 300 nM clonazepam, respectively. Reprinted from Vitorica *et al.* (1990a), with permission.

cell differentiation rather than to synaptogenesis. With the exception of the subplate of the developing cerebral cortex (see below), most authors have noticed that the developmental expression of the GABAR/BZDR is not directly related to synaptogenesis. Some examples are derived from developmental studies on the GABAR and BZDR in the cerebellum (Palacios and Kuhar, 1982; Rotter and

Frostholm, 1986; Frostholm and Rotter, 1987; Zdilar *et al.*, 1992). In rats and mice the cerebellum develops postnatally. The dividing external granule cell layer, from which the granule cells are derived, remains unlabeled with either [³H]FNZ or [³H]MUS. The binding of the latter to the granule cell GABAR occurs for the first time during their migration from the external granule layer to their final position in the granule cell layer, that is before afferent and efferent synaptic contacts are established by these cells. In the adult brain, the predominant binding of [³H]FNZ is to the molecular layer. In contrast, the predominant binding of [³H]MUS is to the granule cell layer. These results show one more example of the localization mismatch between GABAR and BZDR binding sites as mentioned above. The binding of [³H]FNZ to the molecular layer increases dramatically during PD11–PD15. Most likely this binding represents the binding of [³H]FNZ to the α_1 subunit of the GABAR/BZDR. *In situ* hybridization studies (Zdilar *et al.*, 1992) indicate that the expression of the α_1 subunit in Purkinje cells appears very early in development between PD1 and PD5, that is, prior to the establishment of afferent and efferent synaptic connections in these cells. Later, at PD11–PD13 the α_1 mRNA is also expressed in stellate and basket cells. These studies were done with a riboprobe for the α_1 subunit. These results indicate that the expression of the GABAR/BZDR occurs before synaptogenesis. However, the *in situ* hybridization studies by Gambarana *et al.* (1990), using an oligonucleotide for the α_1 subunit, show that the α_1 subunit is synthesized between the second and third postnatal week and therefore the expression of this subunit is most likely related to synapse formation. The discrepancy between the two studies might be explained by the increased sensitivity of the *in situ* hybridization assays done in the former study in which a full-length radiolabeled riboprobe was used instead of the end-labeled 40-mer oligonucleotides used in the latter study.

The relationship between the formation and maintenance of synaptic contacts and GABAR/BZDR expression has also been investigated by using mouse neurological mutants: Purkinje cell degeneration (PKCD), weaver, staggerer and reeler (Rotter *et al.*, 1988; Rotter and Frostholm, 1988) as well as in the rat dystonic mutant (Beales *et al.*, 1990). In the PKCD mutant in which the Purkinje cells have disappeared by PD45, there is a reduction in [³H]FNZ binding in all the cell layers of the cerebellum. However, in the deep cerebellar nuclei, where the axons of the Purkinje cells make GABAergic contacts, there is an increase of [³H]FNZ binding which might be the result of a denervation supersensitivity phenomenon. In the PKCD mutants there is also decreased [³H]MUS binding to the granule cell layer. The weaver mutant, which shows an important loss of granule cells, has increased [³H]FNZ binding and decreased [³H]MUS binding in all layers of the vermis. In the staggerer where the number of granule, Purkinje and Golgi cells as well as the number of synapses between parallel fibers and Purkinje cell dendrites are decreased, the binding of both [³H]FNZ and [³H]MUS is diminished in all the layers. In addition, there is also an increased binding of [³H]FNZ to the deep cerebellar nuclei. In the reeler where all cells are malpositioned and the Purkinje cells are deprived of the afferent basket cell inputs, there is an increase of [³H]FNZ binding to all the layers of the cerebellum including the molecular layer. These results indicate that the expression of the GABAR/BZDRs occurs in the absence of a full complement of GABAergic afferents, however the maintenance of the normal receptor levels depends on the presence of normal synaptic contacts (Rotter *et al.*, 1988; Rotter and Frostholm, 1988). In the dystonic rat mutant there is a concomitant increase of GAD activity in the deep cerebellar nuclei with a decrease in [³H]MUS binding. These results suggest the existence of receptor down regulation in these nuclei (Beales *et al.*, 1990). It seems that the synaptic contacts are involved in the regulation of the normal expression of the receptors. Nevertheless, the cells also express receptors in the absence of synapses.

6.2.4 IMMUNOCYTOCHEMISTRY

Our monoclonal antibody 62-3G1 to β_2 and β_3 subunits of the GABAR/BZDR has been used in several developmental studies of the expression of the GABAR/BZDR. Thus, Huntley *et al.* (1990) have studied the development of the sensory–motor cortex of the macaque monkey. The GABAR/BZDR immunoreactivity with mAb 62-3G1 appeared at ED121 in layers III and IV and subplate. During development there are laminar changes in the distribution of the GABAR, reaching the adult distribution by PD1.5. In the adult the highest level of immunoreactivity occurs in layers II and III (Fig. 6.10). The changes in the laminar distribution of the immunoreactivity parallelled the distribution of the GABA neurons which indicated the existence of early GABA and GABAR interactions. The immunoreactivity in the subplate supports the notion that this layer is a temporary target for ingrowing cortical afferents during early development. Later in development this layer disappears and some cells become interdispersed in the white matter.

In the rat neocortex (Cobas *et al.*, 1991) the GABAR/BZDR immunoreactivity with mAb 62-3G1 appeared first at ED14 in the external primordial plexiform layer. By ED16 the receptor was present in lamina I and subplate. At ED18 the receptor appeared in the lower part of the cortical plate. In the latter, the GABAR/BZDR immunoreactivity and the distribution of GABA neurons show a similar spatiotemporal and inside–out gradient. Nevertheless, the expression of the GABAR/BZDR precedes both the axogenesis of cortical interneurons and the formation of symmetrical synapses. Therefore, the GABA system during early embryogenesis might have a morphogenetic rather than synaptic role. However, in the rat brain subplate as indicated above in the case of the monkey, the GABA system might be involved in transient synaptic connectivity.

In the rat thalamus (Bentivoglio *et al.*, 1991) the GABAR/BZDR immunoreactivity with mAb 62-3G1 undergoes rearrangements during the first postnatal weeks. During this time the levels of GABAR/BZDR decreased in the reticular nucleus while they increased in the dorsal thalamus. The receptor was expressed at ED14, that is as early as the GABAergic innervation of these nuclei. The study, however, does not show whether GABAergic synaptic contacts have been established by that time.

Lauder *et al.* (1986) have studied the development of the GABA neurons in the rat brain by immunocytochemistry with an antibody to GABA. These authors have shown that GABA appeared first at ED13 in the brainstem, mesencephalon and diencephalon. At ED16 the cells were also found in the basal forebrain and cortex. The temporal and spatial distribution of the GABA immunoreactivity was similar to that of the BZDR revealed by radioligand autoradiography (Schlumpf *et al.*, 1983) as discussed earlier. Nevertheless, the GABA immunoreactivity appeared one day earlier than the BZDR. These results also support a morphogenetic and/or trophic function of GABA during embryogenesis as proposed by Chronwall and Wolff (1980). Although some authors favor the notion that GABARs are formed following synaptogenesis (Gambarana *et al.*, 1990; Meinecke and Rakic, 1990), most of the studies discussed earlier indicate that GABAR/BZDR expression does not follow synaptogenesis. In most cases, GABA, GAD and GABAR are expressed earlier than synaptogenesis (Aoki *et al.*, 1989; Fairén *et al.*, 1986). In fact, in certain cases the GABAR function might modulate synaptogenesis as will be discussed below.

6.3 THE GABAR/BZDR AND BRAIN PLASTICITY

We have already mentioned the plasticity of the GABAR/BZDR during the abnormal development of the brain in studies with neurological mutants. Experimental denervation has also revealed plasticity of the GABAR/BZDR. The denervation of the striatonigral fibers results in an increase in the density of type I and a decrease of type II BZDRs in the substantia nigra (Lo *et al.*, 1983). These results suggest a postsynaptic

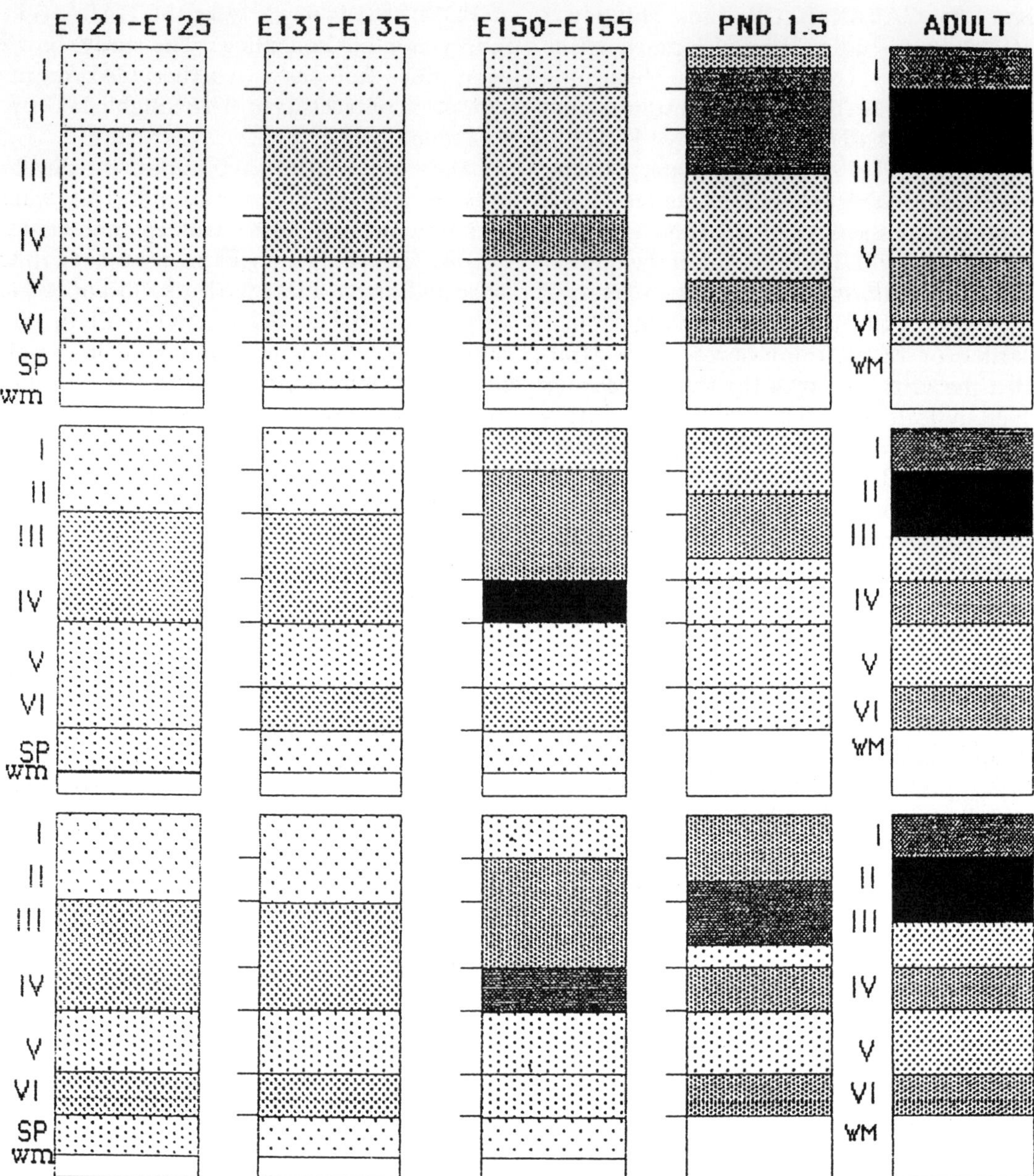

Fig. 6.10. Schematic diagram illustrating the major changes in laminar distribution and staining intensity of receptor immunoreactivity in areas 4, 3a, 3b, 1 and 2 from ED125 to adulthood in the sensory–motor cortex of the macaque monkey. Increasing density of stippling represents increasing intensity of the immunostaining. Note that the laminar boundaries depicted are schematic, and are not meant to reflect changes in laminar thickness with increasing age. Reprinted from Huntley *et al.* (1990), with permission.

localization of BZDR type I and a presynaptic localization of BZDR type II in this structure. They also reveal the possible existence of a phenomenon of denervation supersensitivity as in the case of the PKCD and staggerer mutants already discussed above. In post-mortem human brains from patients with unilateral lesions of the optic nerve, Palacios *et al.* (1987) have found a selective decrease of GABAR/ BZDR in the lateral geniculate body in the layers innervated by the damaged nerve. The previous studies were done by using radioligand autoradiography techniques.

Immunocytochemical techniques with the mAb 62-3G1 to the GABA/BZDR have revealed that in macaque monkey, either intravitreal injection of the sodium channel blocker tetrodotoxin (TTX) or monocular enucleation are followed by a decrease of GABAR/BZDR immunoreactivity in the columns of the visual cortex (areas 17 and 18) dominated by the deprived eye (Fig. 6.11, Hendry *et al.*, 1990). In the same study, similar results were obtained by using [^{3}H]FNZ and [^{3}H]MUS radioligand autoradiography techniques. These results indicate that the density of the GABAR/BZDR in the monkey visual cortex is regulated not only by denervation but also by the electrical activity. Similar activity-dependent regulation of GAD and GABA also occurs in the visual cortex (Hendry and Jones, 1988). These results suggest that decreased GABA-ergic activity in the columns driven by the

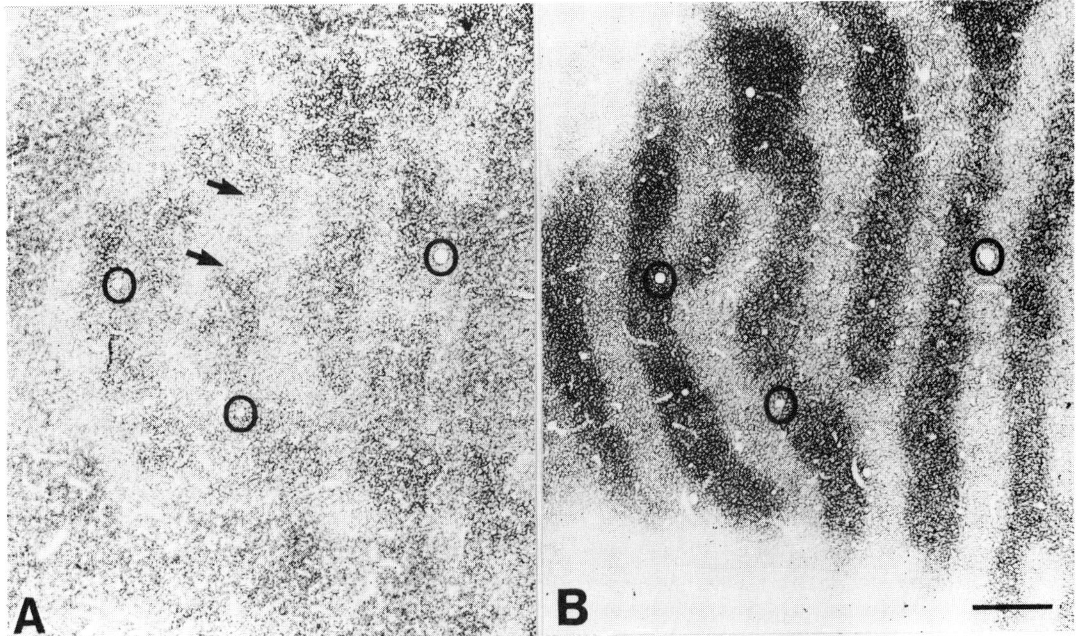

Fig. 6.11.   Photomicrographs of tangential sections through layer IVC of a monkey visual cortex injected intravitreally with TTX 5 days before death. A, GABA$_A$ receptor immunostaining in layer IVC consists of stripes of intense immunostaining that alternate with stripes of lighter immunostaining. Individual stripes pass through both layer IVCβ and the more lightly stained layer IVCα (between arrows). B, Section adjacent to A stained for CO. Intense staining of normal-eye dominance stripes and light staining for injected-eye stripes are evident. Comparison of the profiles of radially oriented blood vessels (circles) shows that the receptor immunostaining in the injected-eye columns is reduced to that in the normal-eye columns. Scale bar, 500 μm. Reprinted from Hendry *et al.* (1990), with permission.

deprived eye might contribute to the functional expansion of the ocular dominance columns driven by the other eye.

In monocularly deprived kittens, changes in the GABAR/BZDR expression was opposite to those observed in monkeys since the density of GABAR increased in area 17 among other regions of the brain (Skangiel-Kramska and Kossut, 1984; Shaw and Cynader, 1988). Therefore, the plastic changes of the GABAR that follow denervation seem to be heterogeneous and adapted to the functional characteristics of the system. This functional heterogeneity might result from a combination of factors: the molecular heterogeneity of the GABAR/BZDR and the differential regulation of the subunit expression as discussed above. No effect on either [^{3}H]FNZ (Shaw *et al.*, 1987) or [^{3}H]MUS (Mower *et al.*, 1988) binding in the cat visual cortex was found after rearing these animals in the dark. Therefore, it seems that the GABAergic system develops normally in the cat visual cortex in the absence of visual input from both eyes. Nevertheless, this phenomenon is not common to all brain structures. Schliebs *et al.* (1986) have reported a decrease in the binding of [^{3}H]FNZ to the lateral geniculate nucleus and superior colliculus of rats reared in the dark.

Evidence of the importance of the role of the GABAR in the synaptic organization of the ocular dominance columns has been provided by Reiter and Stryker (1988). When muscimol is injected in the visual cortex of the cat during the critical period and the eye is monocularly closed, the inputs from the closed eye become dominant. Thus, the less active inputs became dominant when the cortical cell discharges were inhibited by the GABAR/BZDR agonist muscimol. This result was the opposite of what happened when muscimol was not injected. In this case, the open eye was dominant. These results show the pivotal role that the GABAergic transmission, mediated by GABAR/BZDR, has on brain plasticity and synaptogenesis.

6.4   DEVELOPMENT OF THE GABAR/BZDR IN NEURONAL CULTURES

Rat neuronal cultures from cerebral cortex (McCarthy and Harden, 1981; White *et al.*, 1981) and cerebellar granule cells (Meier *et al.*, 1984) as well as neuronal cultures from other species (Huang *et al.*, 1980; Borden *et al.*, 1984; Kuriyama *et al.*, 1987; Mehta and Ticku, 1988) express GABAR/BZDR complexes. Syapin *et al.* (1985) have shown that in embryonic neuronal cultures, most of the BZDR are type II. We have discussed above that during normal development, the early postnatal BZDR is type II. The type I BZDR appears later and increases dramatically during the first and second postnatal weeks coinciding with the expression of the 51 kDa peptide that is PAL by [^{3}H]FNZ as discussed earlier. The development of the GABAR/BZDR complex in primary neuronal cultures from rat brain embryos has been investigated by Vitorica *et al.* (1990b). The receptor complexes of the cultured neurons were photoaffinity labeled with [^{3}H]FNZ, solubilized with detergent and immunoprecipitated with the subunit specific mAb 62-3G1 (to β_2 and β_3) and a polyclonal antibody specific for the α_1 subunit. The results have shown that the cultured neurons express five different photoaffinity labeled peptides of 51, 53, 54, 57 and 59 kDa which are similar in size to the ones found in the brain of newborn rats (Fig. 6.12). However, during *in vitro* development of the neuronal cultures, the transition of the receptors from the embryonic to the mature form does not occur (Fig. 6.8 vs. Fig. 6.13). These findings contrast with the important changes in subunit composition observed during the normal development of the rat brain, mainly regarding the progressive enrichment in the 51 kDa subunit during the early postnatal brain maturation. The neuronal cultures might be missing a cell type and/or soluble factor that regulates the normal expression of the GABAR/BZDR subunits (i.e., the 51 kDa peptide).

Other studies have shown that granule cell cultures from rat cerebellum have high affinity GABAR regardless of the age of the cultures. However, when the cells were cultured during eight days in the presence of 50 μM GABA or muscimol or 150 μM 4,5,6,7-tetrahydroisooxazolo[5-4-c]pyridin-3-ol (THIP) the cultures developed low affinity GABA receptors as well as an increased number of neurites. These effects could be blocked by bicuculline whereas (–)baclofen had no effect (Meier *et al.*, 1984; Hansen *et al.*, 1987). A similar effect on the expression of the low affinity GABAR was produced by taurine (Abraham and Schousboe, 1989). These studies suggest that GABA and taurine exert trophic actions on the neurons, including the induction of the expression of the low affinity GABAR. The trophic

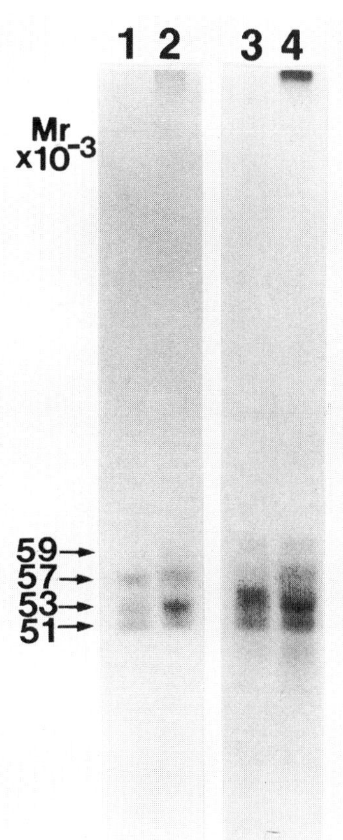

Fig. 6.12.   Fluorographs of the [³H]FNZ photolabeled and immunoprecipitated GABAR/BZDR receptor. Lanes 1 and 2 are immunoprecipitates of newborn rat brain membranes and lanes 3 and 4 show the immunoprecipitated receptor from cells maintained in culture for 14 days. mAb 62-3G1 was used in lanes 1 and 3 and rabbit antiserum A in lanes 2 and 4. Reprinted from Vitorica *et al.* (1990b), with permission.

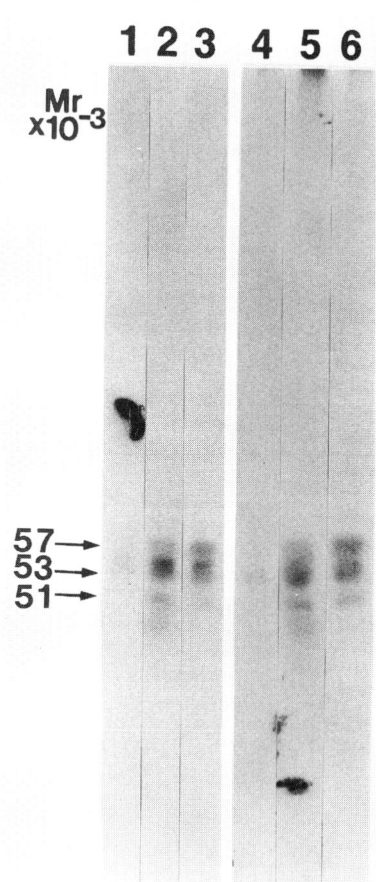

Fig. 6.13.   Fluorographs of the immunoprecipitated [³H]FNZ photoaffinity labeled GABAR/BZDR from rat neuronal cultures. The photoaffinity labeled receptor was solubilized and immunoprecipitated with either mAb 62-3G1 (lanes 1–3) or rabbit antiserum A (lanes 4–6). Cells maintained in culture for 5 days (lanes 1 and 4), 14 days (lanes 2 and 5) and 21 days (lanes 3 and 6) are shown. Reprinted from Vitorica *et al.* (1990b), with permission.

effect results from the binding of GABA and other agonists to the high affinity GABAR. It has been proposed that the hyperpolarization of the membrane which follows the opening of the GABA-gated chloride channel is the signal that mediates the trophic and differentiating actions of GABA both *in vivo* and in culture (Belhage *et al.*, 1990). This hypothesis is consistent with the very early expression of GABA and GABAR during development as indicated above, as well as with the existence of non-vesicular K$^+$-stimulated but Ca^{2+}-independent release of GABA and taurine from the growth cones (Taylor and Gordon-Weeks, 1989; Taylor *et al.*, 1990).

6.5 SUMMARY

The studies on the development of the GABAR/BZDR in the mammalian brain have indicated the following. (1) The expression of GABA, GAD and GABAR/BZDR is an early developmental event that occurs in the embryo before the development of other transmitter–receptor systems and before synaptogenesis. (2) GABA seems to have a trophic and morphogenetic role mediated by GABAR/BZDR during early development. (3) Synaptic connectivity seems to be important for the consolidation and maintenance but not for the initial expression of the GABAR/BZDR. (4) During development, there is differential expression of GABAR/BZDR subunits which seems to be regulated at least in part by cellular interactions and/or trophic factors, electrical activity, GABA and other GABAR/BZDR modulators. The differential expression of various GABAR/BZDR types and subunits during development might reflect the adaptation to the various functional roles that the GABAR/BZDR might play during brain development (i.e., morphogenesis, plasticity, synaptogenesis etc.). (5) There is plasticity in the expression of these receptors that is dependent on neuronal activity and innervation. The GABAergic transmission mediated by GABAR plays an important role in regulating neuronal plasticity and synaptic reorganization in experimental denervation and sensory deprivation.

These results suggest that the GABAR/BZDR might be involved in organizing and maintaining the connectivity and synaptic function of certain brain areas during the normal development, maturation and aging as well as in controlling adaptive synaptic changes in response to injury, sensory deprivation and degenerative neurological disorders.

ACKNOWLEDGEMENT

I thank Ms Laura Morissette for typing the manuscript. The research in my laboratory was supported by grant NS17708 from the National Institute of Neurological Disorders and Stroke.

REFERENCES

Abraham, J.H. and Schousboe, A. (1989) Effects of taurine on cell morphology and expression of low affinity GABA receptors in cultured cerebellar granule cells. *Neurochem. Res.*, **14**, 1031–8.

Aoki, E., Semba, R. and Kashiwamata, S. (1989) When does GABA-like immunoreactivity appear in the rat cerebellar GABAergic neurons? *Brain Res.*, **502**, 245–51.

Beales, M., Lorden, J.F., Walz, E. and Oltmans, G.A. (1990) Quantitative autoradiography reveals selective changes in cerebellar GABA receptors of the rat mutant dystonic. *J. Neurosci.*, **10**, 1874–5.

Belhage, B., Hansen, G.H. and Schousboe, A. (1990) GABA agonist induced changes in ultrastructure and GABA receptor expression in cerebellar granule cells is linked to hyperpolarization of the neurons. *Int. J. Dev. Neurosci.*, **8**, 473–9.

Bentivoglio, M., Spreafico, R., Alvarez-Bolado, G. *et al.* (1991) Differential expression of the GABA$_A$ receptor complex in the dorsal thalamus and reticular nucleus: an immunohistochemical study in the adult and developing rat. *Eur. J. Neurosci.*, **3**, 118–25.

Bloom, F. E. and Iversen, L. L. (1971) Localizing ^{3}H-GABA in nerve terminals of rat cerebral cortex by electron microscopic autoradiography. *Nature*, **229**, 628–30.

Borden, L.A., Czajkowski, C., Chan, C.Y., and Farb, D.H. (1984) Benzodiazepine receptor synthesis and degradation by neurons in culture. *Science*, **226**, 857–60.

Bräestrup, C. and Nielsen, M. (1978) Ontogenic development of benzodiazepine receptors in the rat brain. *Brain Res.*, **147**, 170–3.

Bräestrup, C. and Squires, F. (1977) Specific benzo-diazepine receptors in rat brain characterized by high-affinity [3H]diazepam binding. *Proc. Natl. Acad. Sci. USA*, **74**, 3805–9.

Brüning, G., Bäuer, R. and Baumgarten, H.G. (1990) Postnatal development of [3H]flunitraze-pam and [3H]strychnine binding sites in rat spinal cord localized by quantitative autora-diography. *Neurosci. Lett.*, **110**, 6–10.

Candy, J.M. and Martin, I.L. (1979) The postnatal development of the benzodiazepine receptor in the cerebral cortex and cerebellum of the rat. *J. Neurochem.*, **32**, 655–8.

Carpenter, M.V., Parker, I. and Miledi, F.R.S. (1988) Expression of GABA and glycine receptors by messenger RNAs from the developing rat cere-bral cortex. *Proc. R. Soc. Lond.*, **234**, 159–70.

Chisholm, J., Kellogg, C. and Lippa, A. (1983) Development of benzodiazepine binding sub-types in three regions of rat brain. *Brain Res.*, **267**, 388–91.

Chronwall, B. and Wolff, J.R. (1980) Prenatal and postnatal development of GABA accumulating cells in the occipital neocortex of rat. *J. Comp. Neurol.*, **190**, 187–208.

Cobas, A., Fairén, A., Alvarez-Bolado, G. and Sánchez, M.P. (1991) Prenatal development of the intrinsic neurons of the rat neocortex: a com-parative study of the distribution of GABA immunoreactive cells and the GABA_A receptor. *Neuroscience*, **40**, 375–97.

Coyle, J.T. and Enna, S.J. (1976) Neurochemical aspects of the ontogenesis of GABAenergic neu-rons in the rat brain. *Brain Res.*, **111**, 119–33.

Cutting, G.R., Lu, L., O'Hara, B.F. *et al.*(1991) Clon-ing of the γ-aminobutyric acid (GABA)_pl cDNA: a GABA receptor subunit highly expressed in the retina. *Proc. Natl. Acad. Sci. USA*, **88**, 2673–7.

de Blas, A.L., Kuljis, R.O. and Cherwinski, H.M. (1984) Mammalian brain antigens defined by monoclonal antibodies. *Brain Res.*, **322**, 277–87.

de Blas, A.L., Vitorica, J. and Friedrich, P. (1988) Localization of GABA_A receptor in the rat brain with a monoclonal antibody to the 57 000 M_r peptide of GABA_A receptor/benzodiazepine receptor/Cl⁻ channel complex. *J. Neurosci.*, **8**, 602–14.

Eichinger, A. and Sieghart, W. (1986) Postnatal development of proteins associated with different benzodiazepine receptors. *J. Neurochem.*, **46**, 173–80.

Ewert, M., Shivers, B., Lüddens, H. *et al.* (1990) Subunit selectivity and epitope characterization of monoclonal antibodies directed against the GABA_A/benzodiazepine receptor. *J. Cell Biol.*, **110**, 2043–8.

Ewert, M., de Blas, A.L., Möhler, H. and Seeburg, P.H. (1992) A prominent epitope on GABA_A receptors is recognized by two different mono-clonal antibodies. *Brain Res.*, **569**, 57–62.

Fairén, A., Cobas, A. and Fonseca, M. (1986) Times of generation of glutamic acid decarboxylase immunoreactive neurons in mouse somatosen-sory cortex. *J. Comp. Neurol.*, **251**, 67–83.

Frostholm, A. and Rotter, A. (1987) The ontogeny of [3H]muscimol binding sites in the C57BL/6J mouse cerebellum. *Dev. Brain Res.*, **37**, 157–66.

Gambarana, C., Pittman, R. and Siegel, R.E. (1990) Developmental expression of the GABA_A recep-tor alpha-1 subunit mRNA in the rat brain. *J. Neurobiol.*, **21**, 1169–79.

Garrett, K. M. and Tabakoff, B. (1985) The devel-opment of type 1 and type 2 benzodiazepine receptors in the mouse cortex and cerebellum. *Pharmacol. Biochem. Behav.*, **22**, 985–92.

Garrett, K. M., Saito, N., Duman, R.S. *et al.* (1990) Differential expression of GABA_A receptor subunits. *Mol. Pharmacol.*, **37**, 652–7.

Hansen, G.H., Belhage, B., Schousboe, A. and Meier, E. (1987) Temporal development of GABA ago-nist induced alterations in ultrastructure and GABA receptor expression in cultured cerebellar granule cells. *Int. J. Neurosci.*, **5**, 263–9.

Hendry, S.H.C. and Jones, E.G. (1988) Activity-dependent regulation of GABA expression in the visual cortex of adult monkeys. *Neuron*, **1**, 701–12.

Hendry, S.H.C., Fuchs, J., de Blas, A.L. and Jones, E.G. (1990) Distribution and plasticity of immu-nocytochemically localized GABA_A receptors in adult monkey visual cortex. *J. Neurosci.*, **10**, 2438–50.

Huang, A., Barker, J.L., Paul, S.M. *et al.* (1980) Characterization of benzodiazepine receptors in primary cultures of fetal mouse brain and spinal cord neurons. *Brain Res.*, **190**, 485–91.

Huntley, G.W., de Blas, A.L. and Jones, E.G. (1990) GABA$_A$ receptor immunoreactivity in adult and developing monkey sensory-motor cortex. *Exp. Brain Res.*, **82**, 519–35.

Juiz, J.M., Helfert, R.H., Wenthold, R.J. *et al.* (1989) Immunocytochemical localization of the GABA/benzodiazepine receptor in the guinea pig cochlear nucleus: evidence for receptor localization heterogeneity. *Brain Res.*, **504**, 173–9.

Khrestchatisky, M., MacLennan, A.J., Chiang, M.-Y. *et al.* (1989) A novel α subunit in rat brain GABA$_A$ receptors. *Neuron*, **3**, 745–53.

Klepner, C.A., Lippa, A.S., Benson, P.I. *et al.* (1979) Resolution of two biochemically and pharmacologically distinct benzodiazepine receptors. *Pharmacol. Biochem. Behav.*, **11**, 457–62.

Kuriyama, K., Tomono, S., Kishi, M. *et al.* (1987) Development of gamma-aminobutyric acid (GABA)ergic neurons in cerebral cortical neurons in primary culture. *Brain Res.*, **416**, 7–21.

Lauder, J.M., Han, V.K.M., Henderson, P. *et al.* (1986) Prenatal ontogeny of the GABAergic system in the rat brain: an immunocytochemical study. *Neuroscience,* **19**, 465–93.

Levitan, E.S., Schofield, P.R., Burt, D.R. *et al.* (1988) Structural and functional basis for GABA$_A$ receptor heterogeneity. *Nature*, **335**, 76–9.

Lippa, A.S., Beer, B., Sano, M. C. *et al.* (1981) Differential ontogeny of type 1 and type 2 benzodiazepine receptors. *Life Sci.*, **28**, 2343–7.

Lo, M.M.S., Niehoff, D.L., Kuhar, M.J. and Snyder, S.H. (1983) Differential localization of type I and type II benzodiazepine binding sites in substantia nigra. *Nature*, **306**, 57–60.

Lüddens, H., Pritchett, D.B., Köhler, M. *et al.* (1990) Cerebellar GABA$_A$ receptor selective for a behavioral alcohol antagonist. *Nature*, **346**, 648–51.

Malherbe, P., Sigel, E., Baur, R. *et al.* (1990) Functional characteristics and sites of gene expression of the α$_1$, β, γ$_2$-isoform of the rat GABA$_A$ receptor. *J. Neurosci.*, **10**, 2330–7.

McCabe, R.T. and Wamsley, J.K. (1986) Autoradiographic localization of subcomponents of the macromolecular GABA receptor complex. *Life Sci.*, **39**, 1937–46.

McCarthy, K.D. and Harden, T.K. (1981) Identification of two benzodiazepine binding sites on cells cultured from rat cerebral cortex. *J. Pharmacol. Exp. Ther.*, **216**, 183–91.

Mehta, A.K. and Ticku, M.K. (1988) Developmental aspects of benzodiazepine receptors and GABA-gated chloride channels in primary cultures of spinal cord neurons. *Brain Res.*, **454**, 156–63.

Meier, E., Drejer, J. and Schousboe, A. (1984) GABA induces functionally active low-affinity GABA receptors on cultured cerebellar granule cells. *J. Neurochem.*, **43**, 1737–44.

Meinecke, D.L. and Rakic, P. (1990) Developmental expression of GABA and subunits of the GABA$_A$ receptor complex in an inhibitory synaptic circuit in the rat cerebellum. *Dev. Brain Res.*, **55**, 73–86.

Möhler, H. and Okada, T. (1977) Benzodiazepine receptor: demonstration in the central nervous system. *Science*, **198**, 849–51.

Mower, G.D., Rustad, R. and Frost White, W. (1988) Quantitative comparisons of GABA neurons and receptors in the visual cortex of normal and dark-reared cats. *J. Comp. Neurol.*, **272**, 293–302.

Olsen R.W. and Tobin, A.J. (1990) Molecular biology of GABA$_A$ receptors. *FASEB J.*, **4**, 1469–80.

Palacios, J.M. and Kuhar, M. (1982) Ontogeny of high affinity GABA and benzodiazepine receptors in the rat cerebellum: an autoradiographic study. *Dev. Brain Res.*, **2**, 531–9.

Palacios, J.M., Niehoff, D. L. and Kuhar, M. J. (1979) Ontogeny of GABA and benzodiazepine receptors: effects of Triton X-100, bromide and muscimol. *Brain Res.*, **179**, 390–5.

Palacios, J.M., Wamsley, J.K. and Kuhar, M.J. (1981) High affinity GABA receptors – autoradiographic localization. *Brain Res.*, **222**, 285–307.

Palacios, J.M., Cortés, R. and Probst, A. (1987) Receptor plasticity in the human brain: some autoradiographic studies. *J. Recept. Res.*, **7**, 581–97.

Park, D. and de Blas, A.L. (1991a) Peptide subunits of γ-aminobutyric acid$_A$/benzodiazepine receptors from bovine cerebral cortex. *J. Neurochem.*, **56**, 1972–9.

Park, D. and de Blas, A.L. (1991b) Peptide heterogeneity of GABA$_A$/benzodiazepine receptors in bovine cerebral cortex and cerebellum. *Brain Res.*, **550**, 279–86.

Park, D., Vitorica, J., Tous, G. and de Blas, A.L. (1991) Purification of the γ-aminobutyric acid$_A$/benzodiazepine receptor complex by immunoaffinity chromatography. *J. Neurochem.*, **56**, 1962–71.

Pritchett, D.B., Sontheimer, H., Shivers, B.D. *et al* (1989a) Importance of a novel GABA$_A$ receptor subunit for benzodiazepine pharmacology. *Nature*, **338**, 582–5.

Pritchett, D.B., Lüddens, H. and Seeburg, P. (1989b) Type I and type II GABA$_A$-benzodiazepine receptors produced in transfected cells. *Science*, **245**, 1389–92.

Pritchett, D.B. and Seeburg, P.H. (1990) γ-Aminobutyric acid$_A$ receptor α$_5$-subunit creates novel type II benzodiazepine receptor pharmacology. *J. Neurochem.*, **54**, 1802-4.

Reiter, H.O. and Stryker, M.P. (1988) Neural plasticity without postsynaptic action potentials: less-active inputs become dominant when kitten visual cortical cells are pharmacologically inhibited. *Proc. Natl. Acad. Sci. USA*, **85**, 3623–7.

Richards, J.G., Schoch, P., Haring, P. *et al.* (1987) Resolving GABA$_A$/benzodiazepine receptors: cellular and subcellular localization in the CNS with monoclonal antibodies. *J. Neurosci.*, **7**, 1866–86.

Rotter, A. and Forstholm, A. (1986) Cerebellar benzodiazepine receptor distribution: an autoradiographic study of the normal C57BL/6J and Purkinje cell degeneration mutant mouse. *Neurosci. Lett.*, **71**, 66–71.

Rotter, A. and Frostholm, A. (1988) Cerebellar benzodiazepine receptors: cellular localization and consequences of neurological mutations in mice. *Brain Res.*, **444**, 133–46.

Rotter, A., Gorenstein, C. and Frostholm, A. (1988) The localization of GABA$_A$ receptors in mice with mutations affecting the structure and connectivity of the cerebellum. *Brain Res.*, **439**, 236–48.

Schliebs, R. and Rothe, T. (1988) Development of GABA$_A$ receptor in the cortical visual structures of rat brain. Effect of visual pattern deprivation. *Gen. Physiol. Biophys.*, 7, 281–92.

Schliebs, R., Rothe, T. and Bigl, V. (1986) Dark rearing affects the development of benzodiazepine receptors in the central visual structures of rat brain. *Dev. Brain Res.*, **24**, 179–85.

Schlumpf, M., Richards, J.G., Lichtensteiger, W. and Möhler, H. (1983) An autoradiographic study of the prenatal development of benzodiazepine-binding sites in rat brain. *Neuroscience*, **3**, 1478–87.

Schofield, P.R., Darlison, M.G., Fujita, M. *et al.* (1987) Sequence and functional expression of the GABA$_A$ receptor shows a ligand-gated receptor super-family. *Nature*, **328**, 221–7.

Schwartz, R.D. (1988) The GABA$_A$ receptor-gated ion channel: biochemical and pharmacological studies of structure and function. *Biochem. Pharmacol.*, **37**, 3369–75.

Shaw, C. and Cynader, M. (1988) Unilateral eyelid suture increases GABA$_A$ receptors in cat visual cortex. *Dev. Brain Res.*, **40**, 148–53.

Shaw, C., Aoki, C., Wilkinson, M. *et al.* (1987) Benzodiazepine ([^{3}H]flunitrazepam) binding in cat visual cortex: ontogenesis of normal characteristics and the effects of dark rearing. *Dev. Brain Res.*, **37**, 67–76.

Shaw, C., Needler, M.C., Wilkinson, M. *et al.* (1984) Alterations in the receptor number, affinity and laminar distribution in cat visual cortex during the critical period. *Prog. Neuro-Psychopharmacol. Biol. Psychiat.*, **8**, 627–34.

Shaw, C., Needler, M.C., Wilkinson, M. *et al.* (1986) Modification of neurotransmitter receptor sensitivity in cat visual cortex during the critical period. *Dev. Brain Res.*, **22**, 67–73.

Shivers, B.D., Killisch, I., Sprengel, R. *et al.* (1989) Two novel GABA$_A$ receptor subunits exist in distinct neuronal subpopulations. *Neuron*, **3**, 327–37.

Siegel, R.E. (1988) The mRNA encoding GABA$_A$/benzodiazepine receptor subunits are localized in different cell populations of the bovine cerebellum. *Neuron,* **1**, 579–84.

Sieghart, W. and Drexler, G. (1983) Irreversible binding of [^{3}H]flunitrazepam to different proteins in various brain regions. *J. Neurochem.*, **41**, 47–55.

Sieghart, W. and Karobath, M. (1980) Molecular heterogeneity of benzodiazepine receptor. *Nature*, **286**, 285–7.

Sieghart, W., Mayer, A. and Drexler, G. (1983) Properties of [^{3}H]flunitrazepam binding to different benzodiazepine binding proteins. *Eur. J. Pharmacol.*, **88**, 291–9.

Sigel, E. and Barnard, E.A. (1984) A gamma-aminobutyric acid/benzodiazepine receptor complex from bovine cerebral cortex. Improved purification with preservation of regulatory sites and their interaction. *J. Biol. Chem.*, **259**, 7219–23.

Sigel, E., Baur, R., Trube, G. *et al.* (1990) The effect of subunit composition of rat brain GABA$_A$ receptors on channel function. *Neuron*, **5**, 703–11.

Skangiel-Kramska, J. and Kossut, M. (1984) Increase of GABA receptor binding activity after short lasting monocular deprivation in kittens. *Acta Neurobiol. Exp.*, **44**, 33–9.

Squires, R.F., Saederup, E., Damgaard, I. and Schousbae, A. (1990) Development of benzodiazepine and pictrotopin (*t*-butylbicyclophorothionate) binding sites in rat cerebellar granule cells in culture. *Neurochem.*, **54**, 473–8.

Syapin, P.J., Cole, R., de Vellis, J. *et al.* (1985) Benzodiazepine binding characteristics of embryonic rat brain neurons grown in culture. *J. Neurochem.*, **45**, 1797–801.

Taguchi, J.-I. and Kuriyama, K. (1984) Purification of γ-aminobutyric acid (GABA) receptor from rat brain by affinity column chromatography using a new benzodiazepine 1012-S, as an immobilized ligand. *Brain Res.*, **323**, 219–26.

Taylor, J. and Gordon-Weeks, P.R. (1989) Developmental changes in the calcium dependency of γ-aminobutyric acid release from isolated growth cones: correlation with growth cone morphology. *J. Neurochem.*, **53**, 834–43.

Taylor, J., Docherty, M. and Gordon-Weeks, P.R. (1990) GABAergic growth cones: release of endogenous GABA precedes the expression of synaptic vesicle antigens. *J. Neurochem.*, **54**, 1689–99.

Unnerstall, J.R., Kuhar, M.J., Niehoff, D.L. and Palacios, J.M. (1981) Benzodiazepine receptors are coupled to a subpopulation of GABA receptors: evidence from a quantitative autoradiographic study. *J. Pharmacol. Exp. Ther.*, **218**, 797–804.

Verdoorn, T.A., Draguhn, A., Ymer, S. *et al.* (1990) Functional properties of recombinant rat $GABA_A$ receptors depend upon subunit composition. *Neuron*, **4**, 919–28.

Vitorica, J., Park, D. and de Blas, A.L. (1987) Immunochemical localization of the $GABA_A$ receptor in the rat brain. *Eur.J. Pharmacol.*, **136**, 451–3.

Vitorica, J., Park, D., Chin, G. and de Blas, A.L. (1988) Monoclonal antibodies and conventional antisera to the γ-aminobutyric $acid_A$/benzodiazepine receptor/Cl$^-$ channel complex. *J. Neurosci.*, **8**, 615–22.

Vitorica, J., Park, D., Chin G. and de Blas, A.L. (1990a) Characterization with antibodies of the GABA/benzodiazepine receptor complex during development of the rat brain. *J. Neurochem.*, **54**, 187–94.

Vitorica, J., Park, D. and de Blas, A.L. (1990b) The $GABA_A$/benzodiazepine receptor complex in rat brain neuronal cultures. Characterization by immunoprecipitation. *Brain Res.*, **537**, 209–15.

White, W. F., Dichter, M. A. and Snodgrass, S. N. (1981) Benzodiazepine binding and interactions with the GABA receptor complex in living cultures of rat cerebral cortex. *Brain Res.*, **215**, 162–76.

Wisden, W., Moris, B.J., Darlison, M. G. *et al.* (1988) Distinct $GABA_A$ receptor alpha-subunit mRNAs show differential patterns of expression in bovine brain. *Neuron*, **1**, 937–47.

Wisden, W., Moris, B. J., Darlison, M. G. *et al* (1989) Localization of $GABA_A$ receptor alpha-subunit mRNAs in relation to receptor subtypes. *Mol. Brain Res.*, **5**, 305–10.

Yazulla, S., Studholme, K.M., Vitorica, J. and De Blas, A.L. (1989) Immunochemical localization of $GABA_A$ receptors in goldfish and chicken retinas. *J. Comp. Neurol.*, **280**, 15–26.

Ymer, S., Schofield, P.R., Draguhn, A. *et al.* (1989) $GABA_A$ receptor beta subunit heterogeneity: functional expression of cloned cDNAs. *EMBO J.*, **8**, 1665–70.

Ymer, S., Draguhn, A., Wisden, W. *et al.* (1990) Structural and functional characterization of the γ_1 subunit of GABA/benzodiazepine receptors. *EMBO J.*, **9**, 3261–7.

Zdilar, D., Rotter, A. and Frostholm, A. (1992) Expression of $GABA_A$/benzodiazepine receptor α_1-subunit mRNA and [^{3}H]flunitrazepam binding sites during postnatal development of the mouse cerebellum. *Dev. Brain Res.*, **61**, 63–71.

GABA AND BENZODIAZEPINE RECEPTORS IN THE DEVELOPING VISUAL SYSTEM

Reinhard Schliebs and Thomas Rothe

7.1 INTRODUCTION

For transmitting visual information from the retina via the thalamus (lateral geniculate nucleus, superior colliculus) to the cortex, visual pathways use excitatory synaptic signals (Schiller, 1986). γ-Aminobutyric acid (GABA), acting on both type A and type B receptors, is the major inhibitory transmitter in the visual system, and mediates fast synaptic inhibition by activating a chloride channel. The type A GABA receptor belongs to a super-family of ligand-gated ion channels including glycine, glutamate and nicotinic acetylcholine receptors and is a hetero-oligomeric complex which comprises binding sites for GABA and for its allosteric modulators (benzodiazepines, barbiturates) together with the integral chloride ion channel. The native $GABA_A$ receptor is postulated to have a pentameric structure composed of various combinations of at least four subunits: α, β, γ and δ. Further diversity within each class ($\alpha_1, ..., \alpha_6; \beta_1, ..., \beta_3; \gamma_1, \gamma_2$) has been reported, and thus regional variations in the expression of distinct $GABA_A$ subunits have been observed (for review, see Silvilotti and Nistri, 1991). An additional GABA receptor subunit, termed GABA p_1-receptor, has recently been identified that demonstrates unique pharmacological properties and is primarily expressed in the retina (Cutting *et al.*, 1991).

Baclofen, a GABA receptor agonist, binds selectively to a subpopulation of GABA recognition sites which are termed $GABA_B$ receptors. An unusual characteristic of $GABA_B$ binding is the absolute requirement for Ca^{2+}. The main effect of baclofen via $GABA_B$ receptors is to inhibit presynaptically the release of excitatory transmitters by depressing Ca^{2+} currents probably mediated via a GTP-binding protein, whereas the postsynaptic action of baclofen is an activation of K^+ currents. However, the postsynaptic action of baclofen is nearly negligible when compared to its presynaptic action (Silvilotti and Nistri, 1991).

In most areas of the brain there is a predominance of $GABA_A$ sites over $GABA_B$ ones (exceptionally the molecular layer of cerebellum and interpeduncular nucleus of mesencephalon), although the ratio of $GABA_A$ and $GABA_B$ receptors varies between different brain regions. The functional significance of various ratios of $GABA_A$ and $GABA_B$ sites in the brain is still unclear.

Electrophysiological measurements disclosed an unusual GABA receptor in the visual system of frog and guinea pig. Recordings of

Receptors in the Developing Nervous System Vol. 2: Neurotransmitters. Edited by Ian S. Zagon and Patricia J. McLaughlin. Published in 1993 by Chapman & Hall. ISBN 0 412 49400 0. Vols. 1 and 2 (set) ISBN 0 412 54520 9.

excitatory postsynaptic potentials (EPSPs) in the frog optic tectum by optic nerve stimulation showed that GABA enhances excitatory mechanisms in this brain area (Nistri and Silvilotti, 1985). Application of a very high concentration of GABA in guinea-pig superior colliculus resulted in inhibition, whereas an excitatory effect was observed after application of much lower doses. The sustained excitatory response to GABA displays unusual pharmacological characteristics which suggest the existence of a novel receptor type distinct from the types A and B (for review, see Silvilotti and Nistri, 1991).

The presence of GABAergic local circuit neurons within the visual thalamic nuclei and the visual cortex (VC) of mammals has been disclosed in a number of studies (Ribak, 1978; Sterling and Davis, 1980; McDonald *et al.*, 1981; Hendrickson *et al.*, 1983; Ohara *et al.*, 1983; Somogyi *et al.*, 1983; Fitzpatrick *et al.*, 1984; Montero and Singer, 1984; Penny *et al.*, 1984; Giolli *et al.*, 1985; Gabbott *et al.*, 1986; Somogyi, 1989). GABA-releasing structures are assumed to gate and modify the responses evoked at the postsynaptic cell by the excitatory signals. This is emphasized by the findings that GABA in the visual system is involved in such fundamental phenomena as center-surround inhibition, directional and orientation selectivity, and binocular interaction (Kanno and Okada, 1988; Somogyi, 1989).

Neurotransmitter receptors are one of the decisive links in the chain of synaptic information processing. Due to their adaptive properties in response to altered levels of neurotransmitter as a consequence of changes in neuronal activity (Schwartz *et al.*, 1983), they are of particular importance for functional-adaptive processes occurring during the maturation of the visual system with respect to pattern recognition. GABA$_A$ receptors represent the target structures which receive GABAergic input and are partly associated with benzodiazepine receptors (Stephenson, 1988). The benzodiazepine receptors are thought to play a neuromodulatory role by modifying the activities of a certain population of GABA receptors (Costa, 1988). On the basis of the regulatory properties of benzodiazepines and barbiturates on GABA binding, GABA$_A$ receptors have been further subdivided: GABA$_{A1}$ receptors can be modulated by benzodiazepines and barbiturates, whereas GABA$_{A2}$ receptors are insensitive to the action of these agents (Friedman and Redburn, 1990).

The activation of GABA$_A$ receptors, which are localized mainly on the postsynaptic cell, leads to an inhibition of the excitatory response on the postsynaptic cell. This precise interaction of excitatory and inhibitory signals seems to be a major step in efficient processing of visual information. Thus the question arises as to whether the functional maturation of the GABA receptors in the visual system during an early period of postnatal life is driven by the presence of adequate visual experience. Therefore, in the following section the maturation of GABA$_A$ and benzodiazepine receptors in the visual regions is considered followed by a short description of the effects of visual deprivation on the GABA and benzodiazepine receptor ontogeny.

7.2   GABA AND THE VISUAL SYSTEM

7.2.1   RETINA

GABA is thought to function as a neurotransmitter in the retina. Significant levels of GABA have been detected in the inner plexiform layer where terminals of bipolar, amacrine and ganglion cells are present. There is now evidence that a subset of amacrine cells utilizes GABA as their transmitter (Voaden *et al.*, 1980), and possesses a high affinity uptake and release system for GABA (Brandon *et al.*, 1979). GABAergic amacrine cells constitute a major portion of the inhibitory interneurons, and are reported to make pre- and postsynaptic contacts with bipolar cells and axon terminals and to form synapses onto amacrine cells (including the GABAergic subset) and ganglion cells. They participate in the establishment of receptive field properties like size, velocity and directional selectivity of ganglion cells and seem to be involved in forming both ON and OFF circuitries of light stimulation (e.g. Murashima *et al.*, 1990).

Binding and physiological studies suggest that both GABA$_A$ and GABA$_B$ receptors as well as benzodiazepine receptors are present in the retina, mainly localized within the inner plexiform layer. It is assumed that the p$_1$ GABA receptor subunit recently detected and highly expressed in the retina plays an essential role in retinal neurotransmission (Cutting *et al.*, 1991), but further functional and physiological studies must be awaited. GABA$_A$ receptors may be involved in regulation of both basal and light-dependent acetylcholine release. But GABA$_A$ receptors that are involved in the regulation of acetylcholine release do not seem to be coupled to benzodiazepine and barbiturate modulatory sites (Friedman and Redburn, 1990). GABA$_B$ receptors may participate in the regulation of GABA release in the circuitry for acetylcholine release.

In contrast to the large number of neurochemical and physiological studies of the adult retina, there are fewer data on age-related changes of the GABA system in the mammalian retina (Drago *et al.*, 1989). GABA concentration and glutamic acid decarboxylase (GAD) activity (used as a marker enzyme for GABAergic neurons) in the mouse retina were found to increase from birth until two weeks of age followed by a slight decrease at 20 weeks. GABA and GAD levels reached peaks when the maturation of the retina was complete (Murashima *et al.*, 1990). Similarly, developmental changes in GABA and benzodiazepine binding sites in rat retina have been described (Guarnieri *et al.*, 1982). However, the property of benzodiazepine receptors to be modulated by GABA was steadily lost with age indicating age-dependent functional changes of GABA and benzodiazepine receptors in the retina. It has been shown in the rat retina that GABA$_A$ binding sites increase markedly until postnatal day (PD) 14 followed by a considerable loss of receptor sites until PD 35 by which time the adult value is reached (Fig. 7.1). However, benzodiazepine binding sites rise markedly from PD10 to PD37, persist until PD50 and then slightly decrease until adulthood (Fig. 7.1). The comparison of the developmental appearance of [^{3}H]muscimol and [^{3}H]flunitrazepam binding sites in the rat retina shows that the age-related increase in benzodiazepine binding sites is delayed as compared to the developmental course of GABA$_A$ receptor binding. The distinct temporal appearance of GABA$_A$ and benzodiazepine binding sites might indicate a separate, uncoupled, development of both receptor types or an age-related change in the functional properties of the GABA$_A$–benzodiazepine–chloride ionophore complex. The view of a separate development of GABA$_A$ and benzodiazepine receptors is supported by recent findings in the adult cat retina that a certain number of GABA$_A$ receptors are not coupled to the benzodiazepine binding sites (Friedman and Redburn, 1990). Iontophoretic studies revealed that the action of benzodiazepines in the adult cat retina was selective to ON-type ganglion cells, whereas in kittens aged 7–9 weeks both ON- and OFF-type cells could be influenced by benzodiazepines. This suggests that during postnatal development OFF-type ganglion cells lose GABA receptors that are linked to the benzodiazepine receptor (Robins and Ikeda, 1989). From recent molecular biological studies is it known that different combinations of the various GABA receptor subunits with different sensitivities to GABA and benzodiazepines can exist. Therefore, one might speculate that the age-dependent changes in GABA/benzodiazepine receptor with respect to binding and physiological function might be due to changes in the selective expression and/or suppression of distinct combinations of GABA receptor subunits during the postnatal development. A series of *in situ* hybridization experiments

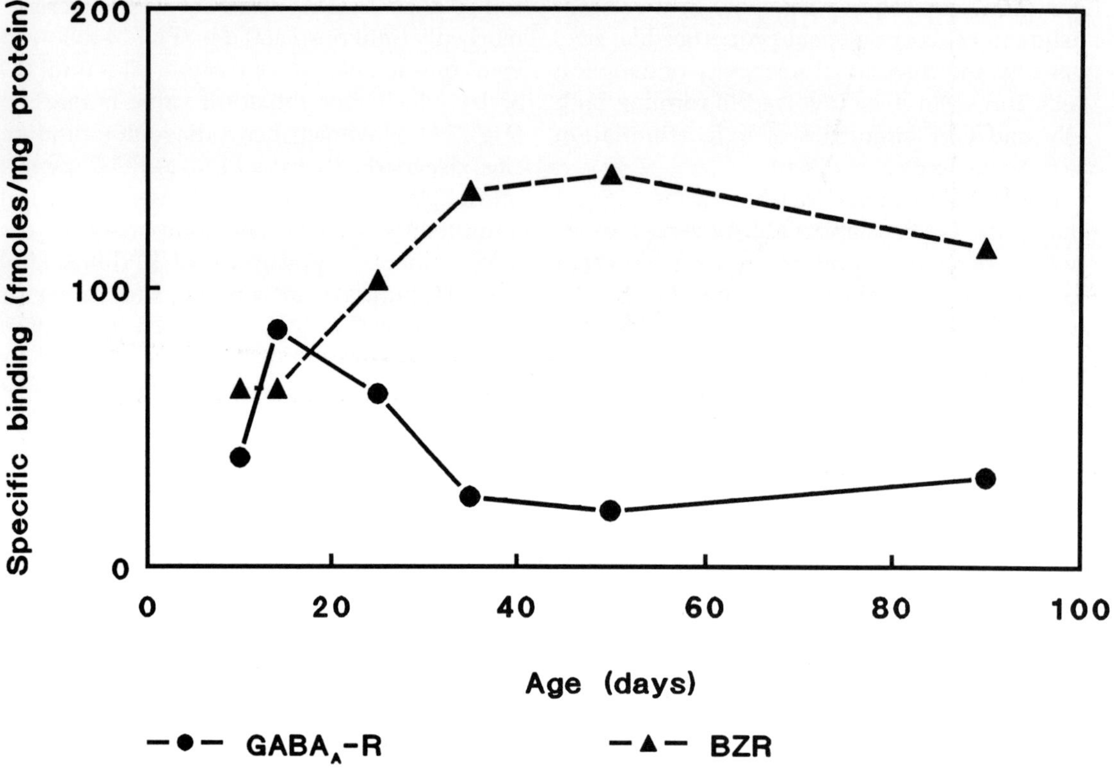

Fig. 7.1. Postnatal development of GABA$_A$ (●) and benzodiazepine (▼) receptor sites in rat retina. Receptor densities were obtained from binding studies in membrane fractions using [^{3}H]muscimol (Schliebs and Rothe, 1988) and [^{3}H]flunitrazepam (Schliebs *et al.*, 1986b) as radioligands. Binding data are expressed as fmol specifically bound radioligand per mg protein content.

performed during the period of early postnatal life should help to elucidate this question.

7.2.2 LATERAL GENICULATE NUCLEUS (LGN)

GABA-mediated inhibition in the LGN has been shown to play a major role in gating, modifying and preserving visual information from the retina. This is emphasized by the high levels of GABA and high activities of GAD found in this region (Ohara *et al.*, 1983; Kanno and Okada, 1988). The dorsal LGN has a laminated structure in many mammals. All layers contain relay cells, local inhibitory interneurons and receive two major excitatory inputs from the retina and the visual cortex

(Somogyi, 1989). Relay neurons in the dorsal LGN receive GABAergic input from axons of the perigeniculate nucleus, and from intrinsic GABAergic interneurons. Electrophysiological experiments demonstrated that GABA$_A$ receptors are involved in such processes like binocular inhibition, center-surround inhibition or spatial frequency tuning difference of X and Y cells.

The postnatal maturation of GABA$_A$ and benzodiazepine receptors in rat LGN is shown in Fig. 7.2. Binding of [^{3}H]muscimol to GABA$_A$ receptors at PD10 is relatively low, increases markedly up to PD30 and then there is a slight decrease until adulthood. In contrast, [^{3}H]flunitrazepam binding at PD10 is relatively high and decreases continuously

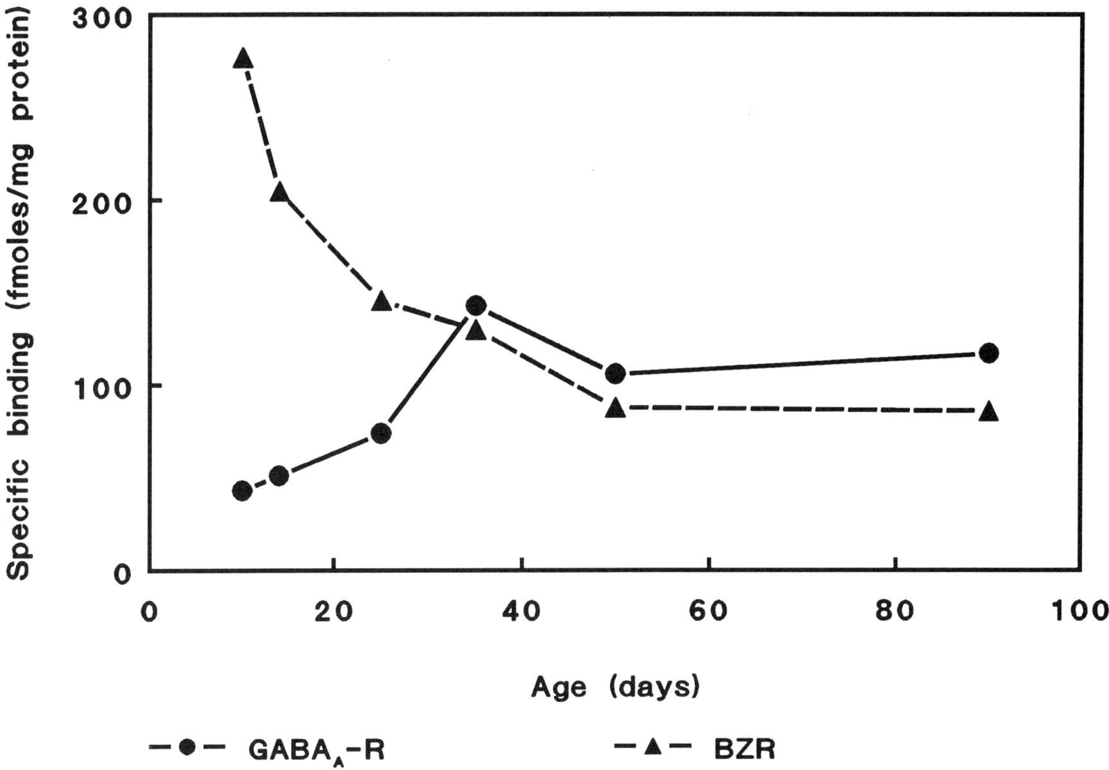

Fig. 7.2.   Postnatal development of GABA$_A$ (●) and benzodiazepine (▲) receptor sites in rat lateral geniculate nucleus. Receptor densities were obtained from binding studies in membrane fractions using [³H]muscimol (Schliebs and Rothe, 1988) and [³H]flunitrazepam (Schliebs *et al.*, 1986b) as radioligands. Binding data are expressed as fmol specifically bound radioligand per mg protein content.

until PD90, when the adult binding level is reached (Fig. 7.2).

One problem in describing developmental patterns of small brain areas is the developmental changes in the reference system itself (Shaw *et al.*, 1984). When [³H]flunitrazepam binding is expressed in terms of total binding (binding sites per LGN), then a continuous increase in benzodiazepine binding is observed (Schliebs *et al.*, 1986b). Apparently, the benzodiazepine binding sites in the LGN develop more slowly than the protein content of tissue. In the rat LGN, GAD activity reached the adult level after two weeks of postnatal life. However, the high affinity uptake of GABA, a marker for GABAergic terminals, showed a marked peak on PD15 as

compared to the adult value (Kvale *et al.*, 1983), indicating that the postnatal ontogeny of GABA$_A$ receptors is delayed in comparison to the development of the presynaptic GABAergic parameters. Similar indications for receptor formation preceding the prior formation of functional synapses were also found in ontogenetic studies in other regions (Brooksbank *et al.*, 1981) or other transmitter systems (Bylund, 1979; Coyle and Yamamura, 1976).

7.2.3   SUPERIOR COLLICULUS

The superior colliculus (SC) receives among others projections from the retina and visual cortex. It is also a laminated structure and the upper half of SC layers responds specifically to visual stimulation and has been reported to contain high concentrations of GABA and GAD activity (Fosse *et al.*, 1989; Kanno and Okada, 1988). Intrinsic GABAergic neurons in the superficial layers of SC are probably involved in shaping receptive fields and in determining orientation selectivity of other SC neurons (Stein and Gallagher, 1981). GABA is mainly localized in interneurons (Houser *et al.*, 1983; Okada, 1974) or in GABAergic fibers from substantia nigra to SC (DiChiara *et al.*, 1979).

[^{3}H]muscimol binding in rat SC rises markedly from PD10 until PD37 followed by a slight decrease until adulthood (Fig. 7.3). By PD9, [^{3}H]flunitrazepam binding in the SC is relatively high. Between PD9 and PD14 binding sites increase slightly and then persist until adulthood (Fig. 7.3). Regardless of minor quantitative differences, the temporal pattern of the development of GABA$_A$ receptors in rat SC correlates rather well with that of benzodiazepine binding sites.

High affinity uptake of GABA in rat SC at birth was already similar to the adult value, but showing a significant peak at PD15. However, GAD activity increased continuously until adulthood and was not fully developed after 30 days (Kvale *et al.*, 1983). In contrast to the finding in the rat LGN, the development of GABA$_A$ binding sites in the SC seems to precede that of GAD activity, which might be partly due to the distinct structural and func-

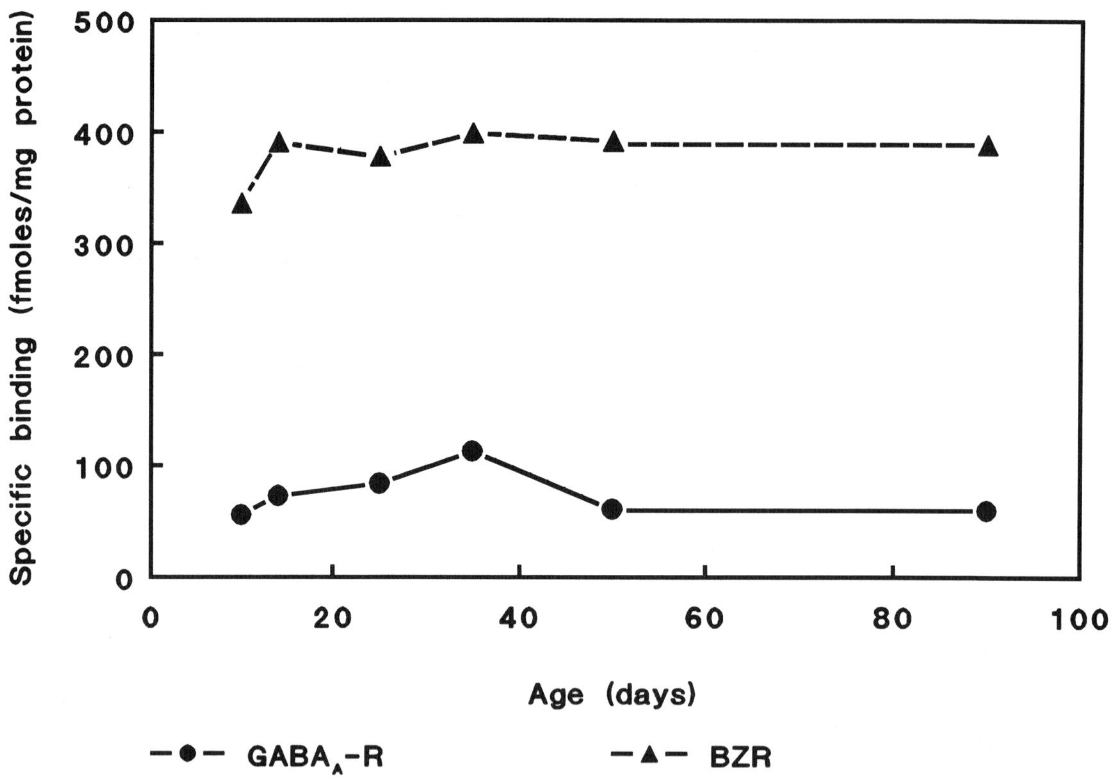

Fig. 7.3.　Postnatal development of GABA$_A$ (●) and benzodiazepine (▲) receptor sites in rat superior colliculus. Receptor densities were obtained from binding studies in membrane fractions using [^{3}H]muscimol (Schliebs and Rothe, 1988) and [^{3}H]flunitrazepam (Schliebs *et al.*, 1986b) as radioligands. Binding data are expressed as fmol specifically bound radioligand per mg protein content.

tional development of SC and LGN in the rat (Kvale *et al.*, 1983).

The ontogeny of GABA$_A$ and benzodiazepine receptors in rat SC does not essentially differ from that in other brain regions suggesting that the development of GABA and benzodiazepine receptors in the SC is not correlated with the development of retinal function and the function of the visual system to respond to light stimuli.

7.2.4   VISUAL CORTEX

In the visual cortex GABA has been suggested to play a role in sharpening orientation and direction selectivity and also in determining the ocular dominance of visual cortical neurons (Mower *et al.*, 1988; Jones, 1990). A number of non-pyramidal cells within the visual cortex use GABA as an inhibitory transmitter. Several types of GABAergic neurons, differing in their synaptic connections, are present in all layers of the visual cortex (Somogyi, 1989). GABAergic nerve terminals are also present in all layers but with higher densities in layers IV, I and VI (Lin *et al.*, 1986). This pattern of GABAergic fibers correlates well with the laminar distribution of benzodiazepine receptors in rat visual cortex, but cannot be easily related to the laminar distribution of GABA$_A$ receptors (Rothe and Schliebs, 1989).

Measurements of the maximum number (B_{max}) of [^{3}H]flunitrazepam binding sites in cat visual cortex showed low values at an early age, rising to a peak in receptor density at about PD60, followed by a decline until adulthood (Shaw *et al.*, 1987). At all ages GABA altered K_d but not B_{max} of benzodiazepine binding.

Similarly, at birth, [^{3}H]muscimol binding sites in cat visual cortex are relatively low but increase continuously until postnatal week 13, after which there is a slight decrease until adulthood (Shaw *et al.*, 1984). However, GAD activity exhibited a gradual increase towards adult levels during the first month, and the adult level was reached during postnatal weeks 5–6 (Fosse *et al.*, 1989), thus indicating that the postnatal ontogeny of GABA$_A$ receptors in cat visual cortex is delayed in comparison to that of the presynaptic GABAergic element.

For both receptor populations, K_d values change during postnatal development. Affinity for [^{3}H]flunitrazepam binding was found to decline with age, whereas GABA$_A$ receptors show an initial increase in affinity until the age of 8 weeks, which was followed by a decline until adulthood (Shaw *et al.*, 1986, 1987). This is in contrast to other studies which failed to detect changes in receptor affinity during development (Brooksbank *et al.*, 1981; Guarnieri *et al.*, 1982; Patel *et al.*, 1980; Schliebs and Rothe, 1988; Skerritt and Johnston, 1982).

In order to correlate developmental changes of B_{max} and K_d Shaw *et al.* (1985) introduced the term of receptor sensitivity which is calculated from the ratio of B_{max} and K_d. The ontogeny of benzodiazepine receptors closely parallels that of GABA$_A$ receptors (Shaw *et al.*, 1984), but peak values differ: benzodiazepine binding sites peak near 60 days whereas GABA binding sites peak near PD95. But using the term of receptor sensitivity, peak receptor sensitivity for both receptor populations occurs near 60 days (Shaw *et al.*, 1986, 1987).

It is noteworthy that the peaks in binding sites of both receptor populations are reached during a period of high cortical plasticity (a critical period, which lasts in the cat until the age of about 3 months). Therefore, the peak in [^{3}H]muscimol binding sites just before the end of the critical period during which the cortex is modified may suggest a role for GABA receptors in altering neuronal response properties (Shaw *et al.*, 1984). The observed increase in receptor number during the critical period followed by a decline from the peak into adulthood, closely parallels the development of the synapse to neuron ratios in cat visual cortex, suggesting that receptors are being added or eliminated together with the

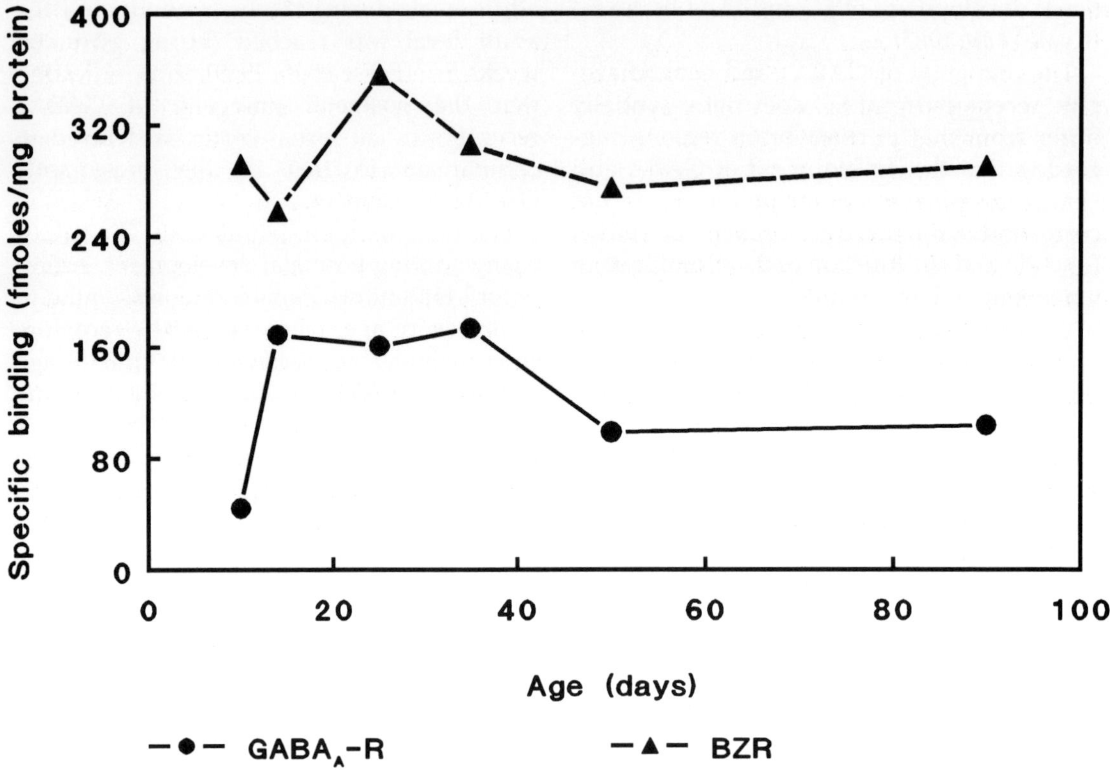

Fig. 7.4.   Postnatal development of GABA$_A$ (●) and benzodiazepine (▲) receptor sites in rat visual cortex. Receptor densities were obtained from binding studies in membrane fractions using [³H]muscimol (Schliebs and Rothe, 1988) and [³H]flunitrazepam (Schliebs *et al.*, 1986b) as radioligands. Binding data are expressed as fmol specifically bound radioligand per mg protein content.

synapses during postnatal development (Shaw *et al.*, 1986).

The laminar distribution pattern of both GABA$_A$ and benzodiazepine receptors in cat visual cortex does not change during the entire postnatal development, in contrast to the postnatal ontogeny of muscarinic acetylcholine receptors, which shows age-dependent laminar alterations of receptor binding during an early period of postnatal life (Shaw *et al.*, 1986; Schliebs and Stewart, 1991).

Biochemical and visual behavioral studies showed that during a certain period of early life the visual system in the rat is also highly sensitive to changes of the visual input (Rothblat *et al.*, 1978; Aurich and Bigl, 1988).

In the rat VC, [³H]muscimol binding rises sharply from PD10 to PD14, remains unchanged until PD37, and then decreases markedly to PD50 when the adult value is reached (Fig. 7.4). This developmental profile is qualitatively similar to that reported previously for the rat cerebral cortex (Aldinio *et al.*, 1980; Patel *et al.*, 1980).

[³H]Flunitrazepam binding to benzodiazepine receptors reached the highest level in the VC on PD25 and then decreased slightly until adulthood (Fig. 7.4). This developmental pattern is very similar to that previously reported in rat cerebral cortex (Candy and Martin, 1979; Mallorga *et al.*, 1980; Regan *et al.*, 1980; Aldinio *et al.*, 1981) and cerebellum (Palacios *et al.*, 1979).

Comparing both receptor populations, the

development of GABA$_A$ receptors precedes that of benzodiazepine receptors, suggesting a distinct maturation of GABA and benzodiazepine receptors in rat VC. This might be partly explained by the finding that in the cortex not all GABA receptors are associated with benzodiazepine receptors. In particular, high affinity GABA$_A$ receptors may not be coupled to benzodiazepine receptors (DeBlas *et al.*, 1988).

As in cat VC, in the rat VC, the peaks in binding sites of both receptor populations are reached during the critical period of development of the rat visual system, which is assumed to last until the age of about 4 weeks (Rothblat *et al.*, 1978).

Regardless of small temporal variations in the developmental profiles, the ontogeny of [^{3}H]muscimol and [^{3}H]flunitrazepam binding sites in rat VC does not essentially differ from that in other non-visual regions. This might indicate that the development of GABA and benzodiazepine binding sites in the visual system is not correlated with the development of retinal function and the functional maturation of the visual system with regard to response to light stimuli. However, it is interesting to note that the development of GABA$_A$ receptors in the rat VC seems to precede that in the subcortical visual regions.

Measurements of GAD activity in individual layers of rat VC indicated a fairly even distribution throughout the cortex, in contrast to the distinct laminar pattern of GABA and benzodiazepine receptors (Rothe and Schliebs, 1989). The activity of GAD in the rat VC increases continuously until PD35 when the adult level is reached, in contrast to the development of high affinity GABA uptake which shows a peak at PD15 (Kvale *et al.*, 1983). This indicates that the ontogeny of GABA$_A$ receptors in rat VC also precedes that of GAD activity, a finding which was also observed for the development of GABA system in rat forebrain (Coyle and Enna, 1976).

In the optic lobe of chicken, a region which corresponds to the VC of mammals, a period of high plasticity was also detected for the development of GABA receptors (Rios *et al.*, 1987). The postnatal development of GABA$_A$ receptor sites in the chick optic lobe shows an increase in binding sites during the first week post-hatching with the highest value at day 6 post-hatching. Between the second and third week GABA$_A$ receptor sites decrease to the adult level (Rios *et al.*, 1987). The descending phase of the developmental course might be explained as a consequence of changes in receptor synthesis or degradation, but it could also result from the elimination of a redundant number of synapses built up during the ascending phase, which is consistent with the hypothesis of selective synaptic stabilization (Changeux and Danchin, 1976). The developmental changes in GABA$_A$ receptor binding in chick optic lobe occur within a period of high cortical plasticity because the effect of dark-rearing on GABA$_A$ receptor binding is maximal at day 6 post-hatching, the time point at which the highest level of GABA$_A$ receptor binding was reached (see also section 7.3).

7.3   GABA AND VISUAL DEPRIVATION

Afferent neuronal activity plays an important role in the development and maturation of the visual system, particularly during a certain period of susceptibility in brain ontogeny. The importance of adequate environmental stimuli for the final tuning of the functional and structural connections during the ontogenetic period of the brain has been strongly emphasized in recent years by a number of electrophysiological, morphological and biochemical studies using the visual system as a model (for reviews, see Singer, 1985; Shaw *et al.*, 1986; Toga, 1987; Hendry and Jones, 1988; Jones, 1990). Light seems to trigger these developmental processes, and once initiated, the completion of the ontogenetic processes can occur without further light input (Mower *et al.*, 1983).

Deprivation of visual stimuli includes such diverse procedures as complete removal of visual input (dark rearing or enucleation) and removal of spatially patterned input by suturing one or both eyelids (monocular or binocular deprivation). Visual deprivation results in abnormalities in visual behavior as well as in electrophysiological and morphological differences in the nuclei of the central visual pathway compared with controls (Singer, 1985).

In cats reared under a normal light/dark cycle, GABA plays a role in determining orientation and direction tuning and ocular dominance. However, in abnormally reared cats GABA seems to play a role in the resultant abnormalities in visual cortical dominance (Burchfiel and Duffy, 1981; Sillito *et al.*, 1981; Mower *et al.*, 1984; Mower and Christen, 1989). In the course of normal development the afferent input is capable of inducing changes in the postsynaptic cell leading to a stabilization of the affected synapses. In this case receptors might play an essential role in mediating neuronal cortical plasticity.

Monocular deprivation of rats from PD10 to PD25 did not affect [^{3}H]muscimol binding in any of the visual regions (retina, LGN, SC, VC) examined (Rothe and Schliebs, 1989). This compares well with similar studies in other species which showed that in cats monocular deprivation also did not affect [^{3}H]muscimol binding in the visual cortex (Mower *et al.*, 1986). These data support the suggestion that physiological light stimulation is not a prerequisite for the correct development of GABA$_A$ receptors in the visual cortex. However, results found in monocularly deprived cats are inconsistent, indicating either no effect (Mower *et al.*, 1986) or an increase in receptor number (Shaw and Cynader, 1988) and changes in GABA$_A$ receptor affinity (Skangiel-Kramska and Kossut, 1984). These conflicting results are assumed to be largely due to differences in preparation and methodology (Shaw and Cynader, 1988). In contrast to the effect of monocular depriva-

tion in cats, a reduction in GABA$_A$ receptor density in the deprived eye columns of layer IV in the visual cortex of monkeys has been detected (Hendry *et al.*, 1990; Jones, 1990) suggesting that GABA$_A$ receptors in adult monkeys undergo a rapid, activity-dependent change in density or structure with monocular deprivation. Thus the effect of monocular deprivation on the development of GABA$_A$ receptors in the visual cortex seems to be species-dependent and might reflect a specific response to the distinct conditions found in area 17 of a particular species (Hendry *et al.*, 1990).

[^{3}H]Flunitrazepam binding to benzodiazepine receptors in the rat LGN, SC, and VC was not affected by monocular deprivation until PD25 (Rothe *et al.*, 1985). Obviously, the development of GABA$_A$ and benzodiazepine receptors in the central visual regions of rats does not need the presence of visual experience, at least until the age of 25 days. This does not compare with previous investigations on the effect of monocular deprivation on alpha- and beta-adrenergic (Aurich *et al.*, 1989), serotoninergic (Aurich *et al.*, 1985), and glutamatergic receptors (Schliebs *et al.*, 1986a), which demonstrated permanent changes in receptor sites particularly in the LGN.

Raising rats in complete darkness from birth until the age of 25 days resulted in decreased [^{3}H]flunitrazepam binding levels in the LGN and in the SC as compared to normally raised control rats, whereas in the retina and VC no effect of dark rearing was detected on benzodiazepine binding (Schiebs *et al.*, 1986b). In the cat VC, however, dark rearing from birth until PD30 resulted in elevated densities of GABA$_A$ and benzodiazepine receptors which were only of transient nature (Shaw *et al.*, 1987; Shaw and Cynader, 1988). In contrast, Mower *et al.* (1988) could not find any differences between normal and dark-reared cats in the number, affinity, or laminar distribution of GABA$_A$ receptors in cat VC. However, dark rearing of chicks resulted in decreased ^{3}H-GABA binding in the optic lobe which was only transient (Rios *et al.*, 1987). But when comparing different species it must be considered that there might be substantial differences in degree and manner in which the visual input can affect visual development.

Total light suppression from birth until adulthood resulted also in increased [^{3}H]serotonin binding in rat SC and VC (Aurich *et al.*, 1985), in contrast to the lack of effect on the development of alpha- and beta-adrenergic receptors in these regions (Schliebs *et al.*, 1982b; Aurich *et al.*, 1989). In the cholinergic system, total visual deprivation led to transient changes of muscarinic acetylcholine receptor binding in the rat SC and in the retina, which seems to reflect alterations in the time course of receptor development rather than permanent changes (Schliebs *et al.*, 1982a). Obviously, the development of various neurotransmitter receptors is differently affected by the complete lack of visual experience indicating that they are involved in different mechanisms of modulating visual information processing.

7.4 SUMMARY AND CONCLUSIONS

The temporal distinct appearance of GABA$_A$ and benzodiazepine receptors in the visual regions suggests that the ontogeny of the GABAergic transmission is controlled in a complex fashion and cannot be easily related to the functional maturation of the visual system. In nearly all visual regions, both receptor populations declined from a peak to a lower binding level in adulthood. The peaks in binding sites of both receptor populations are reached during a period of high cortical plasticity (critical period) and may suggest a role for GABA receptors in altering neuronal response properties. The descending phase of the developmental course might be the consequence of changes in receptor synthesis or elimination of a redundant number of synapses built up during the ascending phase, which is consistent with the hypothesis

of selective synaptic stabilization (Changeux and Danchin, 1976).

Visual deprivation differently affects the ontogeny of GABA receptors in the visual regions of various species reflecting species-dependent differences in which the visual input can influence the maturation of the visual GABAergic system.

Recently, in the visual system some GABA receptor types have been disclosed which showed unusual characteristics with respect to binding and physiology. But whether they will have special functions for visual information processing needs further investigation.

Molecular biological studies revealed a complex pentameric structure of the GABA/ benzodiazepine receptor and a high diversity in each class of subunits. Different combinations of the various GABA receptor subunits with differing sensitivity to GABA and benzodiazepines can exist, but no data on developmental characteristics of the various subunits are available at present. Therefore, one might speculate that the age-dependent changes in GABA/benzodiazepine receptor with respect to binding and physiological function might be due to changes in expression and/or suppression of certain combinations of GABA receptor subunits during postnatal development. *In situ* hybridization experiments performed during the period of early life should help to elucidate this question.

REFERENCES

Aldinio, C., Balzano, M. and Toffano, G. (1980) Ontogenetic development of GABA recognition sites in different brain areas. *Pharmacol. Res. Commun.*, **12**, 495–500.

Aldinio, C., Balzano, M., Savoini, G. *et al.* (1981) Ontogeny of ^{3}H-diazepam binding sites in different rat brain areas. Effect of GABA. *Dev. Neurosci.*, **4**, 461–6.

Aurich, M. and Bigl, V. (1988) A critical period of the development of β-adrenergic receptor binding in the visual system of rat during visual deprivation. *Int. J. Dev. Neurosci.*, **6**, 351–7.

Aurich, M., Schliebs, R. and Bigl, V. (1985) Serotoninergic receptors in the visual system of light-deprived rats. *Int. J. Dev. Neurosci.*, **3**, 285–90.

Aurich, M., Schliebs, R., Stewart, M.G. *et al.* (1989) Adaptive changes in the central noradrenergic system in monocular deprived rats. *Brain Res. Bull.*, **22**, 173–80.

Brandon, C., Lam, D.M.K. and Wu, J.Y. (1979) The γ-amino butyric acid system in the rabbit retina: localization by immunocytochemistry and autoradiography. *Proc. Natl. Acad. Sci. USA*, **76**, 3557–61.

Brooksbank, B.W.L., Atkinson, D.J. and Balasz, R. (1981) Biochemical development of human brain. II. Some parameters of the GABAergic system. *Dev. Neurosci.*, **4**, 188–200.

Burchfiel, J.L. and Duffy, F.H. (1981) Role of intracortical inhibition in deprivation amblyopia: reversal by microiontophoretic bicuculline. *Brain Res.*, **206**, 479–84.

Bylund, D.B. (1979) Regulation of central adrenergic receptors, in *Modulators, Mediators, and Specifiers in Brain Function, Advances in Experimental Medicine and Biology*, Vol. 16 (eds Y.H. Ehrlich, J. Volavka, L.G. Davis and E.G. Brunngraber), Plenum Press, New York, pp. 133–62.

Candy, J.M. and Martin, I.L. (1979) The postnatal development of benzodiazepine receptor in the cerebral cortex and cerebellum of the rat. *J. Neurochem.*, **32**, 655–8.

Changeux, J.P. and Danchin, A. (1976) Selective stabilization of developing synapses as a mechanism for the specification of neuronal networks. *Nature*, **264**, 705–12.

Costa, E. (1988) Polytypic signaling at GABAergic synapses. *Life Sci.*, **42**, 1407–17.

Coyle, J.T. and Enna, S.J. (1976) Neurochemical aspects of the ontogenesis of GABAergic neurons in the rat brain. *Brain Res.*, **111**, 119–33.

Coyle, J.T. and Yamamura, H.I. (1976) Neurochemical aspects of the ontogenesis of cholinergic neurons in the rat brain. *Brain Res.*, **118**, 429–40.

Cutting, G.R., Lu, L., O'Hara, B.F. *et al.* (1991) Cloning of the γ-aminobutyric acid (GABA) ρ_1 cDNA: a GABA receptor subunit highly expressed in the retina. *Proc. Natl. Acad. Sci. USA*, **88**, 2673–7.

deBlas, A.L., Vitorica, J. and Friedrich, P. (1988) Localization of the GABA$_A$ receptor in the rat brain with a monoclonal antibody to the 57,000 M_r peptide of the GABA$_A$ benzodiazepine receptor/Cl$^-$ channel complex. *J. Neurosci.*, **8**, 602–614.

DiChiara, G., Porceddu, M.L., Morelli, M. *et al.* (1979) Evidence for GABAergic projections from the substantia nigra to the ventromedial thalamus and to the superior colliculus of the rat. *Brain Res.*, **176**, 273–84.

Drago, F., Gagliano, C. and Cavaliere, S. (1989) Aging related changes of neurotransmitter in the visual system. *Metab. Pediatr. Syst. Ophthalmol.*, **12**, 21–3.

Fitzpatrick, D., Penny, G.R. and Schmechel, D.E. (1984) Glutamic acid decarboxylase-immunoreactive neurons and terminals in the lateral geniculate nucleus of the cat. *J. Neurosci.*, **4**, 1809–29.

Fosse, V.M., Heggelund, P. and Fonnum, F. (1989) Postnatal development of glutamatergic, GABAergic, and cholinergic neurotransmitter phenotypes in the visual cortex, lateral geniculate nucleus, pulvinar, and superior colliculus in cats. *J. Neurosci.*, **9**, 426–35.

Friedman, D.L. and Redburn, D.A. (1990) Evidence for functionally distinct subclasses of γ-aminobutyric acid receptors in rabbit retina. *J. Neurochem.*, **55**, 1189–99.

Gabbott, P.L.A., Somogyi, J., Stewart, M.G. and Hámori, J. (1986) GABA-immunoreactive neurons in the dorsal lateral geniculate nucleus of the rat: characterization by combined Golgi-impregnation and immunocytochemistry. *Exp. Brain Res.*, **61**, 311–22.

Giolli, R.A., Peterson, C.E., Ribak, C.E. *et al.* (1985) GABAergic neurons comprise a major cell type in rodent visual nuclei: an immunocytochemical study of pretectal and accessory optic nuclei. *Exp. Brain Res.*, **61**, 194–203.

Guarnieri, F., Corda, M.G., Concas, A. *et al.* (1982) Age-related changes of benzodiazepine and GABA binding sites in the rat retina. *Neurobiol. Aging*, **3**, 227–31.

Hendrickson, A.F., Ogren, M.P., Vaughn, J.E. *et al.* (1983) Light and electron microscopic immunocytochemical localization of glutamic acid decarboxylase in monkey geniculate complex: evidence for GABAergic neurons and synapses. *J. Neurosci.*, **3**, 1245–62.

Hendry, S.H.C. and Jones, E.G. (1988) Activity-dependent regulation of GABA expression in the visual cortex of adult monkeys. *Neuron*, **1**, 701–12.

Hendry, S.H.C., Fuchs, J., deBlas, A.L. and Jones, E.G. (1990) Distribution and plasticity of immunocytochemically localized GABA$_A$ receptors in adult monkey visual cortex. *J. Neurosci.*, **10**, 2438–50.

Houser, C.R., Lee, M. and Vaughn, J.E. (1983) Immunocytochemical localization of glutamic acid decarboxylase in normal and deafferented superior colliculus: evidence for reorganization of γ-aminobutyric acid synapses. *J. Neurosci.*, **3**, 2030–42.

Jones, E.G. (1990) The role of afferent activity in the maintenance of primate neocortical function. *J. Exp. Biol.*, **153**, 155–76.

Kanno, S. and Okada, Y. (1988) Laminar distribution of GABA (γ-aminobutyric acid) in the dorsal lateral geniculate nucleus, area 17 and area 18 of the visual cortex, and the superior colliculus of the cat. *Brain Res.*, **451**, 172–8.

Kvale, I., Fosse, V.M. and Fonnum, F. (1983) Development of neurotransmitter parameters in lateral geniculate body, superior colliculus and visual cortex of the albino rat. *Dev. Brain Res.*, **7**, 137–45.

Lin, C.S., Lu, S.M. and Schmechel, D.E. (1986) Glutamic acid decarboxylase and somatostatin immunoreactivities in rat visual cortex. *J. Comp. Neurol.*, **244**, 369–83.

Mallorga, P., Hamburg, M., Tallman, J.F. and Gallager, D.W. (1980) Ontogenetic changes in GABA modulation of brain benzodiazepine binding. *Neuropharmacology*, **19**, 405–8.

McDonald, J.K., Speciale, S.G. and Parnavelas, J.G. (1981) The development of glutamic acid decarboxylase in the visual cortex and the dorsal lateral geniculate nucleus of the rat. *Brain Res.*, **217**, 364–7.

Montero, V. and Singer, W. (1984) Ultrastructure and synaptic relations of neural elements containing glutamic acid decarboxylase (GAD) in the perigeniculate nucleus of the cat. *Exp. Brain Res.*, **56**, 115–25.

Mower, G.D. and Christen, W.G. (1989) Evidence for enhanced role of GABA inhibition in visual cortical dominance of cats reared with abnormal monocular experience. *Dev. Brain Res.*, **45**, 211–18.

Mower, G.D., Christen, W.G. and Caplan, C.J. (1983) Very brief visual experience eliminates plasticity in the cat visual cortex. *Science*, **221**, 178–80.

Mower, G.D., Christen, W.G., Burchfiel, J.L. and Duffy, F.H. (1984) Microiontophoretic bicuculline restores binocular responses to visual cortical neurons in strabismic cats. *Brain Res.*, **309**, 168–72.

Mower, G.D., White, W.F. and Rustad, R. (1986) ³H-muscimol binding of GABA receptors in the visual cortex of normal and monocularly deprived cats. *Brain Res.*, **380**, 253–60.

Mower, G.D., Rustad, R. and White, W.F. (1988) Quantitative comparisons of gamma-aminobutyric neurons and receptors in the visual cortex of normal and dark-reared cats. *J. Comp. Neurol.*, **272**, 293–302.

Murashima, Y.L., Ishikawa, T. and Kato, T. (1990) γ-Aminobutyric acid system in developing and degenerating mouse retina. *J. Neurochem.*, **54**, 893–8.

Nistri, A. and Silvilotti, L. (1985) An unusual effect of γ-aminobutyric acid on synaptic transmission on frog tectal neurones *in vitro*. *Br. J. Pharmacol.*, **85**, 917–22.

Ohara, P.T., Liebermann, A.R., Hunt, S.P. and Wu, J.Y. (1983) Neural elements containing glutamic acid decarboxylase (GAD) in the dorsal lateral geniculate nucleus of the rat: immunocytochemical studies by light and electron microscopy. *Neuroscience*, **8**, 189–211.

Okada, Y. (1974) Distribution of γ-amino butyric acid (GABA) in the layers of superior colliculus of the rabbit. *Brain Res.*, **75**, 362–365.

Palacios, J.M., Niehoff, D.L. and Kuhar, M.J. (1979) Ontogeny of GABA and benzodiazepine receptors: effect of Triton X-100, bromide and muscimol. *Brain Res.*, **179**, 390–5.

Patel, A.J., Smith, R.M., Kingsbury, A.E. *et al.* (1980) Effects of thyroid state on brain development: muscarinic acetylcholine and GABA receptors. *Brain Res.*, **198**, 389–402

Penny, G.R., Conley, M., Schmechel, D.E. and Diamond, I.T. (1984) The distribution of glutamic acid decarboxylase immunoreactivity in the diencephalon of the opposum and rabbit. *J. Comp. Neurol.*, **228**, 38–56.

Regan, J.W., Roeske, W.R. and Yamamura, H.I. (1980) The benzodiazepine receptor: its development and its modulation by γ-amino butyric acid. *J. Pharmacol. Exp. Ther.*, **212**, 137–43.

Ribak, C.E. (1978) Aspinous and sparsely-spinous stellate neurons in the visual cortex of rats contain glutamic acid decarboxylase. *J. Neurocytol.*, **7**, 461–78.

Rios, H., Flores, V. and Fiszer de Plazas, S. (1987) Effects of light- and dark-rearing on the postnatal development of GABA receptor sites in the chick optic lobe. *Int. J. Dev. Neurosci.*, **5**, 319–25.

Robins, J. and Ikeda, H. (1989) Benzodiazepines in the mammalian retina. I. Autoradiographic localisation of receptor sites and the lack of effect on the electroretinogram. *Brain Res.*, **479**, 313–22.

Rothblat, L.A., Schwarts, M.L. and Kasdan, P.M. (1978) Monocular deprivation in the rat: evidence for an age-related defect in visual behavior. *Brain Res.*, **158**, 456–60.

Rothe, T. and Schliebs, R. (1989) Laminar distribution of benzodiazepine receptors in visual cortex of adult rat. *Gen. Physiol. Biophys.*, **8**, 371–80.

Rothe, T., Schliebs, R. and Bigl, V. (1985) Benzodiazepine receptors in the visual structures of monocularly deprived rats. Effect of light and dark adaptation. *Brain Res.*, **329**, 143–50.

Schiller, P.H. (1986) The central visual system. *Vision Res.*, **26**, 1351–86.

Schliebs, R. and Rothe, T. (1988) Development of GABA$_A$ receptors in the central visual structures of rat brain. Effect of visual pattern deprivation. *Gen. Physiol. Biophys.*, **7**, 281–92.

Schliebs, R. and Stewart, M.G. (1991) Laminar postnatal development of muscarinic cholinergic receptors in rat visual cortex and the effect of monocular deprivation. *Neurochem. Int.*, **19**, 143–9.

Schliebs, R., Bigl, V. and Biesold, D. (1982a) Development of muscarinic cholinergic receptor binding in the visual system of monocularly deprived and dark reared rats. *Neurochem. Res.*, **7**, 1181–98.

Schliebs, R., Burgoyne, R.D. and Bigl, V. (1982b) The effect of visual deprivation on β-adrenergic receptors in the visual centres of the rat brain. *J. Neurochem.*, **38**, 1038–43.

Schliebs, R., Kullmann, E. and Bigl, V. (1986a) Development of glutamate binding sites in the visual centres of rat brain. Effect of light deprivation. *Biomed. Biochim. Acta*, **45**, 495–506.

Schliebs, R., Rothe, T. and Bigl, V. (1986b) Dark-rearing affects the development of benzodiazepine receptors in the central visual structures of rat brain. *Dev. Brain Res.*, **24**, 179–85.

Schwartz, J.C., Cortes, C.L., Rose, C. *et al.* (1983) Adaptive changes of neurotransmitter receptor mechanisms in the central nervous system, in *Molecular and Cellular Interactions Underlying Higher Brain Functions, Progress in Brain Research*, Vol. 58 (eds J.–P. Changeux *et al.*), Elsevier, Amsterdam, pp. 117–29.

Shaw, C. and Cynader, M. (1988) Unilateral eyelid suture increases GABA$_A$ receptors in cat visual cortex. *Dev. Brain Res.*, **40**, 148–53.

Shaw, C., Needler, M.C. and Cynader, M. (1984) Ontogenesis of muscimol binding sites in cat visual cortex. *Brain Res. Bull.*, **13**, 331–4.

Shaw, C., Needler, M.C., Wilkinson, M. *et al.* (1985) Modification of neurotransmitter receptor sensitivity in cat visual cortex during the critical period. *Dev. Brain Res.*, **22**, 67–73.

Shaw, C., Wilkinson, M., Cynader, M. *et al.* (1986) The laminar distributions and postnatal development of neurotransmitter and neuromodulator receptors in cat visual cortex. *Brain Res. Bull.*, **16**, 661–71.

Shaw, C., Aoki, C., Wilkinson, M. *et al.* (1987) Benzodiazepine ([^{3}H]flunitrazepam) binding in cat visual cortex: ontogenesis of normal characteristics and the effects of dark rearing. *Dev. Brain Res.*, **37**, 67–76.

Sillito, A.M., Kemp, J.A. and Blakemore, C. (1981) The role of GABAergic inhibition in the cortical effects of monocular deprivations. *Nature (Lond.)*, **291**, 318–20.

Silvilotti, L. and Nistri, A. (1991) GABA receptor mechanisms in the central nervous system. *Progr. Neurobiol.*, **36**, 35–92.

Singer, W. (1985) Central control of developmental plasticity in the mammalian visual cortex. *Vision Res.*, **25**, 389–96.

Skangiel-Kramska, J. and Kossut, M. (1984) Increase of GABA receptor binding activity after short lasting monocular deprivation in kittens. *Acta Neurobiol. Exp.*, **44**, 33–9.

Skerritt, J.H. and Johnston, G.A.R. (1982) Postnatal development of GABA binding sites and their endogenous inhibitors in rat brain. *Dev. Neurosci.*, **5**, 189–97.

Somogyi P. (1989) Synaptic organization of GABAergic neurons and GABA$_A$ receptors in the lateral geniculate nucleus and visual cortex, in *Neural Mechanisms of Visual Perception* (eds D.M.-K. Lam and C.D. Gilbert), Portfolio, Houston, pp. 35–62.

Somogyi, P., Freund, T.F., Wu, J.Y. and Smith, A.D. (1983) The section-Golgi impregnation procedure. Immunocytochemical demonstration of glutamic acid decarboxylase in Golgi impregnated neurons and in their afferent synaptic boutons in the visual cortex in the cat. *Neuroscience*, **10**, 261–94.

Stein, B.E. and Gallagher, H.L. (1981) Maturation of cortical control over superior colliculus cells in cat. *Brain Res.*, **223**, 429–35.

Stephenson, F.A. (1988) Understanding the GABA$_A$ receptor: a chemically gated ion channel. *Biochem. J.*, **249**, 21–32.

Sterling, P. and Davis, T.L. (1980) Neurons in the cat lateral geniculate nucleus that concentrate ^{3}H-γ-amino butyric acid (GABA). *J. Comp. Neurol.*, **192**, 737–49.

Toga, A.W. (1987) The metabolic consequences of visual deprivation in the rat. *Dev. Brain Res.*, **37**, 209–17.

Voaden, M.J., Morjaria, B. and Oraedu, A.C.I. (1980) The localization and metabolism of glutamate, aspartate and GABA in the retina. *Neurochem. Int.*, **1**, 151–65.

RECEPTORS FOR GLUTAMATE AND OTHER EXCITATORY AMINO ACIDS: A CAUSE FOR EXCITEMENT IN NERVOUS SYSTEM DEVELOPMENT

Rebecca M. Pruss

8.1 INTRODUCTION

Excitatory amino acid (EAA) receptors are the major mediators of excitatory transmission in the vertebrate nervous system. The pharmacology and physiology of EAA receptors and EAA associated pathology has received much attention since the discovery in the 1950s of the excitatory and convulsive effects of glutamate (Hayashi, 1954; Curtis *et al.*, 1959). The details of contributions to the understanding of EAA receptors and their roles in the developing and adult nervous system can be found in several excellent reviews (Foster and Fagg, 1984; Mayer and Westbrook, 1987; Monaghan *et al.*, 1989; Wroblewski and Danysz, 1989; Choi and Rothman, 1990; McDonald and Johnston, 1990) and in a series of articles recently compiled by Lodge and Collingridge (1991). The major endogenous activators of EAA receptors are the amino acids, L-glutamate and L-aspartate. However, other sulfur-containing derivatives and analogs of cysteine (cysteic acid, cysteine sulfinic acid, homocysteic acid and homocysteine sulfinic acid) as well as quinolinic acid (a tryptophan metabolite) are also endogenous EAA agonists. Glutamate is the most abundant EAA and fulfills the criteria of a true neurotransmitter: high concentrations of glu-

tamate are stored in vesicles found in nerve terminals; high affinity carriers for acidic amino acids exist on the plasma membranes of neurons and glia and selective high affinity uptake sites for glutamate are found on synaptic vesicles; the enzyme glutaminase which converts glutamine into glutamate is present in nerve terminal mitochondria; exocytotic release of glutamate is Ca^{2+} dependent. In addition, glutamate, although not necessarily the most potent agonist, binds to all EAA receptors. For these reasons EAA receptors are usually simply referred to as glutamate receptors. EAA receptors are found both on neurons and astrocytic glial cells (Barres *et al.*, 1990; Cornell-Bell *et al.*, 1990). EAAs stimulate a variety of responses in these cells through activation of a diverse collection of receptors. Although most EAA receptors are coupled to the activation of cation channels, other EAA receptors are G-protein-coupled receptors. The cation channel-coupled receptors have been termed 'ionotropic' receptors because of their ability to stimulate ion flux whereas the G-protein-coupled receptors have been called 'metabotropic' receptors because they stimulate increases in intracellular second messengers. In these respects, glutamate appears to be similar to the neurotransmitters

Receptors in the Developing Nervous System Vol. 2: Neurotransmitters. Edited by Ian S. Zagon and Patricia J. McLaughlin.
Published in 1993 by Chapman & Hall. ISBN 0 412 49400 0. Vols. 1 and 2 (set) ISBN 0 412 54520 9.

acetylcholine (ACh) and serotonin (5-HT) all of which activate a large number of heterogeneous receptors linked both to ion channels and G proteins. Although some of the excitatory roles of EAA receptors on neurons may be clear, the function of EAA receptors on glial cells is still a mystery.

Many of the structural proteins that form functional EAA-activated ion channels have now been identified by molecular cloning. The cDNA for a metabotropic EAA receptor that is coupled to the activation of phospholipase C has also been cloned. Since the number of EAA receptors appears to be quite large and the functional responses triggered by these receptors quite diverse, attempts have been made to identify selective ligands for the various members of the EAA receptor family. The integration of the pharmacology and the physiology with the structural information gained from the cloning and expression studies will eventually complete our understanding of the roles that EAA receptors play in nervous system development, synaptic plasticity, learning and memory and neurodegenerative processes.

8.2   PHARMACOLOGY

Nature has been generous, though cruel, in providing neurotoxins that are selective ligands for EAA receptor subtypes. These ligands are usually constrained analogs of glutamate that are competitive but more potent agonists of EAA receptors. Their extreme potency as EAA agonists is the reason behind their neurotoxicity since over-activation of EAA receptors leads to cell death by a process termed excitotoxicity. Neurotoxins that activate EAA receptors have been isolated from mushrooms (ibotenic acid), chick peas (β-*N*-oxalylamino-L-alanine), cycad plants (β-*N*-methylamino-L-alanine), seaweed (kainic acid), and plankton (domoic acid) as well as from organisms, such as shellfish, that ingest these food sources. By using the structural leads provided by naturally occurring agonists,

organic chemists have designed synthetic agonists and antagonists of EAA receptors that competed with glutamate for its binding site on the receptors. The structure–activity relationships of these and other compounds have been well described by Watkins *et al.* (1990). The availability of so many ligands and their apparent selectivity has helped in the characterization and classification of EAA receptors. Naturally occurring and endogenous inhibitors of EAA receptors also exist, including arthropod neurotoxins found in the venom of some types of wasps and spiders. These toxins paralyze or kill insects by blocking glutamate receptors that mediate excitatory neurotransmission at the insect neuromuscular junction (Jackson and Usherwood, 1988; Lodge and Johnson, 1990). The active components in these venoms include cysteine-rich polypeptides and acylated polyamines. Although these molecules have effects on vertebrate EAA receptors, pharmacological use of these toxins has been limited by the supply of purified toxins as well as by their complexity and the difficulty inherent in their synthesis. In addition to these neurotoxin antagonists, an endogenous metabolite of tryptophan, kynurenic acid, has been found to inhibit EAA receptors. The fact that kynurenic acid, an EAA receptor antagonist, and quinolinic acid, an EAA receptor agonist, are both metabolites of tryptophan, has led to the speculation that an imbalance in tryptophan metabolism may be involved in some chronic neurodegenerative diseases such as Alzheimer's, Parkinson's, Huntington's and amyotrophic lateral sclerosis. Numerous synthetic compounds like the dissociative anesthetic phencyclidine (PCP) also inhibit EAA receptor activation. Many analogs of these naturally occurring and synthetic receptor antagonists have been used to characterize EAA receptors and identify sites other than the glutamate binding site on EAA receptors. These pharmacological tools have been used to develop a simple pharmacological system for classifying EAA receptors. However, the variety of physiological responses elicited by EAA receptor agonists indicates a greater complexity than can be explained by ligand binding assays. A fuller understanding of the complexity and heterogeneity of EAA receptors is emerging now that some of the receptor proteins have been cloned.

8.2.1 AGONISTS USED TO CLASSIFY NMDA AND NON-NMDA RECEPTORS

Whereas glutamate activates all EAA receptors, a synthetic analog of aspartate, *N*-methyl-D-aspartate (NMDA), has been found to be a selective agonist for a subtype of ionotropic EAA receptor. The receptors that bind and can be activated by NMDA are called NMDA receptors. All other EAA receptors are therefore called 'non-NMDA' receptors. This nomenclature can be confusing since it is meant to refer only to a subtype of glutamate or EAA receptors, not to receptors for other hormones or neurotransmitters. Most naturally occurring neurotoxins appear to be selective agonists for non-NMDA receptors. Non-NMDA receptors are divided into two categories based on the apparent selectivity of two of these neurotoxins, kainate and quisqualate. It was subsequently discovered that quisqualate could activate both an ionotropic and a metabotropic EAA receptor. These two quisqualate receptor subtypes can be distinguished by two synthetic glutamate analogs. AMPA (α-amino-3-hydroxy-5-methyl-4-isoxazole propionic acid) selectively activates quisqualate-activated ion channels whereas [1S,3R]1-amino-1,3-cyclopentanedicarboxylic acid (*trans*-ACPD) selectively activates the metabotropic receptor (Schoepp *et al.*, 1990). The pharmacological classification system and how it evolved, together with the cloned receptor subunits (see later) that appear to correspond to these EAA receptors is shown in Fig. 8.1. In addition to these four receptor subtypes (NMDA, kainate, AMPA and *trans*-ACPD), a fifth type of EAA receptor can be identified using L-2-amino-4-phosphonobutanoic acid (AP4). The AP4 receptor appears to be an autoreceptor on nerve terminals whose function is to inhibit glutamate release. This classification system is given for historical perspective. Although AMPA appears to be a selective ligand for one type of non-NMDA receptor, these receptors also bind kainate (although with lower affinity) and quisqualate and kainate both bind to the same subset of ionotropic EAA receptors. These ligands are classified as receptor agonists because they activate ion currents or stimulate phospholipase C. However, the magnitude of the physiological responses seen with most of these agonists is variable and depends on the cell type or brain region being examined. This variability in response to agonists indicates the existence of receptor subtypes within these broad pharmacological categories.

8.2.2 ANTAGONISTS OF NMDA AND NON-NMDA RECEPTORS

Antagonists have been very valuable tools for studying the pharmacology and physiology of EAA receptors. Although no antagonists of AP4 receptors (the inhibitory autoreceptor) or metabotropic EAA receptors have yet been identified, many antagonists have been identified for ionotropic NMDA and non-NMDA receptors. Pharmacologically, NMDA receptors are the best characterized EAA receptors because of the large number of selective antagonists that affect both the NMDA or glutamate binding site on the receptor or allosteric modulatory sites that control ion channel opening. Competitive antagonists of NMDA receptors bind to the site for glutamate or NMDA but do not activate the ion channel. Although both NMDA and non-NMDA receptors bind glutamate and aspartate, many synthetic analogs of acidic amino acids are selective competitive antagonists of NMDA receptors. Such analogs include 2-amino-5-phosphonovaleric acid (AP5), 2-amino-7-phosphonoheptanoic acid (AP7) and

Pharmacology of EAA Receptors

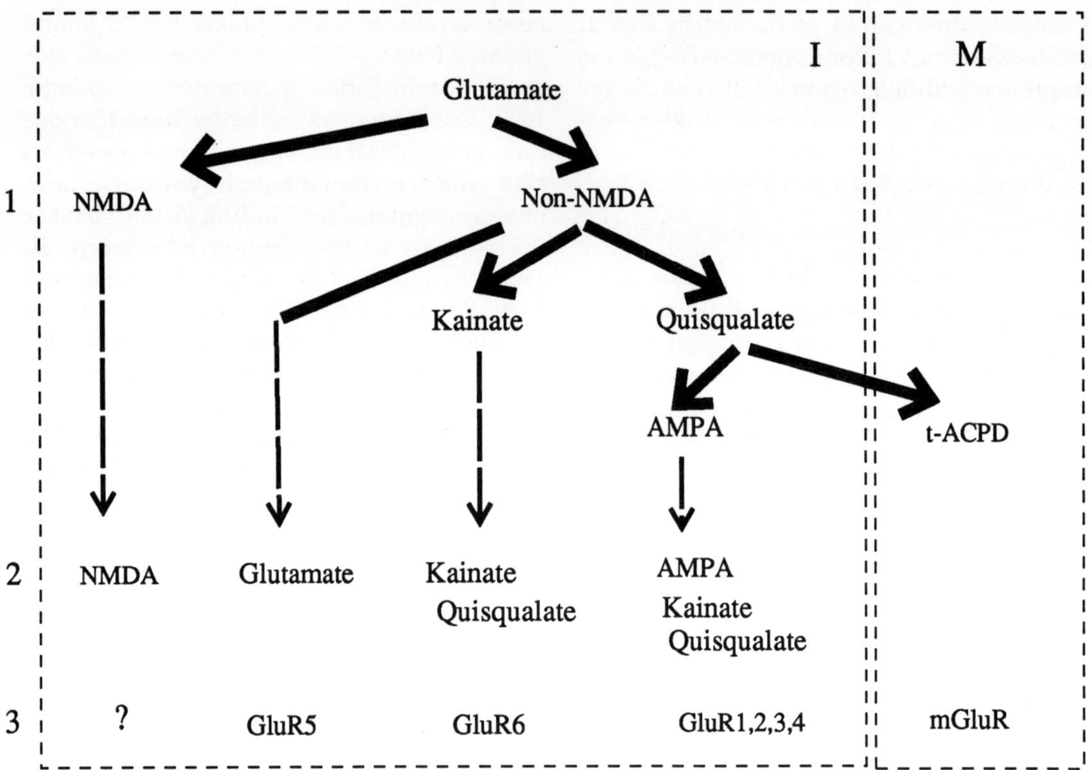

Fig. 8.1.   The classification of EAA receptors based on their pharmacology. High affinity ligand binding is indicated in the area of the diagram labeled 1. The ligands these receptor subtypes actually bind (other than glutamate) is indicated in area 2. The EAA receptor proteins that appear to correspond to these pharmacologically defined receptor subtypes is shown in area 3. Ligand-gated ion channels, or ionotropic receptors are grouped in the box labeled I. Metabotropic receptors, those coupled by G proteins to PI turnover are separated by the box labeled M.

3-(2-carboxy-piperazin-4-yl)propyl-1-phosphonic acid (CPP) (Watkins *et al.*, 1990).

The identification of non-competitive antagonists, those that do not displace glutamate or NMDA from the receptor, have been extremely important in the identification of allosteric, modulatory sites on NMDA receptors. In addition they provide a means of identifying NMDA receptor complexes independent of their EAA binding site. Dissociative anesthetics like PCP have been shown to block ion flux through NMDA receptors. Because the blockade of NMDA receptors by PCP is use-dependent it appears that PCP and its analogs bind to a site within the ion channel of the NMDA receptor complex. MK-801, a high affinity ligand for the PCP binding site on the NMDA receptor, originally synthesized by Merck for use as a neuroprotectant (Wong *et al.*, 1986), has been shown to be a useful tool for characterizing NMDA receptors and responses mediated by NMDA receptors. Divalent cations such as Mg^{2+}, Ni^{2+}, Co^{2+} and Mn^{2+} are also antagonists of NMDA-activated ion channels (Ascher and Nowak, 1988). Their inhibition of Ca^{2+} and Na^+ influx through

NMDA channels can be relieved by depolarization. Since the NMDA receptor is normally blocked by physiological concentrations of Mg^{2+}, activation of NMDA receptors displays voltage dependence, a feature that is unique for a ligand-gated ion channel. Another characteristic of NMDA receptors is their dependence on glycine. Whereas glycine activates receptors coupled to Cl^- channels and, together with γ-aminobutyric acid (GABA), plays an important role as an inhibitory neurotransmitter, the glycine receptor linked to Cl^- channels is blocked by strychnine. By contrast, strychnine has no effect on the ability of glycine to modulate NMDA receptors. Following the discovery that glycine binding is necessary for glutamate to activate NMDA ion channels (Johnson and Ascher, 1987), it was found that kynurenic acid and its more potent analogs antagonize NMDA receptors by displacing glycine from its site on the NMDA receptor complex (Kemp *et al.*, 1988). Although kynurenic acid is also a weak antagonist of non-NMDA receptors, it appears to act as a competitive inhibitor of these receptors since there is no role for glycine in the activation of non-NMDA receptors. Instead, kynurenic acid appears to be a competitive antagonist of non-NMDA receptors. The NMDA receptor is also different from other EAA receptors in its sensitivity to pH (Monaghan and Cotman, 1986; Traynelis and Cull-Candy, 1990) and voltage-independent blockade by Zn^{2+} (Westbrook and Mayer, 1987). In addition, polyamines positively modulate NMDA receptors but not other types of EAA receptors. Although NMDA receptors are generally considered and referred to as a single EAA receptor subtype, there are probably multiple forms of NMDA receptors. Receptor subtypes appear to account for the differential sensitivity of cells with NMDA receptors to the neurotoxic effect of quinolinic acid. In addition, NMDA receptor subtypes can be identified in different brain regions by their preference for agonists or antagonists in ligand binding and receptor autoradiography

studies (Monaghan *et al.*, 1988). A more detailed description of non-competitive NMDA receptor antagonists is given by Lodge and Johnson (1990).

Although NMDA receptors can be antagonized by a variety of different mechanisms, only a few antagonists have been found for non-NMDA receptors and most of these compete for the EAA binding site. Kynurenic acid, as discussed above, is a weak antagonist of non-NMDA as well as NMDA receptors. A class of compounds called quinoxalinediones has been shown to be selective competitive antagonists of kainate, quisqualate and AMPA binding to non-NMDA receptors (Honoré *et al.*, 1988; Sheardown *et al.*, 1990). Some, like 6-nitro,7-cyanoquinoxaline-2,3-dione (CNQX), are also weak inhibitors of NMDA receptors because of their ability to displace glycine from its binding site, others, like 6,7-dinitroquinoxaline-2,3-dione (DNQX) and 2,3-dihydroxy-6-nitro-7-sulfamoyl-benzo(F) quinoxaline (NBQX), have little affinity for the glycine site on NMDA receptors resulting in their increased selectivity for non-NMDA receptors (Watkins *et al.*, 1990). Their potency as non-NMDA receptor antagonists, measured by their ability to displace AMPA and kainate in binding assays, is in the range 10^{-7}–10^{-5} M. Although NBQX appears to have some preference for AMPA binding sites over high affinity kainate binding sites compared to CNQX or DNQX, all three compounds block kainate-, quisqualate-, and AMPA-activated EAA receptors.

8.3 MOLECULAR BIOLOGY

Several strategies have been used to identify and clone cDNAs for EAA receptor proteins. One approach has been to use selective ligands to aid in the purification of EAA receptors or at least their ligand-binding fragments and then use the amino acid sequence of these proteins to design oligonucleotide probes for screening cDNA libraries. By using this strategy and kainate or domoic acid as a ligand two groups have purified and then cloned cDNAs for high affinity kainate binding proteins from frog and chick brain (Gregor *et al.*, 1989; Wada *et al.*, 1989). Both the frog and chick cDNAs code for proteins that are ~450–500 amino acids long. These proteins contain several stretches of hydrophobic amino acids which are predicted to form four transmembrane domains. Although both proteins bind kainate with high affinity and resemble other ligand-gated ion channels, neither the frog nor chick kainate binding protein forms a functional ion channel when they are expressed on their own. Whether these proteins are components of a multimeric complex that is an EAA-activated ion channel remains to be determined.

A second strategy for cloning EAA receptor cDNAs has been to use *Xenopus* oocytes to express mRNA derived from cDNA libraries and assay for the presence of ligand-activated current responses in the oocytes. After successive division of the cDNA library, a single cDNA is isolated that is capable of generating mRNA for a functional receptor. The first cDNA for a protein that forms kainate-activated ion channels and is a member of a large family of ionotropic EAA receptor subunits was cloned from a rat brain cDNA library using this strategy (Hollmann *et al.*, 1989). Oocyte expression was also used successfully to identify and clone cDNAs for a metabotropic EAA receptor (Houamed *et al.*, 1991; Masu *et al.*, 1991).

8.3.1 CLONING OF cDNA FOR SUBUNITS OF IONOTROPIC EAA RECEPTORS

The first subunit for an ionotropic EAA receptor was originally called GluR-K1 because of the large currents elicited in response to kainate compared to other agonists. Using the cDNA sequences corresponding to the protein's transmembrane domains (assumed to be the most highly conserved regions of the

protein) as probes or primers, related proteins have now been identified using low stringency hybridization techniques or the polymerase chain reaction (PCR). Four cDNAs for closely related proteins were isolated: GluR1 (the former GluR-K1), GluR2, GluR3 and GluR4 (also called GluR-A, GluR-B, GluR-C, and GluR-D) (Boulter *et al.*, 1990; Keinänen *et al.*, 1990; Nakanishi *et al.*, 1990). All four of these proteins can be expressed in two different versions, termed flip and flop, depending on the use of alternate exons during mRNA splicing (Sommer *et al.*, 1990). These eight proteins appear to be members of a related family of EAA receptor subunits. All members of this family bind glutamate, kainate, quisqualate and AMPA. Because they bind AMPA with the highest affinity, these receptors have sometimes been referred to as AMPA receptors. Although these subunits have only low affinity for kainate compared to AMPA, kainate is a full agonist for these receptors whereas glutamate, quisqualate and AMPA are variable in their ability to activate the ion channel formed by these subunits.

Additional families of ionotropic EAA receptor subunits have been identified from rat brain cDNA libraries by low stringency hybridization techniques. One receptor subunit called GluR5 was identified based on its 40% sequence identity to GluR4 (GluR1–4 share ~70% sequence identity) (Bettler *et al.*, 1990). Unlike other receptor subunits, GluR5 could only be activated weakly by glutamate and not at all by kainate, quisqualate or AMPA. GluR5 was also found to exist in two forms depending on the presence or absence of a small insert. Recently two additional EAA receptor subunits have been identified that bind glutamate, kainate and quisqualate but not AMPA. One of these subunits, GluR6, was identified by low stringency hybridization using GluR5 as a probe (Egebjerg *et al.*, 1991) whereas the other, KA-1, was identified by PCR using primers based on the sequence of the frog and chick kainate binding proteins (Werner *et al.*, 1991). Both GluR6 and KA-1 have higher affinities for kainate than the other subunits. These receptor subunits may account for some of the high affinity kainate binding sites found in rat brain. When GluR6 was expressed in oocytes, kainate had an EC_{50} of 1 µM for ion current activation compared to an EC_{50} of 35 µM for activation of ion currents when GluR1 was expressed. In ligand binding assays, kainate had a K_d of 5 nM using membranes from mammalian cells expressing the KA-1 subunit. These results are consistent with the receptor binding and autoradiography which indicate that AMPA binding sites and high affinity kainate binding sites are on different proteins and expressed by different cells. However, the results from molecular cloning and expression studies indicate that many, if not all, AMPA binding sites are also low affinity kainate receptors. It remains to be seen whether receptors will be identified that bind AMPA but not kainate. Despite the fact that NMDA receptors are best characterized pharmacologically, none of the EAA receptor subunits that have been identified by molecular cloning have been shown to bind or be activated by NMDA. The presumed association between cloned EAA receptor subunits and pharmacologically characterized EAA receptors is shown in Fig. 8.1.

8.3.2 STRUCTURAL FEATURES OF IONOTROPIC EAA RECEPTOR SUBUNITS

Comparison of the subunits of EAA-activated ion channels with other ligand-gated ion channels such as nicotinic ACh receptors, $GABA_A$ receptors and strychnine-sensitive glycine receptors revealed some structural similarities. The *N*-terminal half of all ligand-gated ion channel proteins, including EAA receptors, is predicted to exist as a large extracellular domain that probably contains the glutamate binding site. The C-terminal half of the molecule contains stretches of hydro-phobic amino acids that are predicted to form four transmembrane alpha helices followed by a short extracellular tail. The most striking difference between EAA receptors and other types of ligand-gated ion channel proteins is their size. EAA receptor subunits are about twice as large as ACh, $GABA_A$ or glycine receptor subunits. EAA receptor proteins are ~900 amino acids long compared to ~450 amino acids for the other types of receptors. Interestingly, the *C*-terminal half of these receptors appears to be similar to the chick and frog kainate binding proteins discussed earlier and the predicted transmembrane domains of all these proteins are homologous. A comparison of the structural features of EAA receptors with other comparable proteins (other ligand-gated ion channels in the case of the ionotropic EAA receptor) is shown in Fig. 8.2.

From studies using radiolabeled GluR probes to map EAA receptor subunit expression by *in situ* hybridization it appears that individual cells may express several different EAA receptor subunits. Based on the structure and stoichiometry of nicotinic ACh receptors, EAA receptors are predicted to be hetero-multimeric complexes of four or five subunits. Therefore, the variations in response to different ligands seen in cells from different parts of the nervous system probably reflect the variation in EAA receptor assembly in different cells. Although it is possible to correlate radioligand binding studies with different GluR subunits, an understanding of the variation in agonist-activated responses will depend on a better understanding of how these receptor subunits combine to form functional glutamate-activated ion channels.

8.3.3 METABOTROPIC EAA RECEPTORS

The cDNA for a metabotropic EAA receptor was cloned independently by two groups using similar oocyte expression strategies (Houamed *et al.*, 1991; Masu *et al.*, 1991). The receptor protein predicted from the cDNA

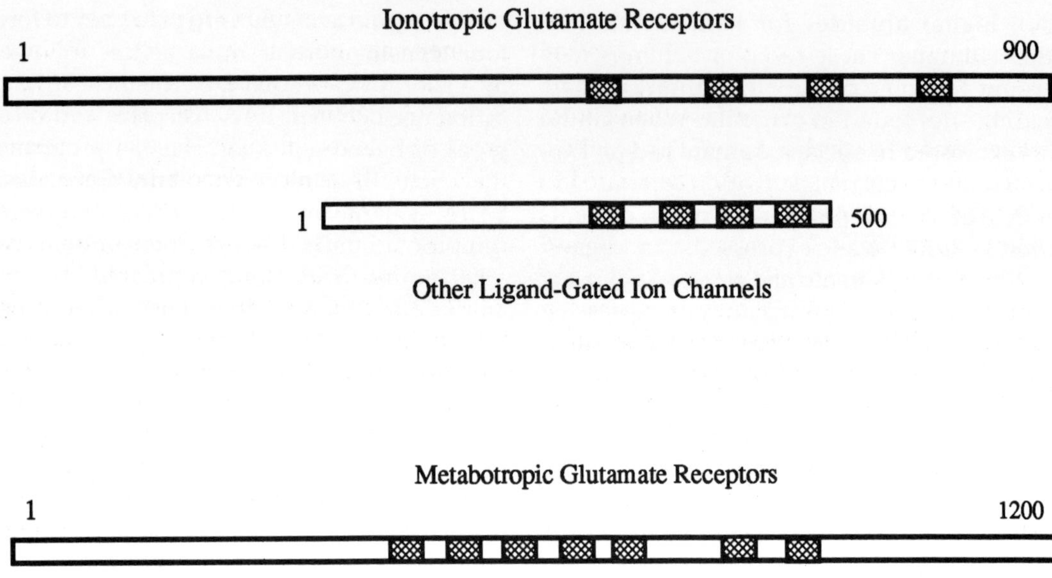

Fig. 8.2. Illustration of the size and structural organization of ionotropic EAA receptor proteins compared to other ligand-gated ion channels (upper) and a metabotropic EAA receptor protein compared to other G-protein-coupled receptors. Hatched boxes indicate regions of hydrophobic amino acids that are predicted to form transmembrane domains. Numbers refer to the approximate length of each protein in amino acid residues.

sequences has seven putative transmembrane domains typical of other members of the family of G-protein-coupled receptors. However, like the ionotropic EAA receptors, the metabotropic EAA receptor is about twice as large as other comparable proteins. Based on the cDNA sequence, the metabotropic EAA receptor is predicted to be ~1200 amino acids long compared to the usual length of ~400–500 amino acids found for other G-protein-coupled receptors. The increased size of the metabotropic receptor is due to long *N*- and *C*-terminal hydrophilic domains that flank the central hydrophobic region. Sequence comparisons show some homology between the extracellular *N*-terminal domains of the metabotropic and ionotropic EAA receptors and these regions have been postulated to be involved in

ligand binding. If this is true, the metabotropic EAA receptor is different from other G-protein-coupled receptors whose ligand binding pocket is predicted to reside within the hydrophobic transmembrane domains. An illustration of the structural characteristics of metabotropic EAA receptors and other G-protein-coupled receptors is shown in Fig. 8.2.

The metabotropic EAA receptor is coupled to phospholipase C activation by a pertussis toxin-sensitive G protein. Activation of phospholipase C results in the hydrolysis of phosphatidylinositol 4,5-bisphosphate (PI) and the production of two second messengers, inositol 1,4,5-trisphosphate (IP3) and 1,2-diacylglycerol (DAG). IP3 stimulates the release of Ca^{2+} from intracellular storage vesicles associated with the endoplasmic reticulum, sometimes referred to as calciosomes. Activation of this response in oocytes expressing the cloned metabotropic EAA receptor occurs in response to micromolar concentrations of glutamate, quisqualate, ibotenate and *trans*-ACPD whereas millimolar concentrations of aspartate, kainate, and AMPA also elicited weak responses.

Many hormones and neurotransmitters that activate G-protein-coupled receptors have been found to bind to a family of receptors coupled to distinct second messenger pathways. For instance, 5-HT and ACh activate several different G-protein-coupled receptors that can either stimulate phospholipase C or inhibit adenylate cyclase. It is probably only a matter of time before additional G-protein-coupled EAA receptors are identified. It will be interesting to discover whether these receptors are coupled to second messengers other than IP3, DAG and the release of intracellular Ca^{2+}.

The identification of a selective agonist and cDNA that can be used as a probe for these proteins provides two valuable tools for studying the expression of metabotropic EAA receptors in the nervous system. Simply measuring PI hydrolysis does not provide rigorous proof of the existence or involvement of a metabotropic EAA receptor. In neurons, increases in intracellular Ca^{2+} due to glutamate-stimulated Ca^{2+} influx can stimulate phospholipase C, a Ca^{2+}-dependent enzyme. This has led to the conclusion by one group that EAA receptors with the pharmacology of NMDA receptors are capable of stimulating PI hydrolysis (Wroblewski *et al.*, 1987). To rule out this possibility, it is, therefore, necessary to measure IP3 formation or increases in intracellular Ca^{2+} in the absence of external Ca^{2+}.

8.4 PHYSIOLOGY

8.4.1 EAA RECEPTORS INCREASE INTRACELLULAR CALCIUM

Regardless of their type, EAA receptors are responsible for increasing intracellular Ca^{2+}. Ionotropic receptors accomplish this by activating plasma membrane ion channels. These channels may be permeable to Ca^{2+} or Na^+ or both Ca^{2+} and Na^+. Receptors linked to Ca^{2+}-permeable channels allow the direct entry of Ca^{2+} whereas receptors linked to Na^+-permeable channels result in depolarization and indirectly stimulate Ca^{2+} influx through voltage-operated Ca^{2+} channels. NMDA receptors have been shown to be highly permeable to Ca^{2+} (MacDermott *et al.*, 1986). Non-NMDA receptors on the other hand display a range of cation permeabilities dependent, at least in part, on the ligand that is used to activate the receptor. Although many electrophysiological studies of non-NMDA receptors failed to demonstrate that these receptors are permeable to Ca^{2+} or other divalent cations, more recent studies using cultured hippocampal or retinal bipolar neurons (Iino *et al.*, 1990; Gilbertson *et al.*, 1991) or oocytes expressing cloned GluRs (Hollmann *et al.*, 1991) have found that non-NMDA receptor-activated ion channels are permeable to Ca^{2+}. Because of the limited number of cell types and GluRs examined it is not clear from electrophysiological studies which cell types or what brain regions will express non-NMDA receptors that can directly gate Ca^{2+}.

To get a better picture of the types of cells that express non-NMDA receptors that are permeable to Ca^{2+}, it is possible to use Co^{2+} uptake as a marker of divalent cation-permeability. Co^{2+}, after precipitation with $(NH_4)_2S$, can be detected as a brown–black deposit in the cell bodies and processes of neurons and astrocytic glial cells that express divalent cation permeable non-NMDA receptors (Pruss *et al.*, 1991). Co^{2+} uptake appears to be a reliable indicator of Ca^{2+} influx although Co^{2+} appears to be somewhat less permeable than Ca^{2+}. Co^{2+} uptake can be used to study the functional properties of non-NMDA receptors on cells in different parts of the nervous system. Co^{2+} permeability appears to require complete activation or opening of the non-NMDA ion channel in response to an agonist. Kainate is able to activate Co^{2+} influx through

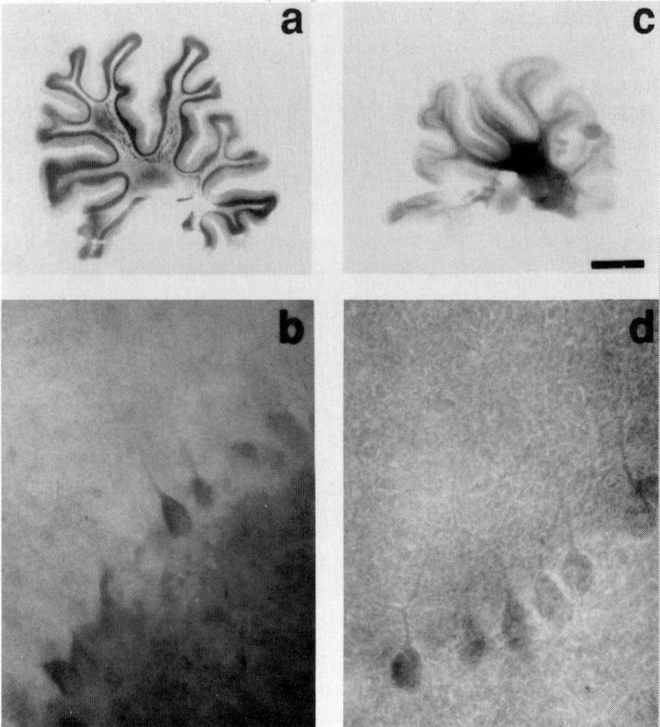

Fig. 8.3.   Different patterns of Co^{2+} uptake are found in adult rat cerebellum after stimulation with kainate (a and b) or glutamate (c and d). Glutamate receptor agonists were added to fresh tissue slices in buffer containing 5 mM $CoCl_2$. After washing and precipitation with $(NH_4)_2S$, the CoS appears as a dark brown or black deposit in cells that have divalent cation-permeable ion channels coupled to non-NMDA receptors. Quisqualate produces a pattern of labeling identical to that seen with glutamate. Kainate causes Co^{2+} uptake into all cells expressing non-NMDA receptors linked to ion channels whereas glutamate activates only a subpopulation of cells. The Co^{2+} uptake in Purkinje neurons is very clear in both kainate and glutamate-treated slices. Co^{2+} is seen in the granule cell layer only after kainate stimulation, but not in response to glutamate. These labeling patterns indicate that granule neurons express different non-NMDA-activated ion channels than Purkinje neurons. Bar is 1.5 mm in (a) and (c) and 20 μm in (b) and (d). For details see Pruss *et al.* (1991).

all non-NMDA receptors made up of subunits that have been shown to bind kainate. Glutamate and quisqualate, which appear to bind to all kainate receptors, and AMPA, which binds to a subset of these receptors, vary in their ability to activate Co^{2+} influx. The pattern of Co^{2+} uptake stimulated by these agonists in adult rat cerebellum and hippocampus is shown in Figs. 8.3 and 8.4. Non-NMDA receptors appear to be able to stimulate divalent cation influx even early in the development of the nervous system since the Co^{2+} uptake pattern in the hippocampus of 5-day-old rats is the same as that seen in the adult (Fig. 8.5). Three kainate receptor subtypes (K1, K2 and K3) can be identified based on the pattern of Co^{2+} uptake. K1 is present on cerebellar granule neurons and is activated by kainate but no other EAA receptor agonist. K2 is activated by kainate and glutamate but not by quisqualate or AMPA and generates a response like that seen in the hippocampal dentate gyrus. K3 is found on Purkinje neurons in the cerebellum and all

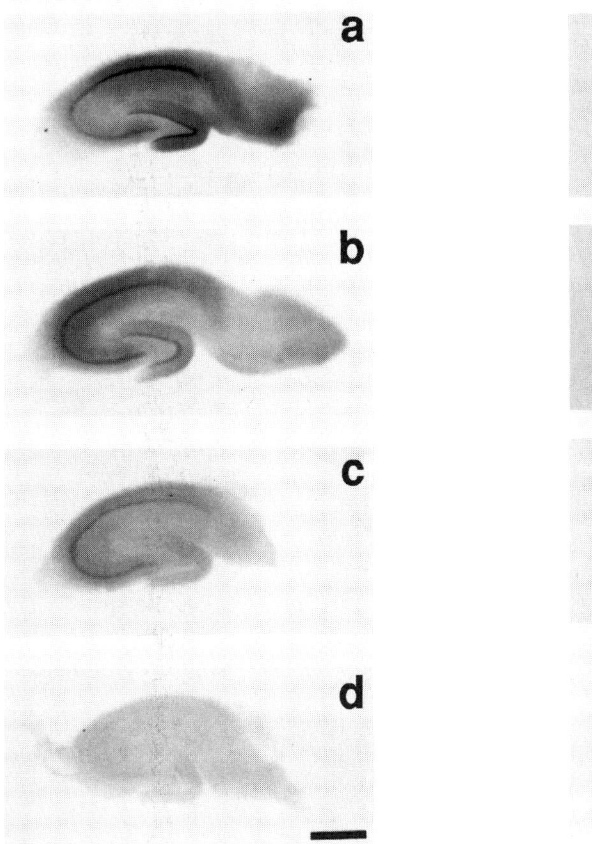

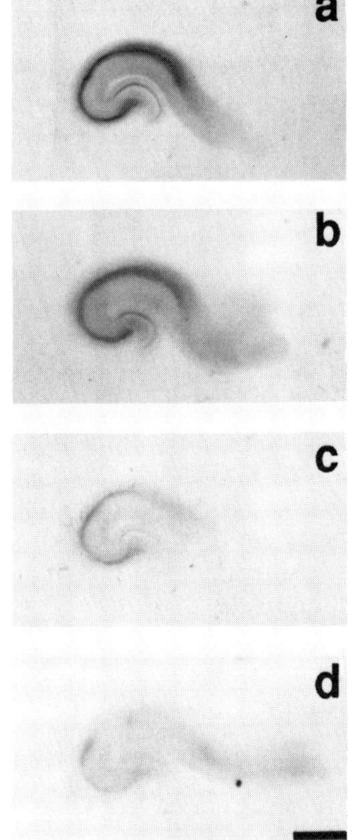

Fig. 8.4.   Uptake of Co^{2+} into slices of adult rat hippocampus. Kainate (a), glutamate (b), and quisqualate (c) were added to fresh tissue slices The pyramidal cell layer is labeled by all three agonists whereas the dentate gyrus does not take up Co^{2+} in response to quisqualate indicating the presence of different non-NMDA receptor-activated ion channels on pyramidal neurons and dentate granule neurons. A control slice is shown in (d); bar is 1 mm. For details see Pruss *et al.* (1991).

Fig. 8.5.   Stimulation of Co^{2+} uptake by non-NMDA receptor agonists in 5-day-old rat hippocampus. Kainate (a), glutamate (b), and AMPA (c) were used to stimulate fresh tissue slices as in Fig. 8.3. A control slice is shown in (d). The same pattern of agonist-activated Co^{2+} uptake is seen in developing hippocampus as in the adult (AMPA stimulates the same pattern of cobalt uptake as quisqualate) indicating full maturation of divalent cation permeable channels linked to non-NMDA receptors even at this age. Bar is 1 mm.

hippocampal pyramidal neurons and is activated by kainate, glutamate, quisqualate, and AMPA. Since Co^{2+} uptake is a reliable indicator of Ca^{2+} permeability, the results of these studies indicate that most if not all non-NMDA receptors are capable of allowing Ca^{2+} to enter cells directly when activated by the appropriate agonist. Glutamate, the endoge-nous agonist, may only be able to stimulate divalent cation influx through a subpopulation of non-NMDA receptors. Non-NMDA receptors are mainly recognized for their permeability to monovalent cations. Activation of non-NMDA receptors can indirectly stimulate Ca^{2+} influx due to depolarization caused by Na^+ influx. Depolarization causes Ca^{2+}

entry by stimulating the opening of voltage operated Ca^{2+} channels and by relieving the NMDA receptors from blockade by Mg^{2+}.

The metabotropic EAA receptor also causes increases in intracellular Ca^{2+}. Like other receptors that stimulate IP3 formation, the metabotropic EAA receptor causes an increase in intracellular Ca^{2+} by releasing Ca^{2+} from intracellular stores. In many cells, the formation of IP3 and the release of intracellular Ca^{2+} stores is invariably linked to the activation of Ca^{2+} influx through a non-selective cation channel in the plasma membrane. The exact mechanism involved in the activation of this plasma membrane channel is not completely understood but may be due to inhibition of a plasma membrane K^+ channel that results in membrane depolarization (Charpak *et al.*, 1990). Alternatively IP3 or a subsequent metabolite of IP3 such as inositol 1,3,4,5-tetrakisphosphate may activate a plasma membrane Ca^{2+} channel (Berridge and Irvine, 1989). Although activation of metabotropic EAA receptors has been shown to decrease Ca^{2+} influx through voltage-operated Ca^{2+} channels, this effect may be mediated by DAG and Ca^{2+} activation of protein kinase C. Activation of metabotropic EAA receptors has been shown to increase the firing rate of neurons. This effect may also be due to inhibition of K^+ channels, as discussed earlier, or it may be an indicator of agonist-activated Ca^{2+} entry which acts to facilitate neuron firing.

8.4.2   CONSEQUENCES OF Ca^{2+} INCREASE

The main function of Ca^{2+} inside cells is the regulation of enzyme activity which is accomplished by Ca^{2+} binding directly to a regulatory domain on an enzyme or by binding to calmodulin (CaM), a calcium-binding protein that can associate with and regulate the activity of a variety of intracellular enzymes (Rasmussen and Means, 1989; Heizmann and Hunziker, 1991). Ca^{2+}-activated enzymes include proteases like calpain, lipases such as phospholipase C and phospholipase A_2, and nucleases. These enzymes participate in catabolic processes but also cause the release of active products such as IP3 or arachidonic acid. Ca^{2+} regulates the level of another second messenger, cyclic AMP, by affecting the activity of adenylate cyclase and cyclic nucleotide phosphodiesterase. Other Ca^{2+}-activated enzymes include nitric oxide (NO) synthase, which forms nitric oxide from arginine. Low concentrations of NO cause vasorelaxation through the activation of guanylate cyclase in vascular smooth muscle cells. However, at high concentrations NO may be neurotoxic due to its ability to inhibit mitochondrial respiration or contribute to the production of reactive free radicals (Dawson *et al.*, 1991). Ca^{2+} also activates protein kinase C and CaM kinases as well as a CaM-dependent phosphatase. These enzymes have important regulatory roles because of their ability to make reversible modifications to proteins. In addition to calmodulin and Ca^{2+}-activated enzymes, Ca^{2+} binds to a large number of proteins whose function may be to sequester intracellular Ca^{2+}. Other Ca^{2+}-binding proteins probably play roles in Ca^{2+}-activated processes, such as cell motility, through their effects on cytoskeleton assembly, and cell and membrane adhesion via annexins and cadherins. Ca^{2+}-binding proteins are also necessary for the fusion of secretory vesicle membranes with the plasma membrane during exocytosis.

Increases in intracellular Ca^{2+} play important roles in the regulation of gene transcription. Increased transcription of mRNA for cellular oncogenes like c-fos and c-jun, and a number of neuropeptides such as enkephalin, vasoactive intestinal peptide, and galanin (Pruss *et al.*, 1985; Pruss and Stauderman, 1988; Rokaeus *et al.*, 1990) occurs in response to increases in intracellular Ca^{2+}. At least two Ca^{2+}-activated protein kinases appear to be able to mediate these transcriptional effects of Ca^{2+}. Protein kinase C activation results in increased formation of c-fos and c-jun heterodimers that act as transcription factors by binding to specific DNA sequences called AP-

1 sites or the TRE (TPA response element) in the promotor regions of genes whose transcription is increased by TPA (12-*O*-tetradecanoylphorbol-13-acetate). CaM kinases have also been shown to mediate Ca^{2+}-activated gene transcription. CaM kinases phosphorylate CREB, a protein that binds to the cyclic AMP response element (CRE). The CRE is a specific sequence in the promoter region of genes that confers inducibility by increased levels of cyclic AMP. Phosphorylation of CREB by either CaM kinases or protein kinase A increases the affinity of CREB for the CRE resulting in a convergence of Ca^{2+} and cAMP signalling pathways (Sheng *et al.*, 1991). The effects of Ca^{2+} on gene transcription are thought to underlie the long-term effects of EAA receptor activation on cell survival and differentiation.

In addition to long-term effects of Ca^{2+} which are dependent on gene transcription, Ca^{2+} triggers acute responses like exocytosis. Trophic or other responses due to EAA receptor activation may be indirectly mediated by the release of growth factors, hormones, neurotransmitters, ATP, ascorbic acid or other factors that influence cell survival, differentiation and function. Whereas the short-term effects of Ca^{2+} on processes like secretion may be triggered by transient elevations in intracellular Ca^{2+}, long-term changes such as increased transcription of genes that code for differentiation-specific products such as neuropeptides appear to require sustained periods of elevated Ca^{2+} (Pruss and Stauderman, 1988). EAA receptor activation may be the means of providing this signal to cells during nervous system development.

In addition to these specific acute and long-term biochemical effects, Ca^{2+} levels play a role in neuronal survival. Although moderate levels of intracellular Ca^{2+} (100–300 nM) appear to enhance neuronal survival, neurotoxicity is associated with high concentrations of intracellular Ca^{2+} (≥ 1 μM) which are sustained for long periods (more than several minutes). This association between excess Ca^{2+} and neurotoxicity has led to the hypothesis that overstimulation of EAA receptors triggers neuronal cell death. Excess release of glutamate can occur as a result of hypoglycemia, hypoxia, uncontrolled seizures, or trauma to the nervous system. Much of the glutamate released during these episodes appears to be due to reversal of the EAA uptake carrier although some is due to Ca^{2+}-dependent release from neurons triggered by EAA receptor activation and/or depolarization (Nicholls and Attwell, 1990). The uncontrolled release of glutamate causes sustained activation of EAA receptors that results in excess Ca^{2+} influx culminating in cell death. This process, termed excitotoxicity, can be provoked by activation of both NMDA and non-NMDA receptors (Meldrum and Garthwaite, 1990). The role of NMDA receptors has received the most attention since these receptors are highly permeable to Ca^{2+} (MacDermott *et al.*, 1986). In models of ischemia and stroke NMDA receptor antagonists have been shown to attenuate neurodegeneration. However, not all cells that are susceptible to excitotoxicity have NMDA receptors. The cerebellar Purkinje cell is one example. Furthermore, quinoxalinediones that are selective antagonists of non-NMDA receptors have been shown to protect against neurodegeneration even when administered up to three hours after an ischemic event. There appear to be at least two forms of neurodegeneration that occur as a result of stroke, ischemia or neurotrauma. First, there is a loss of cells immediately surrounding the site of injury that can be reversed by NMDA receptor antagonists but only within minutes of the injury. Secondarily, there is a loss of cells radiating from the site of injury which is delayed relative to the time of the insult and which can be attenuated by non-NMDA receptor antagonists even hours after the injury. In addition to Ca^{2+}, non-NMDA receptor-activated ion channels are permeable to Na^+. Influx of Na^+, whether alone or in combination with Ca^{2+}, causes cells to swell because of concomitant influx of Cl^- and H_2O. This leads to morphological changes that are different from those seen with Ca^{2+} influx alone. Na^+ influx results in swelling and necrotic cell death whereas Ca^{2+} influx causes the collapse and condensation of the cytoskeleton and other structures inside the cells by a process that, in some cell types, has been called apoptosis. Glutamate and other endogenous EAAs may be able to activate Ca^{2+} entry through NMDA receptors and certain non-NMDA receptor subtypes. Although NMDA receptors are permeable to Ca^{2+}, the passage of Ca^{2+} through the NMDA ion channel may in fact be reduced following ischemia or trauma since the ion channel is blocked by physiological concentrations of Mg^{2+} and acidic pH.

8.5 THE ROLE OF EAA RECEPTORS DURING NERVOUS SYSTEM DEVELOPMENT

8.5.1 EAA RECEPTORS IN THE DEVELOPING NERVOUS SYSTEM

Although most receptor binding and biochemical studies have been done using tissues from rats, other mammalian and non-mammalian species have also been used to study EAA receptors and their roles in the development of the nervous system. Cats and monkeys have been used to study some of the physiological and behavioral effects of EAA receptors. Many cell culture systems and tissue slice preparations come from the mouse as well as rat nervous system, and some studies have been performed measuring EAA receptors in the developing human nervous system. In addition to these mammalian species, EAA receptors and their contribution to nervous system development have been studied using tadpoles and chick embryos and some behavioral studies have been performed using pigeons. In addition to vertebrates, EAA receptors are present, and are important mediators of excitatory neurotransmission in insects. The relative timing of EAA expression and their involvement in vertebrate nervous system development appears to be similar in all species examined.

Many different methods have been used to measure EAA receptors and their roles in nervous system development. In addition to direct quantitation of EAA receptors using radioligand binding and receptor autoradiography, the physiological response to EAA agonists and antagonists has been measured in whole animals and *in vitro* using tissue slices or cell culture. These physiological approaches include measurements of the effects of EAA receptor agonists and antagonists on electrophysiological responses, neurotransmitter release, second messenger production (IP3 and arachidonic acid), uptake or release of $^{45}Ca^{2+}$, changes in intracellular Ca^{2+} using fluorescent dyes, the response of *Xenopus* oocytes following injection of brain mRNA, transcription of mRNA for c-fos of various neuropeptides, the behavior of animals, ability to cause seizures, changes in synaptic responses, neuronal survival both *in vivo* and *in vitro*, and neurite outgrowth using cultured cells. Recently, with the cloning of cDNAs for EAA receptor subunits, the localization and timing of EAA receptor expression in the developing nervous system has been studied by *in situ* hybridization.

EAA receptors appear to be very important in mediating activity-dependent changes in neuronal survival and synaptic plasticity. The role of EAA receptors, particularly NMDA receptors, has been studied in two models of synaptic plasticity: (1) the development of ocular dominance columns in the visual cortex and (2) the phenomenon of long-term potentiation (LTP) in the hippocampus (Collingridge and Singer, 1990). The development of EAA receptors had been correlated with a critical period in the development of the visual system: the time when neuronal activity plays a role in the establishment of ocular dominance columns in the visual cortex and in the determination of retinotectal connections. Although a peak in NMDA receptors is measured in binding assays during this critical period, this peak may be merely correlative rather than causative since NMDA receptor stimulation alone does not induce the development of ocular dominance columns. Furthermore, the critical period can be delayed or prolonged by dark rearing without affecting the peak in the number of NMDA receptors. Even though NMDA receptor activation may not be sufficient to cause the changes in synaptic connections, NMDA receptors appear to be necessary for the changes produced by neuronal activity to occur. During the critical period NMDA antagonists prevent both the disconnection of deprived visual pathways as well as the recovery of stable synapses once the deprived pathway is reactivated.

Another example of synaptic plasticity that involves EAA receptors is LTP, a stimulus-evoked enhancement in synaptic activity that is believed to be related to learning and memory (Nicoll *et al.*, 1988). In the hippocampus and other brain regions where LTP occurs, brief but high frequency stimulation of excitatory inputs leads to an increase in postsynaptic responsiveness that persists for weeks *in vivo*. Simultaneous depolarization accompanied by NMDA receptor activation and increases in intracellular Ca^{2+} in the postsynaptic cell are necessary to induce most (but not all) forms of LTP. The ability to induce LTP in the CA1 region of the hippocampus appears during the second postnatal week in the rat and is correlated with a transient peak in NMDA binding sites measured in this region. The development of LTP in other regions of the hippocampus is also correlated with increases in NMDA receptors. Although the initiation of LTP requires NMDA receptor activation, the maintenance and expression of LTP, that is the increased postsynaptic responsiveness of the neuron, appears to be mediated by non-NMDA receptors since it can be blocked by CNQX but not by NMDA receptor antagonists. Increased PI turnover and the activation of protein kinase C is also correlated with

LTP formation in the hippocampus suggesting a role for metabotropic EAA receptors in the synaptic changes necessary for the development of LTP.

In addition to these specific examples of synapse formation, remodeling and reinforcement, EAA receptor activation stimulates neurite outgrowth *in vitro* and may play a role in neurite extension and sprouting *in vivo* both during development and neuronal regeneration following injury. EAA receptors are particularly abundant in areas of the nervous system where synapses continue to be replaced or remodeled during the life of an animal. These regions include the olfactory bulb where connections with new olfactory receptor neurons occur throughout life, in the hippocampus where synaptic changes involved in learning and memory are continuously being made, and in the cerebellum where synaptic plasticity is necessary to integrate and adapt to changes in sensory, cortical and motor activity. Even in areas of the adult brain that receive little EAA input, there is a transient peak in the expression of EAA receptors that coincides with periods during development when neuronal connections are being established. In the rat cerebellum, although Purkinje neurons express non-NMDA receptors at all stages of development, NMDA receptors are expressed only transiently between the first and third postnatal week (Dupont *et al.*, 1987). This period is correlated with an increased glutamate binding in the molecular layer and the establishment of connections between granule cell parallel fibers and Purkinje cell dendrites (Garcia-Ladona *et al.*, 1991).

Although EAA receptors appear to be present and be involved in excitatory neurotransmission, synapse reinforcement, and postsynaptic responsiveness, all the functional consequences of EAA receptor activation may not develop at the same time as the receptors themselves. Therefore, using EAA-activated responses as an indicator of EAA receptor expression may not be a valid indicator of their presence or absence. For instance, although EAA receptors are expressed very early during development they do not appear to result in excitotoxicity in very young animals and the pattern of receptor-mediated neurotoxicity changes during development (Garthwaite and Garthwaite, 1986; McDonald *et al.*, 1990). Despite the presence of kainate-activated ion channels, kainate is a relatively weak neurotoxin compared to NMDA in young animals whereas the reverse is true in the adult.

In many cases the use of selective antagonists has been used to classify the EAA receptor subtypes responsible for the observed changes in neuronal connections or responses. In other studies the Ca^{2+}-dependence of the EAA-elicited response has been interpreted to mean that NMDA receptors mediate the response. The realization that other EAA receptor subtypes may also directly gate Ca^{2+} may lead to a re-examination of some of the processes believed to be mediated only by NMDA receptors. The importance of non-NMDA receptors and metabotropic EAA receptors in neuronal development and plasticity as well as neurotoxicity should not be underestimated.

8.5.2　LIGAND BINDING AND AUTORADIOGRAPHY

Although functional responses can be used to evaluate and discriminate EAA receptors during different periods of development, their numbers and distribution in nervous system tissue can be measured directly using ligand binding assays (Young and Fagg, 1990). Several different methods are used to quantitate and localize these receptors using radioactive ligands that are either agonists or antagonists of EAA receptors. Binding assays can be performed using either membranes from different regions of the nervous system or tissue sections prepared from animals at different stages of development. Receptor subtypes can be identified and quantitated by measuring the number of [^{3}H]glutamate binding sites that are

displaced by saturating amounts of unlabeled selective ligands. Alternatively, radiolabeled agonists or antagonists that are selective for EAA receptor subtypes can be used. The metabotropic EAA receptor can be labeled using the selective ligand, [^{3}H]-*trans*-ACPD. NMDA receptors can be measured using [^{3}H]CPP, a competitive NMDA receptor antagonist. And non-NMDA receptors can be measured using [^{3}H]CNQX, a non-selective antagonist of these receptors. [^{3}H]AMPA and [^{3}H]kainate are also used to measure non-NMDA receptor subtypes. Whereas [^{3}H]AMPA identifies high affinity AMPA binding sites some or all of these non-NMDA receptors are low affinity receptors for kainate. [^{3}H]kainate, under the conditions used in most binding assays, measures only high affinity kainate binding sites. These high affinity kainate sites may not all be associated with functional EAA receptors; some may be EAA uptake sites or other kinds of binding sites present on neurons and glia (Somogyi *et al.*, 1990; Ortega *et al.*, 1991). To distinguish EAA receptors from their uptake sites, only Na^+-independent binding is measured since the EAA carrier required is dependent on Na^+ which is co-transported with EAAs.

EAA receptors have usually been measured during postnatal development of the rat nervous system although EAA receptor expression occurs much earlier. The density of binding sites increases up to the second or third week postnatally and then declines to adult levels over the next several weeks. Almost all studies using radioligand binding report an 'overshoot' in the number of EAA receptors relative to adult levels that occurs about one to two weeks after birth with the peak in NMDA receptors preceding the peak in non-NMDA receptors. However, the ontogeny of NMDA receptors measured by radioligand binding is dependent on the ligand being used. The binding site for NMDA precedes the development of the strychnine-insensitive [^{3}H]glycine binding site or the binding site for [^{3}H]TCP, an analog of PCP, associated with the NMDA ion channel. Meanwhile, the ontogenies of the glycine site and the TCP binding site associated with the NMDA receptor complex are nearly identical. The transient peak(s) in EAA receptors corresponds to the occurrence of a critical window of nervous system development when synaptic plasticity results in the formation, modification and stabilization of synaptic contacts. This period also corresponds to the time prior to the onset of naturally occurring neuronal cell death. It is possible that the transient peaks in EAA receptors seen by binding assay may be related to a temporary peak in the number of cells expressing EAA receptors. *In situ* hybridization reveals no apparent increases in mRNA for several EAA receptor subunits that might account for the peak in EAA binding sites.

With the cloning of cDNAs for ionotropic and metabotropic EAA receptors their distribution in the rat nervous system has been mapped by *in situ* hybridization. As described earlier, many non-NMDA receptor subunits have now been cloned although at this time no NMDA selective receptor subunit has been identified. Although there are, as yet, no reports of the distribution of mRNA for the metabotropic EAA receptor during nervous system development, its distribution in adult rat hippocampus and cerebellum was reported at the time the sequence of the mRNA and protein was reported (Masu *et al.*, 1991). In the hippocampus, large amounts of mRNA are seen in CA4-dentate gyrus, moderate amounts in CA2–3 and lower levels in CA1. In the cerebellum the mRNA is particularly abundant in Purkinje cells.

The localization of mRNA for subunits of EAA activated ion channels, including specific examination of flip and flop forms, has been studied in both the adult and developing rat nervous system (Bettler *et al.*, 1990; Pellegrini-Giampietro *et al.*, 1991; Monyer *et al.*, 1991; Werner *et al.*, 1991). These studies show that GluR mRNAs vary in their amount and pattern of expression during the development of the nervous system. Some of these subunits have been mapped as early as embryonic day (ED) 10 in the rat where mRNA

for GluRs 4 and 5 were found in postmitotic neurons. The mRNA for all EAA receptor subunits was detected at every stage of development in the rat. The distribution of different GluRs changed during development as did the expression of the flip and flop versions of GluR1–4. The flip versions were almost exclusively expressed before postnatal day (PD) 8 and even after that the flop versions of some GluRs were found only in very restricted patterns only in selected regions of the nervous system. Flip versions of GluRs give greater current responses than the flop versions when the receptor subunits are expressed in mammalian cells (Sommer *et al.*, 1990). The appearance of flop versions occurs about the time of the peak in synapse formation in the rat forebrain (Aghajanian and Bloom, 1967). A switch in EAA receptors to forms that allow less current flow may be important for protecting cells from the neurotoxic consequences of high local concentration of EAAs following synapse formation. The relative abundance of GluRs changes during development: at ED14, GluR2 > GluR4 >GluR1 = GluR3, at PD1 GluR1 > GluR2 > GluR3 >> GluR4, at PD8 GluR1 = GluR3 > GluR2 >> GluR4, and at PD14 GluR1 = GluR2 (due to profoundly increased expression of the flop version of GluR2 in the cerebellum) = GluR3 > GluR4 (where both flip and flop versions are expressed predominantly in cerebellum) (Monyer *et al.*, 1991). In addition to brain, GluR mRNA can be detected in spinal cord, retina, and dorsal root ganglia. For instance, GluR4 was detected in relatively large amounts in retinal ganglion cells and amacrine cells whereas GluR5 mRNA was particularly abundant in dorsal root ganglion neurons. Most GluRs are expressed in large amounts in cerebellum, hippocampus and olfactory bulb. GluR5 is conspicuous by its very restricted pattern of expression in the brain: it is very low in or absent from the hippocampus (while other GluR mRNAs are high) and in the cerebellum GluR5 mRNA is found only in Purkinje neurons. The pattern of EAA subunit expression has yet to provide a clear explanation or understanding of the radio-ligand binding patterns seen with AMPA and kainate or account for changing patterns of susceptibility to EAA-mediated neurotoxicity. Since agonist and antagonist binding as well as ion channel activation will be determined by the combination of subunits expressed, a complete understanding of ligand binding and EAA receptor-mediated functional responses awaits the development of techniques to get stable expression of the receptor subunits in various combinations. Although the possible number of receptors that could be formed from these subunits is quite large, the actual number of combinations will be limited by rules of assembly which are as yet unknown. It will be some time before the receptor subunit combinations can be assigned to functional responses but we are beginning to get a picture of their structure– function relationships.

8.6   SUMMARY

In conclusion, the pharmacology and physiology of EAA receptors has suggested that there are many different receptor subtypes. However, they can be classified in a systematic way on the basis of ligand binding. Cloning and expression studies have just begun to provide information about the structural basis for the sometimes contradictory pharmacology, biochemistry, ligand binding and electrophysiology seen with EAA receptors. The knowledge of the true structural and functional diversity of EAA receptors and their differential distribution in the nervous system will lead to a better understanding of their contribution to nervous system development and function. This information should make it possible to identify and design selective agonists or antagonists of EAA receptors that will be important therapeutically. Since excitatory neurotransmission is of fundamental importance to nervous system function, nature has probably designed multiple receptors to mediate excitatory amino acid-mediated transmission for specialized purposes. Increases in intracellular Ca^{2+} appear to be the primary consequence of EAA receptor activation. The identification of the processes used by neural cells to interpret EAA receptor activation and translate those signals will be the future goal of neurobiologists. The integration of the molecular cloning information with the pharmacology and physiology of EAA receptors will be an exciting challenge that holds the key to our full understanding of the roles of these receptors in nervous system function and dysfunction.

GLUTAMATE RECEPTOR UPDATE

During 1991 and 1992 there was a real explosion in the number of EAA receptors identified, the cDNA for one providing the means to hunt for others like it. At last, several cDNAs for the much sought after NMDA receptor were cloned (Moriyoshi *et al.*, 1991; Meguro *et al.*, 1992; Monyer *et al.*, 1992; Yamazaki *et al.*, 1992b). The four NMDA receptor subunits identified so far can be divided into two families based on their structural similarity: NMDAR1 is only ~20% similar to NMDAR2A, NMDAR2B, and NMDAR2C which are between 55 and 70% similar to each other. Although the mRNAs for these subunits have only been localized in adult rat brain they show interesting differences in their distribution. NMDAR1 and NMDAR2A are distributed throughout the CNS although NMDAR1 mRNA appears to be relatively more abundant than the mRNA for NMDAR2A. NMDAR2B is found mainly in the hippocampus and appears to be absent from cerebellum while NMDAR2C is found only in the cerebellum (Monyer *et al.*, 1992). Two more kainate-selective ion channel subunits have now been described (Bettler *et al.*, 1992; Herb *et al.*, 1992; Morita *et al.*, 1992; Sakimura *et al.*, 1992). Although members of the kainate receptor family do not form functional ion channels on their own, they do when expressed in certain combinations and can even be activated by AMPA (Herb *et al.*, 1992). Mishina and coworkers, who have been cloning cDNAs for EAA receptors expressed in mouse brain

(Sakimura *et al.*, 1990), have also described a cDNA which is structurally related but distinct from previously described NMDA, AMPA or kainate receptor subunits (Yamazaki *et al.*, 1992a). More members of the metabotropic EAA receptor family have also been reported along with descriptions of their coupling to multiple second messenger systems (Aramori and Nakanishi, 1992; Tanabe *et al.*, 1992). Antibodies, as well as *in situ* hybridization, have now been used to map the distribution of the largest metabotropic EAA receptor, mGluRlα, in adult rat brain (Martin *et al.*, 1992). However, no papers have appeared describing the distribution or functional roles of metabotropic EAA receptors in the developing nervous system.

Beyond the identification of additional members of the EAA receptor families by molecular cloning strategies, functional studies have revealed structural elements that determine divalent cation permeability of EAA-activated ion channels. A glutamine residue near the external face of the second putative transmembrane domain of AMPA and kainate receptor subunits appears to play a key role in determining calcium permeability of the ion channel (Hume *et al.*, 1991; Verdoorn *et al.*, 1991). In NMDA receptors an asparagine residue appears to play the same role as glutamine making NMDA channels highly permeable to calcium. An unusual finding is that this strategic glutamine is replaced in some AMPA and kainate receptor subunits by an arginine through the unconventional process of mRNA editing, switching the receptor from a calcium permeable to calcium impermeable ion channel (Sommer *et al.*, 1991). Although the gene for most, if not all, AMPA and kainate receptors codes for a glutamine in this position, (CAG), the adenosine is modified to an inosine resulting in the insertion of an arginine instead of a glutamine residue during protein synthesis. But only in the case of GluR-B, GluR5 and GluR6. And editing appears to be developmentally regulated. Although 100% of the GluR-B mRNA in adult rat brain appears to be edited to the arginine or calcium impermeable form, analysis of rat brain at ED14 and PD0 indicates that ~1% of the total GluR-B message is in the unedited or calcium permeable form (Burnashev *et al.*, 1992). For other EAA subunits that undergo editing the process is not as thorough. In adult rat brain only 40% of the GluR5 mRNA and 75% of the GluR6 mRNA appears in the edited, arginine-containing, calcium impermeable form. It remains to be seen if the editing process shows regional differences either in the adult or developing CNS.

Developmental differences in EAA receptors appear to determine the sensitivity of a cell to ambient levels of glutamate or other endogenous agonists. Variations in splicing and the remarkable process of mRNA editing regulate the magnitude of total current flux and the amount of calcium entry into neurons. EAA receptors expressed early in development, when neurons are dividing, sprouting, migrating, differentiating and establishing synaptic connections, activate greater levels and/or more sustained depolarization and probably allow more calcium entry than adult forms of these receptors. Both the increases in ion flux along with their ability to trigger release of neurotransmitters, neuropeptides and growth factors onto neighboring cells are probably important factors guiding CNS development. However, once synaptic connections have been established, conversion of EAA receptors to lower conductance, less calcium permeable adult forms may be a necessary requirement to prevent excitotoxicity due to increasing ambient levels of glutamate. It will be interesting to see how these developmental changes are triggered and whether a delay or failure in the alteration of EAA receptor subunits plays any role in developmental regulated neuronal cell death in pathological neurodegeneration.

REFERENCES

Aghajanian, G.K. and Bloom, F.E. (1967) The formation of synaptic junctions in developing rat brain: a quantitative electron microscopic study. *Brain Res.*, **6**, 716–27.

Aramori, I. and Nakanishi, S. (1992) Signal transduction and pharmacological characterization of a metabotropic glutamate receptor, mGluR1, in transfected CHO cells. *Neuron*, **8**, 757–65.

Ascher, P. and Nowak, L. (1988) The role of divalent cations in the *N*-methyl-D-aspartate responses of mouse central neurons in culture. *J. Physiol. (Lond.)*, **399**, 247–66.

Barres, B.A., Chun, L.L.Y. and Corey, D.P. (1990) Ion channels in vertebrate glia. *Annu. Rev. Neurosci.*, **13**, 441–74.

Berridge, M.J. and Irvine, R.F. (1989) Inositol phosphates and cell signalling. *Nature*, **341**, 197–205.

Bettler, B., Boulter, J., Hermans-Borgmeyer, I. *et al.* (1990) Cloning of a novel glutamate receptor subunit GluR5: expression in the nervous system during development. *Neuron*, **5**, 583–95.

Bettler, B., Egebjerg, J., Sharma, G. *et al.* (1992) Cloning of a putative glutamate receptor: a low affinity kainate binding subunit. *Neuron*, **8**, 257–65.

Boulter, J., Hollmann, M., O'Shea-Greenfield, A. *et al.* (1990) Molecular cloning and functional expression of glutamate receptor subunit genes. *Science*, **249**, 1033–7.

Burnashev, N., Monyer, H., Seeburg, P.H. and Sakmann, B. (1992) Divalent ion permeability of AMPA receptor channels is dominated by the edited form of a single subunit. *Neuron*, **8**, 189–98.

Charpak, S., Gähwiler, B.H., Do, K.Q. and Knöpfel, T. (1990) Potassium conductances in hippocampal neurons blocked by excitatory amino-acid transmitters. *Nature*, **347**, 765–7.

Choi, D.W. and Rothman, S.M. (1990) The role of glutamate neurotoxicity in hypoxic ischemic neuronal death. *Annu. Rev. Neurosci.*, **13**, 171–82.

Collingridge, G. and Singer, W. (1990) Excitatory amino acid receptors and synaptic plasticity. *Trends Pharmacol. Sci.*, **11**, 290–6.

Cornell-Bell, A.H., Finkbeiner, S.M., Cooper, M.S. and Smith, S.J. (1990) Glutamate induces calcium waves in cultured astrocytes: long-range glial signalling. *Science*, **247**, 470–3.

Curtis, D.R., Phillis, J.W. and Watkins, J.C. (1959) Chemical excitation of spinal neurons. *Nature*, **183**, 611.

Dawson, V.L., Dawson, T.M., London, E. *et al.* (1991) Nitric oxide mediates glutamate neurotoxicity in primary cortical cultures. *Proc. Natl. Acad. Sci. USA*, **88**, 6368–71.

Dupont, J.-L., Gardette, R. and Crepel, F. (1987) Postnatal development of the chemosensitivity of rat cerebellar Purkinje cells to excitatory amino acids. An *in vitro* study. *Dev. Brain Res.*, **34**, 59–68.

Egebjerg, J., Bettler, B., Hermans-Borgmeyer, I. and Heinemann, S. (1991) Cloning of cDNA for a glutamate receptor subunit activated by kainate but not AMPA. *Nature*, **351**, 745–8.

Foster, A.C. and Fagg, G.E. (1984) Acidic amino acid binding sites in mammalian neuronal membranes: their characteristics and relationship to synaptic receptors. *Brain Res. Rev.*, **7**, 103–64.

Garcia-Ladona, F.J., Palacios, J.M., Girard, C. and Gombos, G. (1991) Autoradiographic characterization of [^{3}H]-glutamate binding sites in developing mouse cerebellar cortex. *Neuroscience*, **41**, 243–55.

Garthwaite, G. and Garthwaite, J. (1986) *In vitro* neurotoxicity of excitatory acid analogues during cerebellar development. *Neuroscience*, **17**, 755–67.

Gilbertson, T.A., Scobey, R. and Wilson, M. (1991) Permeation of calcium ions through non-NMDA glutamate channels in retinal bipolar cells. *Science*, **251**, 1613–15.

Gregor, P., Mano, I, Maoz, I. *et al.* (1989) Molecular structure of the chick cerebellar kainate-binding subunit of a putative glutamate receptor. *Nature*, **342**, 689–92.

Hayashi, T. (1954) Effects of sodium glutamate on the nervous system. *Keio J. Med.*, **3**, 183–92.

Heizmann, C.W. and Hunziker, W. (1991) Intracellular calcium-binding proteins: more sites than insights. *Trends Biochem. Sci.*, **16**, 98–103.

Herb, A., Burnashev, N., Werner, P. *et al.* (1992) The KA-2 subunit of excitory amino acid receptors shows widespread expression in brain and forms ion channels with distinctly related subunits. *Neuron*, **8**, 775–85.

Hollmann, M., O'Shea-Greenfield, A., Rogers, S.W. and Heinemann, S. (1989) Cloning by functional expression of a member of the glutamate receptor family. *Nature*, **342**, 643–8.

Hollmann, M., Hartley, M. and Heinemann, S. (1991) Calcium permeability of KA-AMPA-gated glutamate receptor channels: dependence on subunit composition. *Science*, **252**, 851–3.

Honoré, T., Davies, S.N., Drejer, J. *et al.* (1988) Quinoxalinediones: potent competitive non-NMDA glutamate receptor antagonists. *Science*, **241**, 701–3.

Houamed, K.M., Kuijper, J.L., Gilbert, T.L. *et al.* (1991) Cloning, expression and gene structure of a G protein-coupled glutamate receptor from rat brain. *Science*, **252**, 1318–21.

Hume, R.J., Dingledine, R. and Heinemann, S.F. (1991) Identification of a site in glutamate receptor subunits that controls calcium permeability. *Science*, **253**, 1028–31.

Iino, M., Ozowa, S. and Tsuzuki, K. (1990) Permeation of calcium through excitatory amino acid receptor channels in cultured rat hippocampal neurones. *J. Physiol. (Lond.)*, **424**, 151–65.

Jackson, H. and Usherwood, P.N.R. (1988) Spider toxins as tools for dissecting elements of excitatory amino acid transmission. *Trends Neurosci.*, **6**, 278–83.

Johnson, J.W. and Ascher, P. (1987) Glycine potentiates the NMDA response in cultured mouse brain neurons. *Nature*, **325**, 529–31.

Keinänen, K., Wisden, W., Sommer, B. *et al.* (1990) A family of AMPA-selective glutamate receptors. *Science*, **249**, 556–60.

Kemp, J.A., Foster, A.C., Leeson, P.D. *et al.* (1988) 7-Chlorokynurenic acid is a selective antagonist at the glycine modulatory site of the *N*-methyl-D-aspartate receptor complex. *Proc. Natl. Acad. Sci. USA*, **85**, 6547–50.

Lodge, D.W. and Collingridge, G. (eds) (1991) *The Pharmacology of Excitatory Amino Acids*, Elsevier, Cambridge.

Lodge, D. and Johnson, K.M. (1990) Noncompetitive excitatory amino acid receptor antagonists. *Trends Pharmacol. Sci.*, **11**, 81–6.

MacDermott, A.B., Mayer, M.L., Westbrook, G.L. *et al.* (1986) NMDA receptor-activation increases cytoplasmic calcium concentration in cultured spinal cord neurons. *Nature*, **321**, 519–22.

Maguro, H., Mori, H., Araki, K. *et al.* (1992) Functional characterization of a heteromeric NMDA receptor channel expressed from cloned cDNAs. *Nature*, **357**, 70–4.

Martin, L.J., Blackstone, C.D., Huganir, R.L. *et al.* (1992) Cellular localization of a metabotropic glutamate receptor in rat brain. *Neuron*, **9**, 259–70.

Masu, M., Tanabe, Y., Tsuchida, K. *et al.* (1991) Sequence and expression of a metabotropic glutamate receptor. *Nature*, **349**, 760–5.

Mayer, M.L. and Westbrook, G.L. (1987) Permeation and block of *N*-methyl-D-aspartatic acid receptor channels by divalent cations in mouse cultured central neurons. *J. Physiol. (Lond.)*, **394**, 501–27.

McDonald, J.W. and Johnston, M.V. (1990) Physiological and pathophysiological roles of excitatory amino acids during central nervous system development. *Brain Res. Rev.*, **15**, 41–70.

McDonald, J.W., Trescher, W.H. and Johnston, M.V. (1990) The selective ionotropic-type quisqualate receptor agonist AMPA is a potent neurotoxin in immature rat brain. *Brain Res.*, **526**, 165–8.

Meguro, H., Mori, H., Araki, K. *et al.* (1992) Functional characterization of a heteromeric NMDA receptor channel expressed from cloned cDNAs. *Nature*, **357**, 70–4.

Meldrum, B. and Garthwaite, J. (1990) Excitatory amino acid neurotoxicity and neurodegenerative disease. *Trends Pharmacol. Sci.*, **11**, 379–87.

Monaghan, D.T. and Cotman, C.W. (1986) Identification and properties of *N*-methyl-D-aspartate receptors in rat brain synaptic plasma membranes. *Proc. Natl. Acad. Sci. USA*, **83**, 7532–5.

Monaghan, D.T., Olvermann, H.J., Nguyen, L. *et al.* (1988) Two classes of *N*-methyl-D-aspartate recognition sites: differential distribution and differential regulation by glycine. *Proc. Natl. Acad. Sci. USA*, **85**, 9836–40.

Monaghan, D.T., Bridges, R.J. and Cotman, C.W. (1989) The excitatory amino acid receptors: their classes, pharmacology, and distinct properties in the function of the central nervous system. *Annu. Rev. Pharmacol. Toxicol.*, **29**, 365–402.

Monyer, H., Seeburg, P.H. and Wisden, W. (1991) Glutamate-operated channels: developmentally early and mature forms arise by alternative splicing. *Neuron*, **6**, 799–810.

Monyer, H., Sprengel, R., Schoepfor, R. *et al.* (1992) Heteromeric NMDA receptors: Molecular and functional distinction of subtypes. *Science*, **256**, 1217–21.

Morita, T., Sakimura, K., Kushiya, E. *et al.* (1992) Cloning and functional expression of a cDNA encoding the mouse β2 subunit of the kainate-selective glutamate receptor channel. *Mol. Brain Res.*, **14**, 143–6

Moriyoshi, K., Masu, M., Ishii, T. *et al.* (1991) Molecular cloning and characterization of the rat NMDA receptor. *Nature*, **354**, 31–7.

Nakanishi, N., Shneider, N.A. and Axel, R. (1990) A family of glutamate receptor genes: evidence for the formation of heteromultimeric receptors with distinct channel properties. *Neuron*, **5**, 569–81.

Nicholls, D. and Attwell, D. (1990) The release and uptake of excitatory amino acids. *Trends Pharmacol. Sci.*, **11**, 462–8.

Nicoll, R.A., Kauer, J.A. and Malenka, R.C. (1988) The current excitement in long-term potentiation. *Neuron*, **1**, 97–103.

Ortega, A., Eshhar, N. and Teichberg, V.I. (1991) Properties of kainate receptor/channels on cultured Bergman glia. *Neuroscience*, **41**, 335–49.

Pellegrini-Giampietro, D.E., Bennett, M.V.L. and Zukin, R.S. (1991) Differential expression of three glutamate receptor genes in developing rat brain: an *in situ* hybridization study. *Proc. Natl. Acad. Sci. USA*, **88**, 4157–61.

Pruss, R.M. and Stauderman, K.A. (1988) Voltage regulated calcium channels involved in the regulation of enkephalin synthesis are blocked by phorbol ester. *J. Biol. Chem.*, **263**, 13173–8.

Pruss, R.M., Moskal, J.R., Eiden, L.E. and Beinfeld, M. (1985) Specific regulation of VIP biosynthesis by phorbol ester in bovine chromaffin cells. *Endocrinology*, **117**, 1020–6.

Pruss, R.M., Akeson, R.L., Racke, M.M. and Wilburn, J.L. (1991) Agonist-activated cobalt uptake identifies divalent cation-permeable kainate receptors on neurons and glial cells. *Neuron*, **7**, 1–10.

Rasmussen, C.D. and Means, A.R. (1989) Calmodulin, cell growth and gene expression. *Trends Neurosci.*, **11**, 433–8.

Rokaeus, Å., Pruss, R.M. and Eiden, L.E. (1990) Regulation of galanin expression by calcium, protein kinase A and protein kinase C messenger systems in chromaffin cells. *Endocrinology*, **127**, 3096–102.

Sakimura, K., Bujo, H., Kushiya, E. *et al.* (1990) Functional expression from cloned cDNAs of glutamate receptor species responsive to kainate and quisqualate. *FEBS Lett.*, **272**, 73–80.

Sakimura, K., Morita, T., Kushiya, E. and Mishina, M. (1992) Primary structure and expression of the τ2 subunit of the glutamate receptor channel selective for kainate. *Neuron*, **8**, 267–374.

Schoepp, D., Bockaert, J. and Sladeczek, F. (1990) Pharmacological and functional characteristics of metabotropic excitatory amino acid receptors. *Trends Pharmacol. Sci.*, **11**, 508–14.

Sheardown, M.J., Nielsen, E.O., Hansen, A.J. *et al.* (1990) 2,3-Dihydroxy-6-nitro-7-sulfamoyl-benzo-(F)quinoxaline: a neuroprotectant for cerebral ischaemia. *Science*, **247**, 571–4.

Sheng, M., Thompson, M.A. and Greenberg, M.E. (1991) CREB: a Ca^{2+}-regulated transcription factor phosphorylated by calmodulin-dependent kinases. *Science*, **252**, 1427–30.

Sommer, B., Keinänen, K., Verdoorn, T.A. *et al.* (1990) Flip and flop: a cell-specific functional switch in glutamate-operated channels of the CNS. *Science*, **249**, 1580–5.

Sommer, B., Köhler, M., Sprengel, R. and Seeburg, P.H. (1991) RNA editing in brain controls a determinant of ion flow in glutamate-gated channels. *Cell*, **67**, 11–19.

Somogyi, P., Eshhar, N., Teichberg, V.I. and Roberts, J.D.B. (1990) Subcellular localization of a putative kainate receptor in Bergmann glial cells using a monoclonal antibody in the chick and fish cerebellar cortex. *Neuroscience*, **35**, 9–30.

Tanabe, Y., Masu, M., Ishii, T. *et al.* (1992) A family of metabotropic glutamate receptors. *Neuron*, **8**, 169–79.

Traynelis, S.F. and Cull-Candy, S. (1990) Proton inhibition of N-methyl-D-aspartate receptors in cerebellar neurons. *Nature*, **345**, 347–50.

Verdoorn, T.A., Burnashev, N., Monyer, H. *et al.* (1991) Structural determinants of ion flow through recombinant glutamate receptor channels. *Science*, **252**, 1715–18.

Wada, K., Dechesne, C.J., Shimaski, S. *et al.* (1989) Sequence and expression of a frog brain complementary DNA encoding a kainate-binding protein. *Nature*, **342**, 684–9.

Watkins, J.C., Krogsgaard-Larsen, P. and Honoré, T. (1990) Structure–activity relationships in the development of excitatory amino acid receptor agonists and competitive antagonists. *Trends Pharmacol. Sci.*, **11**, 25–33.

Werner, P., Voigt, M., Keinänen, K. *et al.* (1991) Cloning of a putative high-affinity kainate receptor expressed predominantly in hippocampal CA3 cells. *Nature*, **351**, 741–4.

Westbrook, G.L. and Mayer, M.L. (1987) Micromolar concentrations of Zn^{2+} antagonize NMDA and GABA responses of hippocampal neurons. *Nature*, **328**, 640–3.

Wong, E.H., Kemp, J.A., Priestley, T. *et al.* (1986) The anti-convulsant MK-801 is a potent N-methyl-D-aspartate antagonist. *Proc. Natl. Acad. Sci. USA*, **83**, 7104–8.

Wroblewski, J.T., Nicoletti, F., Fadda, E. and Costa, E. (1987) Phencyclidine is a negative allosteric modulator of signal transduction at two subclasses of excitatory amino acid receptors. *Proc. Natl. Acad. Sci. USA*, **84**, 5068–72.

Wroblewski, J.T. and Danysz, W. (1989) Modulation of glutamate receptors: molecular mechanisms and functional implications. *Annu. Rev. Pharmacol. Toxicol.*, **29**, 441–74.

Yamazaki, M., Araki, K., Shibate, A. and Mishima, M. (1992a) Molecular cloning of a cDNA encoding a novel member of the mouse glutamate receptor channel family. *Biochem. Biophys. Res. Commun.*, **183**, 886–92.

Yamazaki, M., Mori, H., Araki, K. *et al.* (1992b) Cloning, expression and modulation of a mouse NMDA receptor subunit. *FEBS Lett.*, **300**, 39–45.

Young, A.B. and Fagg, G.E. (1990) Excitatory amino acid receptors in the brain: membrane binding and receptor autoradiographic approaches. *Trends Pharmacol. Sci.*, **11**, 126–33.

9

NEUROTENSIN RECEPTORS DURING THE DEVELOPMENT OF THE CENTRAL NERVOUS SYSTEM

F. Javier Garcia-Ladona, Guadalupe Mengod and José M. Palacios

During the last two decades increasing evidence has accumulated supporting a role for neuropeptides as classical neurotransmitters or neuromodulators at synapses of the central and peripheral nervous systems (see Polak, 1989 for a review). Although their involvement in brain function is well documented, other possible roles, for example that of neurotrophic factor, have also been proposed based on their mitogenic activity (Hanley, 1985). Studies on neuropeptide receptor changes during brain development are suggestive of such a role. Neurotensin is one of those peptides whose receptors present developmentally regulated patterns in brain. Because of the availability of experimental tools for the analysis of its pre- and post-synaptic components, both at the mRNA and protein levels, the neurotensinergic synapse offers an interesting model to analyse transmitter receptor interactions during development. This chapter reviews the available information on changes in neurotensin receptor (NTR) binding and NTR mRNA levels and their correlation with neurotensin (NT) mRNA during the development of the rat central nervous system (CNS) the species for which most of the available data have been generated.

9.1 NEUROTENSIN AS A NEUROTRANSMITTER

Neurotensin, a 13 amino acid peptide (Fig. 9.1), was first isolated from bovine hypothalamus and later localized in several other regions of the central and peripheral nervous system (Kahn *et al.*, 1980; Jennes *et al.*, 1982). When injected into different brain regions, NT induces several physiological and behavioral effects such as decrease of locomotor activity (Kalivas *et al.*, 1983), hypothermia and antinociception (Kalivas *et al.*, 1982). Radioimmunoassay and immunohistochemical studies demonstrated its synaptosomal localization and uneven regional localization in the brain (Jennes *et al.*, 1982; Kalivas *et al.*, 1982; Inagaki *et al.*, 1982; Roberts *et al.*, 1984). Although the precise molecular mechanisms mediating the actions of the peptide are not fully understood, it has been shown that NT can modulate second messenger pathways (Goedert *et al.*, 1984a; Gilbert and Richelson, 1984; Bozou *et al.*, 1986), inducing increases of cytosolic Ca^{2+} (Sato *et al.*, 1991) and protein phosphorylation via Ca^{2+}/calmodulin dependent protein kinases (Kasckow *et al.*, 1991).

The rat gene coding for a NT precursor has been cloned and sequenced (Kislauskis *et al.*,

Receptors in the Developing Nervous System Vol. 2: Neurotransmitters. Edited by Ian S. Zagon and Patricia J. McLaughlin. Published in 1993 by Chapman & Hall. ISBN 0 412 49400 0. Vols. 1 and 2 (set) ISBN 0 412 54520 9.

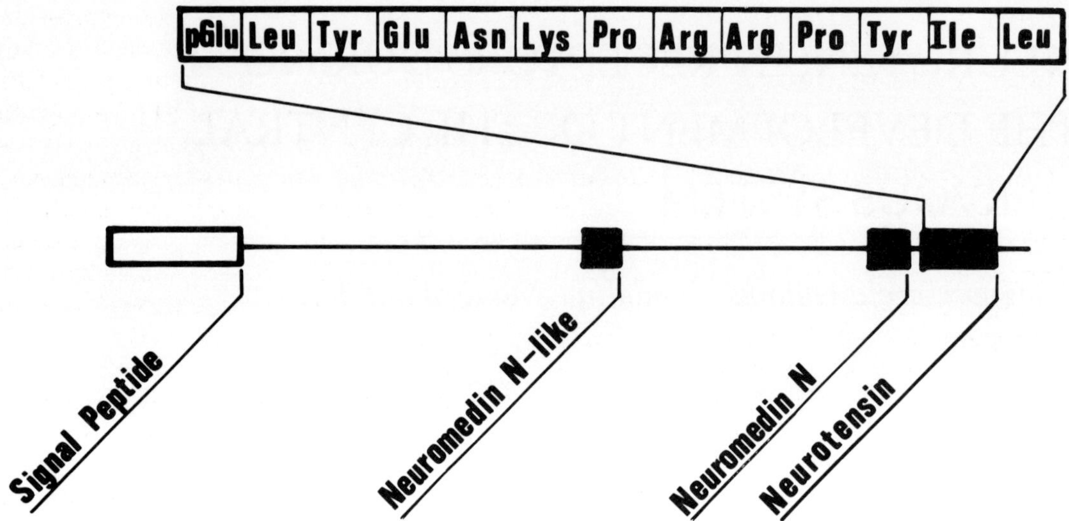

Fig. 9.1.   Neurotensin/neuromedin N gene and neurotensin amino acid sequence.

1988). This gene contains, in addition, sequences coding for a NT-like peptide, neuromedin N (Fig. 9.1). The expression of the NT gene can be induced by several agents such as nerve growth factor (NGF), activators of adenylate cyclase, protein kinase C inhibitors such as staurosporine and lithium (Dobner *et al.*, 1988; Tischler *et al.*, 1991). Nerve cells expressing NT mRNA (Alexander *et al.*, 1989) are detectable only in discrete areas of the adult brain such as hippocampus, striatum, cingulate cortex and hypothalamus (see later).

9.2   NEUROTENSIN BINDING SITES IN THE CNS: CHARACTERIZATION AND DISTRIBUTION

NT actions are mediated by its interaction with specific membrane receptors. Ligand binding studies using either ^{125}I-labeled NT or [^{3}H]NT have allowed the characterization of NT binding sites in brain (for review see Uhl, 1990).

NT binding sites have been detected in brain membrane fractions from several mammalian species. Different authors have reported the presence of at least two types of NT-

binding site exhibiting different affinities for the peptide. NT-binding sites (K_d=3 nM) are present in high density in the frontal cortex, hypothalamus, midbrain, striatum and thalamus, and are less abundant in the hippocampus, olfactory bulb and cerebellum (Uhl and Snyder, 1977; Goedert *et al.*, 1984). Agonists for these high affinity sites are NT itself and xenopsin (an NT analog from *Xenopus*). Displacement data have shown that the partial sequence NT(8–13) is the more potent NT-analog, indicating that the C-terminal part of NT is the active region. Recently, a protein (mol.wt. 72 kDa) containing NT-binding sites (K_d=5 nM) was purified from bovine brain (Mills *et al.*, 1988).

Binding of NT to synaptic membranes has revealed the presence of high and low affinity NT-binding sites (Mazella *et al.*, 1983, 1988). In these studies, the high affinity component is located in a protein of 100 kDa and displays a K_d of 0.10 nM whereas the low affinity component has a K_d around 4 nM. These values are similar to those reported in studies describing only one type of NT binding site. Moreover, the low affinity NT binding site displayed the special characteristic of being sensitive to the antihistaminic agent levo-castine (Mazella *et al.*, 1988). However, a low affinity site insensitive to this agent has also been characterized (Kitabgi *et al.*, 1987) and purified (Mills *et al.*, 1988).

Ligand binding autoradiography has been used to determine the precise anatomical distribution of NT binding sites in CNS. The pharmacological characteristics of NT bind-ing sites detected by autoradiography *in vitro* agree in general with previous results obtain-ed with membrane preparations: the K_d is around 9 nM and NT(8–13) is the more potent fragment of the native peptide inhibiting the binding of radiolabeled NT (Moyse *et al.*, 1987). Although NT binding sites are widely distributed in brain, the highest densities are localized to discrete brain areas.

In adult animals, NT binding sites are specially abundant in parts of the nigrostriatal and mesolimbic systems. High densities of NT binding sites are present in the substantia nigra pars compacta and ventral tegmental area (Young and Kuhar, 1981; Moyse *et al.*, 1987). Several studies indicate that NTR are localized to dopaminergic cell bodies (Palacios and Kuhar, 1981) and can be pre-synaptically located on dopaminergic termi-nals in caudate-putamen, olfactory tubercle and nucleus accumbens (Quirion *et al.*, 1985). High levels of NT binding are also detected in the cingulate cortex, olfactory bulb and olfactory tubercle, Islands of Calleja, vertical limb of diagonal band of Broca and nucleus basalis, amygdala, pre- and postsubiculum, ventral part of dentate gyrus, anterior dorsal thalamic nucleus, suprachiasmatic nucleus, and zona incerta (Young and Kuhar, 1981; Goedert *et al.*, 1984b; Kohler *et al.*, 1985; Moyse *et al.*, 1987).

9.3   CHARACTERISTICS OF NEUROTENSIN RECEPTOR GENE

The isolation and characterization of a gene coding for a rat NTR has been reported by Tanaka *et al.* (1990). This gene, isolated from a rat brain cDNA library by expression cloning in an oocyte system encodes for a 424 amino acid protein. The analysis of its sequence confirms that this NTR belongs to the G-protein-coupled receptor family. It has seven hydrophobic putative transmembrane domains. Northern blot analyses with poly $(A)^+$ mRNA purified from several tissues show that NTR mRNA is relatively abundant in brain whereas the levels in peripheral tissues are only 15% of those observed in the brain.

In oocytes injected with *in vitro* transcribed NTR mRNA, NT elicits a response which is mediated by an endogenous Ca^{2+}-dependent Cl^- channel and results from the activation of G protein coupled to phosphoinositide-Ca^{2+} second messenger systems. This result con-firms NT activation of different second messenger pathways in the cell (Goedert *et al.*, 1984a; Gilbert and Richelson, 1984; Bozou *et al.*, 1986; Sato *et al.*, 1991; Kasckow *et al.*, 1991).

The transfection of COS cells, a monkey kidney cell line, with the NTR gene allowed the pharmacological characterization of this receptor. [³H]NT binds specifically and in a saturable fashion to membrane fractions from these transfected cells. Scatchard plots revealed a single site of binding with a K_d of

Table 9.1.   IC$_{50}$ Values (M) for the displacement of [^{125}I]NT binding to membranes of cDNA NTR transfected cells (a) and [^{3}H]NT binding to rat brain membranes (b) by different peptide analogs of NT. Data from (a) Tanaka *et al.* (1990) and (b) Goedert *et al.* (1984a)

	IC$_{50}$ (M)	
	(a)	(b)
NT(8–13)	2.5×10^{-10}	27×10^{-10}
NT	7.4×10^{-10}	31×10^{-10}
Xenopsin	8.6×10^{-10}	34×10^{-10}
Neuromedin N	$26 \ \times 10^{-10}$	—
NT(9–13)	$240 \ \times 10^{-10}$	603×10^{-9}

0.16 nM, close to the value observed in rat brain membranes. Displacement experiments revealed that among different NT analogs the NT(8–13) was the more potent and effective inhibiting ^{125}I-labeled NT binding, followed by xenopsin that was equipotent to NT. Neuromedin N and NT(9–13) were less effective than NT, and NT(1–8) displayed no activity (Table 9.1), thus confirming that the C-terminal part of NT is more important for the interaction with NTR (Mazella *et al.*, 1989). The antihistaminic levocastine acting on the low affinity binding sites was ineffective in the cloned NTR. Therefore, the cloned NTR is a NT receptor exhibiting the high affinity binding site described earlier. At present, the sequence of NTR reported by Tanaka *et al.* (1990) is the only one available. However, the possibility that there are other subtypes of NTR, with different molecular or pharmacological characteristics, cannot be ruled out. Multiplicity of receptor subtypes for a given neurotransmitter appears to be the rule rather than the exception.

9.4 NEUROTENSIN RECEPTORS DURING THE DEVELOPMENT OF THE CNS

9.4.1 NEUROTENSIN BINDING SITES DURING RAT BRAIN ONTOGENY

The evolution of NTR levels during ontogeny of the rat brain has been studied using membrane binding assays and autoradiographic techniques (Kiyama *et al.*, 1987; Schotte and Laduron, 1987; Palacios *et al.*, 1988).

At least two different patterns of NTR development have been observed in the rat CNS. A 'classical' one, where NTR density progressively increases during embryonic life to reach the adult patterns in the 2nd/3rd postnatal week, for example in the substantia nigra (Fig. 9.2b) and a second one in which

NT binding sites are present at very high levels around the second week of postnatal life (Fig. 9.2a) and then decrease to adult values, for example in the neocortex.

NT binding sites can be detected early in the prenatal period. At gestational day 14 (GD14) they are mainly present in the spinal cord. The density of NTR increases between GD16 and GD18, especially in the developing neocortex where two bands of high density were observed: one in the molecular layer and parts of the cortical plate and the other localized in the more internal part corresponding to layer VI (Figs 9.3 and 9.5). High densities of binding sites were also observed in the developing hypothalamus and substantia nigra. Low levels were found in most of the brainstem nuclei with the exception of the developing nucleus tractus solitarius where a medium density of binding sites was present.

After birth, the pattern of transient expression was particularly dramatic in the neocortex (Kiyama *et al.*, 1987; Palacios *et al.*, 1988) (Figs 9.3 and 9.4). A double band of labeling was seen, with highest densities in the more external (layer I) and more internal layers (layers V and VI) respectively (Fig. 9.5). In other brain areas the distribution of NT binding sites was comparable to the adult pattern. High densities of NT binding sites were also observed in the septum, Islands of Calleja, olfactory bulb, medial habenula and dentate gyrus. At postnatal day (PD)21 the distribution of NT binding sites was that of the adult brain (Fig. 9.6).

9.4.2 THE DEVELOPMENT OF NEUROTENSIN SYSTEM IN OTHER SPECIES

An early and transient expression of NT binding sites has also been reported in human brain. Mailleux *et al.* (1990) have observed high levels of NT binding transiently expressed in the embryonic human inferior

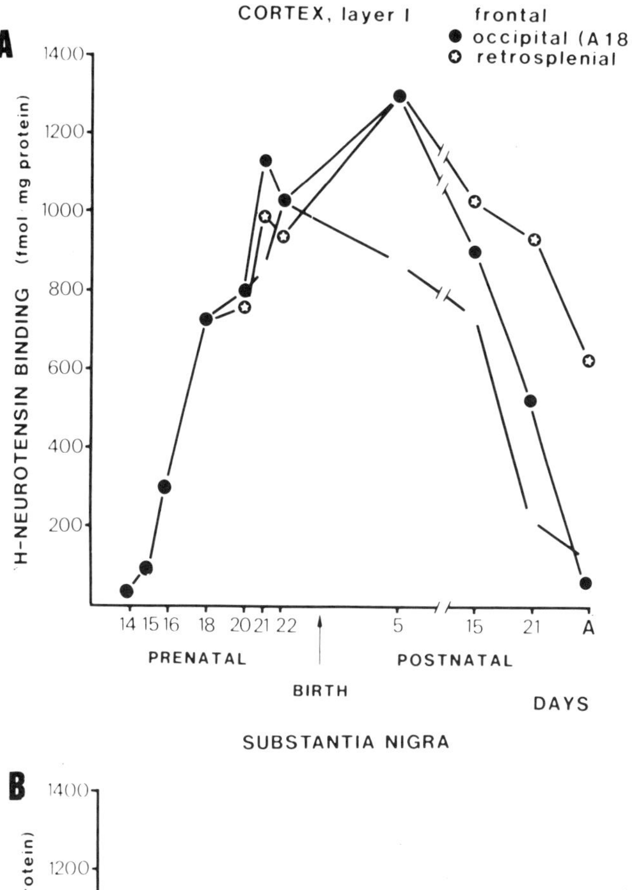

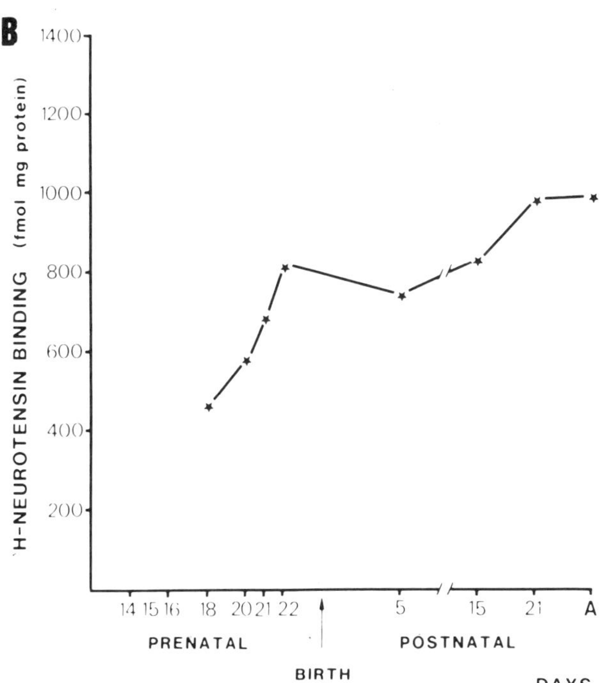

Fig. 9.2. Microdensitometric quantification of [³H]neurotensin binding in (a) layer I of different cortical areas and (b) substantia nigra, during ontogeny.

olive. A different developmental pattern of NT immunoreactivity was described (Mailleux and Vanderhaeghen, 1988) in this nucleus. In our laboratory (Palacios *et al.*, unpublished) we have examined NTR in the human fetus striatum and observed a pattern of distribution comparable to that seen in the adult, although the islands of high density receptor binding were more marked. Goedert *et al.* (1984c) have studied the neuro-

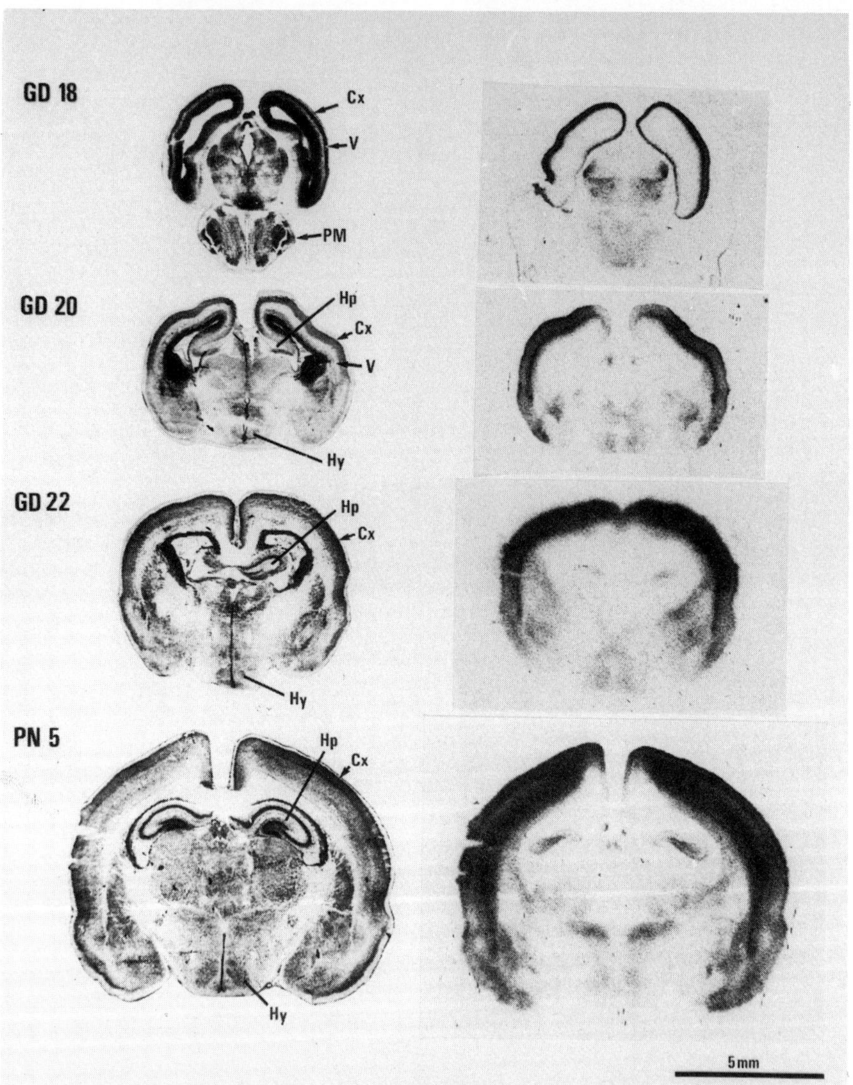

Fig. 9.3. Photomicrographs of coronal sections of rat brain at different ages during the development of the animal (GD18, GD20, GD22, PN5). On the left side sections stained for cresyl violet are shown, on the right side the autoradiograms of sections incubated with [³H] neurotensin. Reproduced from Palacios *et al.* (1988) with permission. Cx, neocortex; Hp, hippocampus; Hy, hypothalamus; PM, pons medulla.

tensinergic system in cat striatum, showing that no postnatal modifications occur in this area as suggested by the developmental pattern of NT immunoreactivity and [³H]NT binding sites. Transient expression of NT binding sites has also been reported in the cat inferior olive (Mailleux *et al.*, 1989). Thus, as in the rat, in other species different regions of the brain present differential patterns of ontogenic changes.

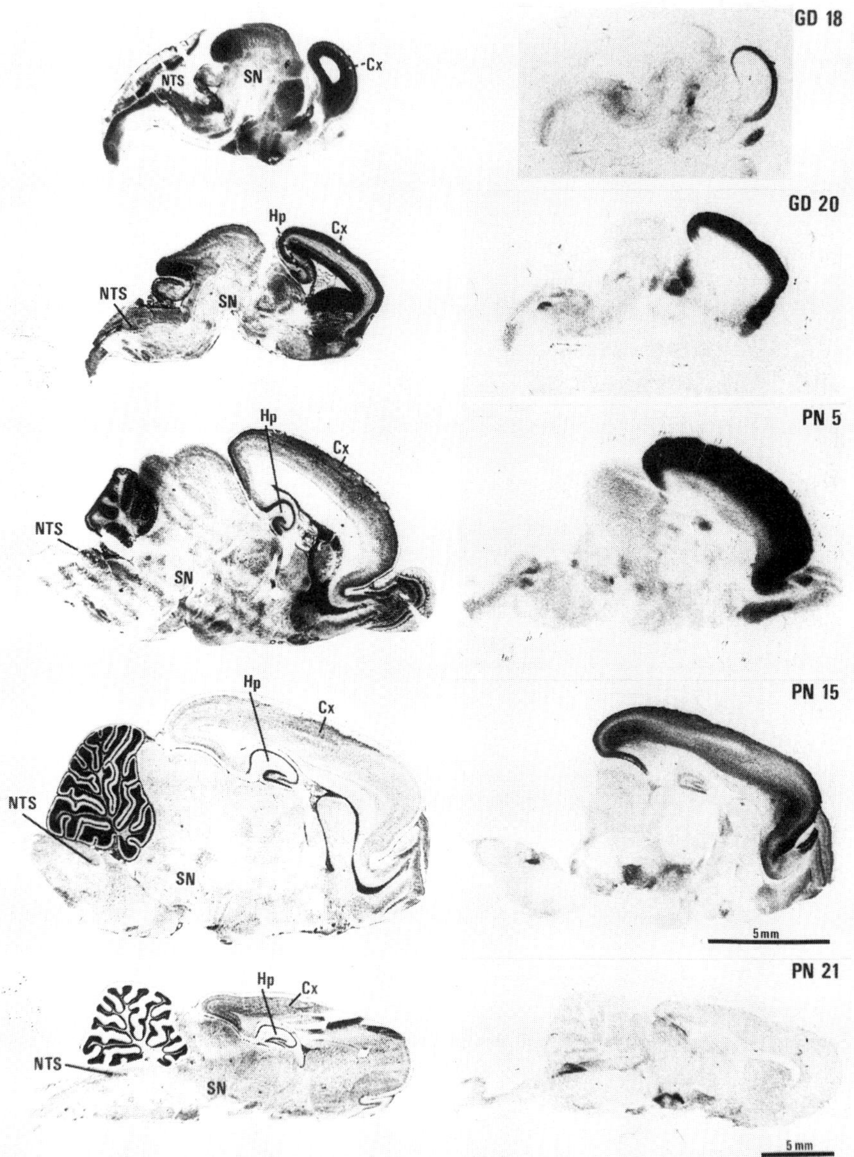

Fig. 9.4.   Photomicrographs of sagittal sections of rat brain and cervical spinal cord at different ages during gestation (GD18, GD20) and early postnatal stages (PN5, PN15, PN21). See legend to Fig. 9.3 for details. NTS, nucleus tractus solitarius; SN, substantia nigra.

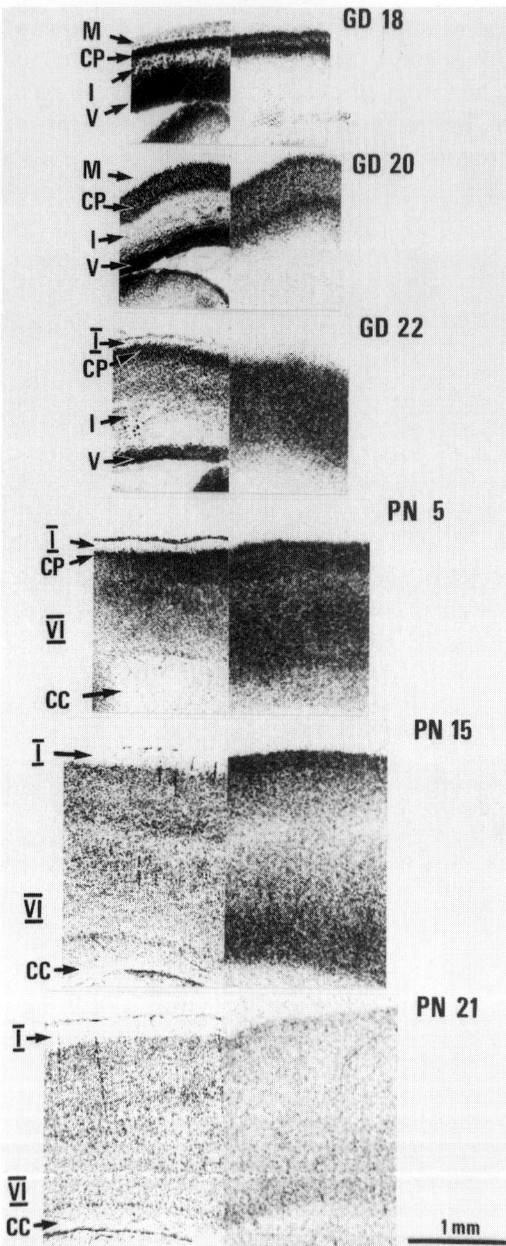

Fig. 9.5. [³H] Neurotensin binding sites in coronal sections of the neocortex of the rat brain at different pre- (GD18, GD20, GD22) and postnatal (PN5, PN15, PN21) ages. See legend to Fig. 9.3 for details. CC, corpus callosum; CP, cortical plate; I, subplate layer; V, VI, layers V and VI.

9.4.3 NEUROTENSIN RECEPTOR mRNA DISTRIBUTION IN DEVELOPING BRAIN: CORRELATION WITH NEUROTENSIN BINDING SITES

Although developmental changes in the density of NT binding sites were observed (Kiyama *et al.*, 1987; Palacios *et al.*, 1988), the mechanisms involved in the regulation of NTR densities which may account for these modifications have not yet been extensively investigated. The recent cloning and sequencing of a NTR gene allows exploration of new aspects of NTR regulation during the development of the CNS. Using *in situ* hybridization histochemistry we have examined the presence of NTR mRNA synthesizing cells in several brain regions during development. The relative content of NTR mRNA at each age is summarized in Table 9.2. NTR mRNA in adult animals was detected in significant amounts only in substantia nigra pars compacta, ventral tegmental area, retrorubral field and medial lemniscus whereas in young animals NTR mRNA was seen in a larger number of brain areas (Table 9.2).

In situ hybridization histochemistry of NTR mRNA revealed two different patterns of expression during the postnatal development of the CNS. In some areas NTR mRNA levels progressively increased after birth, reaching the adult pattern after the third postnatal week (i.e. in substantia nigra pars compacta, ventral tegmental area and nucleus basalis) whereas in other regions it peaked at different times, decreasing later to adult levels (Table 9.2). The peaks were observed at different times.

(a) Cortical areas

The levels of NTR mRNA were very high in cortex of young animals but followed a distinct complex pattern depending on both age and cortical area. The motor area of frontal cortex expressed NTR mRNA at higher levels in the internal layers than in the sensory area, where it was expressed in both

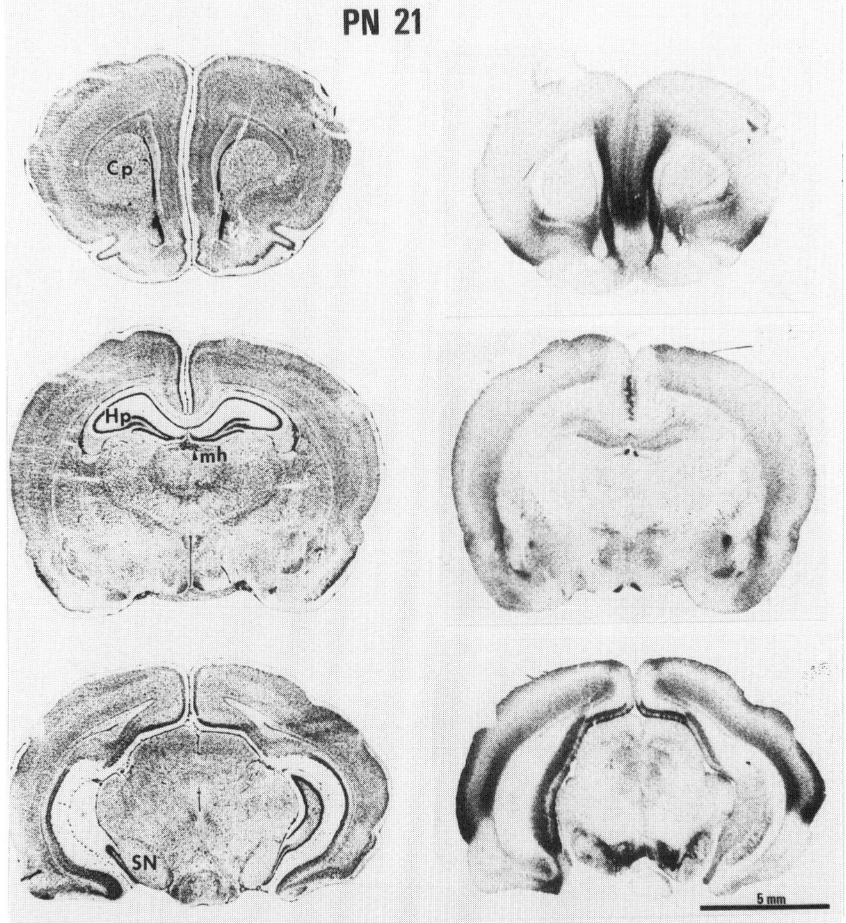

Fig. 9.6.   Coronal sections of the rat brain at PN15. See Fig. 9.3 legend for details. Cp, caudate putamen; mh, medial habenula.

the external and the internal layers (Table 9.2). By PD5, in the anterior part of frontal cortex motor area, the differences in the hybridization signal between the external and the internal layers were more evident, being stronger in the internal layer (Fig. 9.7D,F). However, in the posterior part of the brain the mRNA levels were high in the external layer (Fig. 9.7H,J). In contrast, at this time in the sensory area the highest mRNA levels were always in the internal layers (Fig. 9.7J,L). At PD1 all layers of retrosplenial cortex express NTR mRNA. During development the internal layers dis-

play very low levels of mRNA and by PD10 it is only detectable in the more external layers (Table 9.2). At PD15 moderate levels of mRNA expression are seen (Fig. 9.10F) becoming undetectable by PD20 (Table 9.2).

(b)   Nigrostriatal system and basal forebrain

The dopaminergic efferents are the more important projections originating in substantia nigra and innervating the caudate-putamen, amygdala and globus pallidus. Palacios and

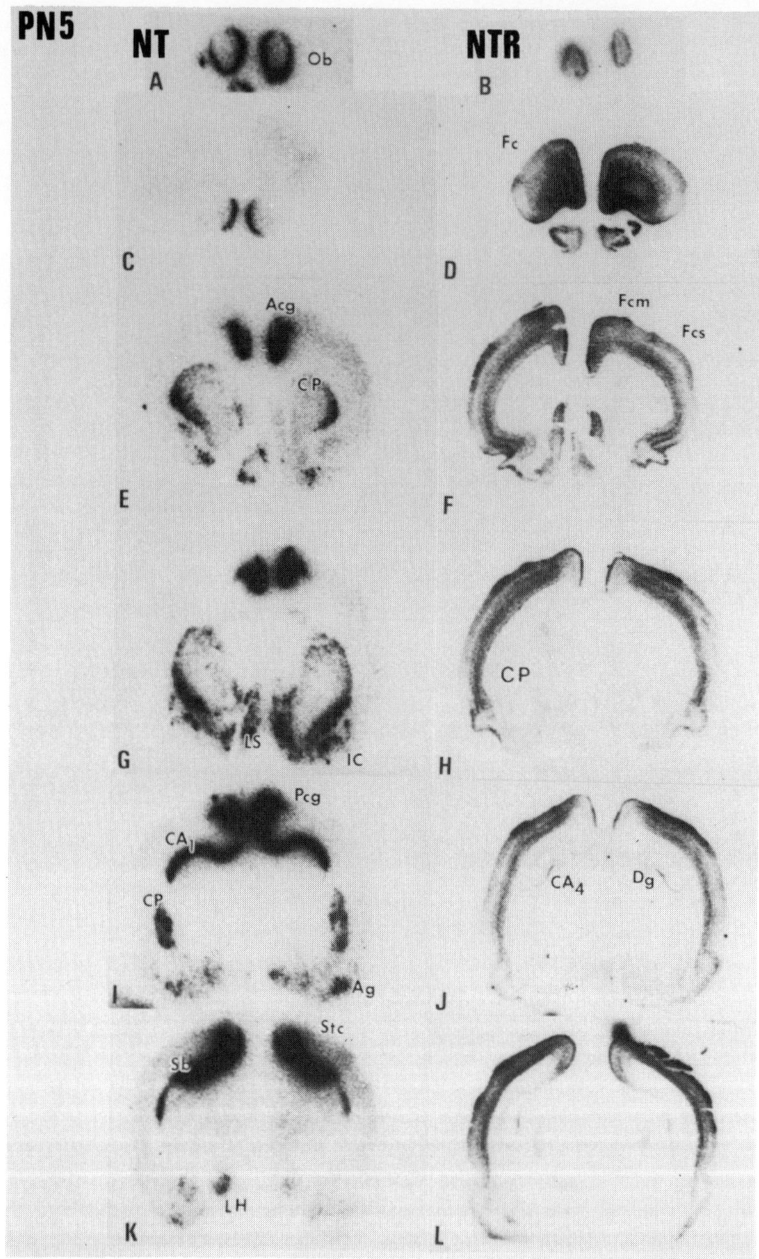

Fig. 9.7. Autoradiographic visualization of mRNA coding for neurotensin (left column) and neurotensin receptor (right column) in the rat brain at PN5. The images are photographs from autoradiograms generated by hybridization of tissue sections with oligonucleotide probes complementary to selected regions of the NT and NTR genes. Dark areas are rich in hybridization signal. Abbreviations: Acg, anterior cingulate cortex; Ag, amygdaloid nuclei; CA1, CA4, fields of Ammon's Horn; CP, cortical plate; Dg, dentate gyrus; Fcm, Fcs, frontal cortex motor and sensory areas; GB, globus pallidus; Hb, habenular nuclei; HG, hypoglossus nucleus; IC, inferior colliculus; LH, lateral hypothalamic area; LS, lateral septum; Ltdg, laterodorsal tegmental nucleus; Ms, medial septum; NB, nucleus basalis; Ob, olfactory bulb; Pcg, posterior cingulate cortex; Po, primary olfactory cortex; Pst, principal sensory trigeminal nucleus; RSS, retrosplenial cortex; SNC, substantia nigra compacta; Sb, subiculum; Stc, striate cortex; VDB, diagonal band of Broca's vertical limb; VMH, ventromedial hypothalamic nucleus; VTA, ventral tegmental area. Reproduced from García-Ladona *et al.* (1991), with permission.

Table 9.2.   Relative levels of NTR mRNA in different rat brain regions during postnatal development

Regions	PD1	PD5	PD10	PD15	PD20	Adult
Nucleus basalis	-	-	+++	++++	+++++	+++++
Substantia nigra	-	-	++	+++	++++	++++
Ventral tegmental area	-	-	++	+++	++++	++++
Olfactory bulb	++	+++	++++	+++++	+++	+
Ammon's Horn	++	++	++	++	-	-
Dentate gyrus granule cells	++	++	++	++	-	-
Hilus	++	++++	+++	++	-	-
Habenular nucleus	-	-	+++	+++	-	-
Nucleus geniculate lateral	+++	+++	+	-	-	-
Neocortex						
Frontal cortex somatosensory internal	+++++	++++++	++	-	-	-
Frontal cortex somatosensory external	+++++	++++	++	-	-	-
Frontal cortex motor internal	++++++	++++++	++	-	-	-
Frontal cortex motor external	++++	+++	++	-	-	-
Anterior cingulate internal	+++	+++	++	-	-	-
Anterior cingulate medium	+++	++++	+++	-	-	-
Anterior cingulate external	+++	+++	+	-	-	-
Retrosplenial internal	+++++	++++	-	-	-	-
Retrosplenial external	+++++	++++	++++	++++	+	-
Striate internal	+++++	+++++	+++	++	-	-
Striate medium	+++++	+++++	+++	++	-	-
Striate external	+++++	+++++	+++	-	-	-

-, no significant amounts were detected.

Kuhar (1981) have shown that NTR binding sites are localized in the dopaminergic cells of substantia nigra, further adding to the wealth of data on NT interactions with the dopaminergic system (Nemeroff, 1980; Govoni *et al.*, 1980; Hökfelt *et al.*, 1984; Shi and Bunney, 1991; Levant *et al.*, 1991). There is increasing evidence for a general presence of NT binding sites in the nigrostriatal and other dopaminergic systems (Hokfelt *et al.*, 1984; Govoni *et al.*, 1980; Levant *et al.*, 1991; Nemeroff *et al.*, 1980; Shi and Bunney, 1991). Autoradiographic studies (Young and Kuhar, 1981; Quirion *et al.*, 1982; Moyse *et al.*, 1987) reported the presence of NT binding sites in basal ganglia and related zones (i.e. caudate-putamen, islands of Calleja, olfactory tubercle, substantia nigra, and ventral tegmental area). However, in adult animals NTR mRNA expression was found mainly in substantia nigra and ventral tegmental area (Fig. 9.11A), whereas in caudate-putamen and islands of Calleja no significant levels were observed in these animals.

During development, significant levels of NTR mRNA were seen in the substantia nigra pars compacta and ventral tegmental area as early as PD10 and increased thereafter (Table 9.2). Although a postnatal increase in [^{3}H]NT binding sites was described at the same period, high levels of binding were found even at younger ages (Palacios *et al.*, 1988) (Figs 9.2B and 9.4).

(c) Hippocampal formation

Intermediate levels of NTR mRNA were observed at PD1 in the Ammon's horn, hilus and granule cells of dentate gyrus (Table 9.2). Between PD5 and PD10 high mRNA levels were observed in the hilus but decreased by PD15 (Fig. 9.10B,C). NT mRNA content decreased during development in the Ammon's horn fields, thus in these areas no detectable amount of NTR mRNA was present in adult animals.

The pattern of expression of NTR mRNA in the hippocampal formation during development agrees with previous data on NT binding sites (Palacios *et al.*, 1988). NTR is mainly expressed until PD15 in the CA4 field of Ammon's horn and infragranular and granular layers of dentate gyrus. During the postnatal period, the dentate gyrus is still developing and, at the time we detected NTR mRNA and NT binding sites (Palacios *et al.*, 1988), granule cells migrate from the hilus towards the granular layer (Bayer, 1980), thus supporting the hypothesis that NT plays a role in the maturation of the dentate gyrus and its related connections.

Autoradiographic studies have shown the presence of high levels of NT binding sites in the subiculum of adult hippocampus (Moyse *et al.*, 1987) which disagrees with *in situ* hybridization data. This mismatch possibly indicates that NT binding sites present in this area are presynaptic receptors in the afferent fibers.

(d) Olfactory system

NTR mRNA was expressed in the first postnatal week (Fig. 9.8B) in the internal granule cell layer and at lower levels in the external plexiform cell layer. It increased until PD15 (Fig. 9.9B) decreasing thereafter (Table 9.2). NTR binding was detected in the plexiform and internal granular layers of adult olfactory system (Young and Kuhar, 1981; Moyse *et al.*,

1987), indicating the existence of NT circuits in the olfactory bulb.

The transient expression of NTR could indicate a neurotrophic effect of NT during the maturation of different CNS regions. This effect could be important for migrating cells in cortical structures as well as granule cells in dentate gyrus. These hypotheses were also suggested by Gonzalez *et al.* (1991) on somatostatin receptors, showing also a changing developmental pattern of somatostatin binding in cortex. Transient expression of NTR in cortex is possibly related to cortical cell death. In fact, such a phenomenon seems to be the cause of observed transient expression of other neuropeptidergic systems in brain cortex (Parnavelas and Cavanaugh, 1988; Fitzpatrick-McElligott *et al.*, 1991).

Another possible explanation for the transient presence of NTR mRNA could be a developmental regulation of NTR while synaptic contacts are taking place in target areas (being NTR presynaptic components) which could also explain the mismatch between the mRNA and the binding site. However, because of the different levels of sensitivity of the techniques used in these studies, caution has to be exerted in interpreting these results. Finally, the possibility of a neuropeptide–receptor interaction during maturation cannot be ruled out as it was proposed for cortical structures (Cowan *et al.*, 1984). For instance, the transient expression of NTR mRNA in hippocampus could be related to neurotensinergic efferent connections from other lymbic structures such as septum and diagonal band given that during development a high expression of NT mRNA was observed in these areas peaking around the second postnatal weak (see below).

Data from lesion studies led to the hypothesis that levels of the agonist may regulate the presence of neurotransmitter receptors in target areas. Surgical or chemical denervation paradigms demonstrated an up-regulation of receptor density in several neurotransmitter systems such as adrenergic receptors in lateral geniculate body (Menkes *et al.*, 1983) and

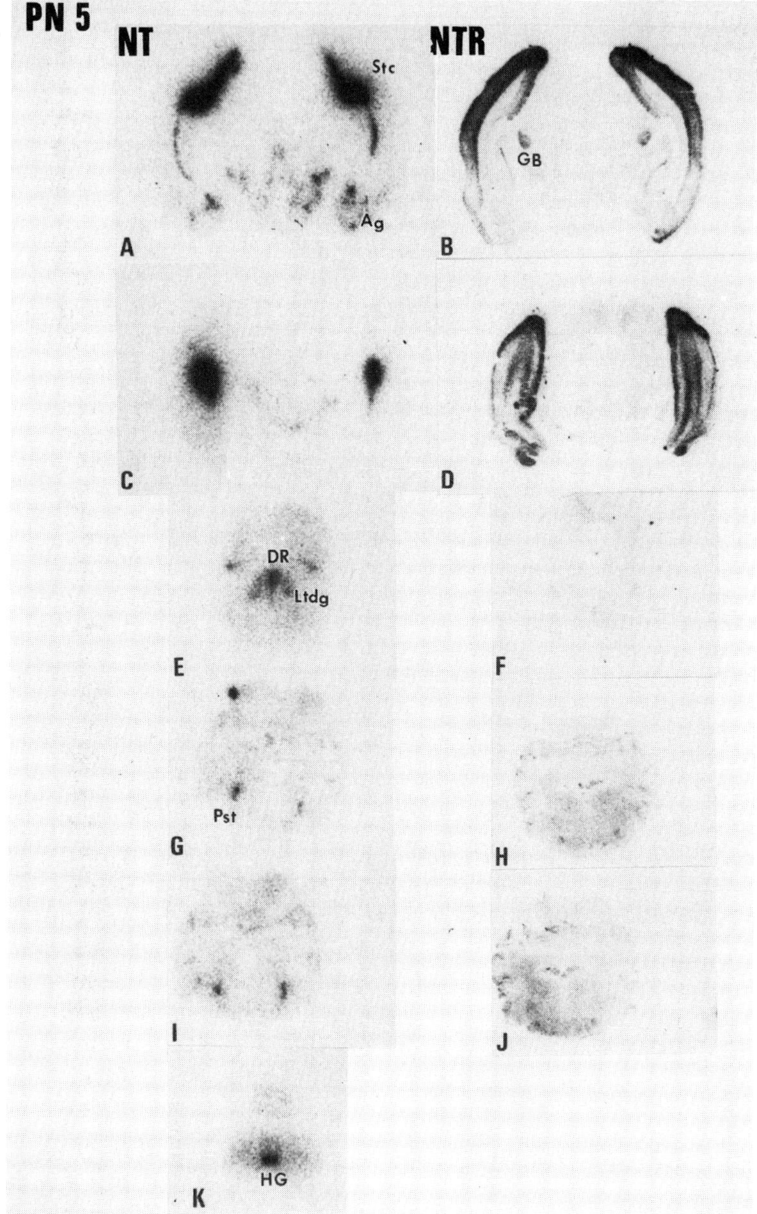

Fig. 9.8.   Postnatal day 5. See legend to Fig. 9.7 for details.

cortex (Minneman *et al.*, 1981), and GABA receptors in substantia nigra (Penney *et al.*, 1981). Similar results were obtained for NTR in substantia nigra after chronic neuroleptic treatments (Uhl and Kuhar, 1984). The possibility that such a regulation occurs during developmental synaptic maturation of CNS cannot be excluded (Hunt, 1988). This possibility is analysed for the neurotensinergic system in this section. A comparison of the genetic

Table 9.3.   Relative levels of NT mRNA in different rat brain areas during postnatal development

Regions	PD1	PD5	PD10	PD15	PD20	Adult
Caudate putamen	+++++	+++++	++++	+++	+++	+++
Islands of Calleja	+++++	+++++	+++	+++	+++	+++
Olfactory bulb	++++++	++++	+++	++	++	+
Subiculum	++++++	+++++	++++	++++	++++	++++
Ammon's Horn	+++	++	++	++	-	-
Dentate gyrus	+++	++	++	++	-	-
Amygdala	++++	++++	+++	++	+	+
Septum	++	++++	+++	+++	++	++
Diagonal band	+	+++	++++	+++	+++	+++
Lateral hypothalamus	+++++	++++	++++	+++	-	-
Inferior olive	-	-	++++	++++	++	++
Dorsal raphe	++++	+++	+++	++	++	+
Lateral tegmental nucleus	+++	++	+	+	+	+
Dorsal parabrachialis nucleus	+++	+++	+++	++	-	-
Ventral parabrachialis nucleus	+++	+++	+++	++	-	-
Dorsal lateral lemniscus	-	+++	++	+	+	+
Dorsal cochlear	-	++++	++++	++++	++++	++++
Lateral vestibular	-	++++	++++	++++	++++	++++
Parvocellular nucleus	-	++++	++++	++++	++++	++++
Solitarius nucleus	-	++++	++++	++++	++++	++++
Spinal trigeminal nucleus	-	-	++++	++++	+++	+++

-, no significant levels were detected.

expression of NTR and NT can help to understand how the presence of the transmitter may control the presence of the receptor during ontogeny.

9.4.4 DEVELOPMENTAL PATTERN OF NEUROTENSIN mRNA LEVELS IN RAT BRAIN

The levels and distribution of NT mRNA during postnatal development in the rat brain have been examined using *in situ* hybridization histochemistry (Sato *et al.*, 1990; Kiyama *et al.*, 1991a,b; our unpublished results). In adult animals, NT mRNA was detected in more regions that NTR mRNA: dorsal and ventral areas of the striatum, nucleus accumbens, Calleja islands, olfactory tubercle, inferior olive, nucleus hypoglossus, nucleus spinalis trigemini, dorsal raphe, cerebellar peduncle (interal tegmental nucleus),

parabranchial nucleus, dorsal cochlear nucleus, vestibular nucleus, nucleus tractus solitarius. During development NT mRNA was also detected in other CNS regions (Table 9.3).

As for NTR, different patterns of NT mRNA content were observed during the postnatal development (Table 9.3). High levels of NT mRNA were observed in the first postnatal days decreasing progressively with time in the intermediate layers of anterior and posterior cyngulate cortex, caudate-putamen, islands of Calleja, olfactory bulb, CA1 field of Ammon's horn, dentate gyrus, amygdala, lateral hypothalamic area (Figs 9.7A–K; 9.9A–I; 9.10A–G), dorsal raphe, lateral tegmental nucleus, dorsoparabranchial nucleus and ventroparabranchial nucleus and inferior olive (Fig. 9.8A–K). Relatively high levels of NT mRNA were found through the developmental period mainly in subiculum, dorsal cochlear nucleus, lateral vestibular nucleus,

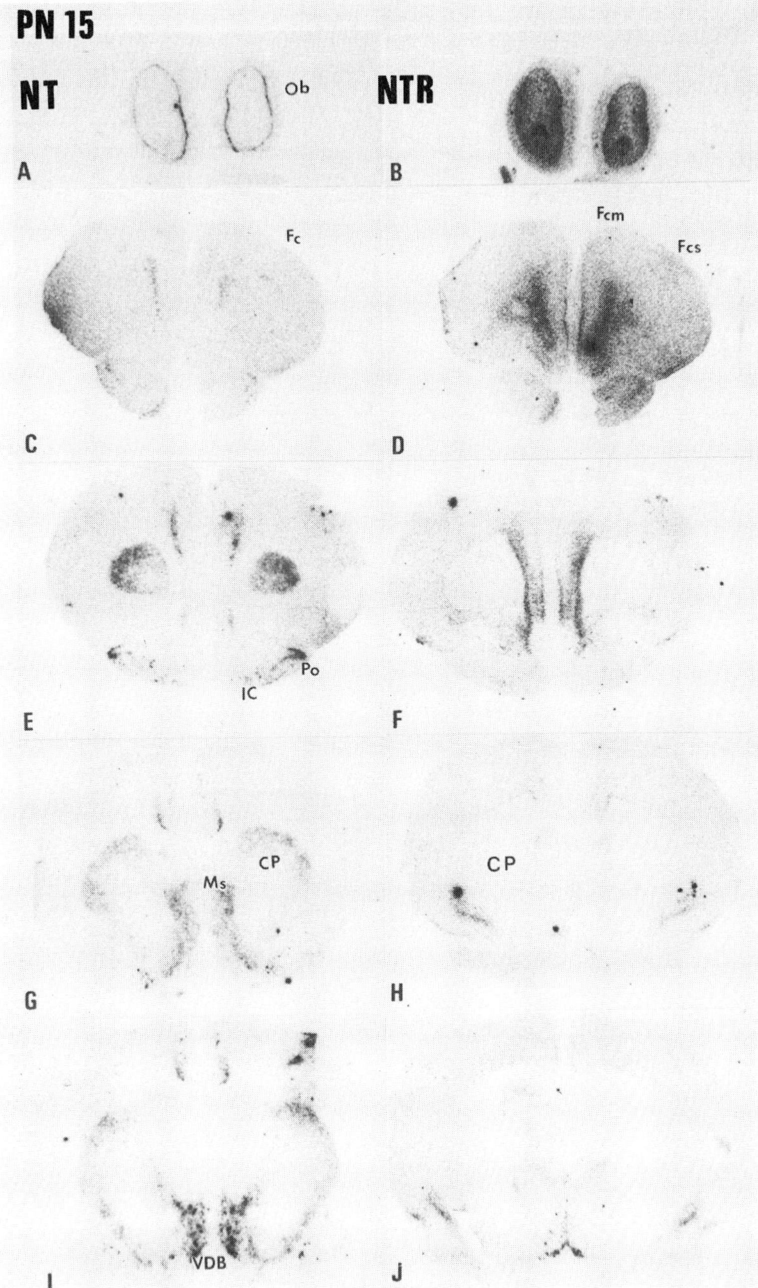

Fig. 9.9.   Postnatal day 15. See legend to Fig. 9.7 for details.

parvocellular nucleus, nucleus solitarius and spinal trigeminal nucleus, although these brainstem nuclei displayed detectable levels of NT mRNA only after PD5. Finally, NT mRNA content peaked at different ages of postnatal development in septum, diagonal band and inferior olive.

In general, hybridization data for NT

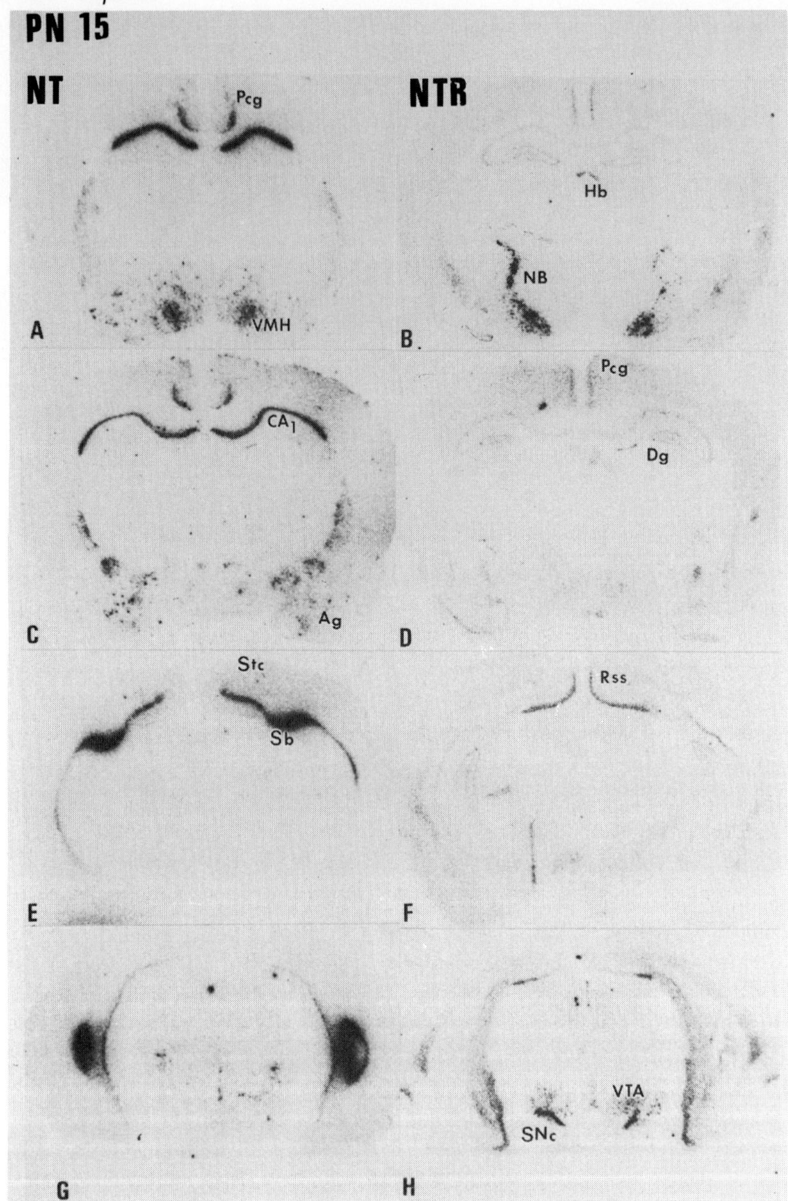

Fig. 9.10.   Postnatal day 15. See legend to Fig. 9.7 for details.

mRNA and immunohistochemical results are in good agreement in adult and developing brain (Hara *et al.*, 1982; Minagawa *et al.*, 1983). However, mismatches between the levels of peptide and mRNA can be observed. Several brain areas displayed high levels of NT mRNA through development although immunopositive cells or fibers have not been described (subiculum, CA1 field of the hippocampus) or are seen only in the early postnatal period (brain stem nuclei). These mismatches between NT and NT mRNA can

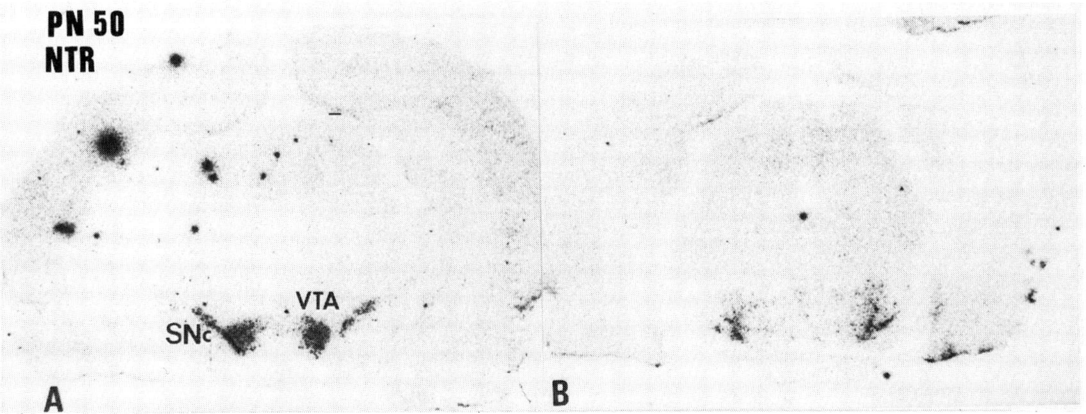

Fig. 9.11.   Visualization of NTR mRNA at postnatal day 50. Patterns at PN50 are similar to adult.

be explained by different mechanisms involving NT gene expression and processing in different steps (Uhl and Nishimori, 1990).

The NT gene can be actively expressed, translated and the peptide transported to distant target areas or turnover rates of both NT and NT mRNA are very slow. As mentioned above, the NT gene expression is regulated by second messenger pathways (Dobner *et al.*, 1988; Tischler *et al.*, 1991) and this could be specially important in the perinatal period. We cannot rule out the possibility that the processing of the peptide is attenuated in the earlier steps of development because the machinery to process the peptide is still immature, as suggested for other systems (Rehfeld, 1990). Thus, although the NT mRNA is present, NT is not easily detectable by immunohistochemistry.

Differential processing of neuropeptide genes is a common characteristic of several peptidergic systems (Uhl and Nishimori, 1990). It was demonstrated that NT gene contains a NT-like sequence as well as the neuromedin N sequences (Dobner *et al.*, 1987; Kislauskis *et al.*, 1988). During development a different processing of the gene coding for NT and neuromedin can be activated at different developmental steps, thus, in adults the expression of NT-containing gene does not necessarily imply NT

production. This mechanism of differential processing has been suggested for NT in a recent study demonstrating different ratios of NT/neuromedin gene products in several brain regions (Kitabgi *et al.*, 1991).

The comparative analysis of the immunohistochemical and autoradiographic studies revealed the existence of several mismatches between the presence of fibers and/or NT-immunopositive cells and NTR. To explain these mismatches between neurotransmitters and receptors, different hypotheses have been proposed (for extensive review see Herkenham, 1987). In the neurotensinergic system, the most striking mismatch is possibly that found in developing neocortex. High levels of NTR mRNA and NT binding sites are observed during early development whereas either NT-immunoreactive fibers or cells, or NT mRNA levels were not detected except in some specific cases. In this case, NT would act as a paracrine peptide by diffusion from distant areas. A direct regulation of NTR by the presence of NT can be expected for instance in the septohippocampal and inferior olive system. In these areas, NT mRNA is present in increasing amounts during the first postnatal week whereas NTR mRNA and NT binding progressively decrease in the same period (Tables 9.2 and 9.3).

9.5 CONCLUDING REMARKS

The understanding of the molecular mechanisms involved in the maturation of the neurotensinergic system during postnatal development of the CNS is still incomplete.

Receptor autoradiography studies (Palacios *et al.*, 1988) have revealed the transient presence of NT binding sites in several brain regions even at the prenatal period. This phenomenon has also been described for other neuropeptides (Gonzalez *et al.*, 1991; Tribollet *et al.*, 1991). Although transient expression for some of the classical neurotransmitters can be the result of simple synaptic modifications occurring during postnatal development, for the case of NT and other neuropetides it may also imply a direct neurotrophic role (Hanley, 1985) during this period of maturation. It is evident that there are mismatches between the presence of the NT and mRNAs (Uhl and Nishimori, 1990) and/or by the transport of the molecule far from the area of synthesis. Mismatches between the presence of NT-immunopositive structures and binding sites or NTR and their respective mRNAs can be explained by the different rate of turnover of both protein and mRNAs.

Whereas in some brain regions the NT and NTR mRNA contents may account for a down-regulation induced by increasing concentration of the neuropeptide (i.e. septo-hippocampal formation), in others, such regulation is not so clear as NT mRNA decreases at the same time as NTR mRNA is up-regulated (striatonigral system). However, further studies should be carried out to verify if lack of detection of NT immunopositive cells or fibers is due to a rapid turnover or transport of the peptide. Also, the obtaining of antibodies raised against synthetic peptides from the NTR protein sequence may help in the elucidation of real mismatches of neurotensinergic system as well as the mechanisms of synaptic maturation during development.

ACKNOWLEDGEMENTS

Part of the studies reviewed in this chapter were carried out at Preclinical Research, Sandoz Pharma Ltd, Basel (Switzerland), and in collaboration with the Institute of Pathology, University of Basel, particularly with Professor A. Probst. We gratefully acknowledge the help of all our colleagues at these institutions. We also thank Dr R.G.W. Gristwood for critical reading of the manuscript, and Ms M.D. Moliné for expert clerical help.

REFERENCES

Alexander, M.J., Miller, M.A., Dorsa, D.M. *et al.* (1989) Distribution of neurotensin/neuromedin N mRNA in rat forebrain: unexpected abundance in hippocampus and subiculum. *Proc. Natl. Acad. Sci. USA*, **86**, 5202–6.

Bayer, S.A. (1980) Development of the hippocampal region in the rat. I. Neurogenesis examined with ^{3}H-thymidine autoradiography. *J. Comp. Neurol.*, **190**, 87–114.

Bozou, J.C., Amar, S., Vincent, J.P. and Kitabgi, P. (1986) Neurotensin mediated inhibition of cyclic AMP formation in neuroblastoma N1E-115 cells: involvement of the inhibitory GTP-binding component of adenylate cyclase. *Mol. Pharmacol.*, **29**, 489–96.

Cowan, W.M., Fawcett, J.W., O'Leary, D.D.M. and Stanfield, B.B. (1984) Regressive events in neurogenesis. *Science*, **225**, 1258–65.

Dobner, P.R., Barber, D.L., Villa-Komaroff, L. and McKiernan, C. (1987) Cloning and sequence analysis of cDNA for the canine neurotensin/neuromedin N precursor. *Proc. Natl. Acad. Sci. USA*, **84**, 3516–20.

Dobner, P.R., Tischler, A.S., Lee, Y.C. *et al.* (1988) Lithium dramatically potentiates neurotensin/neuromedin N gene expression. *J. Biol. Chem.*, **263**, 13983–6.

Fitzpatrick-McElligott, S., Card, J.P., O'Kane, T.M. and Baldino, F. (1991) Ontogeny of somatostatin mRNA-containing perikarya in the rat central nervous system. *Synapse*, **7**, 123–34.

Garcia-Ladona, F.J., Palacios, J.M. and Mengod, G. (1992) Neurotransmitter/receptor mismatches during ontogeny: an in situ hybridization correlative study of the ontogenic expression of neurotensin and neurotensin receptor mRNAs in rat brain. (Submitted.)

Gilbert, J.A. and Richelson, E. (1984) Neurotensin stimulates formation of cyclic GMP in murine neuroblastoma clone N1E-115. *Eur. J. Pharmacol.*, **99**, 245–6.

Goedert, M., Pinnock, R.D, Downes, C.P. *et al.* (1984a) Neurotensin stimulates inositol phospholipid hydrolysis in rat brain slices. *Brain Res.*, **323**, 193–7.

Goedert, M., Pittaway, K., Williams, B.J. and Emson P.C. (1984b) Specific binding of tritiated neurotensin to rat brain membranes: characterization and regional distribution. *Brain Res.*, **304**, 71–81.

Goedert, M., Mantyh, P.W., Emson, P.C. and Hunt, S.P. (1984c) Inverse relationship between neurotensin receptors and neurotensin-like immunoreactivity in the cat striatum. *Nature*, **307**, 543–6.

Gonzalez, B.J., Leroux, P., Bodenant, C. and Vaudry, H. (1991) Ontogeny of somatostatin receptors in the rat somatosensory cortex. *J. Comp. Neurol.*, **305**, 177–88.

Govoni, S., Hong, J.S., Yang, H.Y.T. and Costa, E. (1980) Increase of neurotensin content elicited by neuroleptics in nucleus accumbens. *J. Pharmacol. Exp. Ther.*, **215**, 413–17.

Hanley, M.R. (1985) Neuropeptides as mitogens. *Nature*, **315**, 14–15.

Hara, Y., Shiosaka, S., Senba, E. *et al.* (1982) Ontogeny of the neurotensin-containing neuron system of the rat: immunohistochemical analysis. I. Forebrain and diencephalon. *J. Comp. Neurol.*, **208**, 177–95.

Herkenham, M. (1987) Mismatches between neurotransmitter and receptor localizations in brain: observations and implications. *Neuroscience*, **23**, 1–38.

Hökfelt, T., Everitt, B.J., Theodorson-Norheim, E. and Goldstein, M. (1984) Occurrence of neurotensin-like immunoreactivity in subpopulations of hypothalamic, mesencephalic, and medullary catecholamine neurons. *J. Comp. Neurol.*, **222**, 543–59.

Hunt, S.P. (1988) The development of neurotransmitter receptors, in *The Making of the Nervous System* (eds J.G. Parnavelas, C.D. Stern and R.V. Stirling), Oxford University Press, Oxford, pp. 454–72.

Inagaki, S., Shinoda, K., Kubota, Y. *et al.* (1983) Evidence for the existence of a neurotensin-containing pathway from the endopiriform nucleus and the adjacent prepiriform cortex to the anterior olfactory nucleus and nucleus of diagonal band (Broca) of the rat. *Neuroscience*, **8**, 487–93.

Jennes, L., Stumpf, W.E. and Kalivas, P.W. (1982) Neurotensin: topographical distribution in rat brain by immunohistochemistry. *J. Comp. Neurol.*, **210**, 211–24.

Kahn, D., Abrams, G.M., Zimmerman, E.A. *et al.* (1980) Neurotensin neurons in the rat hypothalamus: an immunocytochemical study. *Endocrinology*, **107**, 47–54.

Kalivas, P.W., Jennes, L., Nemeroff, C.B. and Prange Jr, A.J. (1982) Neurotensin: topographical distribution of brain sites involved in hypothermia and antinociception. *J. Comp. Neurol.*, **210**, 225–38.

Kalivas, P.W., Burgess, S.K., Nemeroff, C.B. and Prange Jr., A.J. (1983) Behavioural and neurochemical effects of neurotensin microinjection into the ventral tegmental area of the rat. *Neuroscience*, **8**, 495–505.

Kasckow, J., Cain, S.T. and Nemeroff, C.B. (1991) Neurotensin effects on calcium/calmodulin-dependent protein phosphorylation in rat neostriatal slices. *Brain Res.*, **545**, 343–6.

Kislauskis, E., Bullock, B., McNeil, S. and Dobner, P.R. (1988) The rat gene encoding neurotensin and neuromedin N. *J. Biol. Chem.*, **263**, 4963–8.

Kitabgi, P., Rostene, W., Dussaollant, M. *et al.* (1987) Two populations of neurotensin binding sites in murine brain: discrimination by the antihistaminic levocastine reveals markedly different radioautographic distribution. *Eur. J. Pharmacol.*, **140**, 285–93.

Kitabgi, P., Masuo, Y., Nicot, A. *et al.* (1991) Marked variations of the relative distributions of neurotensin and neuromedin N in micro-punched rat brain areas suggest differential processing of their common precursor. *Neurosci. Lett.*, **124**, 9–12.

Kiyama, H., Inagaki, S., Kito, S. and Tohyama, M. (1987) Ontogeny of [³H]neurotensin binding sites in the rat cerebral cortex: autoradiographic study. *Dev. Brain Res.*, **31**, 303–6.

Kiyama, H., Emson, P.C., Sato, M. and Tohyama, M. (1991a) The transient appearance of proneurotensin mRNA in the rat hypoglossal nucleus during development. *Dev. Brain Res.* **58**, 293–6.

Kiyama, H., Sato, M., Emson, P.C. and Tohyama, M. (1991b) Transient expression of neurotensin mRNA in the mitral cells of rat olfactory bulb during development. *Neurosci. Lett.*, **128**, 85–9.

Kiyama, H., Shiosaka, S., Sakamoto, N. *et al.* (1986) A neurotensin immunoreactive pathway from the subiculum to the mammillary body in the rat. *Brain Res.*, **375**, 357–9.

Kohler, C., Radesater, A.C., Hall, H. and Winblad, B. (1985) Autoradiographic localization of [³H]neurotensin-binding sites in the hippocampal region of the rat and primate brain. *Neuroscience*, **16**, 577–87.

Levant, B., Bissette, G., Widerlov, E. and Nemeroff, C.B. (1991) Alterations in regional brain neurotensin concentrations produced by atypical antipsychotic drugs. *Regul. Pept.*, **32**, 193–201.

Mailleux, P. and Vanderhaeghen, J.-J. (1988) Transient neurotensin in the human inferior olive during development. *Brain Res.*, **456**, 199–203.

Mailleux, P., Schiffmann, S.N., Halleux, P. *et al.* (1989) Transient neurotensin in the cat inferior olive during development. *Neurochem. Int.*, **14**, 159–61.

Mailleux, P., Pelaprat, D. and Vanderhaeghen, J.-J. (1990) Transient neurotensin high-affinity binding sites in the human inferior olive during development. *Brain Res.*, **508**, 345–8.

Mazella, J., Poustis, C., Labbe, C. *et al.* (1983) Monoiodo-[Trp11]neurotensin, a highly radioactive ligand of neurotensin receptors. *J. Biol. Chem.*, **258**, 3476–81.

Mazella, J., Chabry, J., Kitabgi, P. and Vincent, J.P. (1988) Solubilization and characterization of active neurotensin receptors from mouse brain. *J. Biol. Chem.*, **263**, 144–9.

Mazella, J., Chabry, J. and Vincent, J.P. (1989) Purification of the neurotensin receptor from mouse brain by affinity chromatography. *J. Biol. Chem.*, **264**, 5559–63.

Menkes, D.B., Gallager, D.W., Reinhard, J.F. and Aghajanian, G.K. (1983) Alpha-adrenoceptor denervation supersensitivity in brain: physiological and receptor binding studies. *Brain Res.*, **272**, 1–12.

Mills, A., Demoliou-Mason, C.D. and Barnard, E. (1988) Purification of the neurotensin receptor from bovine brain. *J. Biol. Chem.*, **263**, 13–16.

Minagawa, H., Shiosaka, S., Inagaki, S. *et al.* (1983) Ontogeny of neurotensin-containing neuron system of the rat: immunohistochemical analysis II. Lower brain stem. *Neuroscience*, **8**, 467–86.

Minneman, K.P., Pittman, R.N. and Molinoff, P.B. (1981) β-adrenergic receptor subtypes: properties, distribution and regulation. *Annu. Rev. Neurosci.*, **4**, 419–61.

Moyse, E., Rostène, W., Vial, M. *et al.* (1987) Distribution of neurotensin binding sites in rat brain: a light microscopic radioautographic

study using monoiodo[¹²⁵I]Tyr3-neurotensin. *Neuroscience*, **22**, 525–36.

Nemeroff, C.B. (1980) Neurotensin: perchance an endogenous neuroleptic? *Biol. Psychiatry*, **15**, 283–302.

Palacios, J.M. and Kuhar, M.J. (1981) Neurotensin receptors are located on dopamine-containing neurones in rat midbrain. *Nature*, **294**, 587–9.

Palacios, J.M., Pazos, A., Dietl, M.M. *et al.* (1988) The ontogeny of brain neurotensin receptors studied by autoradiography. *Neuroscience*, **25**, 307–17.

Parnavelas, J.G. and Cavanagh, M.E. (1988) Transient expression of neurotransmitters in developing neocortex. *Trends Pharmacol. Sci.*, **11**, 92–3.

Penney, J.B., Pan, H.S., Young, A.B. *et al.* (1981) Quantitative autoradiography of [³H]muscimol binding in the rat brain. *Science*, **214**, 1036–8.

Polak, J.M. (ed.) (1989) *Regulatory Peptides*, Birkhäuser Verlag, Basel.

Quirion, R., Gaudreau, P., St Pierre, S. *et al.* (1982) Autoradiographic distribution of [³H]neurotensin receptors in rat brain: visualization by tritium sensitive film. *Peptides*, **3**, 757–63.

Quirion, R., Chiueh, C.C., Everist, H.D. and Pert, A. (1985) Comparative localization of neurotensin receptors on nigrostriatal and mesolimbic dopaminergic terminals. *Brain Res.*, **327**, 385–9.

Rehfeld, J.F. (1990) Posttranslational attenuation of peptide gene expression. *FEBS Lett.*, **268**, 1–4.

Roberts, G.W., Woodhams, P.L., Polak, J.M. and Crow, T.J. (1984) Distribution of neuropeptides in the limbic system of the rat: the hippocampus. *Neuroscience*, **11**, 35–77.

Sato, M., Lee, Y., Zhang, J.H. *et al.* (1990) Different ontogenic profiles of cells expressing preproneurotensin/neuromedin N mRNA in the rat posterior cingulate cortex and the hippocampal formation. *Dev. Brain Res.*, **54**, 249–55.

Sato, M., Shiosaka, S. and Tohyama, M. (1991) Neurotensin and neuromedin N elevate the cytosolic calcium concentration via transiently appearing neurotensin binding sites in cultured rat cortex cells. *Dev. Brain Res.*, **58**, 97–103.

Schotte, A. and Laduron, P.M. (1987) Different postnatal ontogeny of two [³H]neurotensin binding sites in rat brain. *Brain Res.*, **408**, 326–8.

Shi, W.X. and Bunney, B.S. (1991) Neurotensin modulates autoreceptor mediated dopamine effects on midbrain dopamine cell activity. *Brain Res.*, **543**, 315–21.

Tanaka, K., Masu, M. and Nakanishi, S. (1990)

Structure and functional expression of the cloned rat neurotensin receptor. *Neuron*, **4**, 847–54.

Tischler, A.S., Ruzicka, L.A. and Dobner, P.R. (1991) A protein kinase inhibitor, staurosporine, mimics nerve growth factor induction of neurotensin/neuromedin N gene expression. *J. Biol. Chem.*, **266**, 1141–6.

Tribollet, E., Goumaz, M., Raggenbass, M. *et al.* (1991) Early appearance and transient expression of vasopressin receptors in the brain of rat fetus and infant. An autoradiographical and electrophysiological study. *Dev. Brain Res.*, **58**, 13–24.

Uhl, G.R. (1990) Neurotensin receptors, in *Handbook of Chemical Neuroanatomy*, Vol.9. (eds A. Bjorklund, T. Hökfelt and M.J. Kuhar), Elsevier Science, Amsterdam, pp. 443–53.

Uhl, G.R. and Kuhar, M. (1984) Chronic neuroleptic treatment enhances neurotensin receptor binding in human and rat substantia nigra. *Nature*, **309**, 350–2.

Uhl, G.R. and Nishimori, T. (1990) Neuropeptide gene expression and neural activity: assessing a working hypothesis in nucleus caudalis and dorsal horn neurons expressing preproenkephalin and preprodynorphin. *Neurobiology*, **10**, 73–99.

Uhl, G.R. and Snyder, S.H. (1977) Neurotensin receptor binding, regional and subcellular distributions favor transmitter role. *Eur. J. Pharmacol.*, **41**, 89–91.

Young, W.S. and Kuhar, M.J. (1981) Neurotensin receptor localization by light microscopic autoradiography in rat brain. *Brain Res.*, **206**, 273–85.

OPIOID RECEPTORS AND THE DEVELOPING NERVOUS SYSTEM

Sandra E. Loughlin and Frances M. Leslie

10.1 INTRODUCTION

Three endogenous opioid peptide families have been identified, including those derived from the proenkephalin, prodynorphin and proopiomelanocortin genes (Imura *et al.*, 1985, for review). Peptides derived from these genes interact with three major classes of opioid receptor, mu, delta and kappa (Leslie, 1987). These receptors are widely distributed throughout the central nervous system and mediate the biological effects of endogenous opioid peptides, as well as of certain non-peptide analgesic drugs such as morphine and ketocyclazocine. In addition, epsilon (Schulz *et al.*, 1979; Chang *et al.*, 1984) and zeta (Zagon *et al.*, 1989) receptor types have been described, which have high affinity and selectivity for the opioid peptides beta-endorphin and enkephalin, respectively.

Both opioid peptides and their receptors appear very early in development, are expressed transiently in a number of brain regions, and reach their mature distributions relatively late in ontogeny (Leslie and Loughlin, 1992). It has, therefore, been suggested that the endogenous opioid systems may mediate basic developmental processes such as cell division and differentiation (Kent *et al.*, 1982). A large body of literature supports this hypothesis (see Zagon and McLaughlin, 1983; McDowell and Kitchen, 1987; Leslie and Loughlin, 1992, for reviews). Clinical data on the subtle, but long-lasting effects of maternal opioid exposure are consistent with a large number of studies demonstrating influences of *in vivo* and *in vitro* administration of opioid agonists and antagonists on developmental processes. Such effects are distinct from those of opioids in the adult nervous system (Olson *et al.*, 1991, for recent review).

Whereas manipulation of opioid systems may modulate developmental processes, such actions may be either direct or indirect and it is not yet clear whether opioids serve as neurotrophic factors. Neurotrophic factors may act by regulating cell division or cell survival, or have effects on process outgrowth and/or synthesis of proteins necessary to differentiated functions of neurons (Hefti *et al.*, 1992). In order to demonstrate that a compound has a neurotrophic role, several criteria must be met, including the presence of the endogenous compound and the receptor to which it binds at the appropriate time and in the appropriate neuronal elements (Fallon and Loughlin, 1992). This chapter describes the ontogeny of endogenous peptide and receptor expression and briefly reviews the evidence for a role of endogenous systems in basic developmental processes.

10.2 ENDOGENOUS OPIOID PEPTIDES

Endogenous opioid peptides are synthesized in the nervous system as large precursors and

Receptors in the Developing Nervous System Vol. 2: Neurotransmitters. Edited by Ian S. Zagon and Patricia J. McLaughlin. Published in 1993 by Chapman & Hall. ISBN 0 412 49400 0. Vols. 1 and 2 (set) ISBN 0 412 54520 9.

are cleaved enzymatically to yield a variety of peptide fragments, at least 17 of which have opioid activity (Douglass *et al.*, 1984; Imura *et al.*, 1985; Hollt, 1986). Three genes encode the opioid peptide precursors, proopiomelano-cortin (POMC), proenkephalin and prodynor-phin (see Hollt, 1991, for review). They are differentially expressed in the nervous system, as well as in a number of peripheral tissues, including endocrine glands. The peripheral expression of opioid peptides is of particular importance during developmental time periods before the blood–brain barrier is effective, since central nervous system onto-geny may be under the influence of circulat-ing opioids.

Immunocytochemical and radioimmunoas-say studies have provided a detailed descrip-tion of the developmental localization of opioid peptide-containing neuronal elements in brain. Since the peptide products of the three genes exhibit many structural similari-ties, detailed anatomical studies on the distri-butions of these systems have relied on the use of multiple antisera directed at specific amino acid sequences. More recently, mole-cular biological techniques have allowed the study of the developmental expression of mRNA for each of the genes. However, both approaches are limited by the sensitivity of assays for detection of low levels of products which may be expressed transiently or in restricted regions.

10.2.1 ENDOGENOUS OPIOID PEPTIDES IN THE ADULT

While recent *in situ* hybridization studies have confirmed the neurochemical identity of many cell groups previously localized by immuno-cytochemistry, the precise distribution of opioid pathways and terminals remains some-what controversial. Combined immunocyto-chemical and retrograde tracing studies have elucidated many of the discrete opioid path-ways. A detailed, critical review of the distri-bution of endogenous opioids is, however,

outside the scope of this chapter. The adult systems are described briefly here and the subsequent discussion concentrates on what is known of the developmental appearance of each system.

(a) Proopiomelanocortin

The POMC gene (Uhler and Herbert, 1983) is expressed in the brain, the pituitary, and a number of peripheral tissues. Expression of the gene is regulated by steroids and neuro-transmitters which modulate c-AMP and intracellular calcium (Hollt, 1991). The POMC precursor is the source of the opioid peptide beta-endorphin, as well as beta-lipotropin, adrenocorticotropic hormone and melano-cyte-stimulating hormones (Mains *et al.*, 1977).

In the central nervous system, beta-endor-phin and other POMC products are present in neurons of the hypothalamus and in the brain-stem nucleus tractus solitarius. POMC neurons in the nucleus tractus solitarius innervate the caudal brainstem, perhaps including the lateral reticular nucleus (Schwartzberg and Nakane, 1982). Processes of POMC cells in the arcuate extend for long distances to innervate many brain regions (Khachaturian *et al.*, 1985a). Many periventricular regions, includ-ing septum, bed nucleus of the stria terminalis, periventricular thalamus and periaqueductal gray are innervated, as well as brainstem reticular and autonomic nuclei. A marked absence of POMC fibers is observed in stria-tum, hippocampus and cortex.

(b) Proenkephalin

The proenkephalin gene (Rosen *et al.*, 1984) is expressed in the pituitary, the adrenal medul-la and a number of other peripheral tissues, as well as the central nervous system. Intra-cellular calcium and factors which modulate adenylate cyclase or protein kinase C regulate gene expression (Hollt, 1991). The opioid peptides met- and leu-enkephalin, and other enkephalin-containing peptides such as met-enkephalin-arg-gly-leu, met-enkephalin-arg-phe, peptide E, peptide F and BAM-22P, are derived from preproenkephalin (Imura *et al.*, 1985).

In contrast to the fairly limited distribution of POMC-containing neurons, the population of cells in the central nervous system expressing proenkephalin is relatively ubiquitous (Khachaturian *et al.*, 1985a; Petrusz *et al.*, 1985; Merchenthaler *et al.*, 1986; Harlan *et al.*, 1987; Fallon and Ciofi, 1990). *In situ* hybridization studies have generally found larger numbers of cell bodies than immunocytochemical studies, probably due to low concentrations of peptide in the normal soma (Schafer *et al.*, 1991). Enkephalinergic systems consist of both long- and short-projecting neurons. Cell bodies are located in most regions of the telencephalon, including the cerebral cortex, olfactory tubercle, amygdala, hippocampus, striatum, septum, bed nucleus of the stria terminalis and preoptic area. Enkephalinergic cells are also present in most hypothalamic nuclei, certain thalamic nuclei, a number of brainstem cell groups including colliculi, periventricular gray, interpeduncular nucleus, reticular and autonomic nuclei, Golgi cells of the cerebellum and spinal cord. Fibers and terminals are observed in a large number of brain structures. Cortical regions are sparsely innervated, with scattered fibers in entorhinal and cingulate cortices, whereas the hippocampus is more densely innervated. Basal telencephalic structures such as the nucleus accumbens, bed nucleus of the stria terminalis, caudate putamen and amygdala contain many enkephalinergic fibers. A dense projection from the caudate putamen to the globus pallidus, substantia nigra and other pallidal structures exists. Most hypothalamic nuclei are moderately innervated, whereas only certain thalamic nuclei, including the medial and lateral geniculate bodies and paraventricular nuclei contain enkephalinergic fibers. Fibers are also observed in the colliculi, central gray, ventral tegmental area, interpe-duncular nuclei, raphe nuclei, locus coeruleus and brainstem motor, autonomic and reticular nuclei.

(c) Prodynorphin

Expression of the prodynorphin, also known as proenkephalin B, gene (Civelli *et al.*, 1985) is greatest in the brain, especially in the striatum, hypothalamus and hippocampus. Little is known of the regulation of expression of prodynorphin, but, like the other opioid genes, it is positively regulated by c-AMP (Hollt, 1991). Prodynorphin is the source of the opioid peptide, dynorphin A, as well as leu-enkephalin and other leu-enkephalin containing peptides including alpha- and beta-neoendorphin (Hollt, 1986).

The majority of the brain regions which contain enkephalin cells and terminals also contain dynorphin (Khachaturian *et al.*, 1983; Morris *et al.*, 1986; Schafer *et al.*, 1988; Fallon and Ciofi, 1990). However, there are significant differences in the distribution of the two opioid peptide systems. In cortex, there is a large number of dynorphin-containing cells and processes, especially in entorhinal and cingulate cortices. Hippocampal dynorphin terminals are of much greater density than enkephalin and are distributed to all hippocampal subfields except CA1. In the basal telencephalon, dynorphin is distributed more sparsely in the bed nucleus of the stria terminalis. Dynorphin cells and fibers are densely localized to the ventral pallidum and largely avoid the enkephalin-rich globus pallidus. In the hypothalamus, dynorphin cells and fibers are found in most of the nuclei which contain enkephalin, with minor differences in distribution. The distribution in thalamus is restricted to the same nuclei as enkephalin, including geniculate bodies and paraventricular nuclei. In the substantia nigra, dynorphin and enkephalin are differentially distributed such that the substantia nigra pars reticulata contains dynorphin, whereas enkephalin is localized to the substantia nigra pars compacta. Dynorphin is not present in the interpeduncular nucleus, but is distributed similarly to enkephalin in brainstem and spinal cord.

(d) Co-localization

The localization of the three endogenous opioid peptide families is, thus, quite extensive. The beta-endorphin system is organized such that a limited number of cell bodies, located in discrete nuclei, have long projections to large portions of the basal forebrain, diencephalon and brainstem, particularly periventricular regions. In contrast, the enkephalin and dynorphin systems are characterized by a diffuse distribution of cell bodies throughout the neuraxis, with projections to the majority of brain systems. Whereas enkephalin and dynorphin are co-localized in a few structures, they are largely present in separate populations of neurons and fibers which are co-distributed (Fallon and Ciofi, 1990). It is likely that the enkephalin and dynorphin inputs may converge on individual target cells to modulate their output. In some of the regions where all three peptide systems are co-distributed, it is possible that neurons may be bathed in a complex mixture of opioid peptides.

(e) Central nervous system opioids in non-neuronal cells

In addition to its neuronal expression, there is evidence suggesting that proenkephalin is expressed in glial cells in the central nervous system (CNS). Proenkephalin mRNA has been localized to astrocytes in culture (Vilijn *et al.*, 1989; Hauser *et al.*, 1990; Stiene-Martin *et al.*, 1991b) and cultured astrocytes have been shown to release enkephalin (Batter *et al.*, 1991). Although enkephalin immunoreactivity within glia of the adult CNS has not been observed, immunoreactive glia have been detected in developing cerebellum (Zagon and McLaughlin, 1990).

10.2.2 ONTOGENY OF ENDOGENOUS OPIOID PEPTIDES

All three opioid genes are expressed early in embryonic development (Rosen and Polakiewicz, 1989), as are their peptide products (see Khachaturian *et al.*, 1985a; McDowell and Kitchen, 1987, for reviews). In addition to the CNS expression of these opioid peptides, the prenatal brain may be exposed to circulating endogenous opioids from peripheral sources. These include POMC products from the pituitary (Schwartzberg and Nakane, 1982; Khachaturian *et al.*, 1983) and ovaries (Shaha *et al.*, 1984), proenkephalin products from the adrenal medulla (Keshet *et al.*, 1989; Coulter *et al.*, 1990) and prodynorphin products from the posterior pituitary derived from hypothalamic magnocellular neurons (Watson *et al.*, 1982).

By embryonic day 17 (ED17), the rat brain contains 20% of adult levels of a 'morphine-like' substance (Garcin and Coyle, 1976). The opioid peptides met-enkephalin, dynorphin and beta-endorphin are detectable in mouse brain by ED11.5, before any opioid binding sites are detectable (Rius *et al.*, 1991a). In early postnatal ontogeny, the relative distributions of the opioid peptides and receptors suggest that no simple relationships can be defined between any one peptide and any one receptor type (Loughlin *et al.*, 1985). Therefore, the developmental appearance of each peptide family will be discussed and subsequently the ontogeny of each receptor type will be reviewed.

(a) Proopiomelanocortin

POMC mRNA has been detected in the rat central nervous system by ED10.5 (Elkabes *et al.*, 1989). Although levels are still low at birth and during the first postnatal week, they are higher at postnatal day 8 (PD8) than during the next two weeks. In the fourth postnatal week, there is a rapid increase to adult levels (Rosen and Polakiewicz, 1989). Studies on the distribution of beta-endorphin immunoreactivity have shown a similar biphasic appearance of the peptide (Bayon *et al.*, 1979; Bloom *et al.*, 1980). Beta-endorphin is present by ED16 in most brain regions at approximately 2% of adult levels. The concentration of peptide relative to brain protein increases little after ED18 (Ng *et al.*, 1984). Within some brain regions, peptide content remains level or decreases slightly until PD14, then increases to adult levels (Kapcala, 1983). Immunoreactive cells and fibers are visible in hypothalamus in the second embryonic week (Schwartzberg and Nakane, 1982; Khachaturian *et al.*, 1983; Elkabes *et al.*, 1989; Rius *et al.*, 1991a), with earliest visualization of hypothalamic cells by ED10.5 and hypothalamic fibers by ED11.5 (Elkabes *et al.*, 1989). Immunoreactive fibers are present in mesencephalon by ED11.5, in spinal cord by ED12.5 and the adult pattern of terminal arborization is recognizable by ED13.5–15.5 (Elkabes *et al.*, 1989).

Transient expression of POMC peptides has been reported in spinal cord from ED 14–PD28 (Haynes *et al.*, 1982) and in the early postnatal forebrain germinal zone (Kent *et al.*, 1982; Loughlin *et al.*, 1991). This transient POMC expression may be in both neuronal and non-neuronal cells. POMC peptide expression in adult peripheral nervous system may reappear following injury (Berry and Haynes, 1989). It has been suggested that POMC peptides may modulate cellular proliferation or guide migration of postmitotic cells (Loughlin *et al.*, 1991) or may regulate the transition from the mitotic cycle to the onset of differentiation (Berry and Haynes, 1989).

(b) Proenkephalin

Proenkephalin mRNA is detectable in rat brain by ED11–12 (Keshet *et al.*, 1989), with enkephalin immunoreactivity visible by ED 11.5 (Rius *et al.*, 1991a). By ED13, both met- and leu-enkephalin like immunoreactivity are present in multiple forms (Dahl *et al.*, 1982).

Enkephalin immunoreactive fibers are present in hindbrain by ED15 and throughout the neuraxis by ED17–18 (Palmer *et al.*, 1982). Although enkephalin is present in many brain regions by ED16, it is present in much lower concentrations than beta-endorphin (Bayon *et al.*, 1979). Unlike beta-endorphin, the developmental distribution of enkephalin does not resemble that of the adult. Met- and leu-enkephalin levels increase independently in postnatal ontogeny, suggesting that differential processing occurs or that leu-enkephalin is a significant product of prodynorphin (Patey *et al.*, 1980).

The ontogeny of striatal enkephalin has been examined by a number of investigators. In fetal pig, hippocampal mRNA increases linearly, whereas striatal mRNA peaks in midgestation, then declines (Pittius *et al.*, 1987). Proenkephalin mRNA is detectable in rat striatum by ED16, followed by the appearance of immunoreactivity by ED17–18. This appearance follows the gradient of cell generation (Tecott *et al.*, 1989). Schwartz and Simantov (1988) have reported that striatal enkephalin mRNA peaks at PD2, then decreases to embryonic levels by PD7, increasing in postnatal weeks 2–4 to adult levels. This developmental profile is similar to that reported for enkephalin immunoreactivity (Bayon *et al.*, 1979). In contrast, Rosen and Polakiewicz (1989) have reported a peak at PD8–10, followed by a 50-fold decrease at PD28.

Hippocampal enkephalin immunoreactivity contrasts with the early appearance of striatal enkephalin in that fibers are first seen at postnatal stages (Gall *et al.*, 1984). At ED 16, enkephalin immunoreactivity is detectable in hippocampal extracts, but at very low levels (Bayon *et al.*, 1979). It should be noted that the hippocampus is a late developing structure, with cell division continuing into postnatal stages.

A transient appearance of enkephalin immunoreactivity has been reported in the cerebellar germinal zone (Zagon *et al.*, 1985). While enkephalin mRNA was not detected in the neurons of the external granular layer, Purkinje cells and some Golgi cells express enkephalin mRNA (Osborne *et al.*, 1991). Levels of expression have been shown to follow the maturational gradient of these cells. In the retina, enkephalin mRNA is present during development, but not in the adult (Isayama and Zagon, 1991). Enkephalin mRNA is also observed in mesodermally derived cells before differentiation (Keshet *et al.*, 1989). It has, thus, been suggested that enkephalin may be an important regulator of cell division or differentiation (Zagon and McLaughlin, 1993).

(c) Prodynorphin

Less is known of the developmental appearance of prodynorphin in the central nervous system. As the distribution parallels that of enkephalin, many similarities exist in the developmental appearance of enkephalin and dynorphin systems in brain, especially in striatum (Pittius *et al.*, 1987; Rosen and Polakiewicz, 1989). Thus, in the fetal pig, prodynorphin mRNA increases linearly in hippocampus, whereas in the striatum it peaks at midgestation, then declines (Pittius *et al.*, 1987). In mouse brain, dynorphin immunoreactivity is first seen at ED11.5 (Rius *et al.*, 1991b) and prodynorphin products have been detected in rat brain at ED20 (Zamir *et al.*, 1985). Prodynorphin mRNA is first detectable in rat striatum at ED15 (Alvarez-Bolado *et al.*, 1990). By the day of birth, striatal dynorphin immunoreactivity has a patchy appearance (Loughlin *et al.*, 1985). Striatal prodynorphin mRNA levels have been reported to peak at PD8–10, then decrease 50-fold by PD28 (Rosen and Polakiewicz, 1989). Hypothalamic and medullary prodynorphin mRNA levels show similar postnatal increases followed by decreases. The postnatal appearance of prodynorphin mRNA has also been seen in cortical regions with iso- and allocortices expressing this gene in the first postnatal week and the hippocampus at PD14 (Alvarez-Bolado *et*

al., 1990; Sato *et al.*, 1991). Dynorphin immunoreactivity, in contrast, is seen at PD6 (Gall 1984).

(d) Opioid peptide processing

As has been noted, the interpretation of studies on the developmental appearance of opioid peptides is complicated by cross-reactivity caused by similarities in peptide structure. In addition, synthesis, degradation and post-translational processing of the peptides might affect the results obtained. Post-translational processing of propeptide precursors giving rise to active peptide products has been shown to occur early in ontogeny (Seizinger *et al.*, 1984a,b; McMillen *et al.*, 1988; Coulter *et al.*, 1990; Sei and Dores, 1990; Hylka and Thommes, 1991; Quinlan and Alessi, 1991; Rius *et al.*, 1991b). Opioid precursors may be differentially processed at different developmental stages, such that the ratio of long- to short-chain peptides decreases as maturation occurs. The degradative enzyme, neutral endopeptidase, which is also known as enkephalinase, has been detected in fetal brain (Patey *et al.*, 1980). Thus, mechanisms for peptide processing, release and degradation are apparent early in development.

10.3 OPIOID RECEPTORS

The receptors to which both exogenous and endogenous opioid compounds bind have been intensively studied. The pharmacology of the opioid receptor types is reviewed here followed by their localization in adult brain. Subsequent sections discuss the developmental appearance of each type and review the literature suggesting that early receptors may differ in their pharmacological characteristics from those of the adult.

10.3.1 PHARMACOLOGY

Three major classes of opioid receptor, mu, delta and kappa have been identified within the mammalian CNS and periphery (Leslie, 1987). An epsilon receptor, with high affinity and selectivity for beta-endorphin, has also putatively been identified from functional (Schulz *et al.*, 1981) and radioligand binding (Nock *et al.*, 1990) studies, although this is somewhat controversial (reviewed by Leslie, 1987). Most recently, a zeta receptor, with high affinity and selectivity for enkephalin peptides, has been identified in immature tissues (Zagon *et al.*, 1990, 1991). The properties and possible functional roles of this receptor are described by Zagon and McLaughlin (1993) and will not be addressed further at this point.

The ontogeny of opioid receptors has been examined primarily by radioligand binding. Although membrane binding and autoradiographic studies have yielded generally similar findings (McDowell and Kitchen, 1987; Leslie and Loughlin, 1992), the use of quantitative autoradiography has provided a higher degree of anatomical resolution such that regional and temporal changes within a single brain structure can be studied in great detail. Although early reports of opioid receptor ontogeny involved the use of radioligand binding conditions which were not highly selective for a given receptor type (Clendennin *et al.*, 1976; Coyle and Pert, 1976; Kirby, 1981), later studies have discriminated the developmental appearance of multiple receptor types (reviewed by Leslie and Loughlin, 1992).

Recent functional and radioligand binding data suggest that mu, delta and kappa opioid receptors may consist of multiple subtypes. Early evidence of mu receptor heterogeneity was provided by the work of Pasternak *et al.* (1980), who demonstrated a differential appearance of high and low affinity morphine binding sites in developing brain. The ontogeny of these two affinity states correlated with the functional expression of morphine analgesia and respiratory depression, respectively. Although later studies have provided additional support for the concept of inde-

pendent mu_1 and mu_2 receptors (Pasternak and Wood, 1986), the differential developmental appearance of these sites has not been examined in further detail and will not be discussed here. In most recent studies described in this review, mu receptor ontogeny has been examined using [^{3}H]DAGOL, a ligand with equal affinity for mu_1 and mu_2 sites (Clark *et al.*, 1988).

An increasing body of evidence has indicated heterogeneity of kappa receptors within mammalian CNS (reviewed by Unterwald *et al.*, 1991). Although some investigators have reported the existence of as many as four receptor subtypes (Rothman *et al.*, 1990; Kinouchi and Pasternak, 1991), there is presently strongest evidence for the subclassification of $kappa_1$ and $kappa_2$ sites (Zukin *et al.*, 1988). Whereas most early studies of kappa receptor ontogeny used binding conditions which were somewhat selective for the $kappa_1$ site, we have recently compared the ontogeny of $kappa_1$ and $kappa_2$ subtypes in rat brain (Leslie and Smith, 1992) (section 10.3.3).

Most recently, functional and radioligand binding studies have suggested the possible existence of more than one type of delta receptor (Negri *et al.*, 1991; Wild *et al.*, 1991). As yet, however, radioligand binding conditions have not been described which clearly discriminate these putative subtypes. In the present review, delta receptors will therefore be discussed as a single class.

10.3.2 OPIOID RECEPTOR DISTRIBUTION IN THE ADULT

Quantitative autoradiography studies have shown that mu, delta and kappa opioid binding sites are present in most regions of the CNS (Khachaturian *et al.*, 1985a; Leslie and Loughlin, 1992, for review). They are heterogeneously distributed such that each type has a distinct localization. Whether or not these sites represent functional receptors remains to be determined, but a large number of studies have indicated that opioids exert receptor-mediated effects within many systems of adult and developing brain (DeVries *et al.*, 1991; Olson *et al.*, 1991; also see below).

We review here the distribution of mu, delta and kappa receptors in the brain of the adult rat, the most studied species. Descriptions for forebrain areas are drawn from a number of recent mapping studies which report largely similar distributions (Kornblum *et al.*, 1987a; Mansour *et al.*, 1987, 1991; Sharif and Hughes, 1989).

(a)　Mu binding sites

High densities of mu receptors are present in a number of telencephalic structures. Binding in cerebral cortex is moderately dense, but exhibits regional and laminar heterogeneity. In many cerebral cortical regions, binding is greater in layers I, III and IV or V. Within the hippocampus, mu receptors are especially dense in the dentate granule cell layer and are also dense in the pyramidal cell layer and stratum lacunosum moleculare. Olfactory bulb, accessory olfactory bulb and accessory olfactory nucleus exhibit high levels of binding which correspond to specific cell layers. The distribution of mu binding sites in olfactory tubercle is patchy and probably corresponds to a dense accumulation of receptors over striatal compartments within sparse background labeling. This patchy distribution of mu binding is also present in the lateral nucleus accumbens, with more homogeneous labeling in medial nucleus accumbens. The patchy distribution of mu binding in the caudate putamen is superimposed on a moderately dense matrix. The patches of dense labeling correspond to the striosomes, which are an important aspect of the compartmental organization of the caudate putamen (Graybiel, 1984). In the adult rat, binding is sparse in the globus pallidus and entopeduncular nucleus (Herkenham and McLean, 1986), as it is in the lateral septum. Moderate levels of binding are present in the medial septum and bed

nucleus of the stria terminalis. Certain nuclei of the amygdala contain significant populations of mu receptors.

Mu binding sites are present in moderate to dense concentration in many thalamic nuclei, but not in ventroposterior, parafascicular and reticular nuclei. Binding is especially dense in the medial habenula and subfornical organ, whereas the lateral habenula and zona incerta have no binding sites. In the hypothalamus, mu receptors are most densely localized in suprachiasmatic, dorsomedial, ventromedial and supramammillary nuclei and in anterior and lateral hypothalamic areas.

Mu labeling is moderately dense in the mesencephalon, with a laminar distribution in superior colliculus and dense labeling in the inferior colliculus and periaqueductal gray. The substantia nigra pars compacta and portions of the pars reticulata and the ventral tegmental area contain moderate densities of mu receptors. Dense labeling is also seen in the medial terminal nucleus and several raphe nuclei.

Brainstem nuclei in the pons and medulla also contain mu binding sites, with highest densities in the parabrachial, raphe magnus and spinal trigeminal nuclei, nucleus of the solitary tract and dorsal motor nucleus of the vagus (Xia and Haddad, 1991). Binding sites are also present in locus coeruleus. In contrast, the cerebellum is largely devoid of mu binding sites in the adult.

The superficial layer and the medial portion of lamina V of the spinal cord show moderate densities of mu binding sites (Morris and Herz, 1987; Besse *et al.*, 1990).

(b)　Kappa binding sites

Early studies on the distribution of kappa sites in rat brain used conditions which were somewhat selective for kappa$_1$ receptors, which have high affinity and selectivity for the opioid peptide, dynorphin A, and for arylacetamides such as U50,488 (Kornblum *et al.*, 1987a; Mansour *et al.*, 1987; Leslie and

Loughlin, 1992). More recently, it has been shown that only 10% of rat brain kappa sites are of the kappa$_1$ type (Hurlbut *et al.*, 1988, 1992; Unterwald *et al.*, 1991). The majority of non-mu, non-delta sites in this species have a distinct pharmacological profile, with high affinity for certain benzomorphan drugs (Chang *et al.*, 1984; Zukin *et al.*, 1988; Hurlbut *et al.*, 1988). Although it has been suggested that these represent beta-endorphin-selective epsilon sites (Nock *et al.*, 1990), these have been more commonly designated as kappa$_2$ (Zukin *et al.*, 1988; Attali and Vogel, 1990). Kappa$_1$ and kappa$_2$ sites are found in different brain regions and are differentially distributed in the rat and guinea-pig brain. These differences are thought to account for the complex pharmacological effects of kappa opioids.

In the rat brain, kappa$_1$ sites are sparse in neocortical structures, though dense binding is observed in the claustrum and endopiriform nucleus. Binding is strikingly low in hippocampus. Kappa$_1$ sites are heterogeneously distributed in moderate densities in the olfactory tubercle and are present in caudate putamen and nucleus accumbens, with greater density in the medial-dorsal caudate putamen, fundus striata and medial accumbens. Certain nuclei of the amygdala contain moderate densities of kappa$_1$ sites, including basolateral and medial nuclei. Low densities of binding sites are found in certain thalamic nuclei, especially the midline nuclei, whereas the majority of thalamic nuclei contain no kappa$_1$ sites. Binding in hypothalamus is moderately dense, especially in the lateral hypothalamic area and ventromedial nucleus. In the mesencephalon, binding sites are of greatest density in superior colliculus, interpeduncular nucleus and central gray.

Kappa$_2$ sites are present in many brain structures in higher concentrations than kappa$_1$ sites. In rat brain there are marked similarities between the distribution of these sites and that of the mu receptor (Unterwald *et al.*, 1991; Leslie and Smith, 1992). However, recent pharmacological and developmental data indicate that, though distributed in the same brain regions as mu sites, these kappa$_2$ sites represent a distinct receptor population (Unterwald *et al.*, 1991; Leslie and Smith, 1992; also see below).

The distribution of kappa$_2$ sites in adult rat brain has recently been described (Unterwald *et al.*, 1991). Neocortical kappa$_2$ binding sites are diffusely distributed throughout layers I–VI. In the hippocampus, binding sites are present in moderate density in the pyramidal cell layer and the granular cells of the dentate gyrus. Moderately high densities of kappa$_2$ sites are observed in the caudate putamen, nucleus accumbens and olfactory tubercle. Dense binding is present in amygdala, especially in basolateral and medial nuclei. In thalamus, kappa$_2$ sites are moderately dense in lateral, posterior and midline nuclear groups, and particularly dense in the habenula. Diffuse labeling is seen throughout the hypothalamus. In mesencephalon, binding is present in superior colliculus, medial geniculate, central gray, substantia nigra pars reticulata and interpeduncular nucleus. A moderate level of binding is also present in the locus coeruleus, with a low density in the cerebellum.

(c) Delta binding sites

Delta binding sites are also present in high densities in rat brain, particularly in the telencephalon. In the adult cortex, delta binding is relatively diffuse, with moderate density in deep and superficial layers and lower levels of binding in intermediate layers. Delta receptors in hippocampus are present in low concentrations with no marked regional or laminar heterogeneity. In olfactory bulb, the densest labeling is in the external plexiform layer, with moderate levels in granule and internal plexiform layers. Diffuse delta binding of moderate density is present in olfactory tubercle and nucleus accumbens. Binding in the caudate putamen is equivalent in density to that of mu receptors overall but the pattern of labeling differs in that a diffuse distribution of sites increases in density from medial to lateral. The globus pallidus, endopeduncular nucleus and septal nuclei show very low levels of delta binding whereas the bed nucleus of the stria terminalis and the nucleus of the diagonal band show moderate levels. Moderate levels of delta binding sites are also present in those nuclei of the amygdala which express mu sites.

The thalamus and hypothalamus show very low levels of delta binding, except in the ventromedial nucleus. Binding is also sparse throughout the mesencephalon except for moderate concentrations of receptors in the interpeduncular nucleus. Delta sites are diffusely distributed in the brainstem, at levels 2–40 times lower than mu concentrations (Xia and Haddad, 1991). In the spinal cord, some delta binding sites are detectable in the superficial layers of the dorsal horn (Besse *et al.*, 1990).

10.3.3 ONTOGENY OF OPIOID RECEPTORS

Early studies on the developmental appearance of opioid receptors in brain using relatively non-selective antagonists showed that binding sites are detectable by ED14–16 (Coyle and Pert, 1976; Clendennin *et al.*, 1976; Kirby, 1981). They follow a complex course of development with significant regional variation in time of appearance. The ontogeny of opioid binding sites in brain has been reviewed in detail by McDowell and Kitchen (1987) and Leslie and Loughlin (1992) and is summarized briefly here. In rat and mouse, mu and kappa receptors appear in embryonic brain, whereas delta receptors appear later (Leslie *et al.*, 1982; Tavani *et al.*, 1985; Kitchen *et al.*, 1990; Rius *et al.*, 1991a; Leslie and Loughlin, 1992). The time course of appearance in human brain follows a similar profile, with mu and kappa sites appearing earlier than delta (Magnan and Tiberi, 1989).

(a) Mu binding sites

In the mouse brain, mu binding sites are detectable by ED12.5 (Rius *et al.*, 1991a). In the rat brain, they are present by ED14 (Kent *et al.*, 1982; Kornblum *et al.*, 1987a; Leslie and Loughlin, 1992). Relative to brain protein, mu binding sites decline in concentration during the first postnatal week (Spain *et al.*, 1985; Petrillo *et al.*, 1987; Attali and Vogel, 1990), increase during the second and third weeks, then decline to adult levels. Quantitative autoradiography studies have demonstrated a similar developmental profile (Kent *et al.*, 1982; Unnerstall *et al.*, 1983; Kornblum *et al.*, 1987a). Autoradiograms showing the postnatal development of mu receptor binding in selected regions are shown in Figs. 10.1 and 10.2.

Neocortical labeling of mu sites is first detected at ED20, heterogeneously distributed in both cortical layers and regions. Binding increases throughout early postnatal weeks, especially during the third week. Hippocampal binding is diffuse at birth, becomes laminated by the end of the first postnatal week and achieves near adult levels during the second week. In olfactory bulb and anterior olfactory nucleus, mu binding sites are present in high concentrations at birth. The density of binding decreases markedly in the external plexiform layer between PD2 and PD9, when it begins to increase in other layers. It has been suggested that this developmental profile corresponds to the maturation of specific cell types (Unnerstall *et al.*, 1983).

Prenatally, mu receptor binding is first detected in the striatal anlage at ED14 (Kent *et al.*, 1982; see below). By ED16, diffuse labeling is apparent throughout the developing caudate putamen and nucleus accumbens. Binding remains diffuse until 1–2 days after birth, when it begins to assume the adult patchy distribution (Kent *et al.*, 1982; Murrin and Ferrer, 1984; Kornblum *et al.*, 1987a). Throughout the next two postnatal weeks,

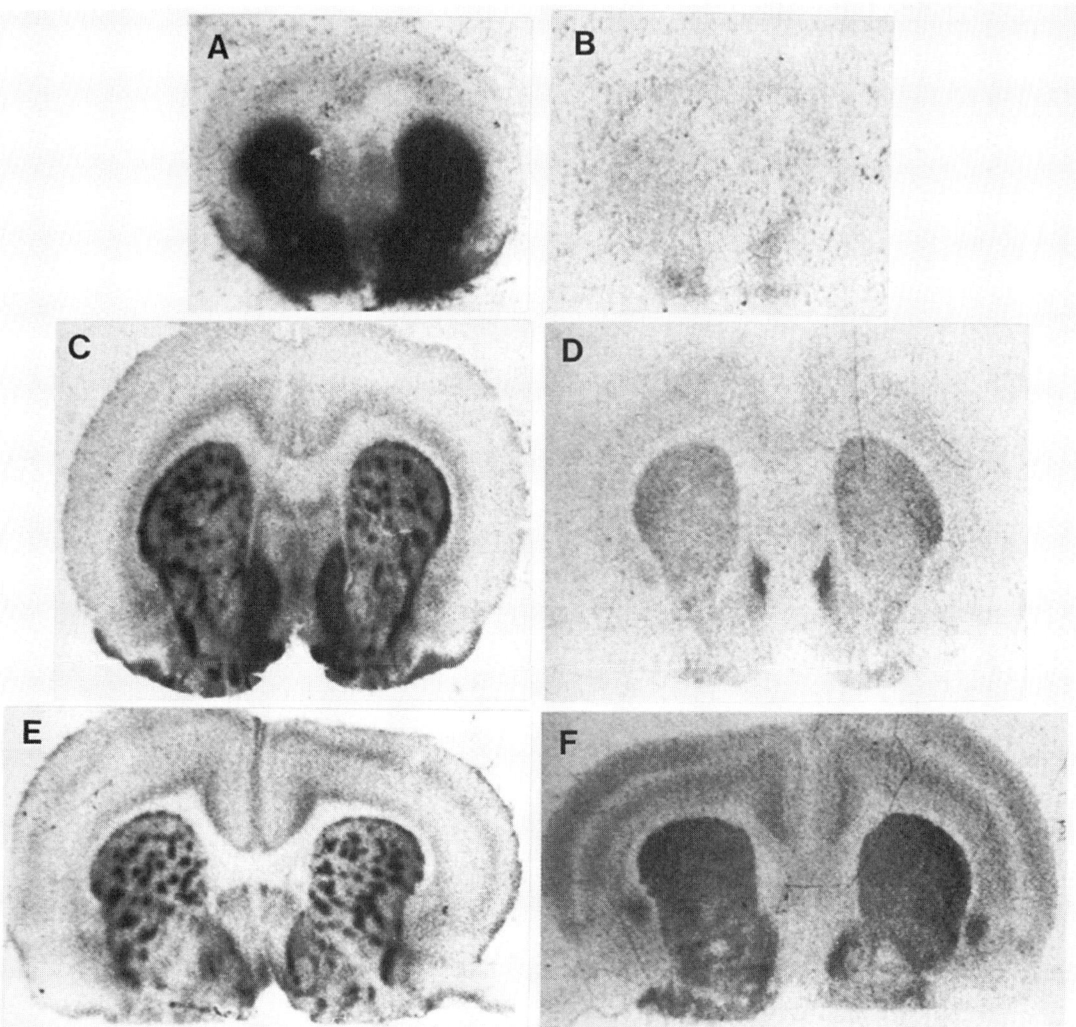

Fig. 10.1.   Postnatal development of mu (A, C, E) and delta (B, D, F) binding sites in forebrain. Autoradiograms showing the distribution of mu and delta receptors at PD1 (A, B), PD9 (C, D) and PD17 (E, F) were generated by incubation with [³H]DAGOL and [³H]DADLE under conditions selective for mu and delta binding sites, respectively.

mu binding site density decreases markedly in the surrounding matrix while remaining constant in the patches (Fig. 10.1). In olfactory tubercle and lateral nucleus accumbens, diffuse binding of moderately high density is present at birth which becomes patchy as it decreases to adult levels (Fig. 10.1). In medial nucleus accumbens, binding decreases but remains diffuse (Fig. 10.1). Mu binding sites in the globus pallidus are also dense at birth and decrease to very low levels thereafter. In contrast, binding in the amygdala is present at birth and increases over the next three weeks.

The ontogeny of mu binding sites in the thalamus is largely postnatal, appearing during the second week in adult patterns and increasing to adult levels in the third week

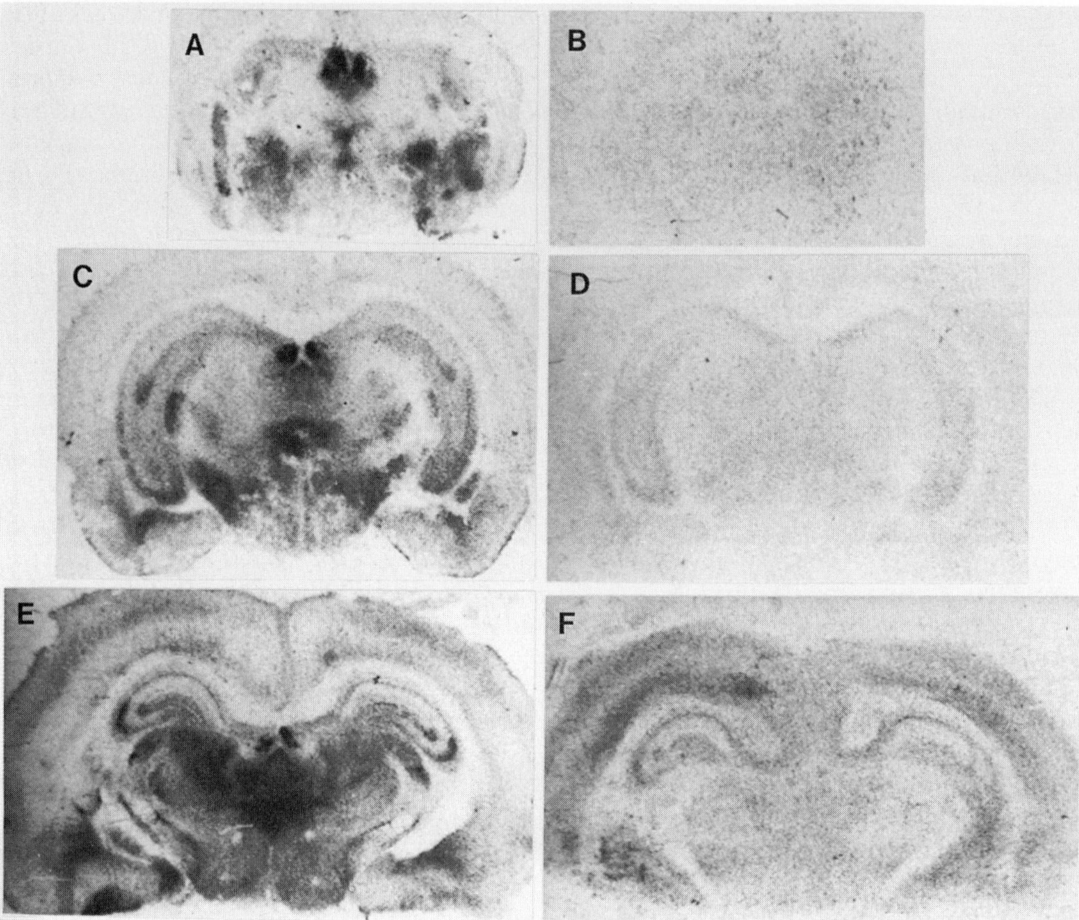

Fig. 10.2.   Postnatal development of mu (A, C, E) and delta (B, D, F) binding sites at the level of the posterior diencephalon. Autoradiograms showing the distribution of mu and delta receptors at PD1 (A, B), PD9 (C, D) and PD17 (E, F) were generated by incubation with [^{3}H]DAGOL and [^{3}H]DADLE under conditions selective for mu and delta binding sites, respectively.

(Fig. 10.2). In the hypothalamus, mu receptors are present in most regions at PD2 and adult levels are reached by the second postnatal week. In the mesencephalon, mu receptors are diffusely distributed throughout the superior colliculus during the first postnatal week, increasing in density and becoming laminated over the next two weeks. Mu sites are detectable in the substantia nigra early in postnatal development and decline two-fold to adult concentrations. Mu sites are also present in brainstem nuclei at birth, especially in

cardiorespiratory nuclei (Xia and Haddad, 1991). Densities increase to adult levels in early postnatal weeks, maturing ahead of forebrain sites. Binding sites in spinal cord achieve adult distribution and densities by the second postnatal week (Besse *et al.*, 1990).

(b)   Kappa binding sites

Recent studies have shown that both kappa$_1$ and kappa$_2$ sites are present in rat brain, with the kappa$_2$ site predominating (Hurlbut *et al.*,

1988; Zukin *et al.*, 1988; Unterwald *et al.*, 1991). In most developmental studies to date, binding conditions have been used which label predominantly kappa$_1$ sites. Minor discrepancies in reported results may reflect labeling of different proportions of the kappa$_2$ receptor population.

In membrane binding studies, kappa$_1$ sites in brain are present in significant densities at birth, and exhibit a small increase in density during the subsequent two postnatal weeks (Spain *et al.*, 1985; Petrillo *et al.*, 1987; Kitchen *et al.*, 1990). Kappa binding sites in spinal cord increase significantly during the first two postnatal weeks, then decline to adult levels (Allerton *et al.*, 1989; Attali and Vogel, 1990).

In autoradiographic studies, a significant density of kappa$_1$ sites is detectable at birth in many brain regions (Kornblum *et al.*, 1987a). In neocortex, kappa receptors do not reach significant levels until PD12, increasing slightly to adult levels in the next few days. The higher levels of kappa binding present in claustrum and endopiriform nucleus appear earlier, suggesting that levels of detection may not be sensitive enough for binding in neocortex to be seen earlier. In hippocampus, kappa receptors are detectable by the end of the second postnatal week.

Binding sites in olfactory tubercle and nucleus accumbens are present at birth and remain constant in density throughout postnatal development. Kappa sites are also present in caudate putamen at birth. In globus pallidus, kappa sites are dense in the first postnatal week, peaking in the second week and decreasing to adult levels by PD17. Kappa sites in amygdala appear differentially across nuclei, with basolateral sites present by PD5 and other nuclei appearing later.

Kappa receptors are not detectable in thalamus until PD6, with sites appearing first in ventral and midline nuclei, and later in others. Within hypothalamus, receptors are present in supraoptic and paraventricular nuclei by PD12 and by PD17 in other nuclei. As for mu sites, kappa binding in superior colliculus is diffuse during the first week after birth, becoming laminated by PD14. Kappa sites are detectable early in the postnatal development of the substantia nigra and remain constant thereafter. Little is known of the ontogeny of kappa sites in brainstem. In spinal cord, kappa binding sites are present in greater density and distribution at PD9–16 than in the adult (Allerton *et al.*, 1989).

Whereas earlier autoradiographic studies have used binding conditions which were somewhat selective for kappa$_1$ sites, binding conditions have recently been used which label both populations of kappa receptors (Leslie and Smith, 1992). In the presence of mu and delta blockers, [^{3}H]diprenorphine labels kappa$_1$ and kappa$_2$ sites with equal affinity (Hurlbut *et al.*, 1988, 1992). By examining [^{3}H]diprenorphine binding in the absence and presence of the kappa$_1$-selective agonist, U50,488, the ontogeny of the two kappa subtypes in rat brain has been compared (Fig. 10.3). Using this approach we have determined that the ontogeny of the kappa$_1$ subtype is largely similar to that described above using more selective labeling conditions. We have also shown that kappa$_2$ sites are present in significant densities at birth in many brain regions, including caudate putamen, nucleus accumbens, olfactory tubercle and habenula (Figs 10.4 and 10.5).

Although there are many similarities in the anatomical distributions of mu and kappa$_2$ sites in adult rat brain (Unterwald *et al.*, 1991; Leslie and Smith, 1992), these receptors exhibit significant differences in their ontogeny. In nucleus accumbens and olfactory tubercle, densities of kappa$_2$ sites are not significantly altered during postnatal development, whereas mu receptor densities decline significantly. In the caudate putamen, kappa$_2$ receptor densities remain constant in the matrix compartment throughout postnatal development, while increasing significantly in the patches. In contrast, mu receptor density remains constant in the patches, while declining significantly in the matrix during the first two postnatal

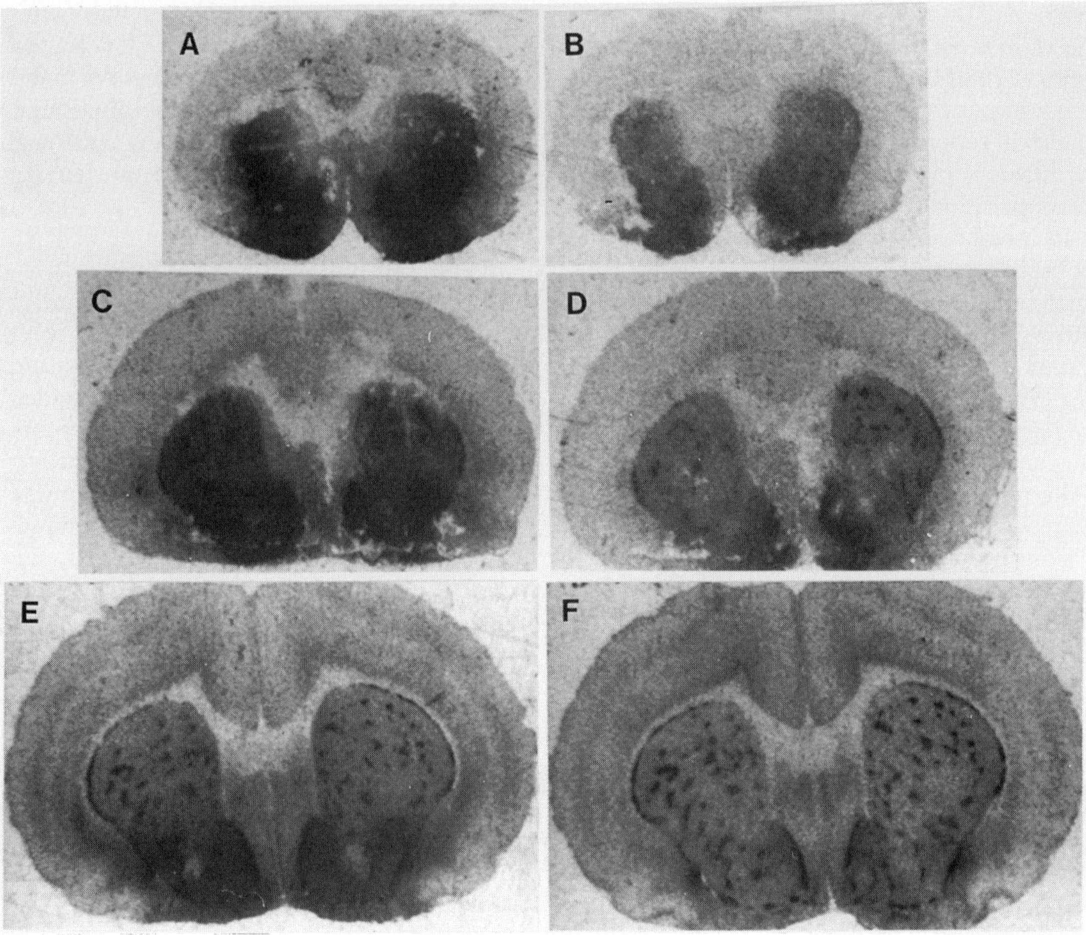

Fig. 10.3.   Postnatal development of kappa binding sites in forebrain. The total population of kappa binding sites was localized with [^{3}H]diprenorphine in A, C and E. Sequential sections (B, D, F) were incubated with [^{3}H]diprenorphine in the presence of U50,488 to block binding to kappa$_1$ sites, and thus represent binding to kappa$_2$ sites. Sections A, B are from PD0, C, D are from PD7 and E, F are from PD21 brains.

weeks. Similar differential developmental profiles for mu and kappa$_2$ receptor populations are observed within the habenula. Such data provide strong evidence that these receptors, though distributed in many similar brain regions, are differentially modulated.

(c)   Delta binding sites

The ontogeny of delta binding sites lags significantly behind that of mu and kappa sites (Figs 10.1 and 10.2). By the end of the first postnatal week, a low density of sites is detectable (Spain *et al.*, 1985; Kornblum *et al.*, 1987a; Milligan *et al.*, 1987; Szucs and Coscia, 1990). These sites are discretely localized to a limited number of regions, particularly nucleus accumbens and olfactory tubercle (Kornblum *et al.*, 1987a). A major increase in density occurs in the second postnatal week, followed by a steady increase which is greater than the rate of increase of protein. It has been

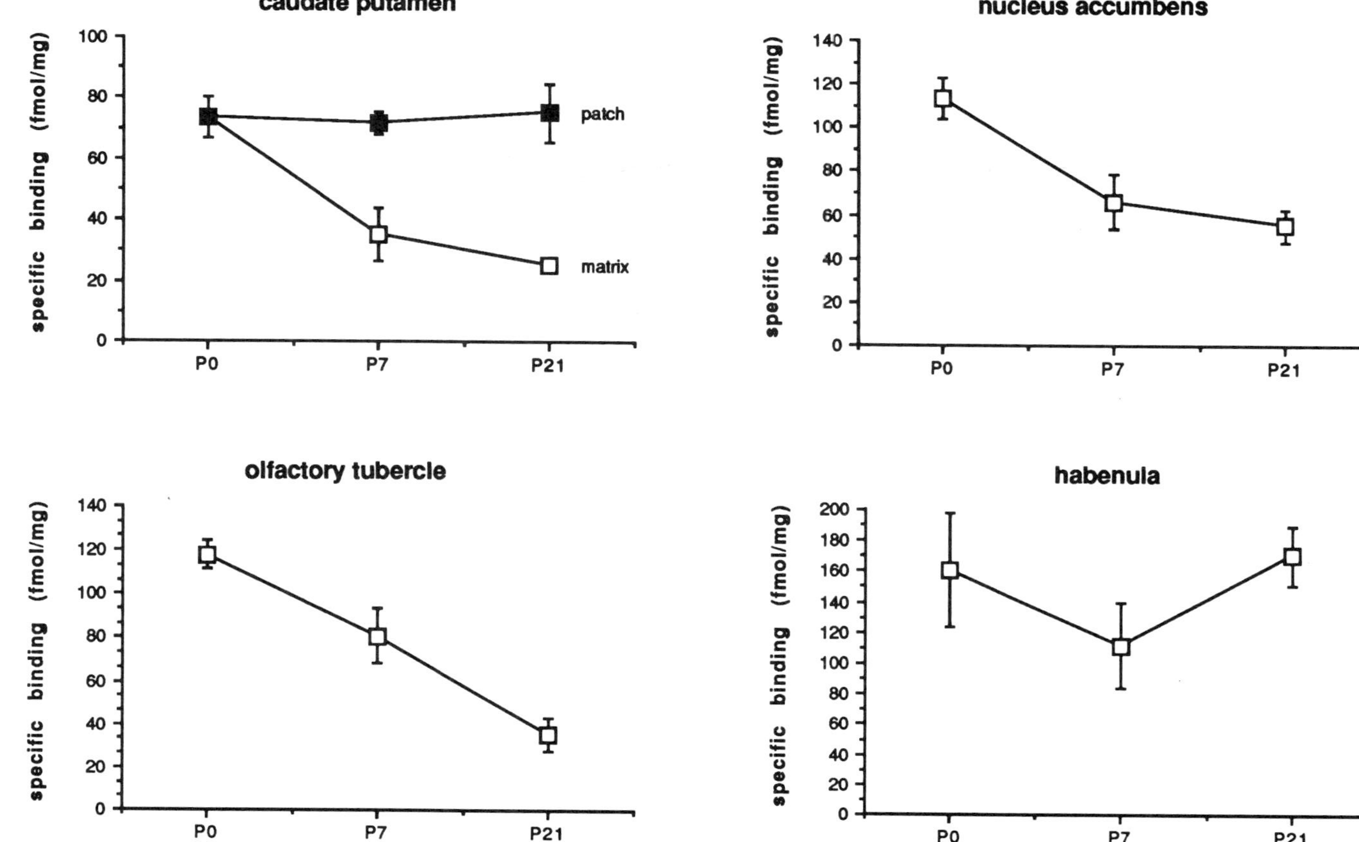

Fig. 10.4. Quantification of postnatal development of mu binding sites in four brain regions. Specific binding (fmol/mg tissue) was determined at PD0, PD7, and PD21 for each region.

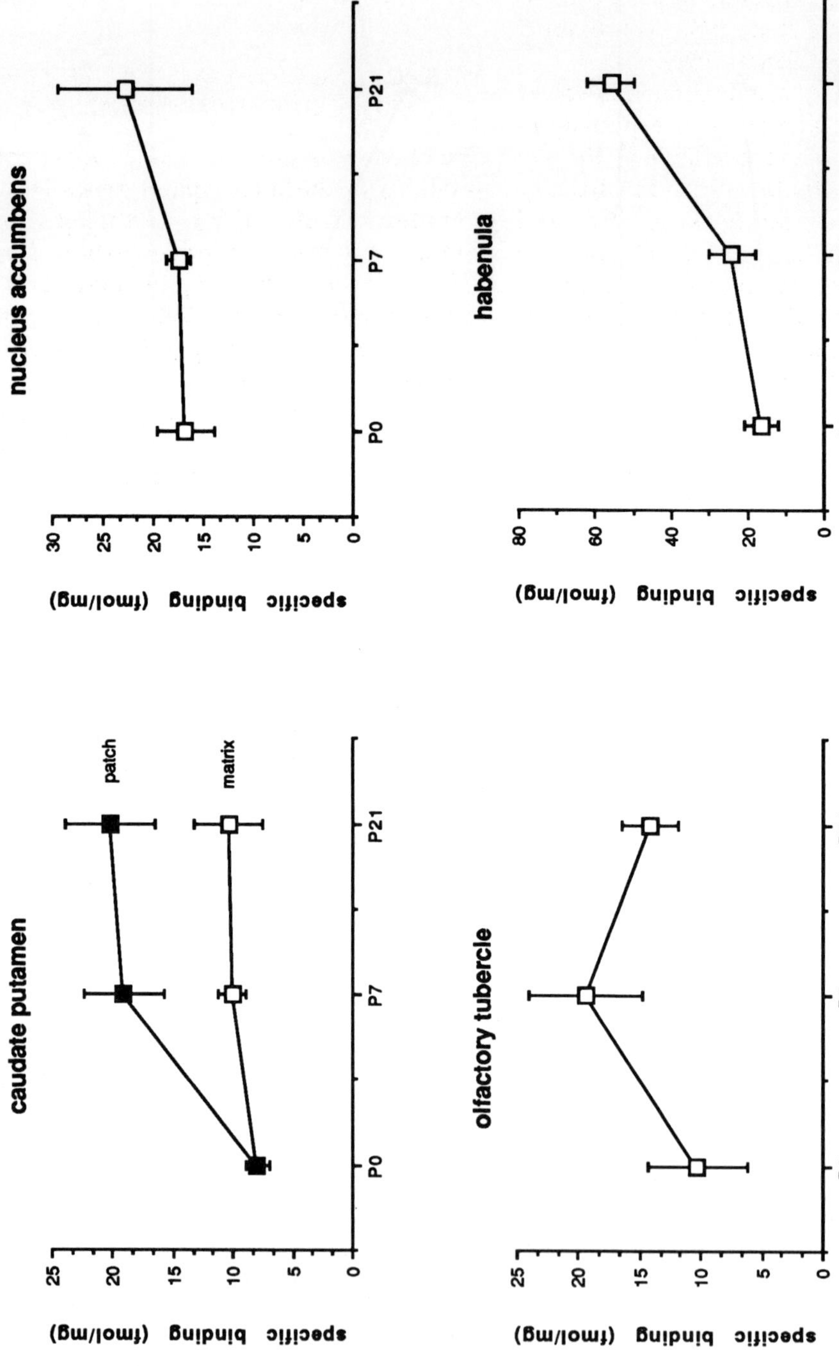

Fig. 10.5. Quantification of postnatal development of kappa$_2$ binding sites in four brain regions. Specific binding (fmol/mg tissue) was determined at PD0, PD7, and PD21 for each region. Note that, though the distribution of kappa$_2$ binding sites resembles that of mu sites, the time course of appearance is different (see text).

suggested that a change in affinity of the delta receptor occurs during ontogeny (Spain *et al.*, 1985), but other investigators have not found such changes (Szucs *et al.*, 1987; Kornblum *et al.*, 1987a; Szucs and Coscia, 1990).

Delta binding sites are detectable in cortex during the first postnatal week, especially in deeper layers and in hippocampus by the end of the second postnatal week. In olfactory bulb, receptors appear between PD9 and PD13, whereas in olfactory tubercle they are present at birth. Both increase throughout development. Some delta labeling is present in the nucleus accumbens at birth, but binding does not reach adult concentrations until PD20. Within the caudate putamen, delta receptors are detectable during the second postnatal week (Fig. 10.1). In globus pallidus, delta sites are present by PD9, at slightly above adult levels, whereas amygdala sites are detectable by the time of birth and increase gradually to adult levels.

Delta receptors are sparse in thalamus, hypothalamus and mesencephalon, except for moderate concentrations in the interpeduncular nucleus (Fig. 10.2). In the brainstem, the ontogeny of delta sites lags significantly behind mu sites, with no detectable binding present at PD1 (Xia and Haddad, 1991). Densities increase differentially across nuclei, showing increases of 2–6-fold in the first postnatal weeks.

10.3.4 OPIOID RECEPTOR EXPRESSION IN GLIA

Whereas glial cells appear to express opioid peptides (see above) and opioid modulation of glial growth has been documented (see below), the expression of opioid receptors by glial cells remains controversial. C_6 glioma cells exhibit binding sites for opioid compounds which may be $kappa_2$ sites (Barg and Simantov, 1991). A number of studies have failed to detect binding sites in primary glial cultures derived from the CNS (Hendrickson and Lin, 1980; Vaysse *et al.*, 1990), and others have detected opioid binding sites whose affinity is similar to sites on neurons (Maderspach and Solomonia, 1988).

10.3.5 BINDING OF ENDOGENOUS LIGANDS: BETA-ENDORPHIN BINDING

The pharmacology of the endogenous peptide products of the three opioid genes is quite complex (Hollt, 1986). Although some endogenous opioid peptides exhibit higher affinity for a given receptor, none exhibits absolute receptor specificity (Leslie, 1987). Furthermore, different peptide products of the same precursor may have strikingly different pharmacological profiles (Hollt, 1986). In view of this, it is not surprising that no strict relationship has been established between the distributions of any opioid peptide and receptor in adult or developing brain (Loughlin *et al.*, 1985; McDowell and Kitchen, 1987; Mansour *et al.*, 1991).

In order to address functionally relevant opioid systems, it is important to examine the binding and activity profiles of the endogenous peptides, as well as of synthetic opioids. McDowell and Kitchen (1987) have reviewed the ontogeny of binding of a number of endogenous opioid peptides, and Zagon and McLaughlin (1993) have discussed the binding of met-enkephalin, and we have examined the ontogeny of ^{125}I-labeled beta-endorphin binding sites in rat brain (Kornblum *et al.*, 1989; Loughlin *et al.*, 1992b).

In adult rat, ^{125}I-labeled beta-endorphin binding is extensively distributed throughout the brain in a pattern consistent with that of a composite of mu and delta receptor distributions. Figure 10.6 shows that this labeling is differentially displaced by mu-, delta- and kappa-selective compounds. Co-incubation of ^{125}I-labeled beta-endorphin with the mu-selective agonist, D-pro[4]-morphiceptin, results in a pattern of labeling that is similar to that of delta receptors. In particular, binding to the patch compartment of the caudate putamen is significantly blocked, leaving diffuse residual

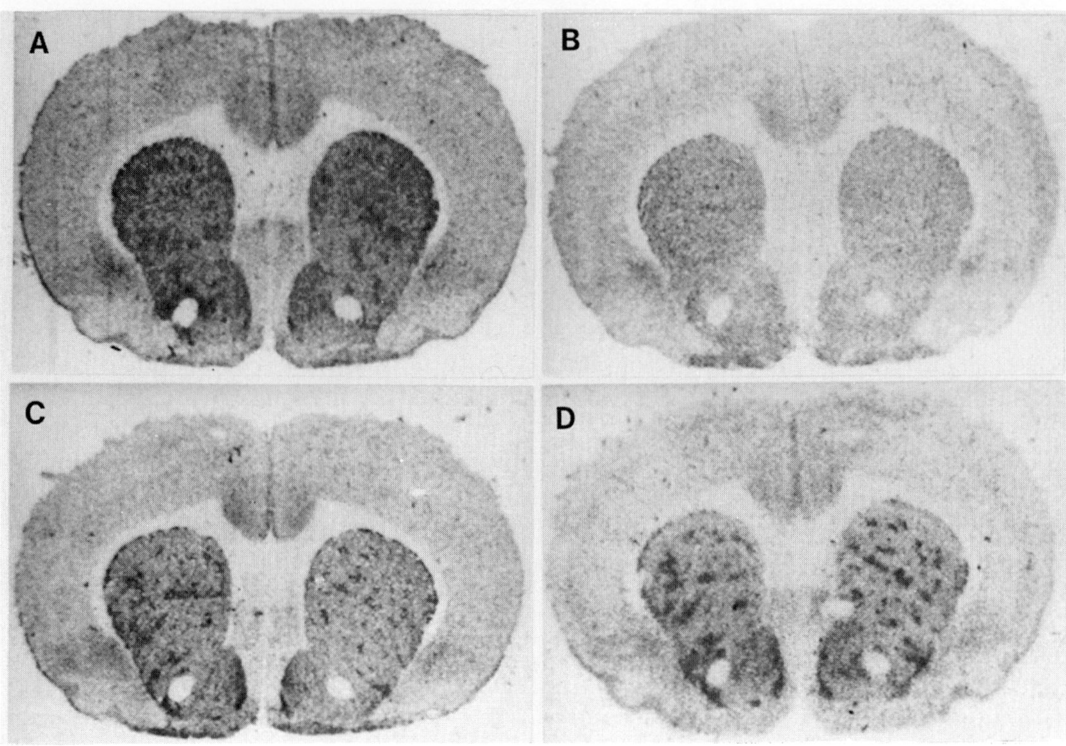

Fig. 10.6. Autoradiograms of ^{125}I-labeled beta-endorphin binding sites in adult rat forebrain. Sections were incubated with ^{125}I-labeled beta-endorphin without blockers (A), and in the presence of D-pro^4-morphiceptin (B), U50,488 (C) or DPDPE (D) to block binding to mu, kappa$_1$ or delta binding sites, respectively. Co-incubation with unlabeled beta-endorphin or naloxone blocked all binding (not shown).

labeling. In contrast, co-incubation of radioligand with the delta-selective agonist, DPDPE, produces a pattern of labeling similar to that of mu receptors, with a striking patchy distribution in the caudate putamen. Although the kappa$_1$-selective agonist, U50,488, does produce some decrease in binding of ^{125}I-labeled beta-endorphin, this inhibition is inconsistent and not regionally selective. Taken together, these data indicate that in adult rat brain the majority of ^{125}I-labeled beta-endorphin binding is to mu and delta receptor populations. This conclusion, consistent with that of Schoffelmeer *et al.* (1990), does not provide evidence for high affinity labeling of a selective epsilon receptor population.

In fetal forebrain, the localization of beta-endorphin binding sites is very similar to that previously shown with mu selective ligands (Figs 10.7 and 10.8). At ED17, mu selective ligands block all but a small population of sites in the ventral forebrain, which is consistent with previous studies suggesting that few delta sites are present before birth (Fig. 10.7). The kappa$_1$-selective compound U50, 488 produces very little, if any, inhibition of binding (Fig. 10.7). These data suggest that, in embryonic brain, ^{125}I-labeled beta-endorphin predominantly labels mu receptor sites.

In both forebrain and hindbrain periventricular regions, ^{125}I-labeled beta-endorphin binding sites are present as early as ED14 (Fig. 10.8). At ED15, dense labeling is present

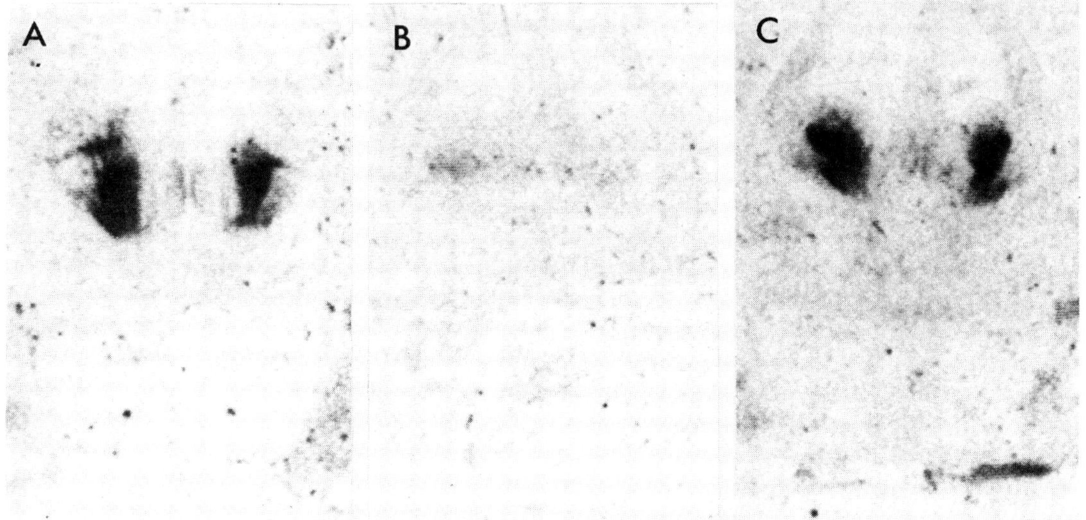

Fig. 10.7. Autoradiograms of ^{125}I-labeled beta-endorphin binding sites in ED17 rat brain. Sections through the developing telencephalon were incubated with ^{125}I-labeled beta-endorphin alone (A), or in the presence of D pro^4-morphiceptin (B), or U50,488 (C) to block binding to mu, and kappa$_1$ binding sites, respectively. Co-incubation with unlabeled beta-endorphin or naloxone blocked all binding (not shown).

in the striatal anlage, with some labeling in the developing cerebral vesicles. Binding sites are detectable in the forebrain germinal zone, especially at its distal poles. Throughout the extent of the midbrain and hindbrain, binding is detectable in the region which surrounds the ventricles. By ED17, this hindbrain labeling is very dense. These data indicate that mu opioid receptors are present at an appropriate time and location to influence the generation and differentiation of many cell groups.

10.4 DEVELOPMENTAL ROLE

It has been suggested that brain opioid binding sites present at early developmental stages differ from those of the adult with regard to selectivity (Barg and Simantov, 1991), molecular weight (McLean *et al.*, 1989), coupling to regulatory proteins (Milligan *et al.*, 1987; Szucs *et al.*, 1987), sensitivity to cation effects (Oetting *et al.*, 1987; Szucs *et al.*, 1987) and association with other opioid receptors (Schoffelmeer *et al.*,

1990). A number of *in vitro* studies, however, suggest that opioid binding sites in developing brain represent functional receptors which bear many similarities to those characterized in adult brain (Chneiweiss *et al.*, 1988; DeVries *et al.*, 1990; Van Vliet *et al.*, 1990; Vaysse *et al.*, 1990; Stiene-Martin and Hauser, 1990, 1991; Hauser and Stiene-Martin, 1991; Stiene-Martin *et al.*, 1991a; Smith *et al.*, 1992a).

A number of reviews (Tempel, 1992; Hammer and Hauser, 1992; Barr, 1992; Kuhn *et al.*, 1992) have recently discussed the effects of opioids on developmental processes. Since more than one in a thousand people in the United States have been exposed to heroin or methadone *in utero* (Tempel, 1992), it is critical to understand the role of opioids in developmental processes. A number of studies support the conclusion that opioids modulate neurogenesis (Kornblum *et al.*, 1987b; Schmahl *et al.*, 1989; Vertes *et al.*, 1982; Zagon and McLaughlin, 1993) and it has been suggested that endogenous opioid systems

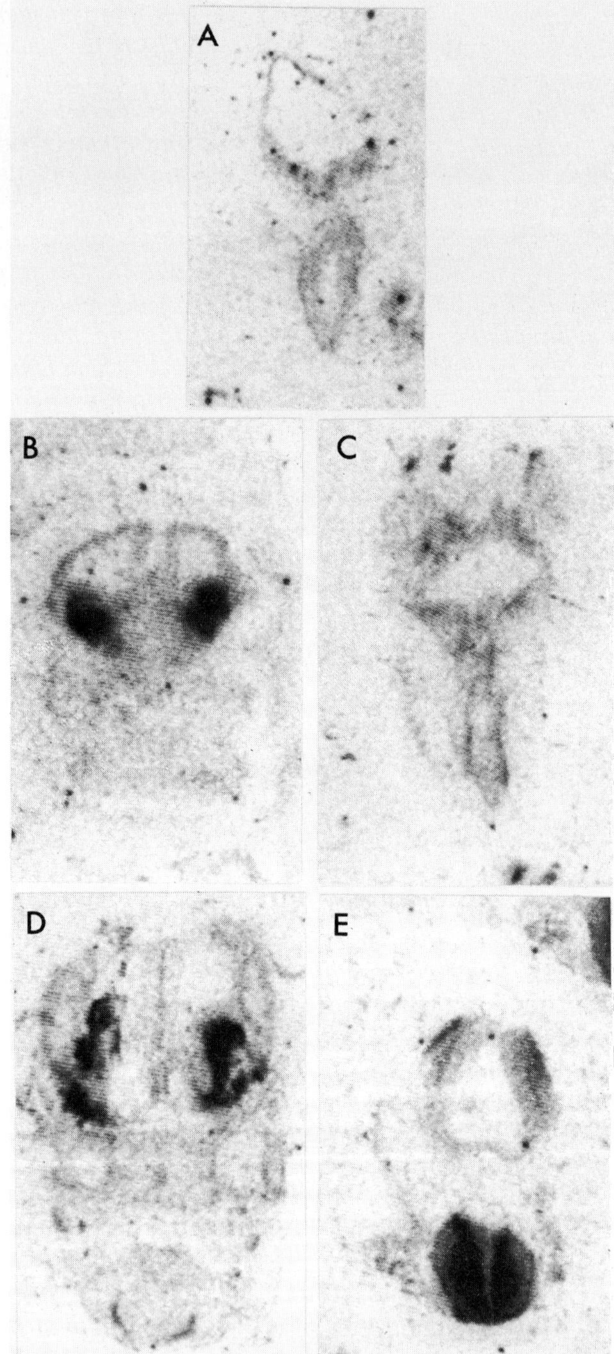

Fig. 10.8.   Prenatal appearance of ^{125}I-labeled beta-endorphin binding sites in ED14 (A), ED15 (B,C), and ED17 (D,E) rat brain. Co-incubation with unlabeled beta-endorphin blocked all binding (not shown).

may affect neuronal migration (Kent *et al.*, 1982; Loughlin *et al.*, 1991). Administration of opioids alters dendritic growth and spine formation (Hauser *et al.*, 1989) and a number of studies support the hypothesis that opioids are regulators of glial growth (Stiene-Martin and Hauser, 1991). Opioid peptides also regulate growth-related enzymes (Bartolome *et al.*, 1986, 1987, 1989).

Elucidating the effects of exogenous opioid administration is complicated by the fact that perinatal administration of opioids affects endogenous opioid systems in complex ways (Chapter 12), both at the level of peptide expression (Uhl *et al.*, 1988; Tiong and Olley, 1988) and receptor expression (e.g. Tsang and Ng, 1980; Bardo *et al.*, 1981; Hammer *et al.*, 1991; Tempel, 1991). Opioids also modulate many other neurotransmitter or hormonal systems (Tempel *et al.*, 1990; Mess *et al.*, 1989). Thus, although the evidence to date favors the conclusion that opioids have important developmental functions, the precise effects and the mechanisms of action of opioids in ontogeny remain to be elucidated.

10.4.1 COMPLEXITY OF INTERACTION BETWEEN OPIOIDS AND OTHER NEUROTRANSMITTER SYSTEMS

(a) Role of opioids in nigrostriatal system development and plasticity

Effects of opioids in developmental processes have been studied in whole brain, neocortex and cerebellum (Zagon and McLaughlin, 1993). However, in the nigrostriatal system, opioid peptides and receptors are present very early and in high density. The nigrostriatal system thus offers a good model in which to study the role of opioids in development, especially with regard to interactions with this well-studied monoaminergic system (Fallon and Loughlin, 1985). In the nigrostriatal system, dopaminergic cells in the substantia nigra innervate striatal regions, including the caudate putamen and nucleus accumbens,

very early in embryonic development (Specht *et al.*, 1981; Voorn *et al.*, 1988). Striatal cells reciprocally innervate the substantia nigra. Opioid peptides and receptors are associated with the nigrostriatal system at all levels and are present in early embryonic development, at the time of neurogenesis and throughout the time of process outgrowth. Beta-endorphin binding sites surround the forebrain germinal zone and are present in caudate putamen during striatal neurogenesis (Loughlin *et al.*, 1992b). They are also detectable in the midbrain regions in which substantia nigra cells develop. Mu binding sites are dense throughout the caudate putamen prenatally and are later lost in the matrix compartment (Kornblum *et al.*, 1987a). Kappa binding sites are present in early postnatal development of the nigrostriatal system (Kornblum *et al.*, 1987a; Leslie and Smith, 1992). A transient population of beta-endorphin immunoreactive cells is present in the forebrain germinal zone which sends processes to the lateral edge of the developing caudate putamen (Loughlin *et al.*, 1991). Proenkephalin and prodynorphin mRNAs are detectable in caudate putamen by ED17 (Tecott *et al.*, 1989) and ED15 (Alvarez-Bolado *et al.*, 1990), respectively.

Since mu receptors in caudate putamen develop their characteristic patchy appearance in registration with the developing dopaminergic innervation (Murrin and Ferrer, 1984), it has been suggested that their expression is under the control of the dopaminergic system and that dopaminergic cues may be necessary for development of normal patch-matrix organization (Fishell and van der Kooy, 1987; Gerfen *et al.*, 1987). Indeed, it has been shown that loss of dopaminergic innervation modulates caudate putamen opioid binding sites (Moon-Edley and Herkenham, 1984; Eghbali *et al.*, 1987; Sirinathsinghji and Dunnett, 1989; Smith *et al.*, 1992b; Loughlin *et al.*, 1992a). The time course of these effects is complex. While a significant loss of binding sites occurs shortly after lesions, much greater losses occur in the ensuing months. Interestingly, the secondary loss of opioid receptors which occurs six months after dopaminergic lesions is reversed by transplants of fetal mesencephalon (Sirinathsinghji and Dunnett, 1989; Smith *et al.*, 1991). Sustained loss of striatal dopaminergic innervation decreases dynorphin mRNA (Jiang *et al.*, 1990), and increases enkephalin mRNA in lesion (Mochetti *et al.*, 1985; Young *et al.*, 1986) and mutant (Loughlin *et al.*, 1992c) models, but such changes are not always correlated with receptor down-regulation (Smith *et al.*, 1992b). Pharmacological manipulation of these systems produces many of the same results as lesions (Chapter 12). Whether such changes in opioid systems occur in Parkinson's disease, in which the dopaminergic cells innervating the caudate putamen die, is controversial (Loughlin *et al.*, 1992a, for review). It is tempting to speculate that decreases in opioid binding sites might underlie some of the changes in affect which occur in Parkinson patients. Increases in enkephalin immunoreactivity have been correlated with severity of symptoms in monkeys exposed to MPTP (Dacko and Schneider, 1991).

Striatal opioids, in turn, are important modulators of dopaminergic function. Opioids modulate dopamine effects on adenylate cyclase in adult and developing striatum (Schoffelmeer *et al.*, 1986, 1987; Milligan *et al.*, 1987; Chneiweiss *et al.*, 1988; DeVries *et al.*, 1990; Van Vliet *et al.*, 1990). Dopamine release from striatal slices is itself modulated by kappa opioids (Mulder *et al.*, 1984; Schoffelmeer *et al.*, 1988; Werling *et al.*, 1988) and this effect is first seen in slices obtained from ED17 embryos (DeVries *et al.*, 1990). We have recently shown that kappa$_1$ receptor agonists inhibit dopamine release from dissociated cultures of ED14 substantia nigra cells by eight days *in vitro* (Smith *et al.*, 1992a).

While a number of investigators have examined the effects of opioids on nigrostriatal plasticity and dopaminergic function, the role of endogenous opioid systems in basic developmental processes remains to be elucidated. Schmahl *et al.* (1989) have shown that opioid antagonist treatment increases mitogenesis in the forebrain germinal zone in 4–12-week-old rats. Such effects probably reflect modulation of glial populations. The availability of well-characterized *in vitro* models will allow the precise determination of whether opioids are capable of influencing the development of neurons in this system.

(b) Role of opioid peptides and receptors in developing CNS

Do endogenous opioids act as neurotrophic factors to mediate basic developmental processes? We have proposed that, in order to be considered as neurotrophic factors, endogenous opioids must meet several criteria, including the presence of endogenous opioids and their receptors at developmentally relevant times and in the appropriate neuronal elements (Fallon and Loughlin, 1992). They might act by regulating cell division or cell survival, or have effects on process outgrowth and/or synthesis of proteins necessary to differentiated functions of neurons (Hefti *et al.*, 1992). Opioid receptors and their endogenous ligands are present early in the development of a number of systems, including the nigrostriatal system, at a time when neurogenesis, neuronal migration, process outgrowth and synaptogenesis are occurring. Opioids and opioid antagonists have been shown to modulate a number of developmental markers (Zagon and McLaughlin, 1993; Chapter 12), including DNA synthesis, enzyme activity, receptor expression, peptide expression, dendritic morphology and cell survival. Whether these are direct or indirect effects remains to be determined. Opioids clearly affect the growth of glial cells (Hauser and Steine-Martin, 1991) and glia may synthesize characterized growth factors, including transforming growth factor alpha (Loughlin *et al.*, 1989, 1992d). Thus, opioids may influence development via interactions with characterized growth factors of glial origin. Interestingly, Schofield *et al.* (1989) have described an opioid binding protein which displays significant homology to a group of cell adhesion and growth factor molecules. Whether opioids might compete for binding to such growth factor receptors *in vivo* remains to be determined.

ACKNOWLEDGEMENTS

The research described here was supported by NIH grants NS 26761 to SEL and NS 19319 to FML. The authors also wish to thank the American Parkinson Disease Association Southern California Chapter for their generous support.

REFERENCES

Allerton, C.A., Smith, J.A.M., Hunter, J.C. *et al.* (1989) Correlation of ontogeny with function of [^{3}H]U69593 labelled kappa opioid binding sites in the rat spinal cord. *Brain Res.*, **502**, 149–57.

Alvarez-Bolado, G., Fairen, A., Douglass, J. and Naranjo, J.R. (1990) Expression of the prodynorphin gene in the developing and adult cerebral cortex of the rat: an in situ hybridization study. *J. Comp. Neurol*, **300**, 287–300.

Attali, B. and Vogel, Z. (1990) Characterization of kappa opiate receptors in rat spinal cord-dorsal root ganglion co-cultures and their regulation by chronic opiate treatment. *Brain Res.*, **517**, 182–8.

Bardo, M.T., Bhatnagar, K.P. and Gebhart, F.F. (1981) Differential effects of chronic morphine and naloxone on opiate receptors, monoamines and morphine-induced behaviors in preweanling rats. *Dev. Brain Res.*, **4**, 139–47.

Barg, J. and Simantov, R. (1991) Transient expression of opioid receptors in defined regions of developing brain: are embryonic receptors selective? *J. Neurochem.*, **57**, 1978–84.

Barr, G.A. (1992) Behavioral effects of opiates during development, in *Development of the Central Nervous System: Effects of Alcohol and Opiates* (ed. M.W. Miller), Wiley-Liss, New York, pp. 221–54.

Bartolome, J.V., Bartolome, M.B., Daltner, L.A. *et al.* (1986) Effects of β-endorphin on ornithine decarboxylase in tissues of developing rats: a potential role for this endogenous neuropeptide in the modulation of tissue growth. *Life Sci.*, **38**, 2355–62.

Bartolome, J.V., Bartolome, M.B., Harris, E.B. and Schanberg, S.M. (1987) N-alpha-acetyl-β-endorphin stimulates ornithine decarboxylase activity in preweanling rat pups: opioid- and non-opioid-mediated mechanisms. *J. Pharmacol. Exp. Ther.*, **240**, 895–9.

Bartolome, J.V., Bartolome, M.B., Harris, E.B. *et al.* (1989) Regulation of insulin and glucose plasma levels by central nervous system beta endorphin in preweanling rats. *Endocrinology*, **124**, 2153–8.

Batter, D.K., Vilijn, M.-H. and Kessler, J. (1991) Cultured astrocytes release proenkephalin. *Brain Res.*, **563**, 28–32.

Bayon, A., Shoemaker, W.J., Bloom, F.E. *et al.* (1979) Perinatal development of the endorphin- and enkephalin-containing systems in the rat brain. *Brain Res.*, **179**, 93–101.

Berry, S. and Haynes, L.W. (1989) The opiomelanocortin peptide family: neuronal expression and modulation of neural cellular development and regeneration in the central nervous system. *Comp. Biochem. Physiol.*, **93A**, 267–72.

Besse, D., Lombard, M.C., Zajac, J.M. *et al.* (1990) Pre- and post-synaptic distribution of mu, delta and kappa opioid receptors in the superficial layers of the cervical dorsal horn of the rat spinal cord. *Brain Res.*, **521**, 15–22.

Bloom, F., Bayon, A., Battenberg, E. *et al.* (1980) Endorphins: developmental, cellular and behavioural aspects. *Adv. Biochem. Psychopharmacol.*, **22**, 619–32.

Chang, K.-J., Blanchard, S.G. and Cuatrecasas, P. (1984) Benzomorphan sites are ligand recognition sites of putative E-receptors. *Mol. Pharmacol.*, **26**, 484–8.

Chneiweiss, H., Glowinski, J. and Premont, J. (1988) Mu and delta opioid receptors coupled negatively to adenylate cyclase on embryonic neurons from the mouse striatum in primary cultures. *J. Neurosci.*, **8**, 3376–82.

Civelli, O., Douglass, J., Goldstein, A. and Herbert, E. (1985) Sequence and expression of the rat prodynorphin gene. *Proc. Natl. Acad. Sci. USA*, **82**, 4291–5.

Clark, J.A., Houghten, R. and Pasternak, G.W. (1988) Opiate binding in calf thalamic membranes: a selective mu$_1$ binding assay. *Mol. Pharmacol.*, **34**, 308–17.

Clendennin, N.J., Petraitis, M. and Simon, E.J. (1976) Ontological development of opiate receptors in rodent brain. *Brain Res.*, **118**, 157–60.

Coulter, C.L., Browne, C.A. and McMillen, I.C. (1990) The molecular weight profile of enkephalin-containing peptides in the sheep adrenal gland changes during development. *Endocrinology*, **127**, 330–6.

Coyle, J.T. and Pert, C.B. (1976) Ontogenetic development of [³H]naloxone binding in rat brain. *Neuropharmacology*, **15**, 555–60.

Dacko, S. and Schneider, J.S. (1991) Met-enkephalin immunoreactivity in the basal ganglia in symptomatic and asymptomatic MPTP-exposed monkeys: correlation with degree of parkinsonian symptoms. *Neurosci. Lett.*, **127**, 49–52.

Dahl, J.L., Epstein, M.L., Silva, B.L. and Lindberg, I. (1982) Multiple immunoreactive forms of [met] and [leu] enkephalin in fetal and neonatal rat brain and in rat gut. *Life Sci.*, **31**, 1853–6.

De Vries, T.J., Hogenboom, F., Mulder, A.H. and Schoffelmeer, A.N.M. (1990) Ontogeny of mu, delta and kappa opioid receptors mediating inhibition of neurotransmitter release and adenylate cyclase activity in rat brain. *Dev. Brain Res.*, **54**, 63–9.

De Vries, T.J., Van Vliet, B.J., Hogenboom, F. *et al.* (1991) Effect of chronic prenatal morphine treatment on mu-opioid receptor-regulated adenylate cyclase activity and neurotransmitter release in rat brain slices. *Eur. J. Pharmacol.*, **208**, 97–104.

Douglass, J., Civelli, O. and Herbert, E. (1984) Polyprotein gene expression: generation of diversity of neuroendocrine peptides. *Annu. Rev. Biochem.*, **53**, 665–715.

Eghbali, M., Santoro, C., Paredes, W. *et al.* (1987) Visualization of multiple opioid-receptor type in rat striatum after specific mesencephalic lesions. *Proc. Natl. Acad. Sci. USA*, **84**, 6582–6.

Elkabes, S., Loh, Y.P., Nieburgs, A. and Wray, S. (1989) Prenatal ontogenesis of pro-opiomelanocortin in the mouse central nervous system and pituitary gland: an *in situ* hybridization and immunocytochemical study. *Dev. Brain Res.*, **46**, 85–95.

Fallon, J.H. and Ciofi, P. (1990) Dynorphin-containing neurons, in *Handbook of Chemical Neuroanatomy*, Vol.9. *Neuropeptides in the CNS*, Part II (eds A. Bjorklund, T. Hökfelt, and M.J. Kuhar), Elsevier, Amsterdam, pp. 1–130.

Fallon, J.H. and Loughlin, S.E. (1985) The substantia nigra, in *The Rat Nervous System* (ed. G. Paxinos), Academic Press, New York, pp. 353–74.

Fallon, J.H. and Loughlin, S.E. (1992) Functional implications of the anatomical localization of neurotrophic factors, in *Neurotrophic Factors* (eds S.E. Loughlin and J.H. Fallon), Academic Press, New York (in press).

Fishell, G. and van der Kooy, D. (1987) Pattern formation in the striatum: developmental changes in the distribution of striatonigral neurons. *J. Neurosci.*, **7**, 1969–78.

Gall, C. (1984) Ontogeny of dynorphin-like immunoreactivity in the hippocampal formation of the rat. *Brain Res.*, **307**, 327–31.

Gall, C., Brecha, N., Chang, K.-J. and Karten, H.J. (1984) Ontogeny of enkephalin-like immunoreactivity in the rat hippocampus. *Neuroscience*, **11**, 359–80.

Garcin, F. and Coyle, J.T. (1976) Ontogenetic development of [³H]naloxone binding and endogenous morphine-like-factor in rat brain, in *Opiates and Endogenous Opioid Peptides* (ed. H.W. Kosterlitz), Elsevier, North Holland, Amsterdam, pp. 267–73.

Gerfen, C.R., Baimbridge, K.G. and Thibault, J. (1987) The neostriatal mosaic: III. Biochemical and developmental dissociation of patch-matrix mesostriatal systems. *J. Neurosci.*, **7**, 3935–44.

Graybiel, A.M. (1984) Correspondence between the dopamine islands and striosomes of the mammalian striatum. *Neuroscience*, **13**, 1157–87.

Hammer, R.P. and Hauser, K.F. (1992) Consequences of early exposure to opioids on cell proliferation and neuronal morphogenesis, in *Development of the Central Nervous System: Effects of Alcohol and Opiates* (ed. M.W. Miller), Wiley-Liss, New York, pp. 319–39.

Hammer, R.P., Seatriz, J.V. and Ricalde, A.R. (1991) Regional dependence of morphine-induced mu-opiate receptor down-regulation in perinatal rat brain. *Eur. J. Pharmacol.*, **209**, 253–6.

Harlan, R.E., Shivers, B.D., Romano, G.J. *et al.* (1987) Localization of preproenkephalin mRNA in the rat brain and spinal cord by *in situ* hybridization. *J. Comp. Neurol.*, **258**, 159–84.

Hauser, K.F., McLaughlin, P.J. and Zagon, I.S. (1989) Endogenous opioid systems and the regulation of dendritic growth and spine formation. *J. Comp. Neurol.*, **281**, 13–22.

Hauser, K.F., Osborne, J.G., Stiene-Martin, A. and Melner, M.H. (1990) Cellular localization of proenkephalin mRNA and enkephalin peptide products in cultured astrocytes. *Brain Res.*, **522**, 347–53.

Hauser, K.F. and Stiene-Martin, A. (1991) Characterization of opioid-dependent glial development in dissociated and organotypic cultures of mouse central nervous system: critical periods and target specificity. *Dev. Brain Res.*, **62**, 245–55.

Haynes, L.W., Smyth, D.G. and Zakarian, S. (1982) Immunocytochemical localization of beta-endorphin (lipotropin c-fragment) in the developing rat spinal cord and hypothalamus. *Brain Res.*, **232**, 115–28.

Hefti, F., Denton, T.L., Knusel, B. and Lapchak, P.A. (1992) Neurotrophic factors: what are they and what are they doing? in *Neurotrophic Factors* (eds S.E. Loughlin and J.H. Fallon), Academic Press, New York (in press).

Hendrickson, C.M. and Lin, S. (1980) Opiate receptors in highly purified neuronal cell populations isolated in bulk from embryonic chick brain. *Neuropharmacology*, **19**, 731–9.

Herkenham, M. and McLean, S. (1986) Mismatches between receptor and transmitter localizations in the brain, in *Quantitative Receptor Autoradiography* (eds C. Boast, E.W. Snowhill and C.A. Altar), Alan Liss, New York, pp. 131–71.

Hollt, V. (1986) Opioid peptide processing and receptor selectivity. *Annu. Rev. Pharmacol. Toxicol.*, **26**, 59–77.

Hollt, V. (1991) Opioid peptide genes: structure and regulation, in *Neurobiology of Opioids* (eds O.F.X. Almeida and T.S. Shippenberg), Springer-Verlag, Berlin, Heidelberg, pp. 11–51.

Hurlbut, D.E., Broide, R.S. and Leslie, F.M. (1988) Evidence for kappa receptor heterogeneity. *Soc. Neurosci. Abstr.*, **14**, 701.

Hurlbut, D.E., Broide, R.S. and Leslie, F.M. (1992) Pharmacological characteristics of kappa₂ opioid binding sites in guinea-pig and rat adult brain. *Brain Res.* (submitted).

Hylka, V.W. and Thommes, R.C. (1991) Avian beta-endorphin: alterations in immunoreactive forms in plasma and pituitary of embryonic and adult chickens. *Comp. Biochem. Physiol.*, **100C**, 643–8.

Imura, I., Kato, Y., Nakai, Y. *et al.* (1985) Endogenous opioids and related peptides: from molecular biology to clinical medicine. *J. Endocrinol.*, **107**, 147–57.

Isayama, T. and Zagon, I.S. (1991) Localization of preproenkephalin A mRNA in the neonatal rat retina. *Brain Res. Bull.*, **27**, 805–8.

Jiang, H.-K., McGinty, J.F. and Hong, J.S. (1990) Differential modulation of striatonigral dynorphin and enkephalin by dopamine receptor subtypes. *Brain Res.*, **507**, 57–64.

Kapcala, L.P. (1983) Discordant changes in immunoreactive ACTH and beta-endorphin in rat brain and pituitary during development. *Clin. Res.*, **31**, 401A.

Kent, J.L., Pert, C.B. and Herkenham, M. (1982) Ontogeny of opiate receptors in rat forebrain: visualization by in vitro autoradiography. *Dev. Brain Res.*, **2**, 487–504.

Keshet, E., Polakiewicz, R.D., Itin, A. *et al.* (1989) Proenkephalin A is expressed in mesodermal lineages during organogenesis. *EMBO J.*, **8**, 2917–23.

Khachaturian, H., Alessi, N.E., Munfakh, N. and Watson, S.J. (1983) Ontogeny of opioid and related peptides in the rat CNS and pituitary: an immunocytochemical study. *Life Sci.*, **33**, 61–4.

Khachaturian, H., Alessi, N.E., Lewis, M.E. *et al.* (1985a) Development of hypothalamic opioid neurons: a combined immunocytochemical and [³H]thymidine autoradiographic study. *Neuropeptides*, **4**, 477–80.

Khachaturian, H., Lewis, M.E., Schafer, M.K.-H. and Watson, S.J. (1985b) Anatomy of the CNS opioid systems. *Trends Neurosci.*, **8**, 111–19.

Kinouchi, K. and Pasternak, G.W. (1991) Evidence for kappa₁ opioid receptor multiplicity in the guinea pig cerebellum. *Eur. J. Pharmacol.*, **207**, 135–41.

Kirby, M.L. (1981) Development of opiate receptor binding in rat spinal cord. *Brain Res.*, **205**, 400–4.

Kitchen, I., Kelly, M. and Viveros, P.M. (1990) Ontogenesis of kappa opioid receptors in rat brain using [³H]U-69593 as a binding ligand. *Eur. J. Pharmocol.*, **175**, 93–6.

Kornblum, H.I., Hurlbut, D.E. and Leslie, F.M. (1987a) Postnatal development of multiple opioid receptors in rat brain. *Dev. Brain Res.*, **37**, 21–41.

Kornblum, H.I., Loughlin, S.E. and Leslie, F.M. (1987b) Effects of morphine on DNA synthesis in neonatal rat brain. *Dev. Brain Res.*, **31**, 45–52.

Kornblum, H.I., Loughlin, S.E., Fallon, J.H. and Leslie, F.M. (1989) Developmental appearance of opioid receptors in embryonic and neonatal rat brain using [¹²⁵I] beta-endorphin: an autoradiographic study. *Adv. Biosci.*, **75**, 277–80.

Kuhn, C.M., Windh, R.T. and Little, P.J. (1992) Effects of perinatal opioid addiction on neurochemical development of the brain, in *Development of the Central Nervous System: Effects of Alcohol and Opiates* (ed. M.W. Miller), Wiley-Liss, New York, pp. 341–61.

Leslie, F.M. (1987) Methods used for the study of opioid receptors. *Pharmacol. Rev.*, **39**, 197–249.

Leslie, F.M. (1992) Neurotransmitters as neurotrophic factors, in *Neurotrophic Factors* (eds S.E. Loughlin and J.H. Fallon), Academic Press, New York (in press).

Leslie, F.M. and Loughlin, S.E. (1992) Development of multiple opioid receptors, in *Development of* the Central Nervous System: Effects of Alcohol and Opiates (ed. M.W. Miller), Wiley-Liss, New York, pp. 255–83.

Leslie, F.M. and Smith, T.E. (1992) Autoradiographic localization of kappa$_2$ opioid binding sites in adult and developing rat brain. *Brain Res.* (submitted).

Leslie, F.M., Tso, S. and Hurlbut, D.E. (1982) Differential appearance of opiate receptor subtypes in neonatal rat brain. *Life Sci.*, **31**, 1393–6.

Loughlin, S.E., Massamiri, T.R., Kornblum, H.I. (1985) Postnatal development of opioid systems in rat brain. *Neuropeptides*, **5**, 469–72.

Loughlin, S.E., Annis, C.M., and Twardzik, D.R. (1989) Growth factors in opioid rich brain regions: distribution and response to intrastriatal transplants. *Adv. Biosci.*, **75**, 403–6.

Loughlin, S.E., Kornblum, H.I., Massamiri, T. and Leslie, F.M. (1991) Transient appearance of beta-endorphin immunoreactive cells within the germinal zone of neonatal rat forebrain. *Int. J. Dev. Neurosci.*, **9**, 493–500.

Loughlin, S.E., An, A. and Leslie, F.M. (1992a) Opioid receptor changes in weaver mouse striatum. *Brain Res.*, **585**, 149–55.

Loughlin, S.E., Kornblum, H.I. and Leslie, F.M. (1992b) Development of [^{125}I] beta endorphin binding in pre- and post-natal rat brain (in preparation).

Loughlin, S.E., Reid, S. and Leslie, F.M. (1992c) Opioid peptide expression is changed in weaver mutant mouse striatum: an in situ hybridization study. *Abstr. Soc. Neurosci.*, **282**, 12.

Loughlin, S.E., Annis, C.M., Gentry, L. *et al.* (1992d) Plasticity of transforming growth factor alpha expression in rat striatum: effects of transplants. *Brain Res. Bull.* (submitted).

Maderspach, K. and Solomania, R. (1988) Glial and neuronal opioid receptors: apparent positive cooperativity observed in intact cultured cells. *Brain Res.*, **441**, 41–7.

Magnan, J. and Tiberi, M. (1989) Evidence for the presence of mu- and kappa- but not of delta-opioid sites in the human fetal brain. *Dev. Brain Res.*, **45**, 275–81.

Mains, R.E., Eipper, E.A. and Ling, N. (1977) Common precursor to corticotropins and endorphins. *Proc. Natl. Acad. Sci. USA*, **74**, 3014–18.

Mansour, A., Khachaturian, H., Lewis, M.E. *et al.* (1987) Autoradiographic differentation of mu, delta, and kappa opioid receptors in the rat forebrain and midbrain. *J. Neurosci.*, **7**, 2445–64.

Mansour, A., Schafer, M.K.-H., Newman, S.W. and Watson, S.J. (1991) Central distribution of opioid receptors: a cross-species comparison of the multiple opioid systems of the basal ganglia, in *Neurobiology of Opioids* (eds O.F.X. Almeida and T.S. Shippenberg), Springer-Verlag, Berlin, Heidelberg, pp. 169–83.

McDowell, J. and Kitchen, I. (1987) Development of opioid systems: peptides, receptors and pharmacology. *Brain Res. Rev.*, **12**, 397–421.

McLean, S., Rothman, R.B., Chuang, D.-M. *et al.* (1989) Cross-linking of [^{125}I]beta-endorphin to mu-opioid receptors during development. *Dev. Brain Res.*, **45**, 283–9.

McMillen, I.C., Mercer, J.E. and Thorburn, G.D. (1988) Pro-opiomelanocortin mRNA levels fall in the fetal sheep pituitary before birth. *J. Mol. Endocrinol.*, **1**, 141–5.

Merchenthaler, I., Maderdrut, J.L., Altschuler, R.A. and Petrusz, P. (1986) Immunocytochemical localization of proenkephalin-derived peptides in the central nervous system of the rat. *Neuroscience*, **17**, 325–48.

Mess, B., Ruzsas, C. and Hayashi, S. (1989) Impaired thyroid function provoked by neonatal treatment with drugs affecting the maturation of monoaminergic and opioidergic neurons. *Exp. Clin. Endocrinol.*, **94**, 73–81.

Milligan, G., Streaty, R.A., Gierschik, P. *et al.* (1987) Development of opiate receptors and GTP-binding regulatory proteins in neonatal rat brain. *J. Biol. Chem.*, **262**, 8626–30.

Mochetti, I., Guidotti, A., Schwartz, J.P. and Dosta, E. (1985) Reserpine changes the dynamic state of enkephalin stores in rat striatum and adrenal medulla by different mechanisms. *J. Neurosci.*, **5**, 3379–85.

Moon-Edley, S. and Herkenham, M. (1984) Comparative development of striatal opiate receptors and dopamine revealed by autoradiography and histofluorescence. *Brain Res.*, **305**, 27–42.

Morris, B.J., Haarmann, I., Kempter, B. *et al.* (1986) Localization of prodynorphin messenger RNA in rat brain by in situ hybridization using a synthetic oligonucleotide probe. *Neurosci. Lett.*, **69**, 104–8.

Morris, B.J. and Herz, A. (1987) Distinct distribution of opioid receptor types in rat lumbosacral spinal cord. *Naunyn-Schmiedebergs Arch. Pharmacol.*, **336**, 240–3.

Mulder, A.H., Burger, D.M., Wardeh, G. *et al.* (1991) Pharmacological profile of various κ-agonists at κ-, μ- and δ-opioid receptors mediating presynaptic inhibition of neurotransmitter release in rat brain. *Br. J. Pharmacol.*, **102**, 518–22.

Murrin, L.C. and Ferrer, J.R. (1984) Ontogeny of the rat striatum: correspondence of dopamine terminals, opiate receptors and acetylcholinesterase. *Neurosci. Lett.*, **47**, 155–60.

Negri, L., Potenza, R.L., Corsi, R. and Melchiorri, P. (1991) Evidence for two subtypes of delta opioid receptors in rat brain. *Eur. J. Pharmacol.*, **196**, 335–6.

Ng, T.B., Ho, W.K.K. and Tam, P.P.L. (1984) Brain and pituitary beta-endorphin levels at different developmental stages of the rat. *Int. J. Protein Res.*, **24**, 141–6.

Nock, B., Giordano, A.L., Cicero, T.J. and O'Conor, L.H. (1990) Affinity of drugs and peptides for U-69, 593-sensitive and insensitive kappa opiate binding sites: the U-69, 593-insensitive site appears to be the beta endorphin-specific epsilon receptor. *J. Pharmacol. Exp. Ther.*, **254**, 412–19.

Oetting, G.M., Szucs, M. and Coscia, C.J. (1987) Differential ontogeny of divalent cation effects on rat brain delta-, mu-, and kappa-opioid receptor binding. *Dev. Brain Res.*, **31**, 223–7.

Olson, G.A., Olson, R.D. and Kastin, A.J. (1991) Endogenous opiates: 1990. *Peptides*, **12**, 1407–32.

Osborne, J.G., Kindy, M.S. and Hauser, K.F. (1991) Expression of proenkephalin mRNA in developing cerebellar cortex of the rat: expression levels coincide with maturational gradients in Purkinje cells. *Dev. Brain Res.*, **63**, 63–9.

Palmer, M.R., Miller, R.J., Olson, L. and Sieger, A. (1982) Prenatal ontogeny of neurons with enkephalin-like immunoreactivity in the rat central nervous system: an immunohistochemical mapping investigation. *Med. Biol.*, **60**, 2–88.

Pasternak, G.W. and Wood, P.L. (1986) Multiple mu opiate receptors. *Life Sci.*, **38**, 135.

Pasternak, G.W., Zhang, A. and Tecott, L. (1980) Developmental differences between high and low affinity opiate binding sites: their relationship to analgesia and respiratory depression. *Life Sci.*, **27**, 1185–90.

Patey, G., De La Baume, S., Gros, C. and Schwartz, J.-C. (1980) Ontogenesis of enkephalinergic systems in rat brain: post-natal changes in enkephalin levels, receptors and degrading enzyme activities. *Life Sci.*, **27**, 245–52.

Petrillo, P., Tavani, A., Verotta, D. *et al.* (1987) Differential postnatal development of mu-, delta- and kappa-opioid binding sites in rat brain. *Dev. Brain Res.*, **31**, 53–8.

Petrusz, P., Merchenthaler, I. and Maderdrut, J.L. (1985) Distribution of enkephalin-containing neurons in the central nervous system, in *Handbook of Chemical Neuroanatomy*, Vol. 4. *GABA and neuropeptides in the CNS*, Part I (eds A. Bjorklund and T. Hökfelt), Elsevier, Amsterdam, pp. 273–334.

Pittius, C.W., Ellendorff, F., Hollt, V. and Parvizi, N. (1987) Ontogenetic development of proenkephalin A and proenkephalin B messenger RNA in fetal pigs. *Exp. Brain Res.*, **69**, 208–12.

Quinlan, P.E. and Alessi, N.E. (1991) Characterization of beta-endorphin-related peptides in the caudal medulla oblongata and hypothalamus of the prenatal, postnatal and adult rat. *Dev. Brain Res.*, **62**, 1–5.

Rius, R.A., Barg, J., Bem, W.T. *et al.* (1991a) The prenatal developmental profile of expression of opioid peptide and receptors in the mouse brain. *Dev. Brain Res.*, **58**, 237–41.

Rius, R.A., Chikuma, T. and Lo, Y.P. (1991b) Prenatal processing of pro-opiomelanocortin in the brain and pituitary of mouse embryos. *Dev. Brain Res.*, **60**, 179–85.

Rosen, H. and Polakiewicz, R. (1989) Postnatal expression of opioid genes in rat brain. *Dev. Brain Res.*, **46**, 123–9.

Rosen, H., Douglass, J. and Herbert, E. (1984) Isolation and characterization of the rat proenkephalin gene. *J. Biol. Chem.*, **259**, 14309–13.

Rothman, R.B., Bykov, V., De Costa, B.R. *et al.* (1990) Interaction of endogenous opioid peptides and other drugs with four kappa opioid binding sites in guinea pig brain. *Peptides*, **11**, 311–31.

Sato, M., Morita, Y., Saika, T. *et al.* (1991) Localization and ontogeny of cells expressing preprodynorphin mRNA in the cerebral cortex. *Brain Res.*, **541**, 41–9.

Schafer, M.K.-H., Herman, J.P., Day, R. *et al.* (1988) The distribution of prodynorphin mRNA throughout the rat brain: a semi-quantitative mapping study. *Soc. Neurosci. Abst.*, **14**, 545.

Schafer, M.K.-H., Day, R., Watson, S.J. and Akil, H. (1991) Distribution of opioids in brain and peripheral tissues, in *Neurobiology of Opioids* (eds O.F.X. Almeida and T.S. Shippenberg), Springer-Verlag, Berlin, Heidelberg, pp. 53–71.

Schmahl, W., Funk, R., Miaskowski, U. and Plendl, J. (1989) Long-lasting effects of naltrexone, an opioid receptor antagonist, on cell proliferation in developing rat forebrain. *Brain Res.*, **486**, 297–300.

Schoffelmeer, A.N.M., Hansen, H.A., Stoof, J.C. and Mulder, A.H. (1986) Blockade of D-2 dopamine receptors strongly enhances the potency of enkephalins to inhibit dopamine-sensitive adenylate cyclase in rat neostriatum: involvement of delta- and mu-opioid receptors. *J. Neurosci.*, **6**, 2235–9.

Schoffelmeer, A.N.M., Hogenboom, F. and Mulder, A.H. (1987) Inhibition of dopamine-sensitive adenylate cyclase by opioids: possible involvement of physically associated mu- and delta-opioid receptors. *Naunyn-Schmiedebergs Arch. Pharmacol.*, **335**, 278–84.

Schoffelmeer, A.N.M., Rice, K.C., Jacobson, A.E. *et al.* (1988) Mu-, delta- and kappa-opioid receptor-mediated inhibition of neurotransmitter release and adenylate cyclase activity in rat brain slices: studies with fentanyl isothiocyanate. *Eur. J. Pharmacol.*, **154**, 169–78.

Schoffelmeer, A.N.M., Yao, Y.-H., Gioannini, T.L. *et al.* (1990) Cross-linking of human [125I] beta-endorphin to opioid receptors in rat striatal membranes: biochemical evidence for the existence of a mu/delta opioid receptor complex. *J. Pharmacol. Exp. Ther.*, **253**, 419–26.

Schofield, P.R., McFarland, K.C., Hayflick, J.S. *et al.* (1989) Molecular characterization of a new immunoglobulin superfamily protein with potential roles in opioid binding and cell contact. *EMBO J.*, **8**, 489–95.

Schulz, R., Faase, E., Wuster, M. and Herz, A. (1979) Selective receptors for beta-endorphin on the rat vas deferens. *Life Sci.*, **24**, 843–50.

Schulz, R., Wuster, M. and Herz, A. (1981) Pharmacological characterization of the epsilon-opiate receptor. *J. Pharmacol. Exp. Ther.*, **216**, 604–6.

Schwartz, J.P. and Simantov, V. (1988) Developmental expression of proenkephalin mRNA in rat striatum and in striatal cultures. *Dev. Brain Res.*, **40**, 311–14.

Schwartzberg, D.G. and Nakane, P.K. (1982) Ontogenesis of adrenocorticotropin-related peptide determinants in the hypothalamus and pituitary gland of the rat. *Endocrinology*, **110**, 855–64.

Sei, C.A. and Dores, R.M. (1989) Changes in the processing of pro-dynorphin end products in the substantia nigra during neonatal development. *Peptides*, **11**, 89–94.

Seizinger, B.R., Grimm, C. and Herz, A. (1984a) Evidence for a differential postnatal development of proenkephalin B (=prodynorphin)-derived opioid peptides in the rat hypothalamus. *Endocrinology*, **115**, 926–35.

Seizinger, B.R., Hollt, V. and Herz, A. (1984b) Postnatal development of beta-endorphin-related peptides in rat anterior and intermediate pituitary lobes: evidence for contrasting develop-ment of proopiomelanocortin processing. *Endocrinology*, **115**, 136–42.

Shaha, C., Margioris, A., Liotta, A.S. *et al.* (1984) Demonstration of immunoreactive beta-endo-phin- and gamma₃-melanocyte-stimulating hormone-related peptides in the ovaries of neonatal, cyclic and pregnant mice. *Endocrinology*, **115**, 378–84.

Sharif, N.A. and Hughes, J. (1989) Discrete mapping of brain mu and delta opioid receptors using selective peptides: quantitative autora-diography species differences and comparison with kappa receptors. *Peptides*, **10**, 499–522.

Sirinathsinghji, D.J.S. and Dunnett, S.B. (1989) Disappearance of the mu opiate receptor patches in the rat neostriatum following lesion-ing of the ipsilateral nigrostriatal dopamine pathway with 1-methyl-4-phenylpyridinium ion (MPP+): restoration by embryonic nigral dopamine grafts. *Brain Res.*, **504**, 115–20.

Smith, J.A.M., Loughlin, S.E. and Leslie, F.M. (1991) Long-term changes in striatal opioid binding sites after 6-hydroxydopamine (6OHDA) lesion of substantia nigra. *Soc. Neurosci. Abst.*, **183**, 3.

Smith, J.A.M., Loughlin, S.E., and Leslie, F.M. (1992a) Kappa opioid inhibition of [³H] dop-amine release from embryonic rat ventral mes-encephalic culture. *Mol. Pharmacol.*, **42**, 575–83.

Smith, J.A.M., Leslie, F.M., Broide, R.S. and Loughlin, S.E. (1992b) Alterations in striatal opioid peptide and receptor expression follow-ing 6-hydroxydopamine lesions of substantia nigra. *Brain Res.* (submitted).

Spain, J.W., Roth, B.L. and Coscia, C.J. (1985) Differential ontogeny of multiple opioid recep-tors (mu, delta, and kappa). *J. Neurosci.*, **5**, 584–8.

Specht, L.A., Pickel, V.M., Joh, T.H. *et al.* (1981) Light-microscopic immunocytochemical localiza-tion of tyrosine hydroxylase in prenatal rat brain. I. Early ontogeny. *J. Comp. Neurol.*, **199**, 233–53.

Stiene-Martin, A. and Hauser, K.F. (1990) Opioid-dependent growth of glial cultures: suppression of astrocyte DNA synthesis by met-enkephalin. *Life Sci.*, **46**, 91–8.

Stiene-Martin, A. and Hauser, K.F. (1991) Glial growth is regulated by agonists selective for multiple opioid receptor types in vitro. *J. Neurosci. Res.*, **29**, 538–48.

Stiene-Martin, A., Gurwell, J.A. and Hauser, K.F. (1991a) Morphine alters astrocyte growth in primary cultures of mouse glial cells: evidence for a direct effect of opiates on neural matu-ration. *Dev. Brain Res.*, **60**, 1–7.

Stiene-Martin, A., Osborne, J.G. and Hauser, K.F.

(1991b) Co-localization of proenkephalin mRNA using cRNA probes and a cell-type-specific immunocytochemical marker for intact astrocytes in vitro. *J. Neurosci. Methods*, **36**, 119–26.

Szucs, M. and Coscia, C.J. (1990) Evidence for delta-opioid binding and GTP-regulatory proteins in 5-day-old rat brain membranes. *J. Neurochem.*, **54**, 1419–25.

Szucs, M., Spain, J.W., Oetting, G.M. *et al.* (1987) Guanine nucleotide and cation regulation of μ, δ, and κ opioid receptor binding: evidence for differential postnatal development in rat brain. *J. Neurochem.*, **48**, 1165–70.

Tavani, A., Robson, L.E. and Kosterlitz, H.W. (1985) Differential postnatal development of mu, delta, and kappa opioid binding sites in mouse brain. *Dev. Brain Res.*, **23**, 306–9.

Tecott, L.H., Rubenstein, J.L.R., Paxinos, G. *et al.* (1989) Developmental expression of proenkephalin mRNA and peptides in rat striatum. *Dev. Brain Res.*, **49**, 75–86.

Tempel, A. (1991) Visualization of mu opiate receptor downregulation following morphine treatment in neonatal rat brain. *Dev. Brain Res.*, **64**, 19–26.

Tempel, A. (1992) Regulation of the opioid system by exogenous drug administration, in *Development of the Central Nervous System: Effects of Alcohol and Opiates* (ed. M.W. Miller), Wiley-Liss, New York, pp. 285–318.

Tempel, A., Kessler, J.A. and Zukin, R.S. (1990) Chronic naltrexone treatment increases expression of preproenkephalin and preprotachykinin mRNA in discrete brain regions. *J. Neurosci.*, **10**, 741–7.

Tiong, G.K.L. and Olley, J.E. (1988) Effects of exposure in utero to methadone and buprenorphine on enkephalin levels in the developing rat brain. *Neurosci. Lett.*, **93**, 101–6.

Tsang, D. and Ng, S.C. (1980) Effect of antenatal exposure to opiates on the development of opiate receptors in rat brain. *Brain Res.*, **188**, 199–206.

Uhl, G.R., Ryan, J.P. and Schwartz, J.P. (1988) Morphine alters preproenkephalin gene expression. *Brain Res.*, **459**, 391–7.

Uhler, M. and Herbert, E. (1983) Complete amino acid sequence of mouse proopiomelanocortin derived from the nucleotide sequence of proopiomelanocortin cDNA. *J. Biol. Chem.*, **258**, 257–61.

Unnerstall, J.R., Molliver, M.E., Kuhar, M.J. and Palacios, J.M. (1983) Ontogeny of opiate binding sites in the hippocampus, olfactory bulb and other regions of the rat forebrain by autoradiographic methods. *Dev. Brain Res.*, **7**, 157–69.

Unterwald, E.M., Knapp, C. and Zukin, R.S. (1991) Neuroanatomical localization of kappa$_1$ and kappa$_2$ opioid receptors in rat and guinea pig brain. *Brain Res.*, **562**, 57–65.

Van der Kooy, D. and Fishell, G. (1987) Neuronal birthdate underlies the development of striatal compartments. *Brain Res.*, **401**, 155–61.

Van Vliet, B.J., Mulder, A.H. and Schoffelmeer, A.N.M. (1990) μ-Opioid receptors mediate the inhibitory effect of opioids on dopamine-sensitive adenylate cyclase in primary cultures of rat neostriatal neurons. *J. Neurochem.*, **55**, 1274–80.

Vaysse, P.J.-J., Zukin, R.S., Fields, K.L. and Kessler, J.A. (1990) Characterization of opioid receptors in cultured neurons. *J. Neurochem.*, **55**, 624–31.

Vertes, Z., Melegh, G.Y., Vertes, M. and Kovacs, S. (1982) Effect of naloxone and d-met^2-pro^5-enkephalinamide treatment on the DNA synthesis in the development rat brain. *Life Sci.*, **31**, 119–26.

Vilijn, M.-H., Das, B., Kessler, J.A. and Fricker, L.D. (1989) Cultured astrocytes and neurons synthesize and secrete carboxypeptidase E, a neuropeptide–processing enzyme. *J. Neurochem.*, **53**, 1487–93.

Voorn, P., Kalsbeek, B., Jorritsma-Byham, B. and Groenewegen, H.J. (1988) The pre-and postnatal development of the dopaminergic cell groups in the ventral mesencephalon and the dopaminergic innervation of the striatum of the rat. *Neuroscience*, **25**, 857–87.

Watson, S.J., Akil, H., Fischli, W. *et al.* (1982) Dynorphin and vasopressin: common localization in magnocellular neurons. *Science*, **216**, 85–7.

Werling, L.L., Frattali, A., Portoghese, P.S. *et al.* (1988) Kappa receptor regulation of dopamine release from striatum and cortex of rats and guinea pigs. *J. Pharmacol. Exp. Ther.* **246**, 282–6.

Wild, K.D., Vanderah, T., Mosberg, H.I. and Porreca, F. (1991) Opioid delta receptor subtypes are associated with different potassium channels. *Eur. J. Pharmacol.*, **193**, 135–6.

Xia, Y. and Haddad, G.G. (1991) Ontogeny and distribution of opioid receptors in the rat brainstem. *Brain Res.*, **549**, 181–93.

Young, W.S., Bonner, T.I. and Brann, M.R. (1986) Mesencephalic dopamine neurons regulate the expression of neuropeptide mRNAs in the rat forebrain. *Proc. Natl. Acad. Sci. USA*, **83**, 9827–31.

Zagon, I.S., Rhodes, R.E. and McLaughlin, P.J. (1985) Distribution of enkephalin immunoreactivity in germinative cells of developing rat cerebellum. *Science*, **227**, 1049–51.

Zagon, I.S., Goodman, S.R. and McLaughlin, P.J. (1989) Characterization of zeta (ζ): a new opioid receptor involved in growth. *Brain Res.*, **482**, 297–305.

Zagon, I.S., Gibo, D. and McLaughlin, P.J. (1990) Expression of zeta, a growth-related opioid receptor, in metastatic adenocarcinoma of the human cerebellum. *J. Natl. Cancer Inst.*, **82**, 325–7.

Zagon, I.S., Gibo, D.M. and McLaughlin, P.J. (1991) Zeta (ζ), a growth-related opioid receptor in developing rat cerebellum: identification and characterization. *Brain Res.*, **551**, 28–35.

Zagon, I.S. and McLaughlin, P.J. (1983) Increased brain size and cellular content in infant rats treated with an opiate antagonist. *Science*, **221**, 1179–80.

Zagon, I.S. and McLaughlin, P.J. (1990) Ultrastructural localization of enkephalin-like immunoreactivity in developing rat cerebellum. *Neuroscience*, **34**, 479–89.

Zagon, I.S. and McLaughlin, P.J. (1993) Opioid factor receptor in the developing nervous system, in *Receptors in the Developing Nervous System*, Vol. 1, (eds I.S. Zagon and P.J. McLaughlin), Chapman & Hall, London, pp. 39–62.

Zamir, N., Quirion, R. and Segal, M. (1985) Ontogeny and regional distribution of proenkephalin- and prodynorphin-derived peptides and opioid receptors in rat hippocampus. *Neuroscience*, **15**, 1025–34.

Zukin, R.S., Eghbali, M., Olive, D. *et al.* (1988) Characterization and visualization of rat and guinea pig brain κ opioid receptors: Evidence for κ_1 and κ_2 opioid receptors. *Proc. Natl. Acad. Sci. USA*, **85**, 4061–5.

SIGMA RECEPTORS, PCP RECEPTORS AND THE DEVELOPING NERVOUS SYSTEM

Edythe D. London and Stephen R. Zukin

11.1 INTRODUCTION

The existence of σ (sigma) receptors was hypothesized on the basis of work by Martin *et al.* (1976), who studied the physiological and behavioral responses to opioid drugs in the chronic spinal dog. Based on the responses to the prototypical ligands, morphine, ketocyclazocine, and *N*-allylnormetazocine (NANM, SKF 10047), this work led to the hypothesis of multiple opioid receptors (μ, κ and σ). Although this landmark work has greatly advanced the field of opiate research, it was later found that the responses to NANM were non-opioid, inasmuch as they were not reversed by naltrexone (Vaupel, 1983). Furthermore, benzomorphan opioids which show high affinity for σ receptors also interact with receptors for phencyclidine (PCP). Therefore, it is likely that canine delirium, which was originally attributed to interactions with the σ 'opioid' receptor, were due to activity at the PCP receptor.

This chapter gives an overview of current knowledge about the pharmacology and physiological importance of the σ and PCP receptors. The role of these receptors in the developing brain is highlighted.

11.2 IDENTIFICATION AND CHARACTERIZATION OF SIGMA AND PCP RECEPTORS

11.2.1 PHARMACOLOGICAL PROPERTIES

The first report of the biochemical identification of σ binding sites appeared about a decade ago, when racemic [³H]N-allylnormetazocine ([³H]SKF-10047) was used to label 'etorphine-inaccessible sites' in guinea-pig brain (Su, 1981, 1982). The sites showed very low affinity for naloxone and some opioid agonists, including etorphine and morphine, but high affinity for benzomorphan opioids, including pentazocine, NANM and cyclazocine. The receptors showed stereoselectivity, favoring the dextrorotatory over the levorotatory isomers of benzomorphan opioid drugs. Although the D_2 dopamine receptor antagonist, haloperidol, was the most potent inhibitor of σ receptor binding, σ sites were unlike dopamine receptors in that they showed higher affinity for the levorotatory enantiomer of butaclamol than for the dextrorotatory isomer. PCP showed activity at the σ sites, but it had about one-tenth the potency of (+)NANM.

High affinity binding sites with pharma-

Receptors in the Developing Nervous System Vol. 2: Neurotransmitters. Edited by Ian S. Zagon and Patricia J. McLaughlin. Published in 1993 by Chapman & Hall. ISBN 0 412 49400 0. Vols. 1 and 2 (set) ISBN 0 412 54520 9.

cological properties of a PCP receptor were observed in rat brain (Vincent *et al.*, 1979; Zukin and Zukin, 1979), and then characterized (e.g., Quirion *et al.*, 1981; Hampton *et al.*, 1982; Vignon *et al.*, 1982, 1983; Vincent *et al.*, 1983; Zukin *et al.*, 1983; Sircar and Zukin, 1985). The PCP receptor is selective for drugs capable of eliciting PCP-like behavioral effects, and does not show high affinity for other categories of hallucinogens, other drugs, or neurochemicals (reviewed by Zukin and Zukin, 1988). One of the earliest reports on the identification of the PCP receptor noted that NANM inhibited binding to the receptor, with a potency that was about one-third that of PCP (Zukin and Zukin, 1979). Observations that specific [³H]PCP binding can be inhibited by cyclazocine and related benzomorphan opioids (Zukin and Zukin, 1981; Zukin *et al.*, 1983) suggested that the PCP and the σ receptor were the same entity. Despite the fact that benzomorphans do interact with the PCP receptor, subsequent studies have shown that σ receptors are distinct from PCP receptors anatomically, ontogenetically and phylogenetically (Vu *et al.*, 1990).

11.2.2 RADIOLIGAND SPECIFICITY (FIG. 11.1)

Since the initial description of specific σ receptor binding, there have been attempts to develop selective, high affinity radioligands for assay. Early studies of this receptor used racemic mixtures of benzomorphan opioids as radioligands. After the first studies with (±)NANM (Su, 1981, 1982), other assays utilized the benzomorphan (±)ethylketocyclazocine (Tam, 1983). The naloxone-inaccessible binding of this radioligand showed stereoselectivity, favoring the dextrorotatory isomers of benzomorphans, a finding that was consistent with earlier studies of the σ receptor. Based on these observations, subsequent assays used (+)-[³H]NANM (Tam and Cook, 1984; Tam 1985; Su *et al.*, 1988b; McCann and Su, 1990) or (+)-[³H]ethylketocyclazocine (Tam, 1985) as radioligands for σ receptor binding assays. Other studies took advantage of the very high affinity of haloperidol for the σ receptor, and used [³H]haloperidol as the radioligand in the presence of either (+)butaclamol or spiperone to block binding of the radioligand to D₂ dopamine receptors (Tam

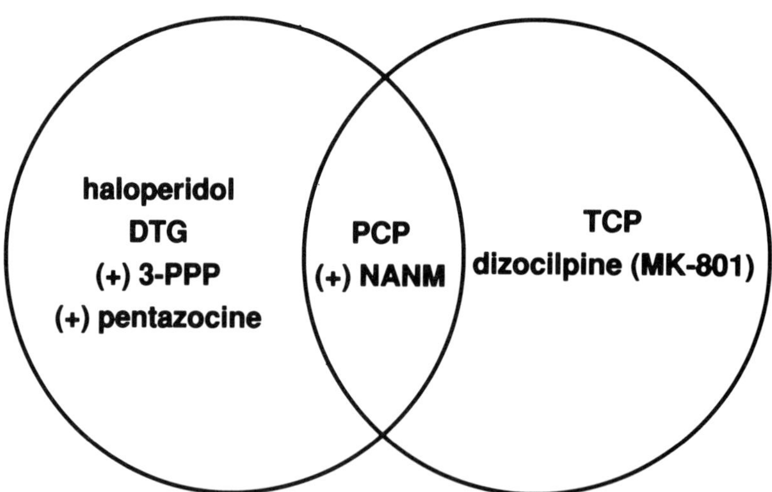

Fig. 11.1. Diagram depicting selectivity of ligands for the σ receptor (left circle) and PCP receptor (right). Area of overlap indicates ligands that interact with both receptors. Abbreviations: DTG, 1,3-di(2-tolyl)guanidine; (+)3-PPP, (+)-3-(3-hydroxyphenyl)-*N*-(1-propyl)piperidine; PCP, phencyclidine; (+)NANM, (+)*N*-allylnormetazocine; TCP, [1-(2-thienyl) cyclohexyl] piperidine.

and Cook, 1984; Su *et al.*, 1988b; McCann and Su, 1990; Vu *et al.*, 1990). Although not as potent at σ sites as haloperidol, the ligand (+)-3-(3-hydroxyphenyl)-*N*-(1-propyl)piperidine ((+)-[³H]3-PPP) has also been used (Largent *et al.*, 1984, 1986a), and it apparently shows higher selectivity for σ sites than many of the other radioligands used previously. Also notable is the finding that [³H]dextromethorphan and σ ligands bind to at least one common high affinity site (Musacchio *et al.*, 1989a,b). The observation that certain symmetrically substituted guanidines can inhibit the binding of (+)-[³H]NANM to membranes from guinea-pig brain led to the development of 1,3-di(2-[5-³H]tolyl)guanidine ([³H]DTG) as a radioligand for the σ receptor (Weber *et al.*, 1986). This ligand had the advantage over the benzomorphans of being highly selective for the σ site, and it yielded a high degree of specific binding in preparations of guinea-pig brain. Most recently, the synthesis of optically pure (+)-[³H]pentazocine, a potent and selective ligand for σ receptors has offered the promise of being a valuable radioligand (de Costa *et al.*, 1989).

[³H]Phencyclidine, the prototypical ligand of the PCP receptor, has proven less than optimally selective for this receptor. Although it displays high affinity for the PCP receptor (K_d = 0.15–0.25 μM) (Zukin and Zukin, 1979; Vincent *et al.*, 1983), it also has significant, albeit lower affinity for σ receptors (IC_{50} values about 1–2 μM against several tritiated ligands for σ receptors (Tam, 1983; Weber *et al.*, 1986)). PCP and related compounds also inhibit dopamine uptake (Vignon *et al.*, 1988), and inhibit the binding of [³H]mazindol to sites on dopamine transporters (Kuhar *et al.*, 1990). *N*-(1-[2-thienyl]cyclohexyl)[³H]piperidine ([³H]TCP), a thienyl derivative of [³H]PCP, displays higher affinity and specificity for the PCP receptor than does PCP itself (Sircar and Zukin, 1985). The most potent and selective ligand currently available for the PCP receptor is dizocilpine (MK-801) (Wong *et al.*, 1986, 1988; Foster and Wong, 1987).

11.2.3 BIOCHEMICAL STUDIES OF SIGMA RECEPTORS

Some of the earliest studies of sigma receptors indicated a protein nature of the sites. When membranes from guinea-pig brain were treated with heat, σ binding, assayed with [³H]NANM or [³H]dextromethorphan, was reduced (Su, 1982; Craviso and Musacchio, 1983a,b). Binding also displayed a marked dependence on pH, with optima achieved above pH 8.0, when either [³H]dextromethorphan or (+)-[³H]-3-PPP were used as radioligands (Craviso and Musacchio, 1983a; Largent *et al.*, 1987). Furthermore, treatment of membranes with proteases reduced the binding of [³H]NANM (Su, 1982) and [³H]dextromethorphan (Craviso and Musacchio, 1983a).

The development of an azido derivative of DTG, [³H]azido-DTG, has allowed photoaffinity labeling and molecular weight determination of a σ receptor protein in guinea-pig brain (Kavanaugh *et al.*, 1988). Irradiation of the ligand–receptor complex with light of 336 nm wavelength produced covalent binding of the azido probe to the receptor; and sodium dodecyl sulfate (SDS)–polyacrylamide gel electrophoresis demonstrated that labeling was to a single polypeptide with a molecular weight of 29 kDa. Pharmacological characterization indicated a ligand specificity consistent with that of the sigma receptor. Similarly, [³H]azido-DTG labels a polypeptide of 29 kDa in NCB-20 cells, which bear σ receptors (Largent *et al.*, 1986a; Adams *et al.*, 1987; Kushner *et al.*, 1988).

In contrast to the findings regarding σ receptors in guinea-pig brain and NCB-20 cells, [³H]azido-DTG labeled two polypeptides, with molecular weights of 18 and 21 kDa in PC12 cells, suggesting heterogeneity of σ receptors (Hellewell and Bowen, 1990). Further evidence for heterogeneity from this study included a reduced affinity for dextrorotatory benzomorphan opioids, as compared to the affinity of the classical σ receptor characterized in the guinea-pig brain. The concept of heterogeneity of σ receptors was consistent with findings of Bowen *et al.* (1989), who noted that irradiation of rat brain membranes with light of 254 nm, a treatment that alters ultraviolet absorbing residues in proteins, had differential effects on the binding of (+)-[^{3}H]NANM as compared to the binding of (+)-[^{3}H]-3-PPP and [^{3}H]DTG. Irradiation increased the binding of [^{3}H]NANM, but decreased binding of the non-benzomorphan radioligands. In treated membranes, DTG and (+)-3-PPP were less potent inhibitors of (+)-[^{3}H]NANM binding than in unirradiated membranes; however, the potencies of the benzomorphans as inhibitors were relatively unaffected. Furthermore, in unirradiated membranes, DTG and haloperidol were competitive inhibitors of (+)-[^{3}H]-3-PPP binding whereas the benzomorphan opioids, (+)NANM and (+)pentazocine, were uncompetitive inhibitors.

Electrophysiological and receptor binding studies in intact NCB-20 cells also indicated the existence of more than one type of σ receptor (Wu *et al.*, 1990). Whole cell voltage clamp studies demonstrated that σ ligands caused an apparent inward current, which was due to blockade of a tonic, outward potassium current. Although the rank order of drugs in producing this effect generally resembled the order of potencies of these drugs in inhibiting binding to σ receptors reported previously, several compounds showed reverse stereoselectivity. Binding assays performed on intact cells under conditions similar to the electrophysiological studies (e.g., Dulbecco's phosphate-buffered saline instead of 50 mM Tris-HCl) revealed the existence of two populations of sites. The high affinity sites showed a pharmacological selectivity similar to that seen previously in membranes from NCB-20 cells (Largent *et al.*, 1986a; Kushner *et al.*, 1988), but the low affinity sites showed a rank order of potencies that correlated well with potencies in the voltage clamp assays, including absence of the usual stereoselectivity for benzomorphan opioids and 3-PPP.

11.2.4 BIOCHEMICAL CHARACTERIZATION OF PCP RECEPTORS

The PCP receptor is proteinaceous in nature, as it is inactivated by heat, and is destroyed by proteases, such as trypsin, pronase, and papain (Zukin and Zukin, 1979; Zukin *et al.*, 1983; Vignon *et al.*, 1982). Pretreatment with *N*-ethylmaleimide or iodoacetamide inhibits activity, suggesting the importance of sulfhydryl groups at or near the binding site, or at a site that influences the tertiary structure of the receptor protein (Zukin and Zukin, 1979).

Photoaffinity labeling of PCP receptors with the photoactive PCP derivative azido-[^{3}H]PCP, followed by SDS–polyacrylamide gel electrophoresis, has been used to determine the polypeptides comprising the binding site. When such a procedure is carried out with membrane homogenates, a number of polypeptides are labeled (Haring *et al.*, 1986, 1987; Sorensen and Blaustein, 1986). However, photoaffinity labeling carried out in lipid vesicles reconstituted with solubilized and partially purified, active PCP receptors results in labeling of only two polypeptides (M_r 98 000 and 59 000), both of which have also been identified in membrane homogenates (Scheideler and Zukin, 1990). Such data suggest that the PCP receptor comprises multiple protein subunits. However, azido-[^{3}H]dizocilpine was reported to label a single polypeptide of M_r 120 000 (Sonders *et al.*, 1990), in agreement with the target size of M_r 118 000 derived from radiation inactivation of the [^{3}H]TCP binding site (Honoré *et al.*, 1989).

11.2.5 ENDOGENOUS LIGANDS

Several research groups have reported preliminary findings on endogenous substances that were extracted from brain, and which appeared to act as ligands for σ and PCP receptors. Acid extraction of porcine brain yielded two factors, termed α- and β-endopsychosin, representing material which selectively inhibited binding to PCP and σ receptors, respectively

(Quirion *et al.*, 1984; Contreras *et al.*, 1987; DiMaggio *et al.*, 1988). α-Endopsychosin was characterized as having a molecular weight of about 3000 Da. It was regionally co-localized in brain with PCP receptors, and it mimicked the effects of PCP upon turning behavior when injected unilaterally into the substantia nigra of the rat. In contrast, β-endopsychosin inhibited the binding of (+)-[³H]NANM and (+)-[³H]-3-PPP without affecting binding to the PCP receptor, opioid receptors or dopamine receptors. Both endogenous factors were sensitive to protease, consistent with a peptide nature. Extracts of human brain have also been found to contain substances that selectively inhibit binding of radioligands to the σ and PCP receptors, as well as factors which inhibit binding to both sites (Zhang *et al.*, 1988). At least two factors, with molecular weights of approximately 700 Da, were partially purified. One of these factors showed activity at σ and PCP receptors whereas another appeared to have preferential activity at the PCP receptor.

Extracts of guinea-pig brain contained two endogenous factors, termed 'sigmaphins', which interacted preferentially with σ receptors (Su *et al.*, 1986). These factors had activity at opioid receptors, but were almost inactive in PCP receptor binding assays. Treatment with trypsin largely reduced inhibitory potencies of the sigmaphins, and, therefore, suggested a protein composition.

Still another factor, isolated from bovine brain, reportedly mimicked the effects of PCP-like drugs upon spontaneous and NMDA-evoked neurotransmitter release (Zukin *et al.*, 1987). The dose–response curves for the endogenous factor were parallel to those of PCP.

Despite these intriguing reports, there has been no report of a definitive characterization of an endogenous ligand for the σ receptor or the PCP receptor. In the case of putative endogenous ligands for the PCP receptor, it is possible that these factors were endogenous substances that antagonize the function of the NMDA receptor at sites distinct from the PCP receptor.

11.3 DISTRIBUTION OF SIGMA AND PCP RECEPTORS IN THE BRAIN

The earliest studies of σ receptors, performed by incubating racemic [³H]NANM or dextro-rotatory isomers of [³H]NANM or [³H]ethylketocyclazocine with suspensions from grossly dissected areas of guinea-pig brain, demonstrated the highest level of binding in the brainstem and midbrain, followed by the cerebellum (Su, 1982; Tam, 1985); lower levels of binding were seen in the striatum and cortex. Subsequent autoradiographic studies *in vitro* with (+)-[³H]-3-PPP and (+)-[³H]NANM demonstrated discrete localizations of σ receptors to many limbic, brainstem and cerebellar regions (Largent *et al.*, 1984, 1986b; Gundlach *et al.*, 1985). Particularly high densities of σ sites were observed in the pyramidal cell layer of the hippocampus, hypothalamus, zona incerta, subiculum, superficial layer of the cingulate cortex, Purkinje cell layer of the cerebellum, and pontine and cranial nerve nuclei. A similar distribution was seen when [³H]DTG was used as the radioligand (Weber *et al.*, 1986). Sigma receptors have also been labeled *in vivo* in mouse brain using (+)-[³H] NANM and [³H]haloperidol, with a distribution pattern that was consistent with *in vitro* biochemical and autoradiographic studies (Weissman *et al.*, 1990a).

As was the case for the σ receptor, the earliest reports on identification of the PCP receptor provided information about the distribution of the receptor in brain (Vincent *et al.*, 1979; Zukin and Zukin, 1979). With [³H]PCP as the radioligand in homogenates of regions of rat brain, specific binding was high in telencephalic areas, including hippocampus, caudate putamen and neocortex; levels of binding were considerably lower in the cerebellum and medulla/pons. Subsequent autoradiographic studies with [³H]PCP and [³H]TCP provided more detailed anatomical information, but generally confirmed the reports of dense labeling of cortical areas and the hippocampal formation, with lower levels of binding in the brainstem and cerebellum (Quirion *et al.*, 1981; Sircar and Zukin, 1985; Largent *et al.*, 1986b). Thus, despite the fact that some benzomorphans and PCP bind to both σ and PCP receptors, the distribution of specific binding using the prototypic ligands for the respective receptors ([³H]NANM for the σ receptor, [³H]PCP for the PCP receptor) showed neuroanatomical divergence. Anatomical studies were also among the key findings that linked the PCP receptor functionally to the *N*-methyl-D-aspartate (NMDA) subset of glutamate receptors in brain. In this regard, autoradiographic studies of [³H]TCP binding showed a high concordance with the binding of [³H]glutamate, assayed under conditions that select for NMDA receptors (Maragos *et al.*, 1986).

11.4 FUNCTIONAL RELEVANCE OF SIGMA AND PCP RECEPTORS

11.4.1 SIGMA RECEPTOR

It has been hypothesized that the σ receptor was the biochemical entity in brain that mediated the psychotomimetic effects of benzomorphan opioids. Evidence for this hypothesis derived, in part, from the fact that the prototypic σ receptor ligand NANM produced psychotomimetic effects in human volunteers (Keats and Telford, 1964) and delirium in the chronic spinal dog (Martin *et al.*, 1976). Racemic NANM was used in both studies. Notably, behavioral effects in the human study were not challenged with an opioid antagonist. Therefore, it is possible that psychotomimesis resulted from interactions with opioid receptors. Some opioids produce naltrexone-reversible psychotomimetic effects (Jasinski *et al.*, 1968; Martin, 1983; Pfeiffer *et al.*, 1986; Musacchio, 1990), which may be attributable to interactions with κ opioid receptors (Pfeiffer *et al.*, 1986; Su, 1991). In this regard, levorotatory benzomorphan opioids are potent ligands for the κ opioid receptor (Su, 1985). Another major

consideration is the fact that NANM interacts with the PCP receptor; therefore, canine delirium in the original report on σ effects may have reflected an interaction of NANM with the PCP receptor.

The aforementioned links between the σ receptor and psychosis seem weak in light of developments over the past 15 years. Nonetheless, the σ hypothesis generated interest in developing σ antagonists, which might prove to be potential agents for the treatment of psychotic disorders. A consistent finding that fueled the search for a σ ligand which would be an antipsychotic drug was the observation that haloperidol was an extremely potent σ receptor ligand. More recent studies demonstrated that several new drugs, with varying affinities for different receptors, and antipsychotic activity in preclinical (and in some cases clinical) trials, have affinity for the σ receptor as a common feature (Largent *et al.*, 1988; Su *et al.*, 1988b). Furthermore, the observation that σ receptors show a lower density in samples of cerebral cortex from patients who died with schizophrenia than in age-matched controls is additional evidence favoring the 'σ hypothesis of psychosis' (Weissman *et al.*, 1990b).

Aside from the potential role of the σ receptor in psychotic disorders and in the development of potential psychotherapeutic medications, the σ receptor has been implicated in numerous functions of the brain and other organ systems (reviewed by Walker *et al.*, 1990; Su, 1991). For example, σ receptors appear to have a role in motor function. Evidence for this view derives from observations that microinjection of σ receptor ligands into the red nucleus produced dystonic effects in the rat, and that circling behavior was induced by intranigral administration of the selective σ receptor ligand di-*o*-tolylguanidine (DTG) (Weber *et al.*, 1986; Walker *et al.*, 1988). It also seems that σ receptors influence the firing pattern of dopaminergic neurons of the substantia nigra (Engberg and Wikström, 1991). The finding that progesterone and other steroids interact with σ receptors in the brain and spleen has led to the hypothesis that this receptor may represent a link between the nervous, endocrine and immune systems (Su *et al.*, 1988a).

11.4.2 PCP RECEPTOR

Numerous lines of evidence indicate that the PCP receptor represents a binding site within the ion channel gated by the NMDA receptor. These include anatomical co-localization of the two sites, as well as correlation of potencies of PCP receptor ligands in PCP receptor binding assays with potencies in use- and voltage-dependent non-competitive antagonism of NMDA receptor-mediated conductances, and sensitivity of specific binding of PCP receptor radioligands to stimulation by NMDA receptor agonists and the co-agonist glycine and to inhibition by NMDA receptor antagonists (reviewed in Javitt and Zukin, 1989b). This functional relationship between PCP and NMDA receptors has several important consequences for performance of PCP receptor binding studies. The NMDA receptor is governed by a multi-state mechanism in which binding of at least one molecule of agonist is required for detection of PCP receptor radioligand binding to the closed conformation of the channel, and binding of two molecules of agonist and of glycine is required for detection of PCP receptor radioligand binding to the open conformation (Javitt *et al.*, 1990). Furthermore, PCP receptor radioligands manifest different binding characteristics depending on whether the channel is in the agonist-associated closed conformation or in the open conformation. Radioligand access to the PCP receptor via the open channel gives rise to a fast exponential of association in kinetic binding assays, or to an apparent high affinity component of binding in saturation studies. By contrast, radioligand access via the agonist-associated, closed channel gives rise to a slow exponen-

tial of association or to an apparent low affinity component of binding. The fast component of binding serves as a biochemical marker of activated NMDA receptor channels (Javitt and Zukin, 1989a). Therefore, it is only in the presence of saturating concentrations of L-glutamate (or NMDA) and glycine that binding properties of PCP receptor ligands can be accurately assessed based upon a simple linear bimolecular model. Under other incubation conditions, total radioligand binding will detect a combination of closed and activated channels, binding may not reach equilibrium for up to 24 h, and only kinetic assay designs, with appropriate non-linear weighted curve-fitting of data, will yield accurate binding parameters. Another important consequence is that the total level of radioligand binding to the PCP receptor, encompassing both fast and slow or high and low affinity components, increases with increasing channel activation (Javitt and Zukin, 1989a,b; Javitt *et al.*, 1990). Thus, an apparent developmental alteration in total receptor density could in principle reflect an altered proportion of activated NMDA channels whereas receptor density actually is unchanged. In fact, the issue of whether developmental alterations in expression and assembly of NMDA receptor subunits may result in alterations in sensitivity to activation has not yet been addressed.

Recently, an NMDA receptor has been cloned and sequenced (Moriyoshi *et al.*, 1991). It consists of 938 amino acids with a calculated molecular mass of 105 500 (M_r value of 105.5 kDa). This single protein appears to contain the PCP receptor, as (+)dizocilpine potently inhibits steady-state currents induced by 100 mM NMDA in *Xenopus* oocytes in which the receptor is expressed.

Another study reported the cloning of a glutamate-binding protein of M_r 57 020 (Kumar *et al.*, 1991). When this protein is combined with several other distinct subunits, not yet cloned, full NMDA receptor-channel function, including PCP receptor binding, is reconstituted. The issue of whether the native PCP receptor corresponds to either of these entities remains to be determined.

The PCP receptor has also been implicated in psychosis. The similarity between the effects of NANM and PCP in the 'canine delirium' model (Vaupel, 1983) suggests that the psychotomimetic effects of σ opioids in the human could be mediated at the PCP receptor. The PCP–NMDA hypothesis of schizophrenia postulates an endogenous deficit in NMDA receptor-mediated neuro-transmission in this disorder. It is based on several observations. Single doses of PCP (which would yield serum and brain concentrations at which the PCP receptor would be the predominant molecular target), given to normal human volunteers induces a transient psychotomimetic state incorporating positive and negative symptoms, formal thought disorder, and neuropsychological deficits closely resembling those seen in schizophrenia; the same low dose produces prolonged exacerbation of disease-specific symptoms and signs in schizophrenic subjects. Furthermore, schizophrenia-like abnormalities in event-related potentials may be induced by PCP-like agents, whereas administration of large oral doses of the NMDA receptor co-agonist glycine have been reported to produce clinical improvements in some schizophrenic patients (reviewed by Javitt and Zukin, 1991).

11.5 SIGMA AND PCP RECEPTORS IN THE DEVELOPING BRAIN

Ontogenetic studies of σ and PCP receptors have been performed in rats of several strains. Although methodological differences (binding conditions, radioligands) confound direct comparisons in some cases, the findings generally confirm a separation between these two historically linked but functionally distinct receptors.

11.5.1 ONTOGENY OF THE SIGMA RECEPTOR

Three studies have assessed σ receptor binding in the brains of rats at different ages. In the first of these studies (Majewska *et al.*, 1989), σ receptor binding parameters were assayed in crude, washed membranes from whole brains of Fischer-344 rats. The radio-ligand was [^{3}H]haloperidol in the presence of spiperone to inhibit radioligand binding to dopamine receptors. Samples were taken from rats at 1, 7, 14, 30, 90 days, 6 months and 1 year postnatally. Because the mass of the rat brain increases through the fourth postnatal month (Rosenberg and Stern, 1966), ages prior to 4 months represent developmental periods. The maximal survival time for the Fischer-344 rats was reported to be approximately 35 months, with a mean survival of 29 months (Coleman *et al.*, 1977). Therefore, rats of this strain can be considered mature, but not senescent at one year of age (London *et al.*, 1981). The results demonstrated no alteration in binding affinity or in the density of σ receptors, expressed per mg protein, in the rat brain from the early postnatal period through the first postnatal year. The lack of postnatal development of σ receptors, as compared to the documented changes in other neurotransmitter receptors, and the fact that σ receptors have high densities in peripheral organs, such as the liver (Samovilova *et al.*, 1988) and immune (Su *et al.*, 1988b; Wolfe *et al.*, 1988) and endocrine tissues (Wolfe *et al.*, 1989) suggested a fundamental role for the σ receptor beyond a function in the brain (Majewska *et al.*, 1989).

Paleos *et al.* (1990) used (+)-[^{3}H]-3-PPP to label σ receptors in the brains of Sprague–Dawley rats taken on prenatal day 2, postnatal days 1, 6 and 28, and adulthood (defined by a body weight of 225 g). The findings indicated a decline in receptor density, expressed as fmol bound per mg protein, from the prenatal period to the early postnatal period, followed by another substantial decline at postnatal day 28, with no subsequent change. Receptor affinity was lower on prenatal day 2 and on postnatal day 1 than at later ages. These findings confirmed the lack of a developmental increase, when results were viewed as a function of protein concentration, although the developmental decline noted here presented an inconsistency. This finding reflected a marked developmental increase in protein content of the brains of the Sprague–Dawley rats assayed. In fact, when σ receptor binding was reported as total pmol bound per brain, there was an increase in receptor binding throughout development.

Another study of σ receptors in the brains of Sprague–Dawley rats used [^{3}H]DTG as the radioligand (Matsumoto *et al.*, 1989). The results indicated an age-related loss of receptor density and affinity, when animals 2–3 months old were compared with rats at 5–6 months of age. The loss of receptors seemed to have functional importance in terms of affecting movement and posture, as unilateral microinjection of DTG into the substantia nigra produced fewer contralateral turns in the older animals. Furthermore, the older animals showed less pronounced postural changes in response to the unilateral injection of DTG into the red nucleus. As no age-related loss of σ receptor binding in brain during the first postnatal year of life was seen in Fischer-344 rats, it is possible that genetic strain may be an important determinant in effects of age on the σ receptor. In this regard, strain differences in age effects on other neurochemical systems in brain have been reported (Waller *et al.*, 1983).

11.5.2 ONTOGENY OF THE PCP RECEPTOR

Seven studies have addressed PCP receptor binding as a function of development in the rat. Of these, two (Majewska *et al.*, 1989; Paleos *et al.*, 1990) have directly compared σ and PCP receptors, whereas two others (McDonald *et al.*, 1990; Tremblay *et al.*, 1990)

have compared development of PCP receptors with development of other components of the NMDA receptor complex.

Sircar and Zukin (1983) characterized prenatal development of PCP receptors labeled with [^{3}H]PCP using a filtration assay of whole brain homogenates derived from Blue Spruce hooded rats from gestational day 13 until parturition, and from adult rats (120 days old). For each age group, Scatchard analysis of binding was carried out with radioligand concentrations ranging from 1 to 500 nM. The key observations were that although K_d values remained unchanged throughout development, B_{max} values (expressed per mg protein) demonstrated marked developmental enhancement. Thus, between gestational days 13 and 15, B_{max} represented 10–15% of adult levels. Beginning at gestational day 17, a rapid and marked increase in B_{max} resulted in the attainment of adult levels by gestational day 21. In order to clarify the functional development of this receptor, stereospecificity of [^{3}H]PCP binding was assayed at each age point by comparing the potencies of the potent PCP-like drug dexoxadrol and its 100-fold less active enantiomer levoxadrol for inhibition of specific binding of radioligand. Stereospecificity was not observed until gestational day 19, suggesting a developmental transition to a functionally adult type of PCP receptor at that point in development.

Majewska *et al.* (1989) undertook a comparative study of the ontogeny of σ receptors labeled with [^{3}H]haloperidol (see above) and PCP receptors labeled with [^{3}H]TCP in Fischer-344 rats. PCP binding assays were accomplished by filtration assays, using radioligand concentrations ranging from 0.5 to 80 nM. From postnatal day 1 to 1 year, K_d values of PCP receptor binding were reported not to change significantly (5.6–15.9 nM). By contrast B_{max} values revealed age-dependent increases, reaching adult levels by 14 days.

Paleos *et al.* (1990) addressed the comparative ontogeny of PCP and σ receptors (see above) in Sprague–Dawley rats. This ontogenetic study was the first to address directly the issue of the dependence of PCP receptor binding on the endogenous concentrations of NMDA receptor agonist (glutamate) and co-agonist (glycine), which were measured and determined to be above the concentrations giving maximal enhancement of [³H]TCP binding throughout the developmental phases examined. Glutamate, 100 μM, was also added to each assay tube. Specific binding of 1 nM [³H]TCP from 2 days prior to parturition to 28 days postnatally increased monotonically, reaching adult levels by postnatal day 21. Scatchard analysis of [³H]TCP binding was conducted at five developmental points from prenatal day 2 to adulthood. Whereas K_d values were stable over this development range (8–10 nM), B_{max} values increased from 350 fmol/mg protein at prenatal day 2 to 1865 fmol/mg protein at adulthood; adult levels were attained by postnatal day 28. Densities of PCP receptors increased more rapidly than developmental increases in brain protein or gross brain mass. Pharmacological selectivity and stereospecificity of the PCP receptor were examined at postnatal day 6, and were found to manifest adult characteristics.

Shinohara *et al.* (1989) examined the ontogeny of [³H]TCP binding in forebrains derived from male Wistar rats from gestational day 18 to postnatal day 50. PCP receptors were also characterized autoradiographically in slide-mounted 10 μM coronal sections, that were exposed to tritium-sensitive film for 4 weeks, or were wiped off and counted by scintillation spectrometry after incubation with [³H]TCP. Assays were carried out at ten age points ranging from gestational day 18 to postnatal day 50. No alteration in K_d of [³H]TCP binding (13.6–14.5 nM) was observed during development. However, B_{max} increased sigmoidally from 183 to 556 fmol/mg protein during this period. This was the first study to investigate whether the sensitivity of PCP receptor binding to inhi-

bition by the direct NMDA receptor antagonist D-(–)2-amino-5-phosphonovaleric acid (D(–)AP5) undergoes developmental alterations. No significant change in the IC₅₀ value for D(–)AP5 was observed between postnatal day 7 (376 ± 56 μM) and day 50 (441 ± 11 μM). The autoradiographic portion of the study confirmed the increase in PCP receptor density between postnatal days 7 and 50. Regional localization of binding did not change significantly between these time points, but the distribution of binding in hippocampus at postnatal day 7 was more homogeneous than at postnatal day 50. These investigators also addressed the developmental profile of forebrain concentrations of glutamate and glycine (expressed as nmol/mg protein). Glutamate content declined prenatally, and then increased progressively. Glycine content was stable between gestational day 18 and postnatal day 3, and then decreased. However, these amino acid changes were not sufficient to account for the observed alterations in [³H]TCP binding parameters.

Tremblay *et al.* (1990) studied the postnatal development of [³H]TCP binding sites in hippocampal specimens derived from Wistar rats in both autoradiographic and membrane preparations. They compared the developmental pattern of the PCP receptor with that of the [³H]glycine site of the NMDA receptor complex. Furthermore, they addressed the question of whether there is a developmental alteration in the sensitivity of [³H]TCP binding to added NMDA. Scatchard analysis of [³H]TCP binding was carried out at nine age points ranging from birth to 80 days postnatally. Values of K_d were stable during development, with B_{max} values increasing monotonically 2.4-fold. When incubations were carried out in the presence of 100 μM NMDA, similar increases in B_{max} values of [³H]TCP binding, without significant alterations in K_d values, were observed at all developmental stages. The same was true in the presence of 5 μM glycine, which further potentiated NMDA-stimulated [³H]TCP binding at all developmental stages, and in the presence of 2 mM Mg^{2+}, which inhibited PCP receptor binding to a similar extent at all developmental stages examined. Autoradiograms indicated that density of hippocampal PCP receptors increased progressively from postnatal day 4 to adulthood, with the increase most prominent in strata oriens and radiatum of CA1 and in the molecular layer of the dentate gyrus. By postnatal day 22, the anatomical distribution appeared identical to that in the adult. By contrast, strychnine-insensitive [^{3}H]glycine sites, with an anatomical distribution very similar to that of [^{3}H]TCP sites, reached adult levels of binding at postnatal day 10. It is of interest that developmental patterns of [^{3}H]TCP and [^{3}H]glycine sites in hippocampus are distinct not only from each other, but also from that of the agonist recognition site of the NMDA receptor (Tremblay *et al.*, 1988), which displays a rapid increase after birth to levels exceeding those in the adult, with a decrease after the third postnatal week to adult levels. It is likely that future molecular studies of the NMDA receptor complex will reveal developmental alterations in subunit expression, as have been detected for GABA receptors (Sato and Neale, 1989) and nicotinic acetylcholine receptors (Schuetze, 1986).

The issue of comparative development of multiple receptor domains of the NMDA complex was directly addressed by McDonald *et al.* (1990). Quantitative autoradiography in hippocampal sections derived from Sprague–Dawley rats was carried out at postnatal days 1, 5, 7, 10, 14, 21, 28 and 90 using [^{3}H]glutamate under conditions directing the radioligand to NMDA recognition sites, [^{3}H]TCP to label the PCP receptor, and [^{3}H]glycine to label the glycine recognition site of the NMDA receptor complex. In all of six hippocampal regions examined, ontogenetic profiles of binding to the PCP and glycine sites were very similar, progressively increasing in density and attaining adult levels by postnatal days 14–21. By contrast, NMDA recognition sites underwent much more rapid increases in density early in postnatal development, and were significantly higher than adult levels in one or more hippocampal regions studied at time points between postnatal days 10 and 28.

Morin *et al.* (1989) undertook a study of the development of [^{3}H]dizocilpine binding sites in Wistar rats at five developmental points from 3 to 20 days postnatally, and in tissue homogenates derived from adult hippocampus, cortex and brainstem. Single-point determinations indicated progressive increases in total [^{3}H]dizocilpine binding (expressed as fmol/mg protein) in all three regions between postnatal days 3 and 10. Levels statistically indistinguishable from adult values were attained by day 10 in hippocampus and cortex. In brainstem, levels significantly greater than those found in adults were apparent from day 3, reaching a maximum of nearly four times the adult level by day 15, and remaining significantly greater than the adult level at day 20. Scatchard analyses were carried out in extensively washed cortical membranes derived from 7-day-old and adult rats, with the goal of eliminating the influences of endogenous glutamate and glycine upon PCP receptor binding. Under these conditions, saturation data could be accurately fit only by a biphasic binding model encompassing distinct low (K_d = 109–185 nM) and high affinity (K_d = 4–6.9 nM) components. A developmental increase in B_{max} was observed only for the high affinity component of binding. The final component of the study was an examination of the abilities of 10 μM glutamate, 100 μM D-(–)AP5, 10 μM glycine, 50 μM PCP and 100 mM Mg^{2+} to modulate specific binding of 3 nM (^{3}H)dizocilpine in homogenates derived from cortex and hippocampus of 7-day-old and adult rat brain. The most significant finding was that in both brain regions the competitive NMDA antagonist D-(–)AP5 failed to reduce [^{3}H]dizocilpine binding in neonatal brain, although it did so robustly in adult brain.

11.6 SUMMARY AND CONCLUSIONS

It is clear from the above discussion that PCP and σ receptors follow sharply distinct courses of development. All studies of PCP receptor ontogeny have revealed patterns of progressive developmental increases in density in forebrain areas, without significant alterations in affinity. There is some disagreement among investigators concerning the developmental stage at which adult levels of PCP receptor binding are attained. The discrepancies may be due to differences in strains of rat, radioligand, or different binding assay methods. By contrast, all studies of σ receptor ontogeny fail to find any developmental increase in receptor density.

The PCP receptor is an integral component of the NMDA receptor complex, and functional coupling between the PCP and NMDA domains appears to exist throughout development. Nonetheless, the transient increases in density of NMDA recognition sites to significantly above adult levels have no parallel in the development of forebrain PCP receptors.

No study of PCP receptor ontogeny has to date addressed the issue of what proportions of the detected radioligand binding are to activated (open) as opposed to resting (closed) conformations of the NMDA receptor-channel complex. It has been shown that under any conditions other than the presence of saturating concentrations of L-glutamate and glycine, radioligand binding to the PCP receptor of the adult rat brain detects a combination of open and closed channels, manifested as apparent high and low affinity binding in saturation studies (Javitt and Zukin, 1989a) or as fast and slow exponentials of association and dissociation in kinetic paradigms (Javitt and Zukin, 1989b). Only kinetic assays are capable of assessing biochemically the crucial question of whether the degree of activatability of NMDA receptors undergoes developmental alteration. The NMDA receptor, encompassing a PCP receptor domain, probably represents only one form of the native receptor. It is likely that the divergent developmental patterns of PCP as opposed to NMDA recognition sites will ultimately be demonstrated to result from developmental alterations in subunit expression and assembly.

REFERENCES

Adams, J.T., Keana, J.F.W. and Weber, E (1987) Labeling of the haloperidol-sensitive sigma receptor in NCB-20 hybrid neuroblastoma cells with [^{3}H]-(1,3)-di-ortho-tolyl-guanidine and identification of the binding subunit by photoaffinity labeling with [^{3}H]-*m*-azido-di-ortho-tolyl-guanidine. *Soc. Neurosci. Abstr.*, **13**, 1703.

Bowen, W.D., Hellewell, S.B. and McGarry, K.A. (1989) Evidence for a multi-site model of the rat brain σ receptor. *Eur. J. Pharmacol.*, **163**, 309–18.

Coleman, G. L., Barthold, L.S., Osbaldiston, G.W. *et al.* (1977) Pathological changes during aging in barrier-reared Fischer–344 male rats. *J. Gerontol.*, **32**, 258–78.

Contreras, P.C., DiMaggio, D.A. and O'Donohue, T.L. (1987) An endogenous ligand for the sigma opioid binding site. *Synapse*, **1**, 57–61.

Craviso, G.L. and Musacchio, J.M. (1983a) High-affinity dextromethorphan binding sites in guinea pig brain. I. Initial characterization. *Mol. Pharmacol.*, **23**, 619–28.

Craviso, G.L. and Musacchio, J.M (1983b) High-affinity dextromethorphan binding sites in guinea pig brain. II. Competition experiments. *Mol. Pharmacol.*, **23**, 629–40.

de Costa, B.R., Bowen, W.D., Hellewell, S.B. *et al.* (1989) Synthesis and evaluation of optically pure [^{3}H]-(+)-pentazocine, a highly potent and selective radioligand for σ receptors. *FEBS Lett.*, **251**, 53–8.

DiMaggio, D.A., Contreras, P.C and O'Donohue, T.L. (1988) Biological and chemical characterization of the endopsychosins: distinct ligands for PCP and sigma sites, in *Sigma and Phencyclidine-like Compounds as Molecular Probes in Biology* (eds E.F. Domino and J.-M. Kamenka), NPP Books, Ann Arbor, pp. 157–71.

Engberg, G. and Wikström, H. (1991) σ-receptors: implication for the control of neuronal activity of nigral dopamine-containing neurons. *Eur. J. Pharmacol.*, **201**, 199–202.

Foster, A.C. and Wong, E.H.F. (1987) The novel anticonvulsant MK801 binds to the activated state of the *N*-methyl-D-aspartate receptor in rat brain. *Br.J. Pharmacol.*, **91**, 403–9.

Gundlach, A.L., Largent, B.L. and Snyder, S.H. (1985) Phencyclidine and σ opiate receptors in brain: biochemical and autoradiographical differentiation. *Eur.J. Pharmacol.*, **113**, 465–6.

Hampton, R.Y., Medzihradsky, F., Woods, J.H. *et al.* (1982) Stereospecific binding of [³H]-phencyclidine in brain membranes. *Life Sci.*,**30**, 2147–54.

Haring, R.H., Kloog, Y. and Sokolovsky, M. (1986) Identification of polypeptides of the phencyclidine receptor of rat hippocampus by photoaffinity labeling with [³H]azidophencyclidine. *Biochemistry*, **25**, 612–20.

Haring, R.H., Kloog, Y., Kalir, A. *et al.* (1987) Binding studies and photoaffinity labeling identify two classes of phencyclidine (PCP) receptors in rat brain. *Biochemistry*, **26**, 5854–61.

Hellewell, S.B. and Bowen, W.D. (1990) A sigma-like binding site in rat pheochromocytoma (PC12) cells: decreased affinity for (+)-benzomorphans and lower molecular weight suggest a different sigma receptor form from that of guinea pig brain. *Brain Res.*, **527**, 244–53.

Honoré, T., Drejer, J., Nielsen, E.Ø. *et al.* (1989) Molecular target size analyses of the NMDA-receptor complex in rat cortex. *Eur. J. Pharmacol.*, **172**, 239–47.

Jasinski, D.R., Martin, W.R. and Sapira, J.D. (1968) Antagonsim of the subjective, behavioral, pupillary, and respiratory depressant effects of cyclazocine by naloxone. *Clin. Pharmacol. Ther.*, **9**, 215–22.

Javitt, D.C. and Zukin, S.R. (1989a) Biexponential kinetics of [³H]MK-801 binding: evidence for access to closed and open *N*-methyl-D-aspartate receptor channels. *Mol. Pharmacol.*, **35**, 387–93.

Javitt, D.C. and Zukin, S.R. (1989b) Interaction of [³H]MK-801 with multiple states of the *N*-methyl-D-aspartate receptor complex of rat brain. *Proc. Natl. Acad. Sci. USA*, **86**, 740–4.

Javitt, D.C. and Zukin, S.R. (1991) Recent advances in the phencyclidine model of schizophrenia. *Am. J. Psychiatry*, **148**, 1301–8.

Javitt, D.C., Frusciante, M.J. and Zukin, S.R. (1990) Rat brain *N*-methyl-D-aspartate receptors require multiple molecules of agonist for activation. *Mol. Pharmacol.*, **37**, 603–7.

Kavanaugh, M.P., Tester, B.C., Scherz, M.W. *et al.* (1988) Identification of the binding subunit of the σ-type opiate receptor by photoaffinity labeling with 1-(4-azido-2-methyl[6-³H]phenyl)-3-(2-methyl[4,6-³H]phenyl) guanidine. *Proc. Natl. Acad. Sci. USA.*, **85**, 2844–8.

Keats, A. and Telford, J. (1964) Narcotic antagonists as analgesics. Clinical aspects, in *Molecular Modification in Drug Design, Advances in Chemistry*, 45th edn (ed R.F. Gould), American Chemical Society, Washington, DC., pp 170–6.

Kuhar, M.J., Boja, J.W. and Cone, E.J. (1990) Phencyclidine binding to striatal cocaine receptors. *Neuropharmacology*, **29**, 295–7.

Kumar, K.N., Tilakaratne, N., Johnson, P.S. *et al.* (1991) Cloning of cDNA for the glutamate-binding subunit of an NMDA receptor complex. *Nature*, **354**, 70.

Kushner, L., Zukin, S.R. and Zukin, R.S. (1988) Characterization of opioid, σ, and phencyclidine receptors in the neuroblastoma-brain hybrid cell line NCB-20. *Mol. Pharmacol.*, **34**, 689–94.

Largent, B.L., Gundlach, A.L. and Snyder, S.H. (1984) Psychotomimetic opiate receptors labelled and visualized with (+)-[³H]-3-(3hydroxyphenyl)-*N*-(1-propyl)piperidine. *Proc. Natl. Acad. Sci. USA.*, **81**, 4983–7.

Largent, B.L., Gundlach, A.L. and Snyder, S.H. (1986a) σ receptors on NCB-20 hybrid neurotumor cells labeled with (+)[³H]SKF 10,047 and (+)[³H]3-PPP. *Eur. J. Pharmacol.*, **124**, 183–7.

Largent, B.L., Gundlach, A.L. and Snyder, S.H. (1986b) Pharmacological and autoradiographic discrimination of sigma and phencyclidine receptor binding sites in brain with (+)-[³H]SKF 10047, (+)-[³H]-3-[3-hydroxyphenyl]-*N*-(1-propyl) piperidine and [³H]-1-[1-(thienyl)-cycloexyl] piperidine. *J. Pharmacol. Exp. Ther.*, **238**, 739–48.

Largent, B.L., Wikström, H., Gundlach, A.L. *et al.* (1987) Structural determinants of σ receptor affinity. *Mol. Pharmacol.*, **32**, 772–84.

Largent, B.L., Wikström, H., Snowman, A.M. *et al.* (1988) Novel antipsychotic drugs share high affinity for σ receptors. *Eur. J. Pharmacol.*, **155**, 345–7.

London, E.D., Nespor, S.M., Ohata, M. *et al.* (1981) Local cerebral glucose utilization during development and aging of the Fischer-344 rat. *J. Neurochem.*, **37**, 217–21.

Majewska, M.D., Parameswaran, S. and London, E.D. (1989) Divergent ontogeny of sigma and phencyclidine binding sites in the rat brain. *Dev. Brain Res.*, **47**, 13–18.

Maragos, W.F., Chu, D.C.M., Greenamyre, J.T. *et al.* (1986) High correlation between the localization of [³H]TCP binding and NMDA receptors. *Eur.J. Pharmacol.*, **123**, 173–4.

Martin, W.R., Eades, C.G., Thompson, J.A. *et al.* (1976) The effects of morphine- and nalorphine-like drugs in the non-dependent and morphine-dependent chronic spinal dog. *J. Pharmacol. Exp. Ther.*, **197**, 517–32.

Martin, W.R. (1983) Pharmacology of opioids. *Pharmacol. Rev.*, **35**, 283–323.

Matsumoto, R.R., Bowen, W.D. and Walker, J.M. (1989) Age-related differences in the sensitivity of rats to a selective sigma ligand. *Brain Res.*, **504**, 145–8.

McCann, D.J. and Su, T.-P. (1990) Haloperidol-sensitive (+)[^{3}H]SKF-10,047 binding sites (σ sites) exhibit a unique distribution in rat brain subcellular fractions. *Eur. J. Pharmacol.*, **188**, 211–18.

McDonald, J.W., Johnston, M.V. and Young, A.B. (1990) Differential ontogenic development of three receptors comprising NMDA receptor/channel complex in the rat hippocampus. *Exp. Neurol.*, **110**, 237–47.

Morin, A.M., Hattori, H., Wasterlain, C.G. *et al.* (1989) [^{3}H]MK-801 binding sites in neonate rat brain. *Brain Res.*, **487**, 376–9.

Moriyoshi, K., Masu, M., Ishii, T. *et al.* (1991) Molecular cloning and characterization of the rat NMDA receptor. *Nature*, **354**, 31–7.

Musacchio, J.M. (1990) The psychotomimetic effects of opiates and the σ receptor. *Neuropsychopharmacology*, **3**, 191–200.

Musacchio, J.M., Klein, M. and Canoll, P.D. (1989a) Dextromethorphan and sigma ligands: common sites but diverse effects. *Life Sci.*, **45**, 1721–32.

Musacchio, J.M., Klein, M. and Paturzo, J.J. (1989b) Effects of dextromethorphan site ligands and allosteric modifiers on the binding of (+)-[^{3}H]3-(3-hydroxyphenyl)-*N*-(1-propyl)piperidine. *Mol. Pharmacol.*, **35**, 1–5.

Paleos, G.A., Yang, Z.W. and Byrd, J.C. (1990) Ontogeny of PCP and sigma receptors in rat brain. *Dev. Brain Res.*, **51**, 147–52.

Pfeiffer, A., Brantl, V., Herz, A. *et al.* (1986) Psychotomimesis mediated by κ opiate receptors. *Science*, **233**, 774–6.

Quirion, R., Hammer, R.P., Herkenham, M. *et al.* (1981) Phencyclidine (angel dust)/σ 'opiate' receptor: visualization by tritium-sensitive film. *Proc. Natl. Acad. Sci. USA*, **78**, 5881–5.

Quirion, R., DiMaggio, D.A., French, E.D. *et al.* (1984) Evidence for an endogenous peptide ligand of the phencyclidine receptor. *Peptides*, **5**, 967–73.

Rosenberg, A. and Stern, N. (1966) Changes in sphingosine and fatty acid components of the gangliosides in developing rat and human brain. *J. Lipid Res.*, **7**, 122–31.

Samovilova, N.N., Nagornaya, L.V. and Vinogradov, V.A. (1988) (+)-[^{3}H]SK&F 10,047 binding sites in rat liver. *Eur. J. Pharmacol.*, **147**, 259–64.

Sato, T.N. and Neale, J.H. (1989) Type I and type II γ-aminobutyric acid/benzodiazepine receptors: purification and analysis of novel receptor complex from neonatal cortex. *J. Neurochem.*, **52**, 1114–22.

Scheideler, M.A. and Zukin, R.S. (1990) Reconstitution of solubilized delta-opiate receptor binding sites in lipid vesicles. *J. Biol. Chem.*, **265**, 15176–82.

Schuetze, S. (1986) Embryonic and adult acetylcholine receptors: molecular basis of developmental changes in ion channel properties. *Trends Neurosci.*, **9**, 346–8.

Shinohara, K., Nishikawa, T., Ishii, T. *et al.* (1989) Embryonic and postnatal development of *N*-(1-[2-thienyl]cyclohexyl)[^{3}H]piperidine binding sites in rat forebrain homogenates and slices. *Neurosci. Lett.*, **107**, 307–12.

Sircar, R. and Zukin, S.R. (1983) Ontogeny of sigma opiate/phencyclidine-binding sites in rat brain. *Life Sci.*, **33**, 255–8.

Sircar, R. and Zukin, S.R. (1985) Quantitative localization of [^{3}H]TCP binding in rat brain by light microscopy autoradiography. *Brain Res.*, **344**, 142–5.

Sonders, M.S., Barmettler, P., Lee, J.A. *et al.* (1990) A novel photoaffinity ligand for the phencyclidine site of the *N*-methyl-D-aspartate receptor labels a M_r 120,000 polypeptide. *J. Biol. Chem.*, **265**, 6776–81.

Sorensen, R.G. and Blaustein, M.P. (1986) m-Azido-phencyclidine covalently labels the rat brain PCP receptor, a putative K channel. *J. Neurosci.*, **6**, 3676–81.

Su, T.-P. (1981) Psychotomimetic opioid binding: specific binding of [^{3}H]SKF-10047 to etorphine-inaccessible sites in guinea-pig brain. *Eur. J. Pharmacol.*, **75**, 81–2.

Su, T.-P. (1982) Evidence for sigma opioid receptor: binding of [^{3}H]SKF-10047 to etorphine-inacessible sites in guinea-pig brain. *J. Pharmacol. Exp. Ther.*, **223**, 284–90.

Su, T.-P. (1985) Further demonstration of kappa opioid binding sites in the brain: evidence for heterogeneity. *J. Pharmacol. Exp. Ther.*, **232**, 144–8.

Su, T.-P. (1991) σ receptors: putative links between nervous, endocrine and immune systems. *Eur. J. Biochem.*, **200**, 633–42.

Su, T.-P., Weissman, A.D. and Yeh, S.-Y (1986) Endogenous ligands for sigma opioid receptors in the brain ('sigmaphin'): evidence from binding assays. *Life Sci.*, **38**, 2199–210.

Su, T.-P., London E.D. and Jaffe, J.H. (1988a) Steroid binding at σ receptors suggests a link between endocrine, nervous and immune systems. *Science*, **240**, 219–21.

Su, T.-P., Schell, S.E., Ford-Rice, F.Y. *et al.* (1988b) Correlation of inhibitory potencies of putative antagonists for σ receptors in brain and spleen. *Eur. J. Pharmacol.*, **148**, 467–70.

Tam, S.W. (1983) Naloxone-inacessible σ receptor in rat central nervous system. *Proc. Natl. Acad. Sci. USA*, **80**, 6703–7.

Tam, S.W. (1985) (+)–[^{3}H]SKF 10,047, (+)-[^{3}H]ethylketocyclazocine, μ, κ, δ and phencyclidine binding sites in guinea pig brain membranes. *Eur. J. Pharmacol.*, **109**, 33–41.

Tam, S.W. and Cook, L. (1984) σ opiates and certain antipsychotic drugs mutually inhibit (+)-[^{3}H]SKF 10047 and [^{3}H]haloperidol binding in guinea pig brain membranes. *Proc. Natl. Acad. Sci. USA*, **81**, 5618–21.

Tremblay, E., Roisin, M.P., Represa, A. *et al.* (1988) Transient increased density of NMDA binding sites in the developmental rat hippocampus. *Brain Res.*, **461**, 393–6.

Tremblay, E., Roisin-Lallemand, M.-P. and Ben-Ari, Y. (1990) Developmental study of [^{3}H]TCP and [^{3}H]glycine binding sites in the rat hippocampus. *Dev. Brain Res.*, **57**, 21–8.

Vaupel, D.B. (1983) Naloxone fails to antagonize the σ effects of PCP and SKF 10,047 in the dog. *Eur. J. Pharmacol.*, **92**, 269–74.

Vignon, J., Vincent, J.P., Bidard, J.N. *et al.* (1982) Biochemical properties of the brain phencyclidine receptor. *Eur. J. Pharmacol.*, **81**, 531–43.

Vignon, J., Chicheportiche, R., Chicheportiche, M. *et al.* (1983) [^{3}H]TCP: a new tool with high affinity for the PCP receptor in rat brain. *Brain Res.*, **280**, 194–7.

Vignon, J., Cerruti, C., Chaudieu, I. *et al.* (1988) Interaction of molecules in the phencyclidine series with the dopamine uptake system: correlation with their binding properties to the phencyclidine receptor. Binding properties of [^{3}H]BTCP, a new PCP analog, to the dopamine uptake complex, in *Sigma and Phencyclidine-like Compounds as Molecular Probes in Biology* (eds E.F. Domino and J.-M. Kamenka), NPP Books, Ann Arbor, pp. 199–208.

Vincent, J.-P., Kartalovski, B., Geneste, P. *et al.* (1979) Interaction of phencyclidine ('angel dust') with a specific receptor in brain membranes. *Proc. Natl. Acad. Sci. USA*, **76**, 4678–82.

Vincent, J.-P., Bidard, J. -N., Lazdunski, M. *et al.* (1983) Identification and properties of phencyclidine-binding sites in nervous tissues. *Fed. Proc.*, **42**, 2570–3.

Vu, T.H., Weissman, A.D. and London, E.D. (1990) Pharmacological characteristics and distributions of sigma and phencyclidine binding sites in the animal kingdom. *J. Neurochem.*, **54**, 598–604.

Walker, J.M., Matsumoto, R.R., Bowen, W.D. *et al.* (1988) Evidence for a role of haloperidol-sensitive σ-'opiate' receptors in the motor effects of antipsychotic drugs. *Neurology*, **38**, 961–5.

Walker, J.M., Bowen, W.D., Walker, F.O. *et al.* (1990) Sigma receptors: biology and function. *Pharmacol. Rev.*, **42**, 355–402.

Waller, S.B., Ingram, D.K., Reynolds, M.A. *et al.* (1983) Age and strain comparisons of neurotransmitter synthetic enzyme activities in the mouse. *J. Neurochem.*, **41**, 1421–8.

Weber, E., Sonders, M., Quarum, M. *et al.* (1986) 1,3-Di(2-[5-^{3}H]tolyl) guanidine: a selective ligand that labels σ-type receptors for psychotomimic opiates and antipsychotic drugs. *Proc. Natl. Acad. Sci. USA*, **83**, 8784–8.

Weissman, A.D., Broussolle, E.P. and London , E.D. (1990a) In vivo binding of [^{3}H]*d-N*-allylnormetazocine and [^{3}H]haloperidol to sigma receptors in the mouse brain. *J. Chem. Neuroanat.*, **3**, 347–54.

Weissman, A.D., Casanova, M.F., Kleinman, J.E. *et al.* (1990b) Selective loss of cerebral cortical sigma, but not PCP binding sites in schizophrenia. *Biol. Psychiatry*, **29**, 41–54.

Wolfe, S.A., Kulsakdinun, C., Battaglia, G. *et al.* (1988) Initial identification and characterization of sigma receptors on human peripheral blood leukocytes. *J. Pharmacol. Exp. Ther.*, **247**, 1114–19.

Wolfe, S.A., Culp, S.G. and De Souza, E.B. (1989) σ-Receptors in endocrine organs: identification, characterization, and autoradiographic localization in rat pituitary, adrenal, testis, and ovary. *Endocrinology*, **124**, 1160–72.

Wong, E.H.F., Kemp, J.A., Priestly, T. *et al.* (1986) The anticonvulsant MK-801 is a potent *N*-methyl-D-aspartate antagonist. *Proc. Natl. Acad. Sci. USA*, **83**, 7104–8.

Wong, E.H.F., Knight, A.R. and Woodruff, G.N. (1988) [^{3}H]MK-801 labels a site on the *N*-methyl-D-aspartate receptor complex in rat brain membranes. *J. Neurochem.*, **50**, 274–81.

Wu, X.-Z., Bell, J.A., Spivak, C.E. *et al.* (1990) Electrophysiological and binding studies on intact NCB-20 cells suggest presence of a low affinity sigma receptor. *J. Pharmacol. Exp. Ther.*, **257**, 351–9.

Zhang, A.-Z., Mitchell, K.N., Cook, L. and Tam, S.W. (1988) Human endogenous brain ligands for sigma and phencyclidine receptors, in *Sigma and Phencyclidine-like Compounds as Molecular Probes in Biology* (eds E.F. Domino and J.-M. Kamenka), NPP Books, Ann Arbor, pp. 335–43.

Zukin, R.S. and Zukin, S.R. (1981) Demonstration of [^{3}H]-cyclazocine binding to multiple opiate receptor sites. *Mol. Pharmacol.*, **20**, 246–54.

Zukin, R.S. and Zukin, S.R. (1988) The σ receptor, in *The Opiate Receptors* (ed. G.W. Pasternak), The Humana Press, Clifton, NJ, pp. 143–63.

Zukin, S.R. and Zukin, R.S. (1979) Specific [^{3}H]phencyclidine binding in rat central nervous system. *Proc. Natl. Acad. Sci. USA*, **76**, 5372–6.

Zukin, S.R., Fitz-Syage, M.L., Nichtenhauser, R. *et al.* (1983) Specific binding of [^{3}H]phencyclidine in rat central nervous tissue: further characterization and technical considerations. *Brain Res.*, **258**, 277–84.

Zukin, S.R., Zukin R.S., Vale, W. *et al.* (1987) An endogenous ligand of the brain σ/PCP receptor antagonizes NMDA-induced neurotransmitter release. *Brain Res.*, **416**, 84–9.

Ann Tempel

12.1 INTRODUCTION

The biochemical demonstration of opioid receptors was first established in 1973 (Pert and Snyder, 1973). Earlier attempts were difficult due to low levels of specific binding and high background binding. Although these early studies identified only a single type of receptor, pharmacological studies had already suggested the existence of multiple opiate receptor types (Martin *et al.*, 1967). Evidence from behavioral, pharmacological and biochemical studies demonstrates the existence of several opiate receptor classes: mu (μ), delta (δ) and kappa (κ) (Lord *et al.*, 1977; Chang and Cuatrecasas, 1979). The sigma (σ) receptor is no longer considered an opioid receptor and it has been suggested that it is the PCP site of the NMDA receptor (for reviews see Snyder, 1984; Zukin and Zukin, 1988). These receptor classes mediate diverse behavioral effects, exhibit different ligand selectivity patterns and have different distributions throughout the central and peripheral nervous systems. The μ receptor is operationally defined as the high affinity site at which morphine-like opiates produce analgesia and a variety of other classical opiate effects. This receptor has been further subdivided into μ_1 and μ_2 (Pasternak *et al.*, 1983). The δ receptor exhibits a higher affinity for the naturally occurring enkephalins (a class of shorter opioid peptides) than for morphine and was originally found in peripheral tissue such as the mouse vas deferens (Lord *et al.*, 1977). The κ receptor is that site at which ketocyclazocine-like opiates produce analgesia, as well as their unique ataxic and sedative effects (Martin *et al.*, 1976). It is also defined as a receptor highly selective for dynorphin (a 17-amino acid opioid peptide). Some researchers have demonstrated the existence of several κ receptor subtypes as well (Attali *et al.*, 1982; Gouarderes *et al.*, 1983; Zukin *et al.*, 1988). Actions at all three of these sites are reversible by the opioid antagonist naloxone or naltrexone with increasing doses required going from μ to δ to κ receptors. The complex neuropharmacological actions of a given opioid would appear to reflect its interaction at a combination of these and other opioid receptor sites with varying potencies.

One of the aims in anatomically differentiating the opioid receptor types is to gain insight into the functional significance of these receptors. Some of the possible functions are briefly outlined by Mansour *et al.* (1987) and Herkenham and McLean (1986). High densities of μ and κ opioid receptors have been localized to areas involved in processing of all types of sensory information, i.e. visual (superior colliculus, lateral geniculate, optic tract), auditory (inferior colliculus and medial genic-

Receptors in the Developing Nervous System Vol. 2: Neurotransmitters. Edited by Ian S. Zagon and Patricia J. McLaughlin.
Published in 1993 by Chapman & Hall. ISBN 0 412 49400 0. Vols. 1 and 2 (set) ISBN 0 412 54520 9.

ulate), olfactory (olfactory bulb and amygdala) and nociceptive (thalamus, central gray, spinal cord). In contrast, δ receptors appear to be localized primarily to more recently developed areas of the forebrain, which may indicate a need for more specialized receptors or neuromodulators. High levels of κ binding in the hypothalamus and neural lobe of the pituitary suggest that they play a role in neuroendocrine regulation. κ opioid receptors have also been implicated in the regulation of feeding and drinking behaviors suggesting a role in motivational behaviors.

It is now well established that there are three major classes of opioid peptides: β-endorphin, the enkephalins, and dynorphin-related peptides, which serve as the endogenous ligands for the μ, δ and κ receptors. These arise from three different precursor molecules in three independent biosynthetic pathways (for a review see Weber *et al.*, 1983). The first, β-endorphin occurs together with adrenocorticotrophic hormone (ACTH), a peptide which stimulates the adrenal cortex, in the 31 kDa precursor molecule proopiomelanocortin (POMC) (Mains *et al.*, 1977). POMC, β-endorphin, and ACTH are found in highest concentrations in the pars intermedia and pars distalis of the pituitary. Under conditions of severe stress, β-endorphin and ACTH are co-released.

The second class includes methionine- and leucine-enkephalin (Hughes *et al.*, 1975), pentapeptides that arise from pro-enkephalin, an approximately 50 kDa protein that has been identified in the adrenal medulla and in the striatum (Gubler *et al.*, 1982). The pro-enkephalin molecule contains six copies of met-enkephalin and one of leu-enkephalin. The distribution of the enkephalins differs from that of β-endorphin in that they appear to be much more widely distributed throughout the brain.

Dynorphin A, a 17-amino acid peptide that contains the sequence of leu-enkephalin at its N-terminal end, is the most recently discovered opioid peptide (Goldstein *et al.*, 1981). It arises together with dynorphin B, its

1–13 counterpart, from a 4000 dalton peptide which in turn is found in the much larger prodynorphin molecule (James *et al.*, 1982). Dynorphin A and its (1–8) fragment occur in approximately equal concentrations in the brain. Whereas β-endorphin displays equipotency at μ and δ receptors, the enkephalins show a much greater affinity for the δ receptor. It has been speculated that the more stable β-endorphin molecule functions as a neurohormone in both pathways, whereas the enkephalins play a more specific role, acting as neurotransmitters or neuromodulators over shorter distances (Lord *et al.*, 1977). The longer dynorphin forms (1–13 and 1–17) appear to be more κ-selective and are more stable, whereas the shorter fragments (e.g., 1–8) which are potent at both κ and δ receptors and are less stable, are more likely candidates for transmitter-like function (Corbett *et al.*, 1983).

12.2 HISTORY OF SUBSTANCE ABUSE

Descriptions of human neonates undergoing withdrawal following maternal opioid abuse during pregnancy date back to the latter part of the nineteenth century (Goodfriend *et al.*, 1956). Methadone was first synthesized by the Germans during World War II. Favorable preliminary findings by Dole and Nyswander (1965) were followed by the wide-scale use in heroin addiction drug treatment programs (for review, see Hutchings, 1985a). By 1975, there were some 70 000–80 000 heroin addicts in methadone maintenance programs throughout the USA and a significant proportion of these women were of childbearing age. At the time, it was estimated that in the New York City metropolitan area alone, 10 000–12 000 such women were enrolled in methadone programs yet little was known of possible risk to the fetus and the newborn. With the advent of methadone maintenance as an experimental treatment for heroin addiction in the early 1970s, attention turned to the questions of reproductive hazards and developmental toxicity. There are confounding variables in clinical experiments such as women coming from low socioeconomic levels, long histories of drug abuse, poor diets, heavy smokers, use of marijuana, cocaine, barbiturates, tranquilizers or alcohol (for review see Householder *et al.*, 1982; Hutchings and Fifer, 1986).

The type of opioid abused has changed over the years. Until the 1950s, morphine appeared to be the drug of choice but a change to heroin usage was reported by Goodfriend *et al.* (1956). Following Dole and Nyswander's (1965) advocation of the methadone maintenance treatment program as an alternative to heroin dependency, numerous reports have documented methadone-dependent offspring.

In the early 1980s, cocaine became the drug of choice. Cocaine abuse rose sharply during the 1980s to the point that the drug subculture represented a large portion of the population. An estimated 50% or more of pregnancies at some of the inner-city hospitals in New York are cocaine exposed (Dow-Edwards, 1991). Interest in the biological effects of maternal cocaine and/or opioid abuse on the development of the unborn child has been increasing. As a result, several laboratories attempted to develop animal models of prenatal opioid exposure. Laboratory environments would provide controlled experiments alleviating the confounding variables in the clinical studies. However, there are a host of methodological and interpretive problems in animal studies as well (for a comprehensive bibliography see Zagon *et al.*, 1982). Although most studies have used the rat, procedures have differed with respect to strain, dose level, dosing regimen, route of administration, gestational age at treatment and fostering techniques. These factors make it difficult to compare results from one laboratory to another. Because of the importance of these variables in interpretation of data and conclusions drawn, they are briefly discussed below.

12.2.1 SURROGATE FOSTERING

The evidence is sufficiently convincing that prenatal manipulations of the pregnant dam can alter maternal behavior and the mother–infant dyad to produce effects in her offspring, even when treatment is terminated before the last week or so of gestation. Daily drug treatment can disrupt circadian rhythms, cause acute nutritional deficits and either inhibit or interfere with hormones that mediate maternal behavior. Recent studies demonstrate the mother's integral role as a regulator of a host of developmental parameters in her offspring that include activity level, sleep–wake states, milk ingestion, heart rate, oxygen consumption, and growth hormone (see Hofer, 1984 for review). If not included, interpretation of the data will be compromised in that one cannot rule out with any degree of confidence that offspring effects have not been maternally mediated during the postnatal period.

12.2.2 DOSE–RESPONSE RELATIONSHIP

Prenatal drug studies involve two mutually interacting biological systems – the mother and fetoplacental unit. Dose–response relationships are very complex and involve interactive, pharmacological, and toxic effects in the mother and offspring (Hutchings, 1985a). The general principle is that as dose is increased from subpharmacological levels through the pharmacological range of the compound, there is a corresponding increase in toxicity culminating in death. For example, thalidomide at subtoxic, pharmacological levels in the mother, is highly embryotoxic during morphogenesis. Another compound with the same sort of profile and embryotoxic response in humans is the vitamin A derivative, isotretinoin (Hutchings, 1985b). In animal studies, what is frequently seen is that within the pharmacological range of the compound, and at levels that are not toxic to the dam, there is no embryotoxic response. Embryotoxicity is seen only at levels that produce maternal toxicity (this is reviewed and detailed in Hutchings, 1985a). One example in the rat is phencyclidine (PCP), which has been shown to be teratogenic, but only at doses that are highly toxic to the dam. In the case of PCP, Hutchings (1985a) administered two doses that were pharmacologically potent as measured by behavioral effects in the dam (5 or 10 mg/kg) but of relatively low maternal toxicity based on maternal weight gain. Yet they observed normal birthweights and no postnatal behavioral effects on two independent behavioral measures (Hutchings *et al.*, 1984).

Dose–response relationships, though complex, are critical for a meaningful description and understanding of effects. It is helpful that every study includes some measure of maternal dose–response. This is of particular importance when embryotoxicity is found only at doses that produce maternal toxicity. Under these circumstances, it is important to determine whether the effects produced in the offspring are primary effects of the compound or secondary to maternal toxicity.

A variety of drug-induced deficits have been observed in developing rats neonatally addicted to opioids. These deficits have been shown to persist into adulthood (Davis and Lin, 1972; Zagon and McLaughlin, 1977; Freeman, 1980). In general, these deficits have included impaired growth, behavioral, and neuroendocrine responses. This spectrum of drug-related developmental deficits may be a general pattern found in the young animal exposed to neurotropic drugs early in life. Neurotropic drugs such as morphine alter both the levels and the biosynthesis of the catecholamines (Slotkin and Anderson, 1975; McGinty and Ford, 1980; Rech *et al.*, 1980; Slotkin *et al.*, 1982) and such effects may be the neurochemical basis of the alterations and deficits in development (Weiner, 1974).

Little is known about mechanisms regulating the initial expression of opioid receptors. However, receptor development may be influenced quantitatively by a variety of factors in the neuronal environment. It is well established that perinatal exposure to drugs may influence the ontogenesis of brain opioid systems. However, results from a variety of studies are contradictory and complicated apparently due to differences in exposure time, dose and route of administration and the age at which animals are tested. This chapter will attempt to review the studies on prenatal opioid (morphine and cocaine) treatment and the biochemical and cellular alterations in normal development of the central nervous system (CNS) as a result of addiction. Studies in the adult CNS are included for comparison.

12.3 EFFECTS OF CHRONIC ADMINISTRATION IN THE ADULT CNS

12.3.1 OPIOID AGONISTS

A major unresolved issue is whether opioid agonist-induced tolerance and dependence *in vivo* and desensitization *in vitro* are associated with altered receptor numbers. It has been reported that chronic administration of morphine produces either no significant change in the adult CNS in opioid receptor number (Pert *et al.*, 1973; Klee and Streaty, 1976; Hitzemann *et al.*, 1974; Simon and Hiller, 1978; Bardo *et al.*, 1982; Holaday *et al.*, 1982; Perry *et al.*, 1982) or a modest receptor up-regulation (Brady *et al.*, 1989; Rothman *et al.*, 1986). More recent studies report that chronic morphine treatment of rats produces an up-regulation of the low affinity [^{3}H]D-Ala^2, D-Leu^5-enkephalin (DADLE) site in brain (Rothman *et al.*, 1986; Danks *et al.*, 1988; Brady *et al.*, 1989). In contrast, chronic etorphine produces down-regulation of μ and δ receptors *in vivo* (Tao *et al.*, 1988) and chronic enkephalin produces down-regulation of δ receptors *in vivo* (Tao *et al.*, 1987; Steece *et al.*, 1986) and in neurotumor cells (Hazum *et al.*, 1981; Chang *et al.*, 1982; Blanchard *et al.*, 1983). Intrathecal administration of high concentrations of the μ-selective peptide [N-Me-Phe^3, D-

Pro[4]] morphiceptin leads to a reduction in the number of spinal cord, but not brain, μ receptors (Nishino *et al.*, 1990). Only in the neonatal rat, however, is there a correlation between morphine-induced receptor down-regulation and the development of tolerance (Tempel *et al.*, 1988; Tempel, 1991). Although numerous studies have examined the effects of chronic μ and δ opioids on brain receptors, few have addressed the effects of chronic κ agonist administration. Rats treated chronically with κ opioids develop tolerance, as determined in behavioral paradigms measuring analgesia (Bhargava *et al.*, 1989a). Chronic administration of bremazocine leads to a reduction in [3H]bremazocine binding under conditions in which μ and δ binding has been suppressed (Morris and Herz, 1989). Moreover, chronic treatment of rats with U-50,488H (κ opioid) leads to a decrease in [3H]ethylketocyclazocine binding to brain membranes (Bhargava *et al.*, 1989b). Because ethylketocyclazocine is a non-selective μ opioid and because [3H]bremazocine labels non-opioid sites in addition to μ, δ and κ receptors (Zukin *et al.*, 1988), it is not clear whether the observed decreases are in κ opioid receptors.

An important question is why chronic treatment with etorphine and opioid peptides, but not with morphine, produce opioid receptor down-regulation in the adult animal. Several explanations are plausible. For example, it may be that morphine is a partial agonist. Alternatively, interaction of etorphine and enkephalin-like peptides with δ receptors may be critical to long-term changes at the cellular level. Yet another possibility is that opioid peptides bind differently than do opioid narcotic agonists to the same receptor, thereby producing different effects.

12.3.2 CHRONIC OPIOID ANTAGONIST ADMINISTRATION AND THE ADULT CNS

Antagonist-induced opioid supersensitivity and opioid receptor up-regulation are well correlated in the adult CNS. Long-term *in vivo* administration of the opioid antagonist naloxone or naltrexone results in enhanced morphine-induced analgesia (Lahti and Collins, 1978; Tang and Collins, 1978; Tempel *et al.*, 1985; Yoburn *et al.*, 1985) and enhanced effects of morphine on neurons of the locus coeruleus (Bardo *et al.*, 1983b) and the myenteric plexus (Schulz *et al.*, 1979). This functional supersensitivity reflects both an increased number of μ and δ opioid receptors (Zukin *et al.*, 1982; Brunello *et al.*, 1984; Tempel *et al.*, 1985; Danks *et al.*, 1988; Millan *et al.*, 1988; Yoburn *et al.*, 1989) and an increased coupling of receptor to the inhibitory guanyl nucleotide binding protein G_i (Zukin *et al.*, 1982; Tempel *et al.*, 1985). The changes in opioid receptor density vary heterogeneously throughout the brain; highest increases in μ receptor density occur in the nucleus accumbens, amygdala, striatum, layers I and III of the neocortex, and the periaqueductal gray region (Tempel *et al.*, 1984).

Up-regulation of μ opioid receptors has also been observed *in vitro* using explant cultures of fetal mouse spinal cord with attached dorsal root ganglia (DRG) (Tempel *et al.*, 1986). Long-term exposure of the explant cultures to naloxone produces an increase in μ opioid receptor density relative to control cultures, even in the presence of the protein synthesis inhibitor cycloheximide at a concentration that blocks >95% protein synthesis. This finding suggests that antagonist-induced opioid receptor up-regulation does not require the synthesis of new receptor molecules.

12.3.3 REGULATION OF OPIOID PEPTIDE GENE EXPRESSION IN THE ADULT CNS

A number of studies in the adult CNS have shown that chronic opioid drugs regulate opioid peptide levels and that this regulation occurs at the level of gene expression. Chronic naltrexone treatment increases met-enkephalin-like immunoreactivity in the striatum and nucleus accumbens (Tempel *et al.*, 1984). Chronic naltrexone also increases substance P immunoreactivity in the striatum (Tempel *et al.*, 1990) and decreases β-endorphin immunoreactivity in the hypothalamus, thalamus and amygdala (Ragavan *et al.*, 1983). In an effort to determine the mechanism by which opioid antagonists stimulate enkephalin and substance P production, the effects of naltrexone on the mRNA levels of preproenkephalin (PPE) and preprotachykinin (PPT) were studied (Tempel *et al.*, 1990). This study indicated that long-term blockade of opioid receptors by naltrexone leads to large increases in both PPE and PPT mRNA in the striatum. Chronic morphine treatment, on the other hand, decreases striatal PPE mRNA levels (Uhl *et al.*, 1988). These findings suggest that activation or blockade of opioid receptors influence PPE and PPT gene expression.

12.3.4 THE ROLE OF G PROTEINS AND SECOND MESSENGER SYSTEMS IN THE DEVELOPMENT OF TOLERANCE

It has been suggested that opioid-receptor density is controled by either the inhibition or excitation of the cAMP system via G-binding proteins (Costa *et al.*, 1988; Sibley and Lefkowitz, 1985; Mayorga *et al.*, 1989). Guanine nucleotide binding regulator proteins (G proteins) transduce a variety of extracellular signals across the cell membranes into changes in the levels of intracellular second messengers (Gilman, 1987). Members of this protein family are all heterotrimers consisting of α, β and γ subunits. In most biological systems, the GTP-binding α subunits serve as the transducer of signals between receptors and effectors. In contrast, the role of βγ is less well understood. The βγ complex has been demonstrated to facilitate the interaction of the subunits to receptors (Fung, 1983; Florio and Sternwies, 1985) and to mediate the binding of α subunit to membranes (Sternwies, 1986). The α subunit of G protein exhibits variation in structure classifying it into $G\alpha_s$, $G\alpha_i$, $G\alpha_o$, $G\alpha_t$ and $G\alpha_z$ (Stryer, 1986; Gilman, 1987; Provost *et al.*, 1988). $G\alpha_s$ can stimulate the activity of adenylate cyclase whereas $G\alpha_i$ type inhibits adenyl

cyclase activity. The βγ complex does not show any variation in structure and is interchangeable between different Gα proteins.

Acute treatment with opioids inhibits adenylate cyclase, and thereby decreases cAMP levels. Such changes have been described in cultured neuroblastoma × glioma hybrid cells (NG108 cells) (Sharma *et al.*, 1975; Traber *et al.*, 1975) and in brain regions like the locus coeruleus (Duman *et al.*, 1988; Beitner *et al.*, 1989), neostriatum and cerebral cortex (Collier and Roy, 1974; Tsang *et al.*, 1978; Law *et al.*, 1981; Schoffelmeer *et al.*, 1986). In addition, chronic opioid treatment increases levels of $G_{i\alpha}$ and $G_{o\alpha}$, adenylate cyclase activity, cAMP-dependent protein kinase activity in the locus coeruleus (LC) but not in other brain regions (Duman *et al.*, 1988; Nestler and Tallman, 1988; Guitart and Nestler, 1989, 1990a; Nestler *et al.*, 1989; Guitart *et al.*, 1990). It has been proposed that such an upregulated G-protein/cAMP system in the locus coeruleus contributes to opioid tolerance, and dependence in the adult CNS (Nestler *et al.*, 1990a,b; Rasmussen *et al.*, 1990). In addition, exposure of rat spinal cord dorsal rat ganglion co-cultured neurons to κ agonists causes a 60–70% reduction in the $G_{\alpha i}$ subunit (Attali and Vogel, 1989). These results suggest that treatment with opioids alters the G-protein/cAMP levels in different brain regions, however, a causal relationship between opioid receptor density change and G-protein/cAMP levels under such conditions has not yet been demonstrated. Sakellaridis and Vernadakis (1987) have shown in the chick embryo that morphine inhibited forskolin-stimulated adenylate cyclase at embryonic days (ED) 6 and 8, but that morphine had no effect on ED10. Naloxone produced the same effect with the same time schedule. A combination of both morphine and naloxone also decreased the adenylate cyclase activity. These investigators interpreted their findings to indicate that the conventional opioid receptor was not responsible for the observed effects since the inhibitory effects were not naloxone reversible and that naloxone itself produced this effect.

Electrophysiological studies in dorsal rat ganglia spinal cord explant cultures (Crain *et al.*, 1988; Shen and Crain, 1989, 1990) suggest that during sustained morphine exposure, a gradual enhancement of excitatory opioid actions coincides with a reduction in inhibitory effects. These authors hypothesize that the effect of morphine is the result of a balance between inhibitory and excitatory actions. In addition, biochemical studies in the guinea-pig longitudinal muscle with adherent myenteric plexus (Gintzler and Xu, 1991) supports this hypothesis. Their data indicate that different classes of G proteins appear to mediate the opioid enhancement or inhibition of stimulated enkephalin release. In addition, they suggest that a pertussis-toxin (PTX)-sensitive G protein (G_i or G_o) and a cholera-toxin (CTX)-sensitive G protein (G_s) are integral components of the mechanism that mediates opioid inhibition and opioid enhancement, respectively, of evoked enkephalin release. Recently, reciprocal effects of chronic morphine administration on stimulatory (G_s) and inhibitory (G_i) G-protein α subunits were demonstrated in primary cultures of rat striatal neurons (Van Vliet *et al.*, 1991). These data, taken together, suggest that opioid addiction and tolerance involve a crucial balance between inhibitory and stimulatory G proteins. Studies investigating the relationship of G proteins to opiate tolerance are just beginning to emerge. Results from these studies should elucidate cellular mechanisms underlying opiate addiction.

12.4 REGULATION OF OPIOID RECEPTORS IN NEONATAL RAT BRAIN

The first appearance of opioid binding sites in the rat occurs prior to ED14, however, their neuroanatomical distribution is different from that of adult brain (Clendeninn *et al.*, 1976; Coyle and Pert, 1976; Young and Kuhar, 1979; Herkenham and Pert, 1981; for a review see Chapter 10). The different opioid receptor types are expressed at different times during brain development, although in each case the largest increases are observed between 3 and 15 days after birth (Petrillo *et al.*, 1987). Mu opioid receptors are present in the mouse and rat brain at the time of birth and receptor density increases rapidly postnatally (Tavani *et al.*, 1985; Petrillo *et al.*, 1987). By contrast, δ receptors are not detectable until 7–10 days after birth (Tavani *et al.*, 1985; Petrillo *et al.*, 1987; Kent *et al.*, 1982; Leslie *et al.*, 1982; Spain *et al.*, 1985; Tempel *et al.*, 1988), and by 15 days after birth they have reached 50% of adult levels (Tavani *et al.*, 1985). Kappa receptors are present at birth and their density reaches adult levels one week after birth (Leslie *et al.*, 1982; Spain *et al.*, 1985; Barr *et al.*, 1986; Loughlin *et al.*, 1985). These diverse patterns of development suggest that different mechanisms may regulate the expression of the various opioid receptor types.

12.4.1 CHRONIC OPIOID ANTAGONIST ADMINISTRATION

Little is known about the mechanisms regulating the initial expression of opioid receptors. However, receptor development may be influenced quantitatively by a variety of factors in the neuronal environment. Antenatal or perinatal exposure to opioid agonists or antagonists alters the apparent number of opioid receptors in the brain with concomitant changes in antinociceptive responses (Handelmann and Quirion, 1983; Tempel *et al.*, 1988). In addition, stress, pain, and a variety of drugs and toxins alter receptor number (Torda, 1978; Kirby *et al.*, 1982; Watanabe *et al.*, 1983; Moon, 1984), however, the intracellular factors that regulate opioid receptor synthesis, insertion and degradation are unknown.

When naloxone is administered chronically during the prenatal period, the caudal brainstem is more likely than the forebrain to exhibit an increase in opioid receptors (Tsang and Ng, 1980). In contrast, when naloxone is administered chronically during the postnatal period, the rostral brain regions are more likely than the caudal brain regions to exhibit an increase in opioid receptors (Bardo *et al.*, 1982, 1983a). These researchers suggest that opioid receptor systems which are undergoing rapid ontogenetic proliferation may be particularly susceptible to receptor supersensitivity following chronic opioid receptor blockade. Rats injected with naloxone during the first three postnatal weeks showed significant increases in [³H]naloxone binding in the cortex, striatum, hypothalamus and spinal cord. Body and brain weights for naloxone and saline-treated rats were similar indicating that naloxone did not produce a general developmental or nutritional deficit. There were no significant changes in opioid binding one week after cessation of naloxone treatment. These results are similar to what was seen in the adult CNS (Tempel *et al.*, 1984, 1985). Naloxone-treated infants displayed an enhanced response to the anti-nociceptive efficacy of morphine which was also seen in adult rats (Tempel *et al.*, 1985).

12.4.2 CHRONIC AGONIST ADMINISTRATION

The ontogenesis of the opioid system is sensitive to pre- and postnatal exposure to drugs. Results from these studies are contradictory and complicated apparently due to differences in exposure time, dose and route of administration and the age at which animals are tested. In some studies, morphine administration to pups increases μ-ligand binding in the striatum and nucleus accumbens (Handelmann and Quirion, 1983), whereas other studies have failed to find any effect on binding following chronic prenatal morphine treatment (Coyle and Pert, 1976; Bardo *et al.*, 1982). Morphine administration to the dam during pregnancy has also been reported to decrease binding during the first week of life (Kirby and Aronstam, 1983) and increase binding later in life (Iyengar and Rabii, 1982; Tsang and Ng, 1980). Chronic administration of opioid agonists during pre-and/or postnatal development may alter opioid receptor ontogeny, and concomitantly, sensitivity to opioid drugs. Down-regulation of opioid receptors has been demonstrated in whole brain homogenates after chronic morphine treatment (Tempel *et al.*, 1988). One week of prenatal morphine treatment via the dam produced a statistically significant 35% decrease in brain μ-opioid receptors of offspring, on the day of birth. There was no significant change in the affinity or in GABA receptors. Interestingly, this down-regulation was no longer evident by postnatal day (PD) 14. Table 12.1 illustrates the time course of the down-regulation of brain μ-opioid receptors after chronic prenatal treatment with morphine.

Table 12.1.   Changes in brain μ-opioid receptor densities of infant rats after chronic prenatal treatment with morphine

Postnatal day	[³H] DAGO binding		% change
	Control (fmol/mg protein)	Morphine-treated (fmol/mg protein)	
0	(4) 77 ± 2.3	(4) 50 ± 3.5	−35[*]
5	(4) 81 ± 4.3	(4) 59 ± 5.3	−27
14	(3) 81 ± 4.1	(3) 83 ± 4.7	–
28	(4) 132 ± 12.5	(4) 129 ± 11.7	–

[*]Statistically significant difference (two-tailed *t*-test, $P < 0.05$). Pregnant female rats were exposed to morphine. Rat pups were killed by decapitation on the day of parturition (day zero), and 5, 14 and 28 days postnatal. Values were generated by computer-assisted linear regression analysis. Receptor density values are reported as means ± S.E.M. from a minimum of three independent experiments. The number of animals per group for each time point are indicated in parentheses. (Reprinted from Tempel *et al.* (1988), with permission from the publishers.)

(a) Postnatal treatment

Four days of daily postnatal (PD1–4) morphine treatment produced a significant 30% decrease in brain μ-opioid receptors. No change in δ or κ receptors was observed. Further treatment with morphine did not result in any significant changes in μ-opioid receptors relative to saline-treated (control) animals. In fact, the degree of μ-opioid receptor down-regulation diminished over time (Table 12.2) with the age of the animal. At PD14, there was no statistically significant difference in receptor density between brains of control and morphine treated pups, as in adult animals. When rat pups were chronically treated (for 7 or 14 days) with morphine, beginning on PD14 and assayed on either PD22 or PD29, no significant differences in opioid receptors were detected relative to saline treated (control) animals (Tempel *et al.*, 1988). Thus, morphine-induced receptor down-regulation appears to occur only during the first week of life.

In order to visualize the neuroanatomical pattern of opiate receptor changes in specific brain regions, light microscopy receptor autoradiography was carried out on rat brain sections from control and chronic morphine treated pups (Tempel, 1991). In the case of brains exposed to morphine from PD1 to PD4, μ-receptor density on PD5 in the striatal patches was virtually non-existent. There were also significant decreases in the surrounding matrix area (Fig. 12.1A, B). In contrast to what is seen after 4 days of morphine treatment, (PD1–4) 8 days of postnatal (PD 1–8) morphine treatment does not produce a significant loss in striatal μ-opioid receptors (Fig. 12.1C, D). Thus, during a brief period of development, there is a unique plasticity of the immature opioid receptor system that is lost with maturation, however, the significance of these findings is as yet unclear.

Several hypotheses can be proposed to explain our developmental results relative to adult rat CNS data. For this, we reference work on other neurotransmitter systems. In the adrenergic system, it has been shown that agonists produce down-regulation of β-adrenergic receptors. This down-regulation was shown to be associated with an internalization mechanism (Chang and Costa, 1979). Conversely, chronic treatment with β-adrenergic antagonists leads to up-regulation or an increase in β-adrenergic receptor number in adult animals (Galant *et al.*, 1978) and in

Table 12.2. Changes in brain μ-opioid receptor densities of infant rats following chronic postnatal treatment with morphine

Duration of treatment post-parturition (days)	[³H]DAGO binding				
	Control (fmol/mg protein)		Morphine-injected (fmol/mg protein)		% change
4	(6)	81 ± 3.8	(6)	57 ± 3.3	−30[*]
8	(6)	74 ± 8.5	(6)	61 ± 11.5	−18
21	(4)	156 ± 5.7	(4)	135 ± 10.4	−13
28	(3)	132 ± 16.5	(3)	128 ± 8.4	0

[*] Statistically significant difference (two-tailed *t*-test, $P < 0.05$). Neonatal rats were each given one daily s.c. injection of morphine or saline beginning on PD1. Rat pups were killed by decapitation at the times indicated. Values were generated by computer-assisted linear regression analysis. Receptor density values are reported as means ± S.E.M. from a minimum of three experiments. The numbers of animals per group for each time point are indicated in parentheses. (Reprinted from Tempel *et al.* (1988), with permission from the publishers.)

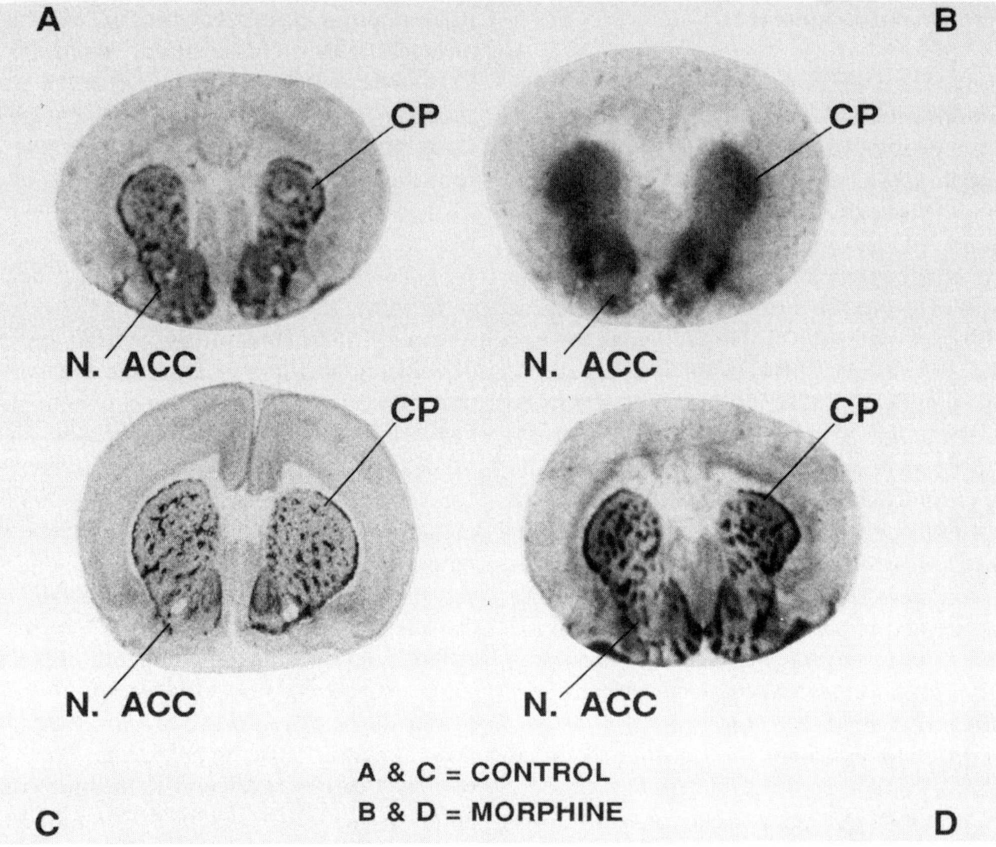

Fig. 12.1. Autoradiographic visualization of alterations in μ-opiate receptor binding at the level of the caudate putamen. Note the lack of μ-receptor 'patches' in the striatum after four days of postnatal morphine treatment (B) relative to control brain (A). Eight days of postnatal morphine treatment (D) induced no significant changes in μ receptor distribution relative to controls (C). Abbreviations; CP, caudate putamen; N. ACC, nucleus accumbens. (Reprinted from Tempel (1991), with permission of the publishers.)

humans (Glaubiger and Lefkowitz, 1977). More recently, several laboratories working with the opioid system have shown that long-term exposure of neurotumor cell lines (Hazum *et al.*, 1981; Chang *et al.*, 1982; Simantov *et al.*, 1982a; Blanchard *et al.*, 1983) and adult rats (Steece *et al.*, 1986) to enkephalin results in a decrease in δ-receptor density. Studies of [^{3}H]DADLE uptake by N4TG1 cells (Hazum *et al.*, 1981; Chang *et al.*, 1982; Simantov *et al.*, 1982b; Blanchard *et al.*, 1983) suggest that enkephalin is internalized via receptor-mediated endocytosis. One hypothesis to explain our findings is that long-

term exposure of the system to opioid agonists or peptides leads to coupling of the receptors to cyclase, followed by internalization. As active receptors disappear from the membrane surface, they would be replaced by inactive or spare receptors. Two possibilities then exist. (1) If the rate of internalization exceeded reactivation, an apparent downregulation in receptor density would be observed. (2) If the rate of reactivation paralleled internalization, no apparent change in receptor density would be observed. If down-regulation involves internalization, degradation of receptors, and recycling or reactivation, it is possible that the internalization process is developed by birth but that the recycling process is not fully developed until 2–3 weeks postnatal. When the recycling process is developed, down-regulation is no longer seen because recycling takes place as quickly as internalization; therefore no difference in receptor number is observed.

Another possibility is that opioids affect second messenger systems. Recent data (section 12.3.4) suggest that the overall effect of morphine is the result of a balance/reciprocal relationship between inhibitory and excitatory G proteins. In the developing CNS, it is not yet known which of the G proteins are present, what proportions they are present in and whether or not their interaction with the receptor and/or cyclase system is operational. All of these factors may be important in elucidating the mechanisms underlying addiction in the developing CNS in comparison to adult CNS tolerance. Future studies in this area will elucidate the exact nature of this interaction.

12.4.3 CHANGES IN OPIOID PEPTIDE SYNTHESIS IN NEONATAL RAT BRAIN

The effect of prolonged opioid use by pregnant women on opioid peptide levels in their newborns has received considerable attention. Opioid drugs cross the placental barrier (Blinick *et al.*, 1975) and their effects on the health of the pregnant woman and developing fetus are great. Pregnancies of opioid-dependent women are often complicated by problems such as toxemia, maternal syphilis and hepatitis and placental problems (Finnegan, 1975; Perlmutter, 1974; Rementeruia and Lotongkhum, 1977; Hans, 1989). Opioid-exposed infants are especially likely to show signs of fetal distress and to be born at low birthweight even when receiving adequate prenatal care (Connaughton *et al.*, 1973; Strauss *et al.*, 1974; Kandall *et al.*, 1975; Hans, 1989). The effects of prenatal opioid exposure on the behavior of newborn infants are also

dramatic. Babies born to morphine- and methadone-dependent mothers have elevated plasma β-endorphin levels for at least 40 days after birth and exhibit behavioral abnormalities (Lesser-Katz, 1982; Bauman and Levine, 1986; Deren, 1986), but fail to exhibit signs of withdrawal (Panerai *et al.*, 1983; Genazzani *et al.*, 1986). More recent studies have shown that during the first week after birth, most opioid-exposed infants show a well-documented neonatal abstinence syndrome that includes a variety of behaviors associated with central and autonomic nervous system hyperarousal such as tremors, hypertonus, hyperactive reflexes, high-pitched crying, poor sleeping and feeding, fever and rapid respiration (Jeremy *et al.*, 1985; Hans, 1989).

Numerous studies have investigated the effects of chronic opioid treatment on brain enkephalin levels, but they have produced conflicting results. Implantation of morphine pellets in mature rats has been reported to reduce brain enkephalin levels after 3 days (Shani *et al.*, 1979) or 11 days (Bergstrom and Terenius, 1983), but to have no effect after 5 days (Childers *et al.*, 1977; Fratta *et al.*, 1977) or 21 days (Shani *et al.*, 1979). Some studies report marked decreases in β-endorphin as well as enkephalin in several discrete areas of the rat brain following treatment in adults with morphine pellets for 30 days (Hollt *et al.*, 1978; Przewlocki *et al.*, 1979). Reduced met-enkephalin levels have also been reported in the hippocampus of the monkey after 10 days of morphine treatment (Elsworth *et al.*, 1986).

In studies of the adult rat CNS, researchers have failed to detect any significant change in opioid peptide levels (Childers *et al.*, 1977; Fratta *et al.*, 1977; Wesche *et al.*, 1977; Shani *et al.*, 1979) in rat brain following chronic morphine treatment. Little is known about the molecular alterations in the functioning of endogenous brain opioid peptide systems as a result of chronic exogenous opioid administration. Neither brain opioid receptors, second messenger systems, nor opioid peptide levels have been shown to vary consistently as a result of drug treatment (Wuster *et al.*, 1983; Akil *et al.*, 1984; Redmond and Krystal, 1984; Collier, 1985; Rothman *et al.*, 1986). In a recent study by Uhl *et al.* (1988), the effect of chronic morphine treatment on opioid peptide gene expression in adult CNS was examined. It was shown that rats made tolerant to morphine via subcutaneous implantation of pellets displayed a significant decrease in striatal preproenkephalin mRNA that persisted during the withdrawal period. In contrast, levels of met-enkephalin were normal at the end of opioid treatment but reduced after withdrawal.

In an effort to gain more insight into these mechanisms clonal cell lines have been studied (Schwartz, 1988). In the NG108-15 neuroblastoma–glioma hybrid cell line, it has been demonstrated that enkephalin peptide content increases after exposure to cAMP (Braas *et al.*, 1983; Yoshikawa and Sabol, 1986). The opioid receptor is known to be linked to adenylate cyclase in an inhibitory fashion (Sharma *et al.*, 1975b). Tolerance and dependence that are produced in animals by chronic exposure to morphine can be reproduced in cells (Sharma *et al.*, 1975a; Sharma *et al.*, 1977). Schwartz (1988) has shown that opioid agonists stimulate proenkephalin synthesis with increases seen in mRNA, precursor forms and enkephalin peptides. However, these effects appear to occur independent of changes in adenylate cyclase activity or cAMP content. It is suggested that proenkephalin synthesis in NG108 cells can be regulated by two different mechanisms, one involving cAMP, and the other regulated by the opioid receptor, but this is yet to be determined.

In the developing CNS, there are very few data and these are complicated to interpret. In a preliminary study, Tempel *et al.* (unpublished) have found significant and contrasting effects of morphine treatment on preproenkephalin mRNA levels. Four days of postnatal (PD1–4) morphine treatment produced a 24% increase in striatal proenkephalin mRNA

levels. However, longer treatments with morphine (PD1–14) decrease striatal preproenkephalin mRNA levels by 39%. This decrease is similar to what was observed in the adult striatum after chronic morphine treatment (Uhl *et al.*, 1988). It is noteworthy that chronic naltrexone treatment produced the exact opposite effect. Four days of postnatal (PD1–4) naltrexone treatment induced a 33% decrease in proenkephalin mRNA levels in striatal tissue whereas 8 days of treatment (PD1–8) produced an increase (+23%) in opioid peptide gene expression. The increase in gene expression is in the same direction, but smaller than that observed in the adult striatum (Tempel *et al.*, 1990). These data suggest that the mechanisms underlying opioid addiction and withdrawal in the developing CNS differ from those in the adult brain. These differences may be due to the interaction of the opioid receptor system with the G-protein/cAMP system.

Future research along these lines will provide valuable information into the mechanisms of addiction and tolerance which will in turn provide insight into the diagnosis and treatment of drug-related effects observed in children born to addicted mothers. Moreover, such studies offer a unique view into the function of the opiate system in normal animals.

12.5 COCAINE AND THE OPIOID SYSTEM

Cocaine use has increased dramatically during the past few years (Clayton, 1985). Recent clinical studies indicate that maternal cocaine use can cause a variety of adverse effects. Cocaine has complex pharmacological effects on the mother and fetus. Cocaine is known to be a potent anorexic agent in both humans and animals and poor nutrition during pregnancy is known to place the infant at risk of developing many physical and functional abnormalities. Many clinical studies report that cocaine produces a premature separation of the placenta from the wall of the uterus with or without premature delivery (Acker *et al.*,

1983; Chasnoff *et al.*, 1985, 1987; Bingol *et al.*, 1987; Ryan *et al.*, 1987; Cherukuri *et al.*, 1988; Critchley *et al.*, 1988; Townsend *et al.*, 1988; Fulroth *et al.*, 1989; Kaye *et al.*, 1989; Valencia *et al.*, 1989).

Several clinical studies report an increase in stillbirth rate in pregnancies complicated by cocaine (Chasnoff *et al.*, 1985; Ryan *et al.*, 1987; Cherukuri *et al.*, 1988; Critchley *et al.*, 1988; Bingol *et al.*, 1989). Neonatal effects include cerebral infarction (Chasnoff *et al.*, 1986), depressed interactive behavior, intrauterine growth retardation, cardiac anomalies as well as fetal deaths (for review see Church *et al.*, 1990). Although these data suggest that cocaine abuse causes problems for the unborn child, there are also contradictory reports. There are reports of cocaine-abusing mothers delivering infants of normal weight and gestational age (Chasnoff *et al.*, 1986, 1987; Madden *et al.*, 1986; Ryan *et al.*, 1987). In addition, the human data are confounded by teenage pregnancy, undernutrition, multiple drug abuse, variability in the concentration of cocaine used and the presence of other chemicals that are sometimes mixed with cocaine (such as heroin, strychnine, talc, amphetamine, etc.). While placental abruption has been identified in one animal study, cocaine generally has no effect on gestational length in the rat (Church *et al.*, 1988; Hutchings *et al.*, 1989; Spear *et al.*, 1989). Animal studies have shown that cocaine reduces litter size but only at the most toxic doses (80–100 mg/kg s.c.) (Church *et al.*, 1988). These doses of cocaine result in a large proportion of maternal lethality and therefore may not be relevant to the human situation.

Although there is a need to establish good animal models for assessment of neurobehavioral and neurochemical consequences of early cocaine exposure, most laboratory studies have focused on maternal toxicity and fetal teratogenicity following gestational cocaine exposure (Mahalik *et al.*, 1980, 1984; Fantel and MacPhail, 1982; Church *et al.*, 1988). Record work has focused on neurobehavioral

and chemical consequences of early cocaine exposure in rodents (Dow-Edwards *et al.*, 1986, 1988; Foss and Riley, 1988; McGivern, 1988; Dow-Edwards, 1989, 1991; Smith *et al.*, 1989; Spear *et al.*, 1989).

The presence of psychoactive drugs can affect any of several neurotransmitter systems which can result in long-term alterations in the structure and function of these systems. Cocaine has been shown to alter activity at the dopaminergic, noradrenergic and serotonergic level. D_1 receptor binding increases by as much as 72% in the caudal portions of caudate nucleus of adult females exposed to cocaine (50 mg/kg) between PD11 and PD20 (Dow-Edwards, 1989). In addition, brain glucose metabolism, an index of functional activity, is differentially affected in males and females postnatally exposed to cocaine. Female treated rats show significantly increased rates of brain functional activity with specific highly dopaminergic regions showing the greatest percentage changes. For example, the cingulate cortex and ventral tegmental area show the largest increases in glucose activity. In contrast to female rats, neonatal cocaine treatment has little effect on male brain glucose metabolism, indicating that females may be more susceptible than males to the toxic effects of cocaine.

A variety of evidence suggests a 'dopamine hypothesis' for the reinforcing properties of cocaine, in particular, the dopamine (DA) mesolimbocortical neurons. D_1 receptors (3H SCH23390) are increased in the caudate of adult females exposed to cocaine between PD 11 and PD20 (for review see Dow-Edwards, 1991). An essential step for understanding the mechanism of action of any drug is to identify its receptor site. Receptor binding studies suggest that the cocaine receptor is the dopamine transporter. Recently, two independent laboratories have cloned the DA transporter (Shimada *et al.*, 1991; Kilty *et al.*, 1991) which should aid the understanding of the pharmacological actions of cocaine. Cocaine first binds to the transporter and blocks DA uptake, which results in a potentiation of dopaminergic neurotransmission in the limbic pathways. However, the existence of another important site or neurotransmitter that is involved in cocaine's actions cannot be ruled out. Serotonin and noradrenergic systems have also been implicated (for a review see Kuhar *et al.*, 1991).

Functional studies of adult rats exposed to cocaine during specific periods of development have shown that several important neuronal circuits are permanently altered by exposure to moderate (not toxic) doses (Dow-Edwards, 1989; Dow-Edwards *et al.*, 1990). Prenatal exposure produces persistent decrements in function in the nigrostriatal pathway which, due to its role in regulation of fine motor coordination, may be involved in the production of tremors seen in newborn humans exposed to cocaine *in utero*. Early postnatal exposure (which approximates the third trimester in human brain development) in the female rat increases function in the mesolimbic dopaminergic system. Since this pathway is involved in the reinforcing effects of cocaine, one might predict that developmental exposure to cocaine could alter drug-seeking behavior. However, the appropriate studies to test this hypothesis have not been carried out.

One of the other systems that may be involved in the reinforcing effects of cocaine is the opioid system. Very few data exist in this area, which is summarized in section 12.5.1.

In the adult CNS, chronic cocaine exposure increases [³H]naloxone binding in regions such as the nucleus accumbens, ventral pallidum and lateral hypothalamus (Hammer, 1989). In addition, cocaine treatment has been shown to increase dynorphin levels (Sivam, 1989; Smiley *et al.*, 1990) with no change in met-enkephalin or substance P levels (Sivam, 1989). These data suggest that cocaine may interact with the κ-opioid receptor, at least, in the adult CNS. Clow *et al.* (1991) examined opiate receptor binding in offspring of cocaine-exposed dams from ED8-20. Gesta-

tional cocaine exposure caused a generalized, dose-dependent increase of [³H]naloxone binding with highest increases in dopaminergic terminal regions as well as limbic and cortical regions. This effect is present 23–24 days after cessation of treatment. The effect is more extensive, and less specific on the opioid system than is seen in the adult after cocaine exposure. However, these studies just touch on the interaction of cocaine and the opioid system. Much more work is needed in order to characterize fully the mechanism of action of cocaine and its interaction with the opioid/dopamine systems.

12.5.1 SECOND MESSENGER SYSTEMS AND COCAINE TREATMENT

To date, there has been very little work on the effects of cocaine treatment on G-protein/second messenger systems. Two research groups have recently begun to explore this interaction in the adult CNS. Nestler *et al.* (1990b) have reported a decrease in $G_{o\alpha}$ and $G_{i\alpha}$ in the A10 DA region and the nucleus accumbens following 14 days of daily cocaine treatment (15 mg/kg). In order to assess further the role of G proteins in cocaine abuse, Steketee *et al.* (1991) examined the effects of PTX administration and ADP-ribosylation of G_i and G_o in these brain regions. The capacity of acute cocaine treatment to increase dopamine release in the nucleus accumbens was significantly increased in rats pretreated 14 days earlier with PTX. These data suggest that injection of PTX into the A10 cell group produces alterations in mesolimbic dopamine function as well as G proteins in cocaine abuse. Although these data may shed light on the mechanism of action of cocaine in the adult CNS, no work has been done in the developing CNS. It is possible that the actions of cocaine in the neonatal CNS may be quite different from those seen in the adult CNS. Future research along these lines is needed to fully comprehend the long-term effects in offspring exposed to cocaine prenatally so that they may best overcome the toxic effects of this drug.

ACKNOWLEDGEMENT

I would like to thank Annette Snoddy for her careful and patient typing of this chapter.

REFERENCES

Acker, D., Sachs, B.P., Tracey, K.J. and Wise, W.E. (1983) Abruptio placentae associated with cocaine use. *Am. J. Obstet. Gynecol.*, **146**, 220–1.

Akil, H., Watson, S.J., Young, E. *et al.* (1984) Endogenous opioids: biology and function. *Annu. Rev. Neurosci.*, **7**, 223–55.

Attali, B., Gouarderes, C., Mazaguil, H. *et al.* (1982) Evidence of multiple kappa binding sites by use of opioid peptides in the guinea pig lumbrosacral spinal cord. *Neuropeptides*, **3**, 53.

Attali, B. and Vogel, Z. (1989) Long-term opiate exposure leads to reduction of the α_i-1 subunit of GTP-binding proteins. *J. Neurochem.*, **53**, 1636–9.

Bardo, M.T., Bhatnagar, R.K. and Gebhart, G.F. (1982) Differential effects of chronic morphine and naloxone on opiate receptors, monoamines and morphine-induced behaviors in preweanling rats. *Dev. Brain Res.*, **4**, 139–47.

Bardo, M.T., Bhatnagar, R.K. and Gebhart, G.F. (1983a) Age-related differences in the effect of chronic administration of naloxone on opiate binding in rat brain. *Neuropharmacology*, **22**, 453–61.

Bardo, M.T., Bhatnagar, R.K. and Gebhart, G.F. (1983b) Chronic naltrexone increases opiate binding in brain and produces supersensitivity to morphine in the locus coeruleus of the rat. *Brain Res.*, **289**, 223–34.

Barr, G.A., Paredes, W., Erickson, K.L. and Zukin, R.S. (1986) κ-opioid receptor-mediated analgesia in the developing rat. *Dev. Brain Res.*, **28**, 145–52.

Bauman, P.S. and Levine, S.A. (1986) The development of children of drug addicts. *Int. J. Addict.*, **21**, 849–63.

Beitner, D.B., Duman, R.S. and Nestler, E.J. (1989) A novel action of morphine in the rat locus coeruleus: persistent decrease in adenylate cyclase. *Mol. Pharmacol.*, **35**, 559–64.

Bergstrom, L. and Terenius, L. (1983) Enkephalin levels decrease in rat striatum during morphine abstinence. *Eur. J. Pharmacol.*, **60**, 349–52.

Bhargava, H.N., Ramarao, P. and Gulati, A. (1989a) Effects of morphine in rats treated chronically with U-50,488H, a kappa opioid receptor agonist. *Eur. J. Pharmacol.*, **162**, 257–64.

Bhargava, H.N., Gulati, A. and Ramarao, P. (1989b) Effect of chronic administration of U-50,488H on tolerance to its pharmacological actions and on multiple opioid receptors in rat brain regions and spinal cord. *J. Pharmacol. Exp. Ther.*, **251**, 21–6.

Bingol, N., Fuchs, M., Diaz, V. *et al.* (1987) Teratogenicity of cocaine in humans. *J. Pediatr.*, **110**, 93–6.

Blanchard, S.G., Chang, K.J. and Cuatrecasas, P. (1983) Characterization of the association of tritiated enkephalin with neuroblastoma cells under conditions optimal for receptor down-regulation. *J. Biol. Chem.*, **258**, 1092–7.

Blinick, G., Inturrisi, C.E., Jerez, E. and Wallach, R.C.(1975) Methadone assays in pregnant women and progeny. *Am. J. Obstet. Gynecol.*, **121**, 617–21.

Braas, K.A., Childers, S.R. and U'Prichard, D.C. (1983) Induction of differentiation increases Met^5-enkephalin and Leu^5-enkephalin content in NG108-15 hybrid cells: an immunocytochemical and biochemical analysis. *J. Neurosci.*, **3**, 1713–27.

Brady, L.S., Herkenham, M., Long, J.B. and Rothman, R.B. (1989) Chronic morphine increases mu-opiate receptor binding in rat brain: a quantitative autoradiographic study. *Brain Res.*, **477**, 382–6.

Brunello, N., Volterra, A., DiGiulio, A.M. *et al.* (1984) Modulation of opioid system in C57 mice after repeated treatment with morphine and naloxone: biochemical and behavioral correlates. *Life Sci.*, **34**, 1669–78.

Chang, D.M. and Costa, E. (1979) Evidence for internalization of the recognition site of beta-adrenergic receptors during receptor subsensitivity induced by (-)isoproterenol. *Proc. Natl. Acad. Sci. USA*, **76**, 3024–8.

Chang, K.J. and Cuatrecasas, P. (1979) Multiple opiate receptors: enkephalins and morphine bind to receptors of different specificity. *J. Biol. Chem.*, **254**, 2610–18.

Chang, K.J., Eckel, R.W. and Blanchard, S.G. (1982) Opioid peptides induce reduction of enkephalin receptors in cultured neuroblastoma cells. *Nature*, **296**, 446–8.

Chasnoff, I.J., Burns, W.J., Schnoll, S.H. and Burns, K.A. (1985) Cocaine use in pregnancy. *N. Engl. J. Med.*, **313**, 666–9.

Chasnoff, I.J., Burns, K.A. and Burns, W.J. (1987) Cocaine use in pregnancy: perinatal morbidity and mortality. *Neurotoxicol. Teratol.*, **9**, 291–3.

Chasnoff, I.J., Bussey, M.E., Sarich, R. and Stack, C.M. (1986) Perinatal cerebral infarction and maternal cocaine use. *J. Pediatr.*, **108**, 456–9.

Cherukuri, R., Minkoff, H., Feldman, J. *et al.* (1988) A cohort study of alkaloidal cocaine ('crack') in pregnancy. *Obstet. Gynecol.*, **72**, 147–51.

Childers, S.R., Simantov, R. and Snyder, S.H. (1977) Enkephalin: radioimmunoassay and radioreceptor assay in morphine dependent rats. *Eur. J. Pharmacol.*, **46**, 289–93.

Church, M.W., Dintcheff, B.A. and Gessner, P.K. (1988) Dose-dependent consequences of cocaine on pregnancy outcome in the Long-Evans Rat. *Neurotoxicol. Teratol.*, **10**, 51–8.

Church, M.W., Overbeck, G.W. and Andrzejczak, A.L. (1990) Prenatal cocaine exposure in the Long-Evans Rat: I. Dose-dependent effects on gestation, mortality, and postnatal maturation. *Neurotoxicol. Teratol.*, **12**, 327–34.

Clayton, R.R. (1985) Cocaine use in the U.S.: in a blizzard or just being snowed, in *Cocaine Use in America: Epidemiologic and Clinical Perspectives, NIDA Monograph* (eds J. Kozel and E.H. Adams), US Dept. Health Human Services, Rockville, MD, Vol. 61, pp. 8–34.

Clendeninn, N.J., Petraitis, M. and Simon, E.J. (1976) Ontological development of opiate receptors in rodent brain. *Brain Res.*, **118**, 157–60.

Clow, D.W., Hammer, R.P., Kirstein, C.L. and Spear, L.P. (1991) Gestational cocaine exposure increases opiate receptor binding in weanling offspring. *Dev. Brain Res.*, **59**, 179–85.

Collier, H.O.J. (1985) A general theory of the genesis of drug dependence by induction of receptors. *Nature*, **205**, 181–3.

Collier, H.O.J., and Roy, A.C. (1974) Morphine-like drugs inhibit the stimulation of E prostaglandins of cyclic AMP formation by rat brain homogenate. *Nature*, **248**, 24–7.

Connaughton, J.F., Finnegan, L.P., Schur, J. and Emich, J.P. (1973) Current concepts in the management of the pregnant opiate addict. *Addictive Dis.*, **2**, 21–36.

Corbett, A.D., Paterson, S.J., McKnight, A.T. *et al.* (1983) Dynorphin 1-8 and dynorphin 1-9 are ligands for the κ-subtype of opiate receptor. *Nature*, **299**, 79–81.

Costa, T., Klinz, F.J., Vachon, L. and Herz, A. (1988) Opioid receptors are coupled tightly to G proteins but loosely to adenylate cyclase in NG108-15 cell membranes. *Mol. Pharmacol.*, **34**, 744–54.

Coyle, J.T. and Pert, C.B. (1976) Ontogenetic development of [^{3}H] naloxone binding in rat brain. *Neuropharmacology*, **15**, 555–60.

Crain, S.M., Shen, K.F. and Chalazonitis, A. (1988) Opioids excite rather than inhibit sensory neurons after chronic opioid exposure of spinal cord-ganglion cultures. *Brain Res.*, **455**, 99–109.

Critchley, H.O.D., Woods, S.M., Barson, A.J. *et al.* (1988) Fetal death in utero and cocaine abuse. Case report. *Br. J. Obstet. Gynaecol.*, **95**, 195–6.

Danks, J.A., Tortella, F.C., Bykov, V. *et al.* (1988) Chronic administration of morphine and naltrexone up-regulate [^{3}H][D-Ala^{2}, D-leu^{5}]enkephalin binding sites by different mechanisms. *Neuropharmacology*, **27**, 965–74.

Davis, M.W. and Lin, C.H. (1972) Prenatal morphine effects on survival and behavior of rat offspring. *Res. Commun. Chem. Pathol. Pharmacol.*, **3**, 205–14.

Deren, S. (1986) Children of substance abusers: a review of the literature. *J. Subst. Abuse Treat.*, **3**, 77–94.

Dole, V.P. and Nyswander, M.A. (1965) Medical treatment for diacetyl-morphine (heroin) addiction. *J. Am. Med. Assoc.*, **193**, 646–50.

Dow-Edwards, D.L. (1989) Long-term neurochemical and neurobehavioral consequences of cocaine use during pregnancy. *Ann. NY Acad. Sci.*, **562**, 280–9.

Dow-Edwards, D.L. (1991) Cocaine effects on fetal development: a comparison of clinical and animal research findings. *Neurotoxicol. Teratol.*, **13**, 347–52.

Dow-Edwards, D.L., Freed, L.A. and Milorat, T.H. (1986) The effects of cocaine on development. *Soc. Neurosci. Abstr.*, **12**, 59.12.

Dow-Edwards, D.L., Fico, T.A. and Hutchings, D.E. (1988) Functional effects of cocaine given during critical periods of development. *Teratology*, **37**, 518.

Dow-Edwards, D.L., Freed, L.A. and Fico, T.A. (1990) Structural and functional effects of prenatal cocaine exposure in adult rat brain. *Dev. Brain Res.*, **57**, 263–8.

Duman, R.S., Tallman, J.F. and Nestler, E.J. (1988) Acute and chronic opiate-regulation of adenylate cyclase in brain: specific effects in locus coeruleus. *J. Pharmacol. Exp. Ther.*, **246**, 1033–9.

Elsworth, J.D., Redmond, D.E. and Roth, R.H. (1986) Effect of morphine treatment and withdrawal on endogenous methionine- and leucine-enkephalin levels in primate brain. *Biochem. Pharmacol.*, **35**, 3415–17.

Fantel, A.G. and MacPhail, B.J. (1982) The teratogenicity of cocaine. *Teratology*, **26**, 17–19.

Finnegan, L.P. (1975) Narcotics dependence in pregnancy. *J. Psychedelic Drugs*, **7**, 299–311.

Florio, V.A. and Sternwies, P.C. (1985) Reconstitution of resolved muscarinic cholinergic receptors with purified GTP-binding proteins. *J. Biol. Chem.*, **260**, 3477–83.

Foss, J.A. and Riley, E.A. (1988) Behavioral evaluation of animals exposed prenatally to cocaine. *Teratology*, **37**, 517.

Fratta, W., Yang, H.Y.T., Hong, J. and Costa, E. (1977) Stability of met-enkephalin content in brain structures of morphine-dependent or footshock stressed rats. *Nature*, **268**, 452–3.

Freeman, P.R. (1980) Methadone exposure in utero: effects on open-field activity in weanling rats. *Int. J. Neurosci.*, **11**, 295–300.

Fulroth, R., Phillips, B. and Durand, D.J. (1989) Perinatal outcome of infants exposed to cocaine and/or heroin in utero. *Am. J. Dis. Child.*, **143**, 905–10.

Fung, B.K.-K. (1983) Characterization of transducin from bovine retinal rod outer segments. I. Separation and reconstitution of the subunits. *Biol. Chem.*, **258**, 10495–502.

Galant, S.P., Durisetti, L., Underwood, S. and Insel, P.A. (1978) Decreased beta-adrenergic receptors on polymorphonuclear leukocytes after adrenergic therapy. *N. Engl. J. Med.*, **299**, 933–6.

Genazzani, A.R., Petraglia, F., Guidetti, R., *et al.* (1986) Neonatal β-endorphin secretion in babies passively addicted to opiates. *Int. Congr. Ser. Excerpta Med.*, **369**, 379–82.

Gilman, A.G. (1987) G proteins: transducers of receptor-generated signals. *Annu. Rev. Biochem.*, **56**, 615–49.

Gintzler, A.R. and Xu, H. (1991) Different G proteins mediate the opioid inhibition or enhancement of evoked [5-methionine] enkephalin release. *Proc. Natl. Acad. Sci. USA*, **88**, 4741–5.

Glaubiger, G. and Lefkowitz, R.J. (1977) Elevated beta-adrenergic receptor number after chronic propranolol treatment. *Biochem. Biophys. Res. Commun.*, **78**, 720–5.

Goldstein, A., Fischli, W., Lowney, L.I. *et al.* (1981) Porcine pituitary dynorphin: complete amino acid sequence of the biologically active heptadecapeptide. *Proc. Natl. Acad. Sci. USA*, **78**, 7219–23.

Goodfriend, M.J., Shey, I.A. and Klein, M.D. (1956) The effects of maternal narcotic addiction on the newborn. *Am. J. Obstet. Gynecol.*, **71**, 29–36.

Gouarderes, C., Attali, B., Audigier, Y and Cros, J. (1983) Interaction of selective mu and delta ligands with the kappa₂ subtype of opiate binding sites. *Life Sci.*, **33**, 175–8.

Gubler, U., Seeberg, P., Hoffman, B.J. *et al.* (1982) Molecular cloning establishes proenkephalin as precursor of enkephalin-containing peptides. *Nature*, **295**, 206–8.

Guitart, X. and Nestler, E.J. (1989) Identification of morphine and cyclic AMP-regulated phosphoproteins (MARPPs) in the locus coeruleus and other regions of rat brain. Regulation by acute and chronic morphine. *J. Neurosci.*, **9**, 4371–87.

Guitart, X. and Nestler, E.J. (1990) Identification of MARPP (14–20), morphine- and cyclic AMP-regulated phosphoproteins of 14–20 kDa, as myelin basic proteins: evidence for their acute and chronic regulation by morphine in rat brain. *Brain Res.*, **516**, 57–65.

Guitart, X., Hayward, M.D., Nissenbaum, L.K. *et al.* (1990) Identification of MARPP-58, a morphine- and cAMP-regulated phosphoprotein of 58 kDa, as tyrosine hydroxylase: evidence for regulation of its expression by chronic morphine in the rat locus coeruleus. *J. Neurosci.*, **10**, 2649–59.

Hammer, R.P. (1989) Cocaine alters opiate receptor binding in critical brain reward regions. *Synapse*, **3**, 55–60.

Handelmann, G.E., and Quirion, R. (1983) Neonatal exposure to morphine increases mu opiate binding in the adult forebrain. *Eur. J. Pharmacol.*, **94**, 357–8.

Hans, S.L. (1989) Development consequences of prenatal exposure to methadone. *Ann. NY Acad. Sci.*, **562**, 195–207.

Hazum, E., Chang, K.J. and Cuatrecasas, P. (1981) Receptor redistribution induced by hormones and neurotransmitters: possible relationship to biological functions. *Neuropeptides*, **1**, 217–30.

Herkenham, M. and Pert, C.B. (1981) Mosaic distribution of opiate receptors, parafascicular projections and acetylcholinesterase in rat striatum. *Nature*, **291**, 415–18.

Herkenham, M. and McLean, S. (1986) Mismatches between receptor and transmitter localizations in the brain, in *Quantitative Receptor Autoradiography* (eds C. Boast, E.W. Altar and C.A. Snowhill), Alan Liss, New York, p. 131–71.

Hitzemann, R., Hitzemann, B. and Loh, H. (1974) Binding of [³H]naloxone in the mouse brain: effect of ions and tolerance development. *Life Sci.*, **14**, 2393–404.

Hofer, M.A. (1984) Relationships as regulators: a psychobiologic perspective on bereavement. *Psychosom. Med.*, **46**, 183–97.

Holaday, J.W., Hitzemann, R.J., Curell, J. *et al.* (1982) Repeated electroconvulsive shock or chronic morphine treatment increases the number of [³H]D-Ala² D-Leu⁵-enkephalin binding sites in rat brain membranes. *Life Sci.*, **31**, 2359–62.

Hollt, V., Przewlocki, R. and Herz, A. (1978) β-Endorphin-like immunoreactivity in plasma, pituitaries and hypothalamus of rats following treatment with opiates. *Life Sci.*, **23**, 1057–66.

Householder, J., Hatcher, R., Burns, W. and Chasnoff, I. (1982) Infants born to narcotic-addicted mothers. *Psychol. Bull.*, **92**, 453–68.

Hughes, J., Smith, T.W., Kosterlitz, H.W. *et al.* (1975) Identification of two related pentapeptides from the brain with potent opiate agonist activity. *Nature*, **258**, 577–80.

Hutchings, D.E. (1985a) Issues of methodology and interpretation in clinical and animal behavioral teratology studies. *Neurobehav. Toxicol. Teratol.*, **7**, 639–42.

Hutchings, D.E. (1985b) Prenatal opioid exposure and the problem of casual inference, in *Current Research on the Consequences of Maternal Drug Use. National Institute on Drug Abuse Research Series* (ed. T.M. Pinkert) DHHS Pub. No. (ADM) 85-1400. Supt of Docs, US Govt Printing Office, Washington, DC, pp. 6–9.

Hutchings, D.E. and Fifer, W.P. (1986) Neurobehavioral effects in human and animal offspring following prenatal exposure to methadone, in *Handbook of Behavioral Teratology* (eds E.P. Riley and C.V. Vorhees), Plenum Press, New York, pp. 141–60.

Hutchings, D.E., Bodnarenko, S.R. and Diaz-DeLeon, R. (1984) Phencyclidine during pregnancy in the rat: effects on locomotor activity in the offspring. *Pharmacol. Biochem. Behav.*, **20**, 251–4.

Hutchings, D.E., Fico, T.A. and Dow-Edwards, D.L. (1989) Prenatal cocaine: maternal toxicity, fetal effects, and locomotor activity in rat offspring. *Neurotoxicol. Teratol.*, **11**, 65–9.

Iyengar, S. and Rabii, J. (1982) Effect of prenatal exposure to morphine on the postnatal development of opiate receptors. *Fed. Proc.*, **41**, 354–7.

James, I.F., Chavkin, C. and Goldstein, C. (1982) Selectivity of dynorphin for kappa opioid receptors. *Life Sci.*, **31**, 1331–4.

Jeremy, R.J. and Hans, S.L. (1985) Behavior of neonates exposed in utero to methadone as assessed on the Brazelton scale. *Infant Behav. Dev.*, **8**, 323–36.

Kandall, S.R., Album, S., Dreyer, E. *et al.* (1975) Differential effects of heroin and methadone on birth weights. *Addictive Dis.*, **2**, 347–55.

Kaye, K., Elkind, L., Goldberg, D. and Tytun, A. (1989) Birth outcomes for infants of drug abusing mothers. *NY State J. Med.*, **89**, 256–61.

Kent, J.L., Pert, C.B. and Herkenham, M. (1982) Ontogeny of opiate receptors in rat forebrain: visualization by *in vitro* autoradiography. *Dev. Brain Res.*, **2**, 487–504.

Kilty, J.E., Lorang, D. and Amara, S.G. (1991) Cloning and expression of a cocaine-sensitive rat dopamine transporter. *Science*, **254**, 578–9.

Kirby, M.L. and Aronstam, R.S. (1983) Levorphanol-sensitive [^{3}H] naloxone binding in developing brainstem following prenatal morphine exposure. *Neurosci. Lett.*, **35**, 191–5.

Kirby, M.L., Gale, T.F. and Mattio, T.C. (1982) Effects of prenatal capsaicin treatment on fetal spontaneous activity, opiate receptor binding, and acid phosphatase in the spinal cord. *Exp. Neurol.*, **76**, 298–308.

Klee, W.A. and Streaty, R.A. (1974) Narcotic receptor sites in morphine-dependent rats. *Nature*, **248**, 61–3.

Kuhar, M.J., Ritz, M.C. and Boja, J.W. (1991) The dopamine hypothesis of the reinforcing properties of cocaine. *Trends Neurol. Sci.*, **14**, 229–302.

Lahti, R.A. and Collins, R.J. (1978) Chronic naloxone results in prolonged increases in opiate binding sites in brain. *Eur. J. Pharmacol.*, **51**, 185–6.

Law, P.Y., Wu, J., Koehler, J. and Loh, H.H.J. (1981) Demonstration and characterization of opiate inhibition of the striatal adenylate cyclase. *J. Neurochem.*, **36**, 1834–6.

Leslie, F.M., Tso, S. and Hurlbut, D.E. (1982) Differential appearance of opiate receptor subtypes in neonatal rat brain. *Life Sci.*, **31**, 1393–6.

Lesser-Katz, M. (1982) Some effects of maternal drug addiction on the neonate. *Int. J. Addict.*, **17**, 887–96.

Lord, J.A.H., Waterfield, A.A., Hughes, J. and Kosterlitz, H.W. (1977) Endogenous opioid peptides: multiple agonists and receptors. *Nature*, **267**, 495–9.

Loughlin, S.E., Massamiri, T., Kornblum, H.I. and Leslie, F.M. (1985) Postnatal development of opioid systems in rat brain. *Neuropeptides*, **5**, 469–72.

Madden, J.D., Payne, T.F. and Miller, S. (1986) Maternal cocaine abuse and effect on the newborn. *Pediatrics*, **77**, 209–11.

Mahalik, M.P., Gautieri, R.F. and Mann, D.E. (1980) Teratogenic potential of cocaine hydrochloride in CS-1 mice. *J. Pharm. Sci.*, **69**, 703–6.

Mahalik, M.P., Gautieri, R.F. and Mann, D.E. (1984) Mechanisms of cocaine-induced teratogenesis. *Res. Commun. Subst. Abuse*, **5**, 279–302.

Mains, R.E., Eipper, B.A. and Ling, N. (1977) Common precursor to corticotropins and endorphins. *Proc. Natl. Acad. Sci. USA*, **74**, 3014–8.

Mansour, A., Khachaturian, H., Lewis, M.E. *et al.* (1987) Autoradiographic differentiation of mu, delta and kappa opioid receptors in the rat forebrain and midbrain. *J. Neurosci.*, **7**, 2445–64.

Martin, W.R., Eades, C.G., Thompson, J.A. *et al.* (1976) The effects of morphine- and nalorphine-like drugs in the nondependent and morphine-dependent chronic spinal dog. *J. Pharmacol. Exp. Ther.*, **197**, 517–32.

Mayorga, L.S., Diaz, R. and Stahl, P.D. (1989) Regulatory role for GTP-binding proteins in endocytosis. *Science*, **244**, 1475–7.

McGinty, J.F. and Ford, D.H. (1980) Effects of prenatal methadone on rat brain catecholamines. *Dev. Neurosci.*, **3**, 224–34.

McGivern, R.F., Raum, W.J., Sokol, R.Z. and Peterson, P. (1988) Long-term effects of prenatal cocaine exposure on scent marking and the HPG axis in male rats. *Teratology*, **37**, 518.

Millan, M.J., Morris, B.J. and Herz, A. (1988) Antagonist-induced opioid receptor upregulation. I. Characterization of supersensitivity to selective mu and kappa agonists. *J. Pharmacol. Exp. Ther.*, **247**, 721–8.

Moon, S.L. (1984) Prenatal haloperidol alters striatal dopamine and opiate receptors. *Brain Res.*, **323**, 109–13.

Morris, B.J. and Herz, A. (1989) Control of opiate receptor number in vivo: simultaneous kappa-receptor down-regulation and mu-receptor up-regulation following chronic agonists/antagonist treatment. *Neuroscience*, **29**, 433–42.

Nestler, E.J. and Tallman, J.F. (1988) Chronic morphine treatment increases cyclic AMP-dependent protein kinase activity in the rat locus coeruleus. *Mol. Pharmacol.*, **33**, 127–32.

Nestler, E.J., Erdos, J.J., Terwilliger, R. *et al.* (1989) Regulation of G-proteins by chronic morphine in the rat locus coeruleus. *Brain Res.*, **476**, 230–9.

Nestler, E.J., Beitner, D.B., Hayward, M. *et al.* (1990a) A general role for adaptations in G-proteins and the cyclic AMP system in mediating the chronic actions of morphine and cocaine on brain function. *Soc. Neurosci. Abstr.*, **16**, 928.

Nestler, E.J., Terwilliger, R.Z., Walker, J.R. *et al.* (1990b) Chronic cocaine treatment decreases levels of the G protein subunits $G_{i\alpha}$ and $G_{o\alpha}$ in discrete regions of rat brain. *J. Neurochem.*, **55**, 1079–82.

Nishino, K., Su, Y.F., Wong, C.S. *et al.* (1990) Dissociation of mu opioid tolerance from receptor down-regulation in rat spinal cord. *J. Pharmacol. Exp. Ther.*, **253**, 67–72.

Panerai, A.E., Martini, A., DiGiulio, A.M. *et al.* (1983) Plasma β-endorphin, β-lipotropin, and met-enkephalin concentrations during pregnancy in normal and drug-addicted women and their newborn. *J. Clin. Endocrinol. Metab.*, **57**, 537–43.

Pasternak, G.W., Gintzler, A.R., Houghton, R.A. *et al.* (1983) Biochemical and pharmacological evidence for opioid receptor multiplicity in the central nervous system. *Life Sci.*, **33** (Suppl 1), 167.

Perlmutter, J.F. (1974) Heroin addiction and pregnancy. *Obstet. Gynecol. Surv.*, **29**, 439–46.

Perry, D.C., Rosenbaum, J.S. and Sadee, W. (1982) In vitro binding of [^{3}H]etorphine in morphine-dependent rats. *Life Sci.*, **31**, 1405–8.

Pert, C.B., Pasternak, G. and Snyder, S.H. (1973) Opiate agonists and antagonists discriminated by receptor binding in brain. *Science*, **182**, 1359–61.

Pert, C.B. and Snyder, S.H. (1973) Opiate receptor demonstration in nervous tissue. *Science*, **179**, 1011–14.

Petrillo, P., Tavani, A., Verotta, D. *et al.* (1987) Differential postnatal development of mu-, delta- and kappa-opioid binding sites in rat brain. *Brain Res.*, **428**, 53–8.

Provost, N.M., Somers, D.E. and Hurley, J.B. (1988) A *Drosophila melanogaster* G protein alpha subunit gene is expressed primarily in embryos and pupae. *J. Biol. Chem.*, **263**, 12070–76.

Przewlocki, R., Hollt, V., Duka, T. *et al.* (1979) Long-term morphine treatment decreases endorphin levels in rat brain and pituitary. *Brain Res.*, **174**, 357–61.

Ragavan, V.V., Wardlaw, S.L., Kreek, M.J. and Frantz, A.G. (1983) Effect of chronic naltrexone and methadone administration on brain immunoreactive beta-endorphin in the rat. *Neuroendocrinology*, **37**, 266–8.

Rasmussen, K., Beitner-Johnson, D.B., Krystal, J.H. *et al.* (1990) Opiate withdrawal and the rat locus coeruleus: behavioral, electrophysiological and biochemical correlates. *J. Neurosci.*, **10**, 2308–17.

Rech, R.H., Lomuscio, G. and Algeri, S. (1980) Methadone exposure in utero: effects on brain biogenic amines and behavior. *Neurobehav. Toxicol.*, **2**, 75–8.

Redmond, D.E. and Krystal, J.H. (1984) Multiple mechanisms of withdrawal from opioid drugs. *Annu. Rev. Neurosci.*, **7**, 443–78.

Rementeria, J.L. and Lotongkhum, K. (1977) The fetus of the drug-addicted woman: conception, fetal wastage, and complications, in *Drug Abuse in Pregnancy and Neonatal Effects* (ed. J.L. Rementeria), Mosby, St Louis, MO, pp. 1–18.

Rothman, R.B., Danks, J.A., Jacobson, A.E. *et al.* (1986) Morphine tolerance increases μ-noncompetitive binding sites. *Eur. J. Pharmacol.*, **124**, 113–19.

Ryan, L., Ehrlich, S. and Finnegan, L. (1987) Cocaine abuse in pregnancy: effects on the fetus and newborn. *Neurotoxicol. Teratol.*, **9**, 295–9.

Sakellaridis, N. and Vernadakis, A. (1987) The chick embryo vs. neural tissue culture as models for the study of opiate neurotoxicity in development, in *Model Systems in Neurotoxicology: Alternative Approaches to Animal Testing* (eds A. Shahar and A.M. Goldberg), Alan R. Liss, New York, pp. 85–100.

Schoffelmeer, A.N.M., Hansen, H.A., Stoff, J.C. and Mulder, A.H. (1986) Blockade of D-2 dopamine receptors strongly enhances the potency of enkephalins to inhibit dopamine-sensitive adenylate cyclase in rat neostriatum: involvement of delta- and mu-opioid receptors. *J. Neurosci.*, **6**, 2235–9.

Schulz, R., Wuster, M. and Herz, A. (1979) Supersensitivity to opioids following the chronic blockade of endorphin action by naloxone. *Naunyn Schmiedebergs Arch. Pharmacol.*, **306**, 93–6.

Schwartz, J.P. (1988) Chronic exposure to opiate agonists increases proenkephalin biosynthesis in NG108 cells. *Mol. Brain Res.*, **3**, 141–6.

Shani, J., Azov, R. and Weissman, B.A. (1979) Enkephalin levels in rat brain after various regimens of morphine administration. *Neurosci. Lett.*, **12**, 319–22.

Sharma, S.K., Klee, W.A. and Nirenberg, M. (1975a) Dual regulation of adenylate cyclase accounts for narcotic dependence and tolerance. *Proc. Natl. Acad. Sci. USA*, **72**, 590–4.

Sharma, S.K., Nirenberg, M. and Klee, W.A. (1975b) Morphine receptors as regulators of adenylate cyclase activity. *Proc. Natl. Acad. Sci. USA*, **72**, 3092–6.

Sharma, S.K., Klee, W.A. and Nirenberg, M. (1977) Opiate-dependent modulation of adenylate cyclase. *Proc. Natl. Acad. Sci. USA*, **74**, 3365–3369.

Shen, K.F. and Crain, S.M. (1989) Dual opioid modulation of the action potential duration of mouse dorsal root ganglion neurons in culture. *Brain Res.*, **491**, 227–42.

Shen, K.F. and Crain, S.M. (1990) Cholera toxin-A subunit blocks opioid excitatory effects on sensory neuron action potentials indicating mediation by G_s-linked opioid receptors. *Brain Res.*, **525**, 225–31.

Shimada, S., Kitayama, S., Lin, C.L. *et al.* (1991) Cloning and expression of a cocaine-sensitive dopamine transporter complementary DNA. *Science*, **254**, 576–8.

Sibley, D.R., and Lefkowitz, R.J. (1985) Molecular mechanisms of receptor desensitization using the beta-adrenergic receptor-coupled adenylate cyclase system as a model. *Nature*, **317**, 124–9.

Simantov, R., Levy, R. and Baram, D. (1982a) Down-regulation of enkephalin (delta) receptors – demonstration in membrane-bound and solubilized receptors. *Biochim. Biophys. Acta*, **721**, 478–84.

Simantov, D., Baram, D., Levy, R. and Hadler, H. (1982b) Enkephalins and α-adrenergic receptors: evidence for both common and differentiable regulatory pathways and down-regulation of the enkephalin receptor. *Life Sci.*, **31**, 1323–6.

Simon, E.J. and Hiller, J.M. (1978) In vitro studies on opiate receptors and their ligands. *Fed. Proc.*, **37**, 141–6.

Sivam, S.P. (1989) Cocaine selectively increases striatonigral dynorphin levels by a dopaminergic mechanism. *J. Pharmacol. Exp. Ther.*, **250**, 818–24.

Slotkin, T.A. and Anderson, T.R. (1975) Sympatho-adrenal development in perinatally addicted rats. *Addict. Dis. Int. J.*, **2**, 243–305.

Slotkin, T.A., Weigle, S.J., Whitmore, W.L. and Seidler, F.J. (1982) Maternal methadone administration: deficient in development of alpha-noradrenergic responses in developing rat brain as assessed by norepinephrine stimulation of ^{33}Pi incorporation into phospholipids in vivo. *Biochem. Pharmacol.*, **31**, 1899–902.

Smiley, P.L., Johnson, M., Bush, L. *et al.* (1990) Effects of cocaine on extrapyramidal and limbic dynorphin systems. *J. Pharmacol. Exp. Ther.*, **253**, 938–43.

Smith, R.F., Mattran, K.M., Kurkjian, M.F. and Kurtz, S.L. (1989) Alterations in offspring behavior induced by chronic prenatal cocaine dosing. *Neurotoxicol. Teratol.*, **11**, 35–8.

Snyder, S.H. (1984) Characterization of opiate binding sites. *Science*, **224**, 22–5.

Spain, J.W., Roth, B.L. and Coscia, C.J. (1985) Differential ontogeny of multiple opioid receptors (mu, delta, and kappa). *J. Neurosci.*, **5**, 584–8.

Spear, L.P., Kirstein, C.L. and Frambes, N.A. (1989) Cocaine effects on the developing central nervous system: behavioral, psychopharmacological and neurochemical studies. *Ann. NY Acad. Sci.*, **562**, 290–307.

Steece, K.A., DeLeon-Jones, F.A., Meyerson, L.R. *et al.* (1986) In vivo downregulation of rat striatal opioid receptors by chronic enkephalin. *Brain Res. Bull.*, **17**, 255–7.

Steketee, J.D., Striplin, C.D., Murray, T.F. and

Kalivas, P.W. (1991) Possible role for G-proteins in behavioral sensitization to cocaine. *Brain Res.,* **545**, 287–91.

Sternwies, P.C. (1986) The purified alpha subunits of G_o and G_i from bovine brain require beta gamma for association with phospholipid vesicles. *J. Biol. Chem.,* **261**, 631–7.

Strauss, M.E., Andresko, M., Stryker, J.C. *et al.* (1974) Methadone maintenance during pregnancy: pregnancy, birth and neonate characteristics. *Am. J. Obstet. Gynecol.,* **120**, 895–900.

Stryer, L. (1986) Cyclic GMP cascade of vision. *Annu. Rev. Neurosci.,* **9**, 87–119.

Tang, A.H. and Collins, R.J. (1978) Enhanced analgesic effects of morphine after chronic administration of naloxone in the rat. *Eur. J. Pharmacol.,* **47**, 473–4.

Tao, P.L., Law, P.Y. and Loh, H.H. (1987) Decrease in delta and mu opioid receptor binding capacity in rat brain after chronic etorphine treatment. *J. Pharmacol. Exp. Ther.,* **240**, 809–16.

Tao, P.L., Chang, L.R., Law, P.Y. and Loh, H. H. (1988) Decrease in delta-receptor density in rat brain after chronic (D-Ala^2, D-Leu^5) enkephalin treatment. *Brain Res.,* **462**, 313–20.

Tavani, A., Robson, L. and Kosterlitz, H.W. (1985) Differential postnatal development of mu, delta, and kappa opioid binding sites in mouse brain. *Dev. Brain Res.,* **23**, 306–9.

Tempel, A. (1991) Visualization of µ opiate receptor downregulation following morphine treatment in neonatal rat brain. *Dev. Brain Res.,* **64**, 19–26.

Tempel, A., Gardner, E.L. and Zukin, R.S. (1984) Visualization of opiate receptor upregulation by light microscopy autoradiography. *Proc. Natl. Acad. Sci. USA,* **81**, 3893–7.

Tempel, A., Gardner, E.L. and Zukin, R.S. (1985) Neurochemical and functional correlates of naltrexone-induced opiate receptor up-regulation. *J. Pharmacol. Exp. Ther.,* **232**, 439–44.

Tempel, A., Crain, S.M., Peterson, E.R. *et al.* (1986) Antagonist-induced opiate receptor upregulation in cultures of fetal mouse spinal cord-ganglion explants. *Brain Res.,* **390**, 287–91.

Tempel, A., Habas J., Paredes, W. and Barr, G.A. (1988) Morphine-induced downregulation of µ-opioid receptors in neonatal rat brain. *Dev. Brain Res.,* **41**, 129–33.

Tempel, A., Kessler, J.A. and Zukin, R.S. (1990) Chronic naltrexone treatment increases expression of preproenkephalin and preprotachykinin mRNA in discrete brain regions. *J. Neurosci.,* **10**, 741–7.

Townsend, R.R., Laing, F.C. and Jeffrey, R.B. (1988) Placental abruption associated with cocaine abuse. *Am. J. Roentgenol.,* **150**, 1339–40.

Traber, J., Gullis, R. and Hamprecht, B. (1975) Influence of opiates on the levels of adenosine 3':5'-cyclic monophosphate in neuroblastoma X glioma hybrid cells. *Life Sci.,* **16**, 1863–8.

Tsang, D. and Ng, S.C. (1980) Effect of antenatal exposure to opiates on the development of opiate receptors in rat brain. *Brain Res.,* **188**, 199–206.

Tsang D., Tan, A.T., Henry, J.L. and Lal, S. (1978) Effect of opioid peptides on L-noradrenaline-stimulated cyclic AMP formation in homogenates of rat cerebral cortex and hypothalamus. *Brain Res.,* **152**, 521–7.

Uhl, G.R., Ryan, J.P. and Schwartz, J.P. (1988) Morphine alters preproenkephalin gene expression. *Brain Res.,* **459**, 391–7.

Valencia, G., McCalla, S., da Silva, M. *et al.* (1989) Epidemiology of cocaine use during pregnancy at Kings County Hospital Center (KCHC). *Pediatr. Res.,* **25**, 265A.

Van Vliet, J., Wardeh, G., Mulder, A.H. and Schoffelmeer, A.N.M. (1991) Reciprocal effects of chronic morphine administration on stimulatory and inhibitory G-protein subunits in primary cultures of rat striatal neurons. *Eur. J. Pharmacol.,* **208**, 341–2.

Watanabe, Y., Shibuya, T., Salafsky, B. and Hill, H.F. (1983) Prenatal and postnatal exposure to diazepam: effects on opioid receptor binding in rat brain cortex. *Eur. J. Pharmacol.,* **96**, 141–4.

Weber, E., Evans, E.J. and Barchas, J.D. (1983) Multiple endogenous ligands for opioid receptors. *Trends Neurosci.,* **6**, 333.

Weiner, N. (1974) Neurotoxicity of opiates during brain development, in *Drugs and the Developing Brain* (eds A. Vernadakis and N. Weiner), Plenum Press, New York, pp. 215–27.

Wesche, D., Hollt, V. and Herz, A. (1977) Radio-immunoassay of enkephalins. Regional distribution in rat brain after morphine treatment and hypophysectomy. *Naunyn-Schmiedebergs Arch. Pharmacol.,* **301**, 79–82.

Wuster, M., Costa, T. and Gramsch, C.H. (1983) Uncoupling of receptors is essential for opiate-induced desensitization (tolerance) in neuroblastoma x glioma hybrid cells NG108-15. *Life Sci.,* **33**, 341–4.

Yoburn, B.C., Goodman, R.G., Cohen, A.C. *et al.* (1985) Increased analgesic potency of morphine and increased brain opioid binding sites in the rat following chronic naltrexone treatment. *Life Sci.,* **36**, 2325–9.

Yoburn, B.C., Kreuscher, S.P., Inturrisi, C.E. and Sierra, V. (1989) Opioid receptor up-regulation and supersensitivity in mice: effect of morphine sensitivity. *Pharmacol. Biochem. Behav.*, **32**, 727–31.

Yoshikawa, K. and Sabol, S.L. (1986) Glucocorticoids and cyclic AMP synergistically regulate the abundance of preproenkephalin messenger RNA in neuroblastoma-glioma hybrid cells. *Biochem. Biophys. Res. Commun.*, **139**, 1–10.

Young, W.S. and Kuhar, M.J. (1979) A new method for receptor autoradiography: [^{3}H] opioid receptors in rat brain. *Brain Res.*, **179**, 255–70.

Zagon, I.S. and McLaughlin, P.J. (1977) The effect of chronic morphine administration on pregnant rats and their offspring. *Pharmacology*, **15**, 302–10.

Zagon, I.S., McLaughlin, P.J., Weaver, D.J. and Zagon, E. (1982) Opiates, endorphins, and the developing organism: a comprehensive bibliography. *Neurosci. Biobehav. Rev.*, **6**, 439–79.

Zukin, R.S. and Zukin, S.R. (1988) The sigma receptor, in *The Opiate Receptors* (ed. G.W. Pasternak), Humana Press, Clifton, NJ, pp. 143–63.

Zukin, R.S., Sugarman, J.R., Fitz-Syaga, M.L. *et al.* (1982) Naltrexone-induced opiate receptor supersensitivity. *Brain Res.*, **245**, 285–92.

Zukin, R.S., Eghbali, M., Olive, D. *et al.* (1988) Characterization and visualization of rat and guinea-pig kappa opioid receptors: evidence for kappa$_1$ and kappa$_2$ receptors. *Proc. Natl. Acad. Sci. USA*, **85**, 4061–5.

INDEX